Data Warehousing and Mining:
Concepts, Methodologies, Tools, and Applications

John Wang
Montclair State University, USA

Volume II

INFORMATION SCIENCE REFERENCE
Hershey · New York

Acquisitions Editor: Kristin Klinger
Development Editor: Kristin Roth
Senior Managing Editor: Jennifer Neidig
Managing Editor: Jamie Snavely
Typesetter: Michael Brehm, Jeff Ash, Carole Coulson, Elizabeth Duke, Jamie Snavely, Sean Woznicki
Cover Design: Lisa Tosheff
Printed at: Yurchak Printing Inc.

Published in the United States of America by
Information Science Reference (an imprint of IGI Global)
701 E. Chocolate Avenue, Suite 200
Hershey PA 17033
Tel: 717-533-8845
Fax: 717-533-8661
E-mail: cust@igi-global.com
Web site: http://www.igi-global.com/reference

and in the United Kingdom by
Information Science Reference (an imprint of IGI Global)
3 Henrietta Street
Covent Garden
London WC2E 8LU
Tel: 44 20 7240 0856
Fax: 44 20 7379 0609
Web site: http://www.eurospanonline.com

Library of Congress Cataloging-in-Publication Data

Data warehousing and mining : concepts, methodologies, tools and applications / John Wang, editor.
p. cm.
Summary: "This collection offers tools, designs, and outcomes of the utilization of data mining and warehousing technologies, such as algorithms, concept lattices, multidimensional data, and online analytical processing. With more than 300 chapters contributed by over 575 experts from around the globe, this authoritative collection will provide libraries with the essential reference on data mining and warehousing"--Provided by publisher.
Includes bibliographical references and index.
ISBN 978-1-59904-951-9 (hbk.) -- ISBN 978-1-59904-952-6 (e-book)
1. Data mining. 2. Data warehousing. I. Wang, John, 1955-
QA76.9.D343D398 2008
005.74--dc22
2008001934

British Cataloguing in Publication Data
A Cataloguing in Publication record for this book is available from the British Library.

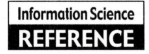

List of Contributors

Abi-Aad, R. / *Concordia University, Canada* 1486
Abidin, Taufik / *North Dakota State University, USA* 3694
Aedo, Ignacio / *Universidad Carlos III de Madrid, Spain* 438
Agrawal, Vikas / *Fayetteville State University, USA* 2201
Ahmed, Mesbah U. / *University of Toledo, USA* 2201
Ahrens, Martin / *Inductis India Pvt. Ltd., India* 2558
Akutsu, Tatsuya / *Kyoto University, Japan* 103
Al-Ahmadi, Mohammad Saad / *Oklahoma State University, USA* 9
Alam, Mansoor / *University of Toledo, USA* 1081
Aler, Ricardo / *Universidad Carlos III de Madrid, Spain* 469
Ali, Shawkat / *Monash University, Australia* 3308
Alvir, Jose Ma. J. / *Pfizer Inc., USA* 3675
Amaral, Luís Alfredo / *University of Minho, Portugal* 880
Amaravadi, Chandra S. / *Western Illinois University, USA* 1689, 2302
Anderson, Marc / *North Dakota State University, USA* 1747
Andreevskaia, A. / *Concordia University, Canada* 1486
Angryk, Rafal / *Tulane University, USA* 2121
Anisimov, Alexander / *Ural State Technical University, Russia* 397
Artz, John M. / *The George Washington University, USA* 3411
Aufaure, Marie-Aude / *Supélec, France* 254
Ayewah, Nathaniel / *Southern Methodist University, Texas, USA* 1400
Azevedo, Paulo Jorge / *Universidade do Minho, Portugal* 1722
Azzag, H. / *Université François-Rabelais de Tours, Laboratoire d'Informatique (EA 2101), France* 1557
Bagchi, Aditya / *Indian Statistical Institute, India* 2338
Bagnall, Anthony / *University of East Anglia, UK* 1320
Bagui, Sikha / *The University of West Florida, USA* 2523
Bandyopadhyay, A.K. / *Jadavpur University, India* 2338
Banerjee, Protima / *Drexel University, USA* 3524
Bapi, Raju S. / *University of Hyderabad, India* 3285
Bapna, Sanjay / *Morgan State University, USA* 1334
Barko, Christopher D. / *University of North Carolina at Greensboro, USA* 2289
Batra, Dinesh / *Florida International University, USA* 280
Baumgartner, Christian / *University for Health Sciences, Medical Informatics and Technology (UMIT), Austria* 1759

Bean, Kathryn / *University of Texas at Dallas, USA* 1400
Bellatreche, Ladjel / *LISI/ENSMA, France* 18
Bennacer, Nacera / *Supélec, France* 3531
Berberidis, Christos / *Aristotle University of Thessaloniki, Greece* 1696, 2993
Besemann, Christopher / *North Dakota State University, USA* 1747
Beynon, Malcolm J. / *Cardiff University, UK* 2943, 3005
Bickel, Steffen / *Humboldt-Universität zu Berlin, Germany* 1454
Bose, Indranil / *The University of Hong Kong, Hong Kong* 1817, 2449, 2762
Botía, Juan A. / *University of Murcia, Spain* 642
Bouet, Marinette / *LIMOS, Blaise Pascal University-Clermont-Ferrand, France* 254
Boujenoui, Ameur / *University of Ottawa, Canada* 3346
Boussaïd, Omar / *University LUMIERE Lyon, France* 254
Braga, Antonio P. / *UFMG, Brazil* 2464
Brathwaite, Charmion / *University of North Carolina at Greensboro, USA* 2856
Brezany, Peter / *University of Vienna, Austria* 755, 913
Bruckner, Robert M. / *Microsoft, USA* 509
Brunzel, Marko / *German Research Center for Artificial Intelligence (DFKI GmbH), Germany* 1978
Bull, Larry / *University of West of England, UK* 1320
Burns, Alex / *The University of Iowa, USA* 2226
Cabrera, Javier / *Rutgers University, USA* 3675
Calero, Coral / *University of Castilla-La Mancha, Spain* 408
Camacho, David / *Universidad Carlos III de Madrid, Spain* 469
Canhoto, Ana Isabel / *Henley Management College, UK* 2316
Cao, Longbing / *University of Technology, Sydney, Australia* 831
Cardell, Nicholas Scott / *Salford Systems, USA* 1519
Caridi, Frank / *Pfizer Inc., USA* 3675
Carvalho, João Álvaro / *Universidade do Minho, Portugal* 2296
Casado, Edilberto / *Gerens Escuela de Gestión y Economía, Peru* 2688
Catarci, Tiziana / *University of Rome "La Sapienza," Italy* 3451
Cawley, Gavin / *University of East Anglia, UK* 1320
Cazier, Joseph A. / *Appalachian State University, USA* 2841
Chahir, Youssef / *Université de Caen, France* 1958
Chakravarthy, Sharma / *The University of Texas at Arlington, USA* 3272
Chan, Chi Kin / *The Hong Kong Polytechnic University, Hong Kong* 2792
Chen, Liming / *Ecole Centrale de Lyon, France* 1958
Chen, Peter P. / *Lousiana State University, USA* 787
Chen, Qiyang / *Montclair State University, USA* 1
Chen, Shaokang / *The University of Queensland, Australia* 3621
Chen, Sherry Y. / *Brunel University, UK* 2273
Chen, Yaohua / *University of Regina, Canada* 2051
Chen, Ye-Sho / *Louisiana State University, USA* 2722
Chen, Ying / *Dalhousie University, Canada* 3176
Chen, Zhengxin / *University of Nebraska at Omaha, USA* 26
Chen, Zhiyuan / *University of Maryland, Baltimore County, USA* 1334
Chesney, David / *Freeman Hospital, UK* 2915

Chesney, Thomas / *Nottingham University Business School, UK* 2915
Cho, Moonjung / *University of Buffalo, USA* 1280
Choi, Kwangmin / *Indiana University, USA* 1784
Chong, P. Pete / *University of Houston-Downtown, USA* 2722
Chu, Feng / *Nanyang Technological University, Singapore* 1706
Chun, Lam Albert Kar / *The University of Hong Kong, Hong Kong* 2762
Chung, Yun-Sheng / *National Tsing Hua University, Taiwan* 1157
Cimpian, Emilia / *University of Innsbruck, Austria* 1037
Cing, Tay Joc / *Nanyang Technological University, Singapore* 3440
Císaro, Sandra Elizabeth González / *Universidad Nacional del Centro de la Provincia de Buenos Aires, Argentina* 75
Conway, Tyrrell / *University of Oklahoma, USA* 1643
Cook, Jack / *Rochester Institute of Technology, USA* 2834
Courant, Jennifer / *Citigroup, USA* 2438
Cuadrado, Juan / *Universidad Carlos III de Madrid, Spain* 469
Cunningham, Colleen / *Drexel University, USA* 787, 1810
Cuzzocrea, Alfredo / *University of Calabria, Italy* 974
Dai, Honghua / *DePaul University, USA* 3557
Dale, Mark S. / *Monash University, Australia* 2772
Daly, Olena / *Monash University, Australia* 336
Damiani, Maria Luisa / *Università di Milano, Italy and Ecole Polytechnique Fédérale, Switzerland* 659
Daneshgar, Farhad / *University of New South Wales, Australia* 2302
Dasu, Tamraparni / *AT&T Labs - Research, USA* 2644
Datta, Amitava / *The University of Western Australia, Australia* 2105
Davies, Simon / *University of Birmingham Research Park, UK* 2915
Davis, Karen C. / *University of Cincinnati, USA* 1606
de Carvalho, André C.P. L. F. / *University of São Paulo, Brazil* 119, 2464
de Souza, Jano Moreira / *Federal University of Rio de Janeiro, Brazil and University of Stuttgart, Germany* 3142
De Vinco, Lara / *Nexera S.c.p.A., Italy* 2067
Dehne, Frank / *Carleton University, Canada* 3176
Delve, Janet / *University of Portsmouth, UK* 2364
Deng, Xiaotie / *City University of Hong Kong, Hong Kong* 1926
Deng, Qun-Shi / *National Chung Cheng University, Taiwan* 1539
Denton, Anne / *North Dakota State University, USA* 1747
Deshpande, Prasad M. / *IBM Almaden Research Center, USA* 179
Díaz, Paloma / *Universidad Carlos III de Madrid, Spain* 438
Dillon, Henry / *Independent Consultant, UK* 2734
Dillon, Tharam S. / *University of Technology, Australia* 485
Ding, Qiang / *Concordia College, USA* 3694
Dodero, Juan Manuel / *Universidad Carlos III de Madrid, Spain* 438
Dreher, Heinz / *Curtin University of Technology, Australia* 1901
Dunham, Margaret H. / *Southern Methodist University, Texas, USA* 1400
Eavis, Todd / *Concordia University, Canada* 3176
El-Ghalayini, Haya / *University of the West of England (UWE), UK* 1068

Emam, Ahmed / *Western Kentucky University, USA* 2900
Enke, David / *University of Missouri – Rolla, USA* 2476
Ezeife, Christie I. / *University of Windsor, Canada* 1335
Fabris, Carem C. / *CPGEI, CEFET-PR, Brazil* 3234
Fadlalla, Adam / *Cleveland State University, USA* 3662
Fang, Qizhi / *Qingdao Ocean University, China* 1926
Felsõvályi, Àkos / *Citigroup, USA* 2438
Fernández-Medina, Eduardo / *Universidad de Castilla-La Mancha, Spain* 679
Ferreira, Pedro Gabriel / *Universidade do Minho, Portugal* 1722
Fovino, Igor Nai / *Joint Research Centre, Italy* 2379
Freitas, Alex / *University of Kent, UK* 119, 3234
Fu, Lixin / *University of North Carolina at Greensboro, USA* 1205, 2402
Fu, Xiuju / *Institute of High Performance Computing, Singapore* 1269
Furtado, Pedro / *University of Coimbra, Portugal* 718
Galitsky, Boris / *Birbeck College University of London, UK* 1714
Gançarski, Pierre / *LSIIT-AFD-Louis Pasteur University, France* 254
Gandy, William / *American Healthways Corp., USA* 1799
Gangopadhyay, Aryya / *University of Maryland, Baltimore County, USA* 1334
Ganguly, Auroop R. / *Oak Ridge National Laboratory, USA* 2618
Gardarin, Georges / *PRiSM Laboratory, France* 622
George, Susan E. / *University of South Australia, Australia* 2517
Ghosh, Sutirtha / *Inductis India Pvt. Ltd., India* 2558
Giudici, Paolo / *University of Pavia, Italy* 364
Goh, John / *Monash University, Australia* 1502
Goh, Liping / *Institute of High Performance Computing, Singapore* 1269
Golfarelli, Matteo / *University of Bologna, Italy* 738
Golovnya, Mikhaylo / *Salford Systems, USA* 1519
Goodhue, Dale L. / *University of Georgia, USA* 2749
Graber, Armin / *BIOCRATES Life Sciences GmbH, Austria* 1759
Greco, Gianluigi / *University of Calabria, Italy* 810
Greenidge, Charles / *University of the West Indies, Barbados* 2263
Griffiths, Benjamin / *Cardiff University, UK* 3005
Gruenwald, Le / *University of Oklahoma, USA* 1643, 2978
Grzes, Marek / *Bialystok Technical University, Poland* 3509
Grzymala-Busse, Jerzy W. / *University of Kansas, USA and Institute of Computer Science, PAS, Poland* 963
Guinot, C. / *CE.R.I.E.S., Unité Biométrie et Epidémiologie, and Université François-Rabelais de Tours, Laboratoire d'Informatique (EA 2101), France* 1557
Gupta, Amar / *University of Arizona, USA* 2618
Gupta, Ashima / *University of Cincinnati, USA* 1606
Guzzo, Antonella / *University of Calabria, Italy* 810
Hamar, Guy B. / *American Healthways Corp., USA* 1799
Hammami, Mohamed / *Faculté des Sciences de Sfax, Tunisia* 1958
Han, Jilin / *University of Oklahoma, USA* 1643
Harper, Franklin Maxwell / *National Civic League, USA* 2427
Harrington, Kara / *University of North Carolina at Greensboro, USA* 2856

Hasan, K. M. Azharul / *University of Fukui, Japan* 3324
Helander, Martin / *Nanyang Technological University, Singapore* 2798
Helen, Wong Oi Ling / *The University of Hong Kong, Hong Kong* 2762
Hernandez-Orallo, Jose / *Technical University of Valencia, Spain* 169
Hernansáez, Juan M. / *University of Murcia, Spain* 642
Hess, Kimberly / *CASA 20th Judicial District, USA* 3067
Hess, Traci J. / *Washington State University, USA* 2185
Higuchi, Ken / *University of Fukui, Japan* 3324
Hippe, Zdzislaw S. / *University of Information Technology and Management, Poland* 963
Hirji, Karim K. / *AGF Management Ltd, Canada* 343
Ho, Tu Bao / *Japan Advanced Institute of Science and Technology, Japan* 3164
Hoffman, Jeff / *The Chubb Group of Insurance Companies, USA* 1888
Hong, Taeho / *Pusan National University, Korea* 1416
Hope, Beverley / *Victoria University of Wellington, New Zealand* 2734
Hsu, D. Frank / *Fordham University, USA* 1157
Hsu, Hui-Huang / *Tamkang University, Taipei, Taiwan* 93
Hsu, Jeffrey / *Fairleigh Dickinson University, USA* 2584
Hsu, Wynne / *National University of Singapore, Singapore* 3477
Hsueh, Hsiang-Yuan / *National Chung Cheng University, Taiwan* 3116
Hu, Xiaohua / *Drexel University, USA* 1810, 3524
Huang, Guan-Shieng / *National Chi Nan University, Taiwan* 1091
Huang, Jie / *University of Texas Southwestern Medical Center, USA* 1400
Huang, Shi-Ming / *National Chung Cheng University, Taiwan* 1539, 3116
Hudson, Laurel R. / *American Healthways Corp., USA* 1799
Hung, GihGuang / *Institute of High Performance Computing, Singapore* 1269
Hung, Wong Cho / *University of Hong Kong, China* 2449
Hutchison, Ron / *The Richard Stockton College of New Jersey, USA* 1747
Hwang, Gwo-Jen / *National University of Tainan, Taiwan, R.O.C.* 2928
Hwang, Sae / *University of Texas at Arlington, USA* 1631
Ines, Li Hoi Wan / *The University of Hong Kong, Hong Kong* 2762
Ito, Takao / *Ube National College of Technology, Japan* 371
Jahangiri, Mehrdad / *University of Southern California, USA* 1250
Jain, Ankur / *Inductis India Pvt. Ltd., India* 2558
Janciak, Ivan / *University of Vienna, Austria* 913
Jatteau, Ganaël / *Université du Québec en Outaouais, Canada* 3346
Jentzsch, Ric / *University of Canberra, Australia* 1435
Jeong, Mina / *Mokpo National University, Korea* 1103
Jespersen, Søren E. / *Linkage, Denmark* 3364
Jiao, Jianxin (Roger) / *Nanyang Technological University, Singapore* 2798
Justis, Robert / *Louisiana State University, USA* 2722
Kalogeras, Ioannis S. / *National Observatory of Athens, Greece* 3645
Kan, Cheng Pui / *University of Hong Kong, China* 2449
Kanellis, Panagiotis / *National & Kapodistrian University of Athens, Greece* 2626
Kantardzic, Mehmed / *University of Louisville, USA* 1573
Kanterakis, Alexandros / *Institute of Computer Science, FORTH, Greece* 2248
Karabatis, George / *University of Maryland, Baltimore County, USA* 1334

Karydis, Ioannis / *Aristotle University of Thessaloniki, Greece* 3586
Katakis, Ioannis / *Aristotle University of Thessaloniki, Greece* 64
Kaur, H. / *Multimedia University, Malaysia* 2506
Kazmierczak, Ed / *University of Melbourne, Australia* 3194
Khan, Latifur / *The University of Texas at Dallas, USA* 2036
Khan, Shiraj / *University of South Florida, USA* 2618
Ki, Lau Wai / *University of Hong Kong, China* 2449
Kido, Takashi / *HuBit Genomix, Inc., Japan* 1674
Kim, Jaehoon / *Sogang University, Korea* 1231
Kim, Jin Sung / *Jeonju University, Korea* 2659
Kim, Sun / *Indiana University, USA* 1784
Kimani, Stephen / *University of Rome "Federico II," Italy* 3451
Kirkman-Liff, Bradford / *Arizona State University, USA* 2494
Knobloch, Bernd / *University of Bamberg, Germany* 449
Koeller, Andreas / *Montclair State University, USA* 350
Koh, Yun Sing / *University of Otago, New Zealand* 3222
Kontio, Juha / *Turku Polytechnic, Finland* 429
Kopanakis, Ioannis / *UMIST Manchester, UK* 2141
Kou, Gang / *University of Nebraska at Omaha, USA* 26
Kretowski, Marek / *Bialystok Technical University, Poland* 3509
Krishna, P. Radha / *Institute for Development & Research in Banking Technology, India* 3285
Kristal, Bruce S. / *Cornell University, USA* 1157
Kumar, Kuldeep / *Bond University, Australia* 1446
Kumar, Navin / *University of Maryland, Baltimore County, USA* 1334
Kumar, Pradeep / *University of Hyderabad, India* 3285
Kundu, Suddha Sattwa / *Inductis India Pvt. Ltd., India* 2558
Kuo, Emily / *Massachusetts Institute of Technology, USA* 3067
Kusiak, Andrew / *The University of Iowa, USA* 2226
LaBrie, Ryan C. / *Seattle Pacific University, USA* 2841
Ladner, Roy / *Naval Research Laboratory, USA* 2121
Laware, Gilbert W. / *Purdue University, USA* 3416
Lawler, James / *Pace University, USA* 2872
Lazarevic, Aleksandar / *University of Minnesota, USA* 2880
Le, D. Xuan / *La Trobe University, Australia* 3386
Le Grand, Bénédicte / *Laboratoire d'Informatique de Paris 6, France* 3531
Lee, Doheon / *Korea Advanced Institute of Science and Technology, Korea* 1103
Lee, JeongKyu / *University of Texas at Arlington, USA* 1631
Lee, Mong Li / *National University of Singapore, Singapore* 3477
Levary, Reuven R. / *Saint Louis University, USA* 2637
Li, Deren / *Wuhan University, China* 1216
Li, Ming / *National Laboratory for Novel Software Technology, China* 2816
Li, Xining / *University of Guelph, Canada* 705
Li, Zhigang / *Southern Methodist University, Texas, USA* 1400
Liabotis, Ioannis / *University of Manchester, UK* 1116
Liao, Shuang-Te / *Ming Chuan University, Taiwan* 1091
Liao, T. Warren / *Louisiana State University, USA* 942

Liberati, Diego / *Consiglio Nazionale delle Ricerche, Italy* 2281
Limanto, Hanny Yulius / *Nanyang Technological University, Singapore* 3440
Lin, Binshan / *Louisiana State University, USA* 1539
Ling, Tok Wang / *National University of Singapore, Singapore* 509
Liu, Chang / *Northern Illinois University, USA* 1
Liu, Han / *University of Toronto, Canada* 509
Liu, Wieguo / *University of Toledo, USA* 2978
Liu, Xiaohui / *Brunel University, UK* 2273
Liu, Xu-Ying / *National Laboratory for Novel Software Technology, China* 2816
Liu, Yanbing / *UEST of China & Chongqing University of Posts and Telecommunications, China* 3033
Liu, Zhigang / *State Key Laboratory of Remote Sensing Science, Beijing Normal University and Wuhan University, China* 1216
Lo, Victor S.Y. / *Fidelity Personal Investments, USA* 2824
Lodi, Stefano / *University of Bologna, Italy* 3451
Long, Lori K. / *Kent State University, USA* 2371
Longo, Giuseppe / *University "Federico II" of Napoli Polo delle Scienze e della Tecnologia, Italy* 2067
Lovell, Brian C. / *The University of Queensland, Australia* 3621
Lu, June / *University of Houston-Victoria, USA* 1
Ludermir, Teresa / *UFPE, Brazil* 2464
Luján-Mora, Sergio / *Universidad de Alicante, Spain* 556, 591
Ma, Z. M. / *Northeastern University, China* 187, 1182
Maffulli, Nicola / *Keele University School of Medicine, UK* 2915
Maniatis, Andreas / *National Technical University of Athens, Greece* 1004
Manolopoulos, Yannis / *Aristotle University of Thessaloniki, Greece* 3212, 3586
Mansmann, Svetlana / *University of Konstanz, Germany* 2164
Marketos, Gerasimos / *University of Piraeus, Greece* 3645
Martakos, Drakoulis / *National & Kapodistrian University of Athens, Greece* 2626
Masud, Mohammad M. / *The University of Texas at Dallas, USA* 2036
Mathieu, Richard / *Saint Louis University, USA* 2637
Mavrogonatos, George / *National Technical University of Athens, Greece* 1004
McClatchey, Richard / *University of the West of England (UWE), UK* 1068
Meng, Jun / *Zhejiang University, China* 1138
Michalarias, Ilias / *National Technical University of Athens, Greece* 1004
Milton, Simon K. / *University of Melbourne, Australia* 3194
Min, Hokey / *University of Louisville-Shelby, USA* 2900
Missaoui, Rokia / *Université du Québec en Outaouais, Canada* 3346
Mitra, Susanta / *International Institute of Information Technology, India* 2338
Mitschang, Bernhard / *University of Stuttgart, Germany* 3142
Mobasher, Bamshad / *DePaul University, USA* 2551, 3557
Mocan, Adrian / *University of Innsbruck, Austria* 1037
Mohammadian, Masoud / *University of Canberra, Australia* 1435
Mohania, Mukesh / *IBM India Research Lab, India* 18
Molluzzo, John C. / *Pace University, USA* 2872
Monteiro, Rodrigo Salvador / *Federal University of Rio de Janeiro, Brazil and University of Stuttgart, Germany* 3142

Montero, Jose D. / *Management Information System Department, Utah State University, USA* 146
Mroczek, Teresa / *University of Information Technology and Management, Poland* 963
Nandkeolyar, Udayan / *University of Toledo, USA* 2201
Nanopoulos, Alexandros / *Aristotle University of Thessaloniki, Greece* 3212, 3586
Naouali, Sami / *Université du Québec en Outaouais, Canada* 3346
Narayanan, L. Venkat / *Satyam Computer Services, Singapore* 1825, 1840
Nassis, Vicky / *La Trobe University, Australia* 485
Natasha, V. / *Multimedia University, Malaysia* 2506
Nayak, Richi / *Queensland University of Technology, Australia and Queensland University, Australia* 1938, 2022, 2697
Neftzger, Amy L. / *American Healthways Corp., USA* 1799
Nemati, Hamid R. / *University of North Carolina at Greensboro, USA* 2289, 2402, 2856
Nguyen, Ha / *Pfizer Inc., USA* 3675
Nguyen, Tho Manh / *Vienna University of Technology, Austria* 755
Nicholson, Scott / *Syracuse University, USA* 2673
Nigro, Héctor Oscar / *Universidad Nacional del Centro de la Provincia de Buenos Aires, Argentina* 75
Nimmagadda, Shastri L. / *Kuwait Gulf Oil Company, Kuwait* 1901
Noriel, Nathaniel B. / *Singapore Institute of Statistics, Singapore* 84
Ntoutsi, Irene / *University of Piraeus, Greece* 228
O'Keefe, Richard / *University of Otago, New Zealand* 3222
Oakley, Peter / *The University Hospital of North Staffordshire, UK* 2915, 3222
Odeh, Mohammed / *University of the West of England (UWE), UK* 1068
Oh, JungHwan / *University of Texas at Arlington, USA* 1631
Ohanekwu, Timothy E. / *University of Windsor, Canada* 1355
Olecka, Anna / *D&B, USA* 1855
Oliveira, Stanley R. M. / *Embrapa Informática Agropecuária, Brazil* 50
Olson, David L. / *University of Nebraska, USA* 1877
Oza, Nikunj C. / *NASA Ames Research Center, USA* 356
Pang, Les / *National Defense University, USA* 389, 2421
Papadopoulos, Apostolos N. / *Aristotle University of Thessaloniki, Greece* 3212
Paquet, Eric / *National Research Council of Canada, Canada* 1623
Park, Seog / *Sogang University, Korea* 1231
Partow-Navid, Parviz / *California State University, Los Angeles, USA* 2888
Pedersen, Torben Bach / *Aalborg University, Denmark* 3364
Pei, Jian / *Simon Fraser University, Canada* 1280
Pelekis, Nikos / *University of Piraeus, Greece and UMIST Manchester, UK* 228, 2141
Peng, Yi / *University of Nebraska at Omaha, USA* 26
Penny, Kay / *Napier University, UK* 2915
Perrizo, William / *North Dakota State University, USA* 3694
Perry, Theodore L. / *American Healthways Corp., USA* 1799
Peter, Hadrian / *University of the West Indies, Barbados* 2263
Petry, Frederick E. / *Tulane University & Naval Research Laboratory, USA* 2121
Pettipher, Mike / *University of Manchester, UK* 1320
Pham, Tho Hoan / *Hanoi University of Education, Vietnam* 3164

Piattini, Mario / *University of Castilla-La Mancha, Spain* 408, 679
Picarougne, F. / *Université François-Rabelais de Tours, Laboratoire d'Informatique (EA 2101), France* 1557
Pontieri, Luigi / *Institute of High Performance Computing and Networks, Italy* 810
Pope, Nigel K. Ll. / *Griffith University, Australia* 1301
Potamias, George / *Institute of Computer Science, FORTH, Greece* 2248
Qin, Qianqing / *Wuhan University, China* 1216
Qu, Zhong / *Chongqing University of Posts and Telecommunications, China* 3493
Radhakrishnan, T. / *Concordia University, Canada* 1486
Raghavan, Vijay / *University of Louisiana at Lafayette, USA* 3085
Rahayu, Wenny / *La Trobe University, Australia* 485, 3386
Raisinghani, Mahesh S. / *University of Dallas, USA* 2468
Rajugan, R. / *University of Technology, Australia* 485
Ramachandran, Karthik / *University of Louisiana at Lafayette, USA* 3085
Ramasamy, Karthikeyan / *Juniper Networks, USA* 179
Ramos, Isabel / *Universidade do Minho, Portugal* 2296
Rao, Ranjan / *Inductis India Pvt. Ltd., India* 2558
Rau-Chaplin, A. / *Dalhousie University, Canada* 3176
Redmond, Richard T. / *Virginia Commonwealth University, USA* 2964
Rizzi, Stefano / *University of Bologna, Italy* 208, 738
Roj, Edward / *Fertilizer Research Institute, Poland* 963
Rosen, Peter A. / *University of Evansville, USA* 9
Rountree, Nathan / *University of Otago, New Zealand* 3222
Rowe, Neil C. / *U.S. Naval Postgraduate School, USA* 1461
Runger, George C. / *Arizona State University, USA* 2494
Rusu, Laura Irina / *La Trobe University, Australia* 530
Saccà, Domenico / *University of Calabria, Italy* 974
Sacharidis, Dimitris / *University of Southern California, USA* 1250
Sadeghian, Pedram / *University of Louisville, USA* 1573
Sadri, Fereidoon / *The University of North Carolina at Greensboro, USA* 2402
Santos, Maribel Yasmina / *University of Minho, Portugal* 880
Saraee, Mohamad / *University of Salford, UK* 1116
Savary, Lionel / *PRiSM Laboratory, France* 622
Savla, Sagar / *The University of Texas at Arlington, USA* 3272
Saygin, Yücel / *Sabanci University, Turkey* 2850
Scheffer, Tobias / *Humboldt-Universität zu Berlin, Germany* 1454
Scholl, Marc H. / *University of Konstanz, Germany* 2164
Schwarz, Holger / *University of Stuttgart, Germany* 3142
Segall, Richard S. / *Arkansas State University, USA* 2088
Seifert, Jeffrey W. / *Congressional Research Service, USA* 3630
Sellis, Timos / *National Technical University of Athens, Greece* 3049
Serafino, Paolo / *University of Calabria, Italy* 974
Serazi, Masum / *North Dakota State University, USA* 3694
Serrano, Manuel / *University of Castilla-La Mancha, Spain* 408
Sethi, Neerja / *Nanyang Technological University, Singapore* 381
Sethi, Vijay / *Nanyang Technological University, Singapore* 381

Shah, Biren / *University of Louisiana at Lafayette, USA* 3085
Shah, Shital / *Rush University Medical Center, Health Systems Management, USA* 2226
Shahabi, Cyrus / *University of Southern California, USA* 1250
Sheta, Walaa M. / *Mubarak City for Scientific Research, Egypt* 1573
Shetty, Sachin / *Old Dominion University, USA* 1081
Shi, Wenzhong / *The Hong Kong Polytechnic University, Hong Kong* 1216
Shi, Yong / *University of the Chinese Academy of Sciences, China and University of Nebraska at Omaha, USA* 26
Shyu, Shyong-Jian / *Ming Chuan University, Taiwan* 1091
Simitisis, Alkis / *National Technical University of Athens, Greece* 3049
Simovici, Dan A. / *University of Massachusetts – Boston, USA* 849
Singh, Manoj K. / *i2 Technologies, USA* 2468
Singh, Rahul / *University of North Carolina at Greensboro, USA* 2964
Skarmeta, Antonio F.G. / *University of Murcia, Spain* 642
Skiadopoulos, Spiros / *National Technical University of Athens, Greece and University of Peloponnese, Greece* 1004, 3049
Skowronski, Boleslaw / *Fertilizer Research Institute* 963
Slusky, Ludwig / *California State University, Los Angeles, USA* 2888
Smith, Edward A. / *University of Arizona and Translational Genomics Research Institute, USA* 2494
Smith, Kate A. / *Monash University, Australia* 2772, 3308
Song, Il-Yeol / *Drexel University, USA* 787
Song, Lei / *University of Guelph, Canada* 705
Song, Min / *Old Dominion University, USA* 1081
Soto, Michel / *Laboratoire d'Informatique de Paris 6, France* 3531
Spaccapietra, Stefano / *Ecole Polytechnique Fédérale de Lausanne, Switzerland* 659
Spiliopoulou, Myra / *Otto-von-Guericke-Universität Magdeburg, Germany* 1987
Sriraam, N. / *Multimedia University, Malaysia* 2506
Staiano, Antonino / *University of Napoli, "Parthenope," Italy* 2067
Stanton, Jeffrey / *Syracuse University, USA* 2673
Steinberg, Dan / *Salford Systems, USA* 1519
Stockinger, Kurt / *University of California, USA* 1590
Studley, Matthew / *University of West of England, UK* 1320
Suh, Woojong / *Inha University, Korea* 1416
Sun, Shixin / *UEST, China* 3033
Sundararaghavan, P. S. / *University of Toledo, USA* 2201
Swierzowicz, Janusz / *Rzeszow University of Technology, Poland* 3611
Tagliaferri, Roberto / *University of Salerno, Italy* 2067
Takaoka, Tadao / *University of Canterbury, New Zealand* 1301
Talburt, John / *University of Arkansas at Little Rock, USA* 3067
Tan, Chew Lim / *National University of Singapore, Singapore* 84
Tang, Hong / *UEST of China & Chongqing University of Posts and Telecommunications, China* 3033
Taniar, David / *Monash University, Australia* 303, 336, 530, 1502, 3386
Techapichetvanich, Kesaraporn / *The University of Western Australia, Australia* 2105
Tekiner, Firat / *University of Manchester, UK* 1320

Templeton, John / *Keele University, School of Medicine, UK* 2915
Testik, Murat Caner / *Cukurova University, Turkey* 2494
Theodoridis, Yannis / *University of Piraeus, Greece* 228, 2141, 3645
Theodoulidis, Babis / *University of Manchester, UK* 1116, 2141
Thorhauge, Jesper / *Conzentrate, Denmark* 3364
Thuraisingham, Bhavani / *The MITRE Corporation, USA and The University of Texas at Dallas* 693, 2036, 3639
Tjioe, Haorianto Cokrowijoyo / *Monash University, Australia* 303
Tjoa, A. Min / *Vienna University of Technology, Austria* 509, 755, 913
Triantafillakis, Aristides / *National & Kapodistrian University of Athens, Greece* 2626
Troutt, Marvin D. / *Kent State University, USA* 2371
Trujillo, Juan / *Universidad de Alicante, Spain* 556, 591, 679
Tsoumakas, Grigorios / *Aristotle University of Thessaloniki, Greece* 64
Tsuji, Tatsuo / *University of Fukui, Japan* 3324
Tsz, Chi King / *University of Hong Kong, China* 2449
Tucker, Travis / *American Healthways Corp., USA* 1799
Tzanis, George / *Aristotle University of Thessaloniki, Greece* 1696, 2993
Vassiliadis, Panos / *University of Ioannina, Greece* 1004, 3049
Vassiliou, Yannis / *National Technical University of Athens, Greece* 1004
Venturini, G. / *Université François-Rabelais de Tours, Laboratoire d'Informatique (EA 2101), France* 1557
Vesonder, Gregg T. / *AT&T Labs - Research, USA* 2644
Viktor, Herna L. / *University of Ottawa, Canada* 1623
Villarroel, Rodolfo / *Universidad Católica del Maule, Chile* 679
Vlahavas, Ioannis / *Aristotle University of Thessaloniki, Greece* 1696
Voges, Kevin E. / *University of Canterbury, New Zealand* 1301
Wang, Baoying / *North Dakota State University, USA* 3694
Wang, Hai / *Saint Mary's University, Canada* 1638, 3027
Wang, Haixun / *IBM, T.J. Watson Research Center, USA* 1280
Wang, Junmei / *National University of Singapore, Singapore* 3477
Wang, Lipo / *Nanyang Technological University, Singapore* 1269, 1706
Wang, Menghao / *UEST of China & Chongqing University of Posts and Telecommunications, China* 3033
Wang, Richard / *Massachusetts Institute of Technology, USA* 3067
Wang, Shouhong / *University of Massachusetts Dartmouth, USA* 1638, 3027
Wang, Wei / *Fudan University, China* 1280
Wangikar, Lalit / *Inductis India Pvt. Ltd., India* 2558
Watkins, Andrew / *Mississippi State University, USA* 3440
Watson, Hugh J. / *University of Georgia, USA* 2749
Weippl, Edgar / *Vienna University of Technology, Austria* 755
Wells, John D. / *Washington State University, USA* 2185
Welzer-Druzovec, Tatjana / *University of Maribor, Slovenia* 3212
Whittley, Ian / *University of East Anglia, UK* 1320
Wickramasinghe, Nilmini / *Cleveland State University, USA* 3662
Wilson, Rick L. / *Oklahoma State University, USA* 9
Wixom, Barbara H. / *University of Virginia, USA* 2749

Wong, T. T. / *The Hong Kong Polytechnic University, Hong Kong* 2704
Wright, Jon R. / *AT&T Labs - Research, USA* 2644
Wu, Desheng / *University of Toronto, Canada,* 1877
Wu, Kesheng / *University of California, USA* 1590
Xiao, Yongqiao / *SAS Inc., USA and Georgia College & State University, USA* 2004, 3252
Yang, Can / *Zhejiang University, China* 1138
Yang, Guizhen / *University at Buffalo, State University of New York, USA* 3252
Yao, James E. / *Montclair State University, USA* 1
Yao, Jenq-Foung (J. F.) / *Georgia College & State University, USA* 2004, 3252
Yao, Yiyu / *University of Regina, Canada* 2051
Yekkirala, Ajay / *North Dakota State University, USA* 1747
Yen, David C. / *Miami University, USA* 3116
Yin, Peng-Yeng / *National Chi Nan University, Taiwan* 1091
Yoo, Illhoi / *Drexel University, USA* 3524
Yoon, Victoria / *University of Maryland Baltimore County, USA* 2964
Yu, Yang / *National Laboratory for Novel Software Technology, China* 2816
Yue, Leung Vivien Wai / *The University of Hong Kong, Hong Kong* 2762
Zaïane, Osmar R. / *University of Alberta, Edmonton, Canada* 50
Zarri, Gian Piero / *University Paris 4/Sorbonne, France* 1376
Zeanah, Jeff / *Z Solutions, Inc., USA* 2566
Zeitouni, Karine / *PRiSM Laboratory, France* 622
Zhan, De-Chuan / *National Laboratory for Novel Software Technology, China* 2816
Zhang, Chengqi / *University of Technology, Sydney, Australia* 831
Zhang, Ji / *University of Toronto, Canada* 509
Zhang, Jianting / *University of New Mexico, USA* 2978
Zhang, Qingyu / *Arkansas State University, USA* 2088
Zhang, Yiyang / *Nanyang Technological University, Singapore* 2798
Zhao, Yan / *University of Regina, Canada* 2051
Zheng, Weimin / *Tsinghua University, China* 1926
Zhou, Zhi-Hua / *National Laboratory for Novel Software Technology, China* 2816
Zhu, Shanan / *Zhejiang University, China* 1138
Zhu, Shanfeng / *City University of Hong Kong, Hong Kong* 1926
Zimbrão, Geraldo / *Federal University of Rio de Janeiro, Brazil* 3142

Contents
by Volume

Volume I

Section 1. Fundamental Concepts and Theories

This section serves as the foundation for this exhaustive reference tool by addressing crucial theories essential to the understanding of data mining and warehousing. Chapters found within these pages provide an excellent framework in which to position data mining and warehousing within the field of information science and technology. Individual contributions provide insight into the critical incorporation of data mining and warehousing in the global community and explore crucial stumbling blocks of this field. Within this introductory section, the reader can learn and choose from a compendium of expert research on the elemental theories underscoring the research and application of data mining and warehousing.

Chapter 1.1. Administering and Managing a Data Warehouse /
James E. Yao, Chang Liu, Qiyang Chen, and June Lu 1

Chapter 1.2. Knowledge Structure and Data Mining Techniques / *Rick L. Wilson, Peter A. Rosen, and Mohammad Saad Al-Ahmadi* 9

Chapter 1.3. Physical Data Warehousing Design /
Ladjel Bellatreche and Mukesh Mohania 18

Chapter 1.4. Introduction to Data Mining Techniques via Multiple Criteria Optimization Approaches and Applications / *Yong Shi, Yi Peng, Gang Kou, and Zhengxin Chen* 26

Chapter 1.5. Privacy-Preserving Data Mining on the Web: Foundations and Techniques /
Stanley R. M. Oliveira and Osmar R. Zaïane 50

Chapter 1.6. Multi-Label Classification: An Overview / *Grigorios Tsoumakas and Ioannis Katakis* 64

Chapter 1.7. Online Data Mining / *Héctor Oscar Nigro and Sandra Elizabeth González Císaro* 75

Chapter 1.8. A Look Back at the PAKDD Data Mining Competition 2006 / *Nathaniel B. Noriel and Chew Lim Tan* 84

Chapter 1.9. Introduction to Data Mining in Bioinformatics / *Hui-Huang Hsu* 93

Chapter 1.10. Algorithmic Aspects of Protein Threading / *Tatsuya Akutsu* 103

Chapter 1.11. A Tutorial on Hierarchical Classification with Applications in Bioinformatics / *Alex Freitas and André C.P.L.F. de Carvalho* 119

Chapter 1.12. Introduction to Data Mining and its Applications to Manufacturing / *Jose D. Montero* 146

Chapter 1.13. Data Warehousing and OLAP / *Jose Hernandez-Orallo* 169

Chapter 1.14. Data Warehousing, Multi-Dimensional Data Models and OLAP / *Prasad M. Deshpande and Karthikeyan Ramasamy* 179

Chapter 1.15. A Literature Overview of Fuzzy Database Modeling / *Z. M. Ma* 187

Chapter 1.16. Conceptual Modeling Solutions for the Data Warehouse / *Stefano Rizzi* 208

Chapter 1.17. Pattern Comparison in Data Mining: A Survey / *Irene Ntoutsi, Nikos Pelekis, and Yannis Theodoridis* 228

Chapter 1.18. Pattern Mining and Clustering on Image Databases / *Marinette Bouet, Pierre Gançarski, and Marie-Aude Aufaure, and Omar Boussaid* 254

Chapter 1.19. Conceptual Data Modeling Patterns: Representation and Validation / *Dinesh Batra* 280

Chapter 1.20. Mining Association Rules in Data Warehouses / *Haorianto Cokrowijoyo Tjioe and David Taniar* 303

Chapter 1.21. Exception Rules in Data Mining / *Olena Daly and David Taniar* 336

Chapter 1.22. Process-Based Data Mining / *Karim K. Hirji* 343

Chapter 1.23. Integration of Data Sources through Data Mining / *Andreas Koeller* 350

Chapter 1.24. Ensemble Data Mining Methods / *Nikunj C. Oza* 356

Chapter 1.25. Evaluation of Data Mining Methods / *Paolo Giudici* 364

Chapter 1.26. Discovering an Effective Measure in Data Mining / *Takao Ito* 371

Chapter 1.27. Data Warehousing and Data Mining Lessons and EC Companies /
Neerja Sethi and Vijay Sethi 381

Chapter 1.28. Best Practices in Data Warehousing from the Federal
Perspective / *Les Pang* 389

Chapter 1.29. Decision Support and Data Warehousing: Challenges of a Global
Information Environment / *Alexander Anisimov* 397

Chapter 1.30. An Experimental Replication with Data Warehouse Metrics /
Manuel Serrano, Coral Calero, and Mario Piattini 408

Chapter 1.31. Data Warehousing Solutions for Reporting Problems / *Juha Kontio* 429

Section 2. Development and Design Methodologies

This section provides in-depth coverage of conceptual architecture, enabling the reader to gain a comprehensive understanding of the emerging technological developments within the field of data mining and warehousing. Research fundamentals imperative to the understanding of developmental processes within information management are offered. From broad examinations to specific discussions on electronic tools, the research found within this section spans the discipline while also offering detailed, specific discussions. Basic designs, as well as abstract developments, are explained within these chapters, and frameworks for implementing secure data warehouses are explored.

Chapter 2.1. A Multi-Agent Approach to Collaborative Knowledge Production /
Juan Manuel Dodero, Paloma Díaz, and Ignacio Aedo 438

Chapter 2.2. A Framework for Organizational Data Analysis and Organizational
Data Mining / *Bernd Knobloch* 449

Chapter 2.3. Rule-Based Parsing for Web Data Extraction / *David Camacho,
Ricardo Aler, and Juan Cuadrado* 469

Chapter 2.4. Conceptual and Systematic Design Approach for XML Document Warehouses /
Vicky Nassis, R. Rajugan, Tharam S. Dillon, and Wenny Rahayu 485

Chapter 2.5. A Framework for Efficient Association Rule Mining in XML Data /
Ji Zhang, Han Liu, Tok Wang Ling, Robert M. Bruckner, and A. Min Tjoa 509

Chapter 2.6. A Methodology for Building XML Data Warehouses / *Laura Irina Rusu,
J. Wenny Rahayu, and David Taniar* 530

Chapter 2.7. Applying UML for Modeling the Physical Design of Data Warehouses / *Sergio Luján-Mora and Juan Trujillo* 556

Volume II

Chapter 2.8. Physical Modeling of Data Warehouses Using UML Component and Deployment Diagrams: Design and Implementation Issues / *Sergio Luján-Mora and Juan Trujillo* 591

Chapter 2.9. GeoCache: A Cache for GML Geographical Data / *Lionel Savary, Georges Gardarin, and Karine Zeitouni* 622

Chapter 2.10. A Java Technology Based Distributed Software Architecture for Web Usage Mining / *Juan M. Hernansáez, Juan A. Botía, and Antonio F.G. Skarmeta* 642

Chapter 2.11. Spatial Data Warehouse Modelling / *Maria Luisa Damiani and Stefano Spaccapietra* 659

Chapter 2.12. Designing Secure Data Warehouses / *Rodolfo Villarroel, Eduardo Fernández-Medina, Juan Trujillo, and Mario Piattini* 679

Chapter 2.13. Privacy-Preserving Data Mining: Development and Directions / *Bhavani Thuraisingham* 693

Chapter 2.14. A Service Discovery Model for Mobile Agent-Based Distributed Data Mining / *Xining Li, Lei Song* 705

Chapter 2.15. Node Partitioned Data Warehouses: Experimental Evidence and Improvements / *Pedro Furtado* 718

Chapter 2.16. Managing Late Measurements in Data Warehouses / *Matteo Golfarelli and Stefano Rizzi* 738

Chapter 2.17. Toward a Grid-Based Zero-Latency Data Warehousing Implementation for Continuous Data Streams Processing / *Tho Manh Nguyen, Peter Brezany, A. Min Tjoa, and Edgar Weippl* 755

Chapter 2.18. Data Warehouse Design to Support Customer Relationship Management Analyses / *Colleen Cunningham, Il-Yeol Song, and Peter P. Chen* 787

Chapter 2.19. An Information-Theoretic Framework for Process Structure and Data Mining / *Gianluigi Greco, Antonella Guzzo, and Luigi Pontieri* 810

Chapter 2.20. Domain-Driven Data Mining: A Practical Methodology / *Longbing Cao and Chengqi Zhang* 931

Chapter 2.21. Metric Methods in Data Mining / *Dan A. Simovici* 849

Chapter 2.22. Mining Geo-Referenced Databases: A Way to Improve Decision-Making /
Maribel Yasmina Santos and Luís Alfredo Amaral 880

Chapter 2.23. Ontology-Based Construction of Grid Data Mining Workflows /
Peter Brezany, Ivan Janciak, and A. Min Tjoa 913

Chapter 2.24. Exploratory Time Series Data Mining by Genetic Clustering /
T. Warren Liao 942

Chapter 2.25. Two Rough Set Approaches to Mining Hop Extraction Data /
Jerzy W. Grzymala-Busse, Zdzislaw S. Hippe, Teresa Mroczek, Edward Roj, and Boleslaw Skowronski 963

Chapter 2.26. Semantics-Aware Advanced OLAP Visualization of Multidimensional
Data Cubes / *Alfredo Cuzzocrea, Domenico Saccà, and Paolo Serafino* 974

Chapter 2.27. A Presentation Model and Non-Traditional Visualization for OLAP /
Andreas Maniatis, Panos Vassiliadis, Spiros Skiadopoulos, Yannis Vassiliou, George Mavrogonatos, and Ilias Michalarias 1004

Chapter 2.28. An Ontology-Based Data Mediation Framework for Semantic
Environments / *Adrian Mocan and Emilia Cimpian* 1037

Chapter 2.29. Engineering Conceptual Data Models from Domain Ontologies:
A Critical Evaluation / *Haya El-Ghalayini, Mohammed Odeh, and Richard McClatchey* 1068

Chapter 2.30. Data Mining of Bayesian Network Structure Using a Semantic Genetic
Algorithm-Based Approach / *Sachin Shetty, Min Song, and Mansoor Alam* 1081

Chapter 2.31. A Bayesian Framework for Improving Clustering Accuracy of
Protein Sequences Based on Association Rules / *Peng-Yeng Yin, Shyong-Jian Shyu, Guan-Shieng Huang, and Shuang-Te Liao* 1091

Chapter 2.32. Improving Classification Accuracy of Decision Trees for Different
Abstraction Levels of Data / *Mina Jeong and Doheon Lee* 1103

Chapter 2.33. Improving Similarity Search in Time Series Using Wavelets /
Ioannis Liabotis, Babis Theodoulidis, and Mohamad Saraee 1116

Chapter 2.34. Cluster-Based Input Selection for Transparant Fuzzy Modeling /
Can Yang, Jun Meng, and Shanan Zhu 1138

Chapter 2.35. Combinatorial Fusion Analysis: Methods and Practices of Combining
Multiple Scoring Systems / *D. Frank Hsu, Yun-Sheng Chung, and Bruce S. Kristal* 1157

Chapter 2.36. Databases Modeling of Engineering Information / *Z. M. Ma* 1182

Volume III

Chapter 2.37. Novel Efficient Classifiers Based on Data Cube / *Lixin Fu* 1205

Chapter 2.38. Partially Supervised Classification: Based on Weighted Unlabeled Samples Support Vector Machine / *Zhigang Liu, Wenzhong Shi, Deren Li, and Qianqing Qin* 1216

Chapter 2.39. Periodic Streaming Data Reduction Using Flexible Adjustment of Time Section Size / *Jaehoon Kim and Seog Park* 1231

Chapter 2.40. Hybrid Query and Data Ordering for Fast and Progressive Range-Aggregate Query Answering / *Cyrus Shahabi, Mehrdad Jahangiri, and Dimitris Sacharidis* 1250

Chapter 2.41. Linguistic Rule Extraction from Support Vector Machine Classifiers / *Xiuju Fu, Lipo Wang, GihGuang Hung, and Liping Goh* 1269

Chapter 2.42. Preference-Based Frequent Pattern Mining / *Moonjung Cho, Jian Pei, Haixun Wang, and Wei Wang* 1280

Section 3. Tools and Technologies

This section presents extensive coverage of the interaction between data mining and warehousing and various tools and technologies that researchers, practitioners, and students alike can implement in their daily lives. These chapters educate readers about fundamental tools such as the Internet and mobile technology, while also providing insight into new and upcoming technologies, theories, and instruments that will soon be commonplace. Within these rigorously researched chapters, readers are presented with countless examples of the tools and technologies essential to the field of data mining and warehousing. In addition, the successful implementation and resulting impact of these various tools and technologies are discussed within this collection of chapters.

Chapter 3.1. Algorithms for Data Mining / *Tadao Takaoka, Nigel K. Ll. Pope, and Kevin E. Voges* 1301

Chapter 3.2. Super Computer Heterogeneous Classifier Meta-Ensembles / *Anthony Bagnall, Gavin Cawley, Ian Whittley, Larry Bull, Matthew Studley, Mike Pettipher, and Firat Tekiner* 1320

Chapter 3.3. Navigation Rules for Exploring Large Multidimensional Data Cubes / *Navin Kumar, Aryya Gangopadhyay, George Karabatis, Sanjay Bapna, and Zhiyuan Chen* 1334

Chapter 3.4. The Use of Smart Tokens in Cleaning Integrated Warehouse Data / *Christie I. Ezeife and Timothy E. Ohanekwu* 1355

Chapter 3.5. An Implemented Representation and Reasoning System for Creating and Exploiting Large Knowledge Bases of "Narrative" Information / *Gian Piero Zarri* 1376

Chapter 3.6. Spatio-Temporal Prediction Using Data Mining Tools / *Margaret H. Dunham, Nathaniel Ayewah, Zhigang Li, Kathryn Bean, and Jie Huang* 1400

Chapter 3.7. Data Mining Using Qualitative Information on the Web / *Taeho Hong and Woojong Suh* 1416

Chapter 3.8. Computational Intelligence Techniques Driven Intelligent Agents for Web Data Mining and Information Retrieval / *Masoud Mohammadian and Ric Jentzsch* 1435

Chapter 3.9. Internet Data Mining Using Statistical Techniques / *Kuldeep Kumar* 1446

Chapter 3.10. Mining E-Mail Data / *Steffen Bickel and Tobias Scheffer* 1454

Chapter 3.11. Exploiting Captions for Web Data Mining / *Neil C. Rowe* 1461

Chapter 3.12. Agent-Mediated Knowledge Acquisition for User Profiling / *A. Andreevskaia, R. Abi-Aad, and T. Radhakrishnan* 1486

Chapter 3.13. Mobile User Data Mining and its Applications / *John Goh and David Taniar* 1502

Chapter 3.14. Mobile Phone Customer Type Discrimination via Stochastic Gradient Boosting / *Dan Steinberg, Mikhaylo Golovnya, and Nicholas Scott Cardell* 1519

Chapter 3.15. Intelligent Cache Management for Mobile Data Warehouse Systems / *Shi-Ming Huang, Binshan Lin, and Qun-Shi Deng* 1539

Chapter 3.16. VRMiner: A Tool for Multimedia Database Mining with Virtual Reality / *H. Azzag, F. Picarougne, C. Guinot, and G. Venturini* 1557

Chapter 3.17. Spatial Navigation Assistance System for Large Virtual Environments: The Data Mining Approach / *Mehmed Kantardzic, Pedram Sadeghian, and Walaa M. Sheta* 1573

Chapter 3.18. Bitmap Indices for Data Warehouses / *Kurt Stockinger and Kesheng Wu* 1590

Chapter 3.19. Indexing in Data Warehouses: Bitmaps and Beyond / *Karen C. Davis and Ashima Gupta* 1606

Chapter 3.20. Visualization Techniques for Data Mining / *Herna L. Viktor and Eric Paquet* 1623

Chapter 3.21. Video Data Mining / *JungHwan Oh, JeongKyu Lee, and Sae Hwang* 1631

Chapter 3.22. Interactive Visual Data Mining / *Shouhong Wang and Hai Wang* 1638

Chapter 3.23. Data Mining in Gene Expression Analysis: A Survey / *Jilin Han, Le Gruenwald, and Tyrrell Conway* 1643

Chapter 3.24. A Haplotype Analysis System for Genes Discovery of Common Diseases / *Takashi Kido* 1674

Section 4. Utilization and Application

This section introduces and discusses a variety of the existing applications of data mining and warehousing that have influenced government, culture, and biology and also proposes new ways in which data mining and warehousing can be implemented in society. Within these selections, particular issues, such as the use of data mining and warehousing in human resources and the incorporation of data analysis techniques into homeland security strategies, are explored and debated. Contributions included in this section provide excellent coverage of today's IT community and insight into how data mining and warehousing impacts the social fabric of our present-day global village.

Chapter 4.1. Strategic Utilization of Data Mining / *Chandra S. Amaravadi* 1689

Chapter 4.2. Biological Data Mining / *George Tzanis, Christos Berberidis, and Ioannis Vlahavas* 1696

Chapter 4.3. Biomedical Data Mining Using RBF Neural Networks / *Feng Chu and Lipo Wang* 1706

Chapter 4.4. Bioinformatics Data Management and Data Mining / *Boris Galitsky* 1714

Chapter 4.5. Deterministic Motif Mining in Protein Databases / *Pedro Gabriel Ferreira and Paulo Jorge Azevedo* 1722

Chapter 4.6. Differential Association Rules: Understanding Annotations in Protein Interaction Networks / *Christopher Besemann, Anne Denton, Ajay Yekkirala, Ron Hutchison, and Marc Anderson* 1747

Chapter 4.7. Data Mining and Knowledge Discovery in Metabolomics / *Christian Baumgartner and Armin Graber* 1759

Chapter 4.8. Comparative Genome Annotation Systems / *Kwangmin Choi and Sun Kim* 1784

Chapter 4.9. The Application of Data Mining Techniques in Health Plan Population Management: A Disease Management Approach / *Theodore L. Perry, Travis Tucker, Laurel R. Hudson, William Gandy, Amy L. Neftzger, and Guy B. Hamar* 1799

Chapter 4.10. Data Mining Medical Digital Libraries / *Colleen Cunningham and Xiaohua Hu* 1810

Volume IV

Chapter 4.11. Data Mining in Diabetes Diagnosis and Detection / *Indranil Bose* 1817

Chapter 4.12. Data Warehousing and Analytics in Banking: Concepts / *L. Venkat Narayanan* 1825

Chapter 4.13. Data Warehousing and Analytics in Banking: Implementation / *L. Venkat Narayanan* 1840

Chapter 4.14. Beyond Classification: Challenges of Data Mining for Credit Scoring / *Anna Olecka* 1855

Chapter 4.15. A TOPSIS Data Mining Demonstration and Application to Credit Scoring / *Desheng Wu and David L. Olson* 1877

Chapter 4.16. The Utilization of Business Intelligence and Data Mining in the Insurance Marketplace / *Jeff Hoffman* 1888

Chapter 4.17. Ontology-Based Data Warehousing and Mining Approaches in Petroleum Industries / *Shastri L. Nimmagadda and Heinz Dreher* 1901

Chapter 4.18. A Study on Web Searching: Overlap and Distance of the Search Engine Results / *Shanfeng Zhu, Xiaotie Deng, Qizhi Fang, and Weimin Zheng* 1926

Chapter 4.19. Data Mining in Web Services Discovery and Monitoring / *Richi Nayak* 1938

Chapter 4.20. A Data Mining Driven Approach for Web Classification and Filtering Based on Multimodal Content Analysis / *Mohamed Hammami, Youssef Chahir, and Liming Chen* 1958

Chapter 4.21. Acquiring Semantic Sibling Associations from Web Documents / *Marko Brunzel and Myra Spiliopoulou* 1987

Chapter 4.22. Traversal Pattern Mining in Web Usage Data / *Yongqiao Xiao and Jenq-Foung (J.F.) Yao* 2004

Chapter 4.23. Facilitating and Improving the Use of Web Services with Data Mining / *Richi Nayak* 2022

Chapter 4.24. E-Mail Worm Detection Using Data Mining / *Mohammad M. Masud, Latifur Khan, and Bhavani Thuraisingham* 2036

Chapter 4.25. User-Centered Interactive Data Mining / *Yan Zhao, Yaohua Chen, and Yiyu Yao* 2051

Chapter 4.26. Advanced Data Mining and Visualization Techniques with Probabilistic Principal Surfaces: Applications to Astronomy and Genetics / *Antonino Staiano, Lara De Vinco, Giuseppe Longo, and Roberto Tagliaferri* 2067

Chapter 4.27. Using Data Mining for Forecasting Data Management Needs / *Qingyu Zhang and Richard S. Segall* 2088

Chapter 4.28. Visual Data Mining for Discovering Association Rules / *Kesaraporn Techapichetvanich and Amitava Datta* 2105

Chapter 4.29. Generalization Data Mining in Fuzzy Object-Oriented Databases / *Rafal Angryk, Roy Ladner, and Frederick E. Petry* 2121

Chapter 4.30. Fuzzy Miner: Extracting Fuzzy Rules from Numerical Patterns / *Nikos Pelekis, Babis Theodoulidis, Ioannis Kopanakis, and Yannis Theodoridis* 2141

Chapter 4.31. Empowering the OLAP Technology to Support Complex Dimension Hierarchies / *Svetlana Mansmann and Marc H. Scholl* 2164

Chapter 4.32. Understanding Decision-Making in Data Warehousing and Related Decision Support Systems: An Explanatory Study of a Customer Relationship Management Application / *John D. Wells and Traci J. Hess* 2185

Chapter 4.33. Statistical Sampling to Instantiate Materialized View Selection Problems in Data Warehouses / *Mesbah U. Ahmed, Vikas Agrawal, Udayan Nandkeolyar, and P. S. Sundararaghavan* 2201

Chapter 4.34. Development of Control Signatures with a Hybrid Data Mining and Genetic Algorithm / *Alex Burns, Shital Shah, and Andrew Kusiak* 2226

Chapter 4.35. Feature Selection for the Promoter Recognition and Prediction Problem / *George Potamias and Alexandros Kanterakis* 2248

Chapter 4.36. Data Warehousing Search Engine / *Hadrian Peter and Charles Greenidge* 2263

Section 5. Organizational and Social Implications

This section includes a wide range of research pertaining to the social and organizational impact of data mining and warehousing around the world. Chapters introducing this section illustrate varying perspectives on organizational data mining, as well as its relationship to cognition. Other contributions discuss the potential of data mining and warehousing for transforming business, government and medicine, as well as providing insight into individual behavior. Particular selections explain the design of a data model for social applications, provide insight into the implications of data mining and warehousing in the banking sector, and explain data mining's use in generating credit scores. The inquiries and methods presented in this section offer insight into the integration of data mining and warehousing in social and organizational settings while also emphasizing the potential for future societal applications.

Chapter 5.1. Data Mining in Practice / *Sherry Y. Chen and Xiaohui Liu* 2273

Chapter 5.2. Model Indentification through Data Mining / *Diego Liberati*................................... 2281

Chapter 5.3. Organizational Data Mining (ODM): An Introduction /
Hamid R. Nemati and Christopher D. Barko.. 2289

Chapter 5.4. Constructionist Perspective of Organizational Data Mining /
Isabel Ramos and João Álvaro Carvalho 2296

Chapter 5.5. The Role of Data Mining in Organizational Cognition /
Chandra S. Amaravadi and Farhad Daneshgar.. 2302

Chapter 5.6. Ontology-Based Interpretation and Validation of Mined Knowledge:
Normative and Cognitive Factors in Data Mining / *Ana Isabel Canhoto*..................................... 2316

Chapter 5.7. Design of a Data Model for Social Network Applications /
Susanta Mitra, Aditya Bagchi, and A.K.Bandyopadhyay 2338

Chapter 5.8. Humanitites Data Warehousing / *Janet Delve*... 2364

Chapter 5.9. Data Mining in Human Resources / *Marvin D. Troutt and*
Lori K. Long... 2371

Chapter 5.10. Privacy Preserving Data Mining, Concepts, Techniques, and
Evaluation Methodologies / *Igor Nai Fovino* 2379

Chapter 5.11. Privacy-Preserving Data Mining and the Need for Confluence of
Research and Practice / *Lixin Fu, Hamid Nemati, and Fereidoon Sadri*....................................... 2402

Chapter 5.12. Data Mining in the Federal Government / *Les Pang* 2421

Volume V

Chapter 5.13. Data Warehousing and the Organization of Governmental Databases / *Franklin Maxwell Harper* 2427

Chapter 5.14. Data Mining and the Banking Sector: Managing Risk in Lending and Credit Card Activities / *Àkos Felsõvályi and Jennifer Courant* 2438

Chapter 5.15. Data Mining for Credit Scoring / *Indranil Bose, Cheng Pui Kan, Chi King Tsz, Lau Wai Ki, and Wong Cho Hung* 2449

Chapter 5.16. Credit Card Users' Data Mining / *André de Carvalho, Antonio P. Braga, and Teresa Ludermir* 2464

Chapter 5.17. Data Mining for Supply Chain Management in Complex Networks / *Mahesh S. Raisinghani and Manoj K. Singh* 2468

Chapter 5.18. Neural Network-Based Stock Market Return Forecasting Using Data Mining for Variable Reduction / *David Enke* 2476

Chapter 5.19. Data Mining and Knowledge Discovery in Healthcare Organizations: A Decision-Tree Approach / *Murat Caner Testik, George C. Runger, Bradford Kirkman-Liff, and Edward A. Smith* 2494

Chapter 5.20. Data Mining Techniques and Medical Decision Making for Urological Dysfunction / *N. Sriraam, V. Natasha, and H. Kaur* 2506

Chapter 5.21. Heuristics in Medical Data Mining / *Susan E. George* 2517

Chapter 5.22. An Approach to Mining Crime Patterns / *Sikha Bagui* 2523

Chapter 5.23. Web Usage Mining Data Preparation / *Bamshad Mobasher* 2551

Chapter 5.24. Classification Of 3G Mobile Phone Customers / *Ankur Jain, Lalit Wangikar, Martin Ahrens, Ranjan Rao, Suddha Sattwa Kundu, and Sutirtha Ghosh* 2558

Chapter 5.25. Impediments to Exploratory Data Mining Success / *Jeff Zeanah* 2566

Section 6. Managerial Impact

This section presents contemporary coverage of the more formal implications of data mining and warehousing, more specifically related to the corporate and managerial utilization of information-sharing technologies and applications, and how these technologies can be facilitated within organizations. Core ideas such as successful data mining in franchise organizations and the use of data analysis to predict customer behavior are discussed throughout these chapters. Contributions within this section seek to answer the fundamental question of data mining and warehousing implementation in organizations: How can particular techniques best be integrated into businesses and what are the potential obstacles to such integration? Particular chapters provide case studies of data mining and warehousing use in business and address some of the most significant issues that have arisen from data mining and warehousing implementation.

Chapter 6.1. Data Mining and Business Intelligence: Tools, Technologies, and Applications / *Jeffrey Hsu* 2584

Chapter 6.2. Data Mining and Decision Support for Business and Science / *Auroop R. Ganguly, Amar Gupta, and Shiraj Khan* 2618

Chapter 6.3. Data Warehousing Interoperability for the Extended Enterprise / *Aristides Triantafillakis, Panagiotis Kanellis, and Drakoulis Martakos* 2626

Chapter 6.4. Data Warehousing and Mining in Supply Chains / *Richard Mathieu and Reuven R. Levary* 2637

Chapter 6.5. Management of Data Streams for Large-Scale Data Mining / *Jon R. Wright, Gregg T. Vesonder, and Tamraparni Dasu* 2644

Chapter 6.6. Customized Recommendation Mechanism Based on Web Data Mining and Case-Based Reasoning / *Jin Sung Kim* 2659

Chapter 6.7. Gaining Strategic Advantage through Bibliomining: Data Mining for Management Decisions in Corporate, Special, Digital, and Traditional Libraries / *Scott Nicholson and Jeffrey Stanton* 2673

Chapter 6.8. Expanding Data Mining Power with System Dynamics / *Edilberto Casado* 2688

Chapter 6.9. Data Mining and Mobile Business Data / *Richi Nayak* 2697

Chapter 6.10. Neural Data Mining System for Trust-Based Evaluation in Smart Organizations / *T. T. Wong* 2704

Chapter 6.11. Data Mining in Franchise Organizations / *Ye-Sho Chen, Robert Justis, and P. Pete Chong* 2722

Chapter 6.12. Translating Advances in Data Mining in Business Operations: The Art of Data Mining in Retailing / *Henry Dillon and Beverley Hope* 2734

Chapter 6.13. Data Warehousing: The 3M Experience / *Hugh J. Watson, Barbara H. Wixom, and Dale L. Goodhue* 2749

Chapter 6.14. Business Data Warehouse: The Case of Wal-Mart / *Indranil Bose, Lam Albert Kar Chun, Leung Vivien Wai Yue, Li Hoi Wan Ines, and Wong Oi Ling Helen* 2762

Chapter 6.15. A Porter Framework for Understanding the Strategic Potential of Data Mining for the Australian Banking Industry / *Kate A. Smith and Mark S. Dale* 2772

Chapter 6.16. Data Mining for Combining Forecasts in Inventory Management / *Chi Kin Chan* 2792

Chapter 6.17. Analytical Customer Requirement Analysis Based on Data Mining / *Jianxin (Roger) Jiao, Yiyang Zhang, and Martin Helander* 2798

Chapter 6.18. Predicting Future Customers via Ensembling Gradually Expanded Trees / *Yang Yu, Chuan Zhan, Xu-Ying Liu, Ming Li, and Zhi-Hua Zhou* 2816

Chapter 6.19. Marketing Data Mining / *Victor S.Y. Lo* 2824

Section 7. Critical Issues

This section addresses conceptual and theoretical issues related to the field of data mining and warehousing, which include the ethical implications of data collection and the numerous approaches adopted by researchers that aid in making data mining and warehousing more effective. Within these chapters, the reader is presented with an in-depth analysis of the most current and relevant conceptual inquires within this growing field of study. Particular chapters address data partitioning, data warehouse refreshment, and mining with incomplete data sets. Overall, contributions within this section ask unique, often theoretical questions related to the study of data mining and warehousing and, more often than not, conclude that solutions are both numerous and contradictory.

Chapter 7.1. Ethics Of Data Mining / *Jack Cook* 2834

Chapter 7.2. Ethical Dilemmas in Data Mining and Warehousing / *Joseph A. Cazier and Ryan C. LaBrie* 2841

Chapter 7.3. Privacy and Confidentiality Issues in Data Mining / *Yücel Saygin* 2850

Chapter 7.4. Privacy Implications of Organizational Data Mining / *Hamid R. Nemati, Charmion Brathwaite, and Kara Harrington* 2856

Chapter 7.5. Privacy in Data Mining Textbooks / *James Lawler and John C. Molluzzo* 2872

Chapter 7.6. Data Mining for Intrusion Detection / *Aleksandar Lazarevic* 2880

Chapter 7.7. E-Commerce and Data Mining: Integration Issues and Challenges / *Parviz Partow-Navid and Ludwig Slusky* 2888

Chapter 7.8. A Data Mining Approach to Formulating a Successful Purchasing Negotiation Strategy / *Hokey Min and Ahmed Emam* 2900

Chapter 7.9. Data Mining Medical Information: Should Artificial Neural Networks Be Used to Analyse Trauma Audit Data? / *Thomas Chesney, Kay Penny, Peter Oakley, Simon Davies, David Chesney, Nicola Maffulli, and John Templeton* 2915

Chapter 7.10. A Data Mining Approach to Diagnosing Student Learning Problems in Sciences Courses / *Gwo-Jen Hwang* 2928

Chapter 7.11. Effective Intelligent Data Mining Using Dempster-Shafer Theory / *Malcolm J. Beynon* 2943

Chapter 7.12. An Intelligent Support System Integrating Data Mining and Online Analytical Processing / *Rahul Singh, Richard T. Redmond, and Victoria Yoon* 2964

Chapter 7.13. A Successive Decision Tree Approach to Mining Remotely Sensed Image Data / *Jianting Zhang, Wieguo Liu, and Le Gruenwald* 2978

Chapter 7.14. Mining for Mutually Exclusive Items in Transaction Databases / *George Tzanis and Christos Berberidis* 2993

Chapter 7.15. Re-Sampling Based Data Mining Using Rough Set Theory / *Benjamin Griffiths and Malcolm J. Beynon* 3005

Chapter 7.16. Data Mining with Incomplete Data / *Hai Wang and Shouhong Wang* 3027

Chapter 7.17. Routing Attribute Data Mining Based on Rough Set Theory / *Yanbing Liu, Shixin Sun, Menghao Wang, and Hong Tang* 3033

Volume VI

Chapter 7.18. Data Warehouse Refreshment / *Alkis Simitisis, Panos Vassiliadis, Spiros Skiadopoulos, and Timos Sellis* 3049

Chapter 7.19. An Algebraic Approach to Data Quality Metrics for Entity Resolution Over Large Datasets / *John Talburt, Richard Wang, Kimberly Hess, and Emily Kuo* 3067

Chapter 7.20. A Hybrid Approach for Data Warehouse View Selection / *Biren Shah, Karthik Ramachandran, and Vijay Raghavan* 3085

Chapter 7.21. A Space-Efficient Protocol for Consistency of External View Maintenance on Data Warehouse Systems: A Proxy Approach / *Shi-Ming Huang, David C. Yen, and Hsiang-Yuan Hsueh* 3116

Chapter 7.22. DWFIST: The Data Warehouse of Frequent Itemsets Tactics Approach / *Rodrigo Salvador Monteiro, Geraldo Zimbrão, Holger Schwarz, Bernhard Mitschang, and Jano Moreira de Souza* 3142

Chapter 7.23. A Hyper-Heuristic for Descriptive Rule Induction / *Tho Hoan Pham and Tu Bao Ho* 3164

Chapter 7.24. Improved Data Partitioning for Building Large ROLAP Data Cubes in Parallel / *Ying Chen, Frank Dehne, Todd Eavis, and A. Rau-Chaplin* 3176

Chapter 7.25. An Ontology of Data Modelling Languages: A Study Using a Common-Sense Realistic Ontology / *Simon K. Milton and Ed Kazmierczak* 3194

Chapter 7.26. Robust Classification Based on Correlations Between Attributes / *Alexandros Nanopoulos, Apostolos N. Papadopoulos, Yannis Manolopoulos, and Tatjana Welzer-Druzovec* 3212

Chapter 7.27. Finding Non-Coincidental Sporadic Rules Using Apriori-Inverse / *Yun Sing Koh, Nathan Rountree, and Richard O'Keefe* 3222

Chapter 7.28. Discovering Surprising Instances of Simpson's Paradox in Hierarchical Multidimensional Data / *Carem C. Fabris and Alex A. Freitas* 3235

Chapter 7.29. Discovering Frequent Embedded Subtree Patterns from Large Databases of Unordered Labeled Trees / *Yongqiao Xiao, Jenq-Foung Yao, and Guizhen Yang* 3252

Chapter 7.30. A Single Pass Algorithm for Discovering Significant Intervals in Time-Series Data / *Sagar Savla and Sharma Chakravarthy* 3272

Chapter 7.31. SeqPAM: A Sequence Clustering Algorithm for Web Personalization / *Pradeep Kumar, Raju S. Bapi, and P. Radha Krishna* 3285

Chapter 7.32. Kernal Width Selection for SVM Classification: A Meta-Learning Approach / *Shawkat Ali and Kate A. Smith* 3308

Chapter 7.33. A Parallel Implementation Scheme of Relational Tables Based on Multidimensional Extendible Array / *K. M. Azharul Hasan, Tatsuo Tsuji, and Ken Higuchi* 3324

Section 8. Emerging Trends

This section highlights research potential within the field of data mining and warehousing while also exploring uncharted areas of study for the advancement of the discipline. Introducing this section are selections providing . Discussions exploring semantic data mining, Web data warehousing and spatio-temporal databases provide insight into forthcoming issues in data mining and warehousing study. These contributions, which conclude this exhaustive, multi-volume set, provide emerging trends and suggestions for future research within this rapidly expanding discipline.

Chapter 8.1. Toward Integrating Data Warehousing with Data Mining Techniques / *Rokia Missaoui, Ganaël Jatteau, Ameur Boujenoui, and Sami Naboulali* 3346

Chapter 8.2. Combining Data Warehousing and Data Mining Techniques for Web Log Analysis / *Torben Bach Pedersen, Jesper Thorhauge, and Søren E. Jespersen* 3364

Chapter 8.3. Web Data Warehousing Convergence: From Schematic to Systematic / *D. Xuan Le, J. Wenny Rahayu, and David Taniar* 3386

Chapter 8.4. Web Technology and Data Warehouse Synergies / *John M. Artz* 3411

Chapter 8.5. Metadata Management: A Requirement for Web Warehousing and KnowledgeManagement / *Gilbert W. Laware* 3416

Chapter 8.6. An Immune Systems Approach for Classifying Mobile Phone Usage / *Hanny Yulius Limanto, Tay Joc Cing, and Andrew Watkins* 3440

Chapter 8.7. User Interface Formalization in Visual Data Mining / *Tiziana Catarci, Stephen Kimani, and Stefano Lodi* 3451

Chapter 8.8. Mining in Spatio-Temporal Databases / *Junmei Wang, Wynne Hsu, and Mong Li Lee* 3477

Chapter 8.9. Algebraic Reconstruction Technique in Image Reconstruction Based on Data Mining / *Zhong Qu* 3493

Chapter 8.10. Evolutionary Induction of Mixed Decision Trees / *Marek Kretowski and Marek Grzes* 3509

Chapter 8.11. Semantic Data Mining / *Protima Banerjee, Xiaohua Hu, and Illhoi Yoo* 3524

Chapter 8.12. Metadata- and Ontology-Based Semantic Web Mining / *Marie Aude Aufaure, Bénédicte Le Grand, Michel Soto, and Nacera Bennacer* 3531

Chapter 8.13. Integrating Semantic Knowledge with Web Usage Mining for Personalization / *Honghua Dai and Bamshad Mobasher* 3557

Chapter 8.14. Mining in Music Databases / *Ioannis Karydis, Alexandros Nanopoulos, and Yannis Manolopoulos* 3586

Chapter 8.15. Multimedia Data Mining Concept / *Janusz Swierzowicz* 3611

Chapter 8.16. Robust Face Recognition for Data Mining / *Brian C. Lovell and Shaokang Chen* 3621

Chapter 8.17. Data Mining and Homeland Security / *Jeffrey W. Seifert* 3630

Chapter 8.18. Homeland Security Data Mining and Link Analysis / *Bhavani Thuraisingham* 3639

Chapter 8.19. Seismological Data Warehousing and Mining: A Survey / *Gerasimos Marketos, Yannis Theodoridis, Ioannis S. Kalogeras* 3645

Chapter 8.20. Realizing Knowledge Assets in the Medical Sciences with Data Mining: An Overview / *Adam Fadlalla and Nilmini Wickramasinghe* 3662

Chapter 8.21. Mining Clinical Trial Data / *Jose Ma. J. Alvir, Javier Cabrera, Frank Caridi, and Ha Nguyen* 3675

Chapter 8.22. Vertical Database Design for Scalable Data Mining / *William Perrizo, Qiang Ding, Masum Serazi, Taufik Abidin, and Baoying Wang* 3694

Preface

In today's data-driven environment, the need for the organization and analysis of a constant flow of information presents a challenge for professionals and researchers. The development, advancement, and implementation of data mining and warehousing techniques have profoundly impacted the ability of researchers and decision makers to identify much more relevant data related to a particular subject, issue, or phenomenon, and, as a result, heighten their understanding and comprehension. With applications in medical research, banking, security, and beyond, data mining and warehousing is both an invisible and powerful force in modern society.

As research projects on data mining and warehousing have grown in both number and popularity, researchers and educators have devised a variety of techniques and methodologies to develop, deliver, and, at the same time, evaluate the effectiveness of their use. The explosion of methodologies in the field has created an abundance of new, state-of-the-art literature related to all aspects of this expanding discipline. This body of work allows researchers to learn about the fundamental theories, latest discoveries, and forthcoming trends in the field of data mining and warehousing.

Constant technological and theoretical innovation challenges researchers in data mining and warehousing to stay abreast of and continue to develop and deliver methodologies and techniques utilizing the latest advancements. In order to provide the most comprehensive, in-depth, and current coverage of all related topics and their applications, as well as to offer a single reference source on all conceptual, methodological, technical, and managerial issues, Information Science Reference is pleased to offer a six-volume reference collection on this rapidly growing discipline. This collection aims to empower researchers, students, and practitioners by facilitating their comprehensive understanding of the most critical areas within this field of study.

This collection, entitled *Data Warehousing and Mining: Concepts, Methodologies, Tools, and Applications* is organized into eight distinct sections, which are as follows: (1) Fundamental Concepts and Theories; (2) Development and Design Methodologies; (3) Tools and Technologies; (4) Utilization and Application; (5) Organizational and Social Implications; (6) Managerial Impact; (7) Critical Issues; and (8) Emerging Trends. The following paragraphs provide a summary of what is covered in each section of this multi-volume reference collection.

Section I, **Fundamental Concepts and Theories**, serves as a foundation for this exhaustive reference tool by addressing crucial theories essential to understanding data mining and warehousing. Opening this elemental section is "Administering and Managing a Data Warehouse" by James E. Yao, Chang Liu, Qiyang Chen, and June Li, which provides a framework for data warehouse design and management while also explaining the pivotal role of data warehouses in shaping business decision making. Similarly, "Conceptual Modeling Solutions for the Data Warehouse" by Stefano Rizzi presents a practical methodology for designing a successful data warehouse. Additional contributions, such as "Knowledge Structure and Data Mining Techniques" by Rick L. Wilson, Peter A. Rosen, and Mohammad Saad Al-Ahmadi, offer a broad overview of procedures and methods essential to data mining. The expertly researched contributions within this section also present an essential introduction to the innumerable applications of data mining and warehousing. "Introduction to Data Mining in Bioinformatics" by Hui-Huang Hsu illustrates how researchers can discover new biological knowledge using data mining techniques, and "Introduction to Data Mining and its Applications to Manufacturing" by Jose D. Montero provides insight into the promising integration of business applications in the field of manufacturing. This foundational section enables readers to learn from expert research on the elemental theories underscoring data mining and warehousing.

Section II, **Development and Design Methodologies**, contains in-depth coverage of conceptual architectures and frameworks, providing the reader with a comprehensive understanding of emerging theoretical and conceptual developments within the field of data mining and warehousing. "A Framework for Organizational Data Analysis and Organizational Data Mining" by Bernd Knobloch offers a methodology for more efficient and beneficial data mining within business organizations. Security, specifically as it relates to the design of data warehouses, is a major concern of researchers and professionals. Within their contribution "Designing Secure Data Warehouses," Rodolfo Villarroel, Eduardo Fernández-Medina, Juan Trujillo, and Mario Piattini compare six different designs for secure data warehouses. From basic designs to abstract development, chapters such as "Metric Methods in Data Mining" by Dan A. Simovici and "An Ontology-Based Data Mediation Framework for Semantic Environments" by Adrian Mocan and Emilia Cimpian serve to expand the reaches of development and design methodologies within the field of data mining and warehousing.

Section III, **Tools and Technologies**, presents extensive coverage of the interaction between various technologies and the field of data mining and warehousing. This symbiotic relationship encourages the advancement and invention of new technologies as well as continual innovation witnessed within the development of new data mining and data warehousing techniques. The rapid expansion of data mining to encompass the extraction of images, audio, and video is evaluated in "Video Data Mining" by JungHwan Oh, JeongKyu Lee, and Sae Hwang. "VRMiner: A Tool for Multimedia Database Mining With Virtual Reality" by Hanene Azzag, Fabien Picarougne, Christiane Guinot, and Gilles Venturini provides a more in-depth analysis of multimedia data mining, explaining how a 3-D method for visualizing multimedia data can be applied to real-world situations. Data mining's role in making technology more efficient is explored in selections such as "Mining E-Mail Data" by Steffen Bickel and Tobias Scheffer and "Mobile User Data Mining and Its Applications" by John Goh and David Taniar. Through these rigorously researched contributions, the reader is provided with countless examples of the up-and-coming tools and technologies that emerge from or can be applied to the multi-dimensional field of data mining and warehousing.

Section IV, **Utilization and Application**, explores the ways in which data mining and warehousing has been adopted and implemented in all facets of society. This collection of innovative research begins with "Strategic Utilization of Data Mining" by Chandra S. Amaravadi, which documents the emergence of data mining as an essential tool in decision support. Specific applications of data warehousing and mining, which are described in selections such as "Biological Data Mining" by George Tzanis, Christos Berberidis, and Ioannis Vlahavas and "The Utilization of Business Intelligence and Data Warehousing and Analytics in Banking: Concepts" by L. Venkat Narayanan relate the importance of data mining and warehousing in specific fields. The essential use and continual evolution of Web data mining is discussed in a set of informative selections, which include "Data Mining in Web Services Discovery and Monitoring" by Richi Nayak and "Traversal Pattern Mining in Web Usage Data" by Yongqiao Xiao and Jenq-Foung Yao. From established applications to forthcoming innovations, contributions in this section provide excellent coverage of today's global community and demonstrate how data mining and warehousing impacts the social, economic, and political fabric of our present-day global village.

Section V, **Organizational and Social Implications**, includes a wide range of research pertaining to the organizational and cultural implications of data mining and warehousing. Introducing this section is "Data Mining in Practice" by Sherry Y. Chen and Xiaohui Liu, a selection that describes how different data mining approaches, such as classification and visualization, make data analysis within organizations more intelligent and automatic. Governmental data mining is explored in "Data Mining in the Federal Government" by Les Pang, which analyzes different federal data mining projects and the knowledge they generated for future organizational implementation. Additional chapters included in this section, such as "Data Mining and Knowledge Discovery in Healthcare Organizations: A Decision-Tree Approach" by Murat Caner Testik, George C. Runger, Bradford Kirkman-Liff, and Edward A. Smith, discuss the use of data mining techniques to both improve and predict an individual's healthcare system use. Organizational data mining and warehousing, however, is not always employed without difficulty, as is illustrated in the concluding chapter of this section, "Impediments to Exploratory Data Mining Success" by Jeff Zeanah. Within this chapter, the author assesses barriers to data mining implementation and provides organizations with guidelines for anticipating and preventing problems. Overall, the discussions presented in this section offer insight into the integration of data mining and warehousing techniques in society and how these techniques can be better structured and implemented in modern-day organizations.

Section VI, **Managerial Impact**, presents contemporary coverage of the more formal implications of data mining and warehousing, which are, more specifically, related to the corporate and managerial utilization of data mining and warehousing within organizations. Core ideas such as integration, evaluation, and potential strategies for increasing the effectiveness of modern organizations are discussed in this collection. "Data Mining and Business Intelligence: Tools, Technologies, and Applications" by Jeffrey Hsu emphasizes the importance of data mining for transforming raw data into useful knowledge that promotes business intelligence. Equally essential to this examination of managerial impact is evaluating the effectiveness of already-implemented data mining and warehousing programs, which is examined at length in chapters such as "The Business Data Warehouse: The Case of Wal-Mart" by Indranil Bose, Lam Albert Kar Chun, Leung Vivien Wai Yue, Li Hoi Wan Ines, and Wong Oi Ling Helen and "Data Warehousing: The 3M Experience" by Hugh J. Watson and Barbara H. Wixom. Considered together, these selections provide a framework for data warehouse design and construction within corporations and also analyze both the successes and pitfalls that such large-scale operations have encountered. This section concludes with a brief overview of data mining for both customer prediction and marketing within a business. Selections such as "Marketing Data Mining" by Victor S.Y. Lo explain how data mining can be used to both explain and predict customer behavior and, therefore, establish long-term customer relationships and increase profit within an organization.

Section VII, **Critical Issues**, presents readers with an in-depth analysis of the more theoretical and conceptual issues within this growing field of study by addressing topics such as the ethics of data mining, mining with incomplete data, and the various theories that have been applied to and derived from the study of data warehousing and mining. Specifically, certain myths and questions regarding how consumer data is extracted and what organizations do with this data are presented in selections such as "Ethical Dilemmas in Data Mining and Warehousing" by Joseph A. Cazier and Ryan C. LaBrie, "Privacy and Confidentiality Issues in Data Mining" by Yücel Saygin, and "Privacy in Data Mining Textbooks" by James Lawler and John C. Molluzzo. Later chapters in this section explore a different aspect of data mining and warehousing—the specific theories, approaches, and algorithms researchers use to both create and analyze data mining and data warehousing tools. "Data Mining with Incomplete Data" by Hai

Wang and Shouhong Wang conceptualizes the common problem of trying to extract information from an incomplete survey data set, while "A Single Pass Algorithm for Discovering Significant Intervals in Time-Series Data" by Sagar Savla and Sharma Chakravarthy introduces a particular algorithm for extracting significant information from time-series data. In all, the theoretical and abstract issues presented and analyzed within this collection form the backbone of revolutionary data mining and warehousing research and inquiry.

The concluding section of this authoritative reference tool, **Emerging Trends**, highlights research potential within the field of data mining and warehousing while exploring uncharted areas of study for the advancement of the discipline. New trends in data mining and warehousing research discussed in this section include pattern mining, which is explored within "Semantic Data Mining" by Protima Banerjee, Xiaohua Hu, and Illhoi Yoo, integrating data mining with data warehousing, which is studied in "Toward Integrating Data Warehousing with Data Mining Techniques" by Rokia Missaoui, Ganaël Jatteau, Ameur Boujenoui, and Sami Naouali, and mining in medical applications, which is discussed by researchers at Pfizer in the selection "Mining Clinical Trial Data." The connection between data analysis and homeland security is explored within "Homeland Security Data Mining and Link Analysis" by Bhavani Thuraisingham and "Data Mining and Homeland Security" by Jeffrey W. Seifert. The latter of these selections documents how "dataveillance," or the monitoring of data related to an individual's activities, can be both a beneficial and controversial addition to national security. The future of data mining and warehousing, with its infinite potential for growth and change, promises to be as influential as its storied history.

Although the contents of this multi-volume book are organized within the preceding eight sections which offer a progression of coverage of the important concepts, methodologies, technologies, applications, social issues, and emerging trends, the reader can also identify specific contents by utilizing the extensive indexing system listed at the end of each volume. Furthermore, to ensure that the scholar, researcher, and educator have access to the entire contents of this multi-volume set, as well as additional coverage that could not be included in the print version of this publication, the publisher will provide unlimited, multi-user electronic access to the online aggregated database of this collection for the life of the edition, free of charge when a library purchases a print copy. In addition to providing content not included within the print version, this aggregated database is also continually updated to ensure that the most current research is available to those interested in data mining and warehousing.

Data mining and warehousing as a discipline has witnessed fundamental changes during the past two decades, allowing researchers and decision makers around the globe to have access to data and information which, two decades ago, was inaccessible. In addition to this transformation, many traditional organizations and business enterprises have taken advantage of the technologies offered by the development of data mining and warehousing in order to expand and augment their existing ability to make the most use of their databases. This has allowed practitioners and researchers to serve their customers, employees, and stakeholders more effectively and efficiently in the modern information age. With continued technological innovations in information and communication technology and with on-going discovery and research into newer and more innovative techniques and applications, the data mining and warehousing discipline will continue to witness an explosion of information within this rapidly evolving field.

The diverse and comprehensive coverage of data mining and warehousing in this six-volume, authoritative publication will contribute to a better understanding of all topics, research, and discoveries in this developing, significant field of study. Furthermore, the contributions included in this multi-volume collection series will be instrumental in the expansion of the body of knowledge in this enormous field, resulting in a greater understanding of the fundamentals while also fueling the research initiatives in emerging fields. We at Information Science Reference, along with the editor of this collection, hope that this multi-volume collection will become instrumental in the expansion of the discipline and will promote the continued growth of data mining and warehousing.

Chapter 2.8
Physical Modeling of Data Warehouses Using UML Component and Deployment Diagrams:
Design and Implementation Issues

Sergio Luján-Mora
University of Alicante, Spain

Juan Trujillo
University of Alicante, Spain

ABSTRACT

Several approaches have been proposed to model different aspects of a Data Warehouse (DW) during recent years, such as the modeling of a DW at the conceptual and logical level, the design of the ETL (Extraction, Transformation, Loading) processes, the derivation of the DW models from the enterprise data models, and customization of a DW schema. At the end of the design, a DW has to be deployed in a database environment, requiring many decisions of a physical nature. However, few efforts have been dedicated to the modeling of the physical design of a DW from the early stages of a DW project. In this article, we argue that some physical decisions can be taken from gathering main user requirements. In this article, we present physical modeling techniques for DWs using the component diagrams and deployment diagrams of the Unified Modeling Language (UML). Our approach allows the designer to anticipate important physical design decisions that may reduce the overall development time of a DW, such as replicating dimension tables, vertical and horizontal partitioning of a fact table, and the use of particular servers for certain ETL processes. Moreover, our approach allows the designer to cover all main design phases of DWs from the conceptual modeling phase to the final implementation. To illustrate our techniques, we show a case study that is implemented on top of a commercial DW management server.

INTRODUCTION

Data warehouses (DW) provide organizations with historical information to support a decision. It is widely accepted that these systems are based on multidimensional (MD) modeling. Thus, research on the design of a DW has been addressed mainly from the conceptual and logical point of view through multidimensional (MD) data models (Abelló, Samos & Saltor, 2001; Blaschka, Sapia, Höfling & Dinter, 1998). However, to the best of our knowledge, there are no standard methods or models that allow us to model all aspects of a DW. Moreover, as most of the research efforts in designing and modeling DWs have been focused on the development of MD data models, the attention to the physical design of DWs from the early stages of a DW project has been very little. Nevertheless, the physical design of a DW is of a vital importance and highly influences the overall performance of the DW (Nicola & Rizvi, 2003) and the following maintenance; moreover, a well-structured physical design policy can provide the perfect roadmap for implementing the whole warehouse architecture (Triantafillakis, Kanellis & Martakos, 2004).

Although in some companies the same employee may take on both the role of DW designer and DW administrator, other organizations may have separate people working on each task. Regardless of the situation, modeling the storage of the data and how it will be deployed across different components such as servers and drives helps to implement and maintain a DW. In traditional software products or transactional databases, physical design or implementation issues are not considered until the latest stages of a software project. Then, if the final product does not satisfy user requirements, designers do a feedback, taking into consideration (or at least bearing in mind) some final implementation issues.

Nevertheless, due to the specific characteristics of DWs, we can address several decisions regarding the physical design of a DW from the early stages of a DW project, with no need to leave them until the final implementation stage. DWs, mainly built for analytical reasons, are queried by final users trying to analyze historical data on which they can base their strategy decisions. Thus, the performance measure for DWs is the amount of queries that can be executed instead of the amount of processes or transactions that it supports. Moreover, the kinds of queries on DWs are demonstrated to be much more complex than the queries normally posed in transactional databases (Kimball, 1996; Poe, Klauer & Brobst, 1998). Therefore, poor performance of queries has a worse impact in DWs than in transactional databases. Furthermore, the set of OLAP (Online Analytical Processing) operations that users can execute with OLAP tools on DWs depends very much on the design of the DW (i.e., on the multidimensional model underneath) (Sapia, 1999; Trujillo, Palomar, Gómez & Song, 2001).

Based on our experience in real-world DW projects, physical storage and query performance issues can be discussed in the early stages of the project. The reason is that in DW projects, final users, analysts, business managers, DW designers, and database administrators participate at least in the first meetings. Therefore, we believe that some decisions on the physical design of DWs can be made in the beginning. Some examples of these decision are as follows: (1) the size and the speed of the hard disk needed to deal with the fact table and the corresponding views; (2) a coherent partitioning of both fact and dimension tables based on data and user requirements; (3) the estimation of the workload needed and the time boundaries to accomplish it. Based on our experience, we believe that making these decisions in the early stages of a DW project will reduce the total development time of the DW.

At this point, we must point out that we are not suggesting that the conceptual modeling of a DW take into account physical issues. Instead, we advocate that the physical aspects and following implementation details from the conceptual modeling of the DW from the early stages of a DW project will benefit the implementation.

In previous works (Luján-Mora & Trujillo, 2003, 2004a), we have proposed a DW development method based on the Unified Modeling Language (UML) (Object Management Group [OMG], 2003) and the Unified Process (UP) (Jacobson, Booch, & Rumbaugh, 1999) to properly design all aspects of a DW. So far, we have addressed modeling of different aspects of a DW by using the UML (Object Management Group [OMG], 2003): MD modeling (Luján-Mora, Trujillo, & Song, 2002a, 2002b; Trujillo et al., 2001), modeling of the ETL processes (Trujillo & Luján-Mora, 2003), and modeling data mappings between data sources and targets (Luján-Mora, Vassiliadis, & Trujillo, 2004). In this article, we complement all of these previous works with a proposal to accomplish the physical design of DWs from the early stages of a DW project. To accomplish these goals, we propose the use of the component diagrams and deployment diagrams of UML. Both component and deployment diagrams must be defined at the same time by DW designers and DW administrators who will be in charge of the subsequent implementation and maintenance. This is due mainly to the fact that while the former know how to design and build a DW, the latter have a better knowledge in the corresponding implementation and the real hardware and software needs for the correct functioning of the DW.

The modeling of the physical design of a DW from the early stages of a DW project with our approach provides us many advantages:

- We address important aspects of the implementation before we start with the implementation process, and therefore, we can reduce the total development time of the DW. This is due mainly to the fact that after the conceptual modeling has been accomplished, we can have enough information to make some decisions regarding the implementation of the DW structures, such as replicating dimension tables or designing the vertical and horizontal partitioning of a fact table.
- We have rapid feedback if there is a problem with the DW implementation, as we easily can track a problem to find out its main reasons.
- It facilitates communication between all people involved in the design of a DW, since all of them use the same notation (based on UML) for modeling different aspects of a DW. Moreover, making sure that the crucial concepts mean the same to all groups and are not used in different ways is critical. In this way, our approach helps to achieve a coherent and consistent documentation during the DW development life cycle.
- It helps us to choose both hardware and software on which we intend to implement the DW. This also allows us to compare and to evaluate different configurations based on user requirements.
- It allows us to verify that all different parts of the DW (e.g., fact and dimension tables, ETL processes, OLAP tools) perfectly fit together.

A short version of this article was presented previously (Luján-Mora & Trujillo, 2004b). In this long version, we have added new stereotypes to consider more physical decisions such as views and indices. Furthermore, we also have included details on the implementation of a case study on a commercial database management server from our component and deployment diagrams, showing the benefit of our approach.

The rest of the article is organized as follows. In Related Work, we briefly comment on other work that addresses the conceptual, logical, and physical design and/or deployment of a DW. In Data Warehouse Design Framework, we briefly introduce our overall method to design all aspects of a DW. In UML Component and Deployment Diagrams, we present main issues that can be specified by using both component and deployment diagrams of UML. In Data Warehouse Physical Design, we describe our approach for using both component and deployment diagrams for the physical design of DWs. In Implementation in Oracle, we provide details on how to use our component and deployment diagrams to implement a DW on a commercial database management server. Finally, we present our conclusions and outline main future work.

RELATED WORK

As this article focuses on the design of DWs and, more specifically, the physical design of DWs, the related work is organized into three subsections about multidimensional modeling, physical design, and implementation of DWs and UML extensibility mechanisms.

Multidimensional Modeling

Several multidimensional (MD) data models have been proposed for DWs. Some of them fall into the logical level (e.g., the well-known star schema by Kimball [1996]). Others may be considered as formal models, as they provide a formalism to consider main MD properties. Blaschka et al. (1998) review the most relevant logical and formal models.

In this subsection, we only make brief reference to the most relevant models that we consider pure conceptual MD models, as this article focuses on the physical design of DWs from the early stages of a DW project. These models provide a high level of abstraction for the main MD modeling properties (e.g., facts, dimensions, classification hierarchies defined along dimensions, the additivity of measures) and are independent of implementation issues. One interesting feature provided by these models is that they provide a set of graphical notations (such as the classical and well-known EER model) that facilitates their use and reading. These are as follows: The Dimensional-Fact (DF) Model by Golfarelli, Maio, and Rizzi (1998); The Multidimensional/ER (M/ER) Model by Sapia, Blaschka, Höfling, and Dinter (1998); The starER Model by Tryfona, Busborg, and Christiansen (1999); the Model proposed by Hüsemann, Lechtenbörger, and Vossen (2000); and The Yet Another Multidimensional Model (YAM2) by Abelló, Samos, and Saltor (2002).

However, none of these approaches for MD modeling considers the design of physical aspects of DWs as an important issue of their modeling, and therefore, they do not solve the problem of physical modeling from the early stages of a DW project.

Physical Design and Implementation Issues of Data Warehouses

Both the research community and companies have devoted few efforts to the physical design of DWs from the early stages of a DW project and incorporate it within a global method that allows designing all main aspects of DWs. In this subsection, we are not presenting research on physical issues of DWs such as new algorithms for defining and managing indices, view materialization, query processing, or performance, as these and other physical aspects of DWs are out of the scope of this article; instead, we concentrate on the modeling of the physical design of DWs from the first stages of a DW project.

Kimball et al. (1998) studied the lifecycle of a DW and proposed a method for the design, development, and deployment of a DW. They discuss the planning of the deployment of a DW and recommend documenting all different deployment strategies. However, they do not provide a standard technique for the formal modeling of the deployment of a DW.

Poe et al. (1998) address the design of a DW from conceptual modeling to implementation. They propose the use of non-standard diagrams to represent the physical architecture of a DW, on the one hand to represent data integration processes and on the other to represent the relationship between the enterprise data warehouse and the different data marts that are populated from it. Nevertheless, these diagrams represent the architecture of the DW from a high level without providing different levels of detail of the subsequent implementation of the DW.

Giovinazzo (2000) discusses several aspects of a DW implementation. Although in this book, other aspects of a DW implementation, such as the parallelism, the partitioning of data in a RAID (Redundant Array of Inexpensive Disk) system, or the use of a distributed database, are addressed, the authors do not provide a formal or standard technique to model all these aspects.

Finally, Rizzi (2003) states that one of the current open problems regarding DWs is the lack of a formal documentation that covers all design phases and provides multiple levels of abstraction (low level for designers and people devoted to the corresponding implementation and high level for final users). The author argues that this documentation is basic for the maintenance and ulterior extension of the DW. In this work, three different detailed levels for DWs are proposed: data warehouse level, data mart level, and fact level. At the first level, the use of the deployment diagrams of UML is proposed to document a DW architecture from a high level of detail. However, these diagrams are not integrated at all with the rest of techniques, models, and/or methods used in the design of other aspects of the DW.

On the other hand, Naiburg and Maksimchuk (2001) have studied the use of UML for the design of databases. Their work is structured around the database design process; therefore, it contains a chapter devoted to database deployment. In this book, it is stated that from the database designers' points of view, in a real database development project, "the biggest benefit in using the UML is the ability to model the tablespaces and quickly understand what tablespaces exist and how tables are partitioned across those tablespaces" (Naiburg & Maksimchuk, 2001). On the other hand, using UML for designing databases has the advantage that a different UML diagram can be used (e.g., package diagram, class diagram, component diagram, and deployment diagram), depending on the particular aspect modeled, and then, many transformations between these diagrams have been widely and recently proposed (Selonen, Koskimies, & Sakkinen, 2003; Whittle, 2000).

Therefore, we argue that there is still a need for providing a standard technique that allows modeling the physical design of a DW from the early stages of a DW project. Another important issue is that this technique is integrated in an overall approach that allows coverage of other aspects of the DW design, such as the conceptual or logical design of the DW or the modeling of ETL processes.

UML Extensibility Mechanism

The UML Extensibility Mechanism package is the subpackage from the UML metamodel that specifies how specific UML model elements are customized and extended with new semantics by using stereotypes, tagged values, and constraints. A coherent set of such extensions, defined for specific purposes, constitutes a UML profile. For example, the UML 1.5 (Object Management Group [OMG], 2003) includes a standard profile for modeling software development processes and another one for business modeling.

A *stereotype* is a model element that defines additional values (based on tagged values), additional constraints, and, optionally, a new graphical representation (an icon); a stereotype allows us to attach a new semantic meaning to a model element. A stereotype is represented either as a string between a pair of guillemots (<< >>) or rendered as a new icon.

A *tagged value* specifies a new kind of property that may be attached to a model element. A tagged value is rendered as a string enclosed by brackets ([ ]) and placed below the name of another element.

Figure 1. Data warehouse design framework

	Source (S) (OLTP, external data, ...)	Integration	Data Warehouse (DW)	Customization	Client (C) (OLAP, data mining, ...)
Conceptual	**SCS** Class diagram *Standard UML*	**DM** Class diagram *Data Mapping Profile*	**DWCS** Class diagram *Standard UML* *Multidimensional Profile*	**DM** Class diagram *Data Mapping Profile*	**CCS** Class diagram *Standard UML* *Multidimensional Profile*
Logical	**SLS** Class diagram *Different data modeling profiles*	**ETL Process** Class diagram *ETL Profile*	**DWLS** Class diagram *Different data modeling profiles*	**Exporting Process** Class diagram *ETL Profile*	**CLS** Class diagram *Different data modeling profiles*
Physical	**SPS** Comp. & deploy. diagrams *Database Deployment Profile*	**Transportation Diagram** Deployment diagram *Database Deployment Profile*	**DWPS** Comp. & deploy. diagrams *Database Deployment Profile*	**Transportation Diagram** Deployment diagram *Database Deployment Profile*	**CPS** Comp. & deploy. diagrams *Database Deployment Profile*

LEGEND: CS: Conceptual Schema, LS: Logical Schema, PS: Physical Schema, Comp. & deploy: Component and deployment

A *constraint* can be attached to any model element to refine its semantics; Warmer and Kleppe (1998) state, "A constraint is a restriction on one or more values of (part of) an object-oriented model or system." In the UML, a constraint is rendered as a string between a pair of braces ({ }) and placed near the associated model element. A constraint on a stereotype is interpreted as a constraint on all types on which the stereotype is applied. A constraint can be defined by means of an informal explanation or by means of OCL (Warmer & Kleppe, 1998; Object Management Group [OMG], 2003) expressions. The OCL is a declarative language that allows software developers to write constraints over object models.

DATA WAREHOUSE DESIGN FRAMEWORK

The architecture of a DW usually is depicted as various layers of data in which data from one layer are derived from data of the previous layer (Jarke, Lenzerini, Vassiliou, & Vassiliadis, 2003). In a previous work (Luján-Mora & Trujillo, 2004a), we have presented a DW development method based on UML (Object Management Group [OMG], 2003) and the UP (Jacobson et al., 1999) that addresses the design and development of both the DW back end and front end. In our approach, we consider that the development of a DW can be structured into an integrated framework with five stages and three levels that define different diagrams for the DW model, as shown in Figure 1 and summarized next.

- **Stages:** We distinguish five stages in the definition of a DW:

- Source, which defines the data sources of the DW, such as OLTP systems, external data sources (syndicated data, census data), and so forth.

- Integration, which defines the mapping between the data sources and the DW.
- Data Warehouse, which defines the structure of the DW.
- Customization, which defines the mapping between the DW and the clients' structures.
- Client, which defines special structures that are used by the clients to access the DW, such as data marts (DM) or OLAP applications.

- **Levels:** Each stage can be analyzed at three different levels or perspectives:

- Conceptual, which defines the DW from a conceptual point of view.
- Logical, which addresses logical aspects of the DW design, such as the definition of the ETL processes.
- Physical, which defines physical aspects of the DW, such as the storage of the logical structures in different disks or the configuration of the database servers that support the DW.

- **Diagrams:** Each stage or level requires different modeling formalisms. Therefore, our approach is composed of 15 diagrams, but the DW designer does not need to define all the diagrams in each DW project; for example, if there is a straightforward mapping between the Source Conceptual Schema (SCS) and the Data Warehouse Conceptual Schema (DWCS), the designer may not need to define the corresponding Data Mapping (DM). In our approach, we use UML (Object Management Group [OMG], 2003) as the modeling language, because it provides enough expressiveness power to address all the diagrams. As UML is a general modeling language, we can use UML extension mechanisms (stereotypes, tag definitions, and constraints) to adapt UML to specific domains. A stereotype is a UML modeling element that extends the UML metamodel in a controlled way (i.e., a stereotype is

Figure 2. From the conceptual to the physical level

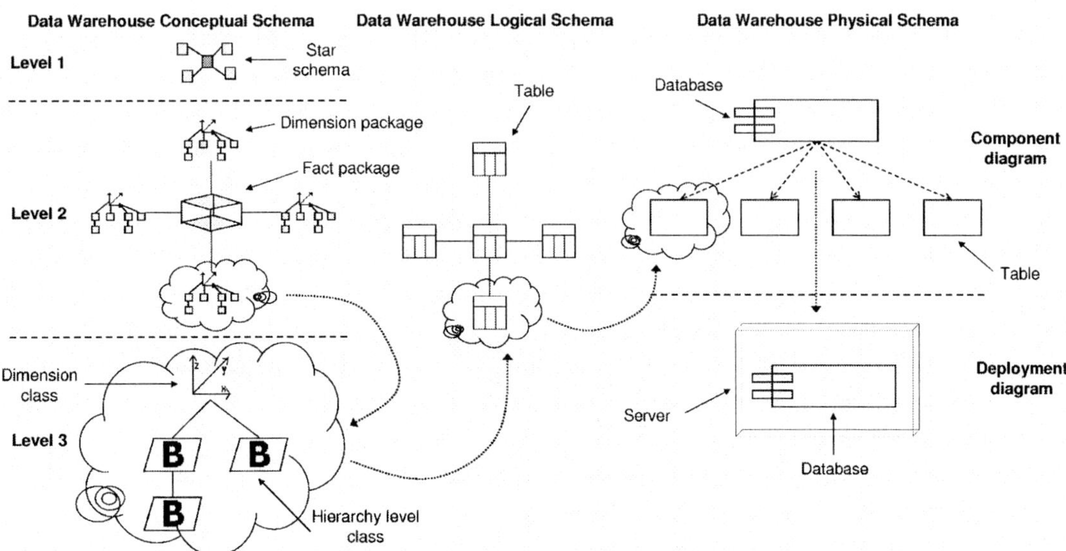

a specialized version of a standard UML element; a tag definition allows additional information about a standard UML element to be specified; and a constraint is a rule that limits the behavior of a UML element). Figure 1 contains the following information for each diagram:

- Name (in bold face): The name we have coined for this diagram.
- UML diagram: The UML diagram we use to model this DW diagram. Currently, we use class, deployment, and component diagrams.
- Profile (in italic font): The dashed boxes show the diagrams where we propose a new profile; in the other boxes, we use a standard UML diagram or a profile from other authors. A profile is an extension to the UML that uses stereotypes, tagged values, and constraints to extend the UML for specialized purposes.

The different diagrams of the same DW are not independent but overlapping. They depend on each other in many ways; therefore, they cannot be created in any order. For example, changes in one diagram may imply changes in another, and a large portion of one diagram may be created on the basis of another diagram. For example, the Data Mapping (DM) is created by importing elements from the Source Conceptual Schema (SCS) and the Data Warehouse Conceptual Schema (DWCS). Moreover, our approach is flexible in the sense that the DW designer does not need to define all the diagrams, but he or she can use what is needed when it is needed and can continue moving forward as necessary.

In previous work, we presented some of the diagrams and the corresponding profiles shown in white dashed boxes in Figure 1: the Multidimensional Profile (Luján-Mora et al., 2002a, 2002b) for the DWCS and the Client Conceptual Schema (CCS); the ETL Profile (Trujillo & Luján-Mora, 2003) for the ETL Process and the Exporting Process; and the Data Mapping Profile (Luján-Mora et al., 2004) for the DM between the SCS and the DWCS and between the DWCS and the CCS. Finally, in light gray dashed boxes, we show the profile we present in this article, the Database Deployment Profile, for modeling a DW at a physical level.

Figure 2 shows a symbolic diagram to summarize our approach and the relationships between the different diagrams (DWCS, DWLS, and DWPS).

- On the left-hand side of this figure, we have represented the DWCS, which is structured into three levels: Level 1 or Model definition; Level 2 or Star schema definition; and Level 3 or Dimension/Fact definition. The different elements drawn in this diagram are stereotyped packages and classes that represent MD concepts. An icon, a new graphical representation, can be associated with a stereotype in UML.
- From the DWCS, we develop the logical model (DWLS, represented in the middle of Figure 2) according to different options, such as ROLAP (Relational OLAP) or MOLAP (Multidimensional OLAP). In this example, we have chosen a ROLAP representation, and each element corresponds to a table in the relational model.
- Finally, from the DWLS, we derive the DWPS, which is represented on the right-hand side of Figure 2. The DWPS shows the physical aspects of the implementation of the DW. This diagram is divided into two parts: the component diagram, which shows the configuration of the logical structures used to store the DW; and the deployment diagram, which specifies different aspects relative to the hardware and software configuration.

Figure 2 shows how our approach allows the designer to trace the design of an element from the conceptual to the physical level. For example, in this figure, we have drawn a cloud around different elements that represent the same entity in different diagrams.

In the following section, we summarize the basic concepts about the UML component and deployment diagrams that we apply for the physical design of DWs in the Data Warehouse Physical Design section.

UML COMPONENT AND DEPLOYMENT DIAGRAMS

According to the UML Specification (Object Management Group [OMG], 2003, p. 3-169), "Implementation diagrams show aspects of physical implementation, including the structure of components and the run-time deployment system. They come in two forms: (1) component diagrams show the structure of components, including the classifiers that specify them and the artifacts that implement them; and (2) deployment diagrams show the structure of the nodes on which the components are deployed."

In the Component Diagram section, we summarize the main concepts about the UML component diagram, whereas in the Deployment Diagram section, we introduce the deployment diagram.

Component Diagram

The UML Specification says, "A component represents a modular, deployable, and replaceable part of a system that encapsulates implementation and exposes a set of interfaces" (Object Management Group [OMG], 2003, p. 3-174). Components represent physical issues such as Enterprise JavaBeans, ActiveX components, or configuration files. A component typically is specified by one or more classifiers (e.g., classes and interfaces) that reside on the component. A subset of these classifiers explicitly defines the component's external interfaces. Moreover, a component also can contain other components. However, a component does not have its own features (attributes and operations).

On the other hand, a component diagram is a graph of components connected by dependency relationships that show how classifiers are assigned to components and how the components depend on each other. In a component diagram (Figure 3), a component is represented using a rectangular box, with two rectangles protruding from the left side.

Figure 3 shows the two different representations of a component and the classifiers it contains:

- On the left-hand side of the figure, the class (Sales) that resides on the component (Facts)

Figure 3. Different component representations in a component diagram

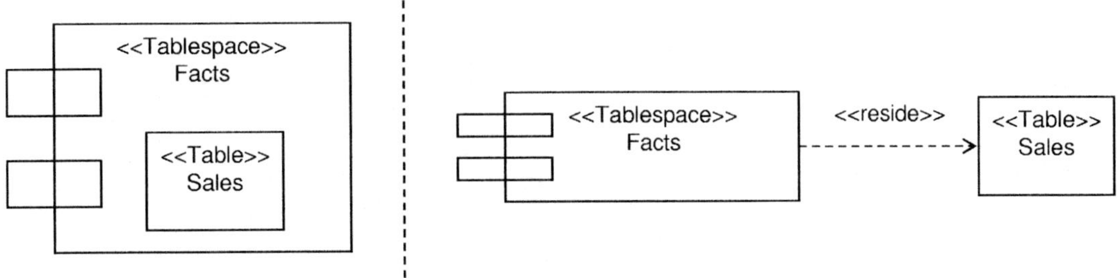

Figure 4. Different node representations in a deployment diagram

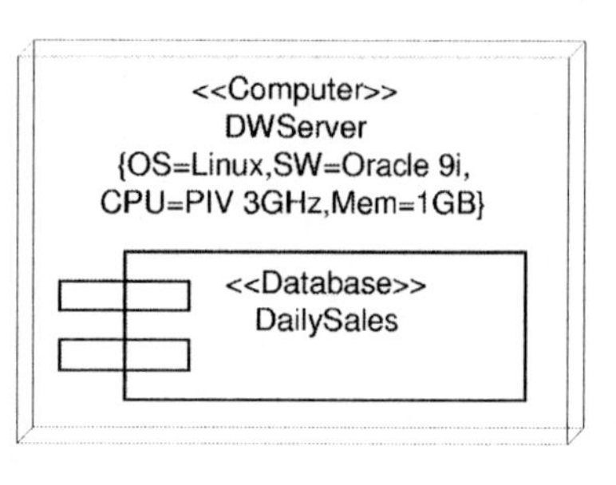

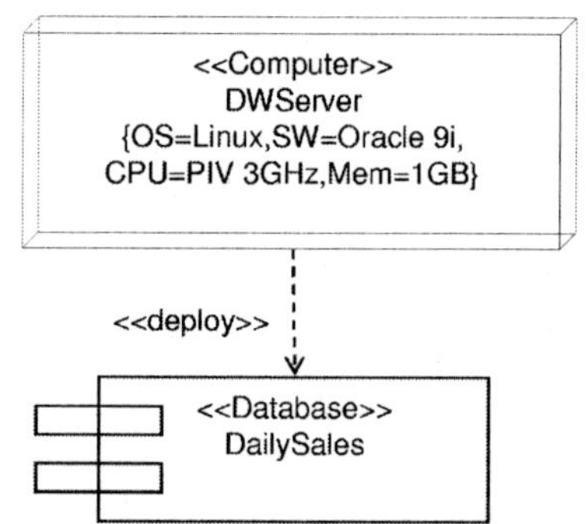

is shown as nested inside the component (this indicates residence and not ownership).
- On the right-hand side of the figure, the class is connected to the component by a <<reside>> dependency.

In these examples, both the component and the class are stereotyped; the component is adorned with the <<Tablespace>> stereotype and the class with the <<Table>> stereotype; these stereotypes are defined by Naiburg and Maksimchuk (2001).

Deployment Diagram

According to the UML Specification, "Deployment diagrams show the configuration of runtime processing elements and the software components, processes, and objects that execute on them" (Object Management Group [OMG], 2003, p. 3-171). A deployment diagram is a graph of nodes connected by communication associations. A deployment model is a collection of one or more deployment diagrams with their associated documentation.

In a deployment diagram, a node represents a piece of hardware (e.g., a computer, a device, an interface) or a software artifact (e.g., Web server, database) in the system, which is represented by a three-dimensional cube. A node may contain components that indicate that the components run or execute on the node.

An association of nodes, which is drawn as a solid line between two nodes, indicates a line of communication between the nodes; the association may have a stereotype to indicate the nature of the communication path (e.g., the kind of channel, communication protocol, or network).

There are two forms of deployment diagram:

1. The descriptor form: Contains types of nodes and components. This form is used as a first-cut deployment diagram during the design of a system, when there is not a complete decision about the final hardware architecture.
2. The instance form: Contains specific and identifiable nodes and components. This form is used to show the actual deployment of a system at a particular site; therefore, it normally is used in the last steps of the implementation activity, when the details of the deployment site are known.

Figure 5. Different levels of detail in a deployment diagram

<<Client>>
WinClient
{OS=Windows XP,
SW=Internet Explorer}
<<HTTP>>
<<Server>>
WinServer
{OS=Windows 2000,
SW=IIS}

<<Client>>
WinClient
<<Ethernet>>
<<Server>>
WinServer
Windows XP
<<TCP/IP>>
Windows 2000
Internet Explorer
<<HTTP>>
IIS

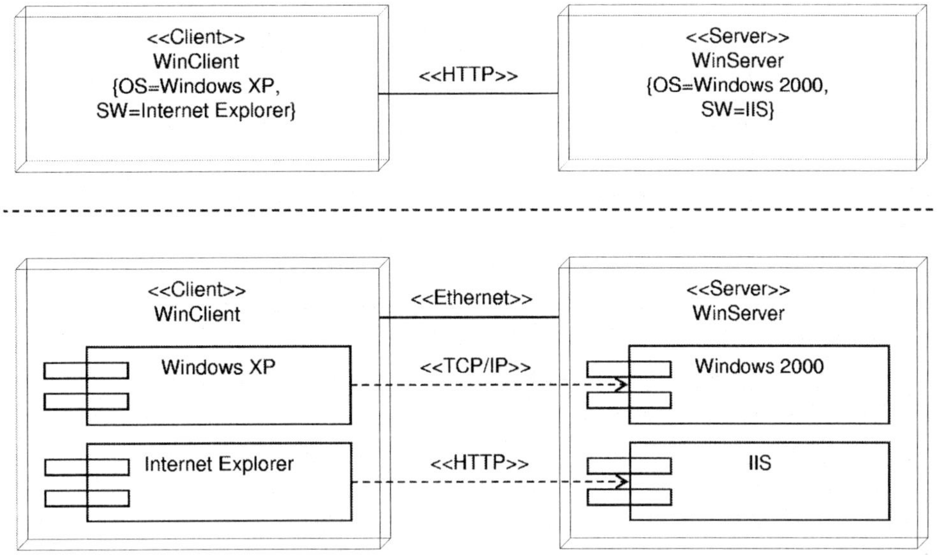

According to Ambler (2002), a deployment diagram normally is used to:

- Explore the issues involved with installing your system into production.
- Explore the dependencies that your system has with other systems that are currently in or planned for your production environment.
- Depict a major deployment configuration of a business application.
- Design the hardware and software configuration of an embedded system.
- Depict the hardware/network infrastructure of an organization.

UML deployment diagrams normally make extensive use of visual stereotypes, because it is easy to read the diagrams at a glance. Unfortunately, there are no standard palettes of visual stereotypes for UML deployment diagrams.

As suggested by Ambler (2002), each node in a deployment diagram may have tens if not hundreds of software components deployed to it; the goal is not to depict all of them but to depict those components that are vital to the understanding of the system.

Figure 4 shows two different representations of a node and the components it contains:

1. On the left-hand side of the figure, the component (DailySales) that is deployed on the node (DWServer) is shown as nested inside the node.
2. On the right-hand side of the figure, the component is connected to the node by a <<deploy>> dependency.

In this example, both the node and the component are stereotyped: the node with the <<Computer>> stereotype and the component with the <<Database>> stereotype. Moreover, the node DWServer contains a set of tagged values (OS, SW, CPU, and Mem) that allow the designer to describe the particular characteristics of the node.

A deployment diagram can be specified at different levels of detail. For example, Figure 5 shows two versions of the same deployment diagram. At the top of Figure 5, the software deployed in

Figure 6. Data warehouse conceptual schema: Level 1

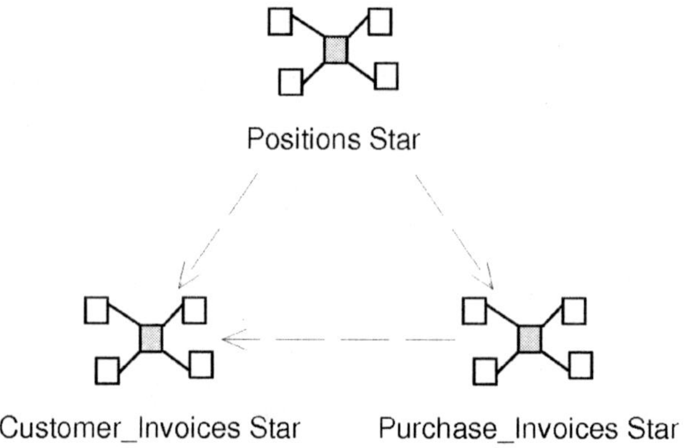

Figure 7. Data warehouse conceptual schema: Level 2 of Customer_Invoices Star

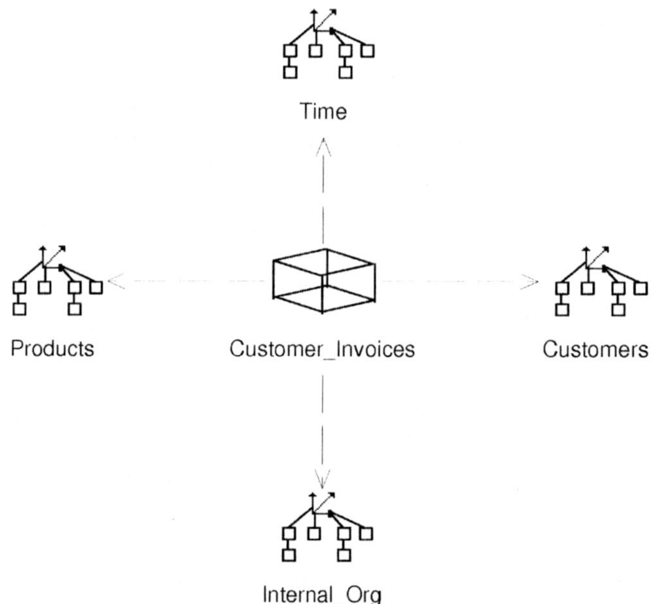

the nodes is specified by means of tagged values. Moreover, the association between the nodes is adorned only with the <<HTTP>> stereotype (HyperText Transfer Protocol), although different protocols can be used in the communication. At the bottom of Figure 5, the software deployed in the nodes is depicted as components, and different stereotyped dependencies (<<TCP/IP>> and <<HTTP>>) indicate how one component uses the services of another component. However, there are more display possibilities; for example, the designer can omit the tagged values in the diagram and capture them only in the supported documentation.

DATA WAREHOUSE PHYSICAL DESIGN

In the Data Warehouse Design Framework section, we briefly describe our design method for DWs. Within this method, we use the component and deployment diagrams to model the physical level of DWs. To achieve this goal, we propose the following five diagrams, which correspond to the five stages presented in the Data Warehouse Design Framework section:

- Source Physical Schema (SPS): Defines the physical configuration of the data sources that populate the DW.
- Integration Transportation Diagram (ITD): Defines the physical structure of the ETL processes that extract, transform, and load data into the DW. This diagram relates the SPS and the next diagram.
- Data Warehouse Physical Schema (DWPS): Defines the physical structure of the DW itself.
- Customization Transportation Diagram (CTD): Defines the physical structure of the exportation processes from the DW to the specific structures employed by clients. This diagram relates the DWPS and the next diagram.
- Client Physical Schema (CPS): Defines the physical configuration of the structures employed by clients in accessing the DW.

The SPS, DWPS, and CPS are based on the UML component and deployment diagrams; whereas, ITD and CTD are based only on the deployment diagrams. These diagrams reflect the modeling aspects of the storage of data (Naiburg & Maksimchuk, 2001), such as the database size, information about where the database will reside (hardware and software), partitioning of the data, and properties specific to the DBMS (Database Management System) chosen.

The five proposed diagrams use an extension of UML that we have called Database Deployment Profile, which is formed by a series of stereotypes, tagged values, and constraints.

Throughout the rest of this article, we use an example to introduce the different diagrams we propose. This example is based partly on the enterprise DW sample database schema from Silverston, Inmon, and Graziano (1997). In this example, final users need a DW in order to analyze the main operations of the company. Because of this, the DW contains information about customers, customer invoices, budget details, products, and suppliers. Moreover, data about the employees, such as salaries, positions, and categories, also are stored in the DW.

The operational data sources are stored in three servers:

1. The sales server, which contains the data about transactions and sales.
2. The CRM (Customer Relationship Management) server, which contains the data about the customers who buy products.
3. The HRM (Human Resource Management) server, which contains data about employees, such as positions and salaries.

Following our approach (Luján-Mora et al., 2002b), we structure the conceptual model into three levels:

Level 1: Model definition. A package represents a star schema of a conceptual MD model. A dependency between two packages at this level indicates that the star schemas share at least one dimension, allowing us to consider conformed dimensions.

Level 2: Star schema definition. A package represents a fact or a dimension of a star schema. A

Figure 8. Data warehouse conceptual schema: Customers dimension

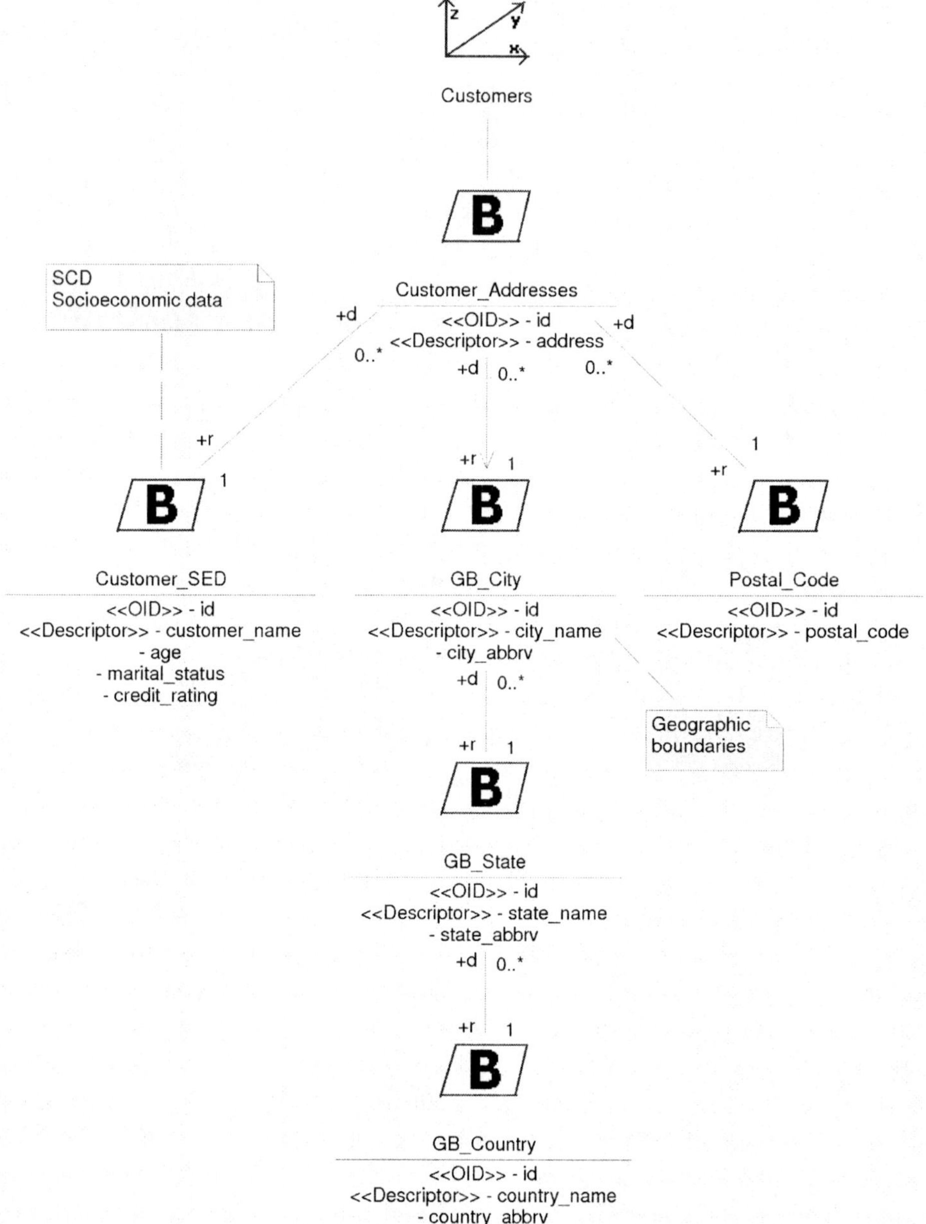

dependency between two dimension packages at this level indicates that the packages share at least one level of a dimension hierarchy.

Level 3: Dimension/fact definition. A package is exploded into a set of classes that represent the hierarchy levels defined in a dimension package or the whole star schema, in the case of the fact package.

Figure 6 shows the first level of the DWCS, which represents the conceptual model of the DW. In our example, the first level is formed by three packages called Customer_Invoices Star, Positions Star, and Purchase_Invoices Star. A dashed arrow from one package to another denotes a dependency between packages (i.e., the packages have some dimensions in common). The direction of the dependency indicates that the common dimensions shared by the two packages were first defined in the package pointed to by the arrow (to start with, we have to choose a star schema to define the dimensions, and then, the other schemas can use them with no need to define them again). If the common dimensions had been defined first in another package, then the direction of the arrow would have been different. In any case, it is better to group together the definition of the common dimensions in order to reduce the number of dependencies. From now on, we will focus our discussion on the Customer_Invoices Star. This star schema represents the invoices belonging to customers.

Figure 7 shows the second level of the DWCS. The fact package Customer_Invoices is represented in the middle of the figure, while the dimension packages are placed around the fact package. As seen in Figure 7, a dependency is drawn from the fact package Customer_Invoices to each one of the dimension packages (Customers, Internal_Org, Products, and Time), because the fact package comprises the whole definition of the star schema and, therefore, uses the definitions of dimensions related to the fact. At level 2, it is possible to create a dependency from a fact package to a dimension package or between dimension packages, but we do not allow a dependency from a dimension package to a fact package, since it is not semantically correct in our technique.

The content of the dimension and fact packages is represented at Level 3. The diagrams at this level are comprised only of classes and associations among them. Figure 8 shows the Level 3 of the Customers dimension package (Figure 7), which contains the definition of the dimension (Customers) and the different hierarchy levels (Customer_Addresses, Customer_SED, GB_City, Postal_Code, GB_City, etc.). GB means geographic boundaries. The hierarchy of a dimension defines how the different OLAP operations (roll up, drill down, etc.) can be applied. In a UML note we highlight that Customer_SED (SED means socioeconomic data) is a Slowly Changing Dimension (SCD), and some kind of solution has to be selected during the implementation (Kimball, 1996).

Figure 9 shows the Level 3 of the Products dimension. This dimension contains two alternative hierarchies: the category of the product (Category) and the supplier of the product (Products_Supplier, City, State, and Country). In the Products_Supplier hierarchy level, msrp means manufacturer's suggested retail price, and uom is the standard unit of measure used for the product.

Figure 10 shows the Level 3 of the Customer_Invoices fact package. In this package, the whole star schema is displayed; the fact class is defined in this package, and the dimensions with their corresponding hierarchy levels are imported from the dimension packages. Because of this, the name of the package where they have been defined previously appears below the package name (e.g., from Products, from Internal_Org). In order to avoid a cluttered diagram, we only show the attributes of the fact class (Customer_Invoices), and we hide the attributes and the hierarchy levels of the dimensions.

Figure 9. Data warehouse conceptual schema: Products dimension

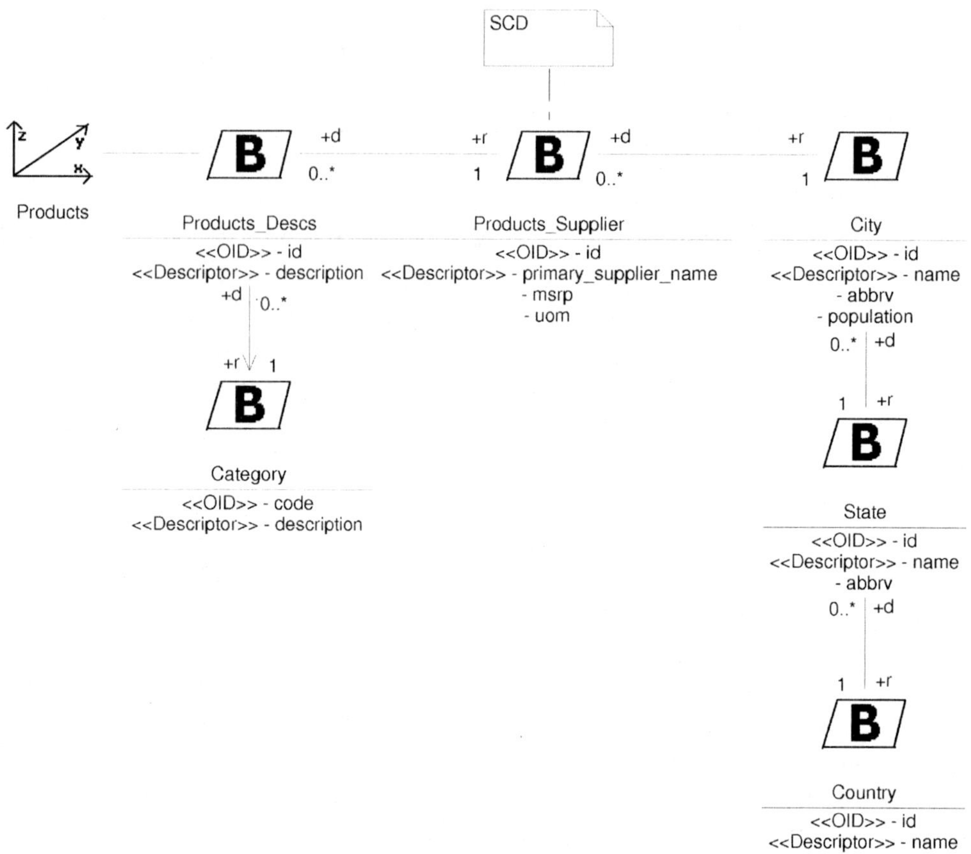

Figure 11 shows the Data Warehouse Logical Schema (DWLS), which represents the logical model of the DW. In this example, a ROLAP system has been selected for the implementation of the DW, which means the use of the relational model in the logical design of the DW. In Figure 11, seven classes adorned with the stereotype <<Table>> are shown: customers, customer_addresses, customer_invoices, internal_org_addresses, geographic_boundaries, product_snapshots, and products.

In the customer_invoices table, the attributes customer_id, bill-to-address, organization_id, org_address_id, and product_code are the foreign keys that connect the fact table with the dimension tables, whereas the attributes quantity, unit_price, amount, and product_cost represent the measures of the fact table. The attribute invoice_date represents a degenerate dimension, whereas the attribute load_date is the date the record was loaded into the DW, which is used in the refresh process. Kimball (1996) coined the term *degenerate dimensions* for data items that perform much the same function as dimensions but are stored in the fact table; however, they are not foreign key links through to dimension tables.

The products and product_snapshots tables contain all the attributes of the different dimension levels (Figure 9) following the star schema approach (Kimball, 1996); some attributes have changed their names in order to avoid repeated names, and some design decisions have been made.

Figure 10. Data warehouse conceptual schema: Customer_Invoices fact

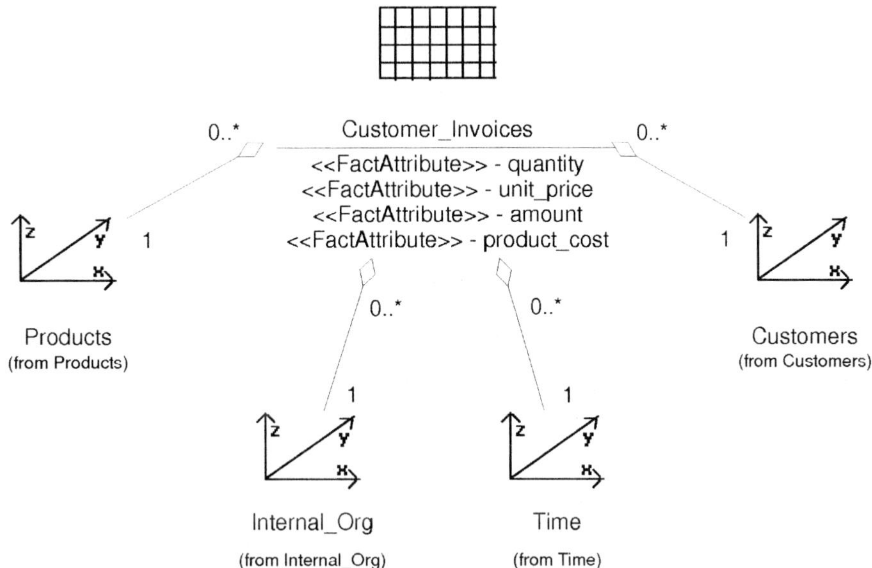

Moreover, we observe that we use UML notes to provide additional information to the diagram.

The following subsections present the five diagrams we propose for the physical design of DWs.

Source Physical Schema

The SPS describes the origins of data of the DW from a physical point of view. Figure 12 shows the SPS of our example, which is formed by three servers called SalesServer, CRMServer, and HRMServer; for each one of them, the hardware and software configuration is displayed by means of tagged values. The first server hosts a database called Sales, whereas the second server hosts a database called Customers.

In our Database Deployment Profile, when the storage system is an RDBMS (Relational Database Management System), we make use of the UML for Profile Database (Naiburg & Maksimchuk, 2001) that defines a series of stereotypes including <<Database>>, <<Schema>>, or <<Tablespace>>. Moreover, we have defined our own set of stereotypes; in Figure 12, we can see the stereotypes <<Server>>, which defines a computer that performs server functions; <<Disk>> to represent a physical disk drive; and <<InternalBus>> to define the type of communication between two elements. In our approach, we represent the configuration parameters of the tablespaces (e.g., size of the tablespace) by means of tagged values; however, these parameters vary greatly, depending on the DBMS, so we only provide a set of common parameters. As UML is extensible, the designer can add additional tagged values as needed to accomplish all the modeling needs of a particular DBMS.

Moreover, whenever we need to specify additional information in a diagram, we make use of the UML notes to incorporate it. For example,

Figure 11. Logical model (ROLAP) of the data warehouse

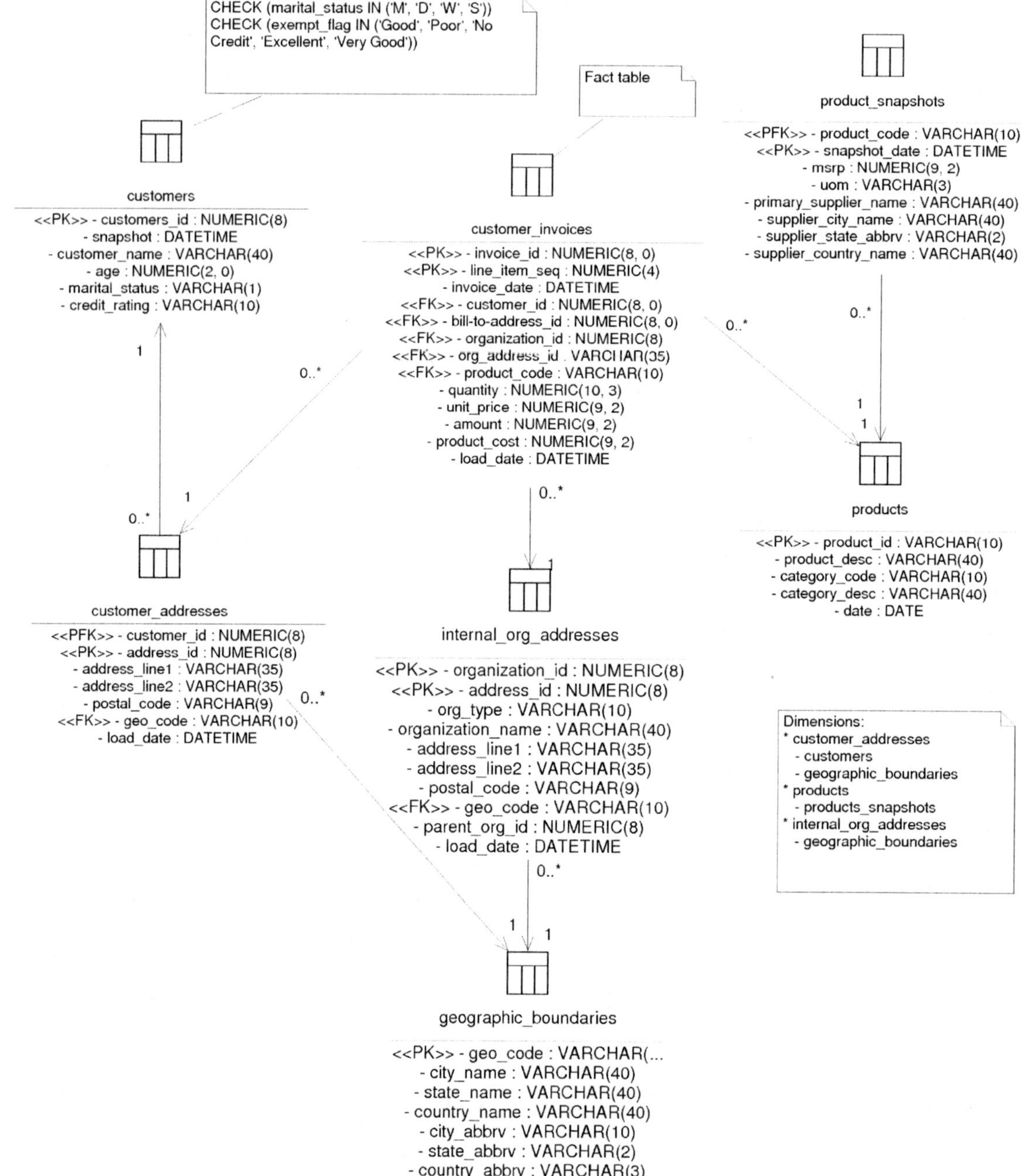

Figure 12. Source physical schema: Deployment diagram

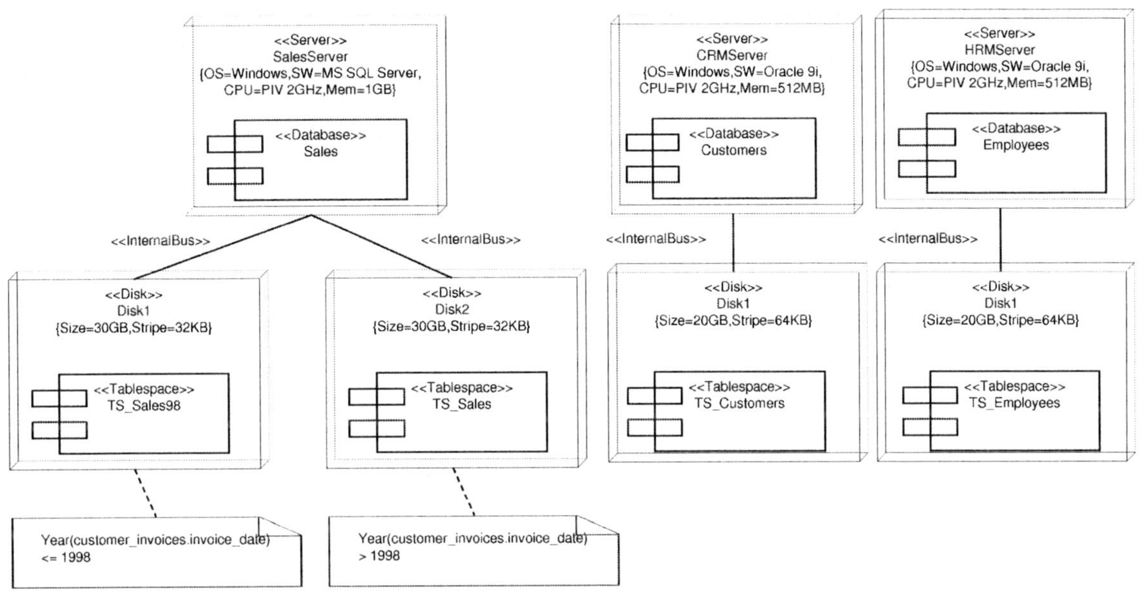

in Figure 12, we have used two notes to indicate how the data are distributed into the two existing tablespaces; the tablespace TS_Sales98 holds data about sales before or in 1998, whereas the tablespace TS_Sales holds the sales after 1998.

Data Warehouse Physical Schema

The DWPS shows the physical aspects of the implementation of the DW. This diagram is divided into two parts: the component diagram and the deployment diagram. In the first diagram, the configuration of the logical structures used to store the DW is shown. For example, in Figure 13, the DW is implemented by means of a database called DWEnt, which is formed by three tablespaces called FACTS, DIMENSIONS, and INDX. The datafiles that the tablespaces use are given, as well: FACTS.ORA, FACTS2.ORA, DIMENSIONS.ORA, DIM-PART2.ORA, and INDX01.DBF.

Figure 14 shows a part of the definition of the tablespaces: the tablespace FACTS hosts the table customer_invoices, and the tablespace DIMENSIONS hosts the dimension tables customers, products, product_snapshots, and the rest of the tables (not shown in the diagram for the sake of simplicity). Below the name of each table, the text (from ROLAP1) is included, which indicates that the tables have been defined previously in a package called ROLAP1 (Figure 11). It is important to highlight that the logical structure defined in the DWLS is reused in this diagram, and therefore, we avoid any possibility of ambiguity or incoherence.

In the second diagram, the deployment diagram, different aspects relative to the hardware and software configuration are specified. Moreover, the physical distribution of the logical structures previously defined in the component diagrams also is represented. For example, in Figure 15, we can observe the configuration of the server that hosts the DW: the server is composed

Figure 13. Data warehouse physical schema: Component diagram (part 1)

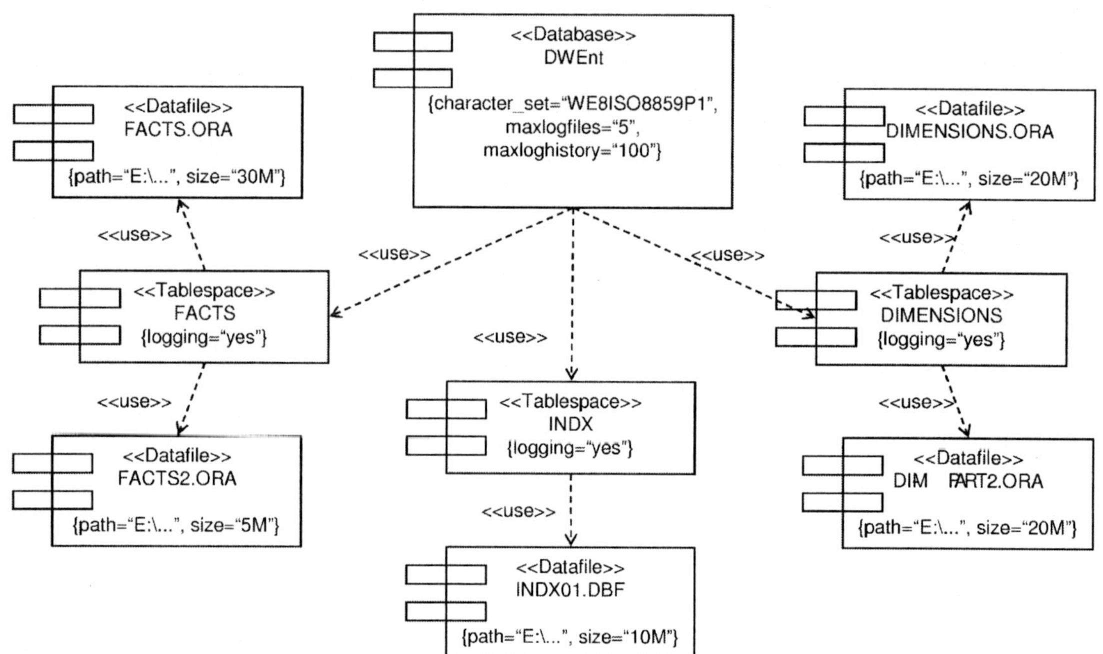

of two disks, one for the operating system (Linux) and the applications (Oracle) and another for the different datafiles (e.g., FACTS.ORA, FACTS2.ORA) that are used by the database (Figure 13).

One of the advantages of our approach is that it allows evaluation and discussion of different implementations during the first stages in the design of a DW. In this way, the designer can anticipate some implementation or performance problems. For example, an alternative configuration of the physical structure of the DW can be established, as shown in Figure 16. In this second alternative, a RAID 0 system has been chosen to host the datafiles that are used by the tablespace FACTS in order to improve the response time of the disk drive and the performance of the system in general. From these two alternative configurations, the DW designer and the DW administrator can discuss the pros and cons of each option.

Integration Transportation Diagram

The ITD defines the physical structure of the ETL processes used in the loading of data in the DW from the data sources. On the one hand, the data sources are represented by means of the SPS, and on the other hand, the DW is represented by means of the DWPS. Since the SPS and the DWPS have been defined previously, in this diagram they are imported.

For example, the ITD for our running example is shown in Figure 17. On the left-hand side of this diagram, different data source servers — SalesServer, CRMServer, and HRMServer — are represented, which have been defined previously in Figure 12; on the right-hand side, the DWServerP, previously defined in Figure 15, is shown. Moreover, the DWServerS, a physical standby database, also is included in the design.

Figure 14. Data warehouse physical schema: Component diagram (part 2)

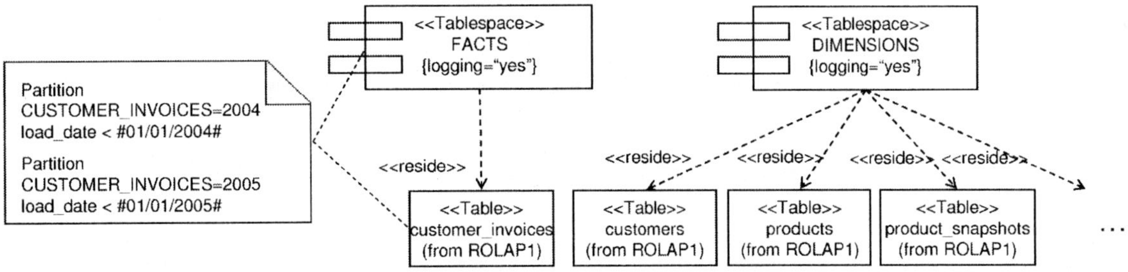

<<Tablespace>>
FACTS
{logging="yes"}
<<Tablespace>>
DIMENSIONS
{logging="yes"}
Partition
CUSTOMER_INVOICES=2004
load_date < #01/01/2004#
Partition
CUSTOMER_INVOICES=2005
load_date < #01/01/2005#
<<reside>>
<<reside>>
<<reside>>
<<reside>>
<<reside>>
<<Table>>
customer_invoices
(from ROLAP1)
<<Table>>
customers
(from ROLAP1)
<<Table>>
products
(from ROLAP1)
<<Table>>
product_snapshots
(from ROLAP1)
...

Figure 15. Data warehouse physical schema: Deployment diagram (version 1)

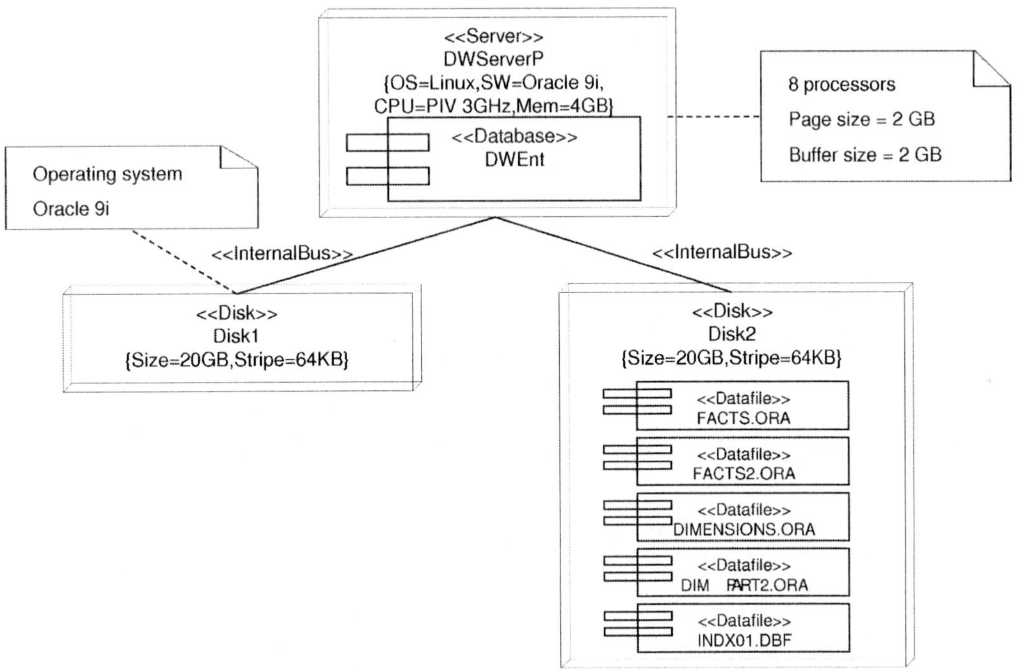

A physical standby database is a byte-by-byte exact copy of the primary database. The primary database records all change and send them to the standby database. A standby database environment is meant for disastrous failures.

In Figure 17, the ETLServer, an additional server that is used to execute the ETL processes, is introduced. This server communicates with the rest of the servers by means of a series of specific protocols: OLEDB to communicate with SalesServer because it uses Microsoft SQLServer and OCI (*Oracle Call Interface*) to communicate with CRMServer, HRMServer, and DWServer because they use Oracle.

Figure 16. Data warehouse physical schema: Deployment diagram (version 2)

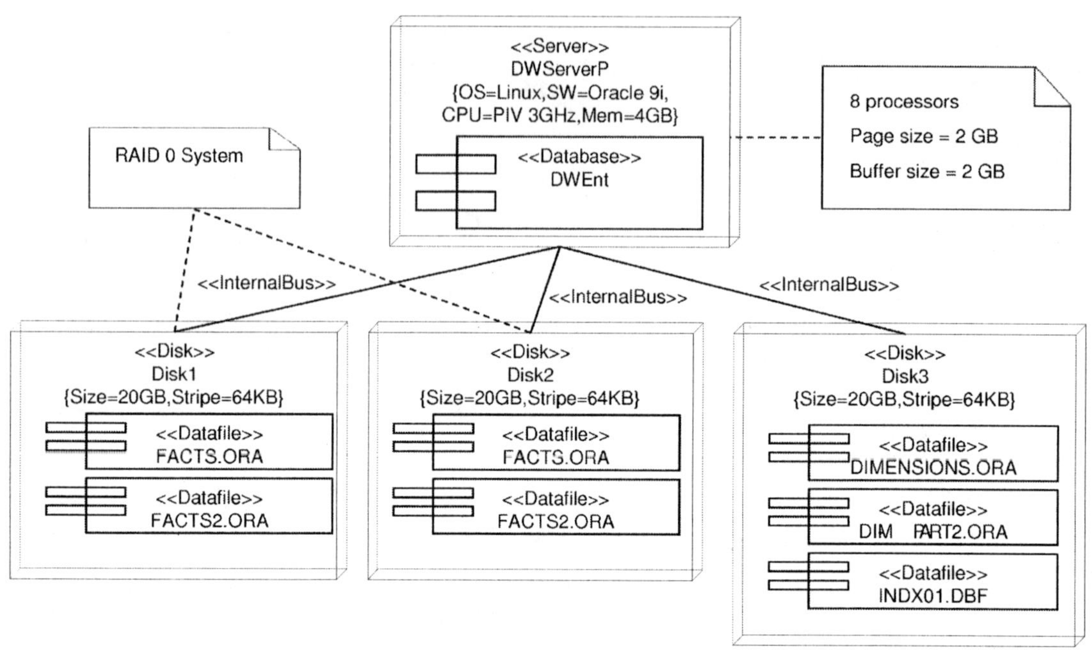

Client Physical Schema

The CPS defines the physical structure of the specific structures that are used by clients to access the DW. Diverse configurations exist that can be used: exportation of data to data marts, use of an OLAP server, and so forth. In our example, we have chosen a client/server architecture, and the same DW server provides access to data for clients. Therefore, we do not need to define a specific structure for clients.

Customization Transportation Diagram

The CTD defines the exportation processes from the DW toward the specific structures used by the clients. In this diagram, the DW is represented by means of the DWPS, and clients are represented by means of the CPS. Since the DWPS and the CPS have been defined previously, in this diagram, we do not have to define them again, but they are directly imported.

For example, in Figure 18, the CTD of our running example is shown. On the left-hand side of this diagram, part of the DWPS, which has been previously defined in Figure 15, is shown; on the right-hand side, three types of clients who will use the DW are shown: a Web client with operating system Apple Macintosh; a Web client with operating system Microsoft Windows; and, finally, a client with a specific desktop application (MicroStrategy) with operating system Microsoft Windows. Whereas both Web clients communicate with the server by means of HTTP, the desktop client uses ODBC (the Open Database Connectivity). In the following section, we explain how to use our component and deployment diagrams to implement a DW on an Oracle database server.

Figure 17. Integration transportation diagram: Deployment diagram

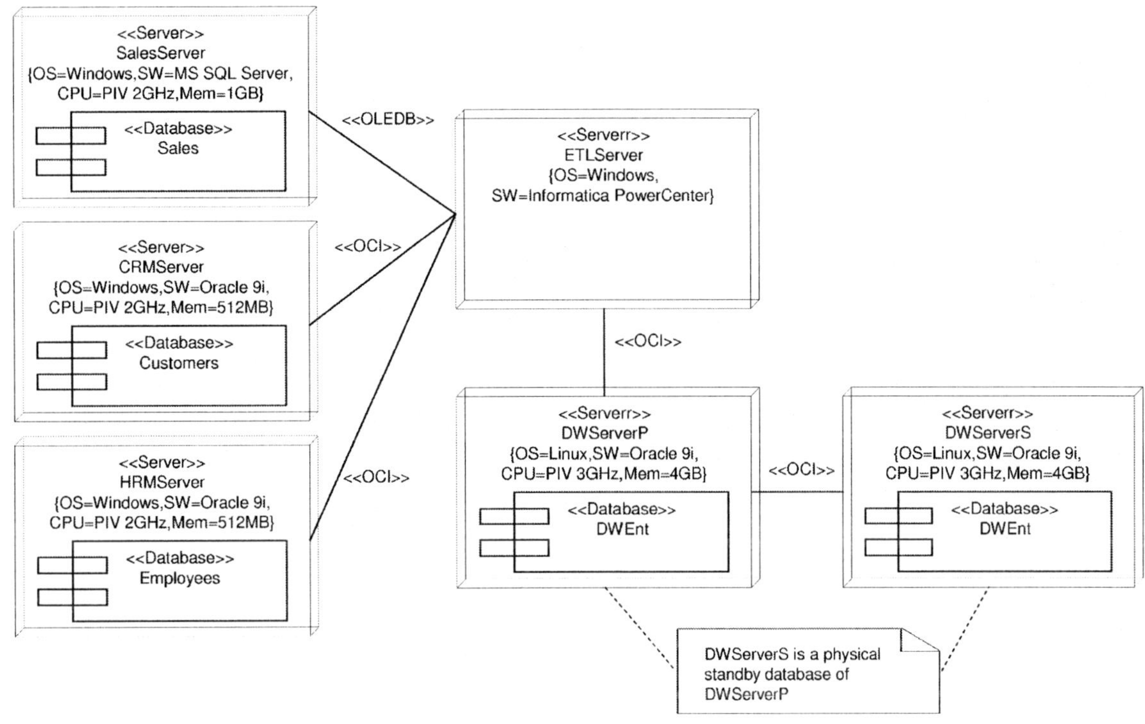

Figure 18. Customization transportation diagram: Deployment diagram

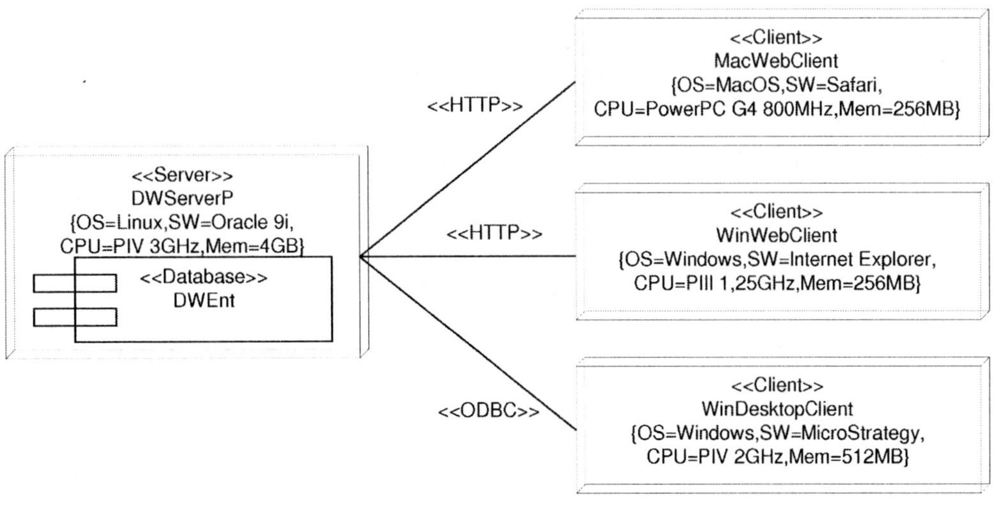

Box 1.

```
CREATE TABLESPACE "FACTS"
LOGGING
DATAFILE 'E:\ORACLE\ORADATA\DWENT\FACTS.ora' SIZE 30M,
'E:\ORACLE\ORADATA\DWENT\FACTS2.ORA' SIZE 5M
EXTENT MANAGEMENT LOCAL SEGMENT SPACE MANAGEMENT AUTO

CREATE TABLESPACE "DIMENSIONS"
LOGGING
DATAFILE 'E:\ORACLE\ORADATA\DWENT\DIMENSIONS.ora' SIZE 20M,
'E:\ORACLE\ORADATA\DWENT\DIM-PART2.ORA' SIZE 20M
EXTENT MANAGEMENT LOCAL SEGMENT SPACE MANAGEMENT AUTO
```

Box 2.

```
CREATE TABLE "SYSTEM"."PRODUCTS" (
"PRODUCT_ID" NUMBER(10) NOT NULL,
"PRODUCT_DESC" VARCHAR2(40) NOT NULL,
"CATEGORY_CODE" VARCHAR2(10) NOT NULL,
"CATEGORY_DESC" VARCHAR2(40) NOT NULL,
"DATE" DATE NOT NULL,
CONSTRAINT "PRODUCT_PK" PRIMARY KEY("PRODUCT_ID"))
TABLESPACE "DIMENSIONS"
PARTITION BY RANGE ("DATE") (PARTITION "PRODUCT=2004"
VALUES LESS THAN (TO_DATE('2004-1-1','YYYY-MM-DD'))
TABLESPACE "DIMENSIONS" ,
PARTITION "PRODUCT=2005"
VALUES LESS THAN (TO_DATE('2005-1-1','YYYY-MM-DD'))
TABLESPACE "DIMENSIONS" );
CREATE INDEX SYSTEM.IDX_PRODUCTS ON SYSTEM.PRODUCTS ("DATE") LOCAL

CREATE TABLE "SYSTEM"."CUSTOMERS" (
"CUSTOMERS_ID" NUMBER(8) NOT NULL,
"SNAPSHOT" DATE NOT NULL,
"CUSTOMER_NAME" VARCHAR2(40) NOT NULL,
"AGE" NUMBER(2) NOT NULL,
"MARITAL_STATUS" VARCHAR2(1) NOT NULL,
"CREDIT_RATING" VARCHAR2(10) NOT NULL,
CONSTRAINT "CUSTOMER_PK" PRIMARY KEY("CUSTOMERS_ID"))
TABLESPACE "DIMENSIONS"

CREATE TABLE "SYSTEM"."CUSTOMER_ADDRESSES" (
"CUSTOMER_ID" NUMBER(8) NOT NULL,
"ADDRESS_ID" NUMBER(8) NOT NULL,
"ADDRESS_LINE1" VARCHAR2(35) NOT NULL,
"ADDRESS_LINE2" VARCHAR2(35) NOT NULL,
"POSTAL_CODE" VARCHAR2(9) NOT NULL,
CONSTRAINT "CUST_ADDRESS_PK" PRIMARY KEY("CUSTOMER_ID", "ADDRESS_ID"),
CONSTRAINT "CUST_ADDRESS_FK" FOREIGN KEY("CUSTOMER_ID")
REFERENCES "SYSTEM"."CUSTOMERS"("CUSTOMERS_ID")
TABLESPACE "DIMENSIONS"
```

Figure 19. Tablespaces definition in Oracle

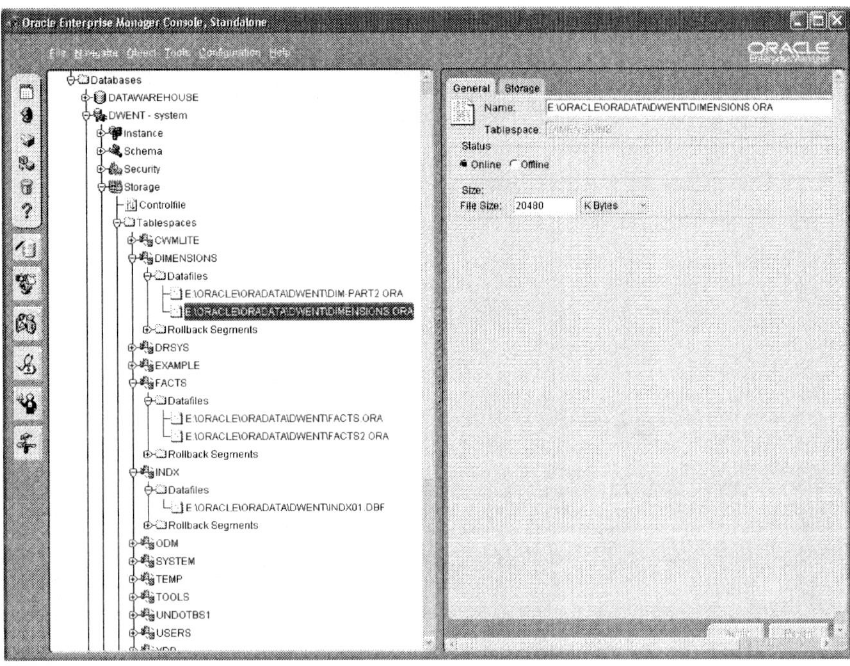

Figure 20. Definition of customer_invoices table in Oracle

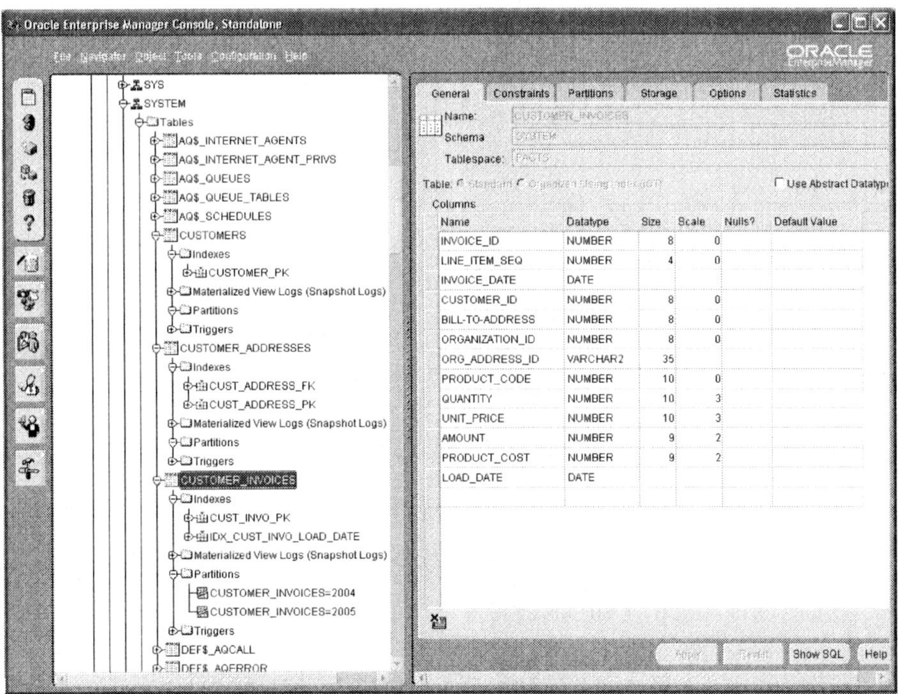

Figure 21. Definition of a partition on customer_invoices table in Oracle

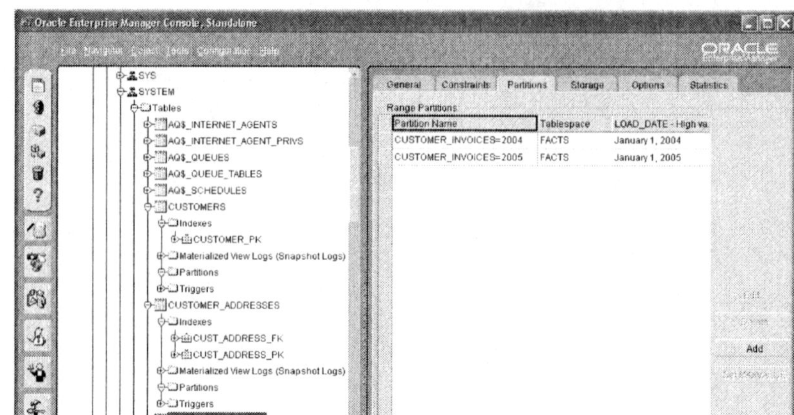

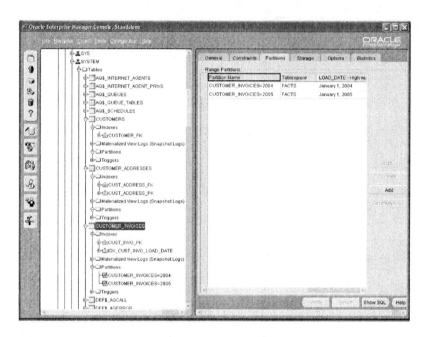

Box 3.

```
CREATE TABLE "SYSTEM"."CUSTOMER_INVOICES" (
"INVOICE_ID" NUMBER(8) NOT NULL,
"LINE_ITEM_SEQ" NUMBER(4) NOT NULL,
"INVOICE_DATE" DATE NOT NULL,
"CUSTOMER_ID" NUMBER(8) NOT NULL,
"BILL-TO-ADDRESS" NUMBER(8) NOT NULL,
"ORGANIZATION_ID" NUMBER(8) NOT NULL,
"ORG_ADDRESS_ID" VARCHAR2(35) NOT NULL,
"PRODUCT_CODE" NUMBER(10) NOT NULL,
"QUANTITY" NUMBER(10, 3) NOT NULL,
"UNIT_PRICE" NUMBER(10, 3) NOT NULL,
"AMOUNT" NUMBER(9, 2) NOT NULL,
"PRODUCT_COST" NUMBER(9, 2) NOT NULL,
"LOAD_DATE" DATE NOT NULL,
CONSTRAINT "CUST_INVO_PK" PRIMARY KEY("INVOICE_ID",
"LINE_ITEM_SEQ"),
CONSTRAINT "CUST_INVO_FK" FOREIGN KEY("CUSTOMER_ID",
"BILL-TO-ADDRESS")
REFERENCES "SYSTEM"."CUSTOMER_ADDRESSES"("CUSTOMER_ID",
"ADDRESS_ID"),
CONSTRAINT "CUST_INVO_FK2" FOREIGN KEY("PRODUCT_CODE")
REFERENCES "SYSTEM"."PRODUCTS"("PRODUCT_ID"),
CONSTRAINT "CUST_INVO_FK3" FOREIGN KEY("ORGANIZATION_ID",
"ORG_ADDRESS_ID")
REFERENCES "SYSTEM"."INTERNAL_ORG_ADDRESSES"("ORGANIZATION_ID",
"ADDRESS_ID"))
TABLESPACE "FACTS"
PARTITION BY RANGE ("LOAD_DATE") (PARTITION
"CUSTOMER_INVOICES=2004"
VALUES LESS THAN (TO_DATE('2004-1-1','YYYY-MM-DD'))
TABLESPACE "FACTS" ,
PARTITION "CUSTOMER_INVOICES=2005"
VALUES LESS THAN (TO_DATE('2005-1-1','YYYY-MM-DD'))
TABLESPACE "FACTS" );
CREATE INDEX SYSTEM.IDX_CUST_INVO_LOAD_DATE ON
SYSTEM.CUSTOMER_INVOICES ("LOAD_DATE") LOCAL
```

IMPLEMENTATION IN ORACLE

Creating a database basically consists of the following steps:

1. Creating the database's datafiles$_{11}$, its control files$_{12}$, and its redo log files$_{13}$.
2. Creating the system tablespaces.
3. Creating the data dictionary (tables, views, etc.).

Creating the Database

Before proceeding, we should point out that the information about the database repository to be created (Figure 13) was passed to our database administrator, who created the database. Following Oracle recommendations, the database creation is an operation that only should be executed by the administrator, who is also responsible for granting database permissions to users. Then, if users need to create several databases or wish to organize a big database consisting of a considerable amount of tables, the DW administrator should organize them by specifying tablespaces and decide which tables to locate in each created tablespace. Therefore, the concept of database for Oracle is at a high administrative level, avoiding programmers and even database designers from creating databases. For this reason, we do not do a deep study on the creating database statement in this section; instead, we will concentrate on the statements and stages that the DW designer has accomplished from our UML component and deployment diagrams described in previous sections. In the following subsections, we will show some SQL sentences automatically generated by the Oracle Enterprise Manager Console tool to implement the DW and some snapshots from the same tool to see the created tablespaces, tables, and indexes.

Creating the Tablespaces

The created database is called DWEnt. The first accomplished task has been to specify the tablespaces where allocating the tables. As seen in Figure 13, we need to create three tablespaces: one for the facts (FACTS), another for the dimensions (DIMENSIONS), and the last for the indexes (INDX). Furthermore, according to the same component diagram, the tablespaces for the facts and the dimensions need to be defined in two datafiles. Datafiles are the logical structure in which Oracle structures a database (i.e., the place where allocating the structures, such as tables, indices, and so forth, defined within a tablespace). Due to the amount of data to be stored in the tablespaces for facts and dimensions, the designer decided to specify two datafiles for each tablespace (Figure 13).

Figure 19 shows the definition of the tablespaces as seen in the Oracle Enterprise Console Manager. In the SQL statements in Box 1, we can see the SQL patterns generated by this management tool for defining tablespaces. We also can notice the datafiles that will be used for the DW to store fact tables, dimension tables, views, and so forth.

Creating the Data Dictionary

Once both tablespaces and datafiles have been created within a database, the next step is to define fact and dimension tables. First of all, in the SQL sentences in Box 2, we can see how we have specified the product dimension table. Apart from the columns, we should point out the fact that the table has been created partitioning it into two different partitions, one for products in year 2004 and another for new products in 2005. Furthermore, due to the fact that a partitioning has been defined, an index for the column in which we define the partitioning (i.e., date) has been created automatically. The SQL statements for the customers and customer_addresses tables are simpler, as no partitioning was defined for these tables in Figure 14. Besides, all dimension tables are defined in the dimension tablespace.

Figure 20 shows the definition of the columns of the fact table customer_invoices in Oracle. Another partitioning has been created in this table. Again, our DW is intended to locate data for every year in each different partition (Figure 14), and therefore, the column in which we base our partition is load_date. Instead of the previous dimension tables, this fact table is defined in the fact tablespace. Moreover, the index creation SQL statement has been changed slightly, because the SQL statement in Box 3 has been generated automatically by the Oracle Enterprise Manager Console, and thus, the index name was specified automatically, based on the database and table names. Then, this index name exceeded the longest index name allowed by Oracle. Therefore, the index name was shortened manually. In Figure 21, we include the definition of the partition on customer_invoices as it is shown in Oracle.

In this section, we have shown how to accomplish the implementation of a DW from our component and deployment diagrams. We believe that the implementation issues considered in our techniques are useful for the final implementation; even more, the DW administrator has implemented the DW according to our physical modeling schema.

CONCLUSION AND FUTURE WORK

After about 15 years of research and development in Data Warehouses (DW), few efforts have been dedicated to the modeling of the physical design of a DW from the early stages of a project. In this article, we have presented an adaptation of the component and deployment diagrams of the Unified Modeling Language (UML), the standard graphical notation for modeling software application needs, for the modeling of the physical design of a DW. This technique is part of our DW engineering process (Luján-Mora & Trujillo, 2004a) that addresses the design and development of both the DW back end and front end. Our method provides a unifying framework that facilitates the integration of different DW models. For example, the DW designers work with the DW administrators to understand the storage needed for the data. Our approach helps the DW designers to coordinate their efforts to include the hardware configuration (servers and drives necessary for the data) as well as the best way to organize the data into the database logical structures (tablespaces and tables).

Moreover, the UML component and deployment diagrams allow the designers to specify hardware, software, and middleware needs for a DW project. The main advantages provided by our approach are as follows:

- It is part of an integrated approach in which we use different diagrams, always following the same standard notation based on UML, for modeling all main aspects of a DW.
- It can be used for the traceability of the design of a DW, from the conceptual model to the physical model.
- It allows reducing the overall development cost as we accomplish implementation issues from the early stages of a DW project. We should take into account that modifying these aspects in subsequent design phases may result in increasing the total cost of the project.
- It supports different levels of abstraction by providing different levels of details for the same diagram.

Since our approach is based on UML, there are different CASE (Computer-Aided Software Engineering) tools that can support our approach, and having the entire design in one language

(UML) breaks down the barriers of communication between the different participants in a DW project.

Regarding future work, we are working on the formal definition of our approach by means of Object Constraint Language (OCL), because the formal definition is a well-defined way to construct a model and avoids confusion and ambiguity. We plan to provide guidelines on developing a DW by means of our DW engineering process. These guidelines will include validation checks to ensure that nothing is missed when going from one step to the next. Finally, we also plan to quantify the actual improvement in the design of a DW due to our approach.

ACKNOWLEDGMENT

We would like to thank Panos Vassiliadis for his helpful comments during the writing of the first version of this article.

This work has been supported partially by the METASIGN project (TIN2004-00779) from the Spanish Ministry of Education and Science.

REFERENCES

Abelló, A., Samos, J., & Saltor, F. (2001). A framework for the classification and description of multidimensional data models. In *Proceedings of the 12th International Conference on Database and Expert Systems Applications (DEXA'01)* (Vol. 2113, pp. 668-677), Munich, Germany. Springer-Verlag.

Abelló, A., Samos, J., & Saltor, F. (2002). YAM2 (Yet another multidimensional model): An extension of UML. In *Proceedings of the International Database Engineering & Applications Symposium (IDEAS'02)* (pp. 172-181), Edmonton, Canada. IEEE Computer Society.

Ambler, S. (2002). *A UML profile for data modeling*. Retrieved June 2003, from http://www.agiledata.org/essays/umlData ModelingProfile.html

Blaschka, M., Sapia, C., Höfling, G., & Dinter, B. (1998). Finding your way through multidimensional data models. In *Proceedings of the 9th International Conference on Database and Expert Systems Applications (DEXA'98)* (Vol. 1460, pp. 198-203), Vienna, Austria. Springer-Verlag.

Giovinazzo, W. (2000). *Object-oriented data warehouse design. Building a star schema*. Upper Saddle River, NJ: Prentice-Hall.

Golfarelli, M., Maio, D., & Rizzi, S. (1998). The dimensional fact model: A conceptual model for data warehouses. *International Journal of Cooperative Information Systems, 7*(2-3), 215-247.

Hüsemann, B., Lechtenbörger, J., & Vossen, G. (2000). Conceptual data warehouse modeling. In *Proceedings of the 2nd International Workshop on Design and Management of Data Warehouses (DMDW'00)* (pp. 6.1-6.11), Stockholm, Sweden.

Jacobson, I., Booch, G., & Rumbaugh, J. (1999). *The unified software development process*. Reading, MA: Addison-Wesley.

Jarke, M., Lenzerini, M., Vassiliou, Y., & Vassiliadis, P. (2003). *Fundamentals of data warehouses* (2nd ed.). Berlin: Springer-Verlag.

Kimball, R. (2002). *The data warehouse toolkit* (2nd ed.). New York: John Wiley & Sons.

Kimball, R., Reeves, L., Ross, M., & Thornthwaite, W. (1998). *The data warehouse lifecycle toolkit*. New York: John Wiley & Sons.

Luján-Mora, S., & Trujillo, J. (2003). A comprehensive method for data warehouse design. In *Proceedings of the 5th International Workshop on Design and Management of Data Warehouses (DMDW'03)* (pp. 1.1-1.14), Berlin, Germany.

Luján-Mora, S., & Trujillo, J. (2004a). A data warehouse engineering process. In *Proceedings of the 3rd Biennial International Conference on Advances in Information Systems (ADVIS'04)* (Vol. 3261, pp. 14-23), Izmir, Turkey. Springer-Verlag.

Luján-Mora, S., & Trujillo, J. (2004b). Modeling the physical design of data warehouses from a UML specification. In *Proceedings of the ACM Seventh International Workshop on Data Warehousing and OLAP (DOLAP 2004)* (pp. 48-57), Washington, DC.

Luján-Mora, S., Trujillo, J., & Song, I. (2002a). Extending UML for multidimensional modeling. In *Proceedings of the 5th International Conference on the Unified Modeling Language (UML'02)* (Vol. 2460, pp. 290-304), Dresden, Germany. Springer-Verlag.

Luján-Mora, S., Trujillo, J., & Song, I. (2002b). Multidimensional modeling with UML package diagrams. In *Proceedings of the 21st International Conference on Conceptual Modeling (ER'02)* (Vol. 2503, pp. 199-213), Tampere, Finland. Springer-Verlag.

Luján-Mora, S., Vassiliadis, P., & Trujillo, J. (2004). Data mapping diagrams for data warehouse design with UML. In *Proceedings of the 23rd International Conference on Conceptual Modeling (ER'04)* (pp. 191-204), Shanghai, China.

Naiburg, E., & Maksimchuk, R. (2001). *UML for database design*. Upper Saddle River, NJ: Addison-Wesley.

Nicola, M., & Rizvi, H. (2003). Storage layout and I/O performance in data warehouses. In *Proceedings of the 5th International Workshop on Design and Management of Data Warehouses (DMDW'03)* (pp. 7.1-7.9), Berlin, Germany.

Object Management Group (OMG). (2003). *Unified modeling language specification 1.5.* Retrieved January 2004, from http://www.omg.org/cgi-bin/doc?formal/03-03-01

Poe, V., Klauer, P., & Brobst, S. (1998). *Building a data warehouse for decision support* (2nd ed.). Upper Saddle River, NJ: Prentice-Hall.

Rizzi, S. (2003). Open problems in data warehousing: Eight years later. In *Proceedings of the 5th International Workshop on Design and Management of Data Warehouses (DMDW'03)*, Berlin, Germany.

Sapia, C. (1999). On modeling and predicting query behavior in OLAP systems. In *Proceedings of the 1st International Workshop on Design and Management of Data Warehouses (DMDW'99)* (pp. 1-10), Heidelberg, Germany.

Sapia, C., Blaschka, M., Höfling, G., & Dinter, B. (1998). Extending the E/R model for the multidimensional paradigm. In *Proceedings of the 1st International Workshop on Data Warehouse and Data Mining (DWDM'98)* (Vol. 1552, pp. 105-116), Singapore. Springer-Verlag.

Selonen, P., Koskimies, K., & Sakkinen, M. (2003). Transformation between UML diagrams. *Journal of Database Management, 14*(3), 37-55.

Silverston, L., Inmon, W., & Graziano, K. (1997). *The data model resource book: A library of logical data models and data warehouse designs*. New York: John Wiley & Sons.

Triantafillakis, A., Kanellis, P., & Martakos, D. (2004). Data warehouse interoperability for the extended enterprise. *Journal of Database Management, 15*(3), 73-84.

Trujillo, J., & Luján-Mora, S. (2003). A UML based approach for modeling ETL processes in data warehouses. In *Proceedings of the 22nd International Conference on Conceptual Modeling (ER'03)* (Vol. 2813, pp. 307-320), Chicago, Illinois, USA. Springer-Verlag.

Trujillo, J., Palomar, M., Gómez, J., & Song, I. (2001). Designing data warehouses with OO conceptual models. *IEEE Computer, Special Issue on Data Warehouses, 34*(12), 66-75.

Tryfona, N., Busborg, F., & Christiansen, J. (1999). starER: A conceptual model for data warehouse design. In *Proceedings of the ACM 2nd International Workshop on Data Warehousing and OLAP (DOLAP'99)* (pp. 3-8), Kansas City, Missouri, USA. ACM.

Warmer, J., & Kleppe, A. (1998). *The object constraint language: Precise modeling with UML*. Reading, MA: Addison-Wesley.

Whittle, J. (2000). Formal approaches to systems analysis using UML: An overview. *Journal of Database Management, 11*(4), 4-13.

This work was previously published in Journal of Database Management, Vol. 17, Issue 2, edited by K. Siau, pp. 12-42, copyright 2006 by IGI Publishing, formerly known as Idea Group Publishing (an imprint of IGI Global).

Chapter 2.9
GeoCache:
A Cache for GML Geographical Data

Lionel Savary
PRiSM Laboratory, France

Georges Gardarin
PRiSM Laboratory, France

Karine Zeitouni
PRiSM Laboratory, France

ABSTRACT

GML is a promising model for integrating geodata within data warehouses. The resulting databases are generally large and require spatial operators to be handled. Depending on the size of the target geographical data and the number and complexity of operators in a query, the processing time may quickly become prohibitive. To optimize spatial queries over GML encoded data, this chapter introduces a novel cache-based architecture. A new cache replacement policy is then proposed. It takes into account the containment properties of geographical data and predicates, and allows evicting the most irrelevant values from the cache. Experiences with the GeoCache prototype show the effectiveness of the proposed architecture with the associated replacement policy, compared to existing works.

INTRODUCTION

The increasing accumulation of geographical data and the heterogeneity of Geographical Information Systems (GISs) make difficult efficient query processing in distributed GIS. Novel architectures (Boucelma, Messid, & Lacroix, 2002; Chen, Wang, & Rundensteiner, 2004; Corocoles & Gonzalez, 2003; Gupta, Marciano, Zaslavsky, & Baru, 1999; Leclercq, Djamal, & Yétongnon, 1999;Sindoni, Tininini, Ambrosetti, Bedeschi, De Francisci, Gargano, Molinaro, Paolucci, Patteri, & Ticca, 2001; Stoimenov, Djordjevic-Kajan,

& Stojanovic, 2000; Voisard & Juergens, 1999; Zhang, Javed, Shaheen, & Gruenwald, 2001) are based on XML, which becomes a standard for exchanging data between heterogeneous sources. Proposed by OpenGIS (2003), GML is an XML encoding for the modeling, transport, and storage of geographical information including both the spatial and non-spatial fragments of geographical data (called features). As stressed in (Savary & Zeitouni, 2003), we believe that GML is a promising model for geographical data mediating and warehousing purpose.

By their nature, geographical data are large. Thus GML documents are often of important size. The processing time of geographical queries over such documents in a data warehouse can become too large for several reasons:

1. The query evaluator needs to parse entire documents to find and extract query relevant data.
2. Spatial operators are not cost effective, especially if the query contains complex selections and joins on large GML documents.

Moreover, computational costs of spatial operators are generally more expensive than those of standard relational operators. Thus, geographical queries on GML documents raise the problem of memory and CPU consumption. To solve this problem, we propose to exploit the specificities of a semantic cache (Dar, Franklin, Jonsson, Srivastava, & Tan, 1996) with an optimized data structure. The proposed structure aims at considerably reducing memory space by avoiding storing redundant values. Furthermore, a new cache replacement policy is proposed. It keeps in cache the most relevant data for better efficiency.

Related works generally focus on spatial data stored in object-relational databases (Beckmann, Kriegel, Schneider, & Seeger, 1990). The proposed cache organizations are better suitable for tuple-oriented data structures (Brinkhoff, 2002). Most cache replacement policies are based on Least Recently Used (LRU) and its variants. Other cache replacement policies proposed in the literature (Arlitt, Friedrich, Cherkasova, Dilley, & Jin, 1999; Cao & Irani, 1997; Lorenzetti & Rizzo, 1996) deal with relational or XML databases, but have not yet investigated the area of XML spatial databases.

The rest of the chapter is organized as follows: The second section gives an overview of related works. In the third section we present our cache architecture adapted for GML geographical data. The fourth section discusses the inference rules of spatial operators and presents an efficient replacement policy for geographical data considering inference between spatial operators. The fifth section shows some results of the proposed cache implementation and replacement policy. Finally, the conclusion summarizes our contributions and points out the main advantages of the proposed GML cache-based architecture.

RELATED WORKS

Cache Replacement Policy

In the literature, several approaches have been proposed for cache replacement policy. The most well known is the Least Recently Used (LRU) (Tanenbaum, 1992). This algorithm replaces the document requested the least recently. Rather at the opposite, the Least Frequently Used (LFU) algorithm evicts the document accessed the least frequently. A lot of extensions or variations have been proposed in the context of WWW proxy caching algorithms. We review some in the sequel.

The LRU-Threshold (Chou & DeWitt, 1985) is a simple extension of LRU in which documents larger than a given threshold size are never cached. The LRU-K (O'Neil, O'Neil, & Weikum, 1993) considers the time of the last K references to a page and uses such information to make page-replacement decisions. The page to be dropped is the one with a maximum backward K-distance for all pages in the buffer. The Log(size)+LRU (Abrams, Standbridge, Adbulla, Williams, & Fox, 1995) evicts the document with the largest log(size), and apply LRU in case of equality. The Size algorithm evicts the largest document. The Hybrid algorithm aims at reducing the total latency time by computing a function that estimates the value of keeping a page in cache. This function takes into account the time to connect with a server, the network bandwidth, the use frequency of the cache result, and the size of the document. The document with the smallest function value is then evicted. The Lowest Relative Value (LRV) algorithm includes the cost and the size of a document in estimating the utility of keeping it in cache (Lorenzetti et al., 1996). LRV evicts the document with the lowest utility value.

One of the most successful algorithms is the Greedy Dual-Size (GD-size) introduced by Cao et al. (1997). It takes into account the cost and the size of a new object. When a new object arrives, the algorithm increases the ranking of the new object by the cost of the removed object. In the same spirit, the Greedy Dual-Size Frequency (GDSF) algorithm proposed by Arlitt et al. (1999) takes into account not only the size and the cost, but also the frequency of accesses to objects. As an enhancement of GDSF, Yang, Zhang, and Zhang introduce the time factor (2003). Combined to the Taylor series, it allows predicting the time of the next access to an object. Thus, it provides a more accurate prediction on future access trends when the access patterns vary greatly. But the main bottleneck of this approach is the time consumption to recalculate the priority of each object.

Spatial Cache Replacement Policy

Most proposed spatial cache replacement policies are based on variants of LRU and are developed in the context of relational databases. In the area of spatial database systems, the effect of other page-replacement strategies has not been investigated except in (Brinkhoff, 2002).

Considering a page managed by a spatial database system, one can distinguish three categories of pages (Brinkhoff, Horn, Kriegel, & Schneider, 1993): directory pages (descriptors), data pages (classical information), and object pages (storing the exact representation of spatial objects). Using the type-based LRU (LRU-T), first the object pages are evicted, followed by the data pages, and finally by the directory pages. Using primitive based LRU (LRU-P), pages are removed from buffer according to their respective priorities. If a tree-based spatial access method is used, the highest priority is accorded from the root to the index directory pages, followed by the data pages, and finally the object pages. Thus, the priority of a page depends on its height in the tree.

Let us recall that in GIS jargon, the MBR of an object is the minimum-bounding rectangle of this object. The area of a page of objects is the minimum rectangle including all MBRs of that page. The margin is the border of an area. Beckmann et al. (1990) and Brinkhoff (2002) define five spatial pages-replacement algorithms based on spatial criteria:

1. **Maximizing the area of a page (A):** A page with a large area should stay in the buffer as long as possible. This result from the observation that the larger is the area, the more frequently the page should be requested.
2. **Maximizing the area of the entries of a page (EA):** Instead of the area of a page, the sum of the area of its entries (spatial objects) is maximized.
3. **Maximizing the margin of a page (M):** The margin of a page p is defined as the margin of the MBR containing all entries of p. The larger a page margin is, the longer it will stay in the buffer.
4. **Maximizing the margin of the entries of a page (EM):** Instead of the margin of

a page p, that of the composing MBRs are considered.

5. **Maximizing the overlaps between the entries of a page (EO):** This algorithm tries to maximize the sum of the intersection areas of all pairs of entries with overlapping MBRs.

As a synthesis, Brinkhoff (2002) proposes a combination of LRU-based and spatial page-replacement algorithms. To evict a document, a set of victim candidates is determined using the LRU strategy. Then, the page to be dropped out of the buffer is selected from the candidate set using a spatial page replacement algorithm. The page dropped by this selection is placed in an overflow buffer, where a victim is evicted using the FIFO strategy. Depending on its spatial and LRU criteria, a requested page found in the overflow buffer is moved to the standard part of the buffer, influencing the size of the candidate set.

Buffer cache techniques are mainly used in spatial database systems in order to optimize queries response time. The work conducted by Brinkhoff uses a spatial index for better management of the buffer cache. However, there is no spatial index for GML documents, as they are encoded in XML. Hence, the spatial criteria mentioned above could not be applied. Other criteria must be considered to handle geographical queries. Moreover, semantic cache gives better performances than page or tuple replacement strategies (Dar et al., 1996), but until now, it has not been really studied for geographical queries where data are stored in XML.

Cache Structure for Geographical Queries in GML

Generally, spatial data consume a lot of memory space. Hence, caching spatial objects has a tendency to flood the available space in cache. For example in a spatial join query, a spatial object A can match with several objects B1, B2, etc. Thence, a same object A can be replicated many times in spatial query results. This may considerably reduce the available space in cache, especially when a large amount of spatial fragments must be stored.

To avoid spatial object replication in cache, we propose a simple data structure, which facilitates object identification and isolation. This structure is divided into two parts. The first is devoted to the non-spatial elements of the geographical data. The second one contains non-redundant spatial fragments of geographical data (i.e., only distinct spatial objects are stored in cache). In semantic cache, the semantic region is divided into two parts (Chidlovskii, Roncancio, & Schneider, 1999): the region descriptor describing each query result stored in cache, and the region content where the data are stored. In the case of geographical queries, we introduce two kinds of region content: the *non-spatial region content*, and the *spatial region content*.

The spatial region content contains non-redundant spatial data of geographical query results, whereas the non-spatial region content contains non-spatial data of geographical query results. These region contents are associated with the geographical region descriptor. It contains information about each geographical query stored in cache. The cache is then divided into two parts (see Figure 1): (i) the *non-spatial part of the cache* composed of the *non-spatial region content* and it associated description contained in the *geographical region descriptor* (see the third section); (ii) *the spatial part of the cache* composed of the *spatial region contents* and it associated description contained in the *geographical region descriptor* (see the third section).

Non-Spatial Part of the Cache

All data are encoded in XML. For simplicity and standard enforcement, we encode XML data as DOM trees. More compact structures are possible,

Figure 1. General cache structure

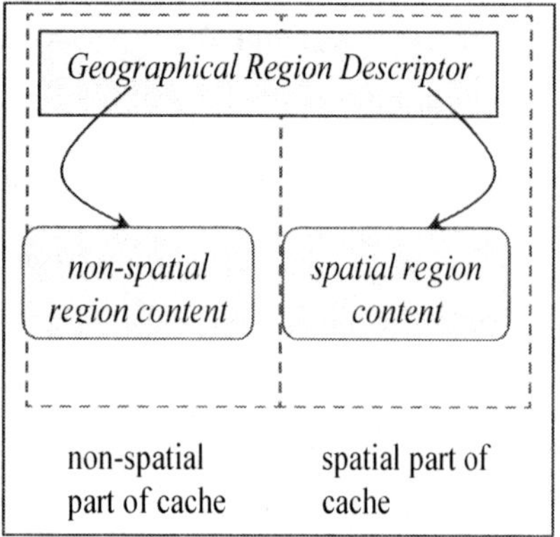

Figure 2. Cache organization

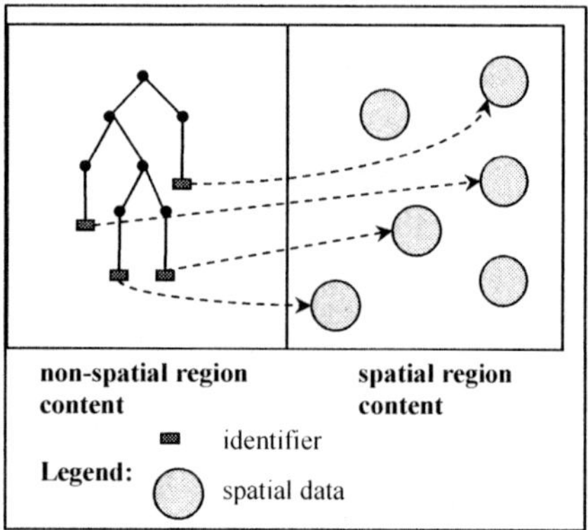

Figure 3. Spatial part of the cache

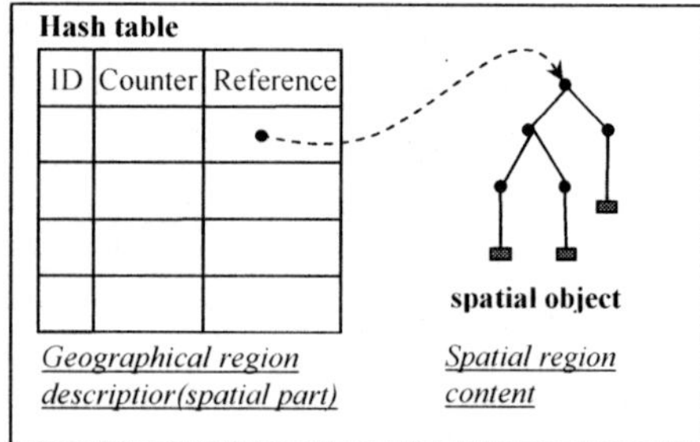

but it would not change the relative results and would require specific developments. Thence, query results are stored in cache as DOM trees that compose the *non-spatial region* content (see Figure 2). To separate non-spatial and spatial data, a generated identifier replaces each GML sub-tree (XML fragment with "gml:" as namespace). The identifier value is used to uniquely retrieve the corresponding spatial fragment stored in the spatial part of the cache structure. For non-spatial data, the *geographical region descriptor* contains classical information about each query (*region formula, replacement values, and pointers*) (Chidlovskii et al., 1999).

The second part of the cache contains a specific structure designed, from one hand, to store only distinct spatial objects, and from the other hand, to quickly retrieve spatial fragments. Each spatial fragment is stored as a DOM tree, since spatial data are in GML. The spatial fragments compose the *spatial region content.* Each spatial fragment is referenced in a hash table. The identifier is used as a key of the hash table. It is composed of a counter value and a reference to the root tree of the spatial fragment (see Figure 3). Thus, each DOM tree corresponding to a spatial fragment can be quickly retrieved in the spatial part of the cache. The counter value indicates the number of references to the spatial fragment from the non-spatial part of the cache. Thence, the entry can be deleted when the counter is 0. The hash table composes the *geographical region descriptor* for spatial data. It contains: identifiers ID, counters and references to spatial fragments (see Figure 3).

Cache Insertion Algorithm

When a geographical query result must be stored in the cache, the spatial and non-spatial fragments have to be identified and stored in the corresponding part of the cache. The insertion algorithm is described in Figure 4. A key issue is the generation of the GML fragment identifier. We simply use the centroïde of the object defined as a point standing for the geometry center. More sophisticated identifiers could be used (e.g., a numerical encoding of the shapes to avoid confusion between geographical objects of same centroïde).

A new DOM tree is created (line 00) for the non-spatial fragment. The XML query result is read homogeneously fragment by fragment. If a non-spatial fragment is read (line 02), then it is placed in the non-spatial part of the cache (line 03). Else, if a spatial fragment is read (line 05), an identifier is automatically generated (line 06). If the identifier already exists in the hash table (line 07), the counter is incremented (line 08). Otherwise, a Dom tree for the spatial fragment is generated (line 11) and inserted into the hash table with the new identifier (line 12). The generated identifier is added in the non-spatial DOM tree (line 14).

Cache Deletion Algorithm

When a geographical query result has to be evicted from the cache, the DOM tree corresponding to the non-spatial part of the result is logically removed. Physically, only the referenced spatial fragments with counter value 0 (no reference) are removed. The algorithm is sketched in Figure 5, which depicts the cache deletion algorithm. A Boolean value defined at line 00 indicates if all counters of a query result have no reference. The DOM tree corresponding to the non-spatial part of the query result is retrieved using the root node (line 01). Then the tree is scanned to get the spatial fragments corresponding to each identifier stored in the tree (line 02). The counter value related to the spatial fragment is decremented (line 03). If this value is equal to 0 (line 04), the DOM tree corresponding to the spatial fragment is removed from the hash table (line 05). Once all identifiers have been scanned, if there is no identifier with associated counter strictly greater than 0 (line 11), the non-spatial tree corresponding to the query result is removed from the cache (line 12).

GEOGRAPHICAL QUERY PROCESSING

Spatial Operators with Inference Rules

In this section, we introduce the spatial operators we use to compose queries. We further detail the inference rules between geographical predicates, which are useful for determining if a query result is included in another one. This is important for semantic cache management.

Spatial Operators and Predicates

For geographical queries, not only conventional comparison operators $\theta = \{\leq, <, =, \neq, >, \geq\}$ must be taken into account, but also spatial operators. To query the database, we use XQuery extended with spatial operators as proposed in GML-QL (Vatsavai, 2002). For illustration purposes, we select the following spatial operators (OpenGIS, 2003):

Figure 4. Cache insertion algorithm

```
Insert_in_Cache(File Query_Result)
{
00 NStree = Create_Dom_Tree()
01 While ReadFile(Query_Result)
02 If (non-spatial data)
03   Insert (NStree, non-spatial data)
04 End If
05 Else If (spatial data)
06   id = Generate_identifier (spatial data)
07   If (HashTable.Contains(id))
08     HashTable.UpdateCounter(id)
09   End If
10   Else
11     Stree = Create_Dom_Tree (spatial
data)
12     Insert_into_HashTable (id, Stree)
13   End Else
14   Insert (NStree, id)
15   End Else If
16 EndWhile
}
```

Figure 5. Cache deletion algorithm

```
Remove_from_Cache (root r)
{
boolean b = true;
NStree = Get_Dom_Tree (r)
For all id in NStree
  HashTable.DecrementsCounter (id))
  If (HashTable.CounterValue (id) == 0)
    HashTable.Remove (id)
  End If
  Else
  b = false
  End Else
End For
If (b)
Remove (NStree)
End If
}
```

- **Overlaps (Path1, Path2):** Determines if two spatial objects specified by the paths Path1 and Path2 are overlapping.
- **Distance (Path1, Path2) θ d:** Determines if the distance between two spatial objects specified by the paths Pahtl and Path2 satisfies the relation θ d, where d is a given distance.
- **Touch (Path1, Path2):** Determines if two spatial objects specified by the paths Path1 and Path2 are touching.
- Intersects (Paht1, Path2): Determines if two spatial objects specified by the paths Path1 and Path2 are intersecting.
- **Within (Path1, Path2):** Determines if the spatial object specified by Path1 is within the spatial object of path Path2.
- **Equals (Path1, Path2):** Determines if the spatial object specified by the path Path1 is "spatially equal" to the spatial object of path Path2.
- **Disjoint (Path1, Path2):** Determines if the spatial object specified by the path Path1 is "spatially disjoint" to the spatial object of path Path2.
- **Crosses (Path1, Path2):** Determines if the spatial object specified by the path Path1 crosses the spatial object of path Path2.
- **Contains (Path1, Path2):** Determines if the spatial object specified by the path Path1 "spatially contains" the spatial object of path Path2.

Some spatial predicates are based on topological relationships and can be deduced from other spatial predicates. This means the results of spatial queries can be contained in results of other queries, with different operators. If inference rules between spatial predicates can be established, then it could be possible to determine whether the result of a geographical query is contained in another one. This feature may allow using the cached results instead of accessing the original document, and thus, optimizing the query.

For example, consider the two GML documents: Feature1.xml in Figure 6, which contains the non-spatial elements T1, T2, T3 and a spatial fragment of type linearGeometry (OpenGIS, 2003) standing for a polyline; and Feature2.xml in Figure 7, which contains the non-spatial fragments N1, N2, and a spatial fragment also of type linearGeometry.

Let R1 and R2 be two GML-QL queries (see Box 1). R1 determines if two objects Feature1 and Feature2 are within a distance closer than 1200. In this case, the query returns the geometry of the two objects and the specified values. R2 determines if two objects Feature1 and Feature2 intersects. In this case, the query returns their geometry with the other specified elements.

Two objects are intersecting if their distance is equals to 0. Hence, the result of R2 is contained in the results of R1 as all the values returned by R2 are included in the values returned by R1, and 1200 > 0. Thus, the result of R2 can be extracted from that of R1 without scanning the database, provided that the result of R2 is kept in cache. As semantic cache replacement policies are based on the semantic of queries, the query-processing algorithm shall determine if a geographical query result is contained in cache. Thus, an originality of semantic caching in geographical databases is the ability to compare geographical predicates.

Spatial Predicate Inference Rules

Let us denote a $\Rightarrow$ b the implication and a $\Leftrightarrow$ b the equivalence inference rule. Let O1 and O2 be two geometric objects. A simple inference rule is, for all real d positive or null: Intersect (O_1, O_2) $\Rightarrow$ (Distance $(O_1, O_2) < d$).

Using the nine operators defined in the third section, more inference rules are given in Figure 8. Notice that, for a couple of objects (O1, O2) verifying a predicate P at level i, if the predicate P implies the predicate Q at level i+1 and if the predicate Q implies the predicate S at level i+2, then the operator P at level i, infers the operator S at level i+2. That is the implication of predicates is transitive. It is important to take into account all rules for better cache management.

In our proposed architecture (Savary et al., 2003), spatial queries are computed using JTS, is a spatial Java API proposed by the OGC (Open Geospatial Consortium) and based on the OpenGIS Simple Feature Specification (OpenGIS, 1999). It implements the spatial operators introduced in the third section. Hence, in this practical context, the use of spatial predicate inference rules makes it possible to determine if a new query is contained or partially contained in cache.

Query Processing

Suppose an input query Q and a set C = $\{C_1, C_2, \ldots, C_K\}$ of K cache queries. Then, according to (Lee & Chu, 1999) there are five general cases of query matching: *exact match*, *containing match*, *contained match*, *overlapping match* and *disjoint match*. Figure 9 illustrates the various cases. The grey box in shop outline stands for the queries results C stored in cache. The white box with full outline represents the new query Q. For an *exact*

Box 1.

```
R1:
For $b in doc("Feature1.xml")/Feature1/Object
For $c in doc("Feature2.xml")/Feature2/Object
Where Distance ($b/gml:LineString/gml:coordinates,$c/gml:LineString/gml:coor-
dinates) < 1200
Return
<Result> { $b/T1 } { $b/T3 }
{$b/gml:LineString/gml:coordinates}
   { $c/N2 }
        {$c/gml:LineString/gml:coordinates}
</Result>
R2:
For $b in doc("Feature1.xml")/Feature1/Object
For $c in doc("Feature2.xml")/Feature2/Object
Where  Intersect  ($b/gml:LineString/gml:coordinates,  $c/gml:LineString/gml:
coordinates)
Return
 <Result>
  { $b/T1 }
 {$b/gml:LineString/gml:coordinates}
  { $c/N2 }
 {$c/gml:LineString/gml:coordinates}
 </Result>
```

Figure 6. The Feature1.xml data

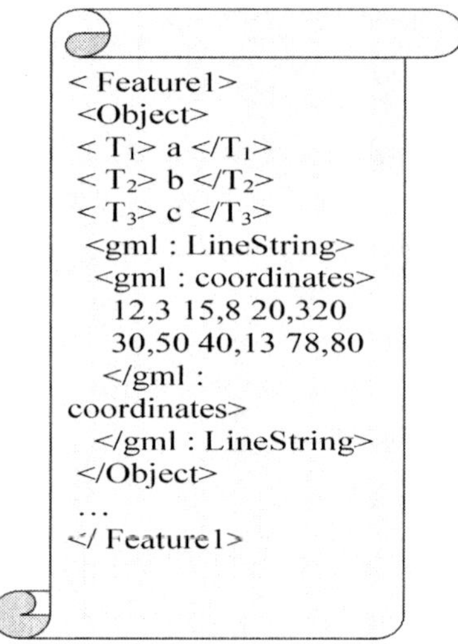

```
< Feature1>
 <Object>
 < T1> a </T1>
 < T2> b </T2>
 < T3> c </T3>
  <gml : LineString>
   <gml : coordinates>
    12,3 15,8 20,320
    30,50 40,13 78,80
   </gml :
coordinates>
  </gml : LineString>
 </Object>
 ...
</ Feature1>
```

Figure 7. The Feature2.xml data

```
< Feature2>
 <Object>
      > i </N1>
      > j </N2>
   <gml : LineString>
    <gml : coordinates>
     6,11 15,13 75,145
     1458, 6321 785,
6831
     45896, 4526
256,745
    </gml :
coordinates>
   </gml : LineString>
 </Object>

</ Feature2>
```

match, the query Q already exists in cache: $\exists C_i \subseteq C \setminus C_i = Q$. For a *containing match*, the query Q is totally contained in the cache: $\exists D \subseteq C \setminus (Q \subset D \wedge D \not\subset Q)$. For a *contained match*, the query Q contains some cache queries: $\exists D \subseteq C \setminus (Q \not\subset D \wedge D \subset Q)$. For an *overlapping match*, the query Q is partially contained in cache: $\exists D \subseteq C \setminus (Q \not\subset D \wedge D \not\subset Q \wedge Q \cap D \neq \emptyset)$. Finally, for a *disjoint match*, the query Q is not contained in cache: $Q \cap C = \emptyset$.

For *exact* and *containing* match, Q can be totally answered from the cache; but for *contained* and *overlapping* match, an access to the database is required in order to answer the *remainder query* (Dar et al., 1996). For geographical queries and especially for join queries, the processing time can be prohibitive. To palliate this disadvantage, we propose to exploit the *geographical region descriptor.*

General Query Processing Scheme

The geographical query processing is depicted in Figure 10. When a new query Q must be processed, the Query Analyzer determines if the query can be answered from the cache, the database or both. Then the Query Rewriter rewrites Q into the *remainder and probe queries*. We introduce two kinds of *probe query*:

- The *elementary probe query* allows retrieving totally or partially the data stored in cache to answer the query Q.
- The *optimized probe query* aims at retrieving the *temporary spatial data* for a further optimization. This will be detailed in the next section.

The *elementary probe query* is performed by the cache Query Executor, with the non-spatial and spatial data stored in cache. In the cases of *contained match, overlapping match,* and *disjoint match,* if an optimization is possible using

Figure 8. Spatial predicates inference rules

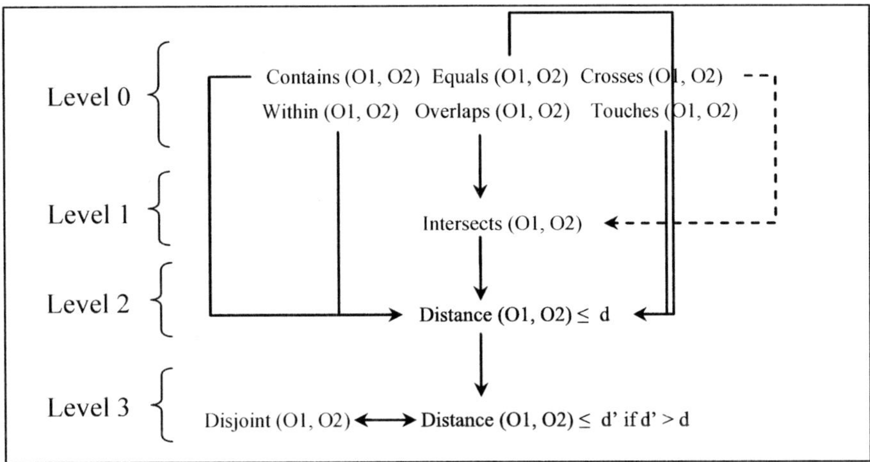

Figure 9. Query match types

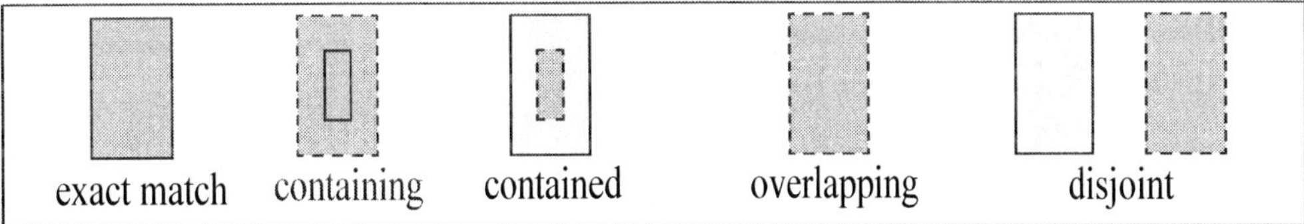

the spatial data stored in cache (as shown in the example in the fourth section), then the required *temporary spatial data* are determined using the *optimized probe query* and forwarded to the Spatial Query Executor. The *temporary spatial data* is composed of *n* documents corresponding to the *n* features implied in the *remainder query*. Each document only contains the geometries of a specific feature. The *remainder query* is split by the Query Decomposer into the *non-spatial remainder query*, and the spatial *remainder query*. The *non-spatial remainder query* is free of spatial predicates and allows interrogating the database. The *database temporary results* (*non-spatial remainder query* results) are then forwarded to the Spatial Query Executor. These *database temporary results* are composed of *n* documents corresponding to the *n* features implied in the *remainder query*. The Spatial Query Executor performs the *remainder spatial query*, which contains the spatial predicates. If the *temporary spatial data* is not empty, an optimization is possible which avoids computing spatial predicates. The result (*remainder query result*) is then sent to the Query Combiner, which merges the *remainder* and *elementary probe query results* into a single GML document R. The query result R of Q is then sent to the cache Replacement Manager. The Cache Replacement Manager determines then if the result has to be placed in cache or not, and which cache queries must be evicted.

Figure 10. Geographical query processing

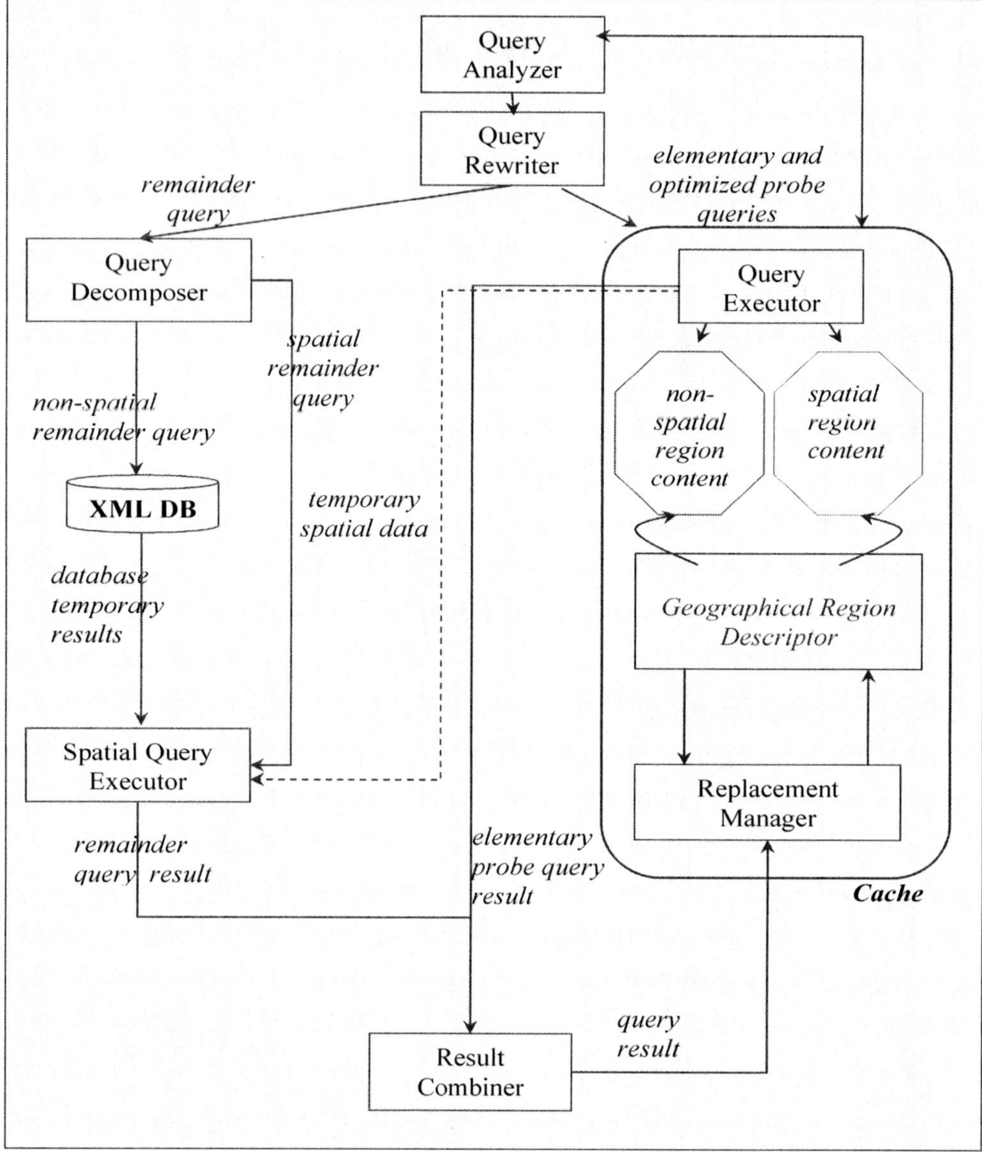

Cache-Based Query Processing and Optimizations

In the cases of *contained match*, *overlapping match*, and *disjoint match*, data, which are not contained in cache, are retrieved from the database. To reduce the I/O and CPU costs, we propose to exploit the spatial part of the cache. The idea is to determine the set of spatial geometries implied in the *remainder query*. If this set is known, then it is possible to directly compare the geometries contained in this set with the geometries contained in the documents of the *database temporary results*. This set of spatial geometries results from the *optimized probe query*. The *optimized probe query* results (*the temporary spatial data*) contain *n* documents corresponding to the *n* features implied in the *remainder query*. Only spatial geometries are stored in these documents. This allows to avoid the process of spatial predicates for the *remainder query result*. Only comparisons with geometry features of *database temporary result* and geometry features of their corresponding *temporary spatial data* are required. Moreover, it makes a better cache management, since the size of those *temporary spatial data* documents is reduced. Thus, performances are improved especially for join queries, where multiple scans are often necessary. For example, if the query Q is the following:

```
For  $b  in  doc("Feature1.xml")/Fea-
    ture1/Object
For  $c  in  doc("Feature2.xml")/Fea-
    ture2/Object
Where  Intersect  ($b/gml:LineString/
    gml:coordinates,  $c/gml:LineS-
    tring/gml:coordinates)
Return <Result> { $b/a } { $c/b } {
    $b/c } </Result>
```

Let C_i be the closest cache query of Q contained in cache, i.e., the query containing all the required entities to answer Q with a minimal number of attributes not contained in Q (not mentioned in the returned result of Q). C_i is found using the inference rules, which have been previously presented (Figure 6). For instance, C_i is found to be the query:

```
For  $b  in  doc("Feature1.xml")/Fea-
    ture1/Object
For  $c  in  doc("Feature2.xml")/Fea-
    ture2/Object
Where Distance ($b/gml:LineString/gml:
    coordinates,  $c/gml:LineString/
    gml:coordinates) < 9
Return
<Result> { $b/e } { $c/f }  </Re-
    sult>
```

Here, the *optimized probe query* contains the *intersect* predicate. It will be applied to the cache query result C_i. The resulting *temporary spatial data* will be composed of two documents containing only geometries of Feature1 and geometries of Feature2 that match. The geometries of these documents will be compared, using their centroïdes, with geometries of those of the *database temporary results*. The required data *a*, *b* and *c* of respectively Feature1 and Feature2 are then retrieved to answer the query Q.

Cache Replacement Policy: B&BGDSF

When a new query result must be stored in a saturated cache, the most irrelevant queries have to be evicted. It is important to take into account the constraints on size but also the cost and frequency of access to spatial objects. Thus, we base our replacement policy on the Greedy Dual-Size Frequency (GDSF) algorithm proposed by Arlitt et al. (1999). This strategy replaces the object with the smallest key value for a certain cost function. When an object i is requested, it is given a priority key K_i computed as follows:

$$K_i = F_i * \frac{C_i}{S_i} + L$$

where: F_i is the frequency usage of the object I; C_i is the cost associated with bringing the object i into cache.; S_i is the size of the object I; and, L is a running age factor that starts at 0 and is updated for each replaced object o to the priority key of this object in the priority queue, i.e., $L = K_o$.

Our cache replacement policy performs in two steps. First, the cost of each query result contained in cache is computed using the GDSF policy. C_i stands for the cost of a geographical query. The computation of C_i for the experimentations is discussed in the experimentation section. Secondly, The Branch and Bound algorithm (Kellerer, Pferschy, & Pisinger, 2004) is used to determine the most irrelevant query result in cache. In the case of spatial cache replacement policy as in the general case, the problem is to determine the optimal set of candidates to be removed. This set must have the minimal cost in cache and the minimal cost out of cache if it must be recomputed, as formalized by equation (i). Secondly, the total size of the set must be equal or greater than the size of the new query result, as formalized by equation (ii) below.

$$MIN(\sum_{i=1}^{n} X_i * C_{CacheCPU} + (1 - X_i) * C_{DiskCPU}) \quad \text{(i)}$$

$$\sum_{i=1}^{n} X_i * S_i \geq T \quad \text{(ii)}$$

In equation (i), X_i is set to 1 if object i is kept in cache, and 0 otherwise, $C_{CacheCPU}$ stands for the cost in CPU of the cached query result and $C_{DiskCPU}$ stands for the cost in CPU to re-compute the query result I from the database. The second constraint concerned the size of the set of cached queries to remove. It is given by equation (ii), where T stands for the size of the new query result, and Si, the size of a cached query result i. This is equivalent to the following expression, which can be resolved using the Branch and Bound algorithm:

$$MAX(\sum_{i=1}^{n} Y_i * (C_{CacheCPU} - C_{DiskCPU})) \quad \text{(i)}$$

where

$$\sum_{i=1}^{n} Y_i * S_i \leq \sum_{i=1}^{n} S_i - T \quad \text{(ii)}$$

PROOF

By replacing X_i by $(1\text{-}Y_i)$ in equation (i), we obtain:

$$\Leftrightarrow MIN(\sum_{i=1}^{n} (C_{CacheCPU} - Y_i * (C_{CacheCPU} - C_{DiskCPU})) \quad \text{(i)}$$

Since $\sum_{i=1}^{n} C_{CacheCPU}$ is a constant, the problem is reduced to the second expression:

$$\Leftrightarrow MIN(-\sum_{i=1}^{n} Y_i * (C_{CacheCPU} - C_{DiskCPU}))$$
$$\Leftrightarrow MAX(\sum_{i=1}^{n} Y_i * (C_{CacheCPU} - C_{DiskCPU})) \quad \text{(i)}$$

$$\Leftrightarrow \sum_{i=1}^{n} (1 - Y_i) * S_i \geq T \Leftrightarrow \sum_{i=1}^{n} Y_i * S_i \leq \sum_{i=1}^{n} S_i - T \quad \text{(ii)}$$

This formalization is generally used to resolve the *knapsack problem*, in operational research (Silvano, 1990). To solve this operational research problem, we choose the Branch and Bound algorithm. The Branch and Bound algorithm take as parameter: the GDSF cost of each cache query result, the size of each query result, and the size of the new query result.

Compared to GDSF, the proposed $B\&B_{GDSF}$ algorithm determines the optimal set of candidates to be replaced by the new query result. Thus, better cache management is done. The results depicted in the experimentations below, show the performances of this algorithm compared to existing ones.

EXPERIMENTATIONS

The Geographical Data Warehouse Architecture

The cache has been implemented in a geographical data warehouse prototype. The architecture of this prototype is depicted in Figure 11.

The geographical data warehouse is entirely built using open sources components based on Java and XML. The data extract from heterogeneous sources are encoded in SOAP (Simple Object Access Protocol)--a standard protocol defined by the W3C - messages and sent to the DSA (Data

Figure 11. Geographical data warehouse architecture

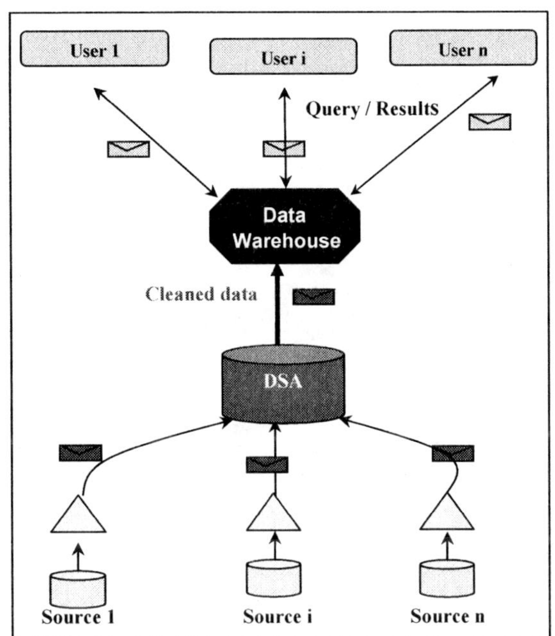

Staging Area) to be processed. In the DSA, the data are integrated using spatial ETL process before it feeds the database of the data warehouse. The XML database eXist has been chosen as the data warehouse manager, since it is an open source and allows querying the non-spatial data using the XQuery language. The geographical queries are computed using the JTS (Java Topology Suite) packages which is OpenGIS compliant. Users queries are sent to the data warehouse embedded in a SOAP message.

The implemented architecture is composed of four levels: sources, the DSA, the data warehouse, and the clients. The communication between these four levels is based on SOAP protocol using the apache tomcat Web application. At the source level, Hibernate 2.0 is used for extracting data from heterogeneous database. Hibernate has been chosen since it allows interrogating any database with the singular HQL language. The data extracted from these sources are sent into a SOAP message to the DSA in order to be cleaned and integrated by the spatial ETL transformation process. Once the data have been prepared, they are sent to the data warehouse in SOAP messages to feed the XML eXist database. By the same way, the user queries are sent into a SOAP message to the data warehouse. The query analyzer then extracts and processes the queries (the fourth section). The user level is composed of a graphical interface, which has been implemented using SWING Java technology. This interface allows users to specify a request and display the queries results. The queries results are encoded in SVG (Scalable Vector Graphic)—the XML format for vector graphic specified by the W3C—embedded in SOAP messages and sent to the users to be graphically displayed.

The Experimentation Parameters and Results

The experimentations were done on a Pentium IV 2.0 GHz with 512 MB of memory. The used dataset is composed of 10000 and 1000 features stored in two GML documents in the eXist (Meier, 2002) XML database. They describe the road networks of Lille (a French town) (Savary & Zeitouni, 2005). Their geometry is of type linearGeometry (OpenGIS, 2003). They have been generated by our ETL process from the source, one was in format shape (the exchange format of ArcGIS software), and the other was stored in the Oracle 9i DBMS which allows spatial data management.

Our cache replacement policy has been compared with GDSF (without Branch and Bound selection), LFU and LRU, which is the most popular in spatial database cache replacement policy. The cache size has been moving from 10 to 100 Mega Bytes. To compare those policies, we have simulated a flow of approximately 70 queries (point, range and join queries), possibly made by different users. We have performed two series of experimentations, consisting in two sets of about 70 GML XQueries. The one (random.gxq file) uses an arbitrary order of queries, while in the second (freq.gxq) the most recent queries are repeated. These two files contain different queries. The evaluation of the first query set is reported in Figures 12, 13, and 16. Whereas the results obtained using freq.gxq file are illustrated in Figures 14, 15, and 17.

Figure 12 shows the hit ratio performances realized with a random dataset queries. Here we can see that in general, GDSF and $B\&B_{GDSF}$, outperform LRU and LFU. When the cache size becomes larger, GDSF roughly scale like $B\&B_{GDSF}$. But with smaller cache size, our $B\&B_{GDSF}$ algorithm performs better than the other cache replacement policies. This is because the Branch and Bound algorithm associated to GDSF only evicts the most irrelevant cache results with smallest cost and which size sum is greater but closest to the new query, which must be put in cache. Thus, the cache is better managed. For example with a cache size equal to 20 Mega Bytes, $B\&B_{GDSF}$ performs about 8% better than GDSF.

Figure 13 shows the cache coverage ratio in percentage. Cache coverage hit shows the effect of partial matching in semantic caching. This information is not readable in traditional cache hit ratio. It is defined as follows (Lee & Chu, 1998):

Given a query set consisting of I queries q_1, ...q_I, let Ai be the number of answers found in the cache for the query q_i, and let Bi be the total number of answers for the query q_i for $1 \leq i \leq I$.

$$CCR = \frac{\sum_{i=1}^{I} QCRi}{I} \quad \text{where } QCRi = \frac{Ai}{Bi} \quad \text{if } Bi > 0.$$

Here again, we can notice that in general GDSF and $B\&B_{GDSF}$ outperform LRU and LFU. But it is interesting to see that compared to Figure 12, when the number of hit ratio is equal for GDSF and $B\&B_{GDSF}$ with a cache size of 50 MB or 60MB, the number of cache coverage ratio is better using the $B\&B_{GDSF}$ algorithm Figure 16 shows the performance *gain* which is defined as follows (Brinkhoff et al., 2002):

$$gain = \frac{disk\_acces\_for\_a\_given\_Policy}{disk\_access\_of\_B\&B_{GDSF}}.$$

where *Policy* stands for one of the three cache replacement policies: GDSF, LRU, and LFU.

The performance gain shows the reduction in the number of disk access using the proposed cache replacement policy, compared to others. In general, we can notice that the gain is positive using our proposed algorithm. However, depending on the cache size, the gain with GDSF is more or less important, especially when the cache size becomes small or large. For example in Figure 12 with a cache size equal to 20 or 100 MB,

Figure 12. Cache hit ratio

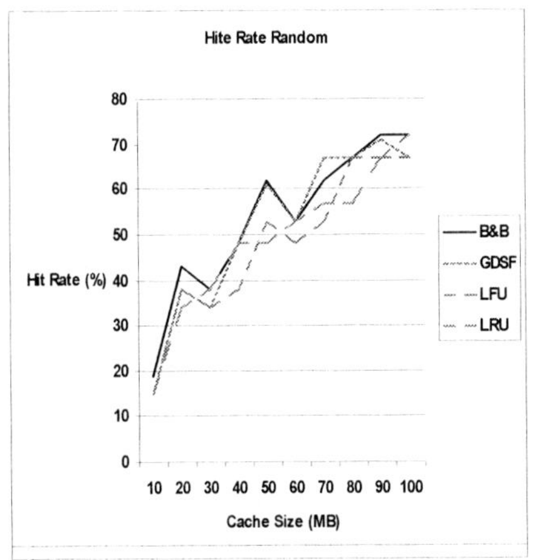

Figure 13. Cache coverage ratio

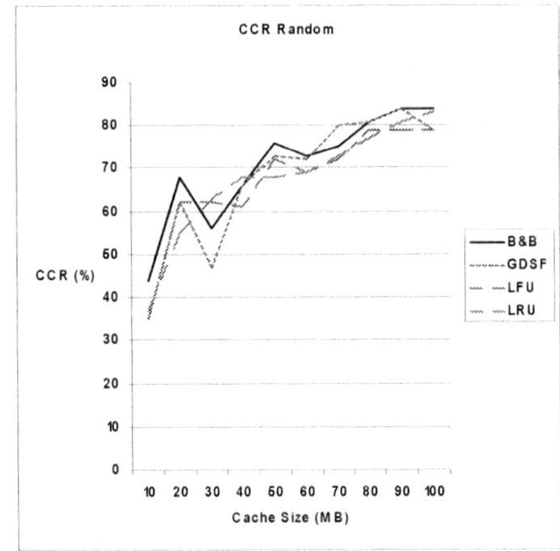

Figure 14. Cache hit ratio

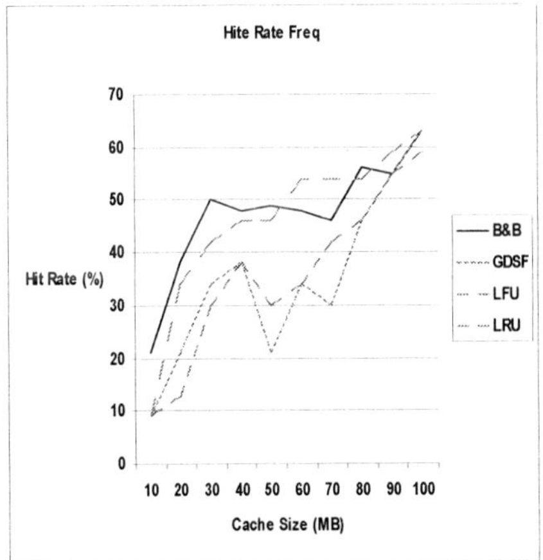

Figure 15. Cache coverage ratio

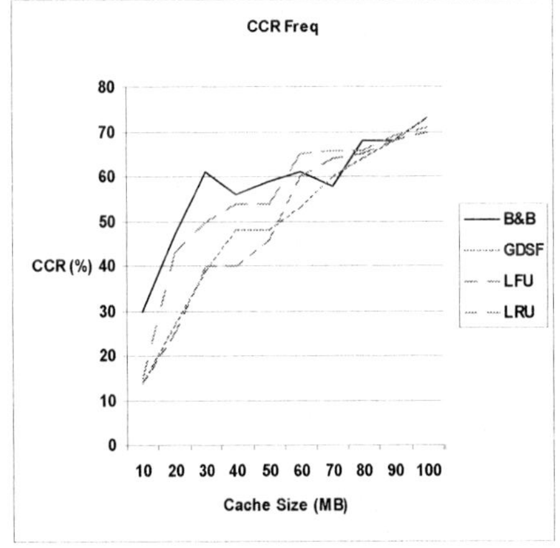

Figure 16. Performance gain

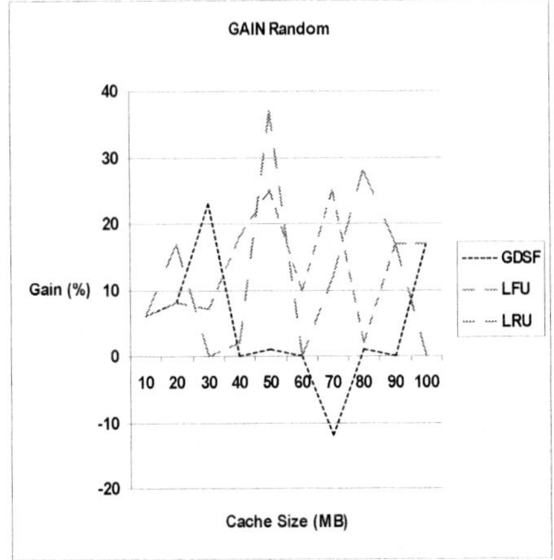

Figure 17. Performance gain

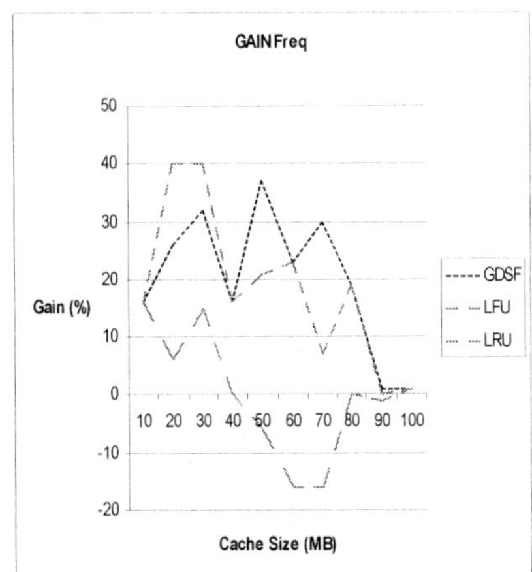

the performances of GDSF are the same or less than $B\&B_{GDSF}$. However, if we look at the gain distribution, the values obtained are about 8 or 18 %. Especially with a cache size of 100 MB, the hit ratio for GDSF and $B\&B_{GDSF}$ are roughly the same, but the gain obtained is different about 18%. These results show that the cache is better managed using $B\&B_{GDSF}$. For a cache size equal to 30 Mega Bytes, we can see that the gain compared to GDSF is roughly equal to 23%. But for a cache size between 40 and 90 MB, the gain compared to GDSF is roughly the same or less than our proposed algorithm. But the Hit Ratio and Cache Coverage Ratio show that better cache management is obtained using the $B\&B_{GDSF}$ algorithm. Moreover, for a cache size of 40 or 60 MB in Figure 12, LRU performs like $B\&B_{GDSF}$; but in Figure 16 the gain obtained compared to LRU is the same or better (for example with a cache size of 40 MB). For a cache size of 50 MB peaks of gain, around 37% with LRU and 25% with LFU are obtained.

Figures 14, 15, and 17 show other experimentations performed with another query set where the most recent queries are repeated. In this case, we notice that LRU outperforms GDSF and LFU, but $B\&B_{GDSF}$ outperforms GDSF, LFU, and LRU, especially when the cache size is smaller (from 10 to 50 MB). This is due to a better selection of cache queries to evict.

These results show that the cache is better managed using the $B\&B_{GDSF}$ compared to GDSF, LFU and LRU cache replacement policies. The cache hit ratio (Figures 12 and 14) show that $B\&B_{GDSF}$ outperforms GDSF, LRU, and LFU; but when the performances of $B\&B_{GDSF}$ are roughly equal to another cache replacement algorithm, the cache coverage ratio (Figures 13 and 15) show that a better cache management is obtained using the proposed algorithm. This is confirmed by the performance gain in disk access presented in Figures 16 and 17.

CONCLUSION

In order to optimize the I/O-performance of a database system, many techniques have been considered. One of the most used in spatial database is the buffering technique associate to the LRU strategy and its variants. Related works generally focus on spatial data stored in object-relational databases. With the popularity of Internet and XML, the OGC proposed the GML semi-structured language, which is an extension of XML for the modeling, transport, and storage of geographical information including both spatial and non-spatial fragments of geographical data. The GML language is more and more used in many distributed architectures. In order to handle efficiently geographical queries stored in GML documents, we have proposed an appropriate cache structure and cache replacement policy. The proposed cache structure aims at considerably reduce the memory used to store geographical queries. Only non redundant spatial fragments of a geographical query are stored. Thus, much more queries can be stored in cache. This architecture is associated to a semantic cache where inference rules between spatial predicates are exploited for a better cache management. Associated with this semantic cache, a new cache replacement policy is proposed. It takes into account the cost implied by the different predicates and gives better performances than existing cache replacement policies like GDSF, LRU, and LFU. The proposed cache replacement policy is based on Branch and Bound and GDSF algorithms. It only evicts the most irrelevant cache results according to the cost and size of data stored in cache. The hit ratio, cache coverage ratio and performance gains show that the proposed $B\&B_{GDSF}$ replacement policy outperforms GDSF, LRU and LFU algorithms. In perspective, this work will be extended to spatiotemporal data stored in semi-structured documents in order to optimize spatiotemporal queries.

REFERENCES

Aboulnaga, A., & Naughton, J. (2000). Accurate estimation of the cost of spatial selections. *IEEE International Conference on Data Engineering (ICDE '00)* (pp. 123-134), San Diego, CA.

Abrams, M., Standbridge, C. R., Adbulla, G., Williams, S., & Fox, E. A. (1995). Caching proxies: Limitations and potentials. *WWW-4*, Boston Conference.

Arlitt, M., Friedrich, R., Cherkasova, L., Dilley, J., & Jin, T. (1999). Evaluating content management techniques for Web proxy caches. In *Proceedings of the Workshop on Internet Server Performance (WISP)*, Atlanta, GA.

Beckmann, N., Kriegel, H. P., Schneider, R., & Seeger, B. (1990). An efficient and robust access method for points and rectangles. In *Proceeding ACM SIGMOD International Conference on Management of Data* (pp. 322-331), Atlantic City, NJ.

Boucelma, O., Messid, M., & Lacroix, Z. (2002). A WFS-based mediation system for GIS interoperability. The *10th ACM International Symposium on Advances in Geographical Information Systems (ACM GIS)*, McLean, VA.

Brinkhoff, T. (2002). A robust and self-tuning page-replacement strategy for spatial database systems. The *8th International Conference on Extending Database Technology (EDBT 2002)* (pp. 533-552), Prague, Czech Republic, 2002. LNCS, Vol. 2287, Springer-Verlag.

Brinkhoff, T., Horn, H., Kriegel, H. P., & Schneider, R. (1993). A storage and access architecture for efficient query processing in spatial databases. In *Proceedings of the 3rd International Symposium on Large Spatial Databases* (pp. 357-376), Singapore. LNCS, Vol. 692, Springer.

Cao, P., & Irani, S. (1997). Cost-aware WWW proxy caching algorithms. *Proceedings of USENIX Symposium on Internet Technologies and Systems (USITS)* (pp. 193-206), Monterey, CA, December.

Chen, L., Wang, S., & Rundensteiner, E. A. (2004). Replacement strategies for XQuery caching systems. *Data and Knowledge Engineering, 49*(2), 145-175.

Chidlovskii, B., Roncancio, C., & Schneider, M. L. (1999). Semantic CacheMechanism for heterogeneous Web querying. In *Proceedings of the 8th World-Wide WebConference (WWW8).*

Chou, H., & DeWitt, D. (1985). An evaluation of buffer management strategies for relational database systems.In *Proceedings of the 11th VLDB Conference.*

Corocoles, J. E., & Gonzalez, P. (2003). Querying spatial resources. An approach to the semantic geospatial Web. *CAISE 2003 Workshop, Web, e-Business, and the Semantic Web (WES: Foundations, Models, Architecture, Engineering and Applications*, LNCS, Springer-Verlag.

Dar, S., Franklin, M. J., Jonsson, B. T., Srivastava, D., & Tan, M. (1996). Semantic data caching and replacement. *Proceedings of the 22nd VLBD Conference*, Bombay, India.

Gupta, A., Marciano, R., Zaslavsky, I., & Baru, C. (1999). Integrating GIS and imagery through XML-based information mediation. In P. Agouris & A. Stefanidis (Eds.), *Integrated spatial databases: Digital images and GIS*, LNCS, Vol. 1737 (available from http://www.npaci.edu/DICE/Pubs/).

Kellerer, H., Pferschy, U., & Pisinger, D. (2004). *Knapsack problems.* Springer.

Leclercq, E., Djamal, B., & Yétongnon, K. (1999). ISIS: A semantic mediation model and an agent-based architecture for GIS interoperability. *Proceedings of the International Database Engineering and Applications Symposium IDEAS* (pp. 81-92), Montreal, IEEE Computer Society.

Lee, D., & Chu, W. (1998). A semantic caching schema for wrappers in Web databases. The *1st ACM International Workshop on Web Information and Data Management (WIDM)*, Washington DC.

Lee, D., & Chu, W. (1999). A semantic caching via query matching for Web sources. In *Proceedings of the 8th ACM International Conference on Information and Knowledge Management (CIKM)*, Kansas City, Missouri, USA.

Lorenzetti, P., & Rizzo, L. (1996). *Replacement policies for a proxy cache*. Technical report, Universita di Pisa.

Luo, O., Naughton, J. F., Krishnamurthy, R., Cao, P., Li, Y. (2000). Active query caching for database {Web} servers. *ACM SIGMOD Workshop on the Web and Databases*, WebDB.

Meier, W. (2002). eXist: An open source native XML database. In A. Chaudri, M. Jeckle, E. Rahm, R. Unland (Eds.), *Web, Web-Services, and Database Systems*. NODe 2002 Web- and Database-Related Workshops. Springer LNCS Series, 2593. Erfurt, Germany.

O'Neil, E. J., O'Neil, P. E., & Weikum, G. (1993). The LRU-K Page replacement algorithm for database disk buffering. In *SIGMOD* (pp. 297-306).

Open GIS Consortium. (1999). Inc., OpenGIS Simple Features Specification for SQL Revision 1.1. OpenGIS Project Document 99-049.

Open GIS Consortium. (2003). Inc., OpenGIS® Geography Markup Language (GML) Implementation Specification, version 3.00, Document 02-023r4.

Pagel, B., Six, H., Toben, H., Widmayer, P. (1993). Towards an Analysis of Range Query Performance in Spatial Data Structures. PODS. 214-221.

Savary, L., & Zeitouni, K. (2003). Spatial data warehouse—A prototype. In The *2nd EGOV International Conference* (pp. 335-340), Prague. LNCS 2739.

Savary, L., & Zeitouni, K. (2005). Automated linear geometric conflation for spatial data warehouse integration process. The *8th AGILE Conference on GIScience*. Estirl, Portugal.

Silvano, M., & Paolo, T. (1990). *Knapsack problems: Algorithms and computer implementations*. John Wiley & Sons.

Sindoni, G., Tininini, L., Ambrosetti, A., Bedeschi, C., De Francisci, S., Gargano, O., Molinaro, R., Paolucci, M., Patteri, P., Ticca, P. (2001). SIT-IN: A real-life spatio-temporal information system. *VLDB 2001* (pp. 711-712).

Stoimenov, L., Djordjevic-Kajan, S., & Stojanovic, D. (2000). Integration of GIS data sources over the Internet using mediator and wrapper technology. *MELECON 2000, 10th Mediterranean Electrotechnical Conference* (Vol. 1, 334-336).

Tanenbaum, A. S. (1992). Modern operating systems. Prentice-Hall.

Theodoridis, Y., & Sellis, T. (1996). A model for the prediction of r-tree performance. In *Proceedings of the 15th ACM Symp. Principles of Database Systems* (pp. 161-171).

Vatsavai, R. R. (2002). GML-QL: A spatial query language specification for GML. *UCGIS Summer*, Athens, Georgia.

Voisard, A., & Juergens, M. (1999). Geographical information extraction: Querying or quarrying? In M. Goodchild, M. Egenhofer, R. Fegeas, C. Kottman (Eds.), *Interoperating geographical information systems*, New York: Kluwer Academic Publishers.

Yang, O., Zhang, H., & Zhang, H. (2003). Taylor series prediction: A cache replacement policy based on second-order trend analysis. The *34th Annual Hawaii International Conference on System Sciences*, Maui, Hawaii.

Zhang, M., Javed, S., Shaheen, A., & Gruenwald, L. (2001). A prototype for wrapping and visualizing geo-referenced data in distributed environments using the XML technology. *ACMGIS at VA* (pp. 27-32).

This work was previously published in International Journal of Data Warehousing and Mining, Vol. 3, Issue 1, edited by D. Taniar, pp. 67-88, copyright 2007 by IGI Publishing, formerly known as Idea Group Publishing (an imprint of IGI Global).

Chapter 2.10
A Java Technology Based Distributed Software Architecture for Web Usage Mining

Juan M. Hernansáez
University of Murcia, Spain

Juan A. Botía
University of Murcia, Spain

Antonio F.G. Skarmeta
University of Murcia, Spain

ABSTRACT

In this chapter we focus on the three approaches that seem to be the most successful ones in the Web usage mining area: clustering, association rules and sequential patterns. We will discuss some techniques from each one of these approaches, and then we will show the benefits of using METALA (a META-Learning Architecture) as an integrating tool not only for the discussed Web usage mining techniques, but also for inductive learning algorithms. As we will show, this architecture can also be used to generate new theories and models that can be useful to provide new generic applications for several supervised and non-supervised learning paradigms. As a particular example of a Web usage mining application, we will report our work for a medium-sized commercial company, and we will discuss some interesting properties and conclusions that we have obtained from our reporting.

INTRODUCTION

When we face the challenge of data recovering from the Internet, or Web mining (WM), we can consider it from two main perspectives: Web resources mining and Web usage mining (WUM). Still, we can split the first one into Web content mining and Web structure mining. While the differences between these resource mining areas sometimes are not clear, WUM is more clearly defined, but it is not isolated either. It aims to describe the behaviour of the users who are surfing the Web. Many techniques and tools have been proposed, giving partial solutions to some of the WUM problems. All the proposed techniques and ideas are showing their importance in many areas, including Web content and Web structure mining.

One of the most accepted definitions for *Web usage mining* is that it is "the application of data mining techniques to large Web data repositories in order to extract usage patterns" (Cooley, Tan & Srivastava, 1999). As we know, Web servers around the world record data about user interaction with the Web pages hosted in the Web servers. If we analyze the Web access logs of different Websites we can know more about the user behaviour and the Web structure, making easier the improvement of the design of the sites among many other applications.

Analyzing data from the access logs can help organizations and companies holding the Web servers to determine the ideal life cycle of their products, the customer needs, effectiveness of new launched products, and more. In short, WUM can support marketing strategies across products over some specific groups of users, and improve the presence of the organization by redesigning the Internet Websites. Even if the organization is based on intranet technologies, WUM can warn of a more effective infrastructure for the organization, and can also warn of improvements and faults of the workgroup communication channels.

This chapter is organized as follows: in the first section, we introduce the three approaches intended for solving the WUM problems and challenges: *clustering, association rules* and *sequential patterns.* We will focus on the techniques that we have integrated in our own architecture, having one technique of each approach. Then, in the second section, we present our software architecture for automated data analysis processes, called *METALA.* We will show how this architecture can be used to support the WUM techniques and models, as well as other data analysis methods like *machine learning,* and how we integrated some algorithms of each one of these three approaches. As an example of a WUM application provided by METALA, in the third section we summarize our work for a medium-sized industrial company, which wanted to know more about the usage of its Website, and about the strengths and flaws of the site and how to improve it. From this work we have proposed some ideas to face the problems we found when applying the WUM techniques. Finally, in the fourth section we give our conclusions and discuss our future work.

WEB USAGE MINING TECHNIQUES

As we have already mentioned, there are three main Web usage mining approaches. They are *clustering, association rules* and *sequential patterns.* Most of the WUM techniques can be included in one of these approaches, and some others may be hybrid. Notice that these techniques are not exclusive to the WUM field: most of them are based on generic algorithms and ideas taken from other more general fields, such as data mining or databases. We first introduce clustering in the first subsection. Then, we overview association rules in the second subsection. We finally discuss sequential patterns in the third subsection.

Clustering

Hartigan (1975) states that Clustering is the "grouping of similar objects from a given set of inputs." That is, given some points in some space, a clustering process groups the points into a number of clusters, where each cluster contains the *nearest* (in some sense) points. This approach is widely used in WUM. The key concept of clustering is the distance measurement used to group the points into clusters. Nevertheless, there are many clustering algorithms and variants of these algorithms.

Nasraoui et al. (2002) introduced the clustering algorithms that we have implemented in our architecture, to be shown in the second section. We have chosen these algorithms for our architecture as they are specifically intended for the WM purpose, and because they have been proved to be efficient during the WM process. All these algorithms are based on the simple K-means algorithm, which must be outlined before we can go further.

The K-means algorithm is one of the best known clustering algorithms. It was first introduced by McQueen (1967). It is an iterative algorithm that divides a set of data samples in a number (*K*, given) of clusters. For this, we first randomly choose the prototypes of each cluster. Then each data sample is classified into the cluster whose prototype is closer to the sample. After that, the prototypes of each cluster are moved to the arithmetic mean of samples in corresponding cluster. As soon as the prototypes stop moving, the algorithm finalizes and the samples remain in the current clusters. The prototypes are also known as the *centroids* or centers of the clusters.

We now review the WM overall process in which we have applied the algorithms of Nasraoui et al. (2002). We start from the logs stored in the Web server, and we must first preprocess the logs in order to limit the noise of the data and to prevent the algorithms from wrong accesses. A good explanation of the preprocess phase can be found in the work of Cooley, Mobasher and Srivastava (1999). For the WM purpose, we must use a discrete version of the K-means basic algorithm, because the centroids (also named *medoids*) are going to be *sessions* of accesses to Web pages (URL accesses) and not just continuous data to be recomputed in each iteration by modifying its numerical values.

As we can see, for the clustering purpose, we consider a *session* as the minimum meaningful data unit. The meaning of the session is the same as the one from the Web servers field: a Web session is a group of URLs accessed by an user during a period of time. There is a session expiration time of the Web server such that if the user stops accessing the Website and exceeds that time, the session ends.

We next define a *distance measurement* between sessions. For this we first identify unique URLs in all the sessions (i.e., a URL that is not repeated in a single session but can appear in other sessions). Then, we can define a distance measurement between sessions by just computing the cosine of the angle between them. That is, if two sessions contain exactly the same URLs, they have a similarity of 1, or 0 otherwise. But this similarity measurement has a drawback: it ignores the URL structure. For example, consider two sessions with the only accesses {/courses/cmsc201} and {/courses/cmsc341}, and also consider the pair {/courses/cmsc341} and {/research/grants}. According to the cosine, the similarity of the sessions will be zero. However, it is clear that the two first sessions are more similar than the other two, because both users seem to be interested in courses. Thus we define a new *syntactic* similarity measurement, at the level of URLs. The new similarity measurement basically measures the overlapping existing between the paths of two URLs. Now, we can define a similarity measurement between user sessions, using this URL level based similarity. In some cases the cosine of the angle of two sessions will give a more intuitive similarity measurement and it will be also used.

Now we can study the algorithms. We first consider the *Fuzzy C-Medoids algorithm* (*FC-Mdd*). Let $X = \{x_i \mid i=1, ..., n\}$ be a set of *n* objects. Let $r(x_i,x_j)$ be the distance from the object x_i to the object x_j. Let $V = \{v_1,v_2,...,v_c\}$, with $v_i \in X$, a subset of X with cardinality *c*. The elements of V are the *centroids*. Let X_c be the set of all subsets V of X with cardinality *c*. Each V represents a specific choice of the prototypes for the *c* clusters in which we wish to partition the data. In the WUM area, X is the total Web sessions set computable from the access log file.

The key concept of the FCMdd algorithm is the possibilistic or fuzzy membership of x_j into the cluster *i*. This membership set is denoted by u_{ij}, and can be heuristically defined in many ways. There are four possible solutions for u_{ij} (two are fuzzy and two are possibilistic). Here we just present one of the fuzzy solutions:

$$u_{ij} = \frac{\left(\frac{1}{r(x_j, v_i)}\right)^{1/(m-1)}}{\sum_{k=1}^{c}\left(\frac{1}{r(x_j, v_k)}\right)^{1/(m-1)}} \tag{1}$$

where the constant $m \in (1, \infty)$ is the "fuzzifier". The larger *m*, the less the difference between the memberships of the different sessions to the different clusters (i.e., the borders among clusters are fuzzier).

With the fuzzy FCMdd algorithm, once we have computed the centroids of the different clusters, we must characterize them in order to interpret the results of the WUM process. This can be simply done by using the weights of the URLs stored in the sessions, which must be placed in the clusters whose centroid is closer. Then, the weights values are computed by dividing the number of times the URL access appears in the cluster by the cardinality of the cluster. Most relevant URLs are then used to tell what the cluster represents.

About the flaws and advantages of the algorithm, we can say that in the worst case, the complexity of FCMdd is $O(n^2)$, due to the step of computing the new centroids. However, it is possible to obtain good results by reducing the amount of candidate objects to be considered when computing the new centroids. If we consider *k* objects, where *k* is a small constant, the complexity reduces to $O(k \times n)$. The problem with the FCMdd algorithm is that it is not *robust*. This means that it is very sensitive to *outliers* (noise in the data) so sometimes it cannot provide good clusters. The problem of handling outliers can be solved with another fuzzy algorithm: *Fuzzy C-Trimmed Medoids Algorithm (FCTMdd)* (also known as *Robust Fuzzy C-Medoids algorithm*). This algorithm can only use the fuzzy version of the membership u_{ij}, stated in equation 1. The "trimmed" values will be those whose distance to the harmonic mean of a subset of objects of *X* is over a specific threshold. We will just keep the first *s* objects obtained by sorting candidate values in ascending order of distance.

The complexity of the robust algorithm is still $O(n^2)$ in the worst case. But like with FCMdd, we can consider just *k* candidate objects to make it almost linear, although we have to add the time of computing the *s* objects. Note that both algorithms may converge to a local minimum. Thus, it is advisable to try many random initializations to increase the accuracy of the results.

There are other clustering techniques applicable in WM, such as the system from Yan et al. (1996) based on the Leader algorithm (Hartigan, 1975), the BIRCH algorithm (Zhang et al., 1997) or the ROCK algorithm (Guha et al., 2000).

Association Rules

The problem of mining association rules between sets of items in large databases was first stated by Agrawal et al. (1993), and it opened a brand new family of techniques in the WUM area. The original problem can be introduced with an example: imagine a market where customers buy different items. They have a "basket" with different acquired products, and we can establish relations between the bought products. Finding all such rules is valuable for cross-marketing, add-on sales, customer segmentation based on buying patterns, and so forth. But the databases involving these applications are very large, so we must use fast algorithms for this task.

If we move the original problem statement to the WM area, we can think of the "baskets" as the users that accessed a Website, the items in the "basket" as the pages of the Website visited and the products as all the pages of the Website. Thus, we can find rules relating the requested URLs of the Website. For example, a rule can say that in 100 cases, 90% of the visitors to the Website of a restaurant who visited the index page also visited both the restaurant menu page and the prices page. The rule should be written:

[index⇒menu,prices](support=100, confidence=90%)

This way the restaurant may know the hit rate of the menu and prices pages from the index page, among other interesting associations.

In this section we just consider the Apriori algorithm (Agrawal & Srikant, 1994), which has been integrated in our architecture. From this algorithm lots of techniques and new algorithms have been proposed, but most of them are just refinements of some of the phases of the basic Apriori algorithm.

The Apriori algorithm is a fast algorithm intended for solving the association rules problem stated above, and it is the algorithm we have included in our METALA architecture for the association rules approach. Before explaining the algorithm, we must state a formal definition of the problem. Let $I=\{i_1,i_2,...,i_m\}$ be a set of literals, called items. In WM, these are the URLs requested by users. Let D be a set of *transactions* (called web sessions in WM), in which each transaction is a set of items so that the set of items T is contained in I $(T \subseteq I)$. There is a unique identifier associated with each transaction. We can say that a transaction T contains X, a set of some of the items of I, if $X \subseteq T$. One *association rule* is an implication of the form $X \rightarrow Y$, where $X \subseteq I$, $Y \subseteq I$, and the intersection $X \cap Y$ is empty. The rule $X \Rightarrow Y$ is true for the set of transactions D with *confidence c* if the c% of the transactions of D containing X also contain Y. The rule $X \Rightarrow Y$ has *support s* in the set of transactions D if the s% of the transactions of D contain $X \cup Y$.

The problem of discovering such association rules can be divided into two subproblems:

1. Finding all the sets of items with a support over a certain threshold. The support for a set of items is defined as the number of transactions containing the set (i.e., the number of transactions in which a group of URLs appears). These sets are called *large itemsets* or *litemsets*.
2. Using the litemsets to generate the desired rules. For example, assume that $ABCD$ and AB are litemsets. We can know if $AB \Rightarrow CD$ by calculating the confidence of the rule, which is the ratio:

   $$\frac{\text{support(ABCD)}}{\text{support(AB)}}.$$

   If this confidence is over a certain threshold, the rule holds, and besides the rule has the minimum support because $ABCD$ is a *large itemset.*

To solve the first subproblem, we can use the algorithm AprioriTid. It passes several times over the data. In the first pass, from the individual items (individual URLs) we get the *large itemsets* (i.e., those with a minimum support previously set; in WM, this is equivalent to setting a threshold, for example 2, and checking which URLs appear in at least two transactions). In each next pass, we start from the set computed in the previous pass. We use this set of candidate itemsets to generate new possible *large itemsets.*

That is, starting from the sets of URL groups with enough support, we build a new set of items made from all the valid combinations of initial URLs (these new combinations are called candidate itemsets). A combination is valid in a transaction if all the subgroups of URLs of the new formed group contain only URLs belonging to such transaction. Once we have checked that some candidates have enough support, the process starts again. It will end when no new sets of large itemsets can be formed.

Once we have got all the large itemsets we can face subproblem 2: getting the association rules existing between the groups of URLs; that is, given a confidence threshold, for each subset $s \subset l$ where s is a subset of items contained in the set of items l, we generate a rule $s \Rightarrow (l - s)$ *if* $support(l) / support(s) \geq confidence$. A fast algorithm to solve this problem is needed (Agrawal & Srikant, 1994).

We now consider the advantages and disadvantages of using Apriori. Its advantages are clear: it solves the problem statement of association rules, and to do it efficiently it takes advantage of the observation that a k-itemset can be frequent only if all its subsets of k-1 items are frequent. Nevertheless, if we consider the computational complexity of the algorithm, it is clear that the cost of the k-th iteration of Apriori strictly depends on both the cardinality of the candidate set C_k and the size of the database D. In fact, the number of possible candidates is, in principle, exponential in the number m of items appearing in the various transactions of D.

Thus, many techniques have been proposed to improve its efficiency, especially the process of counting the support of the candidate itemsets and the process of identifying the frequent itemsets. For the first problem many techniques have been proposed. We can mention hashing (Park et al., 1995), reduction of the number of transactions (Han & Fu, 1995), partitioning (Savasere et al., 1995) sampling (Toivonen, 1996) and dynamic count of itemsets (Brin et al., 1997). The second problem could be stated as follows: we are given m items, and thus there are 2^m potentially frequent itemsets, which form a lattice of subsets over I. However, only a small fraction of the whole lattice is frequent. Most of the algorithms that address this problem only differ in the way they prune the search space to make the candidate generation phase more efficient. And most of these algorithms need to pass more than once over the database, which is too expensive. Zaki et al. (1997) propose a hybrid approach, called *itemset clustering,* which includes several algorithms in which the preprocessed database is scanned just once.

Sequential Patterns

Srivastava et al. (2000) state that the techniques of sequential pattern discovery attempt to find inter-session patterns such that the presence of a set of items is followed by another item in a time-ordered set of sessions or episodes. By using this approach, Web marketers can predict future visit patterns that will be helpful in placing advertisements aimed at certain user groups. Other types of temporal analysis that can be performed on sequential patterns include trend analysis, change point detection, or similarity analysis.

We can find more applications in the Internet area: for example, consider Web hyperlink predicting. If we could know the pages to be requested by the user, we would be able to do precaching with these pages in each connection and therefore speed up the overall browsing process and save bandwidth.

Agrawal and Srikant (1995) considered again the pattern discovery problem (Agrawal et al., 1993) to find an application for sequential pattern discovery. Previously, we have seen that one pattern consisted of an unsorted set of items. Now, the set of items is sorted and the problem consists of, given a sorted sequence, being able to predict a possible continuation of the sequence.

The algorithms proposed by Agrawal and Srikant (1995) have been implemented in the METALA architecture (to be shown in the second section) to have a representation of the sequential patterns approach.

We will see now the new problem statement: given a database D of customer transactions, each transaction has an unique customer identifier, a transaction time and the items acquired in the transaction. An *itemset* is a non-empty set of items, and a *sequence* is an ordered list of *itemsets.* A customer supports a sequence s if s

is contained in the sequence for the client. The *support* for a sequence is defined as the fraction of all customers supporting this sequence. Given *D*, the problem of mining sequential patterns is discovering among all the sequences the *maximal* ones, that is, those sequences that exceed the minimum support threshold and are not contained in any other sequence. Each of these maximal sequences is called *sequential pattern*. The sequence satisfying the minimum support is named *large sequence*.

Three new algorithms were proposed: *AprioriAll, AprioriSome* and *DynamicSome*. All the algorithms have the same phases, and they only differ in the way they perform the *sequence phase* shown below. The phases are the following:

1. *Sort phase:* the original database *D* is sorted using the customer identifier as the primary key and the transaction time as the secondary. This way we get a database of sequences of customers.
2. *Litemset phase:* in this phase we compute the set *L* of all *litemsets*, just like in the same phase of the association rules version of the Apriori algorithm, but only with one difference: we must change the *support* definition. Here is the fraction of customers choosing the itemset in some of their (possibly) several transactions.
3. *Transformation phase:* it is necessary to quickly determine whether a given set of large sequences is contained in a sequence of a customer. Thus, we transform *D* into D_T, for example removing those items that do not belong to any sequence and cannot support anything.
4. *Sequence phase:* using the set of itemsets we compute the desired sequences. Read below.
5. *Maximal phase:* find the maximal sequences from the total of large sequences. For this, starting from the set of large sequences *S*, we must use an algorithm for computing maximal sequences (Agrawal & Srikant, 1995).

The way the *sequence phase* is performed determines the form of the algorithms presented. This phase makes multiple passes over the data. In each pass, we start with an initial set of large sequences. We use this set to generate new potential large sequences, called *candidate* sequences. The AprioriAll algorithm generates the candidates in a similar way as the association rules version of the Apriori algorithm. However, AprioriSome and DynamicSome do it in a different way. They have two distinguished phases: one forward, in which all the large sequences of certain length are computed, and another backwards, where we find the remaining large sequences. Notice that DynamicSome has an additional advantage: it can generate the candidates on-the-fly.

As in the case of the association rules version of the Apriori algorithm, it is not an easy task determining the computational complexity of the presented algorithms. These algorithms have problems when the database of sequences is large or the sequential patterns to be mined are numerous and/or long. There are some other algorithms that try to minimize these problems by reducing the generation of candidate subsequences, for example, Freespan (Han et al., 2000). We can mention some other interesting sequential pattern discovery researches, such as the work of Schechter et al. (1998) or Pitkow and Pirolli (1999).

METALA ARCHITECTURE

METALA is a software architecture that aims to guide the engineering of information systems that support *multi-process inductive learning* (MIL). METALA is defined on the basis of four different layers of abstraction (Figure 1): (1) the object oriented (OO) layer, (2) the middleware layer and (3) the agents layer. The fourth layer is the METALA application.

In the *OO layer,* basic software support tasks are addressed like, for example, training data access and management, machine learning algorithms dynamic behaviour and extensibility mechanisms, learning processes remote monitoring, models management, and so forth. This is the most extensive layer. It defines the ground level tools to work on inductive learning.

The basic unit of learning data in METALA is the *instance.* It is compound by a set of values referring to a particular example in the learning data source. A data source is a set of instances. Instances can be indexed via a cursor named *access.* Data sources can be organized inside a *repository.*

Learning algorithms are seen in the architecture as possible services to offer and use. Dynamic behaviour of any inductive technique is defined via a deterministic finite automata (DFA), as shown in Figure 2.

The most important transition in the DFA is from the *configured* state to the *learning* one by the *learn* token. This token corresponds to a method call in all algorithms. A learning process can be prolonged in time and METALA includes a special mechanism to inform the client when the process has finished, the *listener.* A listener is an object, owned by the client, which receives progress and termination calls from the learning process while the client is doing other tasks. The usual operation has the following sequence: configuring an experiment for a learning technique, launching it (i.e., testing a learning technique with the specified configured parameters), monitoring the progress of the experiment and finally, if the experiment ends successfully, getting the knowledge *model* associated to the experiment for later evaluation and/or utilization. The models are the pieces of knowledge that a concrete learning algorithm induces from data. Models can be stored, recreated and visualized.

METALA is *extendable.* It allows the integration of new techniques by using some basic

Figure 1. Abstraction layers of METALA

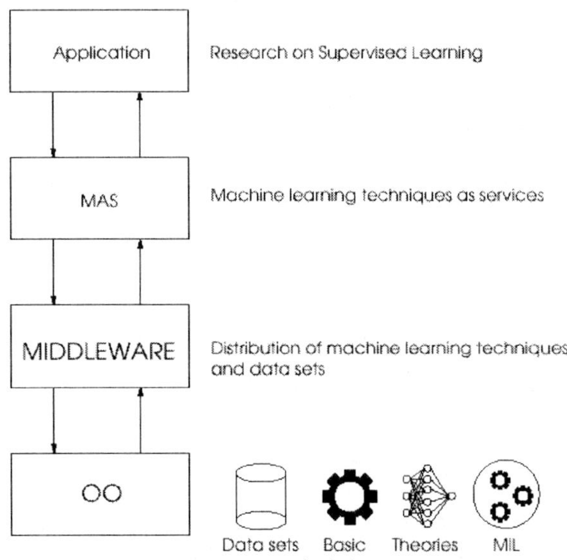

Figure 2. Behaviour that all learning services must show

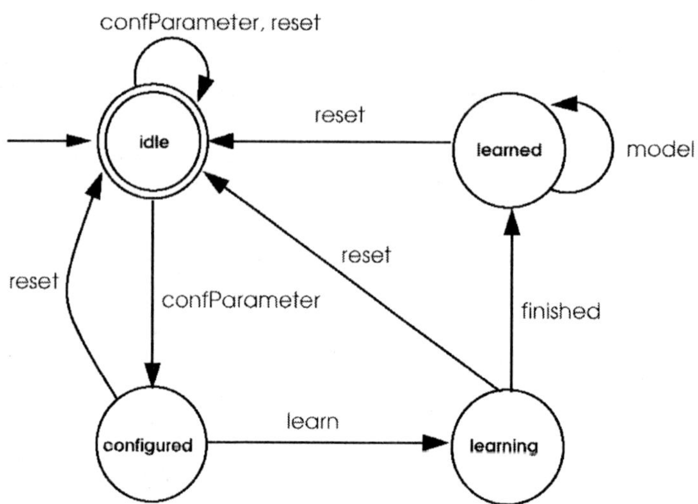

interfaces that all new algorithms must support. Extendability is considered for both algorithms and models. Any new learning technique added to METALA must have a matching model representation.

We now move on the *middleware layer:* it uses the entire framework defined in the OO layer to produce a distributed learning tool. The distribution mechanisms are very technology dependent. The current system is developed with Java and RMI (Sun Microsystems, 1998). With RMI all must be coded in Java, an easy language to learn and use. In this layer we define the *information model* of METALA. It is based on the concept of directory (a hierarchical database that stores data about entities of interest). The directory allows the coordination between the *agents* of the system (read below). It also stores all the needed information about knowledge models, learning techniques and launched experiments.

The *agent layer* is built on the middleware layer. By looking into the directory provided by the information model of METALA, an agent can locate the rest of the agents (and its corresponding services) in the system. In Figure 3, we can see the agent layer of METALA.

There are four kinds of agents. The *user agent* is the front-end to the user. It shows the directory to the user and allows executing and monitoring learning techniques, as well as visualizing and evaluating the obtained models.

The *machine learning agent* (MLA) provides the learning techniques for a specific user and the behaviour of the monitoring task (i.e., which data should be displayed and updated when executing a particular experiment of a particular learning technique). The *directory service agent* (DSA) offers necessary services to access to the directory. Finally, the *data repository agent* (DRA) provides the leaning data needed for executing the learning techniques from the MLA.

In Figure 3 we can see that data may come from different sources: plain text files, relational databases and LDAP databases. Note that in this distributed framework any agent can work in a remote or local manner. Other important features of METALA are the following:

Figure 3. Organization of agents in the METALA architecture

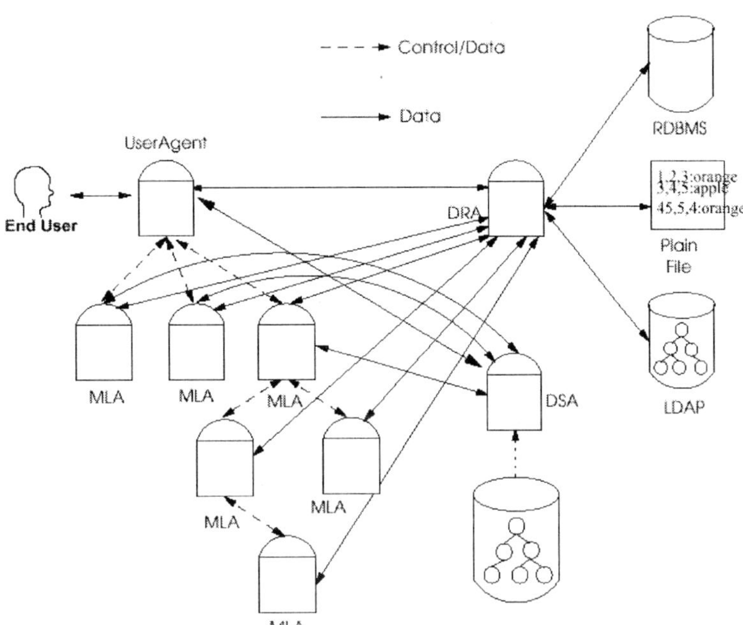

- It is *extendable:* it comes with a methodology to integrate new inductive learning algorithms and models, even new paradigms for learning. Note that METALA was originally intended only for supervised inductive learning. This led us to consider models in a different way, since no inferences are performed by WUM techniques such as clustering or association rules, which are non-supervised ones. Displayed models have more informative power, in this case, than the error estimation measures needed by the supervised learning techniques. We even created a graphical URL recommender tool (Mobasher et al., 2002) integrated into the METALA architecture (Figure 4) based on the information provided by the new WUM models. So the system was positively influenced in the sense that higher software particularization can be done to transform a generic inductive learning tool into a WM tool.
- It is *distributed:* agents can be located at different hosts, to make efficient use of available servers.
- It is *flexible:* all agents can be totally independent and this feature allows an easy adaptation to almost any particular execution condition.
- It is *scalable:* it correctly supports the growth of both pending learning tasks to perform and available servers to manage.
- It is *autonomous:* it is capable of deciding the best way to distribute machine learning tasks among the available learning servers (Botia et al., 2001).
- It is *multi-user:* different users can make use of inductive learning algorithms and the other services that the system offers at the same time; METALA allows management of users and all of them have their own user space to read and write produced results.
- It *is educational:* it is clear that METALA is a good assistance tool for knowing the

Figure 4. URL Recommender Tool (The model used in this case is shown on the left panel. The tool control panel is on the right side. On top of this panel we can see the values of the parameters used for the model of the left side. Below we can select the data source containing the URLs visited by a user. At the bottom we can choose the desired values of support and confidence for getting the recommendations, and the recommendations provided by the tool based on these values and on the observed visited URLs.)

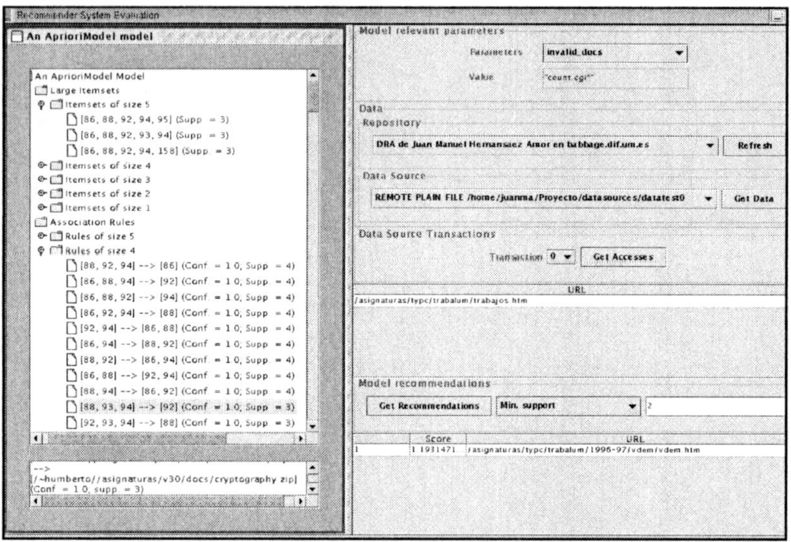

main paradigms of inductive learning. At the present time, with our architecture we can do research on basic techniques like induction of artificial neural networks (Bishop, 1995), decision trees and rules (Quinlan, 1993), genetic algorithms (Goldberg, 1989) and on induction of classification theories by naive bayes (Witten & Frank, 2000). We can also test other high-level techniques that perform what we named *multi-process inductive learning* (MIL), such as boosting (Freund & Schapire, 1996), bagging (Breiman, 1996) and landmarking (Bensusan & Giraud-Carrier, 2000; Fürnkranz & Petrak, 2001). Besides, we have added the capability of researching WM techniques of the most important approaches.

It is also clear that our system is not comparable to others such as SpeedTracer (Wu et al., 1997) in the sense that we are neither looking for a particular technique that we consider the most appropriate for WUM, nor for particular statistical reports. Nowadays, it is commonly accepted that no single supervised machine learning technique behaves better than the rest in any classification and/or regression task (Wolpert, 1996; Wolpert & Macready, 1995). This fact is called the *selective superiority problem* (Brodley, 1993). We believe that we can also apply this fact for non-supervised machine learning techniques, like the WM ones. We offer the possibility of choosing the technique that the user considers more suitable for a particular problem. METALA is extendable and the generated models are usable for creating new applications and tools (see Figure 4 for an example); we can add any learning algorithm that we consider interesting, testing it and generating any reports (see following section) and applications just coding with the provided METALA methodology.

AN APPLICATION EXAMPLE

We now show an example of an application in the real world of some of the WUM algorithms implemented in the METALA architecture. We report here the work we have performed for a medium-size industrial company. This company requested from us a report about its Website. The main targets were to get some valuable information about groups of users and their preferences and about the Website structure. We had at our disposal three years of Web access logs and the *topology* of the Website, which did not change too much during the three years. The Website was structured hierarchically, and under the root node (which was a page including links for selecting language) we found two nodes with exactly the same tree structure, with information pages for the two languages that we could select at the root node.

The WUM algorithms used for the reporting were the association rules Apriori algorithm and the clustering fuzzy C-medoids algorithm. We used the *robust* version of the fuzzy C-medoids algorithm, since it provides more accurate results. We considered unnecessary to use any algorithm of the third WUM approach (sequential patterns), as it was not applicable to this problem.

In order to write the report, we analyzed separately the results for the two algorithms and for the three years. We performed several experiments with all the cases to increase the reliability of the results. Then, we compared and combined the results, removing possibly wrong values (with the Apriori algorithm there are no wrong results, but with the clustering fuzzy robust algorithm, misclassifying objects in the clusters is possible since it contains some random components). Analyzing the results and getting the conclusions useful for the company was not an easy task (see the next subsection for ideas addressing this issue), and it required a lot of human interpretation. Nevertheless, we created a fully qualified report providing interesting results and ideas for improving the presence of the company Website and for easily improving its design and technical structure. The most interesting conclusions can be summed up as follows:

- Thanks to the Apriori algorithm, we knew the hit rate of some pages, on condition that its parent page was visited. We focused on the pages with several links where we realized that only one of them had a low hit rate. For such pages we gave the advice of improving the format of the link to make it more attractive to users. Sometimes pages providing general information presented links where only the first ones were clicked. Hence we proposed distributing the links in a non-sequential manner, to give more hit probability to the links presented at the bottoms of the pages. We also proposed linking these sequential pages to each other to improve the overall hit rate.
- We observed that some pages with a high hit rate in the previous years had a low hit rate with the new naming. Thus, we recommended again changing the name, or providing some new pages with the old name, with new content or directly linking to the pages with the old name, because it seemed that users were not able to reach the newly named pages as before.
- One of the most successful pages of the Website was the chat page. This page provided excellent opportunities to the company to meet the needs of the users, and to meet the requests for information from users about the company products. However, this application was placed in a place that was not

easily reachable (by the obtained rules we realized that the paths traversed to reach the chat were very diverse). Thus we proposed making it accessible from the main browsing bar of the Website. This had a potential risk, however: users using the chat as incidental entertainment. But we considered that this industrial company was specialized enough to not attract such users.

- Analyzing the clusters created by the clustering algorithm, we also realized that most of the users of the chat application used only one of the languages that the Web was intended for, although users in both languages frequently accessed the Website. After checking the chat page, we realized it was written for only one of the languages, and thus we advised redesigning it in order to avoid losing potential customers of the other language.
- Another very important page of the Website was the contact request form. This page put the company directly in contact with its potential customers. We found some clusters profiling users surfing the main pages of the Website, looking for general information about the company and its products. The request form was included in such clusters. Looking in the Apriori results, we discovered that those URLs belonged to long time sessions. Thus, we recommended giving more content to the general browsing pages, providing frames, java applets, other information not directly related to the company, and so forth. The objective was to retain the users surfing the Website long enough to, if she is remotely interested in the company, use the contact request form.
- Evaluating others clusters, we found that some of the products were preferred by users using one language, whereas other products were preferred by the other language users. Thus we recommended that the company study this result, and we advised introducing minor changes to the content of the pages of products to make them more attractive to the users of the other language.
- We also found a case where a page was not being properly accessed, as we found a rule relating very specific URLs but for different languages. After checking the pages, we discovered that some documents were available for the two languages but they were presented in a confusing format, so that users chose the documents randomly. We thus recommended modifying the format.
- Concerning the technical part of the Website, we notice some interesting facts. Some association rules associated URLs being loaded at the same time (for example documents or plugins in a Web page) but the confidence of the rules was not the same when the pages were one in the consequent and the other in the antecedent. Although the difference was not too big, it was significant enough to deserve a study. We discovered that such pages frequently consisted of calling to *Macromedia Flash* documents or to Java applets. So we concluded that some users could not access to these pages and we recommended providing more information about how to get the appropriate plugins, and also designing an alternative version of the pages for the users just interested in the *hard* information content of the pages, or for users that cannot get the plugins. For the users who cannot use the chat, which was a JAVA applet, we advised providing a direct link to the contact request form, to make it easier for them to access the company.
- Some more technical results were provided by the clustering algorithm. We found some clusters relating pages in both languages, in a significant proportion. Since both language versions of the Website contained exactly the same information, these results made no sense for us. To find an explanation we incorporated a statistic analysis process to

the algorithms. This way we could discover that many accesses came from automatic Internet information indexers, such as "robots" and "spiders". These tools index information by traversing the hierarchy of the Web in some order, and thus the results they produce may be difficult to interpret.

Conclusions from the Reporting

We got some interesting ideas from the reporting. The integrated WUM algorithms proved to be useful for the mining purpose, but the results we obtained required a lot of human interpretation. It is necessary to develop automatic mechanisms allowing a fast interpretation of the mining results. As we were writing the report, some repeating tasks were identified:

- For the *Apriori* algorithm, identifying the *interesting* rules was made in the following manner: most of the rules associate index and general-browsing (browsing bars, frames, etc.) pages, so we opt to pay more attention to the more specific URLs, that is, those Web pages located deeper in the Website hierarchy. Thus, the *topology* of the Web site turned out to be very important to identify interesting rules.

Another "rule of thumb" we used in order to identify interesting rules was focusing on the consequent of the rules. The most interesting rules had a more specific URL in the consequent. If a rule was *A→B* and *A* could only be reached from *B,* the rule says nothing because obviously we had to visit B to reach *A*. The rule *B→A* would be more interesting, since it gives us an idea of the possibility of visiting *B* when we are visiting its parent *A,* and thus we would get the hit rate of *B*. Nevertheless, if we study the association rules version of the Apriori algorithm we find that no time order is considered when obtaining the rules; that is, the accessed URLs are stored with no timestamp. So we cannot know if a rule *A→B* indicates *A* occurring before *B* or just the opposite. However, since we have the topology of the Web and the confidence and support of the rule, it is easy to infer time order. Chronological order of the accesses turned out to be a very important issue to be considered, and maybe integrated, in our future work. Taking into account the length of the considered sessions may be an important factor, too.

- For the *robust fuzzy C-medoids* algorithm, the results are the clusters and the weights of the associated URLs. To analyze these results, we prune out those clusters with low cardinality. We set a threshold of how interesting the URLs should be *(score),* and only show those URLs above this threshold. Since the Website was hierarchically structured, we could group the URLs in more general topics by just adding the score of the contained URLs. The most interesting clusters were those with higher cardinality and URL weights, but also those with more specific URLs, the same as before with the association rules.

Analyzing the clusters was easier than analyzing the association rules, but some human interaction was still needed. The clusters with higher cardinality are not always the most interesting ones. Sometimes we look for the *unexpectedness* of the clusters (or, in the case of Apriori, of a rule) and it is not easy checking this. Some interesting work has been made in this sense (Cooley, 2003). Deciding whether a result is unexpected or wrong is difficult, and the topology of the Website, and even the access logs themselves, are needed sometimes.

Finally, the idea of incorporating a kind of analytical tool was interesting. Thanks to the information provided by the statistics analysis we could discover that many accesses recorded in the Web access logs came from "robots" and "spiders" that automatically index information for Web searchers and other Internet applications. Such tools and agents use some kind of algorithms to scan the Website in a prefixed order. This behaviour is not of our interest, since we are looking for understanding and to model the human behaviour while interacting with the Web. We think that the unexpected results that came from these accesses should be pruned out. This gave us the idea of considering the *user agent* field (Luotonen, 1995) of the log in the mining process, because we must determine if the access corresponds to a human user. We also consider incorporating some other fields of the log, as they may provide interesting information. For example, in the event that we do not have at our disposal the topology of the Website, it would be important to use the "Referrer" field of the log, as we may reconstruct the topology of a Website, or at least of a part of it.

CONCLUSION AND FUTURE WORK

In this chapter we have studied one technique from the three most important groups of techniques to perform Web mining on Web usage log files. We have passed through clustering, association rules and sequential patterns. Then we have introduced the tool that we developed to work with some of these algorithms in real world applications. The METALA tool was originally developed to study the generic problem of meta-learning. However, its utility as a tool to manage WUM has been proved in this chapter. Moreover, we also have included in this chapter the results obtained from the application of the tool to a medium-sized company.

Our current and future work consists of migrating the current architecture and information model of METALA to a J2EE (JAVA 2 Enterprise Edition) Application Server and to use Enterprise JavaBeans (Sun Microsystems, 2001), which drastically simplifies the complexity of the current METALA system and offers new capabilities. We are also integrating new Web mining techniques and learning paradigms, researching new Web mining applications and tools using the obtained WUM models, and testing the performance of the new implementation with huge-sized data sources (a critical issue when analyzing Web server logs). Finally, we are researching a way for automatically suggesting to the user the algorithm(s) and configuration parameters that best fit the user needs for a particular problem.

ACKNOWLEDGMENT

This work is supported by the Spanish CICYT through the project TIC2002-04021-C02-02.

REFERENCES

Agrawal, R., & Srikant, R. (1994). Fast algorithms for mining association rules. *Proceedings of the 20th International Conference on Very Large Data Bases* VLDB, (pp. 487-499).

Agrawal, R., & Srikant, R. (1995). Mining sequential patterns. *Eleventh International Conference on Data Engineering,* 3-14.

Agrawal, R., Imielinski, T., & Swami, A.N. (1993). Mining association rules between sets of items in large databases. *Proceedings of the 1993 ACM SIGMOD International Conference on Management of Data,* (pp. 207-216).

Bensusan, H., & Giraud-Carrier, C. (2000). *If you see la Sagrada Familia, you know where you are: Landmarking the learner space.* Technical report.

Bishop, C.M. (1995). *Neural networks for pattern recognition.* Clarendon Press.

Botía, J.A., Skarmeta, A.G., Garijo, M., & Velasco, J.R. (2001). Handling a large number of machine learning experiments in a MAS based system. *Workshop on Multi-Agent Systems Infrastructure and Scalability in Multi-Agent Systems.*

Breiman, L. (1996). Bagging predictors. *Machine Learning, 26*(2), 123-140.

Brin, S., Motwani, R., Ullman, J.D., & Tsur, S. (1997). Dynamic itemset counting and implication rules for market basket data. SIGMOD 1997, *Proceedings ACM SIGMOD International Conference on Management of Data, 5,* (pp. 255-264).

Brodley, C.E. (1993). Addressing the selective superiority problem: Automatic algorithm/model class selection. *Machine Learning: Proceedings of the Ninth International Conference,* (pp. 1-10).

Cooley, R. (2003). The use of Web structure and content to identify subjectively interesting Web usage patterns. *ACM Transactions on Internet Technology, 3*(2), 93-116.

Cooley, R., Mobasher, B., & Srivastava, J. (1999). Data preparation for mining World Wide Web browsing patterns. *Knowledge and Information Systems, 1*(1), 5-32.

Cooley, R., Tan, P-N., & Srivastava, J. (1999). Discovery of interesting usage patterns from Web data. *WEBKDD,* 163-182.

Freund, Y., & Schapire, R.E. (1996). Experiments with a new boosting algorithm. *International Conference on Machine Learning,* 148-156.

Fürnkranz, J., & Petrak, J. An evaluation of landmarking variants. *Proceedings of the ECML/PKDD-01 Workshop: Integrating Aspects of Data Mining, Decision Support and Metalearning,* (pp. 57-68).

Goldberg, D.A. (1989). *Genetic algorithms in search, optimization and machine learning.* Addison-Wesley.

Guha, S., Rastogi, R., & Shim, K. (2000). ROCK: A robust clustering algorithm for categorical attributes. *Information Systems, 25*(5), 345-366.

Han, J., & Fu, Y. (1995, September). Discovery of multiple-level association rules from large databases. *Proceedings of the 1995 International Conference on Very Large Data Bases VLDB'95,* Zurich, Switzerland, 420-431.

Han, J., Pei, J., Mortazavi-Asl, B., Chen, Q., Dayal, U., & Hsu, M-C. (2000). FreeSpan: Frequent pattern-projected sequential pattern mining. *2000 International Conference on Knowledge Discovery and Data Mining.*

Hartigan, J. (1975). *Clustering algorithms.* John Wiley.

Luotonen, A. (1995). The common Logfile format. Available: *http://www.w3.org/Daemon/User/Config/Logging.html.*

McQueen, J. (1967). Some methods for classification and analysis of multivariate observations. *Proceedings of the Fifth Berkeley Symposium on Mathematical Statistics and Probability,* (pp. 281-297).

Mobasher, B., Dai, H., Luo, T., & Nakagawa, M. (2002). Discovery and evaluation of aggregate usage profiles for Web personalization. *Data Mining and Knowledge Discovery, 6,* 61-82.

Nasraoui, O., Krishnapuram, R., Joshi, A., & Kamdar, T. (2002). Automatic Web user profiling and personalization using robust fuzzy relational clustering. E-commerce and intelligent methods. In the series *Studies in fuzziness and soft computing.*

Park, J.S., Chen, M.S., & Yu, P.S. (1995). An effective hash based algorithm for mining association rules. *Proceedings of the 1995 ACM SIGMOD International Conference on Management of Data,* (pp. 175-186).

Pitkow, J.E., & Pirolli, P. (1999). Mining longest repeating subsequences to predict World Wide Web surfing. *USENIX Symposium on Internet Technologies and Systems.*

Quinlan, J.R. (1993). C4.5: Programs for machine learning. *The Morgan Kaufmann series in machine learning.* Morgan Kaufmann.

Savasere, A., Omiecinski, E., & Navathe, S. (1995). An efficient algorithm for mining association rules in large databases. *Proceedings of the 21st VLDB Conference,* (pp. 432-443).

Schechter, S., Krishnan, M., & Smith, M.D. (1998). Using path profiles to predict HTTP requests. *7th International World Wide Web Conference, 30,* 457-467.

Srivastava, J., Cooley, R., Deshpande, M., & Tan, P-N. (2000). Web usage mining: Discovery and applications of usage patterns from Web data. *SIGKDD Explorations, 1*(2), 12-23.

Sun Microsystems. (1998). Java remote method invocation specification. Available: *http://java.sun.com/j2se/1.4.2/docs/guide/rmi/spec/rmiTOC.html.*

Sun Microsystems. (2001). *Enterprise JavaBeans specification, Version 2.0.* Available: *http://java.sun.com/products/ejb/docs.html.*

Toivonen, H. (1996). Sampling large databases for association rules. *Proceedings of the 1996 International Conference on Very Large Data Bases, 9,* (pp. 134-145).

Witten, I.A., & Frank, E. (2000). *Data mining. Practical machine learning tools and techniques with JAVA implementations.* Morgan Kaufmann.

Wolpert, D.H. (1996). The lack of a priori distinctions between learning algorithms. *Neural Computation,* 1341-1390.

Wolpert, D.H., & Macready, W.G. (1995). *No free lunch theorems for search.* Technical report, Santa Fe Institute.

Wu, K., Yu, P.S., & Ballman, A. (1997). SpeedTracer: A Web usage mining and analysis tool. *Internet Computing, 37*(1), 89.

Yan, T., Jacobsen, M., Garcia-Molina, H., & Dayal, U. (1996). From user access patterns to dynamic hypertext linking. *Fifth International World Wide Web Conference,* Paris, France, 1007-1118.

Zaki, M.J., Parthasarathy, S., Ogihara, M., & Li, W. (1997). New algorithms for fast discovery of association rules. *Third International Conference on Knowledge Discovery and Data Mining,* 283-296.

Zhang, T., Ramakrishnan, R., & Livny, M. (1997). BIRCH: A new data clustering algorithm and its applications. *Data Mining and Knowledge Discovery, 1*(2), 141-182.

This work was previously published in Web Mining: Applications and Techniques, edited by A. Scime, pp. 355-372, copyright 2005 by IGI Publishing, formerly known as Idea Group Publishing (an imprint of IGI Global).

Chapter 2.11
Spatial Data Warehouse Modelling

Maria Luisa Damiani
Università di Milano, Italy
Ecole Polytechnique Fédérale, Switzerland

Stefano Spaccapietra
Ecole Polytechnique Fédérale de Lausanne, Switzerland

ABSTRACT

This chapter is concerned with multidimensional data models for spatial data warehouses. Over the last few years different approaches have been proposed in the literature for modelling multidimensional data with geometric extent. Nevertheless, the definition of a comprehensive and formal data model is still a major research issue. The main contributions of the chapter are twofold: First, it draws a picture of the research area; second it introduces a novel spatial multidimensional data model for spatial objects with geometry (MuSD – multigranular spatial data warehouse). MuSD complies with current standards for spatial data modelling, augmented by data warehousing concepts such as spatial fact, spatial dimension and spatial measure. The novelty of the model is the representation of spatial measures at multiple levels of geometric granularity. Besides the representation concepts, the model includes a set of OLAP operators supporting the navigation across dimension and measure levels.

INTRODUCTION

A topic that over recent years has received growing attention from both academy and industry concerns the integration of spatial data management with multidimensional data analysis techniques. We refer to this technology as spatial data warehousing, and consider a spatial data warehouse to be a multidimensional database of spatial data. Following common practice, we use here the term spatial in the geographical sense, i.e., to denote data that includes the description of how objects and phenomena are located on the Earth. A large variety of data may be considered to be spatial, including: data for land use and socioeconomic analysis; digital imagery and geo-sensor data; location-based data acquired through GPS or other positioning devices; environmental phenomena. Such data are collected and possibly marketed by organizations such as public administrations, utilities and other private companies, environmental research centres and spatial data infrastructures. Spatial data warehousing has been recognized as a key technology in enabling the interactive analysis of spatial data sets for decision-making support (Rivest et al., 2001; Han et al., 2002). Application domains in which the technology can play an important role are, for example, those dealing with complex and worldwide phenomena such as homeland security, environmental monitoring and health safeguard. These applications pose challenging requirements for integration and usage of spatial data of different kinds, coverage and resolution, for which the spatial data warehouse technology may be extremely helpful.

Origins

Spatial data warehousing results from the confuence of two technologies, spatial data handling and multidimensional data analysis, respectively. The former technology is mainly provided by two kinds of systems: spatial database management systems (DBMS) and geographical information systems(GIS). Spatial DBMS extend the functionalities of conventional data management systems to support the storage, efficient retrieval and manipulation of spatial data (Rigaux et al., 2002). Examples of commercial DBMS systems are Oracle Spatial and IBM DB2 Spatial Extender. A GIS, on the other hand, is a composite computer based information system consisting of an integrated set of programs, possibly including or interacting with a spatial DBMS, which enables the capturing, modelling, analysis and visualization of spatial data (Longley et al., 2001). Unlike a spatial DBMS, a GIS is meant to be directly usable by an end-user. Examples of commercial systems are ESRI ArcGIS and Intergraph Geomedia. The technology of spatial data handling has made significant progress in the last decade, fostered by the standardization initiatives promoted by OGC (Open Geospatial Consortium) and ISO/TC211, as well as by the increased availability of off-the-shelf geographical data sets that have broadened the spectrum of spatially-aware applications. Conversely, multidimensional data analysis has become the leading technology for decision making in the business area. Data are stored in a multidimensional array (cube or hypercube) (Kimball, 1996; Chaudhuri & Dayla, 1997; Vassiliadis & Sellis, 1999). The elements of the cube constitute the facts (or cells) and are defined by measures and dimensions. Typically, a measure denotes a quantitative variable in a given domain. For example, in the marketing domain, one kind of measure is sales amount. A dimension is a structural attribute characterizing a measure. For the marketing example, dimensions of sales may be: time, location and product. Under these example assumptions, a cell stores the amount of sales for a given product in a given region and over a given period of time. Moreover, each dimension is organized in a hierarchy of dimension levels, each level corresponding to a different granularity for the dimension. For example, *year* is one level of the *time* dimension, while the sequence *day, month, year* defines a simple hierarchy of increasing granularity for the time dimension. The basic operations for online analysis (OLAP operators) that can be performed over data cubes are: *roll-up*, which moves up along one or more dimensions towards more aggregated data (e.g., moving from monthly sales amounts to yearly sales amounts); *drill-down*, which moves down dimensions towards more detailed, disaggregated data and *slice-and-dice*, which performs a selection and projection operation on a cube.

The integration of these two technologies, spatial data handling and multidimensional analysis, responds to multiple application needs.

In business data warehouses, the spatial dimension is increasingly considered of strategic relevance for the analysis of enterprise data. Likewise, in engineering and scientific applications, huge amounts of measures, typically related to environmental phenomena, are collected through sensors, installed on ground or satellites, and continuously generating data which are stored in data warehouses for subsequent analysis.

Spatial Multidimensional Models

A data warehouse (DW) is the result of a complex process entailing the integration of huge amounts of heterogeneous data, their organization into denormalized data structures and eventually their loading into a database for use through online analysis techniques. In a DW, data are organized and manipulated in accordance with the concepts and operators provided by a multidimensional data model. Multidimensional data models have been widely investigated for conventional, non-spatial data. Commercial systems based on these models are marketed. By contrast, research on spatially aware DWs (SDWs) is a step behind. The reasons are diverse: The spatial context is peculiar and complex, requiring specialized techniques for data representation and processing; the technology for spatial data management has reached maturity only in recent times with the development of SQL3-based implementations of OGC standards; finally, SDWs still lack a market comparable in size with the business sector that is pushing the development of the technology. As a result, the definition of spatial multidimensional data models (SMDs) is still a challenging research issue.

A SMD model can be specified at conceptual and logical levels. Unlike the logical model, the specification at the conceptual level is independent of the technology used for the management of spatial data. Therefore, since the representation is not constrained by the implementation platform, the conceptual specification, that is the view we adopt in this work, is more flexible, although not immediately operational.

The conceptual specification of an SMD model entails the definition of two basic components: a set of representation constructs, and an algebra of spatial OLAP (SOLAP) operators, supporting data analysis and navigation across the representation structures of the model. The representation constructs account for the specificity of the spatial nature of data. In this work we focus on one of the peculiarities of spatial data, that is the availability of spatial data at different levels of granularity. Since the granularity concerns not only the semantics but also the geometric aspects of the data, the location of objects can have different geometric representations. For example, representing the location of an accident at different scales may lead to associating different geometries to the same accident.

To allow a more flexible representation of spatial data at different geometric granularity, we propose a SDM model in which not only dimensions are organized in levels of detail but also the spatial measures. For that purpose we introduce the concept of *multi-level spatial measure*.

The proposed model is named MuSD (multigranular spatial data warehouse). It is based on the notions of *spatial fact*, *spatial dimension* and *multi-level spatial measure*. A spatial fact may be defined as a fact describing an event that occurred on the Earth in a position that is relevant to know and analyze. Spatial facts are, for instance, road accidents. Spatial dimensions and measures represent properties of facts that have a geometric meaning; in particular, the spatial measure represents the location in which the fact occurred. A multi-level spatial measure is a measure that is represented by multiple geometries at different levels of detail. A measure of this kind is, for example, the location of an accident: Depending on the application requirements, an accident may be represented by a point along a road, a road segment or the whole road, possibly at different cartographic scales. Spatial measures and dimensions are uniformly represented in terms of the standard spatial objects defined by the Open Geospatial Consortium. Besides the representation constructs, the model includes a set of SOLAP operators to navigate not only through the dimensional levels but also through the levels of the spatial measures.

The chapter is structured in the following sections: the next section, *Background Knowledge*, introduces a few basic concepts underlying spatial data representation; the subsequent section, *State of the Art on Spatial Multidimensional Models*, surveys the literature on SDM models; the proposed spatial multidimensional data model is presented in the following section; and research opportunities and some concluding remarks are discussed in the two conclusive sections.

BACKGROUND KNOWLEDGE

The real world is populated by different kinds of objects, such as roads, buildings, administrative boundaries, moving cars and air pollution phenomena. Some of these objects are tangible, like buildings, others, like administrative boundaries, are not. Moreover, some of them have identifiable shapes with well-defined boundaries, like land parcels; others do not have a crisp and fixed shape, like air pollution. Furthermore, in some cases the position of objects, e.g., buildings, does not change in time; in other cases it changes more or less frequently, as in the case of moving cars. To account for the multiform nature of spatial data, a variety of data models for the digital representation of spatial data are needed. In this section, we present an overview of a few basic concepts of spatial data representation used throughout the chapter.

The Nature of Spatial Data

Spatial data describe properties of phenomena occurring in the world. The prime property of such phenomena is that they occupy a position. In broad terms, a position is the description of a location on the Earth. The common way of describing such a position is through the coordinates of a coordinate reference system.

The real world is populated by phenomena that fall into two broad conceptual categories: entities and continuous fields (Longley et al., 2001). Entities are distinguishable elements occupying a precise position on the Earth and normally having a well-defined boundary. Examples of entities are rivers, roads and buildings. By contrast, fields are variables having a single value that varies within a bounded space. An example of field is the temperature, or the distribution, of a polluting substance in an area. Field data can be directly obtained from sensors, for example installed on satellites, or obtained by interpolation from sample sets of observations.

The standard name adopted for the digital representation of abstractions of real world phenomena is that of *feature* (OGC, 2001, 2003). The feature is the basic representation construct defined in the reference spatial data model developed by the Open Geospatial Consortium and endorsed by ISO/TC211. As we will see, we will use the concept of feature to uniformly represent all the spatial components in our model. Features are spatial when they are associated with locations on the Earth; otherwise they are non-spatial. Features have a distinguishing name and have a set of attributes. Moreover, features may be defined at instance and type level: *Feature instances* represent single phenomena; *feature types* describe the intensional meaning of features having a common set of attributes. Spatial features are further specialized to represent different kinds of spatial data. In the OGC terminology, *coverages* are the spatial features that represent continuous fields and consist of discrete functions taking values over space partitions. Space partitioning results from either the subdivision of space in a set of regular units or cells (*raster* data model) or the subdivision of space in irregular units such as triangles (*tin* data model). The discrete function assigns each portion of a bounded space a value.

In our model, we specifically consider *simple spatial features*. Simple spatial features ("features" hereinafter) have one or more attributes of geometric type, where the geometric type is one of the types defined by OGC, such as point, line and polygon. One of these attributes denotes the position of the entity. For example, the position of the state Italy may be described by a multipolygon, i.e., a set of disjoint polygons (to account for islands), with holes (to account for the Vatican State and San Marino). A simple feature is very close to the concept of entity or object as used by the database community. It should be noticed, however, that besides a semantic and geometric characterization, a feature type is also assigned a coordinate reference system, which is specific for the feature type and that defines the space in which the instances of the feature type are embedded.

More complex features may be defined specifying the topological relationships relating a set of features. Topology deals with the geometric properties that remain invariant when space is elastically deformed. Within the context of geographical information, topology is commonly used to describe, for example, connectivity and adjacency relationships between spatial elements. For example, a road network, consisting of a set of interconnected roads, may be described through a graph of nodes and edges: Edges are the topological objects representing road segments whereas nodes account for road junctions and road endpoints.

To summarize, spatial data have a complex nature. Depending on the application requirements and the characteristics of the real world phenomena, different spatial data models can be adopted for the representation of geometric and topological properties of spatial entities and continuous fields.

STATE OF THE ART ON SPATIAL MULTIDIMENSIONAL MODELS

Research on spatial multidimensional data models is relatively recent. Since the pioneering work of Han et al. (1998), several models have been proposed in the literature aiming at extending the classical multidimensional data model with spatial concepts. However, despite the complexity of spatial data, current spatial data warehouses typically contain objects with simple geometric extent.

Moreover, while an SMD model is assumed to consist of a set of representation concepts and an algebra of SOLAP operators for data navigation and aggregation, approaches proposed in the literature often privilege only one of the two aspects, rarely both. Further, whilst early data models are defined at the logical level and are based on the relational data model, in particular on the star model, more recent developments, especially carried out by the database research community, focus on conceptual aspects. We also observe that the modelling of geometric granularities in terms of multi-level spatial measures, which we propose in our model, is a novel theme.

Often, existing approaches do not rely on standard data models for the representation of spatial aspects. The spatiality of facts is commonly represented through a geometric element, while in our approach, as we will see, it is an OGC spatial feature, i.e., an object that has a semantic value in addition to its spatial characterization.

A related research issue that is gaining increased interest in recent years, and that is relevant for the development of comprehensive SDW data models, concerns the specification and efficient implementation of the operators for spatial aggregation.

Literature Review

The first, and perhaps the most significant, model proposed so far has been developed by Han et al.

(1998). This model introduced the concepts of spatial dimension and spatial measure. Spatial dimensions describe properties of facts that also have a geometric characterization. Spatial dimensions, as conventional dimensions, are defined at different levels of granularity. Conversely, a spatial measure is defined as "a measure that contains a collection of pointers to spatial objects", where spatial objects are geometric elements, such as polygons. Therefore, a spatial measure does not have a semantic characterization, it is just a set of geometries. To illustrate these concepts, the authors consider a SDW about weather data. The example SDW has three thematic dimensions: {temperature, precipitation, time}; one spatial dimension: {region}; and three measures: {region_map, area, count}. While area and count are numeric measures, region_map is a spatial measure denoting a set of polygons. The proposed model is specified at the logical level, in particular in terms of a star schema, and does not include an algebra of OLAP operators. Instead, the authors develop a technique for the efficient computation of spatial aggregations, like the merge of polygons. Since the spatial aggregation operations are assumed to be distributive, aggregations may be partially computed on disjoint subsets of data. By pre-computing the spatial aggregation of different subsets of data, the processing time can be reduced.

Rivest et al. (2001) extend the definition of spatial measures given in the previous approach to account for spatial measures that are computed by metric or topological operators. Further, the authors emphasize the need for more advanced querying capabilities to provide end users with topological and metric operators. The need to account for topological relationships has been more concretely addressed by Marchant et al. (2004), who define a specific type of dimension implementing spatio-temporal topological operators at different levels of detail. In such a way, facts may be partitioned not only based on dimension values but also on the existing topological relationships.

Shekhar et al. (2001) propose a *map cube operator*, extending the concepts of data cube and aggregation to spatial data. Further, the authors introduce a classification and examples of different types of spatial measures, e.g., spatial distributive, algebraic and holistic functions.

GeoDWFrame (Fidalgo et al., 2004) is a recently proposed model based on the star schema. The conceptual framework, however, does not include the notion of spatial measure, while dimensions are classified in a rather complex way.

Pederson and Tryfona (2001) are the first to introduce a formal definition of an SMD model at the conceptual level. The model only accounts for spatial measures whilst dimensions are only non-spatial. The spatial measure is a collection of geometries, as in Han et al. (1998), and in particular of polygonal elements. The authors develop a pre-aggregation technique to reduce the processing time of the operations of merge and intersection of polygons. The formalization approach is valuable but, because of the limited number of operations and types of spatial objects that are taken into account, the model has limited functionality and expressiveness.

Jensen et al. (2002) address an important requirement of spatial applications. In particular, the authors propose a conceptual model that allows the definition of dimensions whose levels are related by a partial containment relationship. An example of partial containment is the relationship between a roadway and the district it crosses. A degree of containment is attributed to the relationship. For example, a roadway may be defined as partially contained at degree 0.5 into a district. An algebra for the extended data model is also defined. To our knowledge, the model has been the first to deal with uncertainty in data warehouses, which is a relevant issue in real applications.

Malinowski and Zimanyi (2004) present a different approach to conceptual modelling. Their SMD model is based on the Entity Relationship modelling paradigm. The basic representation constructs are those of *fact relationship* and *dimension*. A *dimension* contains one or several related levels consisting of entity types possibly having an attribute of geometric type. The *fact relationship* represents an n-ary relationship existing among the dimension levels. The attributes of the *fact relationship* constitute the measures. In particular, a spatial measure is a measure that is represented by a geometry or a function computing a geometric property, such as the length or surface of an element. The spatial aspects of the model are expressed in terms of the MADS spatio-temporal conceptual model (Parent et al., 1998). An interesting concept of the SMD model is that of *spatial fact relationship*, which models a spatial relationship between two or more spatial dimensions, such as that of spatial containment. However, the model focuses on the representation constructs and does not specify a SOLAP algebra.

A different, though related, issue concerns the operations of *spatial aggregation*. Spatial aggregation operations summarize the geometric properties of objects, and as such constitute the distinguishing aspect of SDW. Nevertheless, despite the relevance of the subject, a standard set of operators (as, for example, the operators Avg, Min, Max in SQL) has not been defined yet. A first comprehensive classification and formalization of spatio-temporal aggregate functions is presented in Lopez and Snodgrass (2005). The operation of aggregation is defined as a function that is applied to a collection of tuples and returns a single value. The authors distinguish three kinds of methods for generating the set of tuples, known as *group composition*, *partition composition* and *sliding window composition*. They provide a formal definition of aggregation for conventional, temporal and spatial data based on this distinction. In addition to the conceptual aspects of spatial aggregation, another major issue regards the development of methods for the efficient computation of these kinds of operations to manage high volumes of spatial data. In particular, techniques are developed based on the combined use of specialized indexes, materialization of aggregate measures and computational geometry algorithms, especially to support the aggregation of dynamically computed sets of spatial objects (Papadias, et al., 2001; Rao et al., 2003; Zhang & Tsotras, 2005).

A MULTIGRANULAR SPATIAL DATA WAREHOUSE MODEL: MUSD

Despite the numerous proposals of data models for SDW defined at the logical, and more recently,conceptual level presented in the previous section, and despite the increasing number of data warehousing applications (see, e.g., Bedard et al., 2003; Scotch & Parmantoa, 2005), the definition of a comprehensive and formal data model is still a major research issue.

In this work we focus on the definition of a formal model based on the concept of spatial measures at multiple levels of geometric granularity.

One of the distinguishing aspects of multidimensional data models is the capability of dealing with data at different levels of detail or granularity. Typically, in a data warehouse the notion of granularity is conveyed through the notion of dimensional hierarchy. For example, the dimension *administrative units* may be represented at different decreasing levels of detail: at the most detailed level as municipalities, next as regions and then as states. Note, however, that unlike dimensions, measures are assigned a unique granularity. For example, the granularity of sales may be homogeneously expressed in euros.

In SDW, the assumption that spatial measures have a unique level of granularity seems to be too restrictive. In fact, spatial data are very often available at multiple granularities, since data are collected by different organizations for different purposes. Moreover, the granularity not only regards the semantics (semantic granularity) but also the geometric aspects (spatial granularity) (Spaccapietra et al., 2000; Fonseca et al., 2002). For example, the location of an accident may be modelled as a measure, yet represented at different scales and thus have varying geometric representations.

To represent measures at varying spatial granularities, alternative strategies can be prospected: A simple approach is to define a number of spatial measures, one for each level of spatial granularity. However, this solution is not conceptually adequate because it does not represent the hierarchical relation among the various spatial representations.

In the model we propose, named MuSD, we introduce the notion of *multi-level spatial measure*, which is a spatial measure that is defined at multiple levels of granularity, in the same way as dimensions. The introduction of this new concept raises a number of interesting issues. The first one concerns the modelling of the spatial properties. To provide a homogeneous representation of the spatial properties across multiple levels, both spatial measures and dimensions are represented in terms of OGC features. Therefore, the locations of facts are denoted by feature identifiers. For example, a feature, say p1, of type *road accident*, may represent the location of an *accident*. Note that in this way we can refer to spatial objects in a simple way using names, in much the same way Han et al. (1998) do using pointers. The difference is in the level of abstraction and, moreover, in the fact that a feature is not simply a geometry but an entity with a semantic characterization.

Another issue concerns the representation of the features resulting from aggregation operations. To represent such features at different granularities, the model is supposed to include a set of operators that are able to dynamically decrease the spatial granularity of spatial measures. We call these operators *coarsening operators*. With this term we indicate a variety of operators that, although developed in different contexts, share the common goal of representing less precisely the geometry of an object. Examples include the operators for cartographic generalization proposed in Camossi et al. (2003) as well the operators generating imprecise geometries out of more precise representations (*fuzzyfying* operators).

In summary, the MuSD model has the following characteristics:

- It is based on the usual constructs of (spatial) measures and (spatial) dimensions. Notice that the spatiality of a measure is a necessary condition for the DW to be spatial, while the spatiality of dimensions is optional;
- A spatial measure represents the location of a fact at multiple levels of spatial granularity;
- Spatial dimension and spatial measures are represented in terms of OGC features;
- Spatial measures at different spatial granularity can be dynamically computed by applying a set of coarsening operators; and
- An algebra of SOLAP operators is defined to enable user navigation and data analysis.

Hereinafter, we first introduce the representation concepts of the MuSD model and then the SOLAP operators.

Representation Concepts in MuSD

The basic notion of the model is that of *spatial fact*. A spatial fact is defined as a fact that has occurred in a location. Properties of spatial facts are described in terms of measures and dimensions which, depending on the application, may have a spatial meaning.

A *dimension* is composed of *levels*. The set of levels is partially ordered; more specifically, it constitutes a lattice. Levels are assigned values belonging to *domains*. If the domain of a level consists of features, the level is *spatial*; otherwise it is *non-spatial*. A *spatial measure*, as a dimension, is composed of levels representing different granularities for the measure and forming a lattice. Since in common practice the notion of granularity seems not to be of particular concern for conventional and numeric measures, non-spatial measures are defined at a unique level. Further, as the spatial measure represents the location of the fact, it seems reasonable and not significantly restrictive to assume the spatial measure to be unique in the SDW.

As Jensen et al. (2002), we base the model on the distinction between the intensional and extensional representations, which we respectively call *schema* and *cube*. The schema specifies the structure, thus the set of dimensions and measures that compose the SDW; the cube describes a set of facts along the properties specified in the schema.

To illustrate the concepts of the model, we use as a running example the case of an SDW of road accidents. The *accidents* constitute the spatial facts. The properties of the accidents are modelled as follows: The number of *victims* and the *position* along the road constitute the measures of the SDW. In particular, the position of the accident is a spatial measure. The *date* and the *administrative unit* in which the accident occurred constitute the dimensions.

Before detailing the representation constructs, we need to define the spatial data model which is used for representing the spatial concepts of the model.

The Spatial Data Model

For the representation of the spatial components, we adopt a spatial data model based on the OGC simple features model. We adopt this model because it is widely deployed in commercial spatial DBMS and GIS. Although a more advanced spatial data model has been proposed (OGC, 2003), we do not lose in generality by adopting the simple feature model. Features (simple) are identified by names. Milan, Lake Michigan and the car number AZ213JW are examples of features. In particular, we consider as spatial features entities that can be mapped onto locations in the given space (for example, Milan and Lake Michigan). The location of a feature is represented through a *geometry*. The geometry of a spatial feature may be of type point, line or polygon, or recursively be a collection of disjoint geometries. Features have an application-dependent semantics that are expressed through the concept of *feature type*. Road, Town, Lake and Car are examples of feature types. The *extension* of a feature type, *ft*, is a set of semantically homogeneous features. As remarked in the previous section, since features are identified by unique names, we represent spatial objects in terms of feature identifiers. Such identifiers are different from the pointers to geometric elements proposed in early SDW models. In fact, a feature identifier does not denote a geometry, rather an entity that has also a semantics. Therefore some spatial operations, such as the spatial merge when applied to features, have a semantic value besides a geometric one. In the examples that will follow, spatial objects are indicated by their names.

Basic Concepts

To introduce the notion of schema and cube, we first need to define the following notions: *domain, level, level hierarchy, dimension* and *measure*. Consider the concept of domain. A domain defines the set of values that may be assigned to a property of facts, that is to a measure or to a dimension level. The domain may be single-valued or multi-valued; it may be spatial or non-spatial. A formal definition is given as follows.

Definition 1 (Domain and spatial domain): Let V be the set of values and F the set f features with $F \subseteq V$. A domain Do is single-valued if Do $\subseteq V$; it is multi-valued if Do $\subseteq 2^V$, in which case the elements of the domain are subsets of values. Further, the domain Do is a single-valued spatial domain if Do $\subseteq F$; it is a multi-valued spatial domain if $Do \subseteq 2^F$. We denote with DO the set of domains $\{Do_1, ..., Do_k\}$.

Example 1: In the road accident SDW, the single-valued domain of the property victims is the set of positive integers. A possible spatial domain for the position of the accidents is the set {a4, a5, s35} consisting of features which represent roads. We stress that in this example the position is a feature and not a mere geometric element, e.g., the line representing the geometry of the road.

The next concept we introduce is that of *level*. A level denotes the single level of granularity of both dimensions and measures. A level is defined by a name and a domain. We also define the notion of partial ordering among levels, which describes the relationship among different levels of detail.

Definition 2 (Level): A level is a pair < Ln, Do > where Ln is the name of the level and Do its domain. If the domain is a spatial domain, then the level is spatial; otherwise it is non-spatial.

Let Lv1 and Lv2 be two levels, dom(Lv) the function returning the domain of level Lv, and $\leq_{lv}$ a partial order over V. We say that $Lv1 \leq_{lv} Lv2$ iff for each $v1 \in dom(Lv1)$, it exists $v2 \in dom(Lv2)$ such that $v1 \leq_{lv} v2$. We denote with LV the set of levels. The relationship $Lv1 \leq_{lv} Lv2$ is read: Lv1 is less coarse (or more detailed) than Lv2.

Example 2: Consider the following two levels: L_1=<AccidentAtLargeScale, PointAt1:1'000>, L_2=<AccidentAtSmallScale, PointAt1:50'000>. Assume that Do_1 = PointAt1:1'000 and Do2 = PointAt1:50'000 are domains of features representing accidents along roads at different scales. If we assume that $Do_1 \leq_{lv} Do_2$ then it holds that AccidentAtLargeScale$\leq_{lv}$ AccidentAtSmallScale.

The notion of level is used to introduce the concept of *hierarchy of levels*, which is then applied to define dimensions and measures.

Definition 3 (Level hierarchy): Let L be a set of n levels $L = \{Lv_1, ..., Lv_n\}$. A level hierarchy H is a lattice over L: H =<L, $\leq_{lv}$, Lvtop, Lvbot> where $\leq_{lv}$ is a partial order over the set L of levels, and Lvtop, Lvbot, respectively, the top and the bottom levels of the lattice.

Given a level hierarchy H, the function LevelsOf(H) returns the set of levels in H. For the sake of generality, we do not make any assumption on the meaning of the partial ordering. Further, we say that a level hierarchy is of type *spatial* if all the levels in L are spatial; *non-spatial* when the levels are non-spatial; *hybrid* if L consists of both spatial and non-spatial levels. This distinction is analogous to the one defined by Han et al. (1998).

Example 3: Consider again the previous example of hierarchy of administrative entities. If the administrative entities are described by spatial features and thus have a geometry, then they form a spatial hierarchy; if they are described simply by names, then the hierarchy is non-spatial; if some levels are spatial and others are non-spatial, then the hierarchy is hybrid.

At this point we introduce the concepts of *dimensions*, *measures* and *spatial measures*. *Dimensions* and *spatial measures* are defined as hierarchies of levels. Since there is no evidence that the same concept is useful also for numeric measures, we introduce the notion of hierarchy only for the measures that are spatial. Further, as we assume that measures can be assigned subset of values, the domain of a (spatial) measure is multivalued.

Definition 4 (Dimension, measure and spatial measure): We define:

- A dimension D is a level hierarchy. The domains of the dimension levels are single-valued. Further, the hierarchy can be of type: spatial, non-spatial and hybrid;
- A measure M is defined by a unique level < M, Do >, with Do a multi-valued domain; and
- A spatial measure SM is a level hierarchy. The domains of the levels are multi-valued. Moreover the level hierarchy is spatial.

To distinguish the levels, we use the terms *dimension* and *spatial measure levels*. Note that the levels of the spatial measure are all spatial since we assume that the locations of facts can be represented at granularities that have a geometric meaning. Finally, we introduce the concept of *multigranular spatial schema* to denote the whole structure of the SDW.

Definition 5 (Multigranular spatial schema): A multigranular spatial schema S (schema, in short) is the tuple S =<D1, ..Dn, M1, ...Mm, SM> where:

- Di is a dimension, for each i =1, .., n;
- Mj is a non-spatial measure, for each j =1, .., m; and
- SM is a spatial measure.

We assume the spatial measure to be unique in the schema. Although in principle that could be interpreted as a limitation, we believe it is a reasonable choice since it seems adequate in most real cases.

Example 4: Consider the following schema S for the road accidents SDW:

S =<date, administrativeUnit, victims, location> where:

- *{date, administrativeUnit}* are dimensions with the following simple structure:
  - date =<{year, month } , $\leq_{date}$, month, year> with month $\leq_{date}$ year
  - administrativeUnit =<{municipality, region, state}, $\leq_{adm}$, municipality, state> with municipality $\leq_{adm}$ region $\leq_{adm}$ state;
- victims is a non-spatial measure;
- location is the spatial measure. Let us call M_1 = AccidentAtLargeScale and M_2 = AccidentAtSmallScale, two measure levels representing accidents at two different scales. Then the measure is defined as follows: <{M_1, M_2} $\leq_{pos}$, M_1, M_2> such that $M_1 \leq_{pos} M_2$.

Finally, we introduce the concept of *cube* to denote the extension of our SDW. A cube is a set of cells containing the measure values defined with respect a given granularity of dimensions and measures. To indicate the level of granularity of dimensions, the notion of *schema level* is introduced. A schema level is a schema limited to specific levels. A cube is thus defined as an instance of a schema level.

Definition 6 (Schema level): Let $S = \langle D_1, ..D_n, M_1, ...M_m$, SM> be a schema. A schema level SL for S is a tuple: $\langle DLv_1, ..DLv_n, M_1, ...M_m, Slv \rangle$ where:

- $DLv_i \in$ LevelsOf(D_i), is a level of dimension D_i (for each i = 1, ..., n);
- M_i is a non-spatial measure (for each i =1, …, m); and
- Slv ∈ LevelsOf(SM) is a level of the spatial measure SM

Since non-spatial measures have a unique level, they are identical in all schema levels. The cube is thus formally defined as follows:

Definition 7 (Cube and state): Let $SL = \langle DLv_1, ..DLv_n, M_1, ...M_m, Slv \rangle$ be a schema level.

A cube for SL, C_{SL} is the set of tuples (cells) of the form: <d_1, ..., d_n, m_1, ..., m_m, sv> where:

- d_i is a value for the dimension level DLv_i;
- m_i is a value for the measure M_i; and
- sv is the value for the spatial measure level Slv.

A state of a SDW is defined by the pair <SL, C_{SL}> where SL is a schema level and C_{SL} a cube.

The *basic cube* and *basic state* respectively denote the cube and the schema level at the maximum level of detail of the dimensions and spatial measure.

Example 5: Consider the schema S introduced in example 4 and the schema level <month, municipality, victims, accidentAtlargeScale>. An example of fact contained in a cube for such a schema level is the tuple <May 2005, Milan, 20, A4> where the former two values are dimension values and the latter two values are measure values. In particular, A4 is the feature representing the location at the measure level accidentAtLargeScale.

Spatial OLAP

After presenting the representation constructs of the model, we introduce the spatial OLAP operators. In order to motivate our choices, we first discuss three kinds of requirements that the concept of hierarchy of measures poses on these operators and thus the assumptions we have made.

Requirements and Assumptions

Interrelationship Between Dimensions and Spatial Measures

A first problem due to the introduction of the hierarchy of measures may be stated in these terms: Since a measure level is functionally dependent on dimensions, is this dependency still valid if we change the granularity of the measure? Consider the following example: assume the cube in example 4 and consider an accident that occurred in May 2005 in the municipality of Milan, located in point *P* along a given road, and having caused two victims. Now assume a decrease in the granularity of the position, thus representing the position no longer as a point but as a portion of road. The question is whether the dimension values are affected by such a change. We may observe that both cases are possible: (a) The functional dependency between a measure and a dimension is not affected by the change of spatial granularity of the measure if the dimension value does not depend on the geometry of the measure. This is the case for the dimension *date of accident;* since the date of an accident does not depend on the geometry of the accident, the dimension value does not change with the granularity. In this case we say that the date dimension is *invariant*; (b) The opposite case occurs if a spatial relationships exists between the given dimension and the spatial measure. For example, in the previous example, since it is reasonable to assume that a relationship of spatial containment is implicitly defined between the administrative unit and the accident, if the granularity of position changes, say the position is expressed not by a point but a line, it may happen that the relationship of containment does not hold any longer. In such a case, the value of the dimension level would vary with the measure of granularity. Since this second case entails complex modelling, in order to keep the model relatively simple, we assume that all dimensions are invariant with respect to spatial measure granularity. Therefore, all levels of a spatial measure have the same functional dependency from dimensions.

Aggregation of Spatial Measures

The second issue concerns the operators for the spatial aggregation of spatial measures. Such operators compute, for example, the union and intersection of a set of geometries, the geometry with maximum linear or aerial extent out of a set of one-dimensional and two-dimensional geometries and the MBB (Minimum Bounding Box) of a set of geometries. In general, in the SDW literature these operators are supposed to be applied only to geometries and not to features. Moreover, as previously remarked, a standard set of operators for spatial aggregation has not been defined yet.

For the sake of generality, in our model we do not make any choice about the set of possible operations. We only impose, since we allow representing spatial measures as features, that the operators are applied to sets of features and return a feature. Further, the result is a new or an existing feature, depending on the nature of the operator. For example, the union (or merge) of a set of features, say states, is a newly-created feature whose geometry is obtained from the geometric union of the features' geometries. Notice also that the type of the result may be a newly-created feature type. In fact, the union of a set of states is not itself a state and therefore the definition of a new type is required to hold the resulting features.

Coarsening of Spatial Measures

The next issue is whether the result of a spatial aggregation can be represented at different levels of detail. If so, data analysis would become much more flexible, since the user would be enabled not only to aggregate spatial data but also to dynamically decrease their granularity. To address this requirement, we assume that the model includes not only operators for spatial aggregation but also operators for decreasing the spatial granularity of features. We call these operators *coarsening operators*. As previously stated, coarsening operators include operators for cartographic generalization (Camossi & Bertolotto, 2003) and fuzzyûcation operators. A simple example of fuzzyfication is the operation mapping of a point of coordinates (x,y) into a close point by reducing the number of decimal digits of the coordinates. These operators are used in our model for building the hierarchy of spatial measures.

When a measure value is expressed according to a lower granularity, the dimension values remain unchanged, since dimensions are assumed to be invariant. As a simple example, consider the position of an accident. Suppose that an aggregation operation, e.g., MBB computation, is performed over positions grouped by date. The result is some new feature, say *yearly accidents*, with its own polygonal geometry. At this point we can apply a coarsening operator and thus a new measure value is dynamically obtained, functionally dependent on the same dimension values. The process of grouping and abstraction can thus iterate.

Spatial Operators

Finally, we introduce the Spatial OLAP operators that are meant to support the navigation in MuSD. Since numerous algebras have been proposed in the literature for non-spatial DW, instead of defining a new set of operators from scratch, we have selected an existing algebra and extended it. Namely, we have chosen the algebra defined in Vassiliadis, 1998. The advantages of this algebra are twofold: It is formally defined, and it is a good representative of the class of algebras for cube-oriented models (Vassiliadis, 1998; Vassiliadis & Sellis, 1999), which are close to our model.

Besides the basic operators defined in the original algebra (LevelClimbing, Packing, FunctionApplication, Projection and Dicing), we introduce the following operators: MeasureClimbing, SpatialFunctionApplication and CubeDisplay. The *MeasureClimbing* operator is introduced to enable the scaling up of spatial measures to different granularities; the *SpatialFunctionApplication* operator performs aggregation of spatial measures; *CubeDisplay* simply visualizes a cube as a map. The application of these operators causes a transition from the current state to a new state of the SDW. Therefore the navigation results from the successive application of these operators.

Hereinafter we illustrate the operational meaning of these additional operators. For the sake of completeness, we present first the three fundamental operators of the native algebra used to perform data aggregation and rollup.

In what follows, we use the following conventions: S indicates the schema, and ST denotes the set of states for S, of the form <SL, C> where SL is the schema level <DLv_1, ..., DLv_i, ..., DLv_n, M_1, ..., M_m, Slv> and C, a cube for that schema level. Moreover, the dot notation SL.DLv_i is used to denote the DLv_i component of the schema level. The examples refer to the schema presented in Example 4 (limited to one dimension) and to the basic cube reported in Table 1.

Level Climbing

In accordance with the definition of Vassiliadis, the LevelClimbing operation replaces all values of a set of dimensions with dimension values of coarser dimension levels. In other terms, given a state S=<SL, C>, the operation causes a transition to a new state <SL', C'> in which SL' is the schema level including the coarser dimension level, and C' is the cube containing the coarser values for the given level. In our model, the operation can be formally defined as follows:

Definition 8 (LevelClimbing): The LevelClimbing operator is defined by the mapping: LevelClimbing: ST x D x LV→ ST such that, given a state SL, a dimension D_i and a level lv_i of D_i, LevelClimbing(<SL, Cb>, D_i, lv_i) = <SL', Cb > with lv_i = SL'.Dlv_i.

Example 6: Let SL be the following schema levels: SL= <Month, AccidentPoint, Victims>. Cube 1 in Table 2 results from the execution of Level_Climbing (<SL, Basic_cube>, Time, Year).

Packing

The Packing operator, as defined in the original algebra, groups into a single tuple multiple tuples having the same dimension values. Since the domain of measures is multi-valued, after the operation the values of measures are sets. The new state shows the same schema level and a different cube. Formally:

Definition 9 (Packing): The Packing operator is defined by the mapping: Packing: ST→ ST such that Packing(<SL, C>) = <SL, C'>

Example 7: Cube 2 in Table 3 results from the operation: Pack (SL,Cube1)

Table 1. Cb= Basic cube

Month	Location	Victims
Jan 03	P 1	4
Jeb 03	P 2	3
Jan 03	P 3	3
May 03	P 4	1
Feb 04	P 5	2
Feb 04	P 6	3
Mar 04	P 7	1
May 04	P 8	2
May 04	P 9	3
May 04	P10	1

Table 2. Cube 1

Year	Location	Victims
03	P 1	4
03	P 2	3
03	P 3	3
03	P 4	1
04	P 5	2
04	P 6	3
04	P 7	1
04	P 8	2
04	P 9	3
04	P 10	1

Table 3. Cube 2

year	Location	#Victims
03	{P1,P2,P3,P4}	{4,2,3,1,2,1}
04	{P5,P6,P7,P8,P9,P19}	{3,3,1,3}

FunctionApplication

The FunctionApplication operator, which belongs to the original algebra, applies an aggregation function, such as the standard avg and sum, to the non-spatial measures of the current state. The result is a new cube for the same schema level. Let M be the set of non-spatial measures and AOP the set of aggregation operators.

Definition 10 (FunctionApplication): The FunctionApplication operator is defined by the mapping: FunctionApplication: ST×AOP×M→ ST, such that denoting with op(C, M_i) the cube resulting from the application of the aggregation operator op to the measure Mi of cube C, FunctionApplication(<DLv_1, …, DLv_n, M_1, …M_i, …, M_m, op, M_i) = <SL, C'> with cube C' = op(C, M_i).

SpatialFunctionApplication

This operator extends the original algebra to perform spatial aggregation of spatial measures. The operation is similar to the previous Function-Application. The difference is that the operator is meant to aggregate spatial measure values.

Definition 11 (SpatialFunctionApplication): Let SOP be the set of spatial aggregation operators. TheSpatialFunctionApplication operator is defined by the mapping:

SpatialFunctionApplication: ST×SOP→ ST such that, denoting with op(C, Slv) the cube resulting from the application of the spatial aggregation operator sop to the spatial measure level Slv of cube C, SpatialFunctionApplication(<DLv_1, …, DLv_n, M_1, …, M_m, Slv >, sop) = <SL, C'> with C' = sop(C, Slv).

Table 4. Cube 3

year	# Victims	Location
03	13	Area1
04	10	Area2

Example 8: Cube 3 in Table 4 results from the application of two aggregation operators, respectively on the measures victims and AccidentPoint. The result of the spatial aggregation is a set of features of a new feature type.

Measure Climbing

The MeasureClimbing operator enables the scaling of spatial measures to a coarser granularity. The effect of the operation is twofold: a) it dynamically applies a coarsening operator to the values of the current spatial measure level to obtain coarser values; and b) it causes a transition to a new state defined by a schema level with a coarser measure level.

Defnition 12 (MeasureClimbing): Let COP be the set of coarsening operators. The MeasureClimbing operator is defined by the mapping: MeasureClimbing : ST×COP→ ST such that denoting with:

- op(Slv): a coarsening operator applied to the values of a spatial measure level Slv
- SL =<DLv_1, ..., DLv_i, ..., DLv_n, M_1, ..., M_m, Slv>
- SL' =< DLv_1, ..., DLv_i, ..., DLv_n, M_1, ..., M_m, Slv' >

MeasureClimbing(SL, op)=SL' with Slv' = op(Slv);

Example 9: Cube 4 in Table 5 results from the application of the MeasureClimbing operator to the previous cube. The operation applies a coarsening operator to the spatial measure and thus changes the level of the spatial measure, reducing the level of detail. In Cube 4, "FuzzyLocation" is the name of the new measure level.

DisplayCube

This operator is introduced to allow the display of the spatial features contained in the current cube in the form of a cartographic map. Let MAP be the set of maps.

Defnition 13 (DisplayCube): The operator is defined by the mapping: DisplayCube: ST→MAP so that, denoting with m, a map: DisplayCube(<SL, C>) =m.

As a concluding remark on the proposed algebra, we would like to stress that the model is actually a general framework that needs to be instantiated with a specific set of aggregation and coarsening operators to become operationally meaningful. The definition of such set of operators is, however, a major research issue.

FUTURE TRENDS

Although SMD models for spatial data with geometry address important requirements, such models are not sufficiently rich to deal with more complex requirements posed by innovative applications. In particular, current SDW technology is not able to deal with complex objects. By complex

Table 5. Cube 4

Year	# Victims	FuzzyLocation
03	13	Id
04	10	Id2

spatial objects, we mean objects that cannot be represented in terms of simple geometries, like points and polygons. Complex spatial objects are, for example, continuous fields, objects with topology, spatio-temporal objects, etc. Specific categories of spatio-temporal objects that can be useful in several applications are diverse trajectories of moving entities. A trajectory is typically modelled as a sequence of consecutive locations in a space (Vlachos, 2002). Such locations are acquired by using tracking devices installed on vehicles and on portable equipment. Trajectories are useful to represent the location of spatial facts describing events that have a temporal and spatial evolution. For example, in logistics, trajectories could model the "location" of freight deliveries. In such a case, the delivery would represent the spatial fact, characterized by a number of properties, such as the freight and destination, and would include as a spatial attribute the trajectory performed by the vehicle to arrive at destination. By analyzing the trajectories, for example, more effective routes could be detected. Trajectories result from the connection of the tracked locations based on some interpolation function. In the simplest case, the tracked locations correspond to points in space whereas the interpolating function determines the segments connecting such points. However, in general, locations and interpolating functions may require a more complex definition (Yu et al., 2004). A major research issue is how to obtain summarized data out of a database of trajectories. The problem is complex because it requires the comparison and classification of trajectories. For that purpose, the notion of trajectory similarity is used. It means that trajectories are classified to be the same when they are sufficiently similar. Different measures of similarity have been proposed in the literature (Vlachos et al., 2002). A spatial data warehouse of trajectories could provide the unifying representation framework to integrate data mining techniques for data classification.

CONCLUSION

Spatial data warehousing is a relatively recent technology responding to the need of providing users with a set of operations for easily exploring large amounts of spatial data, possibly represented at different levels of semantic and geometric detail, as well as for aggregating spatial data into synthetic information most suitable for decision-making. We have discussed a novel research issue regarding the modelling of spatial measures defined at multiple levels of granularity. Since spatial data are naturally available at different granularities, it seems reasonable to extend the notion of spatial measure to take account of this requirement. The MuSD model we have defined consists of a set of representation constructs and a set of operators. The model is defined at the conceptual level in order to provide a more flexible and general representation. Next steps include the specialization of the model to account for some specific coarsening operators and the mapping of the conceptual model onto a logical data model as a basis for the development of a prototype.

REFERENCES

Bedard, Y., Gosselin, P., Rivest, S., Proulx, M., Nadeau, M., Lebel, G., & Gagnon, M. (2003). Integrating GIS components with knowledge discovery technology for environmental health decision support. *International Journal of Medical Informatics, 70*, 79-94.

Camossi, E., Bertolotto, M., Bertino, E., & Guerrini, G. (2003). A multigranular spatiotemporal data model. *Proceedings of the 11th ACM International Symposium on Advances in Geographic Information Systems, ACM GIS 2003,* New Orleans, LA (pp. 94-101).

Chaudhuri, S., & Dayal, U. (1997). An overview of data warehousing and OLAP technology. *ACM SIGMOD Record, 26*(1) , 65-74.

Clementini, E., di Felice, P., & van Oosterom, P. (1993). A small set of formal topological relationships suitable for end-user interaction. In *LNCS 692: Proceedings of the 3rd International Symposyium on Advances in Spatial Databases, SSD '93* (pp. 277-295).

Fidalgo, R. N., Times, V. C., Silva, J., & Souza, F. (2004). GeoDWFrame: A framework for guiding the design of geographical dimensional schemas. In *LNCS 3181: Proceedings of the 6th International Conference on Data Warehousing and Knowledge Discovery, DaWaK 2004* (pp. 26-37).

Fonseca, F., Egenhofer, M., Davies, C., & Camara, G. (2002). Semantic granularity in ontology-driven geographic information systems. *Annals of Mathematics and Artificial Intelligence, Special Issue on Spatial and Temporal Granularity, 36*(1), 121-151.

Han, J., Altman R., Kumar, V., Mannila, H., & Pregibon, D. (2002). Emerging scientific applications in data mining. *Communication of the ACM, 45*(8), 54-58.

Han, J. , Stefanovic, N., & Kopersky, K. (1998). Selective materialization: An efficient method for spatial data cube construction. *Proceedings of Research and Development in Knowledge Discovery and Data Mining, Second Pacific-Asia Conference, PAKDD'98* (pp. 144-158).

Jensen, C., Kligys, A., Pedersen T., & Timko, I. (2002). Multidimensional data modeling for location-based services. In *Proceedings of the 10th ACM International Symposium on Advances in Geographic Information Systems* (pp. 55-61).

Kimbal, R. (1996). *The data warehouse toolkit.* New York: John Wiley & Sons.

Longley, P., Goodchild, M., Maguire, D., & Rhind, D. (2001). *Geographic information systems and science.* New York: John Wiley & Sons.

Lopez, I., & Snodgrass, R. (2005). Spatiotemporal aggregate computation: A survey. *IEEE Transactions on Knowledge and Data Engineering, 17*(2), 271-286.

Malinowski, E. & Zimanyi, E. (2004). Representing spatiality in a conceptual multidimensional model. *Proceedings of the 12th ACM International Symposium on Advances in Geographic Information Systems, ACM GIS 2004*, Washington, DC (pp. 12-21).

Marchant, P., Briseboi, A., Bedard, Y., & Edwards G. (2004). Implementation and evaluation of a hypercube-based method for spatiotemporal exploration and analysis. *ISPRS Journal of Photogrammetry & Remote Sensing,* 59, 6-20.

Meratnia, N., & de By, R. (2002). Aggregation and Comparison of Trajectories. *Proceedings of the 10th ACM International Symposium on Advances in Geographic Information Systems, ACM GIS 2002*, McLean, VA (pp. 49-54).

OGC--OpenGIS Consortium. (2001). *OpenGIS® abstract specification, topic 1: Feature geometry (ISO 19107 Spatial Schema).* Retrieved from http://www.opengeospatial.org

OGC—Open Geo Spatial Consortium Inc. (2003). *OpenGIS[a] reference model*. Retrieved from http://www.opengeospatial.org

Papadias, D., Kalnis, P., Zhang, J., & Tao, Y. (2001). Efficient OLAP operations in spatial data warehouses. *LNCS: 2121, Proceedings of the 7h Int. Symposium on Advances in Spatial and Temporal Databases* (pp. 443-459).

Pedersen,T., & Tryfona, N. (2001). Pre-aggregation in spatial data warehouses. *LNCS: 2121, Proceedings. of the 7h Int. Symposium on Advances in Spatial and Temporal Databases* (pp. 460-480).

Rao, F., Zhang, L., Yu,X., Li,Y.,& Chen, Y. (2003). Spatial hierarchy and OLAP-favored search in spatial data warehouse. *Proceedings of the 6th ACM International Workshop on Data Warehousing and OLAP, DOLAP '03* (pp. 48-55).

Rigaux,. P., Scholl, M., & Voisard, A. (2002). *Spatial databases with applications to Gis*. New York: Academic Press.

Rivest, S., Bedard, Y., & Marchand, P. (2001). Towards better support for spatial decision making: Defining the characteristics of spatial on-line analytical processing (SOLAP). *Geomatica, 55*(4), 539-555.

Savary ,L., Wan, T., & Zeitouni, K. (2004). Spatio-temporal data warehouse design for human activity pattern analysis. *Proceedings of the 15th International Workshop On Database and Expert Systems Applications (DEXA04)* (pp. 81-86).

Scotch, M., & Parmantoa, B. (2005). SOVAT: Spatial OLAP visualization and analysis tools. *Proceedings of the 38th Hawaii International Conference on System Sciences.*

Shekhar, S. , Lu. C. T., Tan, X., Chawla, S., & Vatsavai, R. (2001). Map cube: A visualization tool for spatial data warehouse. In H. J. Miller & J. Han (Eds.), *Geographic data mining and knowledge discovery.* Taylor and Francis.

Shekhar, S., & Chawla, S. (2003). *Spatial databases: A tour.* NJ: Prentice Hall.

Spaccapietra, S., Parent, C., & Vangenot, C. (2000). GIS database: From multiscale to multirepresentation. In B. Y.Choueiry & T. Walsh (Eds.), Abstraction, reformulation, and approximation, LNAI 1864. *Proceedings of the 4th International Symposium, SARA-2000,* Horseshoe Bay, Texas.

Theodoratos, D., & Sellis, T. (1999). Designing data warehouses. *IEEE Transactions on Data and Knowledge Engineering, 31*(3), 279-301.

Vassiliadis, P. (1998). Modeling multidimensional databases, cubes and cube operations. *Proceedings of the 10th Scientific and Statistical Database Management Conference (SSDBM '98)* (pp. 53-62).

Vassiliadis, P., & Sellis, T. (1999). A survey of logical models for OLAP databases. *ACM SIGMOD Record, 28*(4), 64-69.

Vlachos, M., Kollios, G., & Gunopulos, D.(2002). Discovering similar multidimensional trajectories. *Proceedings of 18th ICDE* (pp. 273-282).

Wang, B., Pan, F., Ren, D., Cui, Y., Ding, D. et al. (2003). Efficient olap operations for spatial data using peano trees. *Proceedings of the 8th ACM SIGMOD Workshop on Research Issues in Data Mining and Knowledge Discovery* (pp. 28-34).

Worboys, M. (1998). Imprecision in finite resolution spatial data. *GeoInformatica, 2*(3), 257-279.

Worboys, M., & Duckam, M. (2004). *GIS: A computing perspective* (2nd ed.). Boca Raton, FL: CRC Press.

Yu, B., Kim, S. H., Bailey, T., & Gamboa R. (2004). Curve-based representation of moving object trajectories. *Proceedings of the International Database Engineering and Applications Symposium, IDEAS 2004* (pp. 419-425).

Zhang, D., & Tsotras, V. (2005). Optimizing spatial Min/Max aggregations. *The VLDB Journal*, *14*, 170-181.

This work was previously published in Processing and Managing Complex Data for Decision Support Systems, edited by J. Darmont and O. Boussaid, pp. 1-27, copyright 2006 by IGI Publishing, formerly known as Idea Group Publishing (an imprint of IGI Global).

Chapter 2.12
Designing Secure Data Warehouses

Rodolfo Villarroel
Universidad Católica del Maule, Chile

Eduardo Fernández-Medina
Universidad de Castilla – La Mancha, Spain

Juan Trujillo
Universidad de Alicante, Spain

Mario Piattini
Universidad de Castilla – La Mancha, Spain

ABSTRACT

Organizations depend increasingly on information systems, which rely upon databases and data warehouses (DWs), which need increasingly more quality and security. Generally, we have to deal with sensitive information such as the diagnosis made on a patient or even personal beliefs or other sensitive data. Therefore, a final DW solution should consider the final users that can have access to certain specific information. Unfortunately, methodologies that incorporate security are based on an operational environment and not on an analytical one. Therefore, they do not include security into the multidimensional approaches to work with DWs. In this chapter, we present a comparison of six secure-systems design methodologies. Next, an extension of the UML that allows us to specify main security aspects in the multidimensional conceptual modeling is proposed, thereby allowing us to design secure DWs. Finally, we present how the conceptual model can be implemented with Oracle Label Security (OLS10g).

INTRODUCTION

The goal of information confidentiality is to ensure that users can only access information that they are allowed. In the case of *multidimensional* (MD) models, confidentiality is crucial, because very sensitive business information can be discovered by executing a simple query. Sometimes, MD databases and *data warehouses* (DWs) also store information regarding private or personal aspects of individuals; in such cases, confidentiality is redefined as privacy. Ensuring appropriate information privacy is a pressing concern for many businesses today, given privacy legislation such as the United States' HIPAA that regulates the privacy of personal health care information, Gramm-Leach-Bliley Act, Sarbanes-Oxley Act, and the *European Union*'s (EU) Safe Harbour Law.

Generally, information systems security is taken into consideration once the system has been built, is in operation, and security problems have already arisen. This kind of approach — called "penetrate and patch" — is being replaced by methodologies that introduce security in the systems development process. This is an important advance but, unfortunately, methodologies that incorporate security are based on an operational environment and not on an analytical one. If we tried to use the operational environment to process consistent, integrated, well-defined and time-dependent information for purposes of analysis and decision making, we would notice that data available from operational systems do not fulfil these requirements. To solve this problem, we must work in an analytical environment strongly supported by the use of multidimensional models to design a DW (Inmon, 2002).

Several papers deal with the importance of security in the software development process. Ghosh, Howell, and Whittaker (2002) state that security must influence all aspects of design, implementation and software tests. Hall and Chapman (2002) put forward ideas about how to build correct systems that fulfil not only normal requirements but also those pertaining to security. These ideas are based on the use of several formal techniques of requirement representation and a strong correction analysis of each stage. Nevertheless, security in databases and data warehouses is usually focused on secure data storage and not on their design. Thus, a methodology of data warehouse design based on the *Unified Modeling Language* (UML), with the addition of security aspects, would allow us to design DWs with the syntax and power of UML and with new security characteristics ready to be used whenever the application has security requirements that demand them. We present an extension of the UML (*profile*) that allows us to represent the main security information of the data and their constraints in the MD modeling at the conceptual level. The proposed extension is based on the *profile* presented by Luján-Mora, Trujillo, and Song (2002) for the conceptual MD modeling, because it allows us to consider main MD-modeling properties and it is based on the UML. We consider the multilevel security model but focus on considering aspects regarding *read* operations, because this is the most common operation for final user applications. This model allows us to classify both information and users into security classes and enforce mandatory access control. This approach makes it possible to implement the secure MD models with any of the *database management systems* (DBMS) that are able to implement multilevel databases, such as Oracle Label Security (Levinger, 2003) and DB2 Universal Database, UDB (Cota, 2004).

The remainder of this chapter is structured as follows: first, we will briefly analyse each one of the six methodologies that incorporate security into the stages of systems development. The next section summarizes the UML extension for secure data warehouses modeling. Then, we present how the conceptual model can be implemented with a concrete product Oracle*10g* Label Security (OLS*10g*). Finally, we present the main conclusions and introduce our future work.

GENERAL DESCRIPTION OF METHODOLOGIES INCORPORATING SECURITY

The proposals that will be analysed are as follows:

- MOMT: multilevel object-modeling technique (Marks, Sell, & Thuraisingham, 1996);
- UMLSec: secure systems development methodology using UML (Jürgens, 2002);
- Secure database design methodology (Fernandez-Medina & Piattini, 2003);
- A paradigm for adding security into information systems development method (Siponen, 2002);
- A methodology for secure software design (Fernández, 2004); and
- ADAPTed UML: A pragmatic approach to conceptual modeling of OLAP security (Priebe & Pernul, 2001).

We have chosen these six methodologies because the majority of them try to solve the problem of security (mainly confidentiality) from the earliest stages of information systems development, emphasize security modeling aspects, and use modeling languages that make the security design process easier.

Multilevel Object Modeling Technique

Marks, Sell, and Thuraisingham (1996) define *multilevel object-modeling technique* (MOMT) as a methodology to develop secure databases by extending *object-modeling technique* (OMT) in order to be able to design multilevel databases providing the elements with a security level and establishing interaction rules among the elements of the model. MOMT is mainly composed of three stages: the *analysis stage*, the *system design stage*, and the *object design stage*.

UMLSec

Jürgens (2002) offers a methodology to specify requirements regarding confidentiality and integrity in analysis models based on UML. This approach considers an UML extension to develop secure systems. In order to analyse security of a subsystem specification, the behaviour of the potential attacker is modeled; hence, specific types of attackers that can attack different parts of the system in a specific way are modeled.

Secure Database Design

Fernández-Medina and Piattini (2003) propose a methodology to design multilevel databases by integrating security in each one of the stages of the database life cycle. This methodology includes the following:

- a specification language of multilevel security constraints about the conceptual and logical models;
- a technique for the early gathering of multilevel security requirements;
- a technique to represent multilevel database conceptual models;
- a logical model to specify the different multilevel relationships, the metainformation of databases and constraints;
- a methodology based upon the *unified process*, with different stages that allow us to design multilevel databases; and
- a CASE tool that helps to automate multilevel databases analysis and design process.

A Paradigm for Adding Security into IS Development Methods

Siponen (2002) proposes a new paradigm for secure information systems that will help developers use and modify their existing methods as needed. The meta-method level of abstraction offers a perspective on *information systems* (IS) secure development that is in a constant state of emergence and change. Furthermore, developers recognize regularities or patterns in the way problem settings arise and methods emerge.

The author uses the following analytical process for discovering the patterns of security design elements. First, look across information systems software development and information systems security development methodologies in order to find common core concepts (subjects and objects). Second, surface the patterns in existing secure information systems methods resulting in four additional concepts: *security constraints*, *security classifications*, *abuse subjects and abuse scenarios*, and *security policy*. Finally, consult a panel of practitioners for comments about the patterns. This process led to a pattern with six elements. Additional elements can certainly be added to the meta-notation on an ad hoc basis as required.

A Methodology for Secure Software Design

The main idea in the proposed methodology of Fernández (2004) is that security principles should be applied at every development stage and that each stage can be tested for compliance with those principles. The secure software life cycle is as follows: *requirement* stage, *analysis* stage, *design* stage, and *implementation* stage.

- **Requirements stage:** From the use cases, we can determine the needed rights for each actor and thus apply a need-to-know policy. Since actors may correspond to roles, this is now a Role-Based Access Control (RBAC) model.
- **Analysis stage:** We can build a conceptual model where repeated applications of the authorization pattern realize the rights determined from use cases. Analysis patterns can be built with predefined authorizations according to the roles in their use cases.
- **Design stage:** Interfaces can be secured again applying the authorization pattern. Secure interfaces enforce authorizations when users interact with the system.
- **Implementation stage:** This stage requires reflecting the security constraints in the code defined for the application.

ADAPTed UML: A Pragmatic Approach to Conceptual Modeling of OLAP Security

A methodology and language for conceptual modeling of *online analytical processing* (OLAP) security is presented in Priebe and Pernul (2001) by creating a UML-based notation named ADAPTed UML (which uses ADAPT symbols as stereotypes). The security model for OLAP is based on the assumption of a central (administrator-based) security policy. They base the security model on an open-world policy (i.e., access to data is allowed unless explicitly denied) with negative authorization constraints. This corresponds to the open nature of OLAP systems. Also, the authors present a *multidimensional security constraint language* (MDSCL) that is based on *multidimensional expressions* (MDX) representation of the logical OLAP model used by Microsoft.

SUMMARY OF EACH METHODOLOGY'S CONTRIBUTIONS

In Table 1, a synthesis of the contributions is shown, in security terms, made by each one of the analysed methodologies. It is very difficult to develop a methodology that fulfils all criteria and comprises all security. If that methodology was developed, its complexity would diminish its success. Therefore, the solution would be a more complete approach in which techniques and models defined by the most accepted model standards were used. And, if these techniques

Table 1. Contributions made by each one of the analysed methodologies

	Modeling/ Development standard	Technologies	Access control type	Constraints specification	CASE tool support
MOMT	OMT	Databases	MAC	NO	NO
UMLSec	UML patterns	Information systems	MAC (multinivel)	-------	NO
Fernández-Medina & Piattini	UML unified process	Databases	MAC, DAC, RBAC	OSCL (OCL based)	YES
Siponen	---------	Information systems meta-methodology	-------	NO	NO
Fernández	UML patterns	Information systems	Access Matrix RBAC	He refers to OCL as a good solution	NO
ADAPTed UML	ADAPT UML	OLAP	RBAC	MDSCL (MDX-based)	NO

and models could not be directly applied, they must be extended by integrating the necessary security aspects that, at present, are not covered by the analysed methodologies.

UML EXTENSION FOR SECURE MULTIDIMENSIONAL MODELING

In this section, we sketch our UML extension (*profile*) to the conceptual MD modeling of data warehouses. Basically, we have reused the previous profile defined by Lujan-Mora, Trujillo, and Song (2002), which allows us to design DWs from a conceptual perspective, and we have added the required elements that we need to specify the security aspects. Based on Conallen (2000), we define as extension a set of tagged values, stereotypes, and constraints. The tagged values we have defined are applied to certain objects that are especially particular to MD modeling, allowing us to represent them in the same model and on the same diagrams that describe the rest of the system. These tagged values will represent the sensitivity information of the different objects of the MD modeling (fact class, dimension class, base class, attributes, etc.), and they will allow us to specify security constraints depending on this security information and on the values of the attributes of the model. A set of inherent constraints are specified in order to define well-formedness rules. The correct use of our extension is assured by the definition of constraints in both natural language and *Object-Constraint Language* (OCL).

First, we need the definition of some new data types (in this case, stereotypes) to be used in our tagged values definition (see Table 2). All the information surrounding these new stereotypes has to be defined for each MD model, depending on its confidentiality properties and on the number of users and complexity of the organization in which the MD model will be operative.

Next, we define the tagged values of the class, as follows:

(a) **SecurityLevels:** Specifies the interval of possible security level values that an instance of this class can receive. Its type is Levels.

(b) **SecurityRoles:** Specifies a set of user roles. Each role is the root of a subtree of the

Table 2. New stereotypes: Data types

Name	Base class	Description
Level	Enumeration	The type Level will be an ordered enumeration composed by all security levels that have been considered.
Levels	Primitive	The type Levels will be an interval of levels composed by a lower level and an upper level.
Role	Primitive	The type Role will represent the hierarchy of user roles that can be defined for the organization.
Compartment	Enumeration	The type Compartment is the enumeration composed by all user compartments that have been considered for the organization.
Privilege	Enumeration	The type Privilege will be an ordered enumeration composed of all the different privileges that have been considered.
Attempt	Enumeration	The type Attempt will be an ordered enumeration composed of all the different access attempts that have been considered.

general user role hierarchy defined for the organization. Its type is Set(Role).

(c) **Security-Compartments:** Specifies a set of compartments. All instances of this class can have the same user compartments or a subset of them. Its type is Set(Compartment).

(d) **LogType:** Specifies whether the access has to be recorded: none, all access, only denied accesses, or only successful accesses. Its type is Attempt.

(e) **LogCond:** Specifies the condition to fulfil so that the access attempt is registered. Its type is OCLExpression.

(f) **Involved-Classes:** Specifies the classes that have to be involved in a query to be enforced in an exception. Its type is Set(OclType).

(g) **ExceptSign:** Specifies if an exception permits (+) or denies (-) the access to instances of this class to a user or a group of users. Its type is {+, -}.

(h) **ExceptPrivilege:** Specifies the privileges the user can receive or remove. Its type is Set(Privilege).

(i) **ExceptCond:** Specifies the condition that users have to fulfil to be affected by this exception. Its type is OCLExpression.

Figure 1 shows a MD model that includes a fact class (Admission) and two dimensions (Diagnosis and Patient). For example, Admission fact class — stereotype Fact — contains all individual admissions of patients in one or more hospitals and can be accessed by all users who have *secret* (S) or *top-secret* (TS) security labels — tagged-value *SecurityLevels* (SL) of classes, and play health or administrative roles — tagged-value *SecurityRoles* (SR) of classes. Note that the cost attribute can only be accessed by users who play administrative role — tagged-value SR of attributes.

Security constraints defined for stereotypes of classes (fact, dimension, and base) will be defined by using a UML note attached to the corresponding class instance. In this example:

1. The security level of each instance of *Admission* is defined by a security constraint specified in the model. If the value of the *description* attribute of the *Diagnosis_group* to which belongs the *diagnosis* that is related to the *Admission* is *cancer* or *AIDS*, the security level — tagged value *SL* — of this admission will be *top secret*, otherwise

Figure 1. Example of MD model with security information and constraints

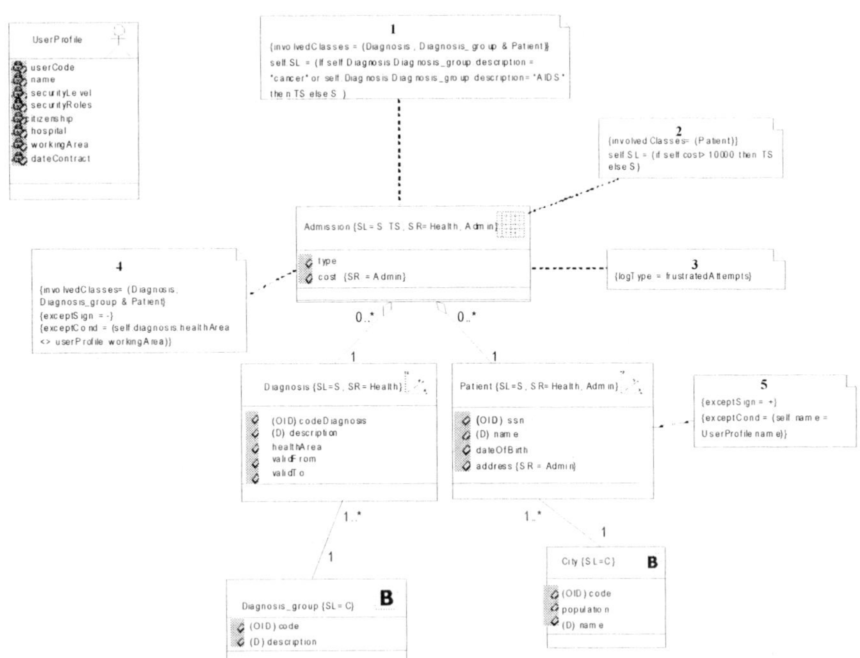

secret. This constraint is only applied if the user makes a query whose the information comes from the *Diagnosis* dimension or *Diagnosis_group* base classes, together with *Patient* dimension — tagged value *involvedClasses*. Therefore, a user who has *secret* security level could obtain the number of patients with *cancer* for each city, but never if information of the *Patient* dimension appears in the query.

2. The security level — tagged value *SL* — of each instance of *Admission* can also depend on the value of the *cost* attribute that indicates the price of the admission service. In this case, the constraint is only applicable for queries that contains information of the *Patient* dimension — tagged value *involved-Classes.*
3. The tagged value *logType* has been defined for the *Admission* class, specifying the value *frustratedAttempts*. This tagged value specifies that the system has to record, for future audit, the situation in which a user tries to access information from this fact class, and the system denies it because of lack of permissions.
4. For confidentiality reasons, we could deny access to admission information to users whose working area is different from the area of a particular admission instance. This is specified by another exception in the *Admission* fact class, considering tagged values *involvedClasses, exceptSign,* and *exceptCond.*
5. Patients could be special users of the system. In this case, it could be possible that patients have access to their own information

Figure 2. Access control mechanism

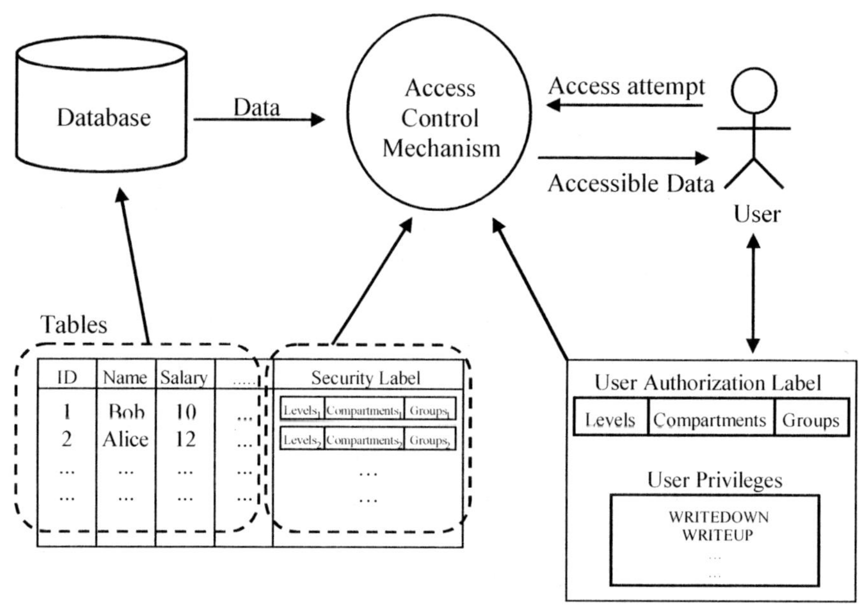

as patients (for instance, for querying their personal data). This constraint is specified by using the *excepSign* and *exceptCond* tagged values in the *Patient* class.

IMPLEMENTING SECURE DWS WITH OLS*10G*

In this section, we present some ideas with regard to how to implement secure DWs with OLS*10g*. We have chosen this model because it is part of one of the most important DBMSs that allows the implementation of label-based databases. Nevertheless, the match between DW conceptual model and OLS*10g* is not perfect. For instance, our general model considers security at the attribute level, and OLS*10g* only supports it at the row level (a coarser granularity access).

OLS*10g* is a component of version 10 of Oracle database management system that allows us to implement multilevel databases. OLS*10g* defines a combined access control mechanism, considering *mandatory access control* (MAC) by using the content of the labels, and *discretionary access control* (DAC), which is based on privileges. This combined access control imposes the rule that a user will only be entitled to access a particular row if he or she is authorized to do so by the DBMS, he or she has the necessary privileges, and the label of the user dominates the label of the row. Figure 2 represents this combined access control mechanism.

According to the particularities of OLS*10g,* the transformation between the conceptual DW model and this DBMS is as follows:

- Definition of the DW schema. The structure of the DW that is composed by fact, dimension, and base classes, including fact attributes, descriptors, dimension attributes, and aggregation, generalization and completeness associations, must be translated into a relational schema. This transformation is similar to the common transformation between conceptual and logical models (see Kimball, Reeves, Ross, & Thornthwaite, 1998).

- Adaptation of the new data types of the UML extension. All new data types (Level, Levels, Role, Compartment, Privilege, and Attempt) are perfectly supported by OLS10g.
- Adaptation of all tagged values that have been defined for the model. *Classes* are now represented as the set of tables of the database. *SecurityLevels, SecurityRoles,* and *SecurityCompartments* must be defined with the following sentences: CREATE_LEVEL, CREATE_GROUP, and CREATE_ COMPARTMENT.
- Adaptation of all tagged values that have been defined for the classes:
  (a) *SecurityLevels, SecurityRoles,* and *SecurityCompartments* are grouped into the security label, with labeling functions. Labeling functions define the information of the security label according to the value of the columns of the row that is inserted or updated.
  (b) *LogType* and *LogCond* are grouped with auditing options.
  (c) InvolvedClasses, Except-Sign, Except-Privilege, and ExceptCond are grouped with SQL predicates.
- Adaptation of all tagged values that have been defined for the attributes. It is important to mention that, in this case, all security tagged values that are defined for each attribute in the conceptual model have to be discarded because OLS*10g* does not support security for attributes (only for rows). This is a limitation of OLS*10g* that has a complex solution, so if it is important to also have security for attributes, another secure DBMS should be chosen.
- Adaptation of security constraints is defined with labeling functions and SQL predicates. The application of labeling functions is very useful in order to define the security attributes of rows and to implement security constraints. Nevertheless, sometimes labeling functions are not enough, being necessary specifying more complex conditions. OLS*10*g provides the possibility of defining SQL predicates together with the security policies. Both labeling functions and SQL predicates will be especially important implementing secure DWs.

We could consider the *Admission* table. This table will have a special column that will store the security label for each instance. For each instance, this label will contain the security information that has been specified in the conceptual model

Figure 3. Security constraints implemented by labeling functions

```
(a) CREATE FUNCTION Which_Cost (Cost: Integer) Return
    LBACSYS.LBAC_LABEL
    As MyLabel varchar2(80);
    Begin
     If Cost>10000 then MyLabel := 'TS::Health,Admin'; else MyLabel :=
       S::Health,Admin';
     end if;
     Return TO_LBAC_DATA_LABEL('MyPolicy', MyLabel);
    End;
(b) APPLY_TABLE_POLICY ('MyPolicy', 'Admission', 'Scheme', ,
    'Which_Cost')
```

in Figure 1 (Security *Level = Secret..TopSecret; SecurityRoles =Health, Admin*). But this security information depends on several security constraints that can be implemented by labeling functions. Figure 3 (a) shows an example by which we implement the security constraints. If the value of *Cost* column is greater than 10000 then the security label will be composed of *TopSecret* security level and *Health* and *Admin* user roles; otherwise, the security label will be composed of *Secret* security level and the same user roles. Figure 3 (b) shows how to link this labeling function with *Admission* table.

According to these transformation rules, the activities for building the secure DW with OLS*10g* are as follows:

- Definition of the DW scheme.
- Definition of the security policy and its default options. When we create a security policy, we have to specify the name of the policy, the name of the column that will store the labels, and finally other options of the policy. In this case, the name of the column that stores the sensitive information in each table, which is associated with the security policy, is *SecurityLabel*. The option *HIDE* indicates that the column *SecurityLabel* will be hidden, so that users will not be able to see it in the tables. The option *CHECK_CONTROL* forces the system to check that the user has reading access when he or she introduces or modifies a row. The option *READ_CONTROL* causes the enforcement of the read access control algorithm for *SELECT, UPDATE* and *DELETE* operations. Finally, the option *WRITE_CONTROL* causes the enforcement of the write access control algorithm for *INSERT, DELETE* and *UPDATE* operations.
- Specification of the valid security information in the security policy.
- Creation of the authorized users and assignment of their authorization information.
- Definition of the security information for tables through labeling functions.
- Implementation of the security constraints through labeling functions.
- Implementation, if necessary, of the operations and control of their security.

Snapshots of Our Prototype from OLS*10g*

In this subsection, we provide snapshots to show how Oracle *10g* works with different secure rules that we have defined for our case study. All these snapshots are captured from the SQL Worksheet tool, a manager tool provided by Oracle to work with Oracle DBMS. Within this tool, we introduce the SQL sentences to be executed in the upper window, and in the lower window, we can see the corresponding answer provided by the server.

First, we have created the database scheme. Then, the security policy, the security information (levels and groups), the users, the labeling functions, the predicates, and the functions have been defined by means of the Oracle policy manager. Finally, in Table 3, we show some inserted rows that will allow us to show the benefits of our approach.

Although the SecurityLabel column is *hidden*, we have shown it for the Admission table, so that we can appreciate that the label for each row is correctly defined, according to the security information and constraints that have been specified in Figure 1.

For the sake of simplicity, we have defined only three users: Bob, who has a "topSecret" security level and who plays the "HospitalEmployee" role; Alice, who has a "Secret" security level and who plays an "Administrative" role; and James, who is a special user because he is a patient (so he will be able to access only his own information and nothing else).

In order to illustrate how the security specifications that we have defined in the conceptual MD modeling (Figure 1) are enforced in Oracle,

Table 3. Rows inserted into the Admission, Patient, Diagnosis and UserProfile tables

ADMISSION

TYPE	COST	SSN	CODE DIAGNOSIS	SECURITY LABEL
Primary	150000	12345678	S1.1	TS::HE, A
Secondary	180000	12345678	S1.2	TS::HE, A
Primary	8000	98765432	D1.1	S::HE, A
Primary	90000	98765432	C1.1	TS::HE, A
Primary	9000	12345678	D1.2	S::HE, A

PATIENT

SSN	NAME	DATE OF BIRTH	ADDRESS	CITY NAME	CITY POPULATION
12345678	James Brooks	12/10/84	3956 North 46 Av.	Florida	15982378
98765432	Jane Ford	10/02/91	2005 Harrison Street	Florida	15982378

DIAGNOSIS

CODE DIAGNOSIS	DESCRIPTION	HEALTH AREA	VALID FROM	VALID TO	DIAGNOSIS GROUP
C1.1	Skin Cancer	Dermatology	01/01/00	01/01/10	Cancer
C1.2	Cervical Cancer	Gynecology	12/10/04	12/10/14	Cancer
D1.1	Diabetes during pregnancy	Gynecology	07/11/03	01/11/13	Diabetes
D1.2	Other diabetes types	Endocrinology	12/12/00	12/12/10	Diabetes
S1.1	Symptomatic infection VIH	Internal medicine	10/11/00	10/11/10	AIDS
S1.2	AIDS related-complex	Internal medicine	11/11/01	11/11/11	AIDS

USERPROFILE

USER CODE	NAME	CITIZENSHIP	HOSPITAL	WORKING AREA	DATE CONTRACT
P000100	James Brooks	Canadian	USA Medical Center		
H000001	Bob Harrison	United States	USA Medical Center	Gynecology	10/12/87
H000002	Alice Douglas	United States	USA Medical Center	Dermatology	11/11/80

we have considered two different scenarios. In the first, we have not implemented the necessary functions to enforce the security rule defined in Note 4 of Figure 1. As it can be observed in Figure 4, the first query, which is performed by Bob (who has a "topSecret" security level and "HospitalEmployee" role), shows all tuples in the database. On the other hand, the second query, performed by Alice (who has a "Secret" security level and an "Administrative" role), does not show the information of patients whose diagnosis is Cancer or AIDS (specified in Note 1 of Figure 1) or whose cost is greater than 10000 (specified in Note 2 of Figure 1).

In the second scenario, we have implemented the security rule that is defined in Note 4 of Figure

Figure 4. First scenario

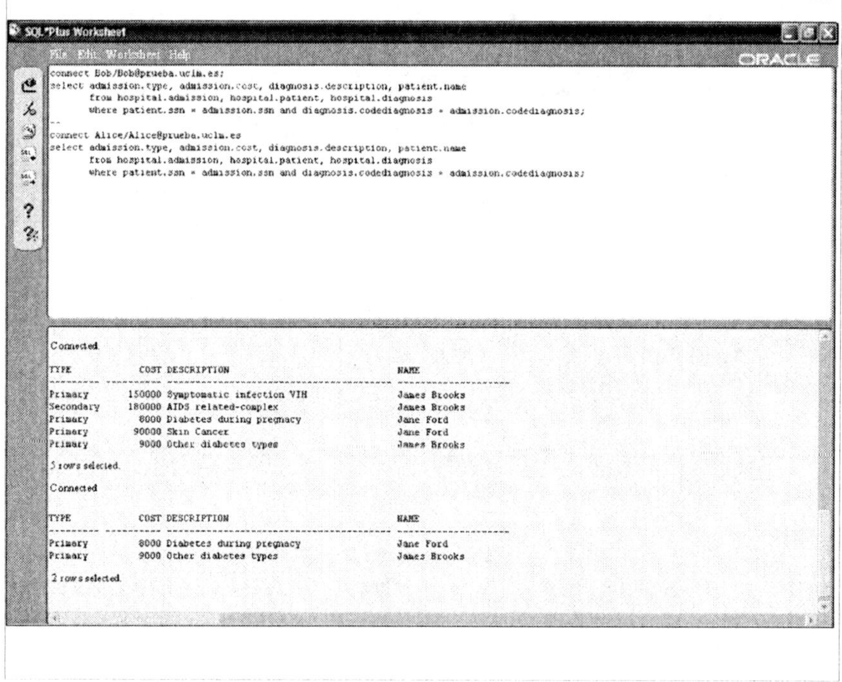

Figure 5. Second scenario

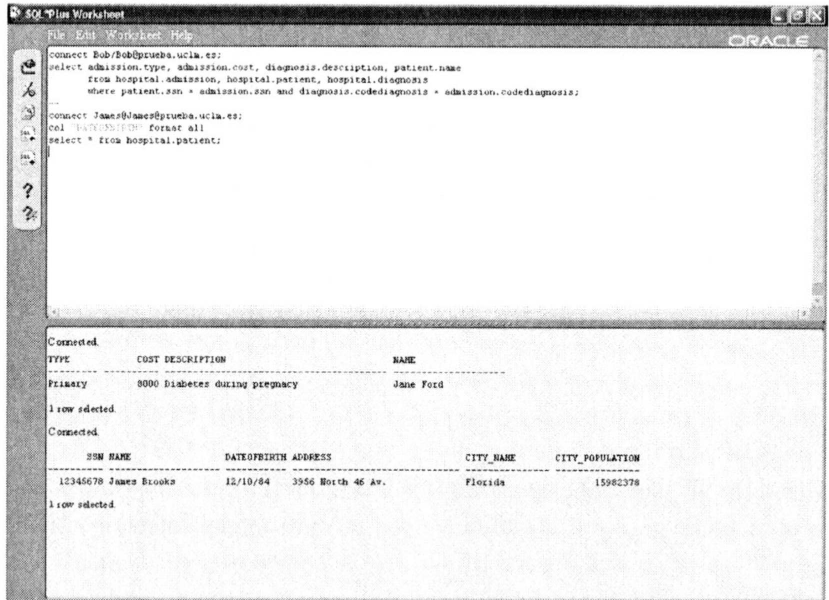

1. As we can observe in Figure 5, the first query, performed by Bob, shows all rows in which Note 4 of Figure 1 is fulfilled (that is to say, when the health area of the patient is the same than the working area of Bob). In the second query, performed by James, only his own information is shown (see Note 5 of Figure 1), hiding the patient information of other patients.

In conclusion, one of the key advantages of the overall approach presented here is that general and important secure rules for DWs that are specified by using our conceptual modeling approach can be directly implemented into a commercial DBMS such as Oracle *10g*. In this way, instead of having partially secure solutions for certain and specific non-authorized accesses, we deal with a complete and global approach for designing secure DWs from the first stages of a DW project. Finally, as we carry out the corresponding transformations through all stages of the design, we can be assured that the secure rules implemented into any DBMS correspond to the final user requirements captured in the conceptual modeling phase.

CONCLUSION

We have made a comparison of methodologies incorporating security in the development of their information systems in order to detect their limitations and to take them as a basis for the incorporation of the security in the uncovered aspects. In this way, we have put forward an extension based on UML as a solution to incorporate security in multidimensional modeling. Our approach, based on a widely accepted object-oriented modeling language, saves developers from learning a new model and its corresponding notations for specific MD modeling. Furthermore, the UML allows us to represent some MD properties that are hardly considered by other conceptual MD proposals. Considering that DWs, MD databases, and OLAP applications are used as very powerful mechanisms for discovering crucial business information in strategic decision-making processes, this provides interesting advances in improving the security of decision-support systems and protecting the sensitive information that these systems usually manage. We have also illustrated how to implement a secure MD model designed with our approach in a commercial DBMS. Our future work will focus on the development of a complete methodology based on UML and the Unified Process in order to develop secure DWs that grant information security and help us to comply with the existing legislation on data protection.

REFERENCES

Conallen, J. (2000). *Building Web applications with UML*. Reading, MA: Addison-Wesley.

Cota, S. (2004). For certain eyes only. *DB2 Magazine, 9*(1), 40-45.

Fernández, E. B. (2004). A methodology for secure software design. The *2004 International Conference on Software Engineering Research and Practice (SERP'04)*, Las Vegas, Nevada.

Fernández-Medina, E., & Piattini, M. (2003, June 21-24). Designing secure database for OLS. In *Proceedings of Database and Expert Systems Applications: 14th International Conference (DEXA 2003)*, Prague, Czech Republic (pp. 130-136). Berlin: Springer-Verlag.

Ghosh, A., Howell, C., & Whittaker, J. (2002). Building software securely from the ground up. *IEEE Software, 19*(1), 14-16.

Hall, A., & Chapman, R. (2002). Correctness by construction: Developing a commercial secure system. *IEEE Software, 19*(1), 18-25.

Inmon, H. (2002). *Building the data warehouse*. New York: John Wiley & Sons.

Jürjens, J. (2002, October 4). UMLsec: Extending UML for secure systems development. In J.

Jézéquel, H. Hussmann & S. Cook (Eds.), *Proceedings of UML 2002 - The Unified Modeling Language, Model Engineering, Concepts and Tools,* Dresden, Germany (pp. 412-425). Berlin: Springer-Verlag.

Kimball, R., Reeves, L., Ross, M., & Thornthwaite, W. (1998). *The data warehousing lifecycle toolkit.* New York: John Wiley & Sons.

Levinger, J. (2003). Oracle label security: Administrator's guide. Release 1 (10.1). Retrieved November 18, 2005, from http://www.oracle-10g-buch.de/oracle_10g_documentation/network.101/b10774.pdf

Luján-Mora, S., Trujillo, J., & Song, I. Y. (2002, September 30-October 4). Extending the UML for multidimensional modeling. In *Proceedings of 5th International Conference on the Unified Modeling Language (UML 2002),* Dresden, Germany (pp. 290-304). Berlin: Springer-Verlag.

Marks, D., Sell, P. , & Thuraisingham, B. (1996). MOMT: A multi-level object modeling technique for designing secure database applications. *Journal of Object-Oriented Programming,* 9(4), 22-29.

Priebe, T., & Pernul, G. (2001, November 27-30). A pragmatic approach to conceptual modeling of OLAP security. In *Proceedings of 20th International Conference on Conceptual Modeling (ER 2001),* Yokohama, Japan (pp. 311-324). Berlin: Springer-Verlag.

Siponen, M. (2002). Designing secure information systems and software (Academic Dissertation). Department of Information Processing Science. University of Oulo, Oulo, Finland.

This work was previously published in Enterprise Information Systems Assurance and Systems Security: Managerial and Technical Issues, edited by M. Warkentin, pp. 295-310, copyright 2006 by IGI Publishing, formerly known as Idea Group Publishing (an imprint of IGI Global).

Chapter 2.13
Privacy-Preserving Data Mining: Developments and Directions

Bhavani Thuraisingham
The MITRE Corporation, USA

ABSTRACT

This article first describes the privacy concerns that arise due to data mining, especially for national security applications. Then we discuss privacy-preserving data mining. In particular, we view the privacy problem as a form of inference problem and introduce the notion of privacy constraints. We also describe an approach for privacy constraint processing and discuss its relationship to privacy-preserving data mining. Then we give an overview of the developments on privacy-preserving data mining that attempt to maintain privacy and at the same time extract useful information from data mining. Finally, some directions for future research on privacy as related to data mining are given.

INTRODUCTION

There has been much interest recently on applying data mining for counter-terrorism applications (see Thuraisingham, 2003a, 2003b). For example, data mining can be used to detect unusual patterns, terrorist activities and fraudulent behavior. While all of these applications of data mining can benefit humans and save lives, there is also a negative side to this technology, since it could be a threat to the privacy of individuals. This is because data mining tools are available on the Web or otherwise, and even naive users can apply these tools to extract information from the data stored in various databases and files, and consequently violate the privacy of individuals. As we have stressed in other papers (see Thuraisingham, 2003a), to carry out effective data mining and extract useful information for counter-terrorism and national security, we need to gather all kinds of information about individuals. However, this information could be a threat to individuals' privacy and civil liberties (Thuraisingham, 2002).

Privacy is getting more attention partly because of counter-terrorism and national security. Recently we have heard a lot about national security in the media. This is mainly because people are now realizing that to handle terrorism, the government may need to collect information about individuals. This is causing a major concern with various civil liberties unions. The challenge is to carry out data mining and yet maintain privacy. This topic is known as privacy-preserving data mining.

This paper discusses developments and directions for privacy-preserving data mining, also sometimes called privacy sensitive data mining or privacy enhanced data mining. We discuss the privacy problem, provide an overview of the developments in privacy-preserving data mining and then discuss some of our research on viewing the privacy problem as an inference problem. In the next section, we first provide an overview of the privacy problem and discuss the connection between the privacy problem and the inference problem. Our research on developing techniques for ensuring privacy follows. This approach is called privacy constraint processing. We also show the connection between privacy-constraint processing and privacy-preserving data mining. Developments in privacy-preserving data mining will be discussed afterwards, along with directions for privacy research.

PRIVACY, DATA MINING AND THE INFERENCE PROBLEM

With the World Wide Web, there is now an abundance of information about individuals that one can obtain within seconds. This information could be obtained through mining or just from information retrieval. Data mining is the process of users posing queries and extracting information previously unknown using machine learning and other reasoning techniques (see Thuraisingham, 1998). Now, data mining is an important technology for many applications. However data mining also causes privacy concerns, as users can now put pieces of information together and extract information that is sensitive or private. Therefore, one needs to enforce controls on databases and data mining tools. That is, while data mining is an important tool for many applications, we do not want the information extracted to be used in an incorrect manner. For example, based on information about a person, an insurance company could deny insurance or a loan agency could deny loans. In many cases these denials may not be legitimate. Therefore, information providers have to be very careful in what they release. Also, data mining researchers have to ensure that privacy aspects are addressed.

We are beginning to realize that many of the techniques that were developed for the past two decades or so on the inference problem can now be used to handle privacy. One of the challenges to securing databases is the inference problem (Air Force Science Board, 1983). Inference is the process of users posing queries and deducing unauthorized information from the legitimate responses that they receive. This problem has been discussed quite a lot over the past two decades (Thuraisingham, 1987; Morgenstern, 1987; Hinke, 1988). However, data mining makes this problem worse. Users now have sophisticated tools that they can use to get data and deduce patterns that could be sensitive. Without these data mining tools, users would have to be fairly sophisticated in their reasoning to be able to deduce information from posing queries to the databases. That is, data mining tools make the inference problem quite dangerous (Clifton & Marks, 1996). While the inference problem mainly deals with secrecy and confidentiality, we are beginning to see many parallels between the inference problem and what we now call the privacy problem.

When the privacy problem is viewed as an inference problem, then we can use the inference controller approach to address privacy. For example, we can develop privacy controllers similar to our approach to developing inference controllers (Thuraisingham, Ford, Collins, & O'Keeffe, 1993). Furthermore, we can also have different degrees of privacy. For example, names and ages together could be less private while names and salaries together could be more private. Names and healthcare records together could be most private. One can then assign some probability or fuzzy value to represent the privacy level of an attribute or collection of attributes. Much work has been carried out on the inference problem in the past. Recently we have conducted some research on applying the techniques for handling the inference problem to handle the privacy problem. Our approach is privacy constraint processing and will be discussed next. Privacy constraint processing can be considered a special case of privacy-preserving data mining. A survey of privacy-preserving data mining is given in the next section.

PRIVACY CONSTRAINT PROCESSING

Overview

In this section we introduce privacy constraints. Essentially, privacy constraints are rules enforced on the data. These rules determine the privacy level of the data. Note that from a database perspective, levels could be assigned to databases, relations, rows, attributes and elements. Our definition of privacy constraints follows along the lines of our work on security constraints (Thuraisingham, 1987, 1990a). Privacy values of the data could take a range of values, including public, semi-public, semi-private and private. Even within a privacy value we could have different levels of privacy, including low-private, medium-private and high-private.

We have defined various types of privacy constraints. We give examples using a medical informatics database. The constraints we have identified include simple constraints, content-based constraints, context or association based constraints, release constraints and event constraints. While we use a relational database to illustrate the concepts, constraints can be defined on object as well as on XML (eXtensible Markup Language) databases.

Simple constraints assign privacy values to attributes, relations or even a database. For example, all medical records are private. Content-based constraints assign privacy values to data depending on content. For example, all financial records are private except for those who are in public office (e.g., the Secretary of Health and Human Services of the United States). Association-based constraints assign privacy values to collections of attributes taken together. For example, names and medical records are private; individually, they are public. That is, one can release names and medical records separately; but one cannot release them together. Furthermore, one has to be careful so that the public user cannot infer medical records for a particular person by posing multiple queries. Event constraints are constraints that change privacy values after an event has occurred. For example, after a patient has been released from a hospital, some information about that patient could be made public, but while in the hospital information about that patient is private. A good example was the sniper shootings that occurred in the Washington D.C. area in Fall 2002. After a victim died, information about that person was released. Until then, the identity of the person was not available to the public.

Finally, release constraints assign privacy values to the data depending on what has already been released. For example, after the medical records have been released, one cannot release any information about the names or social security numbers that can form a link to the medical information.

One could define many more types of privacy constraints. As we explore various applications, we will start defining various classes of constraints. Our main purpose in this chapter is to show how privacy constraints can be processed in a database management system. We call such a system a privacy-enhanced database system. This system is discussed in the next section. A more detailed design of the system and algorithms are given in a paper by Thuraisingham (2003c). Note that some work related to privacy constraint processing is being carried out at IBM under the Hippocratic database project (Agrawal, Kiernan, Ramakrishnan, & Xu, 2002). As we have stated, privacy constraint processing can be considered to be a special case of privacy-preserving data mining. Coming up, we discuss privacy constraints, then privacy-enhanced database systems. The relationship between privacy constraint processing and privacy-preserving data mining will be discussed, and a survey of privacy-preserving data mining will be given.

Specification of Privacy Constraints

In this section we discuss some specific privacy constraints. We need a policy language to specify the constraints. For now, we will specify them using a rule-based format. Here are some examples:

Simple constraints:

Privacy-Level(R, Aj) = L

That is, privacy level of attribute Rj in relation R is L

For example, privacy level of medical records in employee relation is highly private.

Content-based constraints:

COND(Values(S1.B1, S2.B2, - - - - Sm.Bm) → Privacy-Level (R.Aj) = L

That is, if attributes B1, B2, - - - Bm in relations S1, S2 - - - Sm respectively, satisfy some Boolean value, then privacy level of attribute Aj in relation R is L.

For example, if name attribute value of employee relation is Bhavani, then make salary attribute of employee relation of Bhavani semi-private.

Association-based constraints:

Privacy-Level(Together(R1.A1, R2.A2, - - - -Rn.An)) = L

That is, privacy level of attributes A1 in relation R1, attribute A2 in relation R2 - - - attribute An of relation Rn taken together is L.

For example, Employee names and medical records taken together are highly-private.

Event-based constraints:

Event(E) → Privacy-Level(R.Aj) = L

That is, once an event E occurs, then privacy level of attribute Aj of relation R is L.

For example, once the fiscal year ends, financial records of finance relation are made public.

General-released constraints:

Release(R.Ai, L1) →
Privacy-Level(S.Aj) = L2

That is, if any value of attribute Ai of relation R is released to someone who can read information at privacy level L1, then all values of attribute Aj of relation S are assigned privacy level L2.

For example, once any financial record of finance relation is released at privacy-level semi-private, then all salaries also in finance relation can be released at level public.

Individual-released constraints:

Individual-Release (R.Ai, L1) →
Privacy-level(S.Aj) = L2

That is, if an individual financial record of finance relation is released at privacy level semi-private, then the salary value of the corresponding entry in the finance relation is released at the public level.

Aggregate constraints:

Set(S, R) AND Satisfy (S, P) → Privacy-level(S) = L

That is, if R is a relation and S is a set containing tuples of R and S satisfies P, then privacy level of S is L.

For example, if the number of elements in the medical records is more than 10, then the set of medical records is made private.

Level-based constraints:

Level (R. Ai) = L1 → Level(S.Aj) = L2

That is, if the values of attribute Ai of relation R are assigned level L1, then values of attribute Aj of relation S is assigned privacy level L2.

For example, if medical records of relation employee are private, the salaries of relation employee are semi-private.

Logical constraints:

R.Ai → S.Aj

That is, from the value of attribute Ai of relation R we can deduce value of attribute Aj of relation S. For example, from prescription drug information, one can infer information about a person's illness.

In addition to these constraints, one can also have meta constraints that assign values to metadata. An example is that if the value of an attribute is made public, then the attribute itself has to be public. That is, we cannot have a rule that classifies the existence of medical records in a relation employee at the private level and another rule that classifies the medical records at the public level. The constraints are application-specific and are part of the application-specific privacy policy. These constraints are generated by the constraint generator from the specification of the application. Then the consistency checker will check for the consistency of the constraints. Finally, the output will be a set of complete and consistent constraints. In the next section we will see how a database system may process the privacy constraints.

Privacy Enhanced Data Management Systems for Privacy Constraint Processing

Our approach is to augment a database management system (DBMS) with a privacy controller. Such a DBMS is called a privacy enhanced DBMS. The privacy controller will process the privacy constraints. The question is, what are the components of the privacy controller and when do the constraints get processed? We take an approach similar to the approach proposed by Thuraisingham, Ford, Collins and O'Keeffe (1993) for security constraint processing. In our approach, some privacy constraints are processed during database design and the database is partitioned according to the privacy levels. Then, some constraints are processed during database updates. Here, the data is entered at the appropriate privacy levels. Because the privacy values change dynamically, it is very difficult to change the privacy levels of the data in the database in real time. Therefore, some constraints are processed during the query operation. Note that processing constraints in real time will be time consuming. We need some research in this area.

The modules of the privacy enhanced DBMS include the constraint manager, the database designer, the query processor and the update processor. The constraint manager manages the constraints. The database designer processes constraints during database design and assigns levels to the schema. The query processor processes constraints during the query operation and determines what data is to be released. The update processor processes constraints and computes the privacy level of the data. That is, in our approach, some constraints are processed during database design, some during update and some during query operations. Note that in the work we are describing here, instead of sensitivity levels we have privacy levels, and instead of clearance levels we have user roles and credentials; and finally, instead of security constraints we have privacy constraints. We could combine privacy controller with inference controllers to address privacy as well as security. In the following paragraphs we will discuss in more detail the query, update and database design operations.

As stated, some constraints are processed during query operation, some during update and some during database design operation. First consider the query processor. A user poses a query and the query is modified according to the privacy constraints and user's role. For example, if only a physician can read a patient's X-rays, and a secretary queries for the X-rays and billing information, then the query should be modified so that the X-rays are not retrieved. The query modifier will query the constraint manager, retrieve relevant constraints and apply those constraints and modify the query. Sometimes the response may depend on what is previously released. For example, suppose we have a constraint where once the medical records are released then the names cannot be released. In this case, if a user has queried for the medical records and then queries for the names, the release database is examined through the release database manager and the query is modified accordingly. The modified query is sent to the DBMS and the response is obtained. The response manager will also examine the individual release constraints to see whether information about a particular individual has been previously released and determine whether everything in the response should be released. The modified response is given to the user.

One could argue that if the data is entered at the appropriate privacy level, then the query processor is not needed. However, in the real world, privacy policies may change and it may not be feasible to continuously update the database. Therefore we need to process privacy constraints during the query operation. Furthermore, the release constraints are also examined during the query operation.

Next, consider the update processor. In the case of database updates, certain constraints (such as content constraints) are processed by the privacy level computer of the update processor. The privacy level computer will compute the correct privacy level of the data. The data is then entered at the appropriate privacy level. For example, if we have a constraint that states that the medical records for the officers are private, then the privacy level associated with the medical records for the officers are private and entered at the appropriate level.

One of the advantages of computing privacy levels during the update operation is that the query processor may not have to manage all of the constraints, and therefore, the performance of the query processor will improve. However, in the real world the privacy policies may change, and therefore, the query processor has the ability to manage all of the constraints. Thuraisingham, Ford and Collins (1991) have discussed the design of the update processor for processing security constraints in some detail. We are proposing a similar approach for privacy constraint processing.

Next, let us consider the database designer. While certain constraints, such as release constraints, have to be processed during the query operation, and content-based constraints are processed during database update operation, certain other constraints, such as those that assign privacy levels to attributes, are handled during the database design operation. These constraints could be simple constraints or even association constraints. For example, simple constraints may assign privacy levels to, say, medical records. In this case, medical records are assigned privacy value private. Association constraints may assign level private to names and medical records taken together. In this case, while names could be public and medical records could be public, the association between them is private.

The database designer will examine the constraints and essentially assign privacy values to the schemas, such as attributes and relations. In previous work we have discussed algorithms for processing security constraints during database design (Thuraisingham & Ford, 1995). We are proposing a similar approach for handling privacy constraints. That is, the database is partitioned according to the privacy constraints. The only problem here is when the privacy policies change. Then, one needs to redesign a database schema. This is why we enforce all of the constraints during the query operation. One area of research is to develop a scheme to inform the query processor which constraints to examine, as we do not want the query processor to do unnecessary work.

It should be noted that users could collude and obtain information that is sensitive. For example, the system may release X to John and Y to Jane and John and Jane can combine X and Y and infer Z, which may be highly private. Note that we are not enforcing constraints based on user ID. That is, constraints assign privacy levels to the data and if John and Jane are cleared at the private level, then they cannot deduce information at the highly private level, as we would have a constraint that states that X and Y together will be highly private. Now, adapting our approach to an environment that is role-based or user-ID based is more difficult. In addition to enforcing constraints depending on the privacy level, we also need additional constraints that are enforced based on user roles or user IDs.

Privacy Constraint Processing as a Form of Privacy-Sensitive Data Mining

In the previous section we defined privacy constraints and discussed the design of a data management system that processes privacy constraints. Our research has been influenced a great deal by our prior research on the inference problem. Essentially we view the privacy problem as a variation of the inference problem. As stated earlier, there has been a lot of research on privacy-preserving data mining. That is, researchers are developing approaches to carry out successful data mining and at the same time ensure some level of privacy. Next we will survey the developments on privacy-preserving data mining. In this section we will explore the relationship between privacy constraint processing and privacy-preserving data mining.

Essentially, we can view privacy constraint processing as one type of privacy-preserving data mining. The data mining task here is to pose queries to the database and make associations and correlations between the data. The correlations and associations become an issue if they are sensitive, classified or private. Privacy constraints are rules that assign privacy values to the data. Privacy constraint processing essentially prevents an adversary from mining (i.e. posing queries) and extracting associations that are sensitive or private. We take an all-or-nothing approach. That is, all of the data is released or none of the data is released if the association is private. We can adapt the approach proposed here by using some probabilistic reasoning. That is, we can assign probability values, which specify the extent to which the data is private. We can also specify a threshold value. If the computed probability values associated with the data exceed the threshold, then the data is not released. We need more research in this area.

PRIVACY-PRESERVING DATA MINING

As we have mentioned, the challenge is to provide solutions to enhance national security but at the same time ensure privacy. Privacy constraint processing is one approach. The type of data mining that it handles is when users deduce sensitive information from legitimate responses received for the queries. However, there are many outcomes of data mining, such as classification, association, anomaly detection and sequence analysis. We need to develop a comprehensive approach to carry out data mining, extract useful results and still maintain some level of privacy. There is now research at various laboratories on privacy sensitive data mining (Agrawal at IBM Almaden, Gehrke at Cornell University and Clifton at Purdue University). The idea here is to continue with mining but at the same time ensure privacy as much as possible. For example, the approaches examine various data mining techniques, such as cluster formation, decision tree construction and association rule mining, and modify the algorithms so that some level of privacy is preserved (Agrawal & Srikant, 2000; Clifton, 2000; Gehrke, 2002). In this section we discuss the various efforts on privacy sensitive data mining and discuss some directions.

Clifton et al. have proposed various approaches to privacy-sensitive data mining. In one approach they propose the use of the multiparty security policy approach for carrying out privacy-sensitive data mining. In the multiparty approach, the idea is that each party does not know anything except its own inputs and the results. That is, a party does not know about another party's inputs. The solution is to do encryption on a randomly chosen key. Clifton has used this principle and developed privacy-preserving data mining techniques. Essentially he has developed various computations based on the multi-party principle (Clifton, 2003).

Clifton et al. also have proposed a number of other approaches. Kantarcioglu and Clifton (2003) have used a number of tools, including secure multiparty computation, to develop distributed association rule mining algorithms that preserve privacy of the individual sites. The idea behind these algorithms is that for three or more sites, with each site having a private transaction database, the goal is to discover all associations satisfying some given thresholds. Furthermore, no site should be able to learn the contents of a transaction or any rule at any other site. In another approach by Vaidya and Clifton (2003), they develop a version of the K-Means clustering algorithm over vertically partitioned data. The algorithms they have designed essentially follow the standard K-means algorithms. The approximations to the true means are iteratively refined until the improvement in one iteration is below a threshold. At each iteration, every point is assigned to the appropriate cluster, with the minimum distance for each point. As stated by Vaidya and Clifton, once the mappings are known, the local components are computed locally. Then a termination test is carried out. That is, was the improvement to the mean approximation in that iteration below a threshold?

Clifton et al. also have proposed a number of other distributed data mining algorithms. In one approach by Lin and Clifton (2003), they present a secure method for generating an Exception Maximation mixture model (EM) from distributed data sources. Essentially they show that EM cluster modeling can be done without revealing the data points and without revealing which portion of the model came from which site. Clifton et al. have also investigated other approaches for privacy-preserving data mining (Clifton, Kantarcioglu, & Vaidya, 2002).

IBM Almaden was one of the first to develop algorithms for privacy-preserving data mining. In their seminal paper, Agrawal and Srikant (2000) establish this research as an area and discuss various aspects of privacy-preserving data mining and the need for this work. They introduce a quantitative measure to evaluate the amount of privacy offered by a method and evaluate proposed methods against the measure. The approach is to essentially let the users provide a modified value to the sensitive value. They also show a method to reconstruct the original distribution from the randomized data. In another paper by IBM researchers, the authors propose a variation of the results presented by Agrawal and Srikant (Agrawal, & Agrawal, 2001). In particular, they develop optimal algorithms and models for the perturbation approach proposed earlier. Since then, several papers, some of which are variations of the original work, have been published by IBM Almaden Research Center on privacy-preserving data mining.

In addition to Clifton at Purdue and Agrawal at IBM, there are also efforts by Gehrke at Cornell, Kargupta at Baltimore County and Lindell et al. in Israel. In the paper by Evfimievski et al. (2003), the authors present a new formulation of privacy breaches together with an approach called "amplification" for limiting the breaches. They argue that amplification makes it possible to have limits on breaches without knowledge of the original data distribution. They use mining association rules as an example to illustrate their approach (2002). Lindell and Pinkas (2003) show how two parties having private databases can carry out data mining on the union without having to ever know the contents of the other party's database. They show how a decision tree based on their approach can be efficiently computed. Subsequently, they demonstrate their approach on ID3, a well-known decision tree data mining algorithm. Kargupta (2003) considers the problem of computing the correlation matrix from distributed data sets where the owner of the data does not trust the third party who developed the data mining program. They develop a novel approach to compute the correlation matrix for privacy-sensitive data by using a random project-based approach.

Since around the year 2000, numerous papers have been published on privacy-sensitive data mining. We have listed some of the key approaches. Each approach essentially takes a data mining algorithm and shows how it can be modified to take privacy into consideration. Some approaches are based on multiparty computation, some based on perturbation, some based on matrix correlation. We feel that the field, while still not mature, is producing results that can be evaluated using some sort of test bed. We are now ready to develop a test bed with the appropriate data and give feedback to researchers about their approaches. We need a government initiative to develop such a test bed. We need to identify the parameters for the test bed. For example, the computation time could be one parameter. Another parameter could be the number of useful rules produced by the mining algorithms. The usefulness of the rules will have to be determined by the application specialists and the privacy specialists. We need more work to determine the parameters and the type of test bed to be developed. In the meantime, we also need to continue with the research on privacy-preserving data mining. We have come a long way and we still have a lot to do.

DIRECTIONS FOR PRIVACY RESEARCH

We need to attack the privacy problem in many directions. First of all, we need to develop more privacy-sensitive data mining techniques and subsequently explore the foundations. Without a doubt, data mining is a very valuable tool for many applications, including intrusion detection and national security. However, we need to mine the data and extract useful information without violating privacy. As stated previously, we are seeing much progress on privacy-preserving data mining. We still need a lot of research in this area. However, we are also now ready to develop a test bed to analyze the algorithms. We also need to develop taxonomy for the approaches proposed for privacy-preserving data mining. For example, we can group many of the approaches developed so far as belonging to the multi-party-based computation approaches or perturbation-based approaches. There is a lot to do here, and we need government research programs to continue with the research and start evaluation efforts.

We made a lot of progress on security constraint processing in the 1990s. We need to examine the techniques for privacy constraints. We have described system architecture for privacy constraint processing (Thuraisingham, 2003c). IBM's approach is also a very good start. However, we need to specify constraints that are more general and complex, and develop algorithms for processing the constraints. We also need techniques to check the consistency of the privacy constraints. While much of the discussion in this chapter relates to relational databases, the techniques can be adapted for object and XML databases.

Another direction for privacy research is to model the application and reason about the application and detect privacy violations at the application design level. For example, we need to examine the use of conceptual structures and other semantic data models to model the application and capture the privacy constraints. A lot of work was carried out in the 1990s on the use of conceptual structures for handling the inference problem (Thuraisingham, 1991a). We need to examine this work for the privacy problem.

A new direction for privacy is research in trust management and negotiation as well as rights management. The database security community has been carrying out work in this area for the last few years (see, for example, IFIP Database Security Conference Proceedings). We need to see how these techniques can be used for privacy management.

One challenging question for privacy is "is the privacy problem unsolvable?" That is, given a set of privacy constraints and a database, is it possible to determine whether privacy cannot be violated? Back in 1990, we showed that the inference problem was unsolvable (Thuraisingham, 1990b). We need to examine our approach for the privacy problem also. Subsequently we need to analyze the computational complexity of the privacy problem.

One other direction that would be useful is to investigate whether we can use the deductive database approach for privacy management. For example, we have developed logics for multilevel secure databases systems (Thuraisingham, 1991b). Can we develop similar logics for privacy-enhanced database systems? With the deductive database approach, the system may be able to make inferences in a more natural way.

We believe that technology alone will not be sufficient to tackle the privacy problem. We need technologists, policy makers and legal experts to work together. We also need the participation of social scientists to examine the societal impact of data mining. It is not easy for experts in different disciplines to work together and solve problems. However, recently, computer science research is emphasizing interdisciplinary research. As the groups get to know each other and understand the needs and capabilities of each other, we are confident that progress will be made on privacy.

ACKNOWLEDGMENT

I thank the National Science Foundation and the MITRE Corporation for their support to continue my work on data mining, counter-terrorism, information security and privacy.

DISCLAIMER

The views and conclusions expressed in this paper are those of the author and do not reflect the policies or procedures of the National Science Foundation, the MITRE Corporation or the U.S. Government.

REFERENCES

Agrawal, D., & Aggrawal, C. (2001). On the design and quantification of privacy-preserving data mining algorithms. *Proceedings ACM PODS.*

Agrawal, R., & Srikant, R. (2000). Privacy-preserving data mining. *Proceedings of the ACM SIGMOD Conference.*

Agrawal, R., Kiernan, J., Ramakrishnan, S., & Xu, Y. (2002). Hippocratic databases. *Proceedings of the VLDB Conference.*

Air Force Science Board (1983). Multilevel secure database systems. *Air Force Summer Study Report*, Washington D.C.

Clifton, C. (2000). Using sample size to limit exposure to data mining. *Journal of Computer Security, 8*, 120-142.

Clifton, C. (2003). Tools for privacy-preserving distributed data mining. *SIGKDD Explorations, 4*, 28-34.

Clifton, C., Kantarcioglu M., & Vaidya, J. (2002). Defining privacy for data mining. *Proceedings of the Next Generation Data Mining Workshop.*

Clifton, C., & Marks, D. (1996). Security and privacy implications of data mining. *Proceedings of the ACM SIGMOD Conference Workshop on Research Issues in Data Mining and Knowledge Discovery.*

Evfimievski, A., Gehrke, J., & Srikant, R. (2003). Privacy breaches in privacy-preserving data mining. *Proceedings of the ACM PODS Conference.*

Evfimievski, A., Srikant, R., Agrawal, R., & Gehrke, J. (2002). Privacy-preserving mining of association rules. *Proceedings of the ACM SIGKDD Conference.*

Gehrke, J. (2002). Research problems in data stream processing and privacy-preserving data mining. *Proceedings of the Next Generation Data Mining Workshop.*

Hinke, T. (1988). Inference and aggregation detection in database management systems. *Proceedings of the Security and Privacy Conference.*

Kantarcioglu, M., & Clifton, C. (2003). Privacy-preserving distributed mining association rules on horizontally partitioned data. *IEEE Transactions on Knowledge and Data Engineering.*

Kargupta, H. (2003). Privacy sensitive Distributed data mining from multi-party data. *Proceedings of the Security Informatics Symposium.*

Lin, A., & Clifton, C. (2003). Privacy-preserving clustering with distributed EM mixture modeling. *Purdue University Technical Report.*

Lindell Y., & Pinkas, B. (2000). Privacy-preserving data mining. *Proceedings of the Crypto Conference.*

Morgenstern, M. (1987). Security and inference in multilevel database and knowledge base systems. *Proceedings of the ACM SIGMOD Conference.*

Thuraisingham, B. (1987). Multilevel security for relational database systems augmented by an inference engine. *Computers and Security, 6*, 250-66.

Thuraisingham, B. (1990a). Towards the design of a secure data/knowledge base management system. *Data and Knowledge Engineering, 5*, 59-72.

Thuraisingham, B. (1990b). Recursion theoretic properties of the inference problem. *MITRE Report MTP291.*

Thuraisingham, B. (1991a). On the use of conceptual structures to handle the inference problem. *Proceedings of the 1991 IFIP Database Security Conference.*

Thuraisingham, B. (1991b). Nonmonotonic typed multilevel logic for multilevel secure data and knowledge base management system. *Proceedings of the IEEE Computer Security Foundations Workshop.*

Thuraisingham, B. (1998). *Data mining: Technologies, techniques, tools and trends.* CRC Press.

Thuraisingham, B. (2002). Data mining, national security, privacy and civil liberties. *SIGKDD Explorations*, *4*, 1-5.

Thuraisingham, B. (2003a). *Web data mining: Technologies and their applications to business intelligence and counter-terrorism*. Florida: CRC Press.

Thuraisingham, B. (2003b). Data mining for counter-terrorism.. To appear in S. Sivakumar & H. Kargupta (Eds.), *Next generation data mining*. AAAI Press.

Thuraisingham, B. (2003c). Privacy constraint processing in a privacy enhanced database system. *Data and Knowledge Engineering Journal* (forthcoming).

Thuraisingham, B., & Ford, W. (1995). Security constraint processing in a multilevel distributed database management system. *IEEE Transactions on Knowledge and Data Engineering, 7*, 274-93.

Thuraisingham, B., Ford, W., & Collins, M. (1991). Design and implementation of a database inference controller for the update operation, *Proceedings of the IEEE Computer Security Applications Conference*.

Thuraisingham, B., Ford, W., Collins, M., & O'Keeffe, J. (1993). Design and implementation of a database infernce cntroller. *Data and Knowledge Engineering Journal*, *11*, 271-93.

Vaidya J., & Clifton, C. (2003). Privacy-preserving K-means clustering over vertically partitioned data. *Proceedings of the ACM SIGKDD Conference*.

This work was previously published in Journal of Database Management, Vol. 16, No. 1, edited by K. Siau, pp. 75-87, copyright 2005 by IGI Publishing, formerly known as Idea Group Publishing (an imprint of IGI Global).

Chapter 2.14
A Service Discovery Model for Mobile Agent-Based Distributed Data Mining

Xining Li
University of Guelph, Canada

Lei Song
University of Guelph, Canada

ABSTRACT

Mining information from distributed data sources over the Internet is a growing research area. The introduction of mobile agent paradigm opens a new door for distributed data mining and knowledge discovery applications. One of the main challenges of mobile agent technology is how to locate hosts that provide services requested by mobile agents. Traditional service location protocols can be applied to mobile agent systems to explore the service discovery issue. However, because of their architecture deficiencies, they do not adequately solve all the problems that arise in a dynamic domain such as database location discovery. From this point of view, we need some enhanced service discovery techniques for the mobile agent community. This chapter proposes a new model for solving the database service location problem in the domain of mobile agents by implementing a service discovery module based on search engine techniques. As a typical interface provided by a mobile agent server, the service discovery module improves the decision ability of mobile agents with respect to information retrieval. This research is part of the IMAGO system—an infrastructure for mobile agent-based data mining applications. This chapter focuses on the design of an independent search engine, IMAGOSearch, and discusses how to integrate service discovery into the IMAGO system, thus providing a global scope service location tool for intelligent mobile agents.

INTRODUCTION

Mobile agent systems bring forward the creative idea of moving user defined computations–agents towards network resources, and provide a whole new architecture for designing distributed systems. An agent is an autonomous process acting on behalf of a user. A mobile agent roams the Internet to access data and services, and carries out its assigned task remotely. Distributed data mining (DDM) is one of the important application areas of deploying intelligent mobile agent paradigm (Park & Kargupta, 2002; Klusch et al., 2003). Most existing DDM projects focus on approaches to apply various machine leaning algorithms to compute descriptive models of the physically distributed data sources. Although these approaches provide numerous algorithms, ranging from statistical model to symbolic/logic models, they typically consider homogeneous data sites and require the support of distributed databases. As the number and size of databases and data warehouses grow at phenomenal rates, one of the main challenges in DDM is the design and implementation of system infrastructure that scales up to large, dynamic and remote data sources. On the other hand, the number of services that will become available in distributed networks (in particular, on the Internet) is expected to grow enormously. Besides classical services such as those offered by printers, scanners, fax machines, and so on, more and more services will be available nowadays. Examples are information access via the Internet, E-commerce, music on demand, Web services and services that use computational infrastructure that has been deployed within the network. Moreover, the concept of service in mobile agent systems, which will be described in this chapter, has recently come into prominence.

Mobile agents must interact with their hosts in order to use their services or to negotiate services with other agents (Song & Li, 2004). Discovering services for mobile agents comes from two considerations. First, the agents possess local knowledge of the network and have a limited functionality, since only agents of limited size and complexity can efficiently migrate in a network and have little overhead. Hence specific services are required which aim at deploying mobile agents efficiently in the system and the network. Secondly, mobile agents are subject to strong security restrictions, which are enforced by the security manager. Thus, mobile agents should find services that help to complete security-critical tasks, other than execute code that might jeopardize remote servers. Following this trend, it becomes increasingly important to give agents the ability of finding and making use of services that are available in a network (Bettstetter & Renner, 2000).

Some of the mobile agent systems developed in the last few years are Aglets (Lange & Ishima, 1998), Voyager (Recursion Software Inc, 2005), Grasshopper (Baumer et al., 1999), Concordia (Mitsubishi Electric, 1998), and D'Agents (Gray et al., 2000). Research in the area of mobile agents looked at languages that are suitable for mobile agent programming, and languages for agent communication. Much effort was put into security issues, control issues, and design issues. Some state of the art mobile agent systems focus on different aspects of the above issues, for example, Aglets on security, D'Agents on multi-language support, Grosshopper on the implementation of the FIPA (FIPA, 2002) and MASIF (Milojicic et al., 1998) standard. However, few research groups have paid attention to offering an environment to combine the concept of service discovery and mobile agent paradigm. Most existing mobile agent systems require their programmer to specify agent migration itinerary explicitly. This makes mobile agents the weak ability to sense their execution environment and react autonomously to dynamic distributed systems. The objective of our research is to equip mobile agents with system tools such that those agents can search for data sites, move from hosts to hosts, gather information and access databases, carry out complex data mining algorithms, and generate global data model or pattern through the aggregation of the local results.

In this chapter, we propose a new service discovery model DSSEM (discovery service via search engine model) for mobile agents. DSSEM is based on a search engine, a global Web search tool with centralized index and fuzzy retrieval. This model especially aims at solving the database service location problem and is integrated with our IMAGO (intelligent mobile agent gliding on-line) system (Li, 2006). The IMAGO system is an infrastructure for mobile agent applications. It includes code for the IMAGO server—a Multi-threading Logic Virtual Machine, the IMAGO-Prolog—a Prolog-like programming language extended with a rich API for implementing mobile agent applications, and the IMAGO IDE—a Java-GUI-based program from which users can perform editing, compiling, and invoking an agent application. In our system, mobile agents are used to support applications, and service agents are used to wrap database services. Service providers manually register their services in a service discovery server. A mobile agent locates a specific service by submitting requests to the service discovery server with the description of required services. Web pages are used to advertise services. The design goal of DSSEM is to provide a flexible and efficient service discovery protocol in a mobile agent environment.

The rest of the chapter is organized as follows. The following section presents a brief background related to this chapter and discusses the problem of service discovery in mobile agent systems. The third section introduces DSSEM and compares it with several service discovery protocols (SDPs) currently under development. The comparison criteria include functionality, dependency on operating systems and platforms. The fourth section gives an overview of the design of service discovery module and integration with the IMAGO system. Finally, the last section provides some discussion and concluding remarks as well as future work.

BACKGROUND AND MOTIVATION

The general idea of distributed services is that an application may be separated from the resources needed to fulfill a task. These resources are modeled as services, which are independent of the application. Services do not denote software services alone, but any entity that can be used by a person, a program or even another service (Hashman & Knudsen, 2001). Service discovery is a new research area that focuses not just on offering plug and play solutions but aims to simplify the use of mobile devices in a network allowing them to discover services and also be discovered (Ravi, 2001).

In general, the service usage model is role-based. An entity providing a service that can be utilized by other requesting entities acts as a provider. Conversely, the entity requesting the provision of a service is called a requester. To provide its service, a provider in turn can act as a requester making use of other services. To form a distributed system, requesters and providers live on physically separate hosting devices. Providers should, from time to time, advertise services by broadcasting to requesters or registering themselves on third party servers. From requests' point of view, it must be able to:

- Search and browse for services
- Choose the right service
- Utilize the service (Bettstetter & Renner, 2000)

Before a service can be discovered, it should make itself public. This process is called service advertisement. The work can be done when services are initialized, or every time they change their states via broadcasting to anyone who is listening. A service advertisement should consist of the service identifier, plus a simple string saying what the service is, or a set of strings for specifications and attributes. An example is given in Table 1.

Table 1. A typical advertisement of service

Identifier: office-printer-4
Type : printer/postscript/HP20
Speed : 24ppm
Color : yes

There are several ways that a client looks up services it requires. If the client knows the direct address of services, it can make direct requests, or it can listen to broadcasting advertisements and select those it needs. The common method, however, is that the client forms a description of the desired service and asks a known discovery server if there is any service matching the request.

A variety of service discovery protocols (SDPs) are currently under development by some companies and research groups. The most well-known schemes are Sun's Java based Jini™ (Sun, 1999), Salutation (Salutation Consortium, 1998), Microsoft's UPP (Universal Plug and Play, 2000), IETF's draft service location protocol (SLP) (Guttman et al., 1999) and OASIS UDDI (OASIS, 2005). Some of these SDPs are extended and applied by several mobile agent systems to solve the service discovery problem. For example, GTA/Agent (Rubinstein & Carlos, 1998) addresses the service location issue by extending SLP, a simple, lightweight protocol for automatic maintenance and advertisement of intranet services. Though SLP provides a flexible and scalable framework for enabling users to access service information about existence, location, and configuration, it only possesses a local function for service discovery and is not scalable up to global Internet domain because it uses DHCP and multicast instead of a central lookup mechanism. AETHER (Chen, 2000) makes an improvement to Jini by building a dynamic distributed intelligent agent framework. Jini provides a flexible and robust infrastructure for distributed components to find each other on the Internet. However, it relies on the use of standard Java-based interfaces implemented by both the clients and servers in their work. This requires existing systems to be modified for use with Jini, however, a significant amount of the production software currently available around the world is unlikely to be modified. After a study of different SDPs and mobile agent systems that are adopting these methods, we found that several problems cannot be easily solved by the existing protocols due to their limitations.

First of all, most existing works support an attribute-based discovery as well as a simple name lookup to locate a service. Usually, there is only a set of primitive attribute types in the service description, such as string and integer, to characterize a service. Thus, the service discovery process primarily involves activities such as type matching, string comparison, or integer comparison. Here, we define a service description as a sequence of flags or codes that can be multicast to service requesters or be registered on a third-party server for advertisement purposes. Generally speaking, a service description is composed of two parts: property and access. The property of a service description describes the type, characteristics, constraints, and so forth, of a service, which will be published in the service discovery server for advertising purpose. The access of a service is more complicated. It may contain multiple access method tags as there could be multiple ways to invoke a service, for example, using the interface of services, downloading the client-proxy code, locating a database, RPC, RMI or URL location.

For example, Table 2 shows a service description in SLP, where the value of type tag, that is, "service:printer," indicates the property of the service. It also contains some other property tags to describe this resource in detail, such as paper per min. or color-support. In the searching phase, much of the power of SLP derives from its ability to allow exact selection of an appropriate service from among many other advertised services with the same tags. In other words, only the service or services that match the required keywords and attribute values specified by requesters will be found. These keywords and attribute values can be combined into boolean expressions via "AND" and "OR," or common comparison operators "<=," ">," or substring matching. Considering the above example again, the search request from a requester could be "< service:printer, guest, ((name = hp6110) (page per min.>8)) >."

A further step in SDPs development is using eXtensible Markup Language (XML) to describe services. In fact, Web service discovery protocol UDDI, its description language WSDL, as well as the communication protocol SOAP, are all based on XML. In addition, an XML description can be converted to a Java document object model (DOM) so that it can be merged into a service registry system. The example in Table 2 can be described in XML as follows:

```
<description ID="0198">
<type> service: printer </type>
<scope> administration, guest </scope>
<name> hp6110 </name> ......
<usage> //hp6110: 1020/queue1 </usage>
</description>
```

No matter what kind of description format is applied, the lack of rich representation for services has not been changed. The problem arising directly in our project is that these protocols are not adequate to advertise some special services, such as database services. In a DDM environment, data may be stored among physically distributed sites and may be artificially partitioned among different sites for better scalability. Therefore endowing mobile agents with the ability of accessing remote databases is the basis for DDM applications. Obviously, a database system already has a well-defined interface. A data-mining agent requires a way of finding the locations of specific databases and deciding where to move. In this situation, the only way we can accomplish is by registering the database's name and attributes for future discovery. However, for a database service, people care more about the content of the database than its name or structure. Considering an example of a bookstore, before placing an order to the bookstore, customers would like to know if the books they require are available in the store

Table 2. An example of SLP service description

type = service: printer
scope = administrator, guest
name = hp6110
paper per min. = 10
Color-support = yes
usage = //hp6110: 1020/queue1

by checking the summary of all books with some keywords or a fuzzy search criterion. From this point of view, a simple string identifier or XML identifier cannot meet the requirement.

The second problem is ranking. After requesters have searched out all services that may be required, they still need to select the right one for utilization. Just imagine over the entire Internet, tens of thousands of providers could publish their services by their own will. We should be able to know which ones provide the most realistic and highest quality services that users want. Obviously, moving to the hosts one by one to find out the required information is not a wise choice. Therefore, generating a service rank is essential. However, none of the existing SDPs offer such a mechanism for ranking discovered services. They are satisfied with finding a service only, without considering whether the service would be able to serve the requester.

The most significant contribution of our research is that we enrich the service description by using Web pages' URL (later the search engine will index the content referenced by this URL) to replace the traditional string-set service description in mobile agent systems. Because of their specific characteristics, such as containing rich media information (text, sound, image, *etc.*), working with the standard HTTP protocol and being able to reference each other, Web pages may play a key role acting as the template of the service description. On the other hand, since the search engine is a mature technology and offers an automated indexing tool that can provide a highly efficient ranking mechanism for the collected information, it is also useful for acting as the directory server in our model. Of course, DSSEM also benefits from previous service discovery research in selected areas but is endowed with a new concept by combining some special features of mobile agents as well as integrating service discovery tool with agent servers.

DISCOVERY SERVICES VIA SEARCH ENGINE MODEL (DSSEM)

As the most important media type on the Internet today, hypertext Web pages are posted to advertise the information by industries and individuals. Though millions of them are published on the Internet, these Web pages still increase rapidly everyday for a variety of reasons. They are spidered and indexed by commercial search engines such as Google, Yahoo, AltaVista, and so forth. Users can easily find Web pages' locations by submitting the search request to those search engines.

In principle, if we install a lightweight search engine on the service discovery server that could retrieve Web pages posted by service providers and design a Web search interface for the incoming mobile agents, the problems described previously

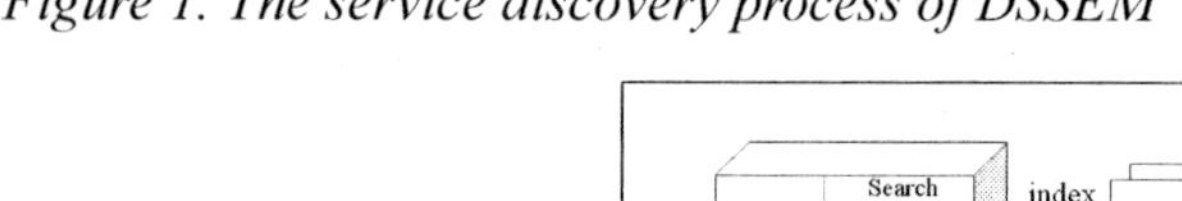

Figure 1. The service discovery process of DSSEM

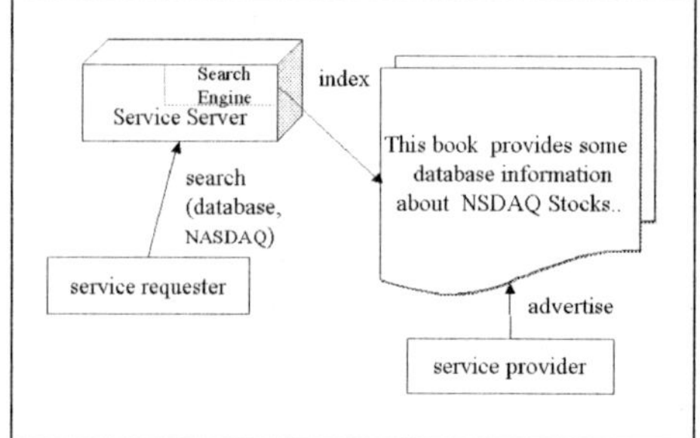

Figure 2. Web representation of database

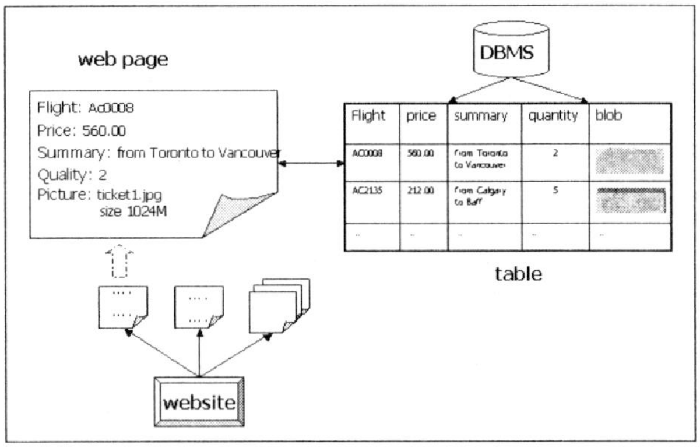

could be solved in an easy way. In this situation, service providers do not need to register the service description on the service discovery server. Instead, they register the URLs of their Web sites that advertise all the information concerning services. As a middleware on the service discovery server, the search engine will periodically retrieve the document indicated in URLs and all of their referencing documents, parse all the tags and words in the documents, and set up the relationship between the keywords and the host address of these service providers.

On the other hand, mobile agents can utilize the system interface by accessing the search engine's database and obtain a destination itinerary that includes a list of ranked host addresses of the service providers. Based on the above discussion, Figure 1 shows the service discovery process of DSSEM.

The current version of DSSEM concentrates on the database service discovery. The database service advertisement information can be easily converted to Web page representation. The specific characteristic of a Web page is that it contains rich media information and flexible layout, and can reference each other. As an example in Figure 2, we can find that a two-dimensional database table can be converted into a one-dimensional Web page. Moreover, some binary data stored in the database such as image can be extracted from higher-dimensional space to a lower-dimensional space as the text representation in the Web page.

To use Web pages as a medium to advertise services for service providers, we should modify the template in the service description of SLP. The remarkable change is that some properties once represented by strings or XML language are now represented as the Web site's home URL. Table

Table 3. An example of service description

type = service: database
name = bookstore
URL = //www.cis.uoguelph.ca/
location(URL)= www.uoguelph.ca
interface = dp_get_set(Handler, 'SQL statement', Result_handler)

3 illustrates a service description template of a bookstore example.

The proposed model is similar to SLP and Jini with respect to the service discovery process; however, it extends those protocols by setting up a centralized, seamless, scalable framework on the Internet. Unlike some multicasting services protocols, the centralized service discovery server makes DSSEM service discovery available on the Internet worldwide. The process of registration is similar to UDDI and the process of discovery is similar to the lookup service in Jini. Besides that, features of mobile agents bring DSSEM other incomparable advantages. First, code mobility is almost impossible in most distributed systems. Therefore a client must download the resource drivers to invoke services. Although RPC or RMI mechanism can help to call services remotely, it might consume tremendous network bandwidth when dealing with services involving a huge amount of data, such as database services. DSSEM goes one step further. It makes agents migrate to the destination hosts and utilize services locally. Secondly, the security issue is seldom considered in current service discovery protocols. However, a mobile agent server requires a strict security concern for authorization and authentication when it accepts the incoming agents and provides them services for utilization.

SERVICE DISCOVERY IN THE IMAGO SYSTEM

IMAGO is a mobile agent system in which agents are programs written in IMAGO Prolog that can move from one host on the Internet to another. Briefly speaking, an agent is characterized as an entity that is mature (autonomous and self-contained), mobile, and bears the mental model of the programmer (intelligent) (Li, 2001; Li & Autran, 2005). From an application point of view, the IMAGO system consists of two kinds of agent servers: stationary server and remote server. The stationary server of an application is the home server where the application is invoked. On the other hand, agents of an application are able to migrate to remote servers. Like a Web server, a remote server must have either a well-known name or a name searchable through the service discovery mechanism. Remote servers should provide services for network computing, resource sharing, or interfaces to other Internet servers, such as Web servers, database servers, and so forth.

In fact, an IMAGO server, no matter stationary or remote, is a multithreading logical virtual machine to host agents and provides a protected agent execution environment. The IMAGO system is portable in the sense that its servers run on virtually all Linux boxes with Gnu C compiler and Posix package. Tasks of an IMAGO server include accepting agents, creating a secure run time environment and supervising agent execution. It must also organize agent migration from or to other hosts, manage communications among agents, authenticate and control access for agent operations, recover agents and the information carried by them in case of network and computer failures and provide some basic services for the agents, such as database service and discovery service.

The architecture of the IMAGO server is shown in Figure 3. In this architecture, the system modules are configured to deal with different tasks. The core module of the IMAGO server is the scheduler. It maintains an agent list where each entry on the list matches different stages of the life cycle of an agent, such as creation, execution, memory management (expansion, contraction, or garbage collection), termination and migration. The agent list is sorted with respect to system-defined priorities. For example, the highest priority is given to agent migration, followed by agent creation and execution, memory manipulation, and finally database service and service discovery.

Figure 3. The Infrastructure of IMAGO System

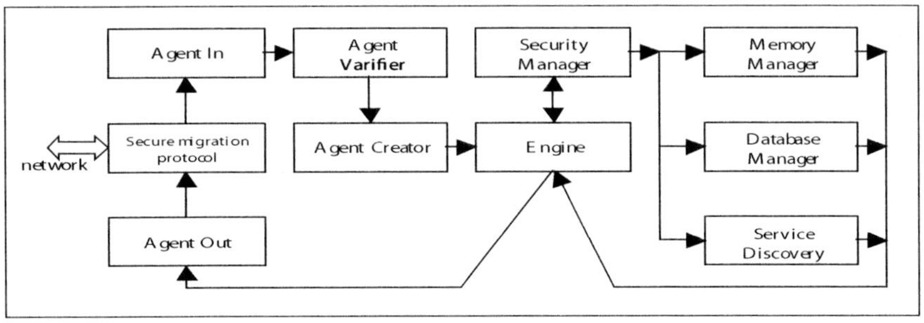

Box 1.

```
dp_connect(URL, DatabaseName, UserName, Password, AccessMode),  //connection
dp_get_set(Handler, 'select .....', ResultHandler),  //data access
dp_disconnect(Handler).                //disconnection
```

In the early phase of system design, database operation becomes the major service to applications in the IMAGO system. Thus, the problem of service discovery focuses on how to find such services effectively. Once a database server has been found, agents may migrate to that remote server and invoke database access locally through built-in primitives.

As an example, the following code shows a typical database access issued by an IMAGO agent (see Box 1).

Before a database service is advertised, the service provider should fill out a registration form and submit the form to an IMAGO service discovery server. The contents of the form include service type, database name, URL of the service provider host, access mode, HTTP URL of the service Web site, interface function, and the verification information. We choose URL as the host address since it is compatible with most commonly used Web browsers and independent of address families (such as IP, IPv6 and IPX).

To illustrate how DSSEM works, Figure 4 shows the steps involved in the service registration and discovery process in our IMAGO system. A service discovery server is called the service location host. In order to gather useful information, the search engine, called IMAGOSearch, should be independently installed on the service location host. This search engine maintains a special database system designed to index Internet addresses (URLs, Usenet, Ftp, image locations, *etc.*). Like traditional search engines, IMAGOSearch consists of three main components: *spider*, *indexer* and *searcher*. They are grouped into two modules, where one module includes spider and indexer, running in the background of a service location host, and the other module is the searcher, running in the foreground to provide discovery services. First, the spider gets the URLs from a URL list that contains initial Web site URLs registered by service providers. The spider then traverses along these URLs in the breadth-first manner and loads the referred hypertext documents into the service

Figure 4. The process of Web search module

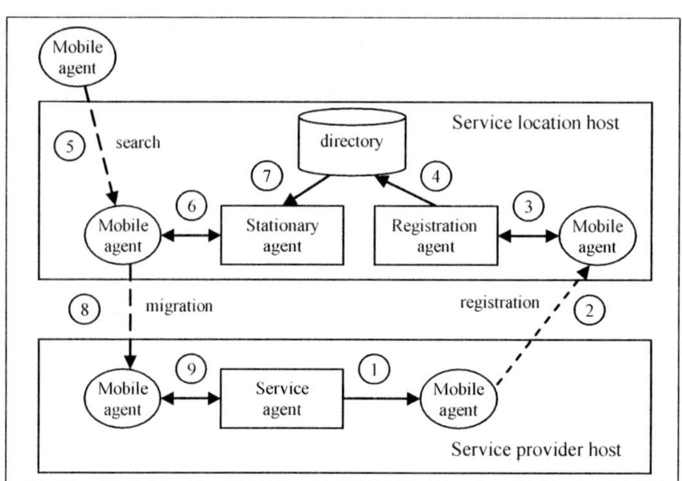

Figure 5. An example of service discovery and data mining process

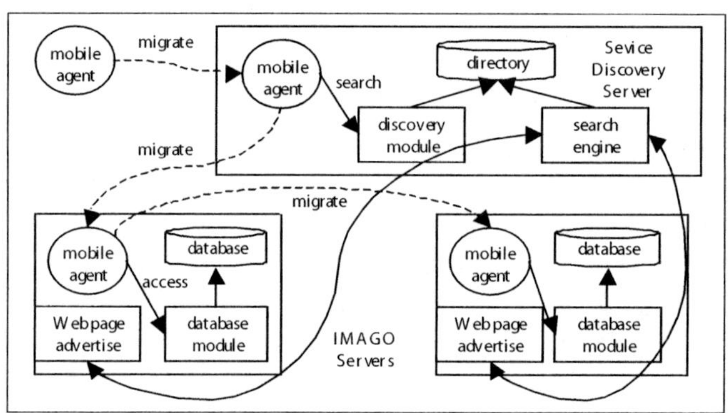

discovery server. The indexer extracts the salient words of these documents. Some related information such as text is also saved into the database for user retrieval. In addition, the indexer looks for URL anchors that reference other documents and adds them to the URL List. Besides considering the weight of each word in the documents (e.g., a word occurrence in the title should be assigned a higher weight than that occurs in the body), IMAGOSearch also pays attention to positions of each word and their relative distance during ranking. The ranking algorithm we use is called the shortest-substring ranking (Charles et al., 2000), which offers a simple way to weight each Web page based on a search criteria and total them up to form Web site ranking. The searcher behaves as a bridge between the IMAGO server and the search engine. It is responsible for accepting the search requests from mobile agents, querying the database, ranking search results and finally returning a destination itinerary.

Box 2.

```
web_search(locate("food," "customer transaction," "imago data server"), 10, R)
```

The application programming interface for mobile agents is a built-in predicate, namely, *web_search(query, number, Result)*, where *query* is a compound term, such as *locate("tsx," "stock transaction," "imago server")*, *number* is an integer indicating the maximum number of results expected, and *Result* is a variable to hold the returned values. When an agent issues a web_search(...) predicate, the agent is blocked and control is transferred to the service discovery module of the hosting IMAGO server. This module will communicate with the searcher, wait for search results, and resume the execution of the blocked agent. For example, suppose a food company wants to analyze the customer transaction records for quickly developing successful business strategies; its DDM agent may move to a known IMAGO service discovery server and then issue a query predicate requesting up to ten possible food industry database locations (see Box 2).

Search result *R* will be delivered to the agent in the form of a list, where list entries are ranked in terms of priorities from high to low. Based on the given itinerary, the mobile agent may travel from host to host to carry out a DDM application. Figure 5 gives an example of service discovery and data mining process.

DISCUSSION AND CONCLUSION

In this chapter, we have discussed the design of a service discovery protocol, namely, DSSEM and its implementation in the IMAGO system. Table 6 (Bettstetter & Renner, 2000) summarizes the main features of selected protocols compared with DSSEM. From implementation point of view, the most critical issue about the performance of search engine is the quality of search results.

However, we cannot make a comparison with other major commercial search engines since they are operating at different levels. Thus, user evaluation is beyond the scope of this chapter. In order to show that our search engine does return useful information, Table 7 gives the experimental results for a query using the keywords "imago lab." The results show that all server URLs have come from reasonably high quality Web sites and, at last check, none were broken links. An R_w value is calculated according to word occurrence, weight and a factor value measuring the distance of keywords by a ranking algorithm (Charles at al., 2000). We define the result that has the highest R_w value as the highest priority and assign it a hundred percent rate, therefore the percentage of other results are rated relatively. Of course a true test of the quality of a search engine would involve extensive experiments, analysis and user evaluation, which is part of our future work.

Aside from the search quality, IMAGOSearch is designed to scale up cost effectively as the size of Web pages grow. Because IMAGOSearch only indexes Web servers registered by IMAGO Server users, we do not need to worry about indexed pages exceeding the maximum size of the database. One endeavour that we are undertaking is to reduce the table redundancy and use the storage efficiently. Our experiment shows that indexing 22,000 different documents consumes only 140Mb disk space. The search time is mostly dominated by the performance of CPU, disk I/O and the underlying database system. When a mobile agent wants to locate certain services, it must first move to the service discovery server, then make a local query and migrate to the destination hosts after obtaining the itinerary. This brings us to the problem that as a central unit, the service discovery server might become a bottleneck, especially when it is handling thousands of Web pages everyday and simultaneously hosting as many incoming mobile agents as possible. A possible solution is to duplicate service discovery servers. Replicas not only make the service discovery mechanism very efficient, but also increase the ability of fault tolerance.

The results of our work are encouraging and further studies in this field are being planned. First, the current implementation of search engine deals with only the AND logical relationship between search strings, it could be enhanced to parse some complex search criteria that combine keywords with Boolean expressions (AND, OR) and conditional expressions (<=, >=, substring match, *etc.*). Secondly, since a database contains multi-dimensional information, how to reflect dimensional relationship by a flat Web page is a big challenge. A possible way to address this issue is to use XML meta-data to describe the database dimension.

ACKNOWLEDGMENT

We would like to express our appreciation to the Natural Science and Engineering Council of Canada for supporting this research.

REFERENCES

Baumer, C., Breugst, M., & Choy, S. (1999). Grasshopper: A universal agent platform based on OMG MASIF and FIPA standards. In *Proceedings of the First International Workshop on Mobile Agents for Telecommunication Applications (MATA'99)* (pp. 1-18).

Bettstetter, C., & Renner, C. (2000). A comparison of service discovery protocols and implementation of the service location protocol. In *Proceedings of the EUNICE 2000, Sixth EUNICE Open European Summer School*, Netherlands.

Charles L., Clarke, A., & Gordon V. (2000). Shortest substring retrieval and ranking. *ACM Transactions on Information Systems, 18*(1), 44-78.

Chen, H. (2000). *Developing a dynamic distributed intelligent agent framework based on Jini architecture*. Master's thesis, University of Maryland, Baltimore County.

FIPA.(2002). *Agent Management Specification.* Retrieved from http://www.fipa.org

Gray, R., Cybenko, G., & Kotz, D. (2002). D'Agents: Applications and performance of a mobile-agent system. *Software—Practice and Experience, 32*(6), 543-573.

Guttman, E., Perkins, C., & Veizades, J. (1999). Service Location Protocol, Version 2, White Paper.

Hashman, S., & Knudsen, S. (2001). *The application of Jini Technology to enhance the delivery of mobile services.* White Paper. Retrieved from http://wwws.sun.com/

John, R. (1999). *UPnP, Jini and Salutaion: A look at some popular coordination framework for future network devices.* Technical Report, California Software Labs.

Klusch, M., Lodi, S., & Moro, G. (2003). The role of agents in distributed data mining: Issues and benefits. In *Proceedings of the IEEE/WIC International Conference on Intelligent Agent Technology (IAT'03)*, Halifax, Canada (pp. 211-217). IEEE Computer Society Press.

Lange, D., & Ishima, M. (1998). *Programming and deploying Java, mobile agents with aglets.* Boston: Addison-Wesley.

Li, X. (2001). IMAGO: A Prolog-based system for intelligent mobile agents. In *Proceedings of Mobile Agents for Telecommunication Applications (MATA'01)* (LNCS 2164, pp. 21-30). Berlin: Springer Verlag.

Li, X. (2006). On the implementation of IMAGO system. *International Journal of Computer Science and Network Security, 6*(2a), 107-118.

Li, X., & Autran, G. (2005). Inter-agent Communication in IMAGO Prolog. (LNAI 3346, pp. 163-180).

Milojicic, D., Breugst, M., & Busse, I. (1998). MASIF: The OMG mobile agent system interoperability facility. In *Proceedings of the Second International Workshop on Mobile Agents* (pp. 50-67).

Mitsubishi Electric ITA. (1998). *Mobile agent computing. A white paper. UDDI White Paper.* Retrieved from http://www.uddi.org

Park, B., & Kargupta, H. (2002). *Distributed data mining: Algorithms, systems, and applications, data mining handbook.* In N. Ye (Ed.), *Data mining handbook* (pp. 341-358). IEA.

Ravi, N. (2001). *Service discovery in mobile environments.* Technical Report. Department of Computer Science and Engineering, University of Texas, Arlington.

Recursion Software Inc. (2005). Voyager Product Documentation. Retrieved from http://www.recursionsw.com/voyager_Documentation.html

Rubinstein, M., & Carlos, O. (1998). Service location for mobile agent system. In *Proceedings of the IEEE/SBT International Telecommunications Symposium (ITS'98)* (pp. 623-626).

Salutation Consortium. (1998) *Salutation Architecture Overview.* White Paper. Retrieved from http:// www.salutation.org/whitepaper/originalwp.pdf

Sun. Technical. (1999). *Jini Architectural overview.* White Paper. Retrieved from http://www.sun.com/jini/

Universal Plug and Play Forum. (2000). *Universal Plug and Play device architecture, Version 0.91.* White Paper.

This work was previously published in Intelligent Information Technologies and Applications, edited by V. Sugumaran, pp. 173-189, copyright 2008 by IGI Publishing, formerly known as Idea Group Publishing (an imprint of IGI Global).

Chapter 2.15
Node Partitioned Data Warehouses:
Experimental Evidence and Improvements

Pedro Furtado
University of Coimbra, Portugal

ABSTRACT

Data Warehouses (DWs) with large quantities of data present major performance and scalability challenges, and parallelism can be used for major performance improvement in such context. However, instead of costly specialized parallel hardware and interconnections, we focus on low-cost standard computing nodes, possibly in a non-dedicated local network. In this environment, special care must be taken with partitioning and processing. We use experimental evidence to analyze the shortcomings of a basic horizontal partitioning strategy designed for that environment, then propose and test improvements to allow efficient placement for the low-cost Node Partitioned Data Warehouse. We show experimentally that extra overheads related to processing large replicated relations and repartitioning requirements between nodes can significantly degrade speedup performance for many query patterns. We analyze a simple, easy-to-apply partitioning and placement decision that achieves good performance improvement results. Our experiments and discussion provide important insight into partitioning and processing issues for data warehouses in shared-nothing environments.

INTRODUCTION

Data Warehouses (DWs) are repositories that typically store large amounts of data that have been extracted and integrated from transactional systems and various other operational sources. Those repositories are useful for online analytical processing (OLAP) and data mining analysis. Typical queries include both standard reporting and ad hoc analysis. They usually are complex and access very large volumes of data, performing time-consuming aggregations. Although data warehouses easily can reach many Giga or Terabytes, users still require fast answers to their analyses. Therefore, performance becomes a major concern in those systems. Although structures such as materialized views and specialized indexes improve response times for predicted queries, parallel processing can be used alone or in conjunction with those structures to offer a major performance boost and to guarantee speedup and scale-up, even for unpredicted ad hoc queries.

Parallel database systems are implemented using one of the following parallel architectures: shared-memory, shared-disk, shared nothing, hierarchical, NUMA (Valduriez & Ozsu, 1999). Each choice has implications for parallel query processing algorithms and data placement. In practice, parallel environments involve several extra overheads related to data and control exchanges between processing units and also concerning storage, so that all components of the system need to be designed to avoid bottlenecks that would compromise the whole processing efficiency. Some parts of the system even may have to account for the aggregate flow into/from all units. For instance, in shared-disk systems, the storage system, including controllers and connections to storage, have to be sufficiently fast in order to handle the aggregate of all accesses without becoming a significant bottleneck for I/O-bound applications. To handle potential bottlenecks, specialized, fast, and fully dedicated parallel hardware and interconnects are required. An attractive alternative is to use a number of low-cost computer nodes in a shared-nothing environment, possibly in a non-dedicated local network, and design the system with special partitioning and processing care. In such an environment, each node has a basic database engine, and the system includes a middle layer providing parallelism to the whole environment. The Node Partitioned Data Warehouse (NPDW) is a generic architecture for partitioning and processing query-intensive data in such an environment. One of the objectives of the Node Partitioned Data Warehouse is to minimize the dependency on very fast, dedicated computing and data exchange infrastructures by optimizing partitioning and making use of replication whenever useful.

DeWitt and Gray (1992) review the major issues in parallel database systems implemented over conventional shared-nothing architectures. One of the major concerns when using such an architecture is to decide how to partition or to cluster relations into nodes, which raises the issue of how to determine the most appropriate partitioning and placement choice for a schema. Data warehouses are a specialized type of database with specific characteristics and requirements that may be useful in the partitioning and placement decision. They are mostly read-only, periodically loaded centralized repositories of data. Replication-related consistency issues are minor when compared to full-blown transactional systems.

The star schema (Kimball, 1996) is part of the typical data organization in a data warehouse, representing a multidimensional logic with a large central fact table and smaller dimension tables. Facts typically are very large relations with hundreds of gigabytes of historical details. Dimen-

Figure 1. Partitioning the star schema

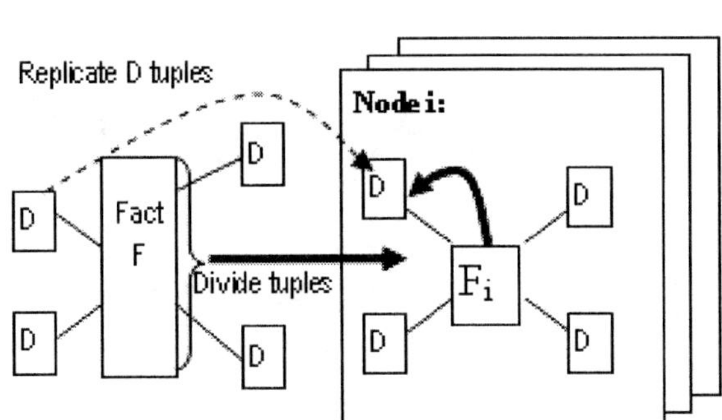

sions are smaller relations identifying entities by means of several descriptive properties.

In that context, a basic placement strategy for the simple star schema replicates dimensions and fully partitions the large central fact horizontally randomly. Figure 1 illustrates the simple placement strategy. The large fact F is partitioned into node fragments F_i, and dimensions D are replicated into all nodes. Very small dimensions even can be cached in memory for faster access and join processing.

An advantage of this simple placement strategy is that it minimizes the amount of data that must be exchanged between nodes, as we shall see further on. Although this strategy can achieve a large speedup when handling simple star schemas with small dimensions, it cannot handle the most severe performance problems in data warehouses, which appear with more complex schemas and access patterns. We show experimentally that the basic partitioning and placement strategy described previously results in severely degraded speedup when applied to those more complex schemas and queries. The schema and query set of the decision support performance benchmark TPC-H (TPC, 1999) is used in our analysis as an example of a complex case. TPC-H is a multidimensional schema and a plausible historical record of a company's business activity and, therefore, a plausible data warehouse. However, its dimensions are not very small, and it contains and queries several large relations, including more than one fact and dimensions. Our objective in this article, therefore, is to analyze this issue experimentally and to propose a generic solution to process efficiently any query. The NPDW partitioning and placement approach takes into consideration the star schema structure but also considers actual data sizes and query workload for a more adaptable partitioning and placement strategy in order to handle more complex data warehouse schemas and queries.

The article is organized as follows. After discussing related work, we describe the Node Partitioned architecture and its basic partitioning and processing strategies. Then we engage in experimental analysis of performance when applying the basic placement strategy and identify the major bottlenecks. We then review the partitioning issue as applied to the NPDW and discuss the modifications we introduced to the basic placement strategy in order to handle complex schemas and queries efficiently. The improved strategy is then evaluated and compared with the basic one. We conclude that the proposed workload-based partitioning and placement together with efficient parallel join processing strategies can prevent very low speedup for most queries.

RELATED WORK

A large amount of work exists on applying parallel and distributed processing techniques to relational database systems. The objective is to apply either inter-query or intra-query parallelism in order to improve performance. DeWitt and Gray (1992) review processing, speedup, and scalability issues in parallel shared-nothing database systems. Parallel and distributed data warehouses share many of the typical performance issues raised for other parallel and distributed databases. They process mostly complex queries over very large data sets and mostly are read-only with only periodic loads, in contrast to transactional processing databases. Two important related and intertwined issues are data allocation (partitioning and placement) and query processing.

Early works on query processing on parallel and distributed databases focus on basic data distribution and processing (placing fragments on sites and processing them) and performance evaluation of the basic schemes (DeWitt & Gray, 1992; Epstein, Stonebraker & Wong, 1978; Yu, Guh, & Zhang, 1987). Numerous researchers focus on efficient processing of queries in parallel and distributed databases in the presence of joins (Chen & Liu, 2000; Khan, McLeod, & Shahabi,

2001; Lee, Park, & Kang, 1999; Liu & Chen, 1996; Sasha & Wang, 1991). Efficient OLAP processing over distributed data is discussed by Akinde and Bohlen (2002). Most query processing algorithms and relevant issues in parallel databases are well reviewed and discussed by Yu and Meng (1998).

Data allocation in parallel and distributed databases has been studied extensively in the context of generic database processing. Early works on data allocation propose alternative algorithms and evaluate them in the presence of transactional and mixed workloads (Apers, 1988; Copeland, Alexander, Boughter, & Keller, 1988; Hua & Lee, 1990; Sacca & Wiederhold, 1985; Yu, Guh, Brill, & Chen, 1989). Livny, Khoshafian, and Boral (1987) compare full partitioning with clustering the relations on a single disk, concluding that partitioning is consistently better for multi-user workloads but can lead to serious performance overhead on complex queries involving joins with high communication overhead. Hua and Lee (1990) propose a solution to data placement using variable partitioning. In their work, the degree of partitioning (n of nodes over which to fragment a relation) is a function of the size and access frequency of the relation. Again, experimental results in Hua and Lee (1990) show that partitioning increases throughput for short transactions, but complex transactions involving several large joins result in reduced throughput with increased partitioning. Williams and Zhou (1998) review five major data placement strategies representing three categories: size-based, access frequency-based, and network traffic based. The study compares the alternative strategies using the TPC-C performance benchmark (TPC, 1999). The authors conclude that the way data are placed in a shared-nothing system can have considerable effect on the performance.

The most promising solutions to extra join overheads that characterize many successful parallel and distributed database systems in shared-nothing environments involve hash partitioning large relations to minimize data exchange requirements (DeWitt & Gerber, 1985; Kitsuregawa, Tanaka, & Motooka, 1983). Parallel hash-join algorithms also are reviewed by Yu and Meng (1998). They consider dynamically partitioning and allocating intervening relation fragments into processors for fast join processing. These strategies typically allocate a hash range to each processor, which builds a hash table, and hashed relation fragments are redirected to the corresponding processor. Other alternative strategies include placement dependency (Liu, Chen, & Krueger, 1996), which uses dependency relationships between relations to collocate fragments for faster processing, and the partition and replicate strategy (Yu, Guh, Brill, & Chen, 1989), which partitions a single relation and replicates all others to process the joins. Algorithms and optimizations for parallel processing of multi-way joins are considered by Sasha, Wang, and Tsong-Li (1991). The authors have proposed algorithms to determine the most efficient join order for multi-way joins over fully partitioned relations in a shared-nothing cluster.

Some of the most promising partitioning and placement approaches focus on query workload-based partitioning choice (Rao, Zhang, & Megiddo, 2002; Zilio, Jhingram, & Padmanabhan, 1994). None of these works considers the specific characteristics of data warehouses, as they are targeted at generic databases. The idea in those works is to use the query workload to determine the most appropriate partitioning attributes, which should be related to typical query access patterns. Rao et al. (2002) optimize the choice of partitioning attributes by predicting the cost of alternative query processing paths that result from different partitioning attribute choices and alternative query processing paths. Heuristic algorithms had to be applied to limit the very large search space of the strategy. The approach has the limitation that it requires tight integration of the placement cost predictor with the query optimizer to evaluate costs of alternative allocations and, therefore, cannot be applied directly in any environment. In contrast, our work in this article and in previous efforts (Furtado, 2004b, 2004c) focuses on evaluating and analyzing generic data partitioning strategies independently of the underlying database server and targeted at node partitioned data warehouses, also improving on basic non-workload-based partitioning (Bernardino, Furtado, & Madeira, 2003). It uses simple workload and schema-based determination of partitioning and replication. Our approach gathers experimental evidence of the factors that should be taken into consideration for a simple and easy-to-apply partitioning and placement strategy to yield almost linear speedup in NPDW.

BASIC PARTITIONING IN NPDW

We already have mentioned the basic partitioning and placement strategy considered for the star schema (Figure 1). Dimensions are replicated, and the fact is partitioned. Heavy replication is feasible in these schemas because the data are not constantly changed, and the only refresh is based on periodically loading new data. The fact can be partitioned using a random or round-robin partitioning strategy. The reason for this placement is to be able to simultaneously parallelize the processing of the largest relation and at the same time process time-consuming operations (e.g., joins and aggregations) locally at each node, therefore minimizing internode communication. In this way, query processing does not become a large burden to the network, and it is less dependent on network bandwidth and data exchange handling issues. We show next why this strategy minimizes internode communication using a simple query example. Each node processes its part of the query independently, so that the system may achieve a speedup that is expected to be near to linear with the number of nodes. Consider a simple OLAP query formulated as:

```
OP(...)
JOIN (F, D1, ..., Dn)
GROUP (G1,..Gm);
```

where OP is an aggregation operator such as SUM, COUNT.

Each node needs to apply exactly the same initial query on its partial data, and the results are merged by applying the same query again at the merging node with the partial results coming from the processing nodes. Figure 2 illustrates this process.

The independent execution of partial joins by nodes is supported by the fact that all but one of the relations in the star (i.e., all dimensions) are replicated into all nodes, which is similar to the rationale of the Partition and Replicate Strategy

Figure 2. Typical OLAP processing in NPDW

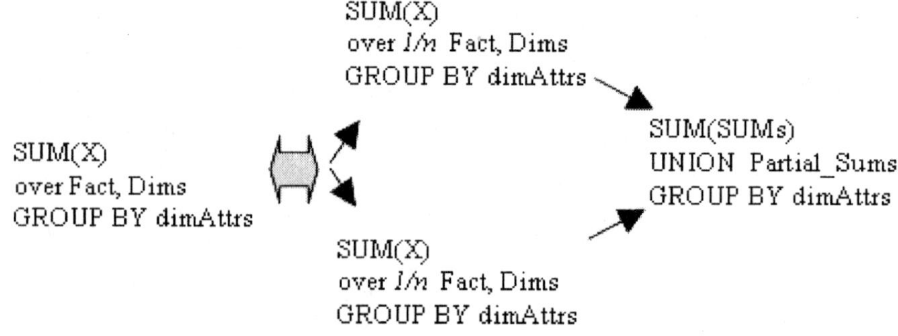

(PRS) (Yu et al., 1989), although in that case, the initial placement of relations was not similar. Considering a single fully-partitioned relation R_i (with partitions R_{ij}) and all the remaining ones (R_l, $l = 1$ *to* n: $l \neq i$) replicated into all nodes, the relevant join property that allows joins to be processed by nodes independently from each other is:

$$R_1 \ ... \ R_i \ ... \ R_n = U_{\text{j over all nodes}} \ R_1 \ ... \ R_{ij} \ ... \ R_n \quad (1)$$

Additionally, even though expression (1) includes a union operator, many other operators denoted here as OP() can be applied before the partial results are collected by some node to process the union of the partial results, due to the property in (2):

$$OP(R_1 \ ... \ R_i \ ... \ R_n) = OP(U_{\text{j over all nodes}} \ R_1 \ ... \ R_{ij} \ ... \ R_n) = OP_y[U_{\text{j over all nodes}} \ OP_x(R_1 \ ... \ R_{ij} \ ... \ R_n)] \quad (2)$$

where OP_x is an operator applied locally at each node and OP_y is a global merge operator. The set OP_x and OP_y replaces OP. Expressions (1) and (2) allow each node to compute part of joins and aggregations over the data it holds, independently from the other nodes. Then the partial results from all nodes are merged in a final step.

BASIC PROCESSING IN NPDW

In this section, we give a brief review of query processing in NPDW and focus only on simple aggregation queries. A detailed study of query processing in NPDW is available elsewhere (Furtado, 2005). Generically, the typical query processing cycle is shown in Figure 3.

Step 1 prepares the node query and merge query components from the original submitted query. Step 2, Send Query, forwards the node query into all nodes in the NPDW, which processes the query locally in Step 3. Each node then sends its partial result into the submitter node, which applies the merge query in Step 5. Step 6 redistributes results into processing nodes, if required (for some queries containing subqueries, in which case more than one processing cycle may be required).

As shown in the example in Figure 3, aggregations over a fact A(F, D_x,...,D_y) can be processed independently in each node, followed by merging (union_all) of the partial result sets and re-applying of the aggregation query over the merged result set. Aggregation primitives are computed at each node. The most common primitives are: LINEAR SUM (LS=SUM(X)); SUM_OF_SQUARES (SS = SUM(X^2)); number of elements (N), extremes MAX and MIN. This step corresponds to applying OP_x in expression (2) of the previous section. Then, the partial results are merged by the union operator in expression (2),

Figure 3. Query processing steps in NPDM

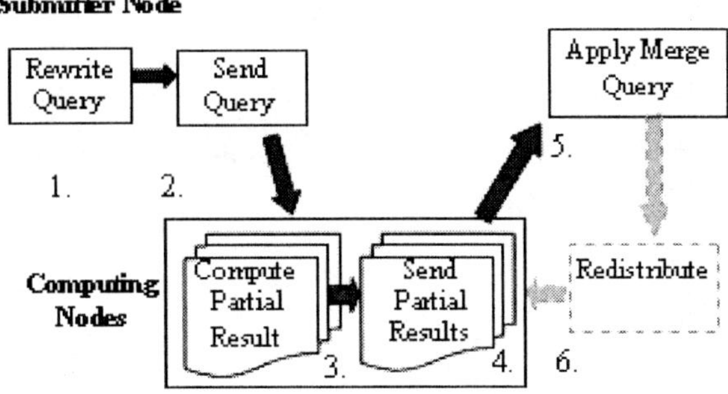

and then a final aggregation function is applied, corresponding to operator OP_y of expression (2). For the most common aggregation operators, the final aggregation function is shown next:

$$COUNT = N = \sum_{all\_nodes} N_{nodei} \quad (3)$$

$$SUM = LS = \sum_{all\_nodes} LS_{nodei} \quad (4)$$

$$AVERAGE = \sum_{all\_nodes} LS_{nodei} / \sum_{all\_nodes} N_{nodei} \quad (5)$$

$$STDDEV = \sqrt{\frac{(\sum SS_{node_i} - \sum LS_{node_i}{}^{2} / N)}{N}} \quad (6)$$

$$MAX = MAX(MAX_{node_i}),$$

$$MIN = MIN(MIN_{node_i}) \quad (7)$$

These expressions show how typical aggregation operators are parallelized by using additive primitives and merge components. For instance, the query transformation step needs to replace each AVERAGE and STDDEV (or variance) expression in the SQL query by a SUM and a COUNT in the first case and by a SUM, a COUNT, and a SUM_OF_SQUARES in the second case in order to determine the local query for each node and by the expressions (3) to (7) in the final merging query.

Figure 4 shows an example of aggregation query processing steps. A query with several aggregation operators is submitted in Step 1. Observe that in Step 2, the query that should be run at nodes is produced by replacing the aggregation operators by the additive primitives used to compute partial node results. The partial results are collected in Step 3 into a memory (the cached table PRqueryX in the example), and in Step 4, the final result is computed by applying a merge query according to expressions (3) to (7).

ANALYZING BASIC PARTITIONING IN NPDW

We are interested in obtaining a generic solution to process efficiently any query in the NPDW, considering more complex and also less denormalized schemas. A schema may comprise several fact tables with many hundreds of millions of rows (corresponding to transaction measurements) and also some large dimensions with millions of rows. It is therefore not unusual for queries to access several large relations, including multiple facts and large dimensions. The star schema itself usually is not isolated but rather included in a larger schema

Figure 4. Basic aggregation query steps

1. Query submission:

```
Select sum(a), count(a), average(a), max(a), min(a),
stddev(a), group_attributes
From fact, dimensions (join)
Group by group_attributes;
```

2. Query rewriting and distribution to each node:

```
Select sum(a), count(a), sum(a x a), max(a), min(a),
group_attributes
From fact, dimensions (join)
Group by group_attributes;
```

3. Results collecting:

```
Create cached table
PRqueryX(node, suma, counta, ssuma, maxa, mina,
group_attributes)
as <insert received results>;
```

4. Results merging:

```
Select sum(suma), sum(counta),
sum(suma)/sum(counta), max( maxa), min(mina)
(sum(ssuma)-sum(suma)²)/sum(counta), group_attributes
From UNION_ALL(PRqueryX), dims (join)
Group by group_attributes;
```

Figure 5. TPC-H schema

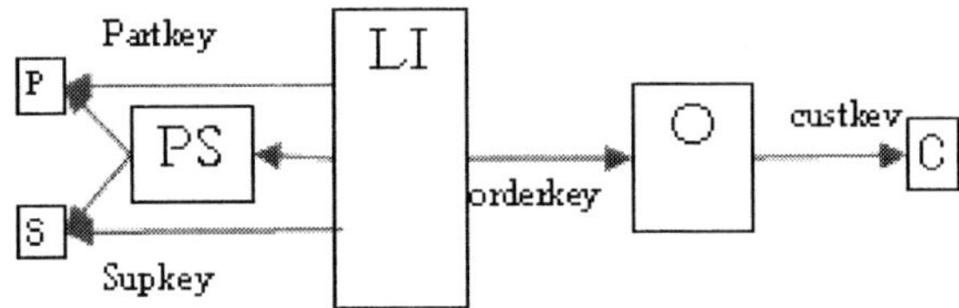

Figure 6. TPC-H speedup on basic placement

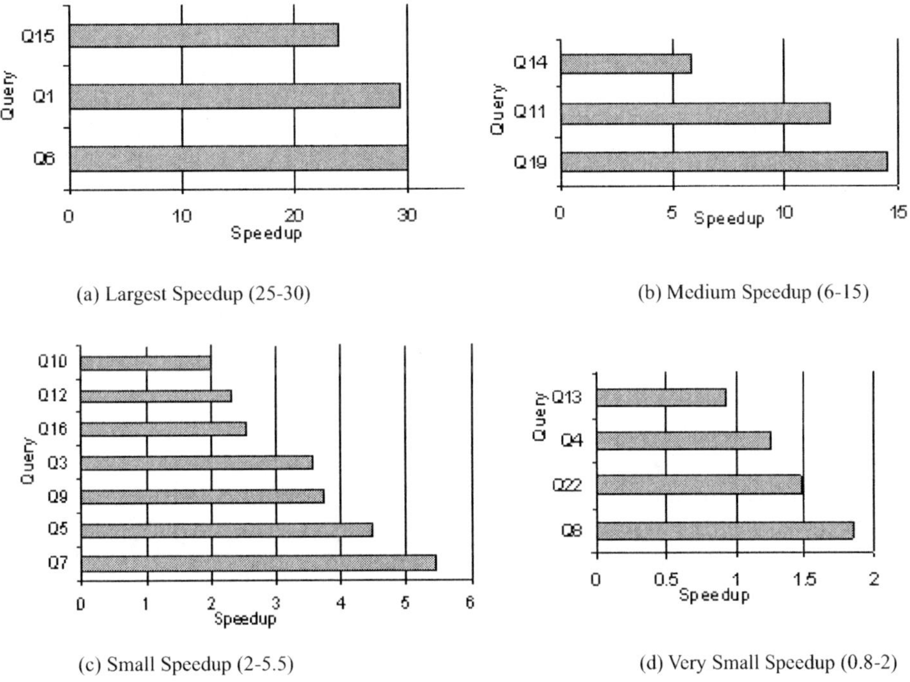

(a) Largest Speedup (25-30)

(b) Medium Speedup (6-15)

(c) Small Speedup (2-5.5)

(d) Very Small Speedup (0.8-2)

comprised of several facts with several conformed dimensions in a bus data warehouse architecture (Kimball, Reeves, Ross, & Thornthwaite, 1998). In this article, we use the TPC-H schema (TPC, 1999) summarized in Figure 5 as a case study. A pure star schema is one with a single large fact and very small dimensions. The major difference of TPC-H, when compared with a pure star schema and query workload, is that TPC-H query set includes frequent joins between two or more large relations. The schema represents ordering and selling activity (LI-lineitem, O-orders, PS-partsupp, P-part, S-supplier, C-customer), where relations such as LI, O, PS, and even P are quite large. There are also two very small relations, NATION and REGION, not depicted in the figure, as they are very small and can be replicated readily into all nodes.

The basic node partitioning strategy determines that dimensions be identified and replicated into all nodes (e.g., P, S, C, and O) and that facts be fully partitioned (PS and LI). As in expression 1 in Figure 5, joins involving only one of the horizontally partitioned relations and any number of replicated relations require no data exchange besides merging partial results, regardless of the partitioning scheme used. Joins between LI and PS incur in repartitioning overheads. In the next section, we will see that this partitioning strategy is inadequate; it does not consider the possibility that partitioned relations may need to be joined, and there are large relations being replicated, which may result in bad performance.

BASIC PARTITIONING RESULTS AND ANALYSIS

In this section, we show results of our experiments on the TPC-H schema (50GB) and query set over 25 nodes. For these experiments, we measured processing and data exchange time on a node with a 32-bit 866 MHz CPU, three IDE hard disks (2x40 GB, 1x80GB), 512 MB of RAM, and a modern DBMS in a 100Mbps switched network. Figure 6 shows the speedup (x-axis) for each of the TPC-H benchmark queries as defined in TPC (1999) (y-axis). The results were clustered according to the amount of speedup achieved for the queries, organized in intervals (25 to 30, 6 to 15, 2 to 5.5, and 0.8 to 2). Observe that the speedup was small for most queries. Given that these experiments concern 25 nodes, a desirable linear speedup would be about 25. The results show that of 19 queries, only a few (3) queries were able to obtain those levels of speedup, and most achieved very low speedup.

The speedup results can be explained if we analyze the relations and queries, as shown in Figure 7 (including fragment sizes when LI and PS are fully partitioned). Table (a) of Figure 7 shows which relations (facts and dimensions in columns) are accessed by each query, organized around the same clusters of Figure 6. Table (b) of Figure 7 shows the size of each relation for the 50GB TPC-H schema used (in gigabytes). Observe that queries with the largest speedup access mostly the LI relation, those with medium speedup access a medium sized-dimension, and most of those with low or very low speedup access the O relation.

Figure 8 shows the node schema resulting from the basic partitioning algorithm. Completely filled boxes represent replicated relations, while partly filled ones represent partitioned relations. In this figure, the relative sizes of the boxes representing relations are indicative of the actual table sizes, and partitioned relations LI and PS

Figure 7. Relations accessed by queries (TPC-H)

	Fact	Dim
Q6	LI	-
Q1	LI	-
Q15	LI	S

	Fact	Dim
Q7	LI	O C S
Q5	LI	O C S
Q9	LI PS	O P S
Q3	LI	O C
Q16	PS	P S
Q12	LI	O
Q10	LI	O C

	Fact	Dim
Q19	LI	P
Q11	PS	S
Q14	LI	P

	Fact	Dim
Q8	LI	O P C S
Q22		OC
Q4	LI	O
Q13		OC
Q21	LI	O S
Q2	PS	P S

(a) Fact and dimensions accessed

	GB
Lineitem (LI)	39/1.56
Partsupp (PS)	3.75/0.15
Orders (O)	9
Part (P)	1
Supplier (S)	0.05
Customer (C)	0.75

(b) Approximate relation sizes (50GB)

Figure 8. Relation layout of nodes in basic partitioning (TPC-H)

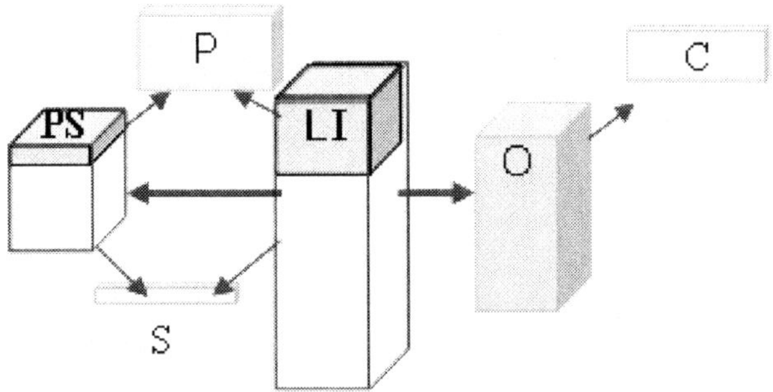

are indicated by showing partitions. Note that replicated relations are now among the largest data sets in each node.

We now review the query patterns that gave rise to each of the categories (large speedup, medium speedup, low speedup, and very low or no speedup) by looking at typical query patterns for each.

In the case of large speedup (25 to 30, Figure 6(a)), nearly linear speedup was obtained for queries exhibiting access patterns such as accessing only LI or LI and the very small relation S. From the relations layout in Figure 8, we can conclude that for these query patterns, each node had to process only about 1/N of the data set (the size of relation LI divided by N nodes), resulting in about N times faster execution. In fact, the speedup can be larger than N, as shown in Figure 7(a), because the performance of algorithms such as hash join degrades as the relation size increases significantly (for lack of memory buffers to optimize the hash join). Query Q15 had slightly less speedup, as it accesses a replicated relation, but a small one (S).

In the case of medium speedup (6 to 15, Figure 6(b)), the reason for the degraded speedup compared to the previous large speedup case is that a typical query in this category accesses, for instance, an LI fragment, together with a replicated relation whose size is of the order of magnitude of the LI fragment (P). Although a large speedup could be achieved by accessing only a fragment of the partitioned relation in each node, that possibility is compromised by the access to a replicated relation with a reasonable size.

In the case of low speedup (2 to 5.5, Figure 6(c)), one of the replicated relations that is accessed (O) is much larger than the LI fragment being accessed (Figure 8). Although we partitioned LI into N fragments in order to have individual nodes processing only 1/N of it, the large size of O caused a small speedup overall.

The relevant pattern when comparing medium to low speedup is that low speedup queries involving the LI relation also involve O, while medium speedup queries involving LI also involve P. The reason for this is that P is much smaller than O and also smaller than LI fragments. The fact that those medium speedup queries have to join LI fragments with full P degrades the speedup, but not as much as joining LI to O. A similar pattern can be seen for queries involving relation PS, whose fragments are very small. Queries joining PS with P have very small speedup, because P is comparatively large, while queries joining PS and S (e.g., query Q11) achieve medium speedup, because S is reasonably small compared to PS.

From these experimental results, we conclude that the basic placement strategy is not adequate to store these complex schemas and query workloads, because, although the facts are partitioned, reasonably large relations such as O and P are replicated into all nodes. We did not take sufficient advantage of the 25 nodes, because we had to deal with full dimensions.

This discussion also can be represented using a mathematical model. Consider relations R_1 and R_2, N nodes and a simplified linear cost model (reviewed in a later section) accounting for processing and joining costs (the only variable considered is the size of the relations). With this model, the cost of processing relation R_i is αR_i (α is a constant factor); the cost of processing a fragment is $\alpha R_i/N$; the cost of joining R_1 to R_2 is $\alpha(R_1 + R_2)$. With N nodes, we would like to process only 1/N of the data in each node, resulting in about N times speedup:

$$\alpha \times \left(\frac{R_1}{N} + \frac{R_2}{N}\right), \alpha\ const = \frac{1}{N}\alpha \times \left(R_1 + R_2\right) \tag{8}$$

However, we have replicated relations, which change the expression to:

$$\alpha \times \left(\frac{R_1}{N} + R_2\right), \alpha\ const \tag{9}$$

The amount of speedup degradation depends on the size of R_2 relative to R_1/N. If replicated relations included in the expression are very large in comparison to fragments, the speedup will be very small.

In this section, we have analyzed experimental results for the basic partitioning approach based on fully partitioning facts and replicating dimensions. We have shown that the approach has important performance limitations, because it does not take into account the actual size of the relations and the query workload. In order to develop an approach that is more adaptable to the actual data sets and query workloads, we must modify the constraint that only the central fact(s) be partitioned and determine the partitioning based on those factors as well. While the rationale for the basic approach was expressed in equations (1) and (2), we review the partitioning and processing issue and relevant costs next, before we propose modifications to the approach that are necessary to solve the performance limitations.

PARTITIONING AND JOIN PROCESSING

In this section, we review the relevant partitioning issue that must be handled in order to improve the speedup of NPDW over the basic strategy. Then, we review relevant cost factors to NPDW. In order to improve the parallelism results significantly, we must partition most relations in NPDW. However, this invalidates expression (1), which allowed joins to be processed independently and in parallel by all nodes and then merged. Instead, repartitioning may become necessary to process many of the joins between partitioned relations. The basic partitioning problem is well described by Liu and Chen (1996). Assuming all queries of the form Q = { target | qualification}, where target is a list of projected attributes and qualification is a list of equi-joined attributes, let a query be Q = {R_1.A, R_2.B | R_1.A = R_2.A $\wedge$ R_2.B = R_3.B}. Then, not all of the three relations can be partitioned, since the first join predicate requires R_2 to be partitioned on A, and the second join predicate requires that R_2 be partitioned on B. We may choose to partition R_1 and R_2 on A and replicate R_3 or to partition R_2 and R_3 on B and replicate R_1. The problem is then to decide which set of relations to partition. Consider that R_1 has a size 12000 and resides in node 1, and that R_2 has size 10000 and resides in node 2. The partition and replicate strategy (PRS) (Yu et al., 1989) involves partitioning R_1 into two fragments, F_{11}, F_{12}, and sending F_{12} to node 2. Relation R_2 is sent to node 1, and then each node processes its part of the join: $R_1 {}_A R_2 = (F_{11} {}_A R_2) \cup (F_{12} {}_A R_2)$ processed in parallel in nodes

1 and 2. Using hash partitioning, it is possible to obtain a more efficient result. We first define hash partitioning.

Definition. A relation R_i with tuples t is *hash partitioned* on attribute A into d disjoint fragments $\{F_{ij}\}$ if (1) $R_i = \cup_i F_{ij}$; (2) $F_{ij} \cap F_{ik} = \emptyset$ for $j \neq k$; and (3)$\forall$ t $\in F_{ij}$, h(t.A) = c_{ij}, where h() is a hash function and c_{ij} is a constant for a given j.

Hash partitioning allows the expression $R_1 \,_A\, R_2$ to be processed faster as $(F_{11} \,_A\, F_{21})$ U $(F_{12} \,_A\, F_{22})$, as fragments are results of hashing the relations such that $F_{11} \,_A\, F_{22} = \emptyset$ and $F_{12} \,_A\, F_{21} = \emptyset$. The join cost at each node is now smaller, as only a fragment of R_2 is joined instead of the whole relation. Communication costs also are reduced, as only a fragment instead of the whole relation R_2 must be moved. On the other hand, hash partitioning R_1 and R_2 introduces some overhead, although R_1 and R_2 can be partitioned simultaneously at each node.

If F_{11} and F_{21} initially are located at node 1 and F_{12}, and F_{22} initially is located at node 2, then the join can proceed without any partitioning and communication costs (except to merge the partial results), as the necessary data are already at the processing nodes. However, if a query joins R_1 to R_2 by another join attribute, or if one of the relations participates in a join with other relation on a different attribute, it is necessary to repartition the fragments and redistribute them before the join can take place. For instance, R_1 can be repartitioned by repartitioning fragments F_{11} and F_{21} in parallel and by exchanging data to build F'_{11} and F'_{21} in nodes 1 and 2, respectively. Repartitioning is not especially costly and, in fact, is typically much faster than partitioning the whole relation at a single node.

A cost model for NPDW should take into account repartitioning as well as other relevant operations incurred when processing queries in that environment.

RELEVANT COSTS IN NPDW

A basic knowledge of the costs that are involved in processing a query over NPDW is useful for the formulation of an effective partitioning strategy. In this section, we briefly discuss the most relevant costs incurred in this context. Given that most relations become partitioned, the main processing costs (listed next) are partitioning, repartitioning, data communication, and local processing.

1. **Partitioning Cost (PC):** Partitioning a relation consists of retrieving the relation from secondary memory, dividing it into fragments by applying a hash function to a join attribute, and assigning buffers for the data to send to other nodes. This involves scanning the relation only once. The partitioning cost is monotonically increasing on the relation size. Since there can be two or more relations to be partitioned, and since they can be processed in parallel in two or more nodes, the partition cost for a given query is the largest partition cost among the nodes participating simultaneously.
2. **Repartitioning Cost (RC):** Repartitioning is similar to partitioning but involves a fragment in each node instead of the whole relation. It is used to reorganize the partitioned relation, hashing on a different equi-join attribute. The fragments resulting from this repartitioning need to be redistributed to other nodes in order to process a hash join.
3. **Data Communication Cost (DC):** The data communication cost is monotonically increasing with the size of the data transferred and equal between any number of nodes. We assume a switched network, as this allows different pairs of nodes to send data simultaneously (with no collisions). This, in turn, allows the repartitioning algorithm to be implemented more efficiently.
4. **Local Processing Cost (LC):** The local processing cost for the join operation typically depends on whether the join is supported by fast access paths such as indexes and the size of

the relations participating in the join. The local processing cost also should account for other operations performed locally. For simplicity, we assume that these costs also increase monotonically on the relation sizes, although, in practice, this depends on several parameters, including memory buffer limitations.

5. **Merging Cost (MC):** The merging cost is related to applying a final query to the collected partial results at the merging node.

We define weighting parameters (Sasha et al., 1991) as a partitioning cost weight, β, and a local processing weight, α, so that β/α denotes the ratio of partitioning costs to local processing costs (e.g., ~2 [Sasha et al., 1991]). Considering large relations with size R_i, N nodes, and the linear cost model described previously, we can obtain a simple expression for the cost of processing joins when repartitioning is required vs. the cost, when the relations are already equi-partitioned. For simplicity, the following expressions consider only two large relations. The fragment size is R_i/N. The join-processing cost for queries requiring the join between equi-partitioned large relations and replicated small relations r_i is:

Join Cost with Equi-partitioned Relations =

$$\alpha \times \left(\frac{R_1}{N} + \frac{R_2}{N} + r_1 + \ldots + r_l \right) \quad (10)$$

The cost when large relations are not equi-partitioned on a switched network includes repartitioning and local processing cost factors with corresponding weights, as shown in (11). The IR symbol in the repartitioning cost factor is an intermediate result from doing independently at all nodes a locally processable part of the joins (those involving replicated and equi-partitioned relations). The IR then must be repartitioned. The value IR/N is the fraction of the IR that is at each node. About 1/N of that fraction (1/N x IR/N) has the correct hash value for the node, therefore requiring no repartitioning.

Join Cost with Repartitioning =

$$\left(\frac{IR}{N} - \frac{IR}{N^2} \right) \times \beta + \alpha \times \left(\frac{R_1}{N} + \frac{R_2}{N} + r_1 + \ldots + r_l \right) \quad (11)$$

The increase in cost of (11) over (10) is therefore

$$\left(\frac{IR}{N} - \frac{IR}{N^2} \right) \times \beta.$$

This overhead depends on the size of IR and is avoided whenever the relations to be joined are equi-partitioned by the appropriate join attribute.

We have already pointed out the reasons for the limitations of a partitioning strategy that does not take into account relation sizes and query workload. The discussion of this section further clarified the costs involved in processing the query and, in particular, joins. In the next section, we propose modifications to NPDW partitioning and join processing that reduce the performance limitations by considering the workload and relation sizes and take full advantage of partitioning options.

PROPOSED MODIFICATIONS TO NPDW

In this section, we propose a partitioning strategy that takes into consideration relation sizes and the query workload to improve query processing performance. From the discussions in the previous sections, it is clear that large relations should be partitioned in NPDW, regardless of whether they are considered facts or dimensions. A small relation in this perspective is one whose contribution to a join processing time cost is expected to be reasonably small either in absolute terms or in comparison with fragment sizes. The placement strategy described next easily can be automated:

1. **Dimensions:** Non-small dimensions simply can be hash partitioned by their primary key. This is because that attribute is expected to be used in every equi-join with facts. The references from fact to dimensions correspond to foreign keys on the fact referencing those equi-join primary keys of the dimensions.
2. **Facts:** The objective is to find the hash-partitioning attribute that minimizes repartitioning costs. A reasonable approximation to this objective is to determine the most frequent equi-join attribute used by the relation. To do this, the placement strategy looks at the frequency of access to other partitioned relations and chooses the most frequent equi-join attribute with those relations as the partitioning attribute.

By collocating relation fragments that are frequent equi-join targets, this simple strategy minimizes repartitioning requirements.

Next, we apply this placement strategy to the TPC-H schema. We arbitrated small as less than 1GB for TPC-H 50GB, so that dimensions C and S are considered small and replicated. This assumption is enough for us to show the advantages of the strategy, but we point out that smaller thresholds allow further speedup in systems with many nodes.

From Figure 7(b), we can see that with our proposed thresholds, S and C are replicated, and P is partitioned by P key. The O relation, if considered a dimension, is immediately key partitioned (orderkey attribute). This relation also could be considered a fact because it links to dimension C, but the resulting partitioning would be the same, if it were considered a fact. Finally, facts should be partitioned according to the most frequent join. From Figure 7(a), we can see that fact LI is joined most frequently to O and, therefore, should be partitioned by O key. Fact PS is partitioned by P key, as shown in Figure 9. This partitioning resulted in two sets of equi-partitioned relations (LI, O), (PS, P) and a set of replicated dimensions (S, C). Joins between relations within each of these sets can be done without repartitioning, and joins with any of the (small) replicated relations also can

Figure 9. Partitioning results with improved algorithm

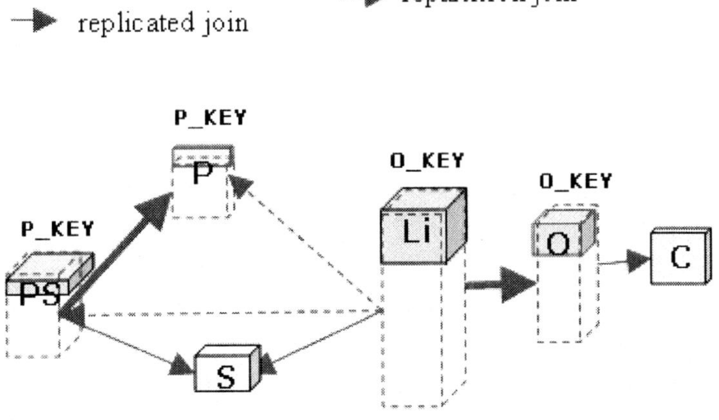

Figure 10. Query graphs

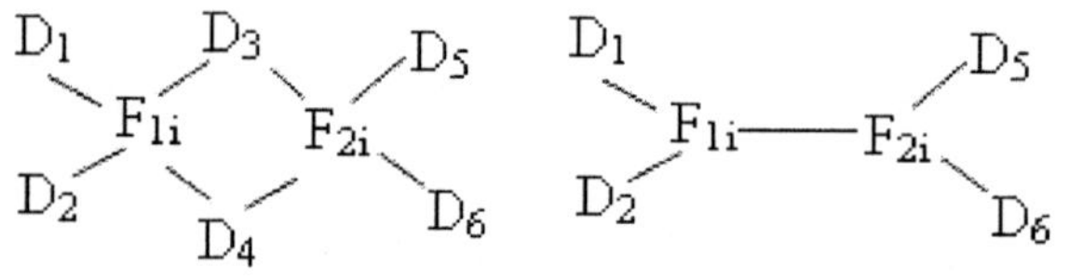

Figure 11. Independent node query processing

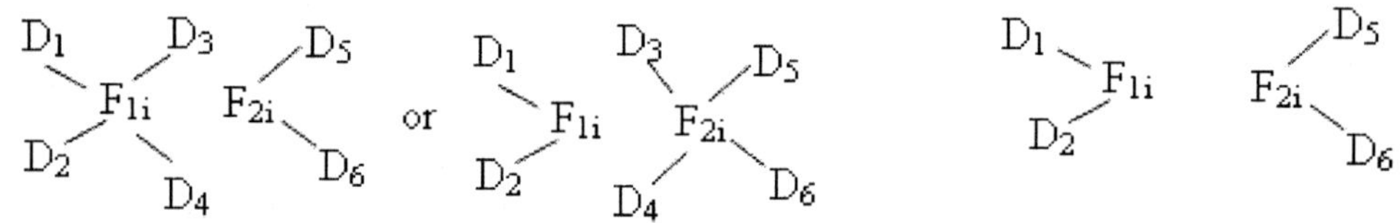

(a) Alternative single node processing of 1st query

(b) Singe node processing of 2nd query

Figure 12. Join processing algorithm

1. Produce Reduced Intermediate Results (IR)

Process in parallel all locally computable joins. In this step apply the selection operators, producing selective intermediate results (IRs);

2. Parallel Hash-join IRs (PHJ-IR)

Determine best join order for the IRs.
For each IR join step
    Repartition intervening IRs by join attribute
    Process the join in parallel at all nodes

be done without any repartitioning. Repartitioning becomes necessary when elements from the previous first and second sets are joined.

Given this mixed hash-partitioning/replicated placement strategy, it is also important to optimize join processing in the presence of non-equi-joins (Furtado, 2004a). Figure 10 shows two examples of query graphs, where vertices correspond to relations and edges are equi-joins. For simplicity, we assume that all equi-joins are on different attributes.

The first step occurs in parallel at all nodes. We take advantage of the replication of (small) dimensions and equi-partitioning to promote partial joins of facts with those relations in individual nodes before repartitioning. The joining with these relations most frequently allows selection conditions to be applied, which results in very good intermediate result selectivity in expression (11). At the end of this phase, we are left with intermediate results for each non equi-partitioned fact.

Figure 11 illustrates the first step applied to the query graphs of Figure 10. Figure 11(a) shows two alternatives, as dimensions D_3 and D_4 can be equi-joined with one or the other fact. The actual join order between each fact and equi-joined dimensions and which of the alternatives in Figure 11(a) is chosen should be determined based on costs.

The intermediate results (IR) now are repartitioned on the equi-join attribute and assigned a join processing node according to the hash value. The rest of the join then is processed in parallel by the nodes. If there are many non-equi-partitioned IRs, the exact order of execution should be determined based on costs. The complete algorithm is given in Figure 12.

After Step 2 of the algorithm, the query processing proceeds at each node. Partial results are then sent to the merging node to compute the final answer. The algorithm in Figure 12 was described sequentially, but the two steps should be pipelined for maximum efficiency. In the next section, we apply this algorithm to the TPC-H experimental setup to compare the performance using the new strategy to that of the basic relation size and workload-insensitive approach.

EXPERIMENTS WITH THE MODIFIED SCHEMA

Figure 13 shows our experimental speedup results considering 25 nodes and the basic TPC-H configuration described before. We also include the results of the basic placement for comparison purposes. The results were clustered using the same groups of Figure 6, according to the amount of speedup achieved for the queries tested in that figure (large, medium, small, and very small speedup). The graphs show for each query (y-axis) the amount of speedup using the basic and improved partitioning (x-axis). Note that most queries exhibit near to linear speedup when the improved strategy is used, even for those queries with small speedup using the basic strategy.

The major improvements in Figure 13(c) have to do with the fact that the O relation is no longer fully replicated, being instead partitioned by the equi-join key to LI. Other queries benefited from PS being equi-partitioned with P. At the same time, queries that were the fastest under the basic placement strategy (Figure 13(a)) are still as fast, because they access partitioned relations in both cases. We also show the importance of applying selection predicates before repartitioning data to process the join in Figure 13(b). Two results are shown for queries Q14 and Q19 joining LI to P. If the whole LI relation is repartitioned, the speedup result will be similar (Q14) or even less (Q19) than the speedup obtained with the basic placement strategy. On the other hand, the dashed lines show that the speedup when the selection predicates are applied to LI before repartitioning is very good. This is the application of the join processing algorithm we presented in the previous section.

While the most important performance issue in the initial replicated placement scheme was the need to process large full relations at each node, repartitioning requirements could be an important overhead in the improved partitioning and placement scheme. However, not only is it possible that many queries do not require repartitioning, but repartitioning also is, in fact, much less expensive than having (possibly very) large relations completely replicated.

Figure 14 shows the runtime cost, measured in elapsed time, for queries that required repartitioning (the cost is measured in runtime). These results compare the total runtime cost (TC) of the improved NPDW scheme vs. the runtime cost of the basic replicated scheme (TC-Replicated), also showing the repartitioning cost for the NPDW scheme. Figure 14(b) shows the cost of repartitioning (in the NPDW scheme) as a percentage of the total costs. The repartitioning cost in Figure 14(a) is clearly a reasonably small fraction of the total runtime cost for most queries, except for Q9. Figure 14(b) shows that the repartitioning cost is about 30% of the total runtime cost for Q9, less than 10% for Q19, and less than 5% for the other two queries, when considering the NPDW scheme,

Figure 13. TPC-H speedup on improved placement

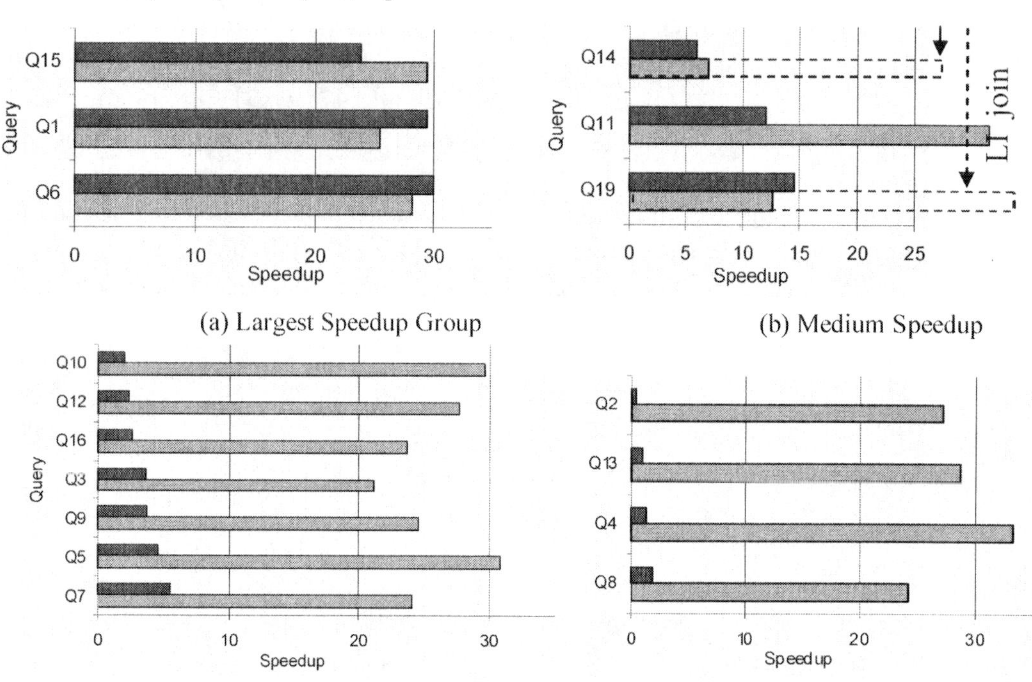

(c) Queries with small speedup using basic partitioning

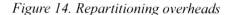

Figure 14. Repartitioning overheads

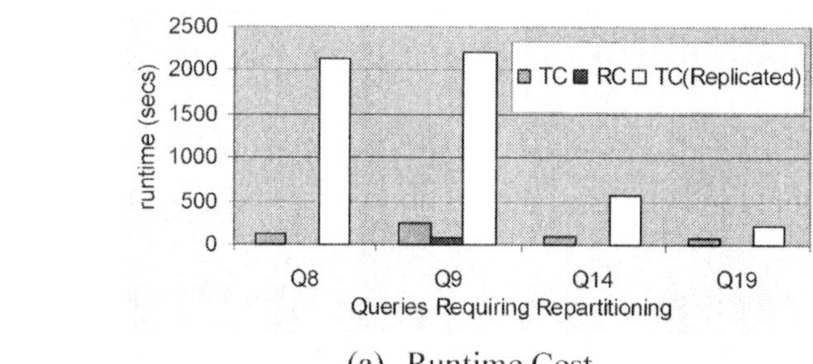

(a) Runtime Cost

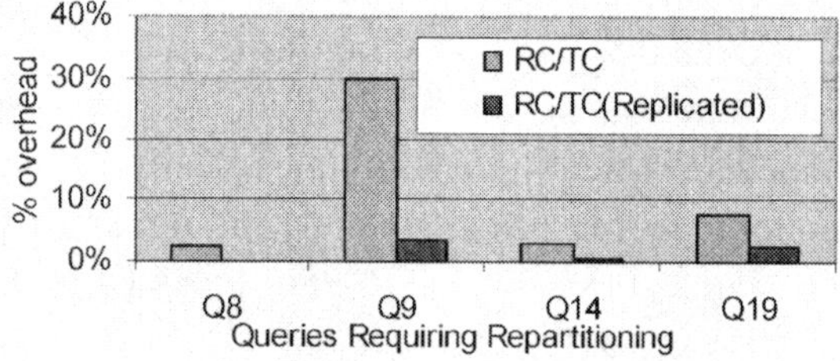

(b) % of Runtime Cost

but a very small fraction when compared with the total cost of the initial replicated scheme (e.g., RC/TC(Replicated) is below 5% even for Q9).

In summary, after applying the simple improved partitioning strategy, the performance results improved significantly. Most queries required no repartitioning at all with the improved placement, and those that did require repartitioning still obtained a speedup that was orders of magnitude faster than the initial replicated placement scheme.

CONCLUSION

In this article, we have discussed partitioning and placement issues in Node Partitioned Data Warehouses. We have reviewed basic placement and processing issues in parallel shared-nothing databases and specifically in the node-partitioned data warehouse, also highlighting the specificities of the data warehouse environment. Considering a simple star schema with a large fact and small dimensions, we discussed a basic and efficient placement strategy with replicated cached dimensions and fully horizontally partitioned fact. We analyzed the shortcomings of such a simple scheme when considering more complex schemas than the single large fact, multiple very small dimensions scenario. Our experimental setup used the schema and query set of the TPC-H decision support benchmark. Our analysis of the results clearly pointed out the limitations of the approach. We have shown that the speedup is totally unsatisfactory for many queries and analyzed why this happens. We explained the problem with the help of access patterns to replicated relations. We proposed a simple and easy-to-apply modification to the basic scheme, which takes into consideration the data warehouse structure with dimensions and facts, query workload, and replication of small relations. It uses a combination of hash, replicated, and repartitioned join to yield very good results. We also discussed important query processing issues related to the strategy. We applied the strategy to TPC-H and observed large improvements in the worst behaving queries. We conclude that the proposed workload-friendly partitioning and placement can prevent very low speedup and provide near to linear speedup for most queries in Node Partitioned Data Warehouses. Future work on this issue includes query optimization in this environment and combining this strategy with bitmap join indexes and materialized views.

ACKNOWLEDGMENTS

I would like to thank the editor in chief, the area editor, and the reviewers for their helpful comments. My thanks are also due to Joao Costa for his help reviewing the text before it was submitted and for his constructive comments.

REFERENCES

Akinde, M., Böhlen, M., Johnson, T., Lakshmanan, L., & Srivastava, D. (2002, March). Efficient OLAP query processing in distributed data warehouses. In *Proceedings of the Eighth Conference on Extending Database Technology* (pp. 20-39), Prague, Czech Republic.

Apers, P. (1988). Data allocation in distributed database systems. *ACM Transactions on Database Systems, 13*(3), 263-304.

Bernardino J., Furtado P., & Madeira H. (2002, July). DWS-AQA: A cost effective approach for very large data warehouses. In *Proceedings of the International Database Engineering and Applications Symposium* (pp. 233-242), Edmonton, Canada.

Chen, H., & Chengwen, L. (2000, July). An efficient algorithm for processing distributed queries using partition dependency. In *Proceedings of the Seventh International Conference on*

Parallel and Distributed Systems (pp. 339-346), Iwate, Japan.

Copeland, G. P., Alexander, W., Boughter, E., & Keller, T. (1988, June). Data placement in bubba. In *Proceedings of the ACM International Conference on Management of Data* (pp. 99-108), Chicago, Illinois, USA.

DeWitt, D., & Gerber, R. (1985, August). Multiprocessor hash-based join algorithms. In *Proceedings of the Eleventh Conference on Very Large Databases* (pp. 151-164), Stockholm, Sweden.

DeWitt, D., & Gray, J. (1992). The future of high performance database processing. *Communications of the ACM, 35*(6), 85-98.

Epstein, R., Stonebraker, M., & Wong, E. (1978, May). Distributed query processing in a relational database system. In *Proceedings of the ACM International Conference on Management of Data* (pp. 169-180), Austin, Texas, USA.

Furtado, P. (2004a, July). Algorithms for efficient processing of complex queries in node-partitioned data warehouses. In *Proceedings of the Eighth International Database Engineering and Applications Symposium* (pp. 117-122), Coimbra, Portugal.

Furtado, P. (2004b, July). Hash-based placement and processing for efficient node partitioned query-intensive databases. In *Proceedings of the Tenth International Conference on Parallel and Distributed Systems* (pp. 127-134), Newport Beach, California, USA.

Furtado, P. (2004c, September). Workload-based placement and join processing in node-partitioned data warehouses. In *Proceedings of the International Conference on Data Warehousing and Knowledge Discovery* (pp. 38-47), Zaragoza, Spain.

Furtado, P. (2005, May). Efficiently processing query-intensive databases over a non-dedicated local network. In *Proceedings of the Nineteenth International Parallel and Distributed Processing Symposium*, Denver, Colorado, USA.

Hua, K. A., & Lee, C. (1990, August). An adaptive data placement scheme for parallel database computer systems. In *Proceedings of the Sixteenth Very Large Data Bases Conference* (493-506), Brisbane, Queensland, Australia.

Khan, L., McLeod, D., & Shahabi, C. (2001). An adaptive probe-based technique to optimize join queries in distributed Internet databases. *Journal of Database Management, 12*(4), 3-14.

Kimball, R. (1996). *The data warehouse toolkit.* New York: J. Wiley & Sons.

Kimball, R., Reeves, L., Ross, M., & Thornthwaite, W. (1998). *The data warehouse life cycle toolkit.* New York: John Wiley & Sons.

Kitsuregawa, M., Tanaka, H., & Motooka, T. (1983). Application of hash to database machine and its architecture. *New Generation Computing, 1*(1), 63-74.

Lee, W., Park, J., & Kang, S. (1999). An asynchronous differential join in distributed data replications. *Journal of Database Management, 10*(3), 3-12.

Liu, C., & Chen, H. (1996a). A hash partition strategy for distributed query processing. In *Proceedings of the International Conference on Extending Database Technology* (pp. 373-387), Avignon, France.

Liu, C., & Chen, H. (1996b, February). A heuristic algorithm for partition strategy in distributed query processing. In *Proceedings of the 1996 ACM Symposium on Applied Computing (SAC)* (pp. 196-200), Philadelphia, Pennsylvania, USA.

Liu, C., Chen, H., & Krueger, W. (1996, February). A distributed query processing strategy using placement dependency. In *Proceedings of the Twelfth International Conference on Data Engineering* (pp. 477-484), New Orleans, Louisiana, USA.

Livny, M., Khoshafian, S., & Boral, H. (1987). Multi-disk management algorithms. *IEEE Data Engineering Bulletin, 9*(1), 24-36.

Rao, J., Zhang, C., Megiddo, N., & Lohman, G. (2002). Automating physical database design in a parallel database. In *Proceedings of the ACM International Conference on Management of Data* (pp. 558-569), Madison, Wisconsin, USA.

Sacca, D., & Wiederhold, G. (1985). Database partitioning in a cluster of processors. *Transactions on Database Systems, 10*(1), 29-56.

Sasha, D., & Wang, T. (1991). Optimizing equijoin queries in distributed databases where relations are hash partitioned. *ACM Transactions on Database System, 16*(2), 279-308.

TPC Benchmark H, Transaction Processing Council. (1999). Retrieved from http://www.tpc.org/

Williams, M., & Zhou, S. (1998). Data placement in parallel database systems. In M. Abdelguerfi & K. Wong (Eds.), *Parallel database techniques* (pp. 203-219). CA: IEEE Computer Society Press.

Wong, E., & Katz, R. H. (1983). Distributing a database for parallelism. In *Proceedings of the ACM International Conference on Management of Data* (pp. 23-29), San Jose, California, USA.

Yu, C., Guh, K., Brill, D., & Chen, A. (1989). Partition strategy for distributed query processing in fast local networks. *IEEE Transactions on Software Engineering, 15*(6), 780-793.

Yu, C. T., & Chang, C. C. (1984). Distributed query processing. *ACM Computing Surveys, 16*(4), 399-433.

Yu, C. T., & Meng, W. (1998). *Principles of database query processing for advanced applications.* San Francisco: Morgan Kaufmann.

Yu, C. T., Keh-Chang, G., Zhang, W., Templeton, M., Brill, D., Chen, A. L. P. (1987). Algorithms to process distributed queries in fast local networks. *IEEE Transactions on Computers, 36*(10), 1153-1164.

Zilio, D., Jhingran, A, & Padmanabhan, S. (1994, November 10). *Partitioning key selection for a shared-nothing parallel database system* (IBM Research Report RC 19820 [87739]). Yorktown Heights, NY: TJ Watson Research Center.

This work was previously published in Journal of Database Management, Vol. 17, Issue 2, edited by K. Siau, pp. 43-61, copyright 2006 by IGI Publishing, formerly known as Idea Group Publishing (an imprint of IGI Global).

Chapter 2.16
Managing Late Measurements in Data Warehouses

Matteo Golfarelli
University of Bologna, Italy

Stefano Rizzi
University of Bologna, Italy

ABSTRACT

Though in most data warehousing applications no relevance is given to the time when events are recorded, some domains call for a different behavior. In particular, whenever late measurements of events take place, and particularly when the events registered are subject to further updates, the traditional design solutions fail in preserving accountability and query consistency. In this article, we discuss the alternative design solutions that can be adopted, in presence of late measurements, to support different types of queries that enable meaningful historical analysis. These solutions are based on the enforcement of the distinction between transaction time and valid time within the schema that represents the fact of interest. Besides, we provide a qualitative and quantitative comparison of the solutions proposed, aimed at enabling well-informed design decisions.

INTRODUCTION

Time is commonly understood as a key factor in data warehousing systems since the decisional process often relies on computing historical trends and on comparing snapshots of the enterprise taken at different moments. Within the multidimensional model, time is usually a dimension of analysis; thus, the representation of the history of fact values across a given lapse of time, at a given granularity, is directly supported. For instance, in a relational implementation for the sales domain, for each day there will be a set of rows in the fact table reporting the values of fact QuantitySold on that day for different products and stores. On the other hand, although the multidimensional model does not inherently represent the history of values for dimensions and their properties, some ad hoc techniques were devised to support the so-called *slowly-changing dimensions* (Kimball, 1996). In both cases, time is commonly meant as valid time in the terminology of temporal databases (Jensen et al., 1994) (i.e., it is meant as the time when an event or change *occurred* in the business domain) (Devlin, 1997). Transaction time, meant as the time when the event or change was *registered* in the data warehouse, is typically given little or no importance since it is not considered to be relevant for decision support.

One of the underlying assumptions in data warehouses is that, once an event has been registered (under the form of a row in the fact table), it is never modified so that the only possible writing operation consists in appending new events (rows) as they occur. While this is acceptable for a wide variety of domains, some applications call for a different behavior. In particular, the values measured for a given event may change over a period of time, to be consolidated only *after* the event has been for the first time registered in the data warehouse. This typically happens when the early measurements made for events may be subject to errors (e.g., the amount of an order may be corrected after the order has been registered) or when events inherently evolve over time (e.g., notifications of university enrollments may be received and registered several days after they were issued).

In this context, if the up-to-date situation is to be made timely visible to the decision makers, past events must be continuously updated to reflect the incoming data. Unfortunately, if updates are carried out by physically *overwriting* past registrations of events, some problems may arise:

- Accountability and traceability require the capability of preserving the exact information the analyst based his or her decision upon. If the old registration for an event is replaced by its latest version, past decisions can no longer be justified.
- In some applications, accessing only up-to-date versions of information is not sufficient to ensure the correctness of analysis. A typical case is that of queries requiring to compare the progress of an ongoing phenomenon with past occurrences of the same phenomenon: since the data recorded for the ongoing phenomenon are not consolidated yet, comparing them with past consolidated data may not be meaningful.

Remarkably, the same problems may arise when events are registered in the data warehouse only once, but with a significant delay with respect to the time when they occurred in the application domain (e.g., there may be significant delays in communicating the daily price of listed shares on the stock market): no update is necessary in this case, yet valid time is not sufficient to guarantee accountability. Thus, in more general terms, we will use term *late measurement* to denote any measurement of an event that is sensibly delayed with respect to the time when the event occurs in the application domain; a late measurement may either imply an update to a previous measurement (as in the case of late corrections to orders) or not (as in the case of shares).

In this article, we discuss and compare the design solutions that can be adopted, in presence of late measurements, to enable meaningful historical analysis aimed at preserving accountability and consistency. These solutions are based on the enforcement of the distinction between transaction time and valid time within the schema that represents the fact of interest.

The rest of the article is organized as follows. In the second and third sections, respectively, we survey the related literature and present the working examples. In the fourth section, we distinguish two possible semantics for facts and give definitions of events and registrations. In the fifth section, we distinguish three basic categories of queries from the point of view of their temporal requirements in presence of late measurements. The sixth and seventh sections introduce, respectively, two classes of design solutions: monotemporal and bitemporal, that are then quantitatively compared in the eighth section. The ninth section concludes by discussing the applicability of the solutions proposed.

RELATED LITERATURE

Several works concerning temporal data warehousing can be found in the literature. Most of them are related to consistently managing updates in dimension tables of relational data warehouses—the so-called *slowly-changing dimensions* (e.g., Letz, Henn, & Vossen, 2002; Yang, 2001). Some other works tackle the problem of temporal evolution and versioning of the data warehouse schema (Bebel, Eder, Koncilia, Morzy, & Wrembel, 2004; Blaschka, Sapia, & Höfling, 1999; Eder, Koncilia, & Morzy, 2002; Golfarelli, Lechtenbörger, Rizzi, & Vossen, 2006; Quix, 1999). All these works are not related to ours since there is no mention of the opportunity of representing transaction time in data warehouses in order to allow accountability and traceability for late measurements.

Devlin (1997) distinguishes between *transient data*, that do not survive updates and deletions, and *periodic data*, that are never physically deleted from the data warehouse. Kimball (1996) introduces two basic paradigms for representing inventory-like information in a data warehouse: the *transactional model*, where each increase and decrease in the inventory level is recorded as an event, and the *snapshot model*, where the current inventory level is periodically recorded. This is then generalized to define a classification of facts based on the conceptual role given to events:

- For a *transactional fact*, each event may either record a single transaction or summarize a set of transactions that occur during the same time interval. Most measures are flow measures (Lenz & Shoshani, 1997): they refer to a time interval and are cumulatively evaluated at the end of that period.
- For a *snapshot fact*, events correspond to periodical snapshots of the fact. Measures are mostly stock measures (Lenz et al., 1997): they refer to an instant in time and are evaluated at that instant.

A similar characterization is proposed by Bliujute, Saltenis, Slivinskas, and Jensen (1998), who distinguish between *state-oriented data* like sales, inventory transfers, and financial transactions, and *event-oriented data*, like unit prices, account balances, and inventory levels. Both distinctions are relevant to our approach and are recalled in the fourth section. Bliujute et al. (1998) also propose a *temporal star schema* that incorporates timestamps into the fact table to model valid time; though such schema is somehow related to the design solutions we propose, it does not take transaction time into consideration and is not analyzed in the light of the late measurements problem.

Pedersen and Jensen (1998) recognize the importance of advanced temporal support in data warehouses, with particular reference to medical applications. Abelló and Martín (2003a) claim that there are important similarities between temporal databases and data warehouses, suggest that both valid time and transaction time should be modeled within data warehouses, and mention the importance of temporal queries. Finally, Abelló and Martín (2003b) propose a storage structure for a bitemporal data warehouse (i.e., one supporting both valid and transaction time). All these approaches suggest the importance of transaction time in data warehouses, but not with explicit reference to the problem of late measurements.

Kimball (2000) raises the problem of late-arriving fact records, generically stating that a bitemporal solution may be useful to cope with them. In the same direction, Bruckner and Tjoa (2002) discuss the problem of data warehouse temporal consistency in consequence of delayed discovery of real-world changes and propose a solution based on transaction time (which they call *revelation time*) and overlapped valid time. Although the article discusses some issues related to late measurements, no emphasis is given to the influence that the semantics of the captured events and the querying scenarios pose on the feasibility of the different design solutions.

WORKING EXAMPLES

In this section, we propose three examples that justify the need for managing late measurements and will be used in the rest of the article to discuss and compare the different design solutions.

In the first example, late measurements (with updates) are motivated by the fact that the represented events inherently evolve over time. Consider a fact modeling the number of enrollments to university degrees; in a relational implementation, a simplified fact table for enrollments could have the following schema (we intentionally do not introduce surrogate keys in the fact table in order to avoid to unnecessarily complicate the examples):

FT_ENROLL(<u>EnrollDate</u>, <u>Degree</u>, <u>AcademicYear</u>, Number)

where EnrollDate is the formal enrollment date (the one reported on the enrollment form). An enrollment is acknowledged by the University secretariat only when the entrance fee is paid; considering the variable delays due to the bank processing and transmitting the payment, the enrollment may be communicated and stored in the university database—and, from there, loaded into the data warehouse—even one month after the enrollment has been done. This is a case of late measurements. Besides: (i) notices of payments for the same enrollment date are spaced out over long periods, and (ii) after paying the fee, students may decide to switch their enrollment from one degree to another. Thus, updates are necessary in order to correctly track enrollments.

The main reason why, in this example, the enrollment date may not be sufficient is related to the soundness of analysis. In fact, most queries on this fact will ask for evaluating the current trend of the number of enrollments as compared to last year. But if the current, partial data on enrollments were compared to the consolidated ones at exactly one year ago, the user would wrongly infer that this year we are experiencing a negative trend for enrollments!

The second example is related to a fact representing the quantities in the lines of orders received by a company selling PC consumables, according to the following schema:

FT_ORDER(<u>OrderNumber</u>, <u>OrderDate</u>, <u>Product</u>, Quantity).

Though the first registration of an order may not involve notable delays, the orders received may be subject to later corrections, which implies late measurements.

The third example, motivated by the delay in communicating information, is that of a fact monitoring the price of listed shares on the stock market:

FT_SHARE(<u>Date</u>, <u>Share</u>, Price).

We assume that this fact is daily fed by importing a file that reports the current quotations; occasional delays in communicating the daily prices will produce late measurements, which in turn will raise problems with justifying the decisions made on previous reports.

EVENTS AND REGISTRATIONS

The aim of this section is to introduce the classification of events and registrations on which we will rely in the next sections to discuss the applicability of the design solutions proposed.

In general terms, the facts to be monitored for decision support fall into two broad categories according to the way they are measured in the application domain. *Flow facts* (called *flow measures* in Lenz et al., 1997) are monitored by collecting their occurrences during a time interval and are cumulatively measured at the end of that period; examples of flow facts are order quantity and number of enrollments. *Stock facts* (called *stock measures* in Lenz et al., 1997) are monitored by periodically sampling and measuring their state; examples of stock facts are the price of a share and the level of a river.

Definition 1 (Event): Given fact F, we call events the results of the monitoring of F. Each event is identified by a set of coordinates, i.e., values for the dimensions of analysis of F. We call the valid time of event e_i the instant vt_i when e_i takes place in the application domain. Event e_i yields a non empty sequence of measurements m_{ij}, $j = 1,...,n_i$ ($n_i \geq 1$).

Each new measurement for an event provides a revised value, typically more accurate than the previous one. Obviously, in order to avoid information loss, each measurement received for an event must be recorded into the data warehouse, which is done when next refresh takes place.

Definition 2 (Registration): Given fact F and a measurement m_{ij} for event e_i, we call registration r_{ij} for m_{ij} the recording[1] of m_{ij} on the data warehouse, done on transaction time tt_{ij} ($tt_{ij} \geq vt_i$). For each i, r_{i1} is called the first registration for e_i and the r_{ij}'s with $j > 1$ (if any) are called the update registrations for e_i. Given $t \geq vt_i$, the current registration for e_i at time t is the one done on transaction time tt_{ij^*} where $j^* = max\{j \mid tt_{ij} \leq t\}$.

With reference to our working examples:

- An event for the (flow) order fact measures the quantity ordered for a given product within a given order issued on a given date (its coordinates). In this case, each event corresponds to a single order line (no aggregation is done) and its valid time is the order date. The first registration of each event is done when the related order is received; an update registration may arise if the ordering customer asks for modifying a quantity in her order.
- An event for the (flow) enrollment fact measures the net number of enrollments made on a given date for a given degree and academic year (its coordinates). In this case, each event aggregates a set of enrollments and its valid time is the enrollment date; after the first registration, a sequence of update registrations is typically made for each event as new data on enrollments made on previous dates are received.
- An event for the (stock) share fact is the observation, made on a given date (valid time), of the price for a given share (date and share are the event coordinates). One single (first) registration is commonly made in this case for each event.

The delay between the time when a measurement is received by the operational database and the transaction time of the corresponding registration in the data warehouse depends on the duration of the *refresh interval* of the data warehouse (i.e., on the time between two consecutive periodical refreshes (typically ranging between 1 and 7 days)). Since, from the point of view of a data warehouse user, a measurement is known only when it is registered, in the following we will assume that *each measurement is synchronous with its registration*.

To clarify this point, Figure 1 shows an example where two events are characterized, respectively, by three and one measurements. The central axis represents the flow of time, and reports the instants when events take place and their measurements are received by the operational database. Such instants are then projected on the other two axes that represent, respectively, valid and transaction time. The instant when an event takes place is its valid time (e.g., vt_1 is the valid time for e_1). Each measurement is registered in the data warehouse not immediately when it is received by the operational database, but when next refresh takes place: for instance, measurement

Table 4. Stock solution for the share fact

Date	Share	Price
Jan. 7, 2006	BigTel	9
Jan. 8, 2006	BigTel	12
Jan. 9, 2006	BigTel	10

Table 5. Flow solution for the share fact

Date	Share	Price
Jan. 7, 2006	BigTel	9
Jan. 8, 2006	BigTel	3
Jan. 9, 2006	BigTel	–2

registration m_{i1} of event ei, under a form depending on the model (flow or stock) adopted for m_{i1};

2. Transaction time is modeled by adding to the fact two new temporal dimensions, used as timestamps to mark the time interval during which each registration is current (*currency interval*);
3. Up-to-date queries are answered by selecting, for each event, the registration that is current now (the one whose currency interval is still open);
4. Rollback queries at time t are answered by selecting, for each event, the registration that was current at t (the one whose currency interval includes t);
5. Historical queries on time interval T are answered by selecting, for each event, the registrations that were current for at least one $t \in T$ (those whose currency interval overlaps with T).

The reason for using two timestamps in a consolidated solution is that each registration records a *state* of the event, which is valid during a time interval, rather than an instant measurement like in a flow solution.

Since in principle these two types of solutions can be combined with either the flow or the stock model for first registrations, four different specific solutions can be distinguished, which we will call *delta-flow, delta-stock, consolidated-flow,* and *consolidated-stock*, respectively. In the following subsections, we will discuss how these solutions are implemented for flow and stock facts subject to late measurements with updates, and for facts subject to late measurements without updates.

Flow Facts with Updates

As seen in the sixth section, a flow fact can be represented within the data warehouse either by flow or stock registrations.

In case of a flow fact represented by flow (first) registrations, the delta solution leads to events, first registrations and update registrations that share the same flow semantics, which means that additivity is preserved for all registrations. Consider for instance the enrollment schema; if a delta-flow solution is adopted, the schema is enriched as follows:

FT_ENROLL(EnrollDate, RegistrDate, Degree, AcademicYear, Number)

where RegistrDate is the dimension added to model transaction time. Table 6 shows a possible set of registrations for a given degree and year, including some positive and negative updates. While each first registration records the exact value of its measurement, each update registration records the difference between its measurement and the previous one.

With reference to these sample data, in the following we report some simple examples of queries of the three types together with their results, and show how they can be computed by aggregating registrations.

1. **q1:** *Daily number of enrollments to electric engineering for academic year 05/06.* This

Table 6. Delta-flow solution for the enrollment fact (update registrations in italics)

EnrollDate	RegistrDate	Degree	AcademicYear	Number
Oct. 21, 2005	Oct. 27, 2005	Elec. Eng.	05/06	5
Oct. 21, 2005	*Nov. 1, 2005*	*Elec. Eng.*	*05/06*	*8*
Oct. 21, 2005	*Nov. 5, 2005*	*Elec. Eng.*	*05/06*	*–2*
Oct. 22, 2005	Oct. 27, 2005	Elec. Eng.	05/06	2
Oct. 22, 2005	*Nov. 5, 2005*	*Elec. Eng.*	*05/06*	*4*
Oct. 23, 2005	Oct. 23, 2005	Elec. Eng.	05/06	3

up-to-date query is answered by summing up Number for all registration dates related to the same enrollment dates, and returns the following result:

EnrollDate	Number
Oct. 21, 2005	11
Oct. 22, 2005	6
Oct. 23, 2005	3

2. **q2:** *Daily number of enrollments to electric engineering for academic year 05/06 as known on Nov. 2.* This rollback query is answered by summing up Number for all registration dates before Nov. 2:

EnrollDate	Number
Oct. 21, 2005	13
Oct. 22, 2005	2
Oct. 23, 2005	3

3. **q3:** *Number of registrations of enrollments to electric engineering received daily for academic year 05/06.* This historical query is answered by summing up Number, for each registration date, along all enrollment dates:

RegistrDate	Number
Oct. 23, 2005	3
Oct. 27, 2005	7
Nov. 1, 2005	8
Nov. 5, 2005	2

For a flow fact represented by flow registrations, also the consolidated solution is possible. In a consolidated-flow solution, the enrollment schema is enriched as follows:

FT_ENROLL(<u>EnrollDate</u>, <u>CurrencyStart</u>, <u>CurrencyEnd</u>, <u>Degree</u>, <u>AcademicYear</u>, Number)

where CurrencyStart and CurrencyEnd delimit the currency interval. Table 7 shows the set of registrations corresponding to those in Table 6: while the first registrations still report the same value for the fact, update registrations now report the exact value of measurements rather than a delta.

Adopting one or the other solution (delta-flow or consolidated-flow) for a flow fact has a deep impact on the response to the workload. For instance it is easy to see that, while in the delta-flow solution queries q_1 and q_2 are answered by accessing several registrations for each event involved, in the consolidated-flow solution they are answered by reading exactly one registration (respectively, the one that is current now and the one that was current on November 2) for each event involved.

In case of a flow fact represented by stock registrations, as seen in the sixth section, a mono-temporal solution leads to an update propagation

Table 7. Consolidated-flow solution for the enrollment fact

EnrollDate	CurrencyStart	CurrencyEnd	Degree	AcademicYear	Number
Oct. 21, 2005	Oct. 27, 2005	Oct. 31, 2005	Elec. Eng.	05/06	5
Oct. 21, 2005	*Nov. 1, 2005*	*Nov. 4, 2005*	*Elec. Eng.*	*05/06*	*13*
Oct. 21, 2005	*Nov. 5, 2005*	—	*Elec. Eng.*	*05/06*	*11*
Oct. 22, 2005	Oct. 27, 2005	Nov. 4, 2005	Elec. Eng.	05/06	2
Oct. 22, 2005	*Nov. 5, 2005*	—	*Elec. Eng.*	*05/06*	*6*
Oct. 23, 2005	Oct. 23, 2005	—	Elec. Eng.	05/06	3

Table 8. Consolidated-stock solution for the share fact

Date	CurrencyStart	CurrencyEnd	Share	Price
Jan. 7, 2006	Jan. 7, 2006	Jan. 11, 2006	BigTel	9
Jan. 7, 2006	*Jan. 12, 2006*	—	*BigTel*	*8.5*
Jan. 8, 2006	Jan. 10, 2006	—	BigTel	12
Jan. 9, 2006	Jan. 10, 2006	Jan. 12, 2006	BigTel	10
Jan. 9, 2006	*Jan. 13, 2006*	—	*BigTel*	*10.5*

problem. This problem also occurs with a bitemporal solution: in fact, since stock registrations are computed by accumulating past flow measurements, each update measurement received for a past event e_i would lead to recording a whole set of update registrations, one for each event with valid time after e_i. Consequently, for a flow fact subject to late registrations with updates, we will not consider stock solutions recommendable.

Stock Facts with Updates

As seen in the sixth section, using flow registrations for a stock fact is not recommendable; thus, we will assume that a stock solution is adopted.

Since both measurements and first registrations have stock semantics, the most immediate choice is the consolidated-stock solution, that gives stock semantics also to update registrations. In the share example, the schema is then enriched as follows:

FT_SHARE(<u>Date</u>, <u>CurrencyStart</u>, <u>CurrencyEnd</u>, <u>Share</u>, Price)

and may be populated for instance by the sample set of registrations reported in Table 8.

An example of up-to-date query on these data is *"find the minimum price of BigTel from Jan. 7 to 9,"* which returns 8.5. A rollback query is *"find the minimum price of BigTel from Jan. 7 to 9, as known on Jan. 10,"* which returns 9. Finally, a historical query is *"find the fluctuation on the price of Jan. 7 for BigTel,"* which returns –0.5 and requires to progressively compute the differences between subsequent registrations. Thus, while up-to-date and rollback queries are very simply answered, historical queries may ask for some computation.

In case of a stock fact, also the delta-stock solution can be applied. See for instance Table 9 that shows the delta solution for the same set of measurements reported in Table 8. In this case, up-to-date and rollback queries that aggregate the fact along valid time would have to be formulated as nested queries relying on different aggregation operators. For instance, the average monthly price for a share is computed by first

Table 9. Delta-stock solution for the share fact

Date	RegistrDate	Share	Price
Jan. 7, 2006	Jan. 7, 2006	BigTel	9
Jan. 7, 2006	*Jan. 12, 2006*	*BigTel*	*–0.5*
Jan. 8, 2006	Jan. 10, 2006	BigTel	12
Jan. 9, 2006	Jan. 10, 2006	BigTel	10
Jan. 9, 2006	*Jan. 13, 2006*	*BigTel*	*0.5*

summing Price along RegistrDate for each Date, then averaging the partial results. On the other hand, a historical query like the one above is very simply answered.

Facts without Updates

In the case of facts where measurements may be delayed but done exactly once for each event, accountability can be achieved, for both flow and stock solutions, by adding a single temporal dimension RegistrDate that models the transaction time. Up-to-date queries are solved without considering transaction times, while rollback queries require to select only the registrations made before a given transaction time. Historical queries make no sense in this context, since only one measurement is made for each event. As a matter of fact, the solution adopted can be considered as a special case of delta solution where no update registrations are recorded.

COMPARISON AND DISCUSSION

This section aims at providing a simple quantitative comparison of the different solutions, in terms of storage space and query performance. Let E, μ, and $|E_q| \leq E$ denote, respectively, the total number of events recorded, the average number of measurements per event, and the number of events involved in query q.

The results are collected in Table 10. The first column of data reports the total number of registrations stored in the fact table (we neglect that, in consolidated solutions, each registration is longer due to the additional timestamp that represents the end of the currency interval). The other three columns report the execution cost for different types of queries, estimated as the number of registrations to be accessed (independently of the execution plans adopted, and assuming that each registration is read only once):

- For up-to-date queries we assume that, at query formulation time, all measurements for the involved events are already available.
- For rollback queries, we consider a border effect related to the distribution along time of the measurements for each event, which reduces by a factor ρ ($0 \leq \rho \leq 1$) the number of registrations to be read (see Figure 2). Such factor heavily depends on the relationship between the width of the time interval defined by the valid times of the involved events, T_q, the relative positioning of the reference time for the query, t, and the average delay of measurements, δ. Figure 3 shows how ρ varies, assuming that measurements delays are normally distributed in time, in function of δ (expressed in numbers of refreshes), when T_q spans 12 refreshes and t falls exactly at the end of T_q.

Figure 2. Distribution of measurements for events; in gray the measurements that are not read by a rollback query with reference time t

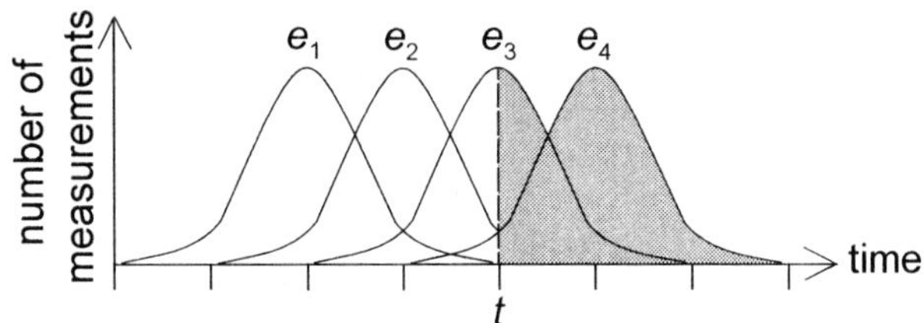

Table 10. Quantitative comparison of the design solutions (UQ, RQ, and HQ stand for up-to-date, roll-back, and historical queries, respectively)

		number of tuples	*UQ*	*RQ*	*HQ*
monotemp.	*flow*	E	$\|E_q\|$	—	—
	stock	E	$\|E_q\|$	—	—
bitemp.	*delta-flow*	μE	$\mu \|E_q\|$	$\rho \mu \|E_q\|$	$\rho' \mu \|E_q\|$
	delta-stock	μE	$\mu \|E_q\|$	$\rho \mu \|E_q\|$	$\rho' \mu \|E_q\|$
	cons.-flow	μE	$\|E_q\|$	$\rho \|E_q\|$	$\rho' \mu \|E_q\|$
	cons.-stock	μE	$\|E_q\|$	$\rho \|E_q\|$	$\rho' \mu \|E_q\|$

Figure 3. Reduction factor in function of the average measurement delay

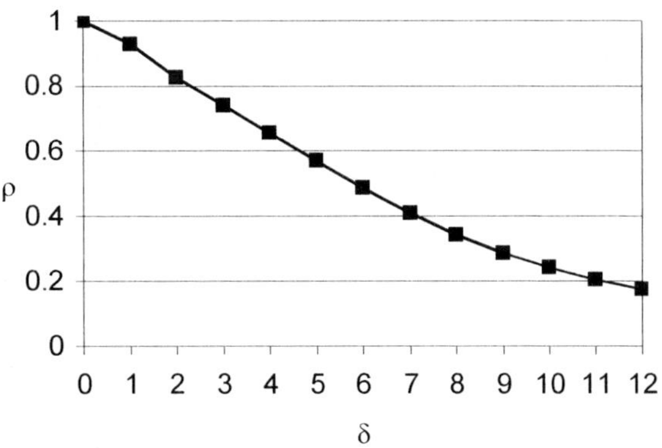

- For historical queries, there still is a reduction factor ρ' that additionally depends on the width of the query reference interval *T*.

Overall, the overhead induced by the design solutions proposed on the query performance and on the storage space heavily depends on the characteristics of the application domain and on the actual workload. In a bitemporal solution, frequent updates determine a significant increase in the fact table size, but this may be due to a wrong choice of the designer, who promoted early registrations of events that are not stable enough to be significant for decision support. The increase in the query response time may be reduced by a proper use of materialized views and indexes: a materialized view aggregating registrations on all transaction times cuts down the time for answering up-to-date queries in delta solutions, while an index on transaction time enables efficient slicing of registrations in consolidated solutions.

CONCLUSION

In this article we have raised the problem of late measurements, meant as retrospective updates to events registered in a data warehouse, and we have shown how conventional design solutions, that only take valid time into account, may fail to provide query accountability and consistency. Then, we have introduced some alternative design solutions that overcome this problem by modeling transaction time as an additional dimension of the fact, and we have discussed their applicability. Table 11 summarizes the results obtained. Most noticeably, the most recommended solutions for a flow and a stock fact with updates are, respectively, consolidated-flow and consolidated-stock. Delta-flow and delta-stock, in fact, create some overhead with up-to-date and rollback queries.

In commercial platforms, late registrations are only partially supported. To the best of our knowledge, the most sophisticated solution is the one adopted by SAP-BW, that supports bitemporal solutions. In particular, BW distinguishes between *cumulative* and *non-cumulative key figures* (corresponding to flow and stock facts). The former are directly modeled in the fact table through a delta-flow solution. The latter can be handled by adopting two different time granularities: at the coarsest one, consolidated values for events are historicized within a support table; at the finest one, delta values are stored within the fact table limitedly to a user-defined time interval. This solution, while guaranteeing good querying performances, limits the expressivity achievable with rollback and historical queries.

We close this section by observing that, in real applications, multiple related facts are normally stored in the same fact table (e.g., in the order example, quantity and unit price for each order line). How do they coexist in presence of late measurements? For simplicity we will reason on the order example, focusing on the event reported below and assuming that (i) on March 17 a correction is received stating that the quantity is not 100 but 110, and (ii) on March 19 another correction is received taking to 1.1 the unit price.

Table 11. Applicability of the design solutions

		without updates		*with updates*	
		flow fact	*stock fact*	*flow fact*	*stock fact*
monotemp.	*flow*	fair, only UQ supported	not recomm.	fair, only UQ supported	not recomm.
	stock	fair, only UQ supported	fair, only UQ supported	not recomm.	fair, only UQ supported
bitemp.	*delta-flow*	good	not recomm.	fair, overhead on UQ and RQ	not recomm.
	delta-stock	good	good	not recomm.	fair, overhead on UQ and RQ
	cons.-flow	—	—	good	not recomm.
	cons.-stock	—	—	not recomm.	good

OrderNumber	OrderDate	Quantity	UnitPrice
11001	Mar. 15, 2007	100	1

We consider three sample cases:

1. Accountability is required for no one of the facts. A monotemporal solution can be adopted; every time a new measurement is made for one of the facts, the related registration is overwritten leaving the other fact unchanged. Thus, on March 20 only one registration is present:

OrderNumber	OrderDate	Quantity	UnitPrice
11001	Mar. 15, 2007	110	1.1

2. Accountability is required for Quantity only. A consolidated-flow solution is adopted; when a new measurement is made for Quantity, an update registration is added that reports the new value for Quantity and the previous value for UnitPrice. If a new measurement is made for UnitPrice, the value of UnitPrice is updated within all the related registrations. Thus, on March 20 we have two registrations present:

Order-Number	Order-Date	Curren-cyStart	Curren-cyEnd	Quan-tity	Unit-Price
11001	Mar. 15, 2007	Mar. 15, 2007	Mar. 16, 2007	100	1.1
11001	Mar. 15, 2007	Mar. 17, 2007	—	110	1.1

3. Accountability is required for both facts. A consolidated-flow/stock solution is adopted; any new measurement for each fact creates a new registration:

Order-Number	Order-Date	Curren-cyStart	Curren-cyEnd	Quan-tity	Unit-Price
11001	Mar. 15, 2007	Mar. 15, 2007	Mar. 16, 2007	100	1
11001	Mar. 15, 2007	Mar. 17, 2007	Mar. 18, 2007	110	1
11001	Mar. 15, 2007	Mar. 19, 2007	—	110	1.1

Remarkably, having two different solutions coexist (like in cases 2 and 3) leads to no additional overhead in query performance, except for some specific historical queries which require to distinguish the updates made to one fact from those made to the other.

REFERENCES

Abelló, A., & Martín, C. (2003a). The data warehouse: An object-oriented temporal database. *Proceedings Jornadas de Ingeniería del Software y Bases de Datos* (pp. 675-684), Alicante, Spain.

Abelló, A., & Martín, C. (2003b). A bitemporal storage structure for a corporate data warehouse. *Proceedings International Conference on Enterprise Information Systems* (pp. 177-183), Angers, France.

Bebel, B., Eder, J., Koncilia, C., Morzy, T., & Wrembel, R. (2004). Creation and management of versions in multiversion data warehouse. *Proceedings ACM Symposium on Applied Computing* (pp. 717-723), Nicosia, Cyprus.

Blaschka, M., Sapia, C., & Höfling, G. (1999). On schema evolution in multidimensional databases. *Proceedings International Conference on Data Warehousing and Knowledge Discovery* (pp. 153-164), Florence, Italy.

Bliujute, R., Saltenis, S., Slivinskas, G., & Jensen, C. S. (1998). Systematic change management in dimensional data warehousing. *Proceedings International Baltic Workshop on Databases and Information Systems* (pp. 27-41), Riga, Latvia.

Bruckner, R., & Tjoa, A. (2002). Capturing delays and valid times in data warehouses--towards timely consistent analyses. *Journal of Intelligent Information Systems, 19*(2), 169-190.

Devlin, B. (1997). Managing time in the data warehouse. *InfoDB, 11*(1), 7-12.

Eder, J., Koncilia, C., & Morzy, T. (2002). The COMET metamodel for temporal data warehouses. *Proceedings International Conference on Advanced Information Systems Engineering* (pp. 83-99), Toronto, Canada.

Golfarelli, M., Lechtenbörger, J., Rizzi, S., & Vossen, G. (2006). Schema versioning in data warehouses: enabling cross-version querying via schema augmentation. *Data and Knowledge Engineering, 59*(2), 435-459.

Jensen, C., Clifford, J., Elmasri, R., Gadia, S. K., Hayes, P. J., & Jajodia, S. (1994). A consensus glossary of temporal database concepts. *ACM SIGMOD Record, 23*(1), 52-64.

Kim, J. S., & Kim, M. H. (1997). On effective data clustering in bitemporal databases. *Proceedings International Symposium on Temporal Representation and Reasoning* (pp. 54-61), Daytona Beach, US.

Kimball, R. (2000). Backward in time. *Intelligent Enterprise Magazine, 3*(15).

Kimball, R. (1996). The data warehouse toolkit. *Wiley Computer Publishing.*

Lenz, H. J., & Shoshani, A. (1997). Summarizability in OLAP and statistical databases. *Proceedings Statistical and Scientific Database Management Conference* (pp. 132-143), Olympia, US.

Letz, C., Henn, E., & Vossen, G. (2002). Consistency in data warehouse dimensions. *Proceedings International Database Engineering and Application Symposium* (pp. 224-232), Edmonton, Canada.

Pedersen, T. B., & Jensen, C. (1998). Research issues in clinical data warehousing. *Proceedings Statistical and Scientific Database Management Conference* (pp. 43-52), Capri, Italy.

Quix, C. (1999). Repository support for data warehouse evolution. *Proceedings International Workshop on Design and Management of Data Warehouses*, Heidelberg, Germany.

Yang, J. (2001). Temporal data warehousing. PhD thesis, Stanford University, UK.

ENDNOTE

1 Importantly, as made clear in the sixth and seventh sections, the value actually stored within a registration is not necessarily the value of m_{ij}, depending on the specific design solution adopted.

This work was previously published in International Journal of Data Warehousing and Mining, Vol. 3, Issue 4, edited by David Taniar, pp.51-67, copyright 2007 by IGI Publishing, formerly known as Idea Group Publishing (an imprint of IGI Global).

Chapter 2.17
Toward a Grid-Based Zero-Latency Data Warehousing Implementation for Continuous Data Streams Processing

Tho Manh Nguyen
Vienna University of Technology, Austria

Peter Brezany
University of Vienna, Austria

A. Min Tjoa
Vienna University of Technology, Austria

Edgar Weippl
Vienna University of Technology, Austria

ABSTRACT

Continuous data streams are information sources in which data arrives in high volume in unpredictable rapid bursts. Processing data streams is a challenging task due to (1) the problem of random access to fast and large data streams using present storage technologies and (2) the exact answers from data streams often being too expensive. A framework of building a Grid-based Zero-Latency Data Stream Warehouse (GZLDSWH) to overcome the resource limitation issues in data stream processing without using approximation approaches is specified. The GZLDSWH is built upon a set of Open Grid Service Infrastructure (OGSI)-based services and Globus Toolkit 3 (GT3) with the capability of capturing and storing continuous data streams, performing analytical processing, and reacting autonomously in near real time to some kinds of events based on a well-established knowledge base. The requirements of a GZLDSWH, its Grid-based conceptual architecture, and the operations of its service are described in this paper. Furthermore, several challenges and issues in building a GZLDSWH, such as the Dynamic Collaboration Model between the Grid services, the Analytical Model, and the Design and Evaluation aspects of the Knowledge Base Rules are discussed and investigated.

INTRODUCTION

We are entering a new area of computing in today's complex world of computational power, very high speed machine processing capabilities, complex data storage methods, next generation telecommunications, new generation operating systems and services, and extremely advanced network services capabilities. At the same time, the number of emerging applications that handle various continuous data streams (Babcock et al., 2002; Chandrasekaran & Franklin, 2002; Lerner & Shasha, 2003; Stonebraker et al., 2003; Widom et al., 2003), such as sensor networks, networking flow analysis, telecommunication fraud detection, e-business, and stock market online analysis is growing.

It is demanding to conduct advanced analysis over fast and huge data streams to capture the trends, patterns, and exceptions. However, fully extracting the latent knowledge inherent within the huge data is still a challenging effort because of the existing insufficient technology. Data streams arrive in high volume and unpredictable rapid bursts and need to be processed continuously.

Processing data streams due to the lack of resources is challenging in the following two respects. On the one hand, random access to fast and large data streams is still impossible in the near future. On the other hand, the exact answers from data streams are often too expensive. Therefore, the approximate query results (Babcock et al., 2002; Dobra et al., 2002; Guha & Koudas, 2002; Kim & Park, 2005; Tucker, Maier, & Sheard, 2003; Widom et al., 2003) are still acceptable, because there is no existing computing capacity powerful enough to produce exact analytical result on continuous data streams.

The significant increased data volume of information manipulated in several domains has affected Data Warehousing (DWH) and Business Intelligence (BI) applications. Data Warehousing and Business Intelligence applications normally are used for strategic planning and decision making. However, existing DWH technologies and tools (e.g., ETL, OLAP) often rely on the assumption that data in the DWH can lag for a tolerable time span (e.g., a few hours) behind the actual operational data, and the decisions are based upon the analytical process on that "window on the past." For many business situations, especially in data stream analysis, this decision-making approach is too slow due to the fast pace of today's business. Today's decisions in the real world thus need more real-time characteristics, and consequently, DWH, BI, ETL tools, and OLAP systems are quickly beginning to incorporate real-time data. A new ETL approach using widely accepted Web technologies recently has been announced (Schlesinger et al., 2005).

Starting from the concept of a Zero-Latency Data Warehouse (ZLDWH) (Bruckner, 2002; Tho & Tjoa, 2003), we extend the system to tackle continuous data streams with the capability of capturing data streams, performing analytical processes, and reacting automatically to some kinds of events based on well-established knowledge.

We do not follow the approximation approach; instead, we capture and store all data streams continuously while performing the analytical processing. Obviously, such systems require a very high computing capacity that is capable of huge storage and computing resources. Fortunately, in the last few years, we have witnessed the emergence of Grid Computing (Foster, Kesselman, & Tuecke, 2001; Joseph & Fellenstein, 2003) as an important new technology accepted by a remarkable number of scientific and engineering fields and by many commercial and industrial enterprises. Grid Computing provides highly scalable, secure, and extremely high performance mechanisms for discovering and negotiating access to remote computing resources in a seamless manner. With the Grid as a flexible, secure, coordinated resource sharing among dynamic collections of individuals, institutions, and resources, our requirements seem to be satisfied. Furthermore, due to unpredictable characteristics of data streams, grid technology is more convincing due to its flexibility.

This paper describes our ongoing work in developing the GZLDSWH built upon a set of OGSA-based Grid services and the GT3 toolkit. The GZLDSWH is composed of several specific grid services for capturing, storing, and performing analysis on continuous data streams and issuing relevant actions or notifications.

This paper is organized as follows. First, we mention other research work related to the paper. Next, the GZLDSWH system overview will be described. We then discuss several approaches in building the GZLDSWH and specify its conceptual architecture. The next section describes the operation of GZLDSWH in conducting analysis processes and reaction on continuous stream. Thereafter, we focus on the main contributions of our paper:

- A model that describes the required dynamic collaboration of the Grid services.
- A grid-based OLAP cube management service.
- The Knowledge Base rules and rule evaluation process.
- A guideline on how to increase performance when nodes fail and need to roll back to previous checkpoints.
- Our ongoing prototype implementation will be described briefly afterwards. Finally, we give our conclusion and mention the future work.

RELATED WORK

ZLDWH requires several extended features compared to the traditional DWH (Bruckner, 2002). First, the traditional batch snapshot approach to extract source data must be replaced by processes that continuously monitor source systems and capture data modifications as they occur and then load those changes into a DWH. Second, not only the continuous data integration but also the real-time automatic analysis engine is necessary to make the DWH more active.

There are several approaches from both academia and the industrial community for realizing zero-latency information systems. Compaq's Zero Latency Enterprise (ZLE) framework (Compaq Corp., 2002) is centered on an operational data store (ODS) as the primary data repository instead of a DWH. NCR's Teradata division has developed the concept of the Active Data Warehouse (Teradata Corp., 2002), which essentially marries the operational data store (ODS) and DWH concepts.

Active rules (Bailey, Poulovassilis, & Wood, 2003; Bertino et al., 2000; Cho et al. 2003) have been widely accepted to achieve the goal of auto-decision making. Research in active databases (ADBs) (Paton & Diaz, 1999) extends the power of active rules to react to events and conditions in the database. Active Data Warehouses (ADWHs) (Thalhammer & Schrefl, 2002) are systems that use event-condition-action rules (ECA) or other event-driven mechanisms in order to carry out routine decision tasks automatically within a DWH environment.

Thalhammer adopted ECA rules to mimic the work of an analyst, so he calls them *analysis rules*. His approach combines ADBs, DWHs, and OLAP to automate decision processes for which well-established decision criteria exist. However, the data integration process is based on traditional batch loading and concerns that the data warehouse state remains constant during a cycle.

Qtool (Bruckner, 2002), Bruckner's solution for zero-latency data warehousing, is based on the continuous data integration using Microsoft Message Queuing (MSMQ) and Teradata Tpump utilities. Although data are loaded and integrated continuously, QTool does not deal with other problems, such as feeding data from heterogeneous data sources, detecting data changes, active rule modeling, or implementing the active decision engine. Obviously, these tools are not designed to support continuous data streams processing.

Data streams have some specific characteristics that make them different from traditional data. They could be infinite, and once a data element has arrived, it should be processed and either archived or deleted (i.e., only a short history can be stored). It is also preferable to process data elements in the order in which they arrive, because sorting even substreams of a limited size is also a blocking operation. So far, research results have been reported for modeling and handling data streams, including algorithms for data stream processing to full-fledged data stream systems.

In continuous query processing, several approximation methods are used for data reduction and synopsis construction, such as sketches (Dobra et al., 2002), random sampling, histograms (Muthukrishnan & Strauss, 2003), and wavelets (Chakrabarti et al., 2001). Some other approximation methods are applied to tackle the blocking operator, such as sliding window (Chandrasekaran & Franklin, 2002), load shedding (Babcock, Datar, & Motwani, 2004), punctuation (Tucker, Maier, & Sheard, 2003). K-Constraints (Babcock & Olston, 2003) are used in clustering and monitoring data streams. Kim and Park (2005) propose an efficient periodic summarization method with a flexible storage allocation to store large volumes of streaming data in stable storage. Other research topics cover data stream management system models, architectures, and related issues such as memory minimization, operator scheduling, query optimization, multiple query, distributed query processing, and so forth (Babcock et al., 2002; Widom et al., 2003).

As another approach, conventional OLAP and data mining models have been extended to deal with data streams, such as multi-dimensional analysis (Han et al., 2002), clustering (Motvani et al., 2000), and classification (Hulten, Spencer, & Domingos, 2001). However, most of the previous approaches on data stream processing focus on approximation methods based on statistical estimations due to the limited storage and computing resources. Our effort, instead, tries to store all data streams and to process them within the grids as if they are stored in one extremely large distributed database.

In recent years, grid computing (Foster & Grossman, 2003; Foster, Kesselman, & Tuecke, 2001; Joseph & Fellenstein, 2003) is emerging as the best solution to the problems posed by the massive computational and data handling requirements. Starting from the concept of linking super computers to benefit from the massive parallelism for computation needs, Grid's focus recently has shifted to more data-intensive applications, where significant processing is conducted on very large amounts of data.

New-generation grid technologies are evolving toward an Open Grid Services Architecture (OGSA) (OGSA, 2003) in which a grid provides an extensible set of services that virtual organizations can aggregate in various ways. The development of the OGSA technical specification is in progress within the Global Grid Forum, covered by the tasks called the Open Grid Services Infrastructure (OGSI) with Globus Toolkit 3 (GT3) (Sotomayor, 2004). The Database Access and Integration Services (DAIS) Group developed a specification for a collection of OGSI-compliant Grid database services. OGSA-DAI Release 3 (GGSA-DAI, 2003), the first reference implementation of the service interface, is already available. So far, only little attention was devoted to knowledge discovery on the Grid. An attempt to design the architecture for performing data mining on the Grid was presented in Cannataro and Talia (2003). The authors present the design of a Knowledge Grid architecture, based on the non-OGSA-based version of the Globus Toolkit, and do not consider any concrete application domain. Moore (2001) presents the concepts of Knowledge-Based Grids, and Roure, Jennings, and Shadbolt1 (2003) explored the Semantic Grid toward a knowledge-centric and metadata-driven computing paradigm. The WP4 of the OGSA-DAI project addresses the design of a distributed query processing service for the grid. Recently, GridMiner (Brezany et al., 2003; Kickinger et al., 2003) has been reported as an evolution of parallel and distributed data mining technology and Grid Database Services development.

Workflow, the coordinated execution of multiple tasks or activities (Marinescu, 2002), can be extended and applied virtually to other areas, from science and engineering to entertainment. Web services already provide mechanisms to handle complex workflows. Since every grid service is a Web service with improved characteristics and services (Sotomayor, 2004) (the converse of this statement is not true), it is possible to adapt the ideas for workflow compositions from Web services and apply them to grid services.

BPEL4WS 1.1 (IBM, 2003) is the actual standard, which describes compositions of Web services. The Grid Services Flow Language (Krishnan, Wagstrom, & Laszewski. 2004) intends to do the same for Grid Services. GSFL is based on the so-called Web Services Flow Language (Leymann, 2003), a predecessor of BPEL4WS, published by IBM. All of these flow language specifications have the same target: describing a business process built up of various Web services. This description then serves as input for a workflow engine like BPWS4J (IBM, 2003) (an engine for BPEL4WS developed by IBM). Such an engine works with the persistent Web services and requires the specification documents of physical Web service URIs. However, in the ZLGDSWH system, the services are transient Grid services, which will be created on demand. Several workflow solutions, such as Triana (Taylor, 2003), Pegasus (Kesselman et al., 2004), GridFlow (Cao et al., 2003), GridPhyN (Deelman et al., 2003), McrunJob (Graham, Evans, & Bertram, 2003) are used to create and manage the Grid computational workflow. However, the above projects did not take into account the automatic collaboration between the Grids services. Because of its automated event-based reaction feature, GZLDSWH requires the higher level of automation in service creation, discovery, invocation, and destruction according to the Grid environments.

THE GZLDSWH OVERVIEW

Starting with the idea of building a Zero-Latency analytical environment dealing with heterogeneous data sources, we extend the system to conduct analysis on continuous data streams. A ZLDWH (Bruckner, 2002; Tho & Tjoa, 2003) aims to significantly decrease the time to react to business events, allowing the organizations to deliver relevant information as fast as possible to applications that need a near real-time action to new information captured by an organization's information system. It enables analysis across corporate data sources while still continuously updating the newly arriving data and notifies the handling of actionable recommendations, alerts, or notifications. In Data Stream applications, events take the form of continuous data streams.

The exact analysis results on these data stream events are very expensive, because they require high computing capacity, which is capable of huge storage and computing resources. Therefore, a Grid-based approach is applied in ZLDWH to tackle the lack of resources for continuous data stream processing. Figure 1 depicts significant phases throughout the overall process of GZLDSWH. Continuous data streams will be captured, cleaned, and stored within the Grids. Whenever the analytical processes need to be executed, immediately after the arrival of new data or based on predefined timely scheduling, the virtual Data Warehouse will be built on the fly (Foster & Grossman, 2003) from data sources stored in the Grid nodes. Obviously, this approach is not concerned with traditional incremental updating issues in Data Warehouse because the virtual DWH is built from scratch using the most current data. Analytical processes then will be executed

Figure 1. The overall process of the grid-enabled zero-latency data warehouse system

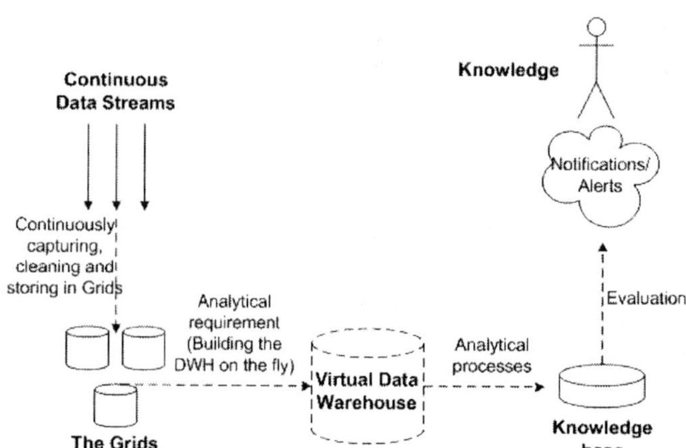

on such virtual DWH, and the results will be evaluated with the use of the Knowledge Base. Finally, dependent on the specification of the Knowledge Base, the system sends notifications, alerts, or recommendations to the users.

There is a significant number of applications in which the conducting of analytical processing on continuous data streams is necessary for detecting trends and abnormal activities. The following scenario describes an example in Mobile Phone Fraud Detection, although the usage of such a system also could be applied in other time-critical decision support applications.

Expert users defined rule specifications for fraud detection (e.g., if an international mobile call lasts more than one hour, it will be considered a fraudulent call). The rule also could be more complex, such as, for example, an international mobile call from Austria to Vietnam of a certain customer that lasts more than 30 minutes will not be considered a fraud if its duration is not more than 1.5 times the customer's average call duration from Europe to Asia within the last three months; otherwise, it is considered a fraudulent call. These rules are stored in the Knowledge Base and are referenced by the Rule Evaluation module before it makes the final decisions. Expert users also could specify how and when the system operates to detect the fraud situation by submitting the pre-defined plan workflow. In this plan, the experts specify the order of module executions and the time point when they should be executed. The whole system operation will then be monitored and controlled by the Workflow Control module, which follows the pre-defined plan.

When the end user makes a phone call, the Call Detail Records (CDRs) are issued continuously as continuous data streams. Because of the special characteristics of continuous data streams, these CDRs must be captured and stored in a timely fashion. The data storage is heterogeneous and geographically distributed in several Grid nodes.

For supporting analytical processing, the virtual DWH is built on the fly from the heterogeneous, distributed data sources as follows. The OLAP server accesses the raw data items from multiple Grid nodes and creates the pre-aggregated data cubes. The Data Mediator allows the OLAP server to access the distributed, heterogeneous data sources as if they were local data sources. In some situations, Data Preprocessing could be necessary to clean the data, standardize the data, or transform data into the common format before storing them into the OLAP cubes.

When the virtual DWH data are available, the OLAP server accepts the queries from Data Analysis or Data Mining tools, executes these

Figure 2. Mobile phone fraud detection

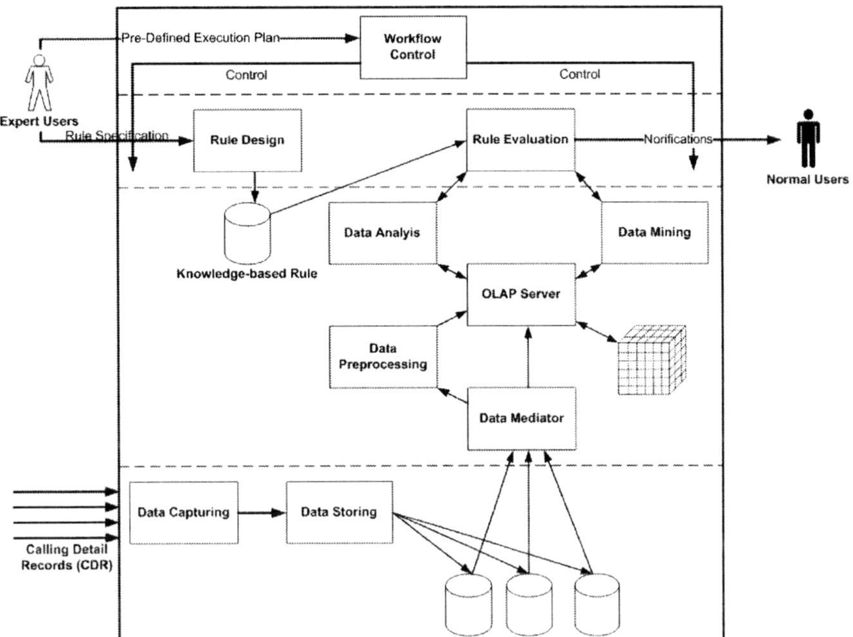

queries and returns the results. To detect a fraud situation, it is necessary to analyze the CDRs at multiple levels with different dimensions and in a variety of time ranges (e.g., to calculate the average international mobile call length from Europe to Asia within the last three months of a certain customer). The OLAP server, therefore, accepts analytical commands such as *drill up, drill down, slide and dice* and performs these operations on the data cubes.

The results are returned to the Rule Evaluation module, which will access the Knowledge Base to evaluate the rules. If some pre-defined criteria in the Knowledge Base are satisfied, the Rule Evaluation module performs suitable actions; for instance, sending of notifications to the users or stopping the telephone service. Particularly, if the rule criteria led to an ambiguous situation, the Rule Evaluation module would issue other analytical queries to further investigate the data.

The previous scenario highlights the following requirements for a GZLDSWH system:

- **Knowledge-Base Rule Support.** The final decisions (e.g., whether a phone call is fraudulent or not) is based on the set of Knowledge-Base rules specified by expert users. The Knowledge-Base Rules preserve the experiences and knowledge to drive the decision support process. It is necessary to develop a component that allows expert users to easily manipulate the existing rules in the Knowledge Base (i.e., insert, delete, or update rules). The autonomous validation and consistency checking of these rules also should be considered. The facilities of auto-inference and auto-learning are also the challenges of knowledge-rule management.
- **Multi-Level Analysis Support.** In order to evaluate the rules, it is necessary to conduct an analytical process on the multi-dimensional historical transaction of customer CDRs in order to identify the customer's pattern. Besides, in some situations, there is an uncertainty or ambiguity in evaluating

the rules during online analysis process. Then, it is necessary to conduct analytical process on the whole CDR at multiple levels in order to define the final decision. CDR streams thus need to be stored without loss within the Grid. DWH repositories and online analytical processing (OLAP) cubes are built on the fly from these Grid nodes' data. The significant parts are (1) the creation and maintenance the OLAP Cube, (2) the OLAP query engine that executes analytical queries on OLAP data, and, optionally, (3) the OLAP Data Mining Engine for execution of the online analytical mining (OLAM) algorithms.

- **Automated Reaction Support.** Whenever the fraud is detected via CDR analysis and rule-based evaluation, the system must have the ability to issue automatically the relevant actions respective to the fraud prevention methods. That could be the alarm or recommendation message sent to the mobile phone customers, or it also could be the service disconnection to prevent more fraudulent activities. However, this automatic reaction must follow the rules stored in the pre-defined Knowledge Base.
- **Distributed, Heterogeneous Data Acquisition.** The ability to use different kinds of data sources and data warehouse repositories is mandatory for Data Grid systems. Therefore, the Data Mediator layer usually is used to access transparently the heterogeneous Grid-based data sources.
- **Transparent User Interface:** The system should be able to provide the users with an interface that is easy and transparent to the Grid, Network, and Location. Customers can access the system in order to register the services or to send feedback from anywhere, using a variety of devices (mobile, PDA, laptop, etc.) without considering the complexity of system architecture. Expert users use the designed interface to specify the rules without considering the Grid structure, network feature, and physical details or location of data sources.
- **Flexibility and Open Architecture.** The essence of Grid is heterogeneity. The GZLDSWH thus should allow the integration of components that are heterogeneous (i.e., written in different programming languages or optimized for different platforms). Technical and contextual changes to the underlying data sources, interface implementations, libraries, and so forth must not affect the module operability. The components and features shall be easily extensible, allowing for plug-ins to be executed. Furthermore, the architecture must be open to integrate specialized third-party toolkits.
- **Grid Environment Information and Monitoring.** The Grid environment always varies significantly during the runtime. Therefore, the system should be able to invoke and execute its component dynamically, depending on the runtime environment. For this purpose, the Grid environment information should be monitored by a special component. It maintains the knowledge about the resources available on the Grid, their capacity, and current utilization. It should be easily queryable, highly scalable, and quickly reactable, even under a heavy load.
- **Dynamic Component Invocation and Execution.** As described in the scenario, the component execution flows are pre-defined by expert users. However, the Grid environment always varies significantly during the runtime. Therefore, the system should be able to invoke and execute its components dynamically, depending on the runtime environment.
- **Integrity and Availability.** Assuming that all Grid nodes are trusted entities, there still is a need to implement integrity checks and address non-availability. If nodes cannot be

trusted, data integrity checks are required. Kamvar et al. (2003) provide an excellent overview of requirements and also show how these issues are addressed in P2P networks. Their solutions also can be adapted to Grid computing, if one is prepared to accept the computational overhead.

There are a lot of other requirements, such as security, high availability, scalability, and performance, but they will be planned for future work.

FRAMEWORK FOR BUILDING A GZLDSWH

Based on the requirements previously discussed, there are three approaches for building the GZLD-SWH system.

Building the Whole System from Scratch

Following this way, we have to develop not only the components of the system but also deal with many other issues in Grid Computing, such as protocol for communication, message passing mechanism, resource management, scheduling, life cycle management, and so forth. The only advantage is that we do not have to obey any specifications and freely apply any container technology, such as J2EE Container or .Net Container, to control and manage our components. However, a lot of work must be done, and the system then cannot be integrated easily with the Grid Community's standard toolkits and specifications. We, thus, did not choose this approach to develop our system.

Open Grid Service Architecture (OGSA)

The second option is to build the system using the Open Grid Service Architecture (OGSA) (OGSA, 2003), which is based on the Open Grid Service Infrastructure specification (OGSI) and Globus Toolkit version 3 (GT3) (2003). During the past years, the Service Oriented Architecture (SOA) gained popularity as a new software engineering paradigm. It arose from the necessity of creating components providing clearly defined small pieces of functionality that later on can be assembled into complex (usually distributed) applications. The Web Services Model follows the SOA and allows applications to communicate using agreed, widespread standards and protocols independent of their implementation and platform. The Open Grid Service Architecture (OGSA) represents the convergence of Web service and Grid computing technologies with the aim of describing the next generation of Grid Architecture in which the components are exchangeable on different layers. Consequently, in OGSA, all kinds of storage and computational resources, components, databases, file systems, and so forth are exposed as services. This approach has the big advantage that the upper-layer components have to be concerned only with a small amount of interfaces, because the implementation is hidden behind the interface.

As mentioned previously, the Open Grid Service Infrastructure (OGSI) defines specifications with many interesting features, such as factory service discovery, instance creation, invocation, lifetime management, notification, and manageability. The Globus Toolkit 3 (GT3) implements most of these OGSI specifications and provides us with the infrastructure components for resource management, monitoring and discovery, security, information services, data and file management, communication, and fault detection. Some features that are still missing are a resource broker, load balancing and scheduling functionality. However, if we develop the system on top of the OGSI and GT3 toolkit, some of our previously mentioned requirements are satisfied, such as Flexibility, Open Architecture, Grid-Network-Location Transparency, and Grid Environment Information. Hence, we chose this approach to develop the GZLDSWH system.

WS-Resource Framework

Around the same time when the OGSI work had been progressing, the Web services architecture evolved, as well; for example, the definition of WSDL 2.0 and the release of new draft specifications, such as WS-Addressing (Globus Alliance, IBM, & HP, 2004). WS-Resource Framework can be viewed as a straightforward refactoring of the concepts and interfaces developed in the OGSI V1.0 specification in a manner that exploits recent developments in Web services architecture. The difference is that WSRF uses different constructs for modeling the stateful resources and the stateless Web services, while OGSI uses the same construct (the service instance). The WS-Resource Framework is still new, and the Globus Toolkit version 4, which integrates this framework, was expected to release at the end of 2004. Therefore, in the future, the GZLDSWH system developed upon OGSI/GT3 should be based on the WS-Resource Framework.

The Grid-Based Conceptual Architecture of GZLDSWH

Each phase of the process in Figure 1 includes several tasks, such as capturing, storing data stream, building OLAP cube, conducting multi-dimensional analysis, and so forth. Sharing the same approach with GridMiner project (Tjoa et al., 2003), each particular task will be realized as a Grid service. The GZLDSWH, thus, is composed of several specific Grid services for capturing, storing, performing analysis on continuous data streams, and issuing relevant actions or notifications reflecting the trends or patterns of the data streams. As we have discussed, these Grid services are built on top of OGSI and the GT3 toolkit and can be grouped into several layers, based on their functionality as described in Figure 3.

The Fabric and Core Layer. The services in this layer are provided by the Globus Toolkit 3, which enables most of the basic operations and communication between the Grid Services. We do not need to develop these fabric services and Grid core services (see the white box in Figure 3).

The Facilities Layer. This layer provides some services for monitoring the Grid environment, scheduling, load-balancing, and so forth. It also provides the transparency of resources, network, and location to the heterogeneous data sources. The services in this layer include the following:

- **System Information Service (SIS).** The SIS is a vital service within every Grid in-

Figure 3. The service components of GZLDSWH

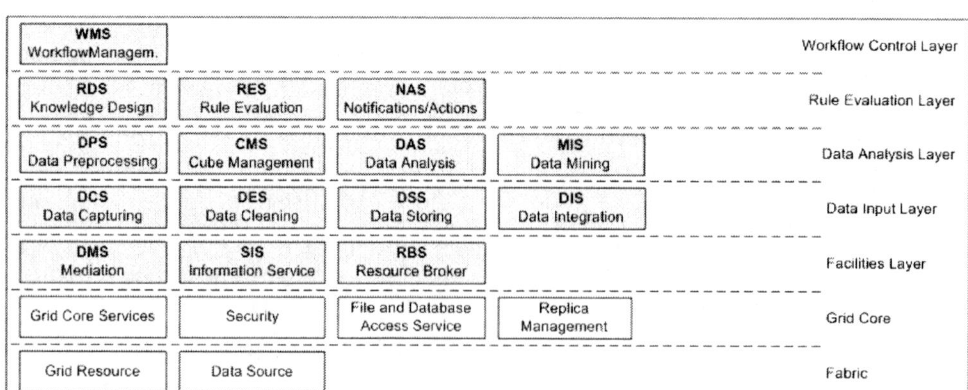

frastructure, providing static and dynamic information on available Grid resources. It is a specialized implementation enabling specific decision-making and monitoring processes.

- **Resource Broker Service (RBS).** The Resource Broker is used to find best-fitting resources for resource allocation, such as matchmaking of the requests and resources. In our system, the Resource Broker is used as a reference for the Workflow Engine in dynamic discovery, creation, binding, and invocation of other available service instances in the runtime.
- **Data Mediation Service (DMS).** The DMS provides a single virtual data source having the same client interfaces as classical Grid data sources, but it integrates data from multiple heterogeneous federated sources from several Grid nodes. This service simplifies access requests and implements transparency for heterogeneous data sources.

The Data Input Layer. This layer provides the services for capturing, cleaning, and storing data within the Grid. The following services belong to this layer:

- **Data Capturing Service (DCS).** The Data Capturing Service works closely with the stream sources, such as sensors, cameras, satellites, and so forth, for capturing data streams in the limited time without data loss.
- **Data Cleaning Service (DES).** In some special situations, the Stream Data should be cleaned prior to being stored in the Grid. Therefore, the Data Cleaning Service is an optional service.
- **Data Storing Service (DSS).** The Data Storing Service resides in each Grid node and is responsible for storing the lossless data streams. This service also could convert the data into an appropriate format for the local data source.
- **Data Integration Service (DIS).** This service is responsible for secure, reliable, efficient management and operation of the necessary data transfers within the Grid environments.

The Data Analysis Layer. The aim of the Data Analysis layer is to support the analytical process on data streams stored somewhere within the Grid. This layer contains the following services:

- **Data Preprocessing Service (DPS).** The Data Preprocessing Service performs several pre-processing activities, such as data cleaning, normalization, selection, reduction, transformation, and so forth, before storing the data into the OLAP cube.
- **OLAP Cube Management Service (CMS).** The OLAP Cube Management Service is one of the major components of the system. This service creates distributed OLAP cubes from several data sources stored at specified Grid nodes. After the initial cube creation, the service can be used for cube interaction and life cycle management.
- **Data Analysis Service (DAS).** The Data Analysis Service is another major component of the system. It works very closely with the OLAP Cube Management Service and performs the analytical process by sending queries and commands such as *drill up*, *drill down*, or *slide and dice*. It thus allows analyses of datasets at different levels of abstraction. The output of the Data Analysis Service is the analysis result, which then will be evaluated by the Knowledge Base for further actions.
- **Data Mining Service (MIS).** The Data Mining Service is created as an extensible framework, providing necessary data mining algorithms and making it convenient for the related application developers to easily plug in their algorithms and tools.

The Rule Evaluation Layer. This layer contains services supporting the rule specification and rule evaluation. The services in this layer provide the interface so that the expert users can interact with the system in order to specify the rules. In addition, normal users can receive notifications, alarms, and recommendations from the system. The following services are available in this layer:

- **Knowledge Base Rule Design Service (RDS).** This service allows expert users to specify the rules stored in the Knowledge Base of the system. The Knowledge Base Rules embed ECA rules, consisting of an event, condition, and action part. However, it carries out complex OLAP analyses on DWH data instead of evaluating simple conditions, as compared to ECA rules in the OLTP system.
- **Rule Evaluation Service (RES).** The RES service takes the analytical results from the DAS, evaluates these results against the Knowledge Base rules, and finally takes suitable actions. It could invoke the NAS to perform suitable actions, if the rule criteria are satisfied. It also could invoke the DAS service or the Data Mining service (MIS) to perform further analytical processes, when the rule criteria are too ambiguous to make the final decision.
- **Notification/Action Service (NAS).** The NAS service manages all possible actions of the GZLDSWH system, such as issuing notifications, alarms, and recommendations to the users to inform them about abnormal trends. In addition, it invokes the Analysis service when receiving the local update message.

The Workflow Control Layer. The Workflow Control Layer contains the Workflow Management Service that controls the dynamic service invocations, executions, and destructions.

- **Workflow Management Service (WMS).** This service allows expert users to specify the logical workflow of system activities and to execute complex and highly dynamic workflows for several heterogeneous Grid services. The WMS service dynamically controls service execution, service termination and communication, and so forth, depending on the Grid environment. Because the Grid environment significantly varies over time, the Dynamic Workflow Control Service is an extremely important component in the system.

THE OPERATION OF GZLDSWH

Within the Grids environment, as described in Figure 4, there are one Master node and several child nodes (Node 1, 2..., Node N). The Master node controls other child nodes to fulfill system activities. The role of these child nodes is to store data within the Grid environment. The Master node, therefore, includes most of the essential services, while the child nodes only contain some data input services and local data update detection services. The Master node also keeps the Grid metadata for Grids management and the Knowledge Base rules for controlling event reaction behavior.

The operation plan of the services in GZLDSWH is specified in the logical workflow, as described in Figure 4, in which the services are arranged in the following logical order. The Data Capturing Service (DCS) receives continuous data streams from stream sources, such as sensor systems, satellites, and so forth. Due to the huge amount of incoming data, the DCS must capture the data timely and invoke available Data Storing Services (DSS) residing at several child nodes for storing data. The DCS also could invoke Data Cleaning Service (DES) in order to clean the data before storing, if necessary.

After storing data at child nodes, the Analysis Service (DAS) at the Master node will be invoked immediately or after a predefined timely schedule, depending on application requirements and performance trade-off. DAS execution will create the virtual Data Warehouse from scratch. For this purpose initially, the DASs available at several local child nodes are invoked. Due to this, the Cube Management Service (CMS) gains the essential raw data at the child nodes in order to build the global OLAP cube. Each child node contains part of the cube; namely, cube chunk. Data then will be integrated into the common format by the Data Integration Service (DIS). Before being stored into the virtual Data Warehouse, data can be passed to the pre-processing phase via the Data Pre-processing Service (DPS). The DPS can perform several tasks, such as data cleaning, data transformation, data normalization, or data reduction. After the global cube is formed, the DAS will perform analysis queries or data mining algorithms (via the equivalent Mining Service [MIS]), based on the data inside the virtual DWH. The analysis results then will be sent to the Rule Evaluation Service (RES).

The RES accesses the Knowledge Base and evaluates the rules. The Knowledge Base rules are provided by the user through the Rule Design Service (RDS). Depending on the rule criteria, the RES could invoke the DAS or MIS for further analysis before issuing the final decision or invoking the Notification-Action Service (NAS) to issue relevant notifications, recommendations, alerts, and so forth to the users. The NAS also could send back other action commands to several Grid child nodes in order to execute several data manipulation operations at the local data sources, such as insert, delete, update, and so forth. Furthermore, the previous analysis process could be executed to answer the analysis queries issued by other applications. Especially if the local data update takes place at the Grid child nodes, the analysis process also could be invoked. Whenever the local data update happens, the NAS at local child node sends the local data updates message events to the NAS of the Master node. The NAS then invokes the DAS, and the Analysis process will be executed.

The invocation between the services previously described is, in fact, more complicated in the dynamic Grid environment because of the computational and networking capabilities and availability of the Grid nodes. Therefore, the Grid services invocation process is monitored strictly by the Resource Broker Service (RRS) and the System Information Service (SIS). These services manage the available resource and find the best fitting resources for resource allocation and dispatch. The role of Workflow Management Service (WMS) is to execute the complex, highly dynamic workflows involving different Grid service instances.

DYNAMIC GRID SERVICE COLLABORATION

As previously mentioned, each OGSI-based service in GZLDSWH is able to perform an individual task within the whole process. Obviously, these services have to collaborate with each other to fulfill the common purpose of the GZLDSWH system.

The Dynamic Service Control Engine (DSCE) has been developed in GridMiner (Kickinger et al., 2003; Kickinger & Brezany, 2004) to control the service execution via the Dynamic Service Control Language (DSCL) document. The exact service handles are specified in the DSCL. Users have to know which service factory to use to perform a certain task. Therefore, the services do not need to communicate with each other. The output of the first service serves as the input of the second service; the output of the second one serves as the input of the third one, and so forth. Thus, no service is aware of other existing services, and each of the services is able to run completely independently.

Figure 4. The operations of GZLDSWH services

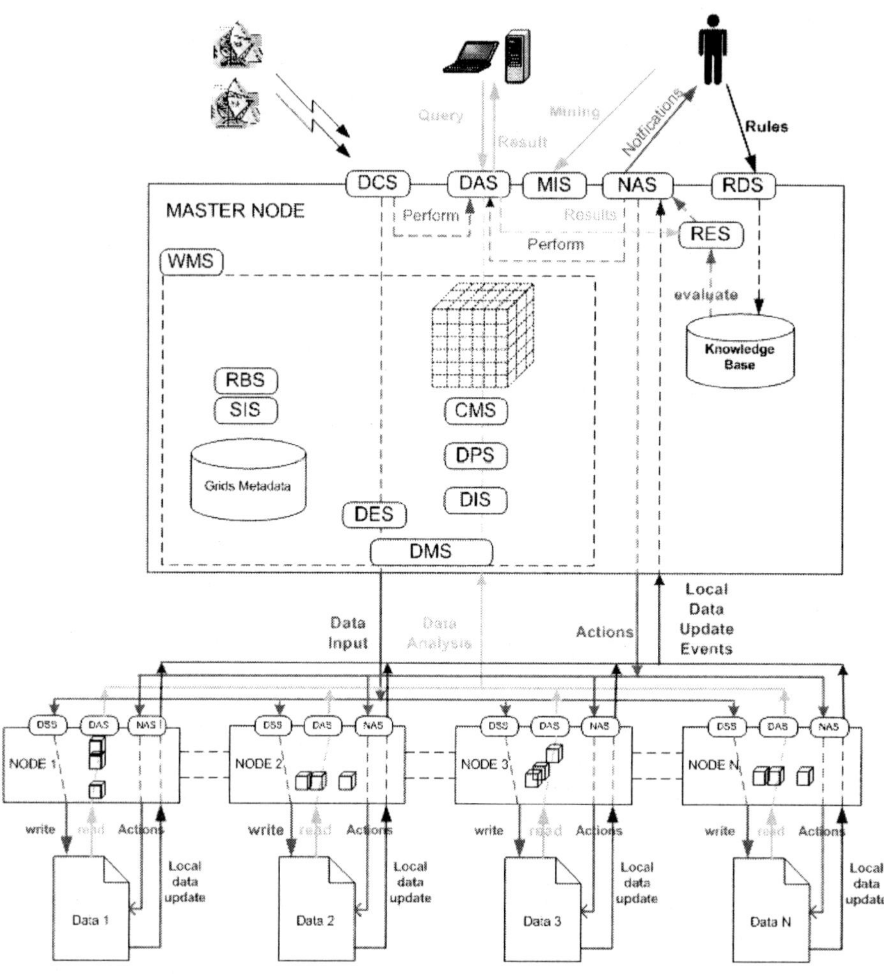

The independence of the various services also allows a parallel execution without any communication overhead. This results in an improvement of performance.

However, in our system, due to the requirement of automated event-based reaction, a service must be able to discover, create, bind, and invoke relevant service instances within the Grid environment. Only the execution flows are specified in advance, in which the services are arranged in the specified logical execution order. During the execution time, the services have their autonomy to discover, create, bind, and invoke relevant physical service instances within the Grid environment, depending on the context at that time. Consider the Telecommunication Fraud Detection scenario, for example. The Call Detail Records (CDRs) are issued continuously as continuous data streams. The Data Capturing Service (DCS) instance is created in the Master Node, and its operations should be executed while the CDRs are still arriving. The DCS instance will invoke the Data Storing Services (DSSs) at several child nodes to store data. However, according to the Grid environment at the runtime, some child nodes are available, and the others could be corrupted or out of resources. The DCS instance thus must be intelligent enough to create and invoke the

Figure 5. The predefine logical workflow of service invocation

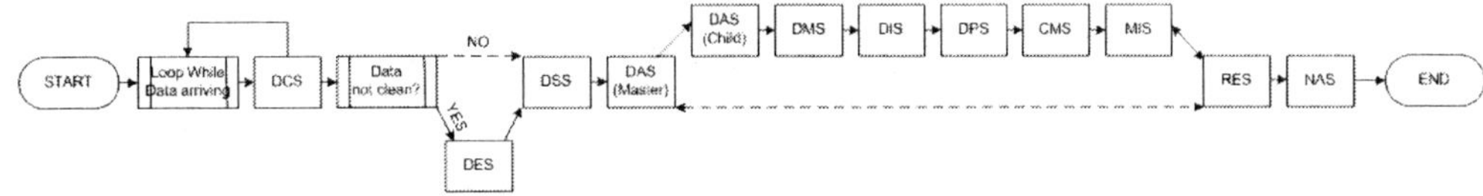

instances at the suitable Grid nodes. If the CDRs data arrives in a burst fashion, the DCS instance has to create many DSS instances in order to store all data in a timely fashion; otherwise, it should destroy some non-used DSS instances to free the resources. The similar situation happens when the DAS instance at Master node decides to create and invoke the DAS instances at the suitable child nodes. Especially if the Rule Evaluation Service (RES) cannot issue the final decision (due to ambiguity between Fraud and non-Fraud call), it has to invoke other analysis services or data mining services for further data analysis instead of invoking the Notification/Action service to alarm the users. Therefore, the dynamic service invocation requirement is extremely important in the GZLDSWH.

To our best knowledge, there are two possible approaches for the service flow execution: centralized control and distributed control. In the former approach, there is a central service control engine that controls all service executions from the start node to the end node of the workflow. The engine itself is responsible for discovering, creating, binding, invoking, and destroying service instances to follow the logical workflow. Thus, the engine must keep the information of the whole workflow and should trace the information of the Grid environment, such as Grid nodes status, resource availability, workloads, and so forth, in order to coordinate the services execution. In the latter approach, there is no such central engine, but each service instance has its own knowledge to invoke the next service instances throughout the workflow. It is not necessary for each service to keep information of the whole workflow. Instead, each service needs to keep only part of the workflow metadata relevant to itself, such as its direct ancestor and descendant services and the Grids environment context at its execution time. That information should be passed to the service as parameters as it is invoked by its ancestor services, and the service will use such information to invoke the next relevant service instances.

Both of the two approaches have advantages and disadvantages. In the centralized approach, the central service control engine, which also could be realized as a service, copes with the coordination between other services. The other services thus only focus on their specific functionality without taking into account the workflow execution. However, it could be the heavy workload for the engine service, if it processes the high complexity workflow or if the number of service instances increases. The distributed control approach, on the other hand, does not have to deal with the bottleneck issue. However, it is more complicated to develop the services, because each service, besides its specific functionality, must be realized as an agent-based solution in order to adapt to flexible service instance invocations. Moreover, the service invocation also would become more complex due to the parameters transferred between the service instances. Further investigation on distributed control approach is out of the scope of this paper. It will be one of our considerations in future work.

In GridMiner (Kickinger et al., 2003; Kickinger & Brezany, 2004), we have developed the Dynamic Service Control Engine (DSCE), which receives the workflow specification document written in DSCL (Dynamic Service Control Language) and executes the workflow by invoking the corresponding Grid service instances specified in the documents. DSCL allows the user to specify variables, workflow structure, and operations to be executed; DSCE will execute the workflow followed by the DSCL document specification. Although the DSCL and DSCE provide some levels of dynamic workflow execution, they still have some limitations. The DSCL does not support branch conditions and loop structures; the DSCE only works with the physical workflow specification document (i.e., the document specifies exactly which factory handle URIs should be invoked to create the instances, which usually are unknown in advance due to the variant Grid environment). We can improve the DSCL and DSCE in GridMiner to support the new requirements in GZLDSWH. The extended DSCL (Tho et al., 2004) will support the condition branches and loop structures as well as allow the references of the service instance handles to be transferred as parameters. The logical workflows are specified with the unknown service instance handles declared as variables. During the execution time, the Re-Writer queries the Resource Broker Service and Information Service to have the relevant dynamic service factory handle references at that time and rewrites the logical workflow to the physical one. The DSCE engine then will invoke these service instances via the reference variables in the physical workflow. That operation will be repeated at each step of the workflow until the whole process is finished.

Extended Dynamic Service Control Language

DSCL is an XML-based language allowing the users to specify the workflow of services activities. It contains exactly two sections:

- **The <variables> section.** All variables must be defined here. The variables either could be the parameters of service calls or the results of service calls. XML Schema Simple Type, Complex Type, and SOAP Arrays Type are supported as variable type.

```
<variables>
   <variable name="iAge">
         <value type="int" 25 </value>
   </variable>
</variables>
```

- The <**composition**> section contains the description of the workflow to be executed. A workflow is comprised of a set of activities that could be classified as *control flow* or *operational.* The control flow activities control the execution of the workflow and, thus, must contain other activities, while the operational activities are the atomic activities that perform operations.

```
<composition>
   control flow activities
   other control flow activities or operational
   activities
   operational activities
</compostion>
```

DSCL allows the users to specify the workflow structure and define the workflow operations.

Workflow Structure

Our DSCL supports four basic execution styles (Sequential, Parallel, Condition Branch, and Loop) by providing several tags; namely, **<Sequence>**, **<Parallel>**, **<Condition>**, and **<Loop>,** respectively. These tags could be nested to realize the complex workflow composition. Figure 6 states an example of a workflow, including all control activities and the respective DSCL document.

```
<composition>
   <sequence>
         activity1
         <condition>
                  cond_var1 = TRUE
                           <loop while cond_var2 =
   TRUE>
```

Figure 6. Composite workflow example

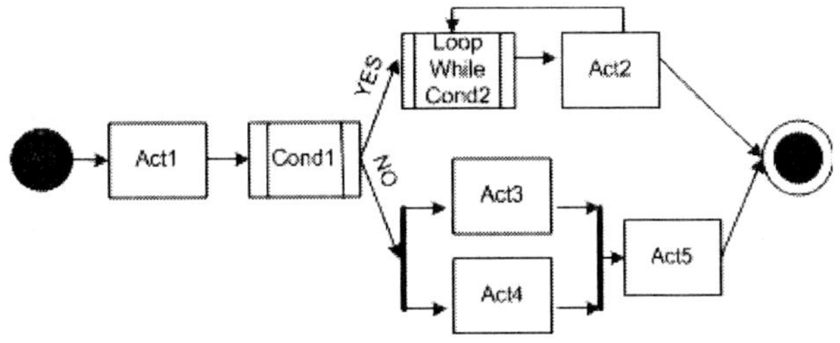


```
                activity2
            </loop>
        cond_var1 = FALSE
        <sequence>
            <parallel>
                activity3
                activity4
            </parallel>
            activity5
        </sequence>
    </condition>
  </sequence>
</compostion>
```

Workflow Operations

Besides the control flow activities, DSCL supports other activities; namely, operational activities. Operational activities perform operations for interacting with the underlying Grid services, such as creating new service instances, destroying instances, invoking operation of services, and querying service data elements. DSCL provides respective tags to specify these operational activities: <**createService**>, <**destroyService**>, <**invoke**>, and <**querySDE**>. The operational activity cannot contain other activities and must have the mandatory attribute; namely, activityID, which is of type DTD. This attribute is necessary in order for the workflow engine to identify the activity.

The major difference between Grid and common Web services is the fact that the Grid service could be either persistent or transient. The persistent service is created and available if its container is running. In contrast, the transient one is created and invoked as and when required and destroyed afterwards. The transient service always is created by its Factory service. The following information is necessary in order to create a new service instance:

1. The location of the factory service
2. Additional service parameters
3. A virtual instance name of the newly created instance

```
<createService activityID ="Act1"
factory-gsh="http://url/serviceFactory"
instance _ name="newInstance1"/>
```

In GZLDSWH, there is a situation where we have more than equivalent factory services at different Grid nodes at the same time. For example, the Data Storing services located at several child nodes when we need to store data stream. In such a situation and in other cases when the Grid environment changes rapidly, the Resource Broker Service decides which Service Factory should be executed to create a new instance according to the availability and resource capacity of the different Grid nodes. DSCL provides the dynamic service creation by allowing Grid service handle references transferred via variables. It is also possible to create the service instance with user-defined parameters via <**parameter**> tag.

```
...
<variables>
  <variable name="factgsh">
      <! Default value of the factory service
  handle>
      <valuetype="string"http:url/serviceFactory
  </value>
  </variable>
...
</variables>
...
<!other services set the value of factgsh, e.g. Resource
  Broker >
...
<! Create the service instance via reference to factory
  handle >
<createService activityID ="Act1"
factory-ref="factgsh"
instance _ name="newInstance1"/>
```

After a service instance is created, it is possible to destroy the instance, invoke its operations, or query its data elements. These activities require the service instance reference in order to identify the relevant instance. We provide three optional attributes; namely, instance-name, instance-gsh, and instance-ref, enabling the reference to a service instance via (1) instance name, (2) Grid service handle, and (3) a variable reference to the instance. The usage of each attribute is optional; however, exactly one of the three attributes must be used together with the activity.

```
...
<destroyService activityID ="Act1"
instance-gsh="http://url/SerInst01"(orinstance-name
  = "Instant1" or instance-ref = "varInstGSH")
...
```

Invoking an operation of a Grid service is similar to invoking a method in common programming language. To invoke an operation of a Grid service, the following information is necessary:

1. **The required Grid service instance.** Referenced by one of the attributes: *instance-name, instance-gsh, instance-ref*
2. **Name of the operation.** Mandatory defined within attribute *operation* and optional attribute *portType*
3. **The required parameters.** Specified in <**parameter**> tags, the parameter is simply a reference to a variable.
4. **Result.** Storing the result of a Grid invocation via <**result**> tag.

```
...
<Invoke activityID ="CleanData01"
instance-gsh="http://url/DES01" (or instance-name
  = "DES01" or instance-ref = "varInstGSH")
operation = "Clean_Data"
<parameter variable= "Data_Strore">
<result variable= "Data_Result">
</Invoke>
...
```

Some of operations do not return the results but store them into so-called service data elements. To allow querying the contents of these elements, DSCL provides the tag <**querySDE**>. This operation requires the reference to Grid service instance and the name of the required service data element (stored in attribute *sdName*).

```
...
<querySDE activityID="act1"
  instance _ name="instance01"
  sdName="value"
  <result variable="var02"/>
</querySDE>
```

Dynamic Workflow Management Service

In GridMiner project (Kickinger & Brezany, 2004), we have developed an engine service called the Dynamic Service Control Engine (DSCE), which processes DSCL documents and controls the service execution in both interactive and batch modes. It provides some interesting features, such as (1) independent processing (without any interaction of the user) of a workflow described in DSCL; (2) the provision of all intermediate results from the services involved; (3) the possibility for a user to stop, cancel, or resume a workflow; and (4) the possibility to change workflow at run time (by stopping the engine, changing the DSCL document, and restarting engine).

Figure 7 describes the Conceptual Architecture of DSCE (Kickinger & Brezany, 2004). The engine is implemented as a stateful, transient OGSI Grid service and has several structured layers. The top layer is the Interface layer, which provides essential operations to control the engine. The Factory interface allows users to create a new DSCE instance for a specific DSCL document via operation *CreateService (DSCLDocument dscl)*. The DSCE engine instance now will be created, and its state will change within its life cycle according to user interactions and the activities execution results. The possible states could be *empty, initialized, running, stopping, waiting, finishing,* or *failure*. The Service interface provides interactive control operations such as *changeWorkflow(), start(), stop(),* and *resume()*, as well as several service data elements containing information about the DSCL document, Workflow state, and activity state.

The middle DSC Engine layer covers the main functionality of DSCE. It controls the whole workflow execution by controlling the execution of activities specified by the DSCL document. First, the DSCL workflow description is parsed; then the network of activities, an internal model of the workflow, is constructed before processing the activities. Such a network of activities describes the dependency between the activities. Each activity could have succeeding and preceding activities. Succeeding activities are executed right after the execution of actual activity is finished. If an activity has more than one successor, all of them will be executed in parallel after the actual activity is finished. Similarly, the activity could not be started until all of its preceding ones are finished. This could happen in some situations like loop or parallel execution. Several internal operations are provided in this layer for managing the workflow, such as *start(), stop(), resume(), reset(), setDSCLWorkflow()*, and so forth, as well as some operations for controlling the activities, such as *startActivity(), EndActivity(), CreateInstanceActivity(), DestroyInstance Activity(), InvokeActivity(), QuerySDE-Actvity(), startNext Activites(), wait-ForPrevious-Activities()*, and so forth. The necessary parameters of all underlying services also are prepared at this layer.

Normally, when a Grid service is developed and implemented, additional stub and proxy classes are generated to hide the complexity of communication between the client and the service. This approach is very common and practically used in all distributed object systems like CORBA, Java RMI, and Web Service. To benefit from this approach, the required services or remote objects must be known at the compilation time. However, this requirement is not satisfied in

Figure 7. Conceptual architecture of DSCE

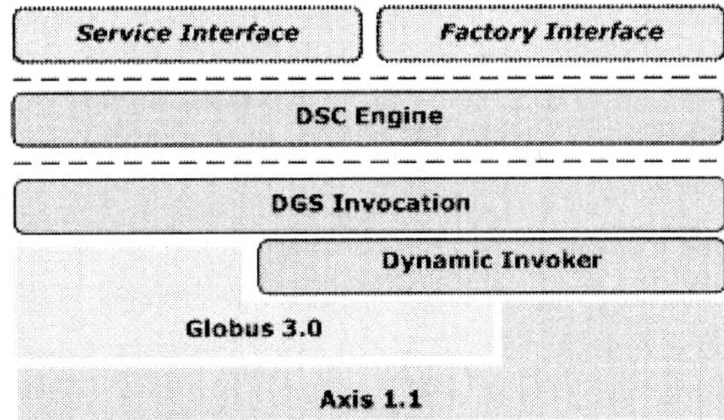

DSCE, because DSCE receives a DSCL workflow description document and will be able to communicate with all services specified within that DSCL. The lowest layer; namely, Dynamic Grid Service Invocation (DGSI), is composed of the DGS Invocation and Dynamic Invoker. It provides classes that allow accessing Grid services and their operations without using common stubs/proxy approach. The Dynamic Invoker, the lowest layer, provides the possibility to invoke any operation on any underlying Grid service. It uses much of ApacheAxis (The Axis Project, 2003), and SOAP engine, which are based on the GT3 toolkit. Dynamic Invoker translates an operation invocation into a SOAP1.1 message and sends it to the corresponding service to invoke specified operations. It provides all necessary marshalling and unmarshalling of arguments, first by fetching the WSDL of the corresponding Grid service (via its handle GSH) and then by setting service port type via *setPortType(String port-TypeName)*, setting operation via *setOperation (String operation Name)*, setting parameters of the operation via *set-Parameters (Object[] params)*. All of information is used to construct an essential SOAP operation call. Finally *invoke()* executes the operation by sending that SOAP message to corresponding services. At the higher layer, the DGS Invocation provides the classes to use stubless operation invocation and to access the functionality of the GT3. It provides three classes; namely, *DGSIService*, *DGSIFactory*, and *DGSI- Listener*, allowing the workflow engine to handle its underlying services, such as creation and destruction of Grid service instance, invocation of operations, querying of service data element, and synchronization of asynchronous service calls.

DSCE suits well in GridMiner, where the interaction role of the user is important. The engine operates, based on the *physical* DSCL document specified by the user (i.e., it only works with the DSCL that specifies exactly the service handles). It does not accept the *logical* workflow, which only specifies the logical name of the required service. In GZLDSWH, we sometimes do not know in advance which service factory should be executed in order to create a new instance. Instead, the decision should depend on the runtime environment. Besides, because of the automated event-based reaction feature of GZLDSWH, a higher level of automation in service invocation engine is required. Therefore, the Dynamic Workflow Management Service in GZLDSWH extends the DSCE with the automatic Workflow Re-Writer ability. Now, the WMS Service will accept the logical DSCL, parse it, and find out logical services (i.e., services that do not have the exact physical factory handle. It then queries the Resource Broker Service to have the relevant physical service factory handle and then rewrites the DSCL with the new factory handle value. It finally passes the rewrite DSCL to the DSCE engine to invoke the services.

Figure 8(a) describes the extended Dynamic Service Control Engine (EDSCE) based on that DSCE engine. This EDSCE engine accepts the logical workflow in Figure 8(b), parses it, and converts to the physical executable workflow, as depicted in Figure 8(c). The significant difference is the values of the parameters that have been supplied after the Re-Writer queries the Resource Broker Services.

THE OLAP CUBE MANAGEMENT SERVICE

As mentioned previously in the overview section, our approach in GZLDWSH is to store all streaming data into Grid nodes and build OLAP cubes from these Grid-based sources prior to executing analytical queries that evaluate the rules. Following this approach, we have to implement the OLAP Cube Management Service that manages the creation, updation, and querying of the associated cube portions distributed over the Grid nodes. The kernel part of this service is the *OLAP* engine. The first prototype of this

Figure 8. The extended service control engine and corresponding DSCL documents

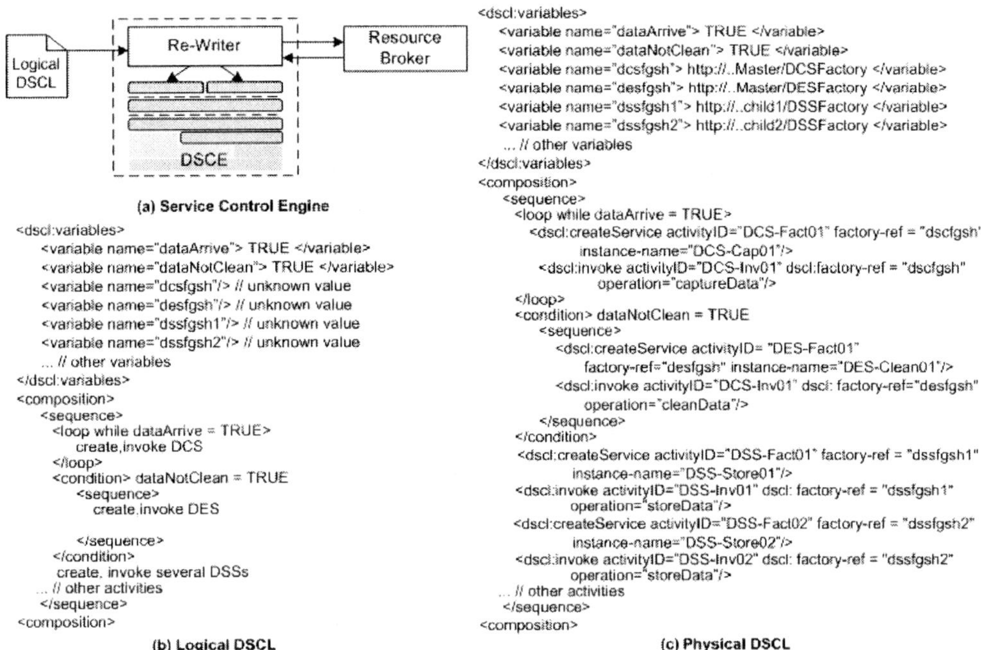

engine (Fiser et al., 2004) has been implemented already in Java.

The OLAP data cube structure consists of an increasing number of chunks, which again consists of a fixed maximum number of measures. A measure is the smallest unit of the cube (one atomic element), and it actually contains just a numeric value. The chunk is a part of the whole cube; it has the same dimensionality like the cube but collects aggregation data at one Grid node. The chunk contains the measures associated with a number of positions of each dimension. Because the amount of memory used by the whole cube usually will be much higher than a system may provide, each chunk offers methods for storing and loading its data onto and from the disk storage. Thus, always only a limited number of chunks is kept within memory at the same time. Storing and loading targeted chunks is called chunk swapping and is a subsystem of the data cube structure implementation. This is similar to paging in modern operating systems, with the distinction that our chunks may grow up to a specific size; hence, the memory resident chunk location table, which is a list of chunks currently resident in memory, varies in size. This is because the aggregation results also are stored within the same data cube.

Special indexing structures and paging mechanisms are necessary to manage the Grid-based OLAP cubes. The index database contains the literal positions, the meta-information, of each dimension and maps unique integer values to position indexes within each dimension. Furthermore, it provides methods for the linearization of multi-dimensional position indexes used for addressing specific measures of the OLAP cube. Several methods are available (e.g., hashing, bit encoded sparse structure [BESS], binary trees, etc.). In order to deal with a huge number of tuples, we need an algorithm, which, on the one hand, is fast and, on the other hand, is not limited to some upper boundary. This is necessary to avoid multiple scans of the source data and allows insertion

Figure 9. OLAP engine architecture

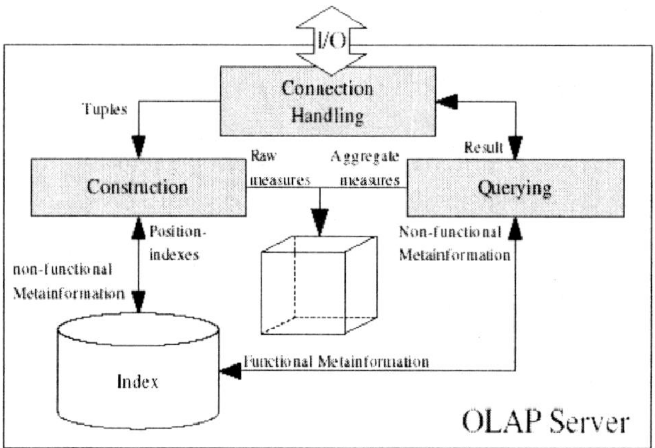

of measure aggregation after cube construction. The method, called Dynamic bit encoding (DBE) (Fiser et al., 2004), was developed for indexing the bit maps of OLAP chunks. DBE is based on BESS with the difference that the bits used for each dimension are kept within bit masks, which are extensible and mutually exclusive to each other from a binary point of view. The position indices are processed by the OR operation using these masks, which results in a linear measure address called the global index.

Figure 9 describes the system architecture of the OLAP engine prototype. The cube construction reads the tuples one after the other, passes over the items to the index database, retrieves their global index, and then passes the (raw) measure and its associated global index to the data cube structure. The Querying functional block is some kind of highly sophisticated, recursively nested loops for aggregation of measures. Because the number of computational operations of nested aggregations depends on the size of the dimensions and thereby on the order in which dimensions are aggregated, the engine uses a kind of query plan optimization to select dimensions in a good way. The procedure of dimension selection is done by traversing a tree, which, in the literature, usually is called the query lattice. The task of aggregation is realized sequentially by loading one chunk after the other and aggregating them one by one. To avoid repeated computation of the same aggregates, they also are stored in the cube structure as if they were raw measures and also get index entries within the index database within appropriate dimensions.

Connection handling is the network interface, which allows user interactions with the system. A typical workflow of the system usage is as follows. After startup, the index tables and the base cube are constructed. This is done upon loading and parsing a structured or semi-structured text file representing database tuples. It is called the origin input stream (see Figure 9). To each position, a unique index value is assigned, and this assignment is kept within the index database. Then all index values from the tuple are merged together using the DBE algorithm. The encoded global index is used to uniquely locate and to store the measure within the cube. After the step of tuples import, the server opens a listening TCP socket and accepts client connections. A simple command language was defined (Fiser et al., 2004) for

Figure 10. Mobile phone fraud detection rule

```
<?xml version="1.0" encodinq="UTF-8"?>
<ReactiveRule name = "Mobile Fraud Analysis">
<variables>
<    variable name="averaqe call lenqth"> </variable>
<    variable name="call lenqth"> </variable>
<    variable name="countrv"> </variable>
    <variable name="area"> </variable>
<    variable name="time_point"> </variable>
<variables>
<Events>
<    Event name = "International lonq call">
      <fact > countrv = "Vietnam"    area = "Asia"
      call length in "current_time" > 30 minutes </fact >
   </Event>
</Events>
<Conditions>
          <fact >
   c       ountrv = "Thailand" or
 c        ountrv = "Indonesia" or countrv ="Malavsia"
           average_call_lenghth in "last 3 hours" > 2 * average_call_lenghth in "last 3 months", area = "Asia"
</fact >
</Conditions>
<Actions>
<    Action name = "Fraud Situation Alarm"/>
</Actions>
</ ReactiveRule>
```

communication between server and client. This is called the control input (output) stream. A client now is able to submit queries. The server supports concurrent sessions, which allows multiple users to log in concurrently.

KNOWLEDGE-BASED RULE DESIGN AND EVALUATION MECHANISM

The Knowledge-Based Rule is the brain of the GZLDSWH and controls how the system reacts automatically to the continuously arriving events, based on the complex incremental multi-dimensional analysis of the collected OLAP data. The rules follow the basic Event-Condition-Action (ECA) rule structure but carry out the complex OLAP analysis instead of evaluating the simple conditions as in ECA rules in OLTP. It is necessary to design a model for maintenance of the Knowledge Base rules that allows users to insert, update, replace, or delete the rules easily. If possible, the rules should be managed in a consistent manner with the option of checking the validation of the rule, avoiding rule conflicts, and maintaining the consistency among the rules.

The mechanism to evaluate the rule is also a challenge, when the events come from different sources within the Grid environment. There are several causes that can trigger a rule. It could be the temporal time events generated from the scheduling service, explicit invocation from another service, or it could be auto-triggered when the Data Service Element (DSE) in another service exists. Right now, we just use a simple XML-based file to manage the rules. The Rule Evaluation Service is invoked explicitly by the Workflow Management Service.

Figure 10 describes a Fraud Detection Rule that monitors the international mobile call. This rule will be triggered when the current international call length to Vietnam of a certain customer is more than 30 minutes. It will check whether the average call length of this customer to other countries in Southeast Asia, such as Thailand, Malaysia, or Indonesia, within the last three hours exceeds twice the average call length to Asia for this customer. If so, the Fraudulent Call alarm will be issued and sent to the customer.

RELIABILITY AND EFFICIENCY

In our architecture, there are two options of how incoming data streams are assigned to nodes in a Grid. First, a master node can select which node on which an incoming data stream will be stored. Second, any node in the Grid may accept an incoming data stream. The node can choose to hand over the data stream to the master node at any point in time.

Grids are characterized by a large number of cooperating nodes. The rationale is that storing of multiple data streams can be handled in parallel and, thus, is more efficient. Amdahl's law (Gene, 1967) describes the speedup of programs, if the performance of some parts (but not all) can be improved. A special case of this improvement is parallel processing. If parts of the code can be executed in parallel, the overall performance will be better.

$1/(S+(1-S)/N)$ shows that speedup depends on how much must be executed sequentially (S in the range of 0.0 to 1.0).

As the size of the Grid (N nodes) increases, reliability of nodes becomes an issue. Even though the mean time to failure (MTTF) of individual nodes decreased over the last years, the size of Grids grows faster, so that the MTTF of the system decreases, if one requires all nodes to be online. Clearly, in a large Grid with N nodes, the setup usually requires only A nodes ($A < N$) to be available for the Grid to work.

If a node receives an incoming data stream and then fails, parts of the data stream might be lost. To avoid inconsistent states, the failed node and possibly some neighboring nodes will need to roll back to the last checkpoint.

Following Elnozahy's models (Elmootazbellah & James, 2004), we look at specific requirements of data streams and how system parameters need to be modified to achieve the primary goal of increasing performance. Each failure causes additional work to be performed; useful work is work that a system performs that will not be lost by a failure. Obviously, the goal is to maximize the ration U of useful work/total work. When serial parts are low, then increasing the number of active nodes A to values near N increases U (Elmootazbellah & James, 2004). Building on Elnozahy's (2004) results, simulations show that checkpoint intervals for Grids with fewer than 4,000 nodes should be greater than 20 minutes. For growing numbers of N, the optimal (maximizing U) interval to set checkpoints is approximately the checkpoint latency, assuming a checkpoint latency of five minutes.

To minimize overhead of creating checkpoints, it is advisable to reduce the length of serial parts in the process of storing data streams. This can be achieved by distributing the stream to several nodes; for instance, 32 KBytes to the first node, the subsequent 32 KBytes to the next node, and so forth. Assuming a data rate of 100 KBit/s the sequential duration of a 32 KByte block is 32 * 1024 * 8 / (1024 * 100) = 2.56 sec.

By adjusting the block length of the stream, the overhead of setting checkpoints can be adjusted. Shorter blocks will reduce the overhead but will increase fragmentation of the stream. Depending on the requirements of retrieving and analyzing the data stream, a decision concerning the tradeoff checkpoint overhead vs. data fragmentation has to be made.

IMPLEMENTATION

In this section, we will describe our ongoing prototype implementation and some experimental performance results. So far, we have developed the prototypes for the Dynamic Service Control Engine (DSCE) and the Sequential OLAP Cube Engine.

Dynamic Service Control Engine Prototype

Figure 11 describes the class architecture of the DSCE prototype, which follows the conceptual architecture mentioned earlier. It is implemented as a stateful and transient OGSA Grid service.

DSCE is a workflow enactment engine that executes a workflow of OGSA Grid Services described by the Dynamic Service Control Language

Figure 11. DCSE class architecture

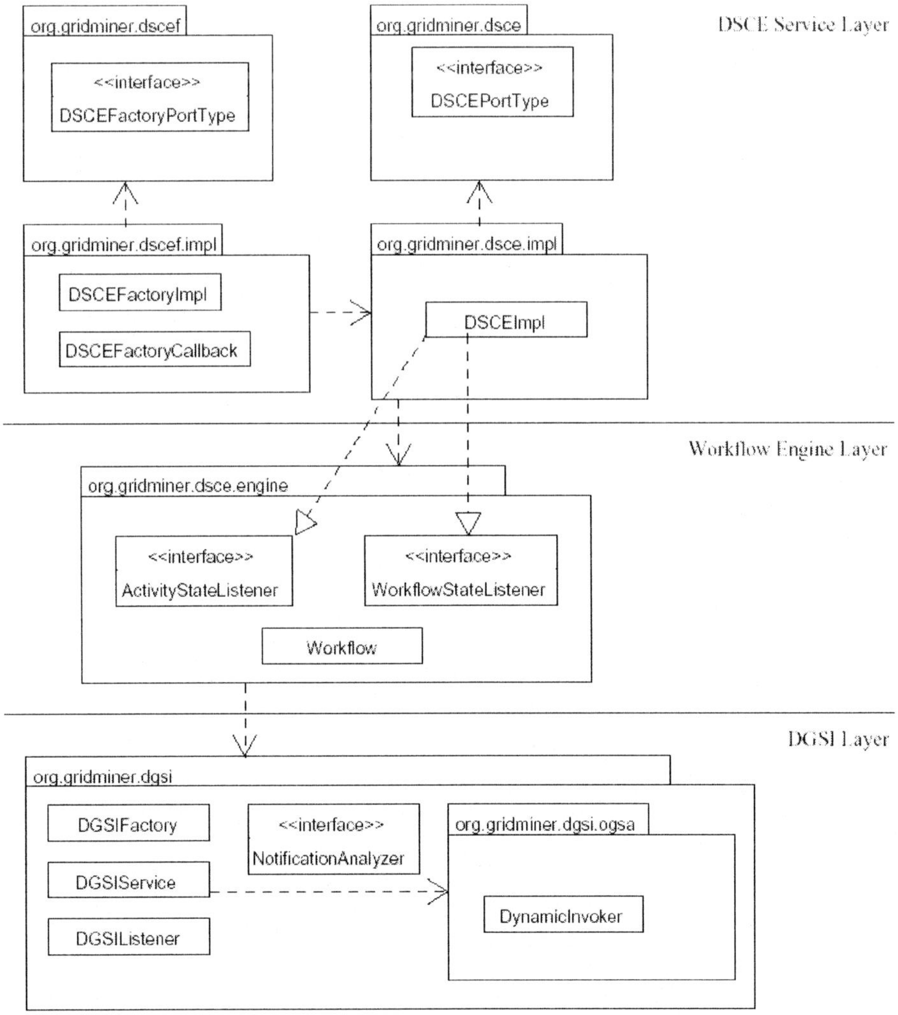

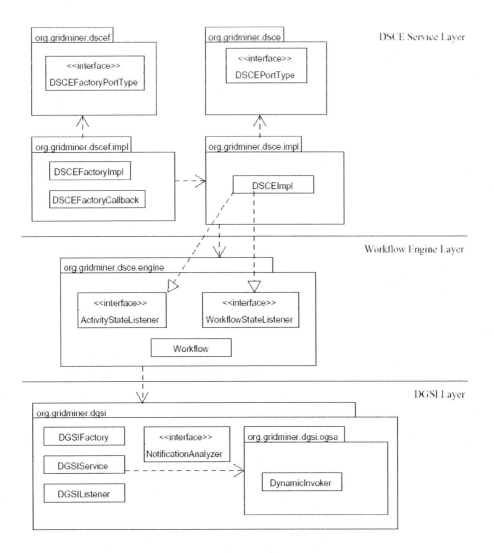

(DSCL). The engine is always in a particular state. Depending on its state, it could accept the client's command, execute the command, and change the state, respectively. This state tells the client about what the engine is doing at the moment and whether it is ready to accept certain commands. The state diagram of the DSCE engine is depicted in Figure 12.

After being created as a service instance, the DSCE engine is in *empty* state. If a client sends a new DSCL workflow description and the engine accepts it, its state will change to *initialized*. The state will change to *running*, if the client starts the workflow and the change of DSCL workflow description is not possible through the change-Workflow() command. If the client needs to change the DSCL description, the engine execution has to be stopped first, and its state will change to *stopping* (during the processing stopping the activities) and *waiting* (when all stated activities have been returned), respectively. The engine could finish without error (*finished* state) or stop the execution with error report (*failure* state).

To execute the performance tests, we define a minimal sub-workflow with no computational cost. This is used as the smallest item throughout all test runs. The sub-workflow consists of the following sequence:

1. Create new *TestGridService* instance
2. Invoke the operation *getFloatValue()*
3. Invoke the operation *setFloatValue(float)*
4. Query the service data element
5. Destroy previously created *TestGridService* instance

To test the workflow engine, 24 different DSCL documents are used (pt1.xml, pt2.xml,..., pt24.xml). DSCL document pt1.xml contains exactly one of the previously defined sub-workflow; DSCL document pt2.xml contains two successive sub-workflows, and so forth. Finally, DSCL document pt24.xml contains 24 successive sub-workflows, which results in 24 × 5 = 120 activities.

A client application (*DSCE-Client*) is implemented to invoke the DSCE service instance, one for each DSCL document. To determine the overhead of the DSCE engine, compared to a direct execution of the workflow, another client has been implemented that executes the sub-

Figure 12. DCSE state diagram

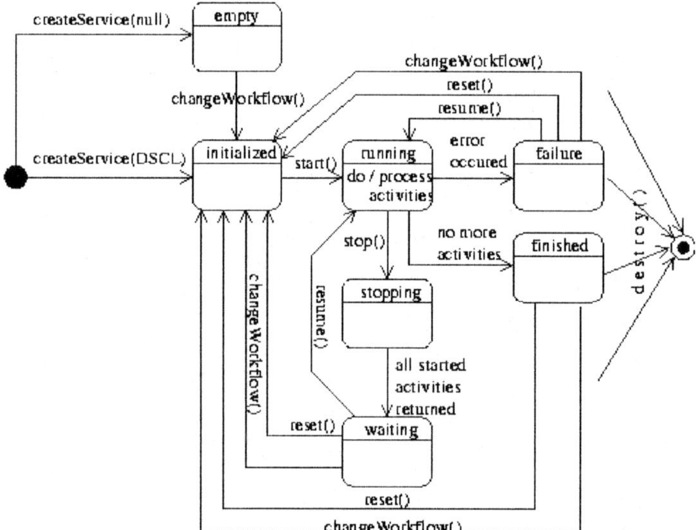

Figure 13. Comparison of overall workflow execution time

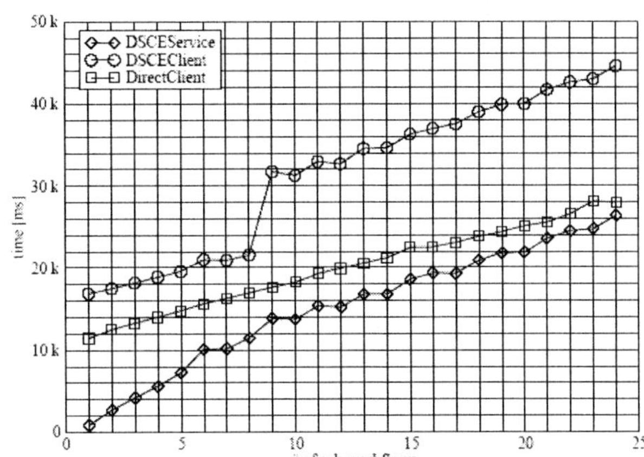

Table 1. Characteristics of the test files used in the test scenarios

	Input I	Input II	Input III
NOL	2000	8000	20000
NODV in column I	100	300	600
NODV in column II	10	15	25
NODV in column III	45	86	125
NODV in column IV	2000	8000	20000
BPCS	90000000	3096000000	37500000000

workflows without usage of DSCE and DSCL (*DirectClient*).

In Figure 13, we can see the execution time of the overall workflow between the two methods: invoke DSCE engine or direct execution of the workflow.

Sequential OLAP Cube Engine

Our prototype implementation of the OLAP engine was tested with three input files (tab separated text files) of different sizes. The test files were structured in a way so that each column represents the values for a dimension, whereby the first column contained the measure values for the cube. Using these input files, we have built three different OLAP cubes with four dimensions, which means that each input file contained data in five columns. Table 1 gives an overview of the characteristics of our test files.

We have treated each test session (which uses a different input file) as a new OLAP scenario and have sent the same queries to the OLAP cube in order to provide an overview of the querying performance relative to the increasing cube size. Each query is represented in a form like [ANY, ANY, ANY, 0], whereby the term ANY indicates that the cube shall be aggregated along this dimension. This means that the costs for a query increases with the increasing number of distinct values in a dimension, when the query value is set to ANY. Figure 14 demonstrates the first performance results, which are more than satisfying for a prototypical sequential OLAP engine. In our future research effort, we plan to investigate the architecture of a parallel and distributed OLAP engine.

Figure 14. Performance results of queri~~ f~~ th~~~ d~ff~~~~t s~enarios

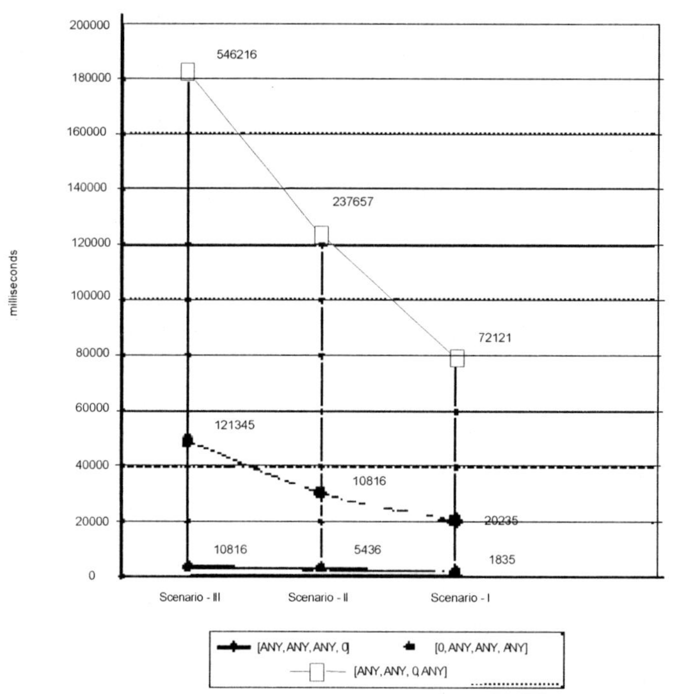

CONCLUSION AND FUTURE WORK

In this paper, we have presented a framework for building the Grid-based Zero-Latency Data Warehouse (GZLDSWH) system for continuous data streams processing and analysis. The GZLDSWH system is built upon the set of OGSI/GT3-based services. Following the predefined reactive rule-based metadata specified by the user, the system can react automatically to continuous data streams in near real time. An adaptive service interaction mechanism is used for the flexible service collaboration. The system then can execute some relevant actions at the data sources or send significant awareness, such as alarms, notifications, recommendations, and so forth to the user.

Our proposed GZLDSWH system is currently an ongoing research, and some of the other issues require further investigation. The major focus of future research should be on solving several significant services within the GZLDSWH, such as Grid Resource Allocation and Scheduling, Heterogeneous Data Mediation and Integration, Grid Data Replica Synchronization (Goel, Sharda & Taniar, 2005), Data Cube Construction and Management, Rule-based Metadata Construction, and Embedding and Evaluation. Other open issues, such as Performance Efficiency in building Data Warehouse on the fly, Distributed Service Control Management, Stream-based Distributed Processing, Heterogeneous Stream, and Availability and Security, also should be considered. In addition, both Grid and Data Stream processing technologies are young and still evolving. The Semantic Grid (Roure, Jennings & Shadboltl, 2003) and Grids services have a role similar to Semantic Web and Web services. Recently, WS-Resource Framework and WS-Notification (Globus Alliance, IBM & HP, 2004) proposals have been announced as an evolution of OGSI with the purpose of effective integration of Grids and Web services standards. The work presented here is closely related to OGSA/OGSI, so it has to adapt to the WS-Resource Framework with suitable modifications. The distributed control of service discovery, creation, invocation, and destroying will be considered as an alternative of collaboration model. The orchestration of Grids or Web services, another approach for solving the workflow problem, also should be investigated further.

ACKNOWLEDGMENT

This work has been partly supported by a Technology Grant South East Asia (No. 322/1/EZA/2002) of the Austrian Council Research and Technology and the ASEA-UNINET.

We are very indebted to Umut Onan and Günter Kickinger for their contribution concerning the implementation aspects. Thanks to Shuaib Karim and Khalid Latif for their English revision efforts.

REFERENCES

Babcock, B., Datar, M., & Motwani, R. (2004). Load shedding for aggregation queries over data streams. In *Proceedings of the International Conference on Data Engineering (ICDE 2004).*

Babcock, B. & Olston, C. (2003). Distributed top-K monitoring. In *Proceedings. of the ACM International Conference on Management of Data (SIGMOD 2003).*

Babcock, B. et al. (2002). Models and issues in data stream systems. In *Proceedings.of the 2002 ACM Symposium on Principles of Database Systems, SIGMOD 2003*, June.

Bailey, J., Poulovassilis, A., & Wood, P. (2003). An event-condition-action language for the semantic Web. *Proceedings of the First International Workshop on Semantic Web and Databases (SWDB' 03).*

Bertino, E. et al. (2000). Trigger inheritance and overriding in an active object database system. *IEEE Transaction on Knowledge and Data Engineering, 12*(4), 588-608.

Bruckner, R. (2002). *Zero-latency data warehousing: Toward an integrated analysis environment with minimized latency for data propagations.* Doctoral thesis, Vienna University of Technology, Vienna.

Cannataro, M. & Talia, D. (2003). The knowledge Grid: An architecture for distributed knowledge discovery. *Communications of the ACM, 46*(1).

Cao, J. et al. (2003). GridFlow: Workflow management for grid computing. In *Proceedings of the Third IEEE/ACM International Symposium on Cluster Computing and Grid.*

Chakrabarti, K. et al. (2001). Approximate query processing using wavelets. *The VLDB Journal, 10.*

Chandrasekaran, S. & Franklin, M. (2002). Streaming queries over streaming data. In *Proceedings of 28th International. Conference. on Very Large Data Bases*, August.

Cho, E. et al. (2003). ARML: An active rule markup language for sharing rules among active. *Information Management System.* Retrieved from *http://www.soi.city.ac.uk/~msch/conf/ruleml/park.pdf*

Compaq Corp. (2002). Compaq global services: Zero latency enterprise. Retrieved from *http://clac.compaq.com/ globalservices/zle/*

Deelman, E. et al. (2003). *Workflow management in GriPhyN: Grid resource management.* Kluwer.

Dobra, A. et al. (2002). Processing complex aggregate queries over data streams. In *Proceedings of the 2002 ACM SIGMOD International Conference on Management of Data.*

Elmootazbellah, N. & James, P. (2004). Checkpointing for peta-scale systems: A look into the future of practical rollback-recovery. *IEEE Transactions of Dependable and Secure Computing, 1*(2), 97-108.

Fiser, B. et al. (2004). On-line analytical processing on large databases managed by computational grids. In *Proceedings of the International Conference on Database and Expert Systems Applications, DEXA04*, Zaragoza.

Foster, I. & Grossman, R. (2003). Data integration in a bandwidth-rich world. *Communications of the ACM, 46*(11).

Foster, I., Kesselman, C., & Tuecke, S. (2001). The anatomy of the grid: Enabling scalable virtual organizations. *International Journal Supercomputer Applications, 15*(3).

Gene, A. (1967). Validity of the single processor approach to achieving large-scale computing capabilities. In *Proceedings of the AFIPS Conference*, (pp. 483-485).

Globus Alliance, IBM, & HP. (2004). The WS-resource framework. Retrieved from *http://www-fp.globus.org/wsrf/ default.asp*

Globus Toolkit. (2003). The globus toolkit. Retrieved from *http://www-unix.globus.org/toolkit/*

Goel, S., Sharda, H., & Taniar, D. (2005). Replica synchronization in grid databases. *International Journal of Web and Grid Services, 1*(1).

Graham, G., Evans, D., & Bertram, I. (2003). Mcrunjob: A high energy physics workflow planner for Grid. *Computing in High Energy and Nuclear Physics.*

Guha, S. & Koudas, N. (2002). Approximating a data stream for querying and estimation: Algorithms and performance valuation. In *Proceedings of the 2002 International Conference on Data Engineering.*

Han, J. et al. (2002). Multi dimensional regression analysis of time-series data streams. In *Proceedings of the 28th VLDB Conference*, Hong Kong.

Hulten, G., Spencer, L., & Domingos, P. (2001). Mining time-changing data streams. In *Proceedings of the Seventh ACM SIGKDD International Conference on Knowledge Discovery and Data Mining, KDD01*, California.

IBM. (2003). The IBM business process execution language for Web services (BPEL4WS). Retrieved from *http://alphaworks.ibm.com/tech/bpws4j*

Joseph, J. & Fellenstein, C. (2003). Grid computing. *Prentice Hall PTR.*

Kamvar, D. et al. (2003). The Eigentrust algorithm for reputation management in P2P networks. In *Proceedings of the 12th International Conference on World Wide Web, WWW2003*, Budapest, Hungary.

Kesselman, C. et al. (2004). Artificial intelligence and grids: Workflow planning and beyond. *IEEE Intelligent System.*

Kickinger, G. & Brezany, P. (2004). The grid knowledge discovery process and corresponding control structures (technical report). Retrieved in March, from *http://www.gridminer.org/publications/gridminer2004-2.pdf*

Kickinger, G. et al. (2003). Workflow management in GridMiner. In *Proceedings of the Third* Cracow Grid Workshop, Cracow, Poland, October 27-29.

Kim, J. & Park, S. (2005). Periodic streaming data reduction using flexible adjustment of time section size. *International Journal of Data Warehousing and Mining, 1*(1), 37-56.

Krishnan, S., Wagstrom, P., & Laszewski, G. (2004). GSFL: A workflow framework for Grid services. Retrieved from *http://www.globus.org/cog/papers/gsfl-paper.pdf*

Lerner, A. & Shasha, D. (2003). The virtues and challenges of ad hoc + streams querying in finance. *Bulletin of the IEEE Computer Society Technical Committee on Data Engineering,*

Leymann, F. (2003). Web services flow language (WSFL 1.0). Retrieved from *http://www-4.ibm.com/software/solutions/webservices/pdf/WSFL.pdf*

Marinescu, D. (2002). *Internet-based workflow management: Toward a semantic Web.* John Wiley.

Moore, R. (2001). Knowledge-based grids. *Technical Report TR-2001-02.* San Diego: Supercomputer Center.

Motvani, R. et al. (2000). Clustering data streams. In *Proceedings of the IEEE Symposium on Foundations of Computer Science, FOCS00,* California.

Muthukrishnan, S. & Strauss, M. (2003). Maintenance of multidimensional histograms. In *Proceedings of the the 23rd Conference FSTTCS 2003,* India.

OGSA. (2003). The globus project: Open grid service architecture. Retrieved from *http://www.globus.org/ogsa/*

OGSA-DAI. (2003). OGSA distributed query processing. Retrieved from *http://www.ogsadai.org/dqp/*

Paton, W. & Diaz, O. (1999). Active database systems. *ACM Computing Surveys, 31*(1).

Roure, D., Jennings, N., & Shadboltl, N. (2003). The semantic grid: A future e-science infrastructure. Retrieved from *http://www.semanticgrid.org/documents/semgrid-journal /semgrid-journal.pdf*

Schlesinger, L. et al. (2005). Supporting the ETL-process by webservice technologies. *International Journal of Web and Grid Services, 1*(1).

Sotomayor, B. (2004). The Globus Toolkit 3 programmer's tutorial. Retrieved from *http://www.casa-sotomayor.net/gt3-tutorial/*

Stonebraker, M. et al. (2003, March). The Aurora and Medusa projects. *Bulletin of the IEEE Computer Society Technical Committee on Data Engineering.*

Taylor, I. (2003). Grid enabling application using triana. In *Proceedings of the Workshop on Grid Applications and Programming Tools.*

Teradata Corp. (2002). Teradata online document. Retrieved from *http://www.teradata.com*

Thalhammer, T. & Schrefl, M. (2002). Realizing active data warehouses with off-the-shelf database technology. *Software Practical Experiment.*

The Axis Project. (2003). Webservices: Axis 1.1. Retrieved from *http://ws.apache.org/axis*

Tho, N. & Tjoa, A. (2003). Zero-latency data warehousing: Continuous data integration and assembling active rules. In *Proceedings of the Fifth International Conference on Information Integration and Web-Based Applications and Services, IIWAS2003,* Jakarta, February.

Tho, N. et al. (2004). Towards service collaboration model in grid-based zero-latency data stream warehouse (GZLDSWH). In *Proceedings of the IEEE International Conference on Service Computing, SCC04,* Shanghai, September.

Tjoa, A. et al. (2003). GridMiner: An infrastructure for data mining on computational grids. In *Proceedings of the APAC Conference and Exhibition on Advanced Computing: Grid Applications and eResearch.*

Tucker, P., Maier, D., & Sheard, T. (2003, March). Applying punctuation schemes to queries over continuous data streams. *Bulletin of the IEEE Computer Society Technical Committee on Data Engineering.*

Widom, J. et al. (2003). Query processing, approximation, and resource management in a data stream management system. In *Proceedings of the First Biennial Conference on Innovative Data Systems Research (CIDR)*, January.

This work was previously published in International Journal of Data Warehousing and Mining, Vol. 1, No. 4, edited by D. Tanier, pp. 22-55, copyright 2005 by IGI Publishing, formerly known as Idea Group Publishing (an imprint of IGI Global).

Chapter 2.18
Data Warehouse Design to Support Customer Relationship Management Analyses

Colleen Cunningham
Drexel University, USA

Il-Yeol Song
Drexel University, USA

Peter P. Chen
Lousiana State University, USA

ABSTRACT

CRM is a strategy that integrates concepts of knowledge management, data mining, and data warehousing in order to support an organization's decision-making process to retain long-term and profitable relationships with its customers. This research is part of a long-term study to examine systematically CRM factors that affect design decisions for CRM data warehouses in order to build a taxonomy of CRM analyses and to determine the impact of those analyses on CRM data warehousing design decisions. This article presents the design implications that CRM poses to data warehousing and then proposes a robust multidimensional starter model that supports CRM analyses. Additional research contributions include the introduction of two new measures, percent success ratio and CRM suitability ratio by which CRM models can be evaluated, the identification of and classification of CRM queries, and a preliminary heuristic for designing data warehouses to support CRM analyses.

INTRODUCTION

It is far more expensive for companies to acquire new customers than it is to retain existing customers. In fact, acquiring new customers can cost five times more than it costs to retain current customers (Massey, Montoya-Weiss & Holcom, 2001). Furthermore, according to Winer (2001), repeat customers can generate more than twice as much gross income as new customers. Companies have realized that instead of treating all customers equally, it is more effective to invest in customers that are valuable or potentially valuable, while limiting their investments in non-valuable customers (i.e., not all relationships are profitable or desirable). As a result of these types of findings as well as the fact that customers want to be served according to their individual and unique needs, companies need to develop and manage their relationships with their customers such that the relationships are long-term and profitable. Therefore, companies are turning to Customer Relationship Management (CRM) techniques and CRM-supported technologies.

In our earlier work (Cunningham, Song, Jung, & Chen, 2003), we defined CRM as a strategy that utilizes organizational knowledge and technology in order to enable proactive and profitable long-term relationships with customers. It integrates the use of knowledge management, or organizational knowledge, and technologies to enable organizations to make decisions about, among other things, product offerings, marketing strategies, and customer interactions. By utilizing a data warehouse, companies can make decisions about customer-specific strategies such as customer profiling, customer segmentation, and cross-selling analysis. For example, a company can use a data warehouse to determine its customers' historic and future values and to segment its customer base. shows four quadrants of customer segmentation: (1) customers that should be eliminated (i.e., they cost more than what they generate in revenues); (2) customers with whom the relationship should be re-engineered (i.e., those that have the potential to be valuable, but may require the company's encouragement, cooperation, and/or management); (3) customers that the company should engage;

Table 1. Customer segments

		Historic Value	
		Low	High
Future Value	High	II. Re-Engineer	IV. Invest
	Low	I. Eliminate	III. Engage

Table 2. Corresponding segmentation strategies

		Historic Value	
		Low	High
Future Value	High	Up-sell & cross-sell activities and add value	Treat with priority and preferential
	Low	Reduce costs and increase prices	Engage customer to find new opportunities in order to sustain loyalty

and (4) customers in which the company should invest (Buttle, 1999; Verhoef & Donkers, 2001). The company then could use the corresponding strategies, as depicted in Table 2, to manage the customer relationships. Table 1 and Table 2 are only examples of the types of segmentation that can be performed with a data warehouse. However, if used, a word of caution should be taken before categorizing a customer into Segment I, because that segment can be further segmented into (a) those customers that serve as benchmarks for more valuable customers, (b) those customers that provide the company with ideas for product improvements or efficiency improvements, and (c) those customers that do not have any value to the company.

It is important to point out that customer segmentation can be further complicated by the concept of extended households. The term *extended household* refers to the relationship that exists between companies (e.g., parent company and subsidiary). The analysis of the relationships that exist between customers (i.e., lines of potential customer influence) is known as household analysis. It is important to understand and manage extended households, because a company's decision to treat a member of one segment potentially could have a negative impact on a related customer. For example, if a customer is in a non-profitable segment, then the company may decide to increase the customer's price. However, if the company is aware that the same non-profitable customer has influence over another customer (e.g., a parent or small business) that is in a more profitable segment, then the company may decide to not increase the customer's price rather than to risk losing both of the customers. Clearly, these social networks of influence are important for companies to identify and manage because of the impact that they can have on the company's ability to retain customers.

Currently, however, there are no agreed upon standardized rules for how to design a data warehouse to support CRM. Yet, the design of the CRM data warehouse model directly impacts an organization's ability to readily perform analyses that are specific to CRM. Subsequently, the design of the CRM data warehouse model contributes to the success or failure of CRM. In fact, recent statistics indicate that between 50% and 80% of CRM initiatives fail due to inappropriate or incomplete CRM processes and poor selection of technologies (Myron & Ganeshram, 2002; Panker, 2002). Thus, the ultimate long-term purpose of our study is to systematically examine CRM factors that affect design decisions for CRM data warehouses in order to build a taxonomy of CRM analyses and to determine the impact of those analyses on CRM data warehousing design decisions.

The taxonomy and heuristics for CRM data warehousing design decisions then could be used to guide CRM initiatives and to design and implement CRM data warehouses. The taxonomy also could be used to customize a starter model for a company's specific CRM requirements within a given industry. Furthermore, that taxonomy also would serve as a guideline for companies in the selection and evaluation of CRM data warehouses and related technologies.

In order to objectively quantify the completeness and suitability of the proposed CRM model (and alternative models), we propose two new metrics: *CRM success ratio* ($r_{success}$) and *CRM suitability ratio* ($r_{suitability}$). The *CRM success ratio* ($r_{success}$) is defined as the ratio of queries that successfully executed to the total number of queries issued against the model. A query is executed successfully if the results that are returned are meaningful to the analyst. The *CRM success ratio* cannot be used only to evaluate our proposed CRM model, but it also can be used to evaluate other CRM data warehouse models, as well. The range of values for $r_{success}$ is between 0 and 1. The larger the value of $r_{success}$, the more successful the model. The following equation defines the CRM success ratio:

$$r_{success} = Q_p / Q_n \quad (1)$$

where Q_p is the total number of queries that successfully executed against the model, and Q_n is the total number of queries issued against the model.

The *CRM suitability ratio* ($r_{suitability}$) is defined as the ratio of the sum of the individual suitability scores to the sum of the number of applicable categories. The following equation defines the CRM suitability ratio:

$$r_{suitability} = (X_iC_i) / N \qquad (2)$$

where N is the total number of applicable analysis criteria, C is the individual score for each analysis capability, and X is the weight assigned to each analysis capability.

The range of values for the $r_{suitability}$ ratio is between 0 and 1, with values closer to 1 being more suitable. Unlike the $r_{success}$ ratio, which can be used to evaluate and compare the richness and completeness of CRM data warehouse models, the $r_{suitability}$ ratio, however, can be used to help companies to determine the suitability of the model based upon the contextual priorities of the decision makers (i.e., based upon the company-specific CRM needs). We utilize the two metrics to evaluate the proposed CRM data warehouse model in our case study implementation.

A brief review of CRM literature is presented in the next section. The section on schema design introduces the analytical CRM analyses requirements that the data warehouse must support as well as provides guidelines for designing the fact tables and the dimensions. The experiment, which is subsequently described with the results in the following section, tests the completeness of the model. The flexibility of the model, the utilization of the CRM analyses, as well as the initial heuristics for designing a CRM data warehouse are presented in the discussion. Finally, the research contributions and future work are discussed in the conclusions

CRM LITERATURE REVIEW

The shift in marketing paradigms from mass marketing to target marketing to the customer-centric one-to-one marketing (known as relationship marketing) is driving CRM (Bose, 2002). Mass marketing is a product-focused approach that allows companies to reach a wide audience with little or no research, irrespective of the consumer's individual needs. Unlike mass marketing, target marketing focuses on marketing to segmented groups that share a similar set of characteristics (e.g., demographic information and purchasing habits). While both approaches are cost-effective, they do not allow for personalization. On the other hand, one-to-one marketing (relationship marketing) enables companies to treat customers individually according to their unique needs. Since not all relationships are profitable or desirable, relationship marketing allows companies to focus on customers that have the best potential lifetime value. In order to identify the appropriate customer-specific approach for managing individual customers, we first must classify customers into one of the four quadrants in Table 1 and subsequently apply the appropriate strategy. In the literature, researchers use the total historical value, total potential future value, and customer lifetime value (CLV). In fact, managing the CLV is essential to the success of CRM strategies (Bose, 2002), because companies that understand and utilize CLV are 60% more profitable than those that do not (Kale, 2004). There are many ways to define and calculate those measures (Hawkes, 2000; Hwang, Jung & Suh, 2004; Jain & Singh, 2002; Rosset, Neumann, Eick & Vatnik, 2003). For the purposes of this article, CLV is the sum of the total historical value and the total potential value for each customer. The following equation defines the *total historical value*:

$$\text{Historical Value} = (\text{Revenue}_j - \text{Cost}_j) \qquad (3)$$

where j is the individual products that the customer has purchased.

In Equation (3), the historical value is computed by summing the difference between the revenue and total cost over every product (j) that the customer has purchased in the past. The cost would include such things as product cost, distribution cost, and overhead cost. Using the same calculation as defined by Hwang et al. (2004), the following equation defines the *potential future value* for a customer:

$$\text{Potential Future Value} = \sum (\text{Probability}_j \times \text{Profitability}_j) \quad (4)$$

where j is the individual products that the customer potentially could purchase.

In Equation (4), the profitability represents the expected revenues minus the sum of the expected costs that would be incurred in order to gain the additional revenues. The probability represents the likelihood that the customer would purchase the product. Thus, the total potential future value would be the sum of individual potential future value of each product that the customer could potentially purchase. The sum of all of the individual customer lifetime values is known as *customer equity* (Rust, Lemon & Zeithaml, 2004).

One of the goals of companies should be to increase their customer equity from one year to the next. By incorporating the ability to compute the CLV into the CRM data warehouse, companies can utilize the CRM data warehouse to determine their customer growth. Additionally, companies can use key performance indicators (KPIs) to identify areas that could be improved. Specific KPIs should relate to the goals of the organization. For example, if a company wants to minimize the number of late deliveries, then an on-time delivery KPI should be selected. Some known KPIs that are relevant to CRM include, but are not limited to, margins, on-time deliveries, late-deliveries, and customer retention rates. Other KPIs that are relevant to CRM include, but are not limited to, marketing cost, number and value of new customers gained, complaint numbers, and customer satisfaction rates (Kellen, 2002).

SCHEMA DESIGN FOR CRM

The first step in any design methodology is to understand the requirements. As such, the minimum requirements for CRM analyses are presented in the CRM analysis requirements section. The specific CRM analysis requirements as well as the need to classify customers according to the four CRM quadrants presented in Table 1 then are used to identify the specific fact tables and dimensions. The heuristics (or guidelines) for modeling the fact tables and dimensions then are explored in the design rationale for the fact tables and design rationale for the dimensions subsections.

CRM Analysis Requirements

The purpose of a data warehouse is not just to store data but rather to facilitate decision making. As such, the first step to designing the schema for the CRM data warehouse is to identify the different types of analyses that are relevant to CRM. For example, some typical CRM analyses that have been identified include customer profitability analysis, churn analysis, channel analysis, product profitability analysis, customer scoring, and campaign management.

In addition to identifying what CRM analyses the data warehouse needs to support, we also must understand how the data analyses are used by the business users. Often, understanding the business use of the data analyses provides additional insights as to how the data should be structured, including the identification of additional attributes that should be included in the model.

Once the specific types of CRM analyses as well as the intended uses of those analyses have been identified, they can be decomposed into the data points that are needed to support the analyses. Moreover, additional data points also can be identified from both experience and literature (Boon, Corbitt, & Parker, 2002; Kellen, 2002; Rust et al., 2004). It should be noted that the additional data points could include non-transactional information such as customer complaints, support calls, and other useful information that is relevant for managing the customer relationships. Furthermore, the non-transactional information could exist in a variety of formats, such as video and graphics (Bose, 2002). Such data formats are beyond the scope of this article. Table 3 identifies the types of analyses that are relevant to CRM as well as some of the data maintenance issues that must be considered. In other words, Table 3 identifies the minimum design requirements for a CRM data warehouse (DW). It should be noted that there is no significance to the order in which the items are listed in Table 3. The design rationale in the following section is based on the minimum design requirements in Table 3.

Design Rationale for the Fact Tables

The model needs to have fact tables that can be used to compute the historical and future values for each customer, because they are used to classify customers. As such, the model consists of a profitability fact table, a future value fact table, a customer service fact table, and various dimensions, which are defined in Table 4. We note that not all of the fact tables and dimensions are included in Figure 1. The profitability fact table includes the attributes (e.g., revenues and all costs—distribution, marketing, overhead, and product) that are required to compute the historical profitability of each transaction in the profitability fact table with the minimum number of joins. That, in turn, improves the performance when querying the data warehouse. Additionally, storing the detailed transactions facilitates the ability to compute the CLV for each customer across each product. Moreover, the model depicted in Figure 1 can be used to calculate KPIs for delivery, such as the number of on-time items and the number of damage-free items. The complement measures are calculated by subtracting the explicitly stored KPI measures from the total quantity. These KPIs are important to track and manage, because they can help organizations to identify internal areas for process improvements and ultimately influence customer satisfaction and possibly customer retention.

The customer service fact table contains information about each interaction with the customer, including the cost of the interaction, the time to resolve the complaint, and a count of customer satisfaction or dissatisfaction. The total historical value of each customer is computed by summing the historical value of each transaction (i.e., the net revenue from the profitability fact table) and then subtracting the sum of the cost of interacting with the customer (i.e., the service cost from the customer service fact table).

In accordance with Equation 4, the future value fact table stores measures that are needed to compute the potential future lifetime value for each customer. For example, among other things, the future value fact table contains the expected gross revenue, costs, expected purchasing frequency, and the probability of gaining additional revenue. It also contains other descriptive attributes that can be used to analyze and categorize the customer's future lifetime value. The customer lifetime value, which is used to classify each customer in one of the four quadrants in Table 1, is computed by summing the historical value for each customer and the future value for each customer.

Design Rationale for the Dimensions

Dimensions are very important to a data warehouse, because they allow the users to easily browse the content of the data warehouse. Special treatment of certain types of dimensions must be taken into consideration for CRM analyses.

Table 3. Minimum design requirements for CRM DWs

No.	Analysis Type/Data Maintenance	Description
3.1	Customer Profitability	Ability to determine profitability of each customer
3.2	Product Profitability	Ability to determine profitability of each product
3.3	Market Profitability	Ability to determine profitability of each market
3.4	Campaign Analysis	Ability to evaluate different campaigns and responses over time
3.5	Channel Analysis	Ability to evaluate the profitability of each channel (e.g., stores, web, and phone)
3.6	Customer Retention	Ability to track customer retention
3.7	Customer Attrition	Ability to identify root causes for customer attrition
3.8	Customer Scoring	Ability to score customers
3.9	Household Analysis	Ability to associate customers with multiple extended household accounts
3.10	Customer Segmentation	Ability to segment customers into multiple customer segmentations
3.11	Customer Loyalty	Ability to understand loyalty patterns among different relationship groups
3.12	Demographic Analysis	Ability to perform demographic analysis
3.13	Trend Analysis	Ability to perform trend analysis
3.14	Product Delivery Performance	Ability to evaluate on-time, late and early product deliveries
3.15	Product Returns	Ability to analyze the reasons for and the impact of products being returned
3.16	Customer Service Analysis	Ability to track and analyze customer satisfaction, the average cost of interacting with the customer and the time it takes to resolve customer complaints
3.17	Up-selling Analysis	Ability to analyze opportunities for customers to buy larger volumes of a product or a product with a higher profitability margin
3.18	Cross-selling Analysis	Ability to identify additional types of products that customers could purchase, which they currently are not purchasing
3.19	Web Analysis	Ability to analyze metrics for web site
3.20	Data Maintenance	Ability to maintain the history of customer segments and scores
3.21	Data Maintenance	Ability to integrate data from multiple sources, including external sources
3.22	Data Maintenance	Ability to efficiently update/maintain data

Table 4. Starter model dimension definitions

Dimension Name	Dimension Definition
Channel Dimension	Stores the different modes for interacting with customers
Customer Dimension	Stores the static information about the customer
Customer Behavior Dimension	Stores the dynamic scoring attributes of the customer
Customer Demographics Dimension	Stores the dynamic demographic characteristics of the customer
CustomerExistence	Tracks the periods in which the customer is a valid
CustomerMarket	Tracks changes in the relationship between the customer and market dimensions
Comments Dimension	Stores the reasons for customer attrition and product returns
Company Representative	Stores the company representatives (sales representatives)
County Demographics Dimension	Stores external demographics about the counties
Extended Household	Represents the fact that the customer may belong to one or more extended households
Market Dimension	The organizational hierarchy and regions in which the customer belongs
Product Dimension	Represents the products that the company sells
ProductExistence	Tracks the periods in which the products are valid
Promotion Dimension	Represents the promotions that the company offers
Prospect	Stores information about prospects
Scenario Dimension	Used to analyze hypothetical up-selling and cross-selling scenarios
Supplier Dimension	Represents the vendors that supply the products
sTime Dimension	The universal times used throughout the schema
Time Dimension	Universal dates used throughout the schema

Each of those dimension types and their special treatments are discussed in the following subsections.

Existence Dimensions and Time

Customer Relationship Management is a process. As with any business process, the CRM process needs to be changed periodically to reflect changes in and additions to the business process (e.g., organizational restructuring due to territory realignments or mergers and acquisitions, new or modified business rules, changes in strategic focus, and modified or new analysis requirements). Thus, time is an inherent part of business systems and must be modeled in the data warehouse. Traditionally, the time dimension primarily participates in a relationship with the fact tables only. Additionally, there are two ways of handling temporal changes: tuple versioning and attribute versioning (Allen & March, 2003). Tuple versioning (or row time stamping) is used in multiple ways to record (1) changes in the active state of a dimension, (2) changes to the values of attributes, and (3) changes in relationships (Todman, 2001). As such, traditional tuple versioning has limitations within the context of CRM. For example, periods of customer inactivity can be determined only by identifying two consecutive tuples where there is a gap in the timestamp. Additionally, queries that involve durations may be spread over many tuples, which would make the SQL statement complex with slow response times (Todman, 2001).

In order to alleviate the issues with traditional time stamping in the context of CRM, each dimension is examined carefully to determine if the dimension (1) contains attributes whose complete set of historical values have to be maintained, or (2) is subject to discontinuous existence (i.e., only valid for specific periods).

If either (1) or (2) is applicable, then a separate dimension is created called an existence dimension. The existence dimensions are implemented as outriggers, and two relationships are created between the time dimension and each outrigger dimension. The two relationships are formed in order to record the date period in which the data instances are valid. In doing so, this facilitates the ability to perform state duration queries and transition detection queries (Todman, 2001). State duration queries contain a time period (start date and end date) in the *where* clause of the query, whereas transition detection queries identify a change by identifying consecutive periods for the same dimension (Todman, 2001).

Careful consideration is given to this step in the design process, because the fact table only can capture historical values when a transaction occurs. Unfortunately, the reality is that there may be periods of inactivity, which would mean that any changes that occur during those periods of inactivity would not be recorded in the data warehouse. This would, in turn, impact the types of analyses that could be done, since one cannot analyze data that one has not recorded.

Mini-Dimensions

If a dimension contains attributes that are likely to change at a different rate than the other attributes within the dimension, then a separate dimension is created as a mini-dimension. The new dimensions are implemented as mini-dimensions as opposed to outriggers in order to allow the user to readily browse the fact table. One benefit of this approach is that the history of the changes in the customer's behavior scores and demographics are stored as part of the fact table, which facilitates robust analyses without requiring the use of Type 1, 2, or 3 techniques (Kimball & Ross, 2002) for the Customer Demographics or Customer Behavior dimensions.

Customer Dimension

The customer must be at the heart of the customer-centric data warehouse. As such, careful attention must be given to the design of the customer dimension, which will force attention on the customer. Direct relationships are formed between the Customer dimension and the Sales Representative, Market, Comment, and Time dimensions in order to allow the user to readily determine the most current values for the sales representative, market, activation date, attrition date, and attrition comments by simply browsing the Customer dimension without having to include a time constraint in the query statement.

Other Dimensions

There is a Household dimension as well as an extended household dimension in order to analyze the potential lines of influence that exists between customers. In accordance with the types of CRM analyses that the data warehouse must support, other dimensions are identified according to the dimensions along which the fact tables are analyzed. For example, other dimensions include the Product, Supplier, Channel, Promotion, Market, and Sales Representative dimensions in order to facilitate the CRM analyses described in the CRM Analysis Requirements section.

As a result of this approach to modeling the dimensions, the only slowly changing dimensions in the model are the County Demographics dimension, the Product dimension, the Supplier dimension, and the Customer dimension.

The model depicted in Figure 1, which is based upon the minimum design requirements and the design rationale presented in this section, is tested to determine its completeness and flexibility for CRM analyses. The experiment that is used to test the model is described in the following section.

EXPERIMENT

The purpose of the experiment is to test the completeness and flexibility of the proposed CRM data warehouse model. Our hypothesis is that the proposed data warehouse starter model has a positive impact on the ability to perform CRM analyses. The implementation, methodology, and selection of the queries that are used in the experiment to test our hypothesis as well as the results are discussed in the specific subsections that follow.

Implementation

We perform a case study to test the validity of our proposed starter model. The proposed CRM data warehouse model is implemented in SQL Server 2000 running on a Windows 2000 server. The hardware computer is a DELL 1650 database server with a single processor and 2.0 MHz. The schema is populated with 1,685,809 rows of data from a manufacturing company.

Methodology

In the experiment, a series of CRM queries are executed against the proposed data warehouse schema. The success rate of the proposed schema is computed as a ratio of the number of successful queries executed divided by the total number of queries used in the investigation. Furthermore, the proposed CRM data warehouse model is tested to determine if it could or could not perform the analyses listed in Table 3. For each analysis in Table 3 that the model could perform, it is given a score of one point; otherwise, the model is given a score of zero points. The sum of the points for the model is computed in order to determine an overall CRM-analysis capability score. The selection of the queries that are used to study the model is discussed in the following section.

Selection of Queries to Test

Since we believe that the proposed data warehouse starter model has a positive impact on the ability to perform CRM analyses, special care was taken in the selection of the queries used for testing in

Figure 1. Proposed CRM data warehouse model

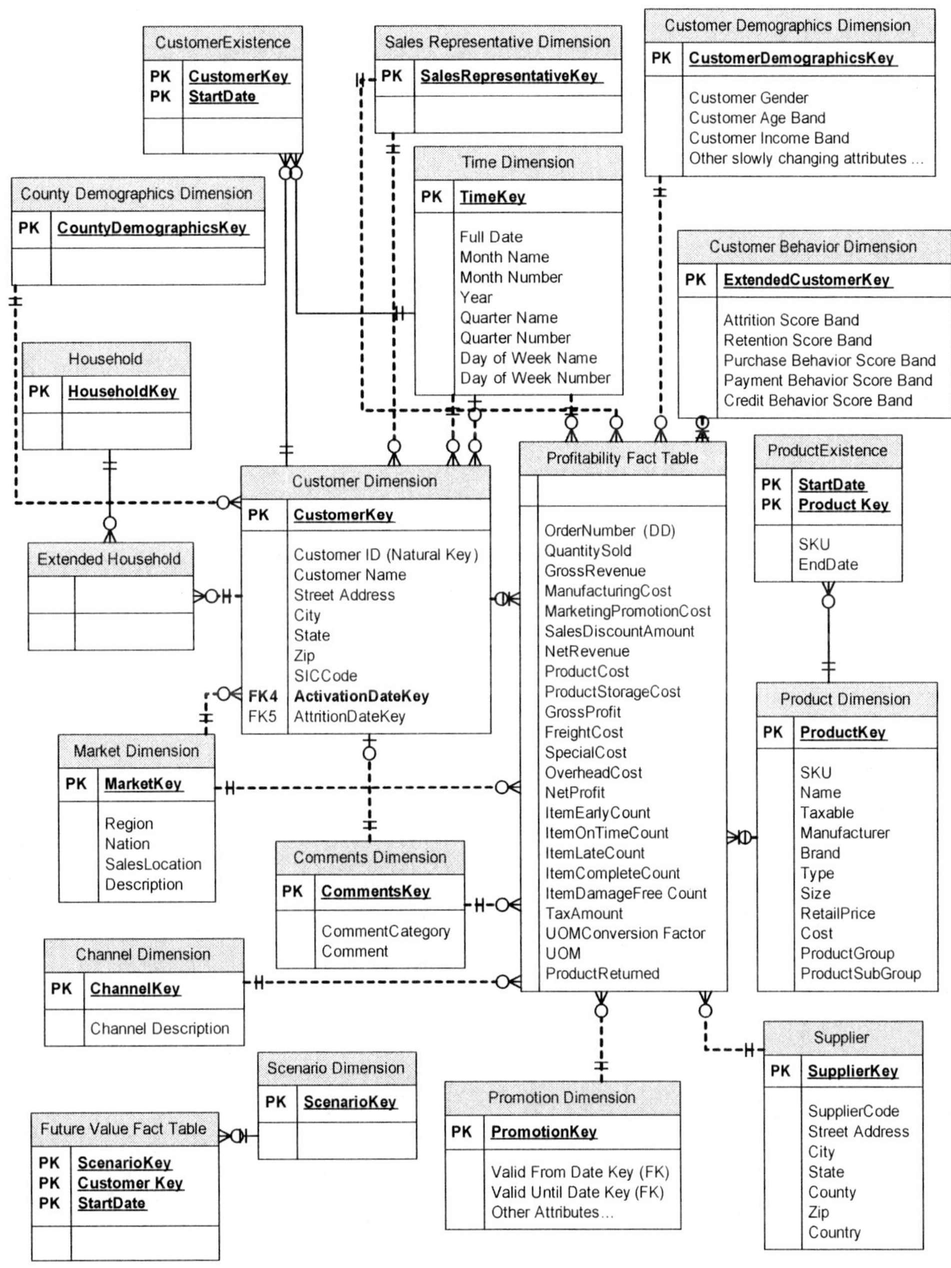

order to avoid any biases in the types of queries used to test the model. Stratified random sampling is used to select the specific queries for the experiment. The stratified random sampling is conducted as follows: (1) representative queries for CRM are gathered from literature and experience; (2) the queries are grouped into categories based upon the nature of the query; (3) within each category, each query is numbered; (4) a random number generator is used to select queries from each category; and (5) the queries whose assigned number corresponds to the number generated by the random number generator are selected. The specific queries that are selected are listed in Table 5. It is important to note that since the queries are randomly selected from a pool of CRM-related queries, it is possible that the $r_{success}$ ratio can be less than one for our proposed model. It is also important to note that the representative CRM queries are queries that equally apply to different industries and not queries that are specific to only one industry. This aspect of the sampling procedure is important in order to make generalizations about the characteristics of the data warehouse schema that should be present in order to perform CRM analyses across different industries.

Results

Our preliminary finding is that the proposed CRM data warehouse model can be used to successfully perform CRM analyses. Based upon the sample queries, our model has a value of 1 and 0.93 for the $r_{success}$ and $r_{suitability}$ ratios, respectively. The individual scores for successfully executing the queries against the model are listed in Table 5. The individual and cumulative scores for the suitability of the proposed CRM data warehouse model are listed in Table 6. It should be noted that there is no significance to the order in which the items are listed in the table.

The scores for items 6.1 through 6.11 in Table 6 are based upon whether or not queries are successfully executed in those categories. The scores for items 14 and 15 are determined while loading data from multiple sources and updating customer scores. Each of the queries that were successfully executed in the experiment is discussed in further detail in the following section in order to highlight the completeness and flexibility of the model.

DISCUSSION

In addition to discussing the completeness and flexibility of the model, this section also presents potential uses of the specific analyses, including KPIs. This section also describes the data quality issues pertaining to CRM data warehouses before presenting a summary of the heuristics for designing data warehouses to support CRM analyses.

Model Completeness and Flexibility

The starter model depicted in Figure 1 can be used for a variety of CRM analyses, including customer profitability analysis, household profitability analysis, demographics profitability analysis, product profitability analysis, channel profitability analysis, and promotion profitability analysis simply by including the appropriate dimensions in the query statement. Furthermore, each query can be modified to include additional measures and descriptions simply by including additional fields from the fact table and the dimensions. Some of those queries are discussed next.

The SQL statement in Figure 2 is used to identify the most profitable customers based upon total revenue and gross margin. By excluding the time dimension, the customer profitability SQL statement identifies the customer's historical lifetime value to the company. This is an important analysis that, in conjunction with the customer's future value and the customer service interaction costs, is used to classify customers in one of the four CRM quadrants (Table 1), which subsequently can be used to determine the appropriate strategy for managing the customer.

The SQL statement in Figure 3 is used to determine the margins for each product and subsequently identifies products that potentially may be eliminated from the company's product line. The ability to be able to determine the lifetime value of each product (irrespective of market) merely by modifying the SQL statement in Figure 3 to exclude the product code further illustrates the flexibility and robustness of the proposed CRM model.

Table 5. Sample CRM analyses

No.	Category	Analysis	Pass	Fail
5.1	Channel Analysis	Which distribution channels contribute the greatest revenue and gross margin?	1	0
5.2	Order Delivery Performance	How do early, on time and late order shipment rates for this year compare to last year?	1	0
5.3	Order Delivery Performance and Channel Analysis	How do order shipment rates (early, on time, late) for this year compare to last year by channel?	1	0
5.4	Customer Profitability Analysis	Which customers are most profitable based upon gross margin and revenue?	1	0
5.5	Customer Profitability Analysis	What are the customers' sales and margin trends?	1	0
5.6	Customer Retention	How many unique customers are purchasing this year compared to last year?	1	0
5.7	Market Profitability Analysis	Which markets are most profitable overall?	1	0
5.8	Market Profitability Analysis	Which products in which markets are most profitable?	1	0
5.9	Product Profitability Analysis	Which products are the most profitable?	1	0
5.10	Product Profitability Analysis	What is the lifetime value of each product?	1	0
5.11	Returns Analysis	What are the top 10 reasons that customers return products?	1	0
5.12	Returns Analysis	What is the impact of the value of the returned products on revenues?	1	0
5.13	Returns Analysis	What is the trend for product returns by customers by product by reason?	1	0
5.14	Customer Attrition	What are the top 10 reasons for customer attrition?	1	0
5.15	Customer Attrition	What is the impact of the value of the customers that have left on revenues?	1	0

The SQL statement in Figure 4 is used to determine and compare Key Performance Indicators (KPIs) for overall on-time, early, and late shipment percentages for different years.

By modifying the statement in Figure 4 to include the Channel Dimension, the performance of each channel from one year to the next is determined. The modified SQL statement can be seen in Figure 5.

The SQL statement in Figure 6 is used to determine the overall profitability of each market. By eliminating the market key from the SQL statement, the profitability for each location is obtained for each location within the organi-

Table 6. Sample suitability for CRM analyses scores

No.	Criteria	Score
6.1	Ability to track retention	1
6.2	Ability to identify root causes for customer attrition	1
6.3	Ability to score customers	1
6.4	Ability to associate customers with multiple extended household accounts	1
6.5	Ability to segment customers into multiple customer segmentations	1
6.6	Ability to maintain the history of customer segments and scores.	1
6.7	Ability to evaluate different campaigns and responses over time	1
6.8	Ability to analyze metrics for website	0
6.9	Ability to understand loyalty patterns among different relationship groups	1
6.10	Ability to perform demographic analysis	1
6.11	Ability to perform trend analysis	1
6.12	Ability to perform customer profitability analysis	1
6.13	Ability to perform product profitability analysis	1
6.14	Ability to integrate data from multiple sources, including external sources	1
6.15	Ability to efficiently update/maintain data.	1
	Total	**14**

Figure 2. Customer profitability analysis query — Which customers are most profitable based upon gross margin and revenue?

```
SELECT b.CustomerKey, b.CustomerName, Sum(a.GrossRevenue) AS TotalRevenue,
Sum(a.GrossProfit) AS TotalGrossProfit, TotalGrossProfit/TotalRevenue AS GrossMargin
FROM tblProfitabilityFactTable a, tblCustomer b
WHERE b.CustomerKey=a.CustomerKey
GROUP BY b.CustomerKey, b.CustomerName
ORDER BY Sum(a.GrossRevenue) DESC;
```

Figure 3. Product profitability analysis query — Which products in which markets are most profitable?

```
SELECT c.Year, b.MarketKey, b.LocationCode, b.Location, b.Description,
b.CompetitorName, d.ProductCode, d.Name, Sum(a.GrossRevenue) AS TotalRevenue,
Sum(a.GrossProfit) AS TotalGrossProfit, TotalGrossProfit/TotalRevenue AS GrossMargin
FROM tblProfitabilityFactTable a, tblMarket b, tblTimeDimension c, tblProductDimension
d
WHERE b.MarketKey=a.MarketKey And a.TimeKey=c.TimeKey And
a.ProductKey=d.ProductKey
GROUP BY c.Year, b.MarketKey, b.LocationCode, b.Location, b.Description,
b.CompetitorName, d.ProductKey, d.ProductCode, d.Name, b.MarketKey
ORDER BY Sum(a.GrossRevenue) DESC;
```

Figure 4. Order delivery performance query — How do early, on time, and late order shipment rates for this year compare to last year?

```
SELECT b.Year, Sum(a.ItemOnTimeCount) AS OnTime, Sum(a.ItemEarlyCount) AS
Early, Sum(a.ItemLateCount) AS Late,
Sum(a.ItemOnTimeCount+a.ItemEarlyCount+a.ItemLateCount) AS TotalCount,
OnTime/Late*100 AS PercentOnTime, Early/TotalCount*100 AS PercentEarly,
Late/TotalCount*100 AS PercentLate
FROM tblProfitabilityFactTable a, tblTimeDimension b
WHERE b.TimeKey = a.TimeKey
GROUP BY b.Year;
```

Figure 5. Order delivery performance and channel analysis query — How do order shipment rates (early, on-time, late) for this year compare to last year by channel?

```
SELECT b.Year, c.ChannelCode, Sum(a.ItemOnTimeCount) AS OnTime,
Sum(a.ItemEarlyCount) AS Early, Sum(a.ItemLateCount) AS Late,
Sum(a.ItemOnTimeCount+a.ItemEarlyCount+a.ItemLateCount) AS TotalCount,
OnTime/Late*100 AS PercentOnTime, Early/TotalCount*100 AS PercentEarly,
Late/TotalCount*100 AS PercentLate
FROM tblProfitabilityFactTable a, tblTimeDimension b, tblChannelDimension c
WHERE b.TimeKey=a.TimeKey And c.ChannelKey = a.ChannelKey
GROUP BY b.Year, c.ChannelCode;
```

Figure 6. Market profitability analysis query — Which markets are most profitable overall?

```
SELECT c.Year, b.MarketKey, b.LocationCode, b.Location, b.Description,
b.CompetitorName, Sum(a.GrossRevenue) AS TotalRevenue, Sum(a.GrossProfit) AS
TotalGrossProfit, TotalGrossProfit/TotalRevenue AS GrossMargin
FROM tblProfitabilityFactTable a, tblMarket b, tblTimeDimension c
WHERE b.MarketKey=a.MarketKey And a.TimeKey=c.TimeKey
GROUP BY c.Year, b.MarketKey, b.LocationCode, b.Location, b.Description,
b.CompetitorName, b.MarketKey
ORDER BY Sum(a.GrossRevenue) DESC;
```

zational hierarchy that is defined in the market dimension.

The SQL statement in Figure 7 demonstrates that by including the product code from the product dimension in the previous SQL statement, the profitability of each product by market is obtained.

The SQL statement in Figure 8 is used to determine the top reasons for product returns. In this case, the basis for the top reasons is merely

Figure 7. Product profitability analysis query — Which products in which markets are most profitable?

```
SELECT c.Year, b.MarketKey, b.LocationCode, b.Location, b.Description,
b.CompetitorName, d.ProductCode, d.Name, Sum(a.GrossRevenue) AS TotalRevenue,
Sum(a.GrossProfit) AS TotalGrossProfit, TotalGrossProfit/TotalRevenue AS
GrossMargin
FROM tblProfitabilityFactTable a, tblMarket b, tblTimeDimension c,
tblProductDimension d
WHERE b.MarketKey=a.MarketKey And a.TimeKey=c.TimeKey And
a.ProductKey=d.ProductKey
GROUP BY c.Year, b.MarketKey, b.LocationCode, b.Location, b.Description,
b.CompetitorName, d.ProductKey, d.ProductCode, d.Name, b.MarketKey
ORDER BY Sum(a.GrossRevenue) DESC;
```

Figure 8. Returns analysis — What are the top reasons that customers return products?

```
SELECT b.CommentsKey, c.ProductCode, c.Name, d.Comment, Sum(a.GrossRevenue)
AS TotalRevenue, Sum(a.GrossProfit) AS TotalGrossProfit,
TotalGrossProfit/TotalRevenue AS GrossMargin, Count(*) AS MembershipCount
FROM tblProfitabilityFactTable a, tblTimeDimension b, tblProductDimension c,
tblCommentDimension d
WHERE a.TimeKey=b.TimeKey And a.ProductKey=c.ProductKey And
a.CommentsKey=d.CommentsKey And a.ProductReturned=Yes
GROUP BY d.CommentsKey, c.ProductCode, c.Name, d.Comment, c.ProductKey
ORDER BY Count(*) DESC, Sum(a.GrossProfit) DESC;
ORDER BY Sum(a.GrossRevenue) DESC;
```

the count of the number of reasons that products are returned. By first grouping the products that are returned according to the reason for their return and the product code and then including the number of returns, revenue, gross profit, and gross margin for each group, the SQL statement in Figure 8 identifies areas upon which the company should improve in order to minimize the number of returns and to improve overall customer satisfaction. Specifically, since companies have limited resources, a company can use the result set to create Pareto charts according to the most frequently occurring problems that have the largest associated gross profits. Management teams then can use the Pareto charts to determine which problems to address first with corrective actions. It should be noted that in order to facilitate quick identification of the most frequently occurring problems that have the largest associated gross profits, the SQL statement in Figure 8 includes an ORDER BY clause.

Furthermore, simply by modifying the SQL statement in Figure 8 to include the year of the transaction from the profitability fact table in the

Figure 9. What is the impact of the value of the returned products on revenues?

```
SELECT b.CommentsKey, c.ProductCode, c.Name, d.Comment, Sum(a.GrossRevenue)
AS TotalRevenue, Sum(a.GrossProfit) AS TotalGrossProfit,
TotalGrossProfit/TotalRevenue AS GrossMargin, Count(*) AS MembershipCount
FROM tblProfitabilityFactTable a, tblTimeDimension b, tblProductDimension c,
tblCommentDimension d
WHERE a.TimeKey=b.TimeKey And a.ProductKey=c.ProductKey And
a.CommentsKey=d.CommentsKey And a.ProductReturned=Yes
GROUP BY d.CommentsKey, c.ProductCode, c.Name, d.Comment, c.ProductKey
ORDER BY Count(*) DESC;
ORDER BY Sum(a.GrossRevenue) DESC;
```

Figure 10. What is the trend for product returns by customers by product by reason?

```
SELECT e.CustomerName, b.Year, b.CommentsKey, c.ProductCode, c.Name,
d.Comment, Sum(a.GrossRevenue) AS TotalRevenue, Sum(a.GrossProfit) AS
TotalGrossProfit, TotalGrossProfit/TotalRevenue AS GrossMargin, Count(*) AS
MembershipCount
FROM tblProfitabilityFactTable a, tblTimeDimension b, tblProductDimension c,
tblCommentDimension d, tblCustomerDimension e
WHERE a.TimeKey=b.TimeKey And a.ProductKey=c.ProductKey And
a.CommentsKey=d.CommentsKey And a.ProductReturned=Yes
GROUP BY e.CustomerName, b.Year, d.CommentsKey, c.ProductCode, c.Name,
d.Comment, c.ProductKey
ORDER BY Count(*) DESC, Sum(a.GrossProfit) DESC;
ORDER BY Sum(a.GrossRevenue) DESC;
```

SELECT clause and the GROUP BY clause, companies can use the results of the modified query to monitor the trend of return reasons over time. Stated differently, companies can use the results of the modified query statement to monitor the impact of the corrective actions over time. Not only can the return analyses be used to monitor the impact of corrective actions, but they also can be used to identify improvement targets, which can be tied to employee (and/or departmental) performance goals.

The SQL statement listed in Figure 9 is used to determine the impact of the returned products on revenues.

The SQL statement listed in Figure 10 is used to identify the trend for product returns by customer, by product, and by reason. The results can be used to identify whether or not a problem is systematic across all customers, many customers, or a few specific customers. This query also can be used to help management make an informed decision with respect to allocating resources to address problems that lead to customers returning products. Additionally, the results can be used by the sales team to gain further insights into why their customers have returned products. The sales team potentially can use that information to

Figure 11. What are the top 10 reasons for customer attrition?

```
SELECT  b.Comment, Count(a.CommentsKey) AS CountReasons
FROM tblCustomer AS a, tblCommentDimension AS b
WHERE a.CommentsKey=b.CommentsKey
GROUP BY  b.Comment
ORDER BY Count(a.CommentsKey) DESC;
```

Figure 12. What is the impact of the value of the customers that have left on revenues?

```
SELECT b.Comment, Count(a.CommentsKey) AS NumberOfTransactions,
Sum(c.GrossRevenue) AS TotalGrossRevenue
FROM tblCustomer AS a, tblCommentDimension AS b, tblProfitabilityFactTable
AS c
WHERE a.CommentsKey=b.CommentsKey AND c.CustomerKey=a.CustomerKey
GROUP BY b.Comment
ORDER BY Count(a.CommentsKey) DESC;
```

work with the customer to resolve the issue(s) in cases where the customer repeatedly has returned products for reasons that cannot be considered the company's mistake. Alternatively, the sales team can use the results to identify accounts that could (should) be charged additional fees if the customer repeatedly returns products.

The SQL statement in Figure 11 is used to identify the top reasons for customer attrition. Figure 12 is used to analyze the impact of customer attrition on the total revenues. By analyzing customer attrition, companies can gain further insights into areas for improvement in order to reduce the attrition rate and thereby improve its overall company value.

Table 7 summarizes some of the possible uses for the CRM analyses that are presented in Table 5.

Data Quality

Given the range of decisions that the CRM data warehouse must be able to support and given the potential impact on companies' profitability, it is imperative that the data are accurate. The data that are used in the CRM analyses originate from disparate data sources that must be integrated into the data warehouse. Under such circumstances, the issue of dirty data arises. Analyzing dirty data, particularly in the context of systems that support corporate decision-making processes (e.g., CRM analyses and subsequent decisions), would result in unreliable results and potentially inappropriate decisions. As such, ensuring data quality within the data warehouse is important to the overall success of subsequent CRM analyses and decisions. Data quality should not be consid-

Table 7. Initial taxonomy of CRM analyses (S = strategic and T = tactical)

#	Decision Class	Category	Analysis	Potential Use(s)	KPI
1	S	Channel Analysis	Which distribution channels contribute the greatest revenue and gross margin?	Resource allocation	
2	S, T	Order Delivery Performance	How do early, on time and late order shipment rates for this year compare to last year?	Setting performance goals	early delivery, on-time delivery, late
3	S	Order Delivery Performance and Channel Analysis	How do order shipment rates (early, on time, late) for this year compare to last year by channel?	Setting performance goals, monitoring trends	early delivery, on-time delivery, late
4	S	Customer Profitability Analysis	Which customers are most profitable based upon gross margin and revenue?	Classify customers	gross margin, revenue
5	S	Customer Profitability Analysis	What are the customers' sales and margin trends?	Classify customers	gross margin, revenue
6	S	Customer Retention	How many unique customers are purchasing this year compared to last year?	Identify the threshold to overcome with new customers	unique customers /year
7	S, T	Market Profitability Analysis	Which markets are most profitable overall?	Setting performance goals, allocate marketing resources	gross margin/market
8	S, T	Market Profitability Analysis	Which products in which markets are most profitable?	Setting performance goals, allocate marketing resources	gross margin/ products/ market
9	S, T	Product Profitability Analysis	Which products are the most profitable?	Managing product cost constraints, identify products to potentially eliminate from product line	gross margin/ product
10	S, T	Product Profitability Analysis	What is the lifetime value of each product?	Managing product cost constraints, identify products to potentially eliminate from product line	gross margin/ product
11	S, T	Returns Analysis	What are the top 10 reasons that customers return products?	Create Pareto charts to identify problems to correct, setting performance goals	count
12	S, T	Returns Analysis	What is the impact of the value of the returned products on revenues?	Create Pareto charts to identify problems to correct, setting performance goals	count, revenue, profit
13	S, T	Returns Analysis	What is the trend for product returns by customers by product by reason?	Create Pareto charts to identify problems to correct, setting performance goals, identify problematic accounts (identify customers that may leave), assess additional service fees	count, revenue, profit
14	S, T	Customer Attrition	What are the top 10 reasons for customer attrition?	Insights for process improvements	attrition rate

ered a one-time exercise conducted only when data are loaded into the data warehouse. Rather, there should be a continuous and systematic data quality improvement process (Lee, Pipino, Strong, & Wang, 2004; Shankaranarayan, Ziad, & Wang, 2003).

One way of minimizing data quality issues is to carefully document the business rules and data formats that then can be used to ensure that those requirements are enforced. Too often, however, thorough documentation of the business rules and data formats is not available in a corporate setting. Therefore, routine data quality audits should be performed on the data in the CRM model in order to identify data quality issues that are not addressed during the ETL process. For example, missing data can be identified during data quality audits and consequently addressed by consulting with a domain expert.

Some forms of dirty data (e.g., outliers) can be identified using data mining techniques and statistical analysis, while other forms of dirty data (e.g., missing data values) are more problematic. Although Dasu, Vesonder, and Wright (2003) assert that data quality issues are application-specific, Kim, Choi, Kim, and Lee (2003) developed a taxonomy of dirty data and identified methods for addressing dirty data based upon their taxonomy. Kim et al. (2003) identified three broad categories of dirty data: (1) missing data; (2) not missing, but wrong data; and (3) not missing and not wrong, but unusable data. They then further decompose each category of dirty data. However, they did not include composite types of dirty data. They provided some suggestions to address the issue of dirty data that can be used to clean dirty data in the CRM model during the ETL process. For example, the use of constraints can be valuable for avoiding instances of missing data (e.g., not null constraints) or incorrect data (e.g., domain ranges and check constraints). It is important to point out that the ability to enforce integrity constraints on inconsistent spatial data (e.g., geographical data such as sales territory alignments) and outdated temporal data is not supported in current database management systems.

Initial Heuristics for Designing CRM Data Warehouses

Once the types of CRM analyses that the data warehouse needs to be able to support have been identified, the data points have been identified, and the granularity has been selected, the next step is designing the data warehouse model to support the analyses that were identified. Based upon our initial findings, Table 8 lists initial heuristics for designing a data warehouse in order to successfully support CRM analyses.

CONCLUSION

In this article, we first present the design implications that CRM poses to data warehousing and then propose a robust multidimensional starter model that supports CRM analyses. Based upon sample queries, our model has a value of 1 and 0.93 for the $r_{success}$ and $r_{suitability}$ ratios, respectively. Our study shows that our starter model can be used to analyze various profitability analyses such as customer profitability analysis, market profitability analysis, product profitability analysis, and channel profitability analysis. In fact, the model has the flexibility to analyze both trends and overall lifetime value of customers, markets, channels, and products simply by including or excluding the time dimension in the SQL statements. Since the model captures rich descriptive non-numeric information that can be included in the query statement, the proposed model can return results that the user easily can understand. It should be noted that such rich information then can be used in data mining algorithms for such things as category labels. As such, we have demonstrated that the robust proposed model can be used to perform CRM analyses.

Table 8. Initial heuristics for designing CRM DWs

#	Heuristic	Benefit
1	Identify the types of CRM analyses, their uses and the data elements required to perform the analyses	The model will be able to support the intended purpose of the analyses
2	Include all attributes required to compute the profitability of each individual transaction in the fact table(s)	The ability to generate a profit & loss statement for each transaction, which can then be analyzed along any dimension
3	Each dimension that will be used to analyze the Profitability fact table should be directly related to the fact table	Provides improved query performance by allowing the use of simplified queries (i.e., support browsing data)
4	Pay careful attention to the Customer dimension	It forces attention to the customer to the center of CRM
5	Create a relationship between the Customer dimension and the Market and Sales Representative dimensions	Provides the ability to quickly determine the current market and Sales Representative for the customer by merely browsing the Customer dimension
6	Include the attrition date and reason for attrition attributes in the Customer dimension	Provides the ability to quickly determine if a customer is no longer a customer by browsing the Customer dimension only
7	Attributes that are likely to change at a different rate than other attributes in the same dimension should be in a separate dimension	Minimize the number of updates
8	Create a separate *existence* dimension for any entity that can have a discontinuous existence	Provides the ability to track the periods in which the instance of the entity is valid (needed to support some temporal queries)
9	Create a separate *existence* dimension for any attribute whose historical values must be kept	Provides the ability to track accurate historical values, even during periods of inactivity
10	Create a relationship between the Time dimension and each *existence* dimension	Provides the ability to perform temporal queries efficiently using descriptive attributes of the Time dimension
11	*Existence* dimensions should be in a direct relationship with their respective original dimensions	
12	There should always be a CustomerExistence dimension	The ability to track and perform analyses on customer attrition
13	If some products are either seasonal or if it is necessary to determine when products were discontinued, then create a ProductExistence dimension	The ability to perform analyses for seasonal and discontinued products
14	There should be a Household dimension and an ExtendedHousehold dimension	Provides the ability to perform Household analyses
15	The organizational hierarchical structure can be contained in one *Market* dimension	Provides the ability to maintain a history of the organizational changes, and the ability to perform analyses according to the organizational structure

Our contributions also include the identification of and classification of CRM queries and their uses, including KPIs; the introduction of a sampling technique to select the queries with which the model is tested; the introduction of two measures (percent success ratio and CRM suitability ratio) by which CRM data warehouse models can be evaluated; and the identification of the initial heuristics for designing a data warehouse to support CRM. Finally, in terms of future work, we plan to classify and test additional CRM analyses, evaluate alternative models using the same set of queries and the $r_{success}$ and $r_{suitability}$ ratios, identify materialized views that are relevant to CRM, and explore CRM query optimization.

ACKNOWLEDGMENT

The research of Peter Chen was partially supported by National Science Foundation grant ITR-0326387 and AFOSR grants F49620-03-1-0238, F49620-03-1-0239, and F49620-03-1-0241.

REFERENCES

Allen, G. N., & March, S. T. (2003). Modeling temporal dynamics for business systems. *Journal of Database Management, 14*(3), 21-36.

Boon, O., Corbitt, B., & Parker, C. (2002, December 2-3). Conceptualising the requirements of CRM from an organizational perspective: A review of the literature. In *Proceedings of 7th Australian Workshop on Requirements Engineering (AWRE2002)*, Melbourne, Australia.

Bose, R. (2002). Customer relationship management: Key components for IT success. *Industrial Management & Data Systems, 102*(2), 89-97.

Buttle, F. (1999). The S.C.O.P.E. of customer relationship management. *International Journal of Customer Relationship Management, 1*(4), 327-337.

Cunningham, C., Song, I-Y., Jung, J. T., & Chen, P. (2003, May 18-21). Design and research implications of customer relationship management on data warehousing and CRM decisions. In M. Khosrow-Pour (Ed.), *Information Technology & Organizations: Trends, Issues, Challenges & Solutions, 2003 Information Resources Management Association International Conference (IRMA 2003)* (pp. 82-85), Philadelphia, Pennsylvania, USA. Hershey, PA: Idea Group Publishing.

Dasu, T., Vesonder, G. T., & Wright, J. R. (2003, August 24-27). Data quality through knowledge engineering. In *Proceedings of SIGKDD '03*, Washington, DC.

Hawkes, V. A. (2000). The heart of the matter: The challenge of customer lifetime value. *CRM Forum Resources, 13*, 1-10.

Hwang, H., Jung, T., & Suh, E. (2004). An LTV model and customer segmentation based on customer value: A case study on the wireless telecommunication industry. *Expert Systems with Applications, 26*, 181-188.

Jain, D., & Singh, S. (2002). Customer lifetime value research in marketing: A review and future direction. *Journal of Interactive Marketing, 16*(2), 34-46.

Kale, S. H. (2004, September/October). CRM failure and the seven deadly sins. *Marketing Management, 13*(5), 42-46.

Kellen, V. (2002, March). *CRM measurement frameworks*. Retrieved July 1, 2002, from http://www.kellen.net/crmmeas.htm

Kim, W., Choi, B., Kim, S., & Lee, D. (2003). A taxonomy of dirty data. *Data Mining and Knowledge Discovery, 7*, 81-99.

Kimball, R., & Ross, M. (2002). *The data warehouse toolkit* (2nd ed.). New York: Wiley Computer Publishing.

Lee, Y. W., Pipino, L., Strong, D. M., & Wang, R. Y. (2004). Process-embedded data integrity. *Journal of Database Management, 15*(1), 87-103.

Massey, A. P., Montoya-Weiss, M. M., & Holcom, K. (2001). Re-engineering the customer relationship: Leveraging knowledge assets at IBM. *Decision Support Systems, 32*(2), 155-170.

Myron, D., & Ganeshram, R. (2002, July). The truth about CRM success & failure. *CRM Magazine, 6*(7). Retrieved July 1, 2002, from http://www.destinationcrm.com/articles/default.asp?ArticleID=2370

Panker, J. (2002, June). Are reports of CRM failure greatly exaggerated? *SearchCRM.com*. Retrieved July 1, 2002, from http://searchcrm.techtarget.com/originalContent/0,289142,sid11_gci834332,00.html

Rosset, S., Neumann, E., Eick, U., & Vatnik, N. (2003). Customer lifetime value models for decision support. *Data Mining and Knowledge Discovery, 7*, 321-339.

Rust, R. T., Lemon, K. N., & Zeithaml, V. A. (2004). Return on marketing: Using customer equity to focus marketing strategy. *Journal of Marketing, 68*(1), 109-139.

Shankaranarayan, G., Ziad, M., & Wang, R. Y. (2003). Managing data quality in dynamic decision environments: An information product approach. *Journal of Database Management, 14*(4), 14-32.

Todman, C. (2001). *Designing a data warehouse*. Upper Saddle River, NJ: Prentice Hall.

Verhoef, P. C., & Donkers, B. (2001). Predicting customer potential value: An application in the insurance industry. *Decision Support Systems, 32*(2), 189-199.

Winer, R. S. (2001). A framework for customer relationship management. *California Management Review, 43*(4), 89-108.

This work was previously published in Journal of Database Management, Vol. 17, Issue 2, edited by K. Siau, pp. 62-84, copyright 2006 by IGI Publishing, formerly known as Idea Group Publishing (an imprint of IGI Global).

Chapter 2.19
An Information-Theoretic Framework for Process Structure and Data Mining

Gianluigi Greco
University of Calabria, Italy

Antonella Guzzo
University of Calabria, Italy

Luigi Pontieri
Institute of High Performance Computing and Networks, Italy

INTRODUCTION

Process mining is a key technology for advanced business process management, aimed at supporting the (re)design phase of process-oriented systems like workflow management (WFM), enterprise resource planning (ERP), customer relationship management (CRM), business to business (B2B), and supply chain management (SCM) systems. In fact, based on the log data—a.k.a. transactional log or audit trail—that are gathered by these systems, process mining techniques (Herbst & Karagiannis, 2000; Schimm, 2003; van der Aalst et al., 2003; van der Aalst, Weijters, & Maruster, 2004) are designed to discover the underlying process model and constraints explaining the episodes recorded. The "mined" model provides the users with a syntectic view of the operations involved in the process, which can be the first step leading to supporting the process with a workflow system. Even if the process is already equipped with a workflow schema, such a model can profitably help in re-engineering that schema.

While the output of process mining techniques has been originally defined to be a *unique* model, recent research (Greco, Guzzo, Pontieri, & Saccà, 2006) has argued the importance of explicitly singling out the variants (i.e., the use cases) for the process at hand. This is particularly useful in the case of a complex process, possibly involving hundreds of activities, for which a unique schema may be an overly-detailed and inappropriate description of the actual process behavior, for it mixes semantically different scenarios. Technically, different variants of a process can be detected by partitioning the traces into clusters so that a different schema is eventually induced for each cluster, by way of traditional process mining algorithms.

However, despite the efforts spent in designing process mining techniques, the actual impact of such techniques in the industry is endangered by some simplifying assumptions. In particular, most of the approaches in the literature, and specifically the ones addressed to the mining of different process variants are *propositional* in that in order to extract a model for the process, they only take account for the sequence of task identifiers associated with logged instances, thereby completely disregarding all *non-structural* information kept by many real systems such as activity executors, time-stamps, parameter values, and various performance data.

Contributions

In this article, a further step toward enhancing the process mining framework is made by presenting an approach to discovering variants for a process by means of a technique for clustering log traces, which takes care of both structural and non-structural aspects. Specifically, beside the list of activity identifiers, each trace is also equipped with a number of *metrics*, which are meant to characterize some performance measures for the enactment of the process such as the total processing time and the quality of the process. These measures are very relevant from a business process intelligence perspective; in fact, the need of getting explanations for why a certain metric has a certain value and for predicting the value of such metrics in forthcoming executions has clearly emerged in advanced frameworks geared for the industry (Casati, Castellanos, Dayal, & Shan, 2005).

In order to take care of the different (both structural and not-structural) execution facets in the clustering, we introduce and discuss an information-theoretic framework that extends previous formalizations in Dhillon, Mallela, & Modha (2003) and Berkhin & Becher (2002), in a way that the structural information as well as each of the performance measures is represented by a proper domain, which is correlated to the

Figure 1. A workflow schema for process HANDLEORDER

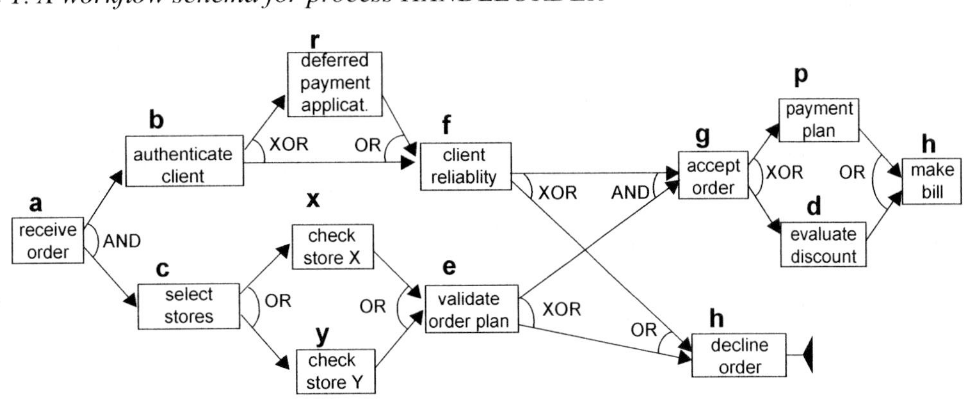

"central domain" of process instances according to the contents of the log.

The clustering of log traces is then performed synergically with the clustering of structural elements and different performance measures. More specifically, our proposal mainly consists in partitioning each of these domains (i.e., traces, activities, and any set of values produced by a single performance metrics) in a way that best fits their mutual correlations--where the minimization of mutual information loss is taken as optimality criterion. In addition, a series of workflow models can be discovered by way of classical process mining techniques in order to effectively describe each cluster of workflow traces.

The simultaneous clustering of *two* different types of objects (short: bi-dimensional *co-clustering*), such as documents and terms in text corpora or genes and conditions in micro-array data, has become a very active research area. In fact, co-clustering techniques turned out to be more effective than traditional clustering approaches in many application contexts, mainly because they enable for partitioning either dimension by way of a proper subset of relevant features (i.e., a cluster in the other dimension). Indeed, such an implicit feature-selection effect, somewhat similar to that of subspace clustering (Parson, Haque, & Liu, 2004), can help in achieving good performances even for sparse and high-dimensional datasets.

However, in order to profitably apply such a co-clustering approach in a process mining setting, we extend it in different respects:

- First, in order to deal with both the structural information (activities) and the performance measures, we consider co-clustering over (at least) three different domains, which has been marginally addressed in the literature. To this purpose, we define an information-theoretic scheme that allows to co-cluster one central domain, collecting workflow traces, and a series of associated domains, corresponding to the set of process activities and to different performance metrics.
- Second, we tackle the co-clustering problem in a scenario where some domains consists of numerical (ordered) values. This problem has not been investigated before and requires the definition of ad-hoc techniques to mine numerical (co)clusters as non-overlapping ranges of values. To this aim, we propose and study an exact (but time-consuming) partitioning method, as well as a faster, greedy, heuristic that provides an approximate solution to the problem.

The approach we devised is illustrated in the article with the help of a formal framework and of a series of algorithms, which are discussed and studied from a theoretical point of view. Moreover, in order to assess the validity of the approach, we implemented all these algorithms and tested them against different data sets.

Plan of the Article

The rest of the article is organized as follows. Section "Formal Framework" proposes the formal framework for process mining based on information-theoretic co-clustering. The following two sections illustrate, respectively, the approach for co-clustering log traces and associated performance metrics, and two alternative methods for partitioning a numerical domain, which can be both exploited in our co-clustering scheme. Subsequently, some results of experimental activity are discussed in Section "Experiments." After presenting a brief overview of related works, we finally draw out our conclusions and outline some directions for further investigation.

FORMAL FRAMEWORK

In this section, we present the formal (information-theoretic) framework for co-clustering log traces based on both structural and non-structural information. Actually, we start by discussing a simple model for processes and log traces.

Workflows: Schemas, Logs, and Metrics Data

Several models and languages have been proposed in the literature to describe process models (van der Aalst & van Hee, 2002). To our aims, elaborate facets of the specifications (e.g., transactional aspects) are not of interest since process mining generally looks at processes from the *control flow* perspective (i.e., by focusing on the activities occurring in the process and the constraints on the precedences among them. Formally, a *workflow schema* is a tuple $\langle D_A, E, a_0, F\rangle$, where D_A is a finite set of *activities*, $E \subseteq (D_A - F)\times(D_A-\{a_0\})$ is a relation of precedences between activities, $a_0 \in D_A$ is the starting activity, $F \subseteq D_A$ is the set of final activities. A workflow schema is usually extended by specifying routing constructs such as parallelism, synchronization, and exclusive choice. More precisely, for any activity a, the split constraint for a is: *(S.i) AND-split* if a activates all of its successor activities, once completed; *(S.ii) OR-split*, if a may activate any number of its successors, once completed; *(S.iii) XOR-split* if a activates exactly one out of all its successor activities, once completed. The join constraint for a is: *(J.i) AND-join* if a can be executed only after all of its predecessors have notified a to start; *(J.ii) OR-join*, if a can be executed as soon as one of its predecessors notifies a to start.

Example 1. As a sample applicative scenario, consider the toy HANDLEORDER *process for handling customers' orders in a business company, which is graphically illustrated in Figure 1. Here, edges clearly represent precedence relationships, while additional constraints are expressed via labels associated with each activity node. For example, task g is an* AND-join *activity, as it must be notified by its predecessors that both the client is reliable and the order can be supplied correctly. Conversely, task c is a* XOR-split *activity, since just one of the stores is checked, based on the product requested in the order—we here assume that the two stores contain different kinds of goods, and that each order can refer to only one product.*

Each time a workflow schema W is enacted in a workflow management system, it produces an *instance* (i.e., a suitable subgraph of $(W.D_A, W.E)$) containing both the initial activity $W.a_0$ and a final activity from $W.F$, that satisfies all the constraints. Actually, many process-oriented systems store partial information about the various instances of a process by tracing events related to the execution of its activities. In particular, the *logs* kept by most such systems simply consist of sequences of event occurrences. By abstracting from their heterogeneity, we simply view a process log as a bag of *traces over* D_A, where each trace s simply is a string over the alphabet D_A representing an ordered sequence of activities occurrences (i.e., $s \in D_A^*$). Given a workflow log, the process mining problem consists in discovering a workflow schema that represents the traces registered in the log in a compact and accurate way. In the following, we introduce some concepts that allow to better characterize the accuracy of a workflow schema in modelling a given workflow log. A trace s is *compliant with* W if the first (resp., last) activity of s is a starting (resp., final) activity w.r.t. to W and there is an instance I of W such that s is a topological sort of I.

Finally, two different functions can be exploited to evaluate the accuracy of W w.r.t. a given log L: *(i) soundness*(W,L), expressing the percentage of instances of W which have some corresponding traces in L, and *(ii) completeness*(W,L), which measures the percentage of traces in L that are compliant with W. These measures can be combined into an overall accuracy score defined as follows: *accuracy*(W,L)= min{ *soundness*(W,L), *completeness*(W,L) }.

Example 2 (...continues): An example log data for process HandleOrder is shown in Table 1, where for each enactment of the process the sequence of executed activities is reported, along with a single performance measure (i.e., the total execution time). Let us now evaluate the accuracy of the workflow schema in Figure 1 with respect to this log. Since all the traces in Table 1 are compliant with the schema, its completeness is maximal (i.e., it is 1). Conversely, the schema admits some spurious execution patterns that never appear in the log, e.g., instances containing both r and p, as well as instances where neither r or d appear. In particular, a half of all possible instances that can be generated by the schema do not correspond to any trace in the log; thus, the soundness of the schema against that log is only 0.5.

The basic idea for improving the quality of the discovered process model is to partition the log into a set of structurally homogeneous clusters, and then derive a different workflow schema for each of these clusters. As a result, we obtain a set of more specific workflow schema that collectively represent the log in a more accurate way, for a lower percentage of spurious instances is modeled.

For instance, in the example log shown in Table 1, we can notice that a, b, c, e, f, g, and h are executed in all the instances of *HandleOrder*, while the other ones appear in a subset of them only. Moreover, every trace containing r also includes p and does not contain d; conversely, d appears any time r does not. Based on these observations, we can recognize two (*structurally*) homogeneous clusters of traces, each one representing a variant of the process: $\{t_1,\ldots t_8\}$ (orders with deferred payment) and $\{t_9,\ldots t_{12}\}$ (orders with immediate payment). By providing each cluster with a specific workflow schema, we can get a more accurate model than that in Figure 1.

The scenario described so far roughly depicts the typical the process-mining setting considered in the literature. However, in several real situations, further data is available in workflow logs, besides pure information on the sequencing of activities. In particular, traces can be associated with some "non-structural" information, involving, for example, data parameters, executors, execution times, and a possibly wide range of additional performance metrics. We believe that such information can be exploited as well in order to extract interesting knowledge on the different ways a workflow process has been executed in the past. For instance, in the example log shown in Table 1, one can note that instances including both r and x take longer than the others, as a result, perhaps, of some additional verification procedure done at store *X* when the ordered goods are being paid in a deferred way.

Throughout the article, we shall focus only on performance metrics that can be modelled as functions that take values from some fixed numerical domain. For instance, for each trace in Table 1 just one additional measure is reported, measuring the total execution time spent for executing the process.

Table 1. A log for process HandleOrder: traces and associated execution time.

ID	task sequence	time
t_1	abcrxefgph	8
t_2	abcrxfegph	9
t_3	acbrxefgph	9
t_4	abcrxefgph	8
t_5	acybrefgph	6
t_6	abcryefgph	4
t_7	abrcyefgph	4
t_8	abcyrfegph	5
t_9	abcyefgdh	3
t_{10}	abcxfyegdh	2
t_{11}	acbfyegdh	1
t_{12}	abfcxegdh	1

More generally, a *process log L* over D_A is a tuple $\langle D_A, D_T, F_{Z^1}, \ldots, F_{Z^N} \rangle$ such that:

- D_T is a set of (identifiers of) traces over D_A,
- F_{Z^i}, with i=1..N, is a function associating a trace in D_T with a real value (expressing a performance measure according to some given metrics), that is, $F_{Z^i} : D_T \mapsto \Re$.

For each performance measure F_{Z^i} with i=1.. N, we define its *active domain*, denoted by D_{Z^i}, as the image of F_{Z^i} over T (i.e., $D_{Z^i} = \{F_{Z^i}(t_j) \mid t_j \in D_T\}$), where for any trace j in T $F_{Z^i}(t_j)$ is the value function F_{Z^i} assigns to t_j.

For instance, in the case of the example process *HandleOrder*, it is $D_T = \{t_1, \ldots, t_{12}\}$, while the active domain of durations is D_{Z^i} ={1,2,3,4,5,8,9}.

It is worth noticing that it would be easy to extend the definition of workflow log previously given in order to take account for an arbitrary number of non-structural information other than pure performance metrics (such as, e.g., executors, parameter values, and time-stamps). However, for the sake of simplicity, we just consider the set of activities as a prototypical example of "categorical domain," yet pinpointing that any such domain can be handled in the same way as the activities' one.

The Information-Theoretic Framework

In this section, we define the basic framework for co-clustering heterogeneous data by specializing and extending the approach in Dhillon et al. (2003).

Based on the data stored in a process log L, we can establish correlations among the *central domain* T, representing the logged process instances, and the *auxiliary* domains D_A, F_{Z^1}, ..., F_{Z^N}. Such correlations can be modelled by considering a random variable T ranging over T, along with a random variable X associated with each auxiliary domain X, where X is a placeholder for any of the auxiliary variables $A, Z^1, \ldots, Z^N$. Then, $p(T,X)$ is to denote the joint-probability distribution between T and X. In particular, $p(t_j,x)$ is the probability for the event that T is the trace t_j *and* X takes the value x.

The problem we want to face is co-clustering the traces in D_T, as well as the structural data encoded in D_A and non-structural data encoded in $D_{Z^1}, \ldots, D_{Z^N}$. More precisely, assume that D_T, D_A and any D_{Z^i} are to be partitioned into, respectively, m^T clusters, say $\hat{D}_T = \{\hat{t}_1, \ldots, \hat{t}_{m^T}\}$, m^A clusters, say $\hat{D}_A = \{\hat{a}_1, \ldots, \hat{a}_{m^A}\}$, and m^i clusters, say $\hat{D}_Z = \{\hat{z}^i_1, \ldots, \hat{z}^i_{m^i}\}$. Then, a *co-clustering* of L is a tuple $\langle C_T, C_A, C_{Z^1}, \ldots, C_{Z^N} \rangle$, where $C_T : D_T \mapsto \hat{D}_T$, $C_A : D_A \mapsto \hat{D}_A$ and, for i=1...N, $C_{Z^i} : D_{Z^i} \mapsto \hat{D}_{Z^i}$, i.e., T (resp., C_A, $C_{Z^1}, \ldots, C_{Z^N}$) is a function that partitions T (resp., D_A, $D_{Z^1}, \ldots, D_{Z^N}$) by assigning its elements to the clusters in $\hat{D}_T$ (resp., $\hat{D}_A$, $\hat{D}_{Z^1}, \ldots, \hat{D}_{Z^N}$).

Notice that, like in Dhillon et al. (2003), for any domain, lower-case letters are used to denote its elements, and upper-case letters to denote the random variable ranging over it; moreover, clusters and clustered random variables are both referred to by hatted letters.

As the effect of co-clustering can be thought of as sort of information compression, the co-clustering problem can be turned into the search of a fixed-size compression scheme that preserves as much as possible of the original mutual information among the domains.

To this aim, for any auxiliary variable $X \in \{A, Z^1, \ldots, Z^N\}$, we can evaluate the *mutual information* $I(T;X)$ between the random variables T and X, ranging over T and X, respectively. As observed in Dhillon et al. (2003), the search of "good" co-clusters along D_T and D_X can be accomplished by minimizing the *loss in mutual information* $\Delta I_X = I(T;X) - I(T_{C_T}; X_{C_X})$ between the random variables T and X and their clustered versions. Indeed, by minimizing the loss in mutual information, one can expect that the random variables T and X retain as much information as possible from the original distributions T and X.

An algorithm for the *bi-dimensional* co-clustering (cf. just one auxiliary domain X is co-clustered with T) is proposed in Dhillon et al. (2003), which relies on expressing ΔI_X as a dissimilarity between the original distribution $p(T,X)$ and a function $q(T,X)$ approximating it: $\Delta I_X = \mathcal{D}(p(T, X) \mid q(T, X))$, where $\mathcal{D}(\cdot\|\cdot)$ denotes the Kullback-Leibler divergence $\mathcal{D}(x, y) = \Sigma_{i,j}\, p(x_i) \cdot \log((p(x_i) / q(y_j))$ and $q(T,X)$ is a function of the form:

$$q(T, X)= p(T, X) \cdot p(T|\hat{T}) \cdot p(X|\hat{X}) \quad (1)$$

which preserves all marginals of $p(T,X)$, and is univocally determined by the co-clustering at hand.

Thus, the bi-dimensional co-clustering problem can be solved by searching the function $q(T,X)$ that is most similar to $p(T,X)$ according to D. To this aim one can express ΔI_X in terms of "individual" loss contributes measuring the (KL-based) dissimilarity between each element in D_X (or, D_T) and the cluster it was assigned to:

$$\Delta I_X = \sum_{x \in D_X} \delta_T(x, C_X(x)) = \sum_{t \in D_T} \delta_X(t, C_T(t)) \quad (2)$$

In fact, the algorithm in Dhillon et al. (2003) searches for the function $q(T,X)$ that is most similar to $p(T,X)$ by an alternate minimization scheme, which considers one dimension (i.e., X or T) per time, and iteratively assigns each element in D_X (resp., D_T) to the cluster that is the "most similar" to it, according to coefficient δ_T (resp., δ_X).

A similar strategy is exploited in our multi-dimensional co-clustering approach, described next, where the case of N>1 is investigated. Notably, in such a case, the co-clustering solution is wanted to guarantee a low value for every pairwise information-loss function.

MULTI-DIMENSIONAL CO-CLUSTERING OF WORKFLOW LOGS

When more than one auxiliary domains are considered, the ultimate goal is to discover a clustering function for each of them, so that a low value for all the pairwise information loss functions is obtained.

A major problem, in this multi-dimensional extension, is that there may well exists two auxiliary dimensions $X, X' \in \{A, Z^1,...Z^N\}$, with $X \neq X'$, such that the best co-clustering solution for the pair D_T and D_X does not conform with the best co-clustering for D_T and $D_{X'}$, as concerns the partitions of traces in D_T.

Our solution to jointly optimize all the pairwise loss functions is to linearly combine them in a global one, by using N+1 weights $\beta_A, \beta_1 .... \beta_N$ with $\beta_A + \Sigma_i \beta_i = 1$ which are meant to quantify the relevance that the corresponding auxiliary domain should have in determining the co-clustering of log data. Therefore, the co-clustering problem for the log L can be rephrased into finding the set of clusters that minimize the quantity:

$$\beta_A \cdot \Delta I_A + \sum_{i=1}^{N} \beta_i \cdot \Delta I_{Z_i} \quad (3)$$

Definition 3: Let $\mathcal{L} = \langle D_A, F_{Z^1}, ..., F_{Z^N} \rangle$ *be a process log, and* $\beta_A, \beta_1 .... \beta_N$ *be real numbers, such that* $\beta_A + \Sigma_i \beta_i = 1$*. Then, a co-clustering* $C = \langle C_T, C_A, C_{Z^1}, ..., C_{Z^N} \rangle$ *of* $\mathcal{L}$ *is optimal if it is a minimum for the function* $\beta_A \cdot \Delta I_A (C_T, C_A) + \Sigma_{i=1..N}\, \beta_i \cdot \Delta I_Z^i (C_T, C_Z^i)$*, i.e., for each co-clustering* C' *of* L*, it holds:* $\beta_A \cdot \Delta I_A (C_T, C_A) + \Sigma_{i=1..N}\, \beta_i \cdot \Delta I_Z^i (C_T, C_Z^i) \geq \beta_A \cdot \Delta I_A (C'_T, C'_A) + \Sigma_{i=1..N}\, \beta_i \cdot \Delta I_Z^i (C'_T, C'^i_Z)$.

In order to compute an optimal co-clustering, an algorithm, called *LogCC*, is, shown in Figure 2. The algorithm takes as input a log $\mathcal{L} = \langle D_A, F_{Z^1}, ..., F_{Z^N} \rangle$, the numbers of clusters $m^T, m^A, m^1, ..., m^N$ required for all domains and (resp., $D_A, D_T, D_{Z^1}, ..., D_{Z^N}$), and the weights $\beta_A, \beta_1, ..., \beta_N$ for the auxiliary domains (resp., $D_A, D_T, D_{Z^1}, ..., D_{Z^N}$).

First, an initial co-clustering $C = \langle C_T^0, C_A^0, C_{Z^1}^0, ..., C_{Z^N}^0 \rangle$ is computed, which is eventually refined in the main loop.

The refinement is done in an alternate manner. Indeed, at each iteration, say s, a (locally) optimal clustering $C_A^{(s)}$ (resp., $C_{Z^i}^{(s)}$) is computed for the activity domain D_A (resp., for each metrics domain

, with $1 \leq i \leq N$), based on the current clustering of the traces in T. Specifically, the cluster assigned to any activity A is computed as follows:

$$C_T^{(S)}(t) = \arg\min_{t \in D_T}(\beta \mathcal{D}\ (p\,(T \mid a) \,\|\, q^{(s-1)}(T \mid a)) \quad (4)$$

where $q(T,A)$ is a distribution approximating $p(T,A)$, and preserving its marginals.

Note that the clustering $C_{Z^i}^{(s)}$, for each (ordered) numerical domain , is computed by function Partition, which splits the domain into a number of intervals, and assigns each of them to one cluster in $\hat{D}_{Z^i}$. The function, which will be discussed later, also takes as input the distribution $q^{(s-1)}$, accounting for the current co-clustering solution. Based on the clusters of A, ,...., , an optimal clustering $C_T^{(S)}$ is computed as follows:

$$C_T^{(S)}(t) = \arg\min_{\hat{t} \in \hat{D}_T} \begin{pmatrix} \beta_A \mathcal{D}(p(A \mid t) \,\|\, q^{(s-1)}(A \mid \hat{t}) + \\ \sum_{i=1}^{N} \beta_i \ \left(p(Z^i \mid t) \,\|\, q^{(s-1)}(A \mid \hat{t})\right) \end{pmatrix} \quad (5)$$

For each trace cluster we eventually derive a workflow schema by way of function MineSchema, implementing some standard process mining algorithm.

Note that when applying LogCC to D_T, D_A, ,...., it is not generally ensured that every information loss function ΔI_X monotonically decreases, for each $X \in \{A, Z^1, \ldots, Z^N\}$. Conversely, the (global) objective function (i.e., $\beta_A \cdot \Delta I_A + \Sigma_{i=1..N}\ \beta_i \cdot \Delta I_i$) is guaranteed to converge to a local optimum, under some technical conditions on the implementation of Partition, which are discussed below.

Definition 4: Let D_Z be a numerical domain, with an associated set $\hat{D}_Z$ of cluster labels, while C_Z: $D_Z \mapsto \hat{D}_Z$ and C_T: $D_T \mapsto \hat{D}_T$ denote two clustering for Z and T, respectively. Moreover, let $q^{C_T,C_Z}(T,Z)$ be a distribution of the form as in Equation 1, entirely determined by T and Z. Then a split function P is a function that computes a new clustering from Z to $\hat{D}_Z$ based on q^{C_T,C_Z}. Moreover, we say that P is loss-safe if, for any D_T, D_Z, C_T, C_Z, it is ΔI_Z (CT, $\mathcal{P}(D_Z, \hat{D}_Z\, q^{C_T,C_Z})) \leq \Delta I_Z(C_T, C_Z)$

*Figure 2. Algorithm **LogCC**: Discovery of co-clusters over workflow traces and correlated data*

Input: a process log $\mathcal{L} = \langle D_T, D_A, \mathcal{F}_{Z^1}, \ldots, \mathcal{F}_{Z^N}\rangle$, real nums $\beta_A, \beta_1, \ldots, \beta_N$, and cardinal nums $m^T, m^A, m^1, \ldots, m^N$

Output: A co-clustering for $\mathcal{L}$, and a set of workflow schemas $\mathcal{W}$;

Extract joint probability functions $p(T,A), p(T,Z^1), \ldots, p(T,Z^N)$, out of $\mathcal{L}$;
Define an arbitrary co-clustering $\langle C_T^0, C_A^0, C_{Z^1}^0, \ldots, C_{Z^N}^0\rangle$ for $\mathcal{L}$;
Compute $q^{(0)}(T,A)$, and $q^{(0)}(T,Z^i)$ for $= 1 \ldots N$;
let $s = 0$, $\Delta I_A^{(0)} = \mathcal{D}(p(T,A) \| q^{(0)}(T,Z^i))$, and $\Delta I_i^{(0)} = \mathcal{D}(p(T,Z^i) \| q^{(0)}(T,Z^i))$;
repeat
  Compute $q^{(s)}(T,A), q^{(s)}(T,Z^i)$ for $i = 1 \ldots N$, and **set** $s = s+1$;
  for each $a \in D_A$ **do** $C_A^{(s)}(a) := \arg\min_{\hat{a} \in \hat{D}_A} \delta_T(a,\hat{a})$;
  for each Z^i **do** $C_{Z^i}^{(s)} :=$ Partition($D_{Z^i}, \hat{D}_{Z^i}, q^{(s-1)}(T,Z^i)$);
  Compute $q^{(s)}(T,A), q^{(s)}(T,Z^i)$ for $i = 1 \ldots N$, and **set** $s = s+1$;
  for each $t \in D_T$ **do** $C_T^{(s)}(t) := \arg\min_{\hat{t} \in \hat{D}_T}\left(\beta_A \delta_A(t,\hat{t}) + \sum_{i=1}^{N} \beta_i \delta_{Z^i}(t,\hat{t})\right)$;
  let $\Delta I_A^{(s)} = \mathcal{D}(p(T,A)^{(s-1)}) \| q(T,A)^{(s-1)}))$;
  let $\Delta I_i^{(s)} = \mathcal{D}(p(T,Z^i)^{(s-1)}) \| q(T,Z^i)^{(t-1)}))$, $\forall Z^i$;
while $\beta_A \Delta I_A^{(s)} + \sum_{i=1}^{N} \beta_i \Delta I_i^{(t)} < \beta_A \Delta I_A^{(s-2)} + \sum_{i=1}^{N} \beta_i \Delta I_i^{(s-2)}$;
$C_T^* := C_T^{(s-2)}$, $C_A^* = C_A^{(s-3)}$, $C_{Z^1}^* = C_{Z^i}^{(s-3)}$, for $= 1 \ldots N$;
for each $\hat{t} \in \hat{D}_T$ **do** $W(\hat{t}) :=$ MineSchema($\{\hat{t} \in \hat{D}_T \mid C_T^*(t) = \hat{t}\}$);
return $\langle C_X^*, C_{Y^1}^*, \ldots, C_{Y^N}^*\rangle$ and $\mathcal{W} = \{W(\hat{t}) \mid \hat{t} \in \hat{D}_T\}$;

Next, we show that any loss-safe implementation of function Partition guarantees convergence to local optima.

Theorem 5: Provided that the split function Partition is loss-safe, algorithm LogCC converges to a local optimum, i.e., to a co-clustering for L such that for any other co-clustering $C' = \langle C'_T, C'_A, C'_{Z^1}, \ldots, C'_{Z^N} \rangle$ *of L it is:*

(a) $\Delta I_X(C_T, C'_X) \geq \Delta I_X(C_T, C_X)\ \forall X \in \{A, Z^1, \ldots, Z^N\}$,

(b) $\beta_A \cdot \Delta I_A(C'_T, C_A) + \Sigma_{i=1..N} \beta_i \cdot \Delta I_Z^i(C'_T, C_Z^i) \geq \beta_A \cdot \Delta I_A(C_T, C_A) + \Sigma_{i=1..N} \beta_i \cdot \Delta I_Z^i(C_T, C_Z^i)$.

Proof The result can be proved by exploiting similar arguments to those in Dhillon et al. (2003), where the proposed bi-clustering algorithm was shown to monotonically decrease the mutual information loss between the two domains considered. In fact, we can compute each pairwise objective function ΔI_X, with $X \in \{A, Z^1, \ldots, Z^N\}$, via a weighted sum of the KL dissimilarity between the distribution $p(X \mid t)$ (resp., $p(T \mid x)$) of each $t \in D_T$ (resp., $x \in D_X$) and the "prototype" $q(X \mid \hat{t}$) (resp., $q(T \mid \hat{x}')$ of the cluster it is currently assigned to—see Equation 2. Basically, this entails that each of the updates carried out according to Equation 4 and Equation 5 cannot yield an increase of the mutual information loss A. Therefore, at each iteration s the it is $\beta_A \Delta I_A^{(s)} \leq \beta_A \Delta I_A^{(s-2)}$. On the other hand, by assuming that algorithm *LogCC* is loss-safe, after any application of it over whatever numerical domain D_{Z^i} it will be $\Delta I_{Z^i}^{(s)} \leq \Delta I_{Z^i}^{(s-2)}$. Putting it all together, after each step of the main loop the following relationship holds:

$$\beta_A \Delta I_A^{(s)} + \sum_{i=1}^{N} \beta_i I_{Z^i}^{(s)} \leq \beta_A \Delta I_A^{(s-2)} + \sum_{i=1}^{N} \beta_i \Delta I_{Z^i}^{(s-2)}$$

Thus, the global objective function is monotonically decreasing, and it eventually converges to a local optimum, because of the structure of the exit condition.

Partitioning Numerical Domains

We next discuss two possible implementations of function partition, computing a proper clustering function C_Z, from a numerical domain $D_Z = \{z_1, \ldots, z_n\}$ to a set of clusters $\hat{D}_Z = \{\hat{z}_1, \ldots, \hat{z}_m\}$.

Differently from the case of generic domains, values in D_Z are disposed in an ordered way (i.e., $z_u < z_v$ for each u, v such that $u < v$) and this ordering must still hold among the clusters (i.e., $\forall \hat{z}_k, \hat{z}_l \in \hat{D}_Z$ s.t. $k < l$). Any value assigned to $\hat{z}_k$ precedes all assigned to $\hat{z}_l$. This requirement clearly makes inappropriate the scheme proposed in Dhillon et al. (2003) and used in algorithm *LogCC* for clustering the "non-numerical" domains D_A and D_T – where elements in the domain are clustered independently of each other.

The problem exposed so far, hereinafter named NDP problem, can be formally defined in terms of an Integer Linear Programming formulation, where the clustering function to compute is encoded by using a boolean variable $w_{i,l}$ for each element z_i and each cluster $\hat{z}_l$, such that $w_{i,l} = 1$ only when z_i is assigned to $\hat{z}_l$.

$$\begin{cases} min\, \mathcal{G}^{C_Z}(D_Z, \hat{D}_Z) = \sum_{l=1}^{m} \sum_{i=1}^{n} w_{i,l} \cdot \delta(z_i, \hat{z}_l) & \\ C_Z(z_i)) = \hat{z}_l \ \ iff \ \ w_{i,l} = 1 & \\ w_{i,l} \in \{0,1\}, \forall i \in [1..n], \ and \ \ \forall l \in [1..m] & (a) \\ \sum_{l=l}^{m} w_{i,l} = 1, \forall i \in [1..n] & (b) \\ w_{i,l} + w_{i+2,l} \leq w_{i+1,l} + 1, \forall l \in [1..m] & (c) \end{cases}$$

ster labels

le q(T, Z)

where $\delta_T(z_i, \hat{z}_l) = p(z_i) \cdot \mathcal{D}(p(T \mid z_i) \| q(T \mid \hat{z}_l))$, as introduced in Equation 2, while $[i..j]$ stands for the range of integer values from i to j.

Notice that the combination of conditions *(a)* and *(b)* leads to a hard clustering of D_Z, while condition *(c)* forces each cluster to encompass an entire range of values in D_Z.

The term $\mathcal{G}^{C_Z}(D_Z, \hat{D}_Z)$, called *partition cost* of C_Z from now on, and used as the objective function of problem NDP, clearly coincides with the mutual information loss ΔI_Z produced by C_Z (along with C_T, here considered as fixed).

An alternative way to encode the clustering function C_Z, making more explicit the requirements on values' contiguousness, is as follows: Let $b_1,...,b_{m-1}$ be $m-1$ boundary indexes such that $1 \leq b_r \leq b_s \leq n+1$ for every r, s, and $1 \leq r < s < m$. A clustering function C_Z can be defined based on these indexes as follows: $C_Z(z_i) = \hat{z}_l$ iff $b_{l-1} \leq i \leq b_l$, where for notation convenience two fixed, additional bounds are considered: $b_0=1$ and $b_m = n+1$. Under these assumptions, problem NDP turns in finding such $m-1$ boundary indexes such that the resulting clustering function C_Z gets the minimum value for:

$$\mathcal{G}^{C_Z}(D_Z, D_Z) = \sum_{z_l \in D_Z} \sum_{i=b_{l-1}}^{b_l - 1} \delta(z_i, z_l) \quad (6)$$

A Dynamic Programming Implementation of Partition: Algorithm NDP-OPT

In order to get a recursive formulation of the problem, let us consider the problem of clustering a range $D_Z^{[u,v]} = \{z_u, ..., z_v\}$ of D_Z (i.e., a contiguous set of values in D_Z), by using a range $\hat{D}_Z^{[r,s]} = \{\hat{z}_r, ..., \hat{z}_s\}$ of $\hat{D}_Z$ (i.e., a contiguous set of clusters in $\hat{D}_Z$).

In order to optimally solve this sub-problem, a suitable function $C_Z^{[u,v]} : D_Z^{[u,v]} \mapsto D_Z^{[r,s]}$ is to be found, such that $\mathcal{G}^{C_Z^{[u,v]}}(D_Z^{[u,v]}, D_Z^{[r,s]})$ is minimal.

In the following theorem, it is shown that such an optimal value for $\mathcal{G}^{C_Z^{[u,v]}}$ can, in fact, be expressed in terms of the optimal solutions of some sub-problems of NDP(D_Z, $\hat{D}_Z$, $q(T, Z)$).

Theorem 7: Let $\mathcal{G}^*(D_Z^{[u,v]}, \hat{D}_Z^{[r,s]})$ *be the minimal partition cost resulting from partitioning* $D_Z^{[u,v]}$ *into the clusters in* $\hat{D}_Z^{[r,s]}$*, with* $l \leq u \leq v \leq n$ *and* $l \leq r \leq s \leq m$*, that is,* $G^*(D_Z^{[u,v]}, \hat{D}_Z^{[r,s]}) = \min_{C_Z[u,v]} G^{C_Z^{[u,v]}}(D_Z^{[u,v]}, \hat{D}_Z^{[r,s]})$. *Then it holds:*

$$\mathcal{G}^*(D_Z^{[u,v]}, \hat{D}_Z^{[r,s]}) = \min_{u \leq \mu \leq v} \begin{bmatrix} \mathcal{G}^*(D_Z^{[u,\mu]}, \hat{D}_Z^{[r,s-1]}) \\ +\mathcal{G}^*(D_Z^{[\mu+1,v]}, \hat{D}_Z^{[s,s]}) \end{bmatrix}$$

where computing the right-side term is trivial as no real partition is required: $\mathcal{G}^*(D_Z^{[\mu+1,v]}, \hat{D}_Z^{[s,s]}) = \sum_{i=\mu+1}^{v} \delta(z_i, \hat{z}_s)$.

Proof Let $C_Z^{[u,v]}$ be a generic function from $D_Z^{[u,v]}$ to $\hat{D}_Z^{[r,s]}$, and $B = \{b_{r-1}, b_r, ..., b_{s-1}, b_s\}$ be the associated set of boundary indexes in $[u.v]$, with $b_{r-1} = u$ and $b_s = v+1$. Then it is: $\mathcal{G}^*(D_Z^{[u,v]}, \hat{D}_Z^{[r,s]}) = \min_B \mathcal{G}^{C_Z^{[u,v]}}(D_Z^{[u,v]}, \hat{D}_Z^{[r,s]})$ $= = \min_B \sum_{l=r}^{s} \sum_{i=b_{l-1}}^{b_l - 1} \delta(i,l) = \min_{b_{s-1}} \left[\min_{B-\{b_{s-1}\}} \left(\sum_{l=r}^{s-1} \sum_{i=b_{l-1}}^{b_l - 1} \delta(i,l) + \sum_{i=b_{s-1}}^{v} (i,l) \right) \right] =$

$$\min_{b_{s-1}} \left[\mathcal{G}^*(D_Z^{[u,b_{s-1}]}, \hat{D}_Z^{[r,s-1]}) + \mathcal{G}^{\varnothing}(D_Z^{[b_{s-1}+1,v]}, \hat{D}_Z^{[s,s]}) \right]$$

Figure 3. Algorithm NDP-OPT*: A dynamic programming implementation of partition*

```
Input:   A domain D_Z, a set of clusters D̂_Z = {ẑ_1, ..., ẑ_m};
Output:  A clustering function from D_Z to D̂_Z;

Compute c_l(i) = Σ_{0<i≤u} δ_T(z_i, ẑ_l), ∀i ∈ [1..n] and ∀l ∈ [1..m];
g(1, i) := c_1(i), ∀i ∈ [1..n];
for s = 2..m do
    for i = 1..n do
        g(s, i) := min_{μ∈[1..i]}( g(s − 1, μ) + c_s(i) − c_s(μ) );
        h(s, i) := arg min_{μ∈[1..i]}( g(s − 1, μ) + c_s(i) − c_s(μ) );
    end for
Let b_m = n + 1, and b_0 = 1;
for l = 1..m do
    Let b_{m−l} = h(m − l + 1, b_{m−l+1});
Set C_Z(z_i) := ẑ_l for each l ∈ [1..m] and for each i ∈ [b_{l−1} + 1..b_l];
return C_Z;
```

The result clearly holds in general, as b_{s-1} is a generic index in $[u.v]$. An exact solution to problem NDP can be computed by the algorithm shown in Figure 3. The core idea is to consider the problem of clustering a range $D_Z^{[u,v]}=\{z_u,\ldots, z_v\}$ of D_Z by only using a range $\hat{D}_Z^{[r,s]}=\{\hat{z}_r,\ldots,\hat{z}_s\}$ of clusters from $\hat{D}_Z$. Based on Theorem 7, this sub-problem can be solved by finding a function $C_Z^{[u,v]}: D_Z^{[u,v]} \mapsto \hat{D}_Z^{[r,s]}$ that minimizes the value $\mathcal{G}^{C_Z^{[u,v]}}(D_Z^{[u,v]},\hat{D}_Z^{[r,s]})$.

Algorithm NDP-OPT is, in fact, a dynamic-programming implementation of Theorem 7. Indeed, at each step, $g(s,i)$ will store the minimal cost of splitting the first i values in D_Z among the first s clusters, that is, $g(s,i)=\mathcal{G}^*(D_Z^{[1,i]},\hat{D}_Z^{[1,s]})$. These costs are computed in increasing order of s and, for each s, in increasing order of i, for $s = 1..m$ and $i = 1..n$. As a basic case, for every range $D_Z^{[1,i]}$ the algorithm stores in $g(1,i)$ the cost that results from assigning all the values in $D_Z^{[1,i]}$ to the sole cluster $\hat{z}_1$. Moreover, in order to efficiently evaluate the cost of assigning all values in $D_Z^{u,v}$ to some cluster $\hat{z}_l$, all prefix sums $c_l(u)=\sum_{0<i\le u}\delta(i,l)$ are preliminary computed—notice that, in fact, it holds: $\mathcal{G}^*(D_Z^{[u,v]},\{z_l\})=c_l(v)-c_l(u-1)$.

In more detail, the optimal split cost $g(s,i)=\mathcal{G}^*(D_Z^{[1,i]},\hat{D}_Z^{[1,s]})$ is singled out by evaluating, for each index $\mu < i$, the value $\mathcal{G}^*(D_Z^{[1,\mu]},\hat{D}_Z^{[1,s-1]})$, stored in $g(s-1, \mu)$, along with $\mathcal{G}^*(D_Z^{[\mu+1,i]},\hat{D}_Z^{[s,s]})$, which can be derived from $c_s(\cdot)$. In addition, the boundary index leading to this optimal split is kept in $h(s, i)$. In the next step, the optimal $m-1$ boundary indexes, which ensure the minimum value (stored in $g(m, n)$) of the whole objective function, are extracted out of table h. Finally, a clustering function is built out of these boundaries, and returned as output.

Proposition 8. Algorithm NDP-OPT *implements a loss-safe partition function.*

Proof The clustering $C_Z^{(s+1)}$ found by algorithm NDP-OPT guarantees the minimal information loss among all possible splits of the numerical domain D_Z taken as input. In particular, the information loss produced by $C_Z^{(s+1)}$ is not higher than that due to the clustering $C_Z^{(s)}$ previously computed in algorithm *LogCC*, and taken as input by *NDP-OPT*.

Theorem 9: Let $n = | D_Z |$, $m = | \hat{D}_Z |$, *with* $1 < m < n$ *and* $P_{T,Z}$ *be the size of the contingency table over* D_Z *and* D_T, *that is, the number of non-zero values in the distribution* $p(T, Z)$. *Then algorithm NDP-OPT computes an exact solution to problem NDP in* $O(n^2 \times m + P_{T,Z})$ *steps.*

Proof The correctness of algorithm NDP-OPT directly descends form observing that each term $g(s, i)$ coincides with the partial solution $\mathcal{G}^*(D_Z^{[1,i]},\hat{D}_Z^{[1,s]})$, righly computed according to the recursive formulation given in Theorem 7. As concerns the computation time, it is clear all the coefficients;

$$\delta_T(z_i,\hat{z}_l)=p(z_i)\cdot\mathcal{D}(p(T\mid z_i)\,\|\,q(T\mid\hat{z}_l)),$$

as well as the associated partial sums $c_l(\cdot)$, can be computed in $O(P_{T,Z} + n\times m)$. All basic costs $g(1,\cdot)$ can be computed in constant time, while computing of both $g(\cdot,\cdot)$ and $h(\cdot,\cdot)$ takes $O(n^2\times m)$. Finally, $O(n+m)$ time suffices to reconstruct both the optimal cluster boundaries, out of table $h(\cdot,\cdot)$, and the final clustering C_Z.

A serious drawback of algorithm NDP-OPT is the quadratic dependency on $| D_Z |$, which makes it unviable in many real-world applications. Hence, we next investigate an alternative, greedy, approach to NDP, which consumes linear time in $| D_Z |$ only.

A Greedy Implementation of Partition: Algorithm NDP-GR

The algorithm in Figure 4 computes a clustering function C_Z from a numerical domain D_Z to a set of clusters $\hat{D}_Z$, both given as input. In this case, however, a binary split procedure over both D_Z

and $\hat{D}_Z$ is performed, where every range $D_Z^{[u,v]} = \{z_u,..., z_v\}$ of values from D_Z is associated with a suitable range $\hat{D}_Z^{[r,s]} = \{\hat{z}_r,...,\hat{z}_s\}$ of clusters from $\hat{D}_Z$. In general, this corresponds to keeping all the values of $D_Z^{[u,v]}$ in a preliminary cluster that will be eventually split into the real clusters $\hat{z}_r,...,\hat{z}_s$. As a special case, associating $D_Z^{[u,v]}$ with a singleton cluster range $\hat{D}_Z^{[r,r]}$ simply means that all the values in $D_Z^{[u,v]}$ are assigned to $\hat{z}_r$. For any of such interval pairs $D_Z^{[u,v]}$ and $\hat{D}_Z^{[r,s]}$, we keep trace of the associated index ranges by storing a tuple $\langle u, v, r, s\rangle$ in R.

In more detail, the algorithm starts with only one cluster, which encompasses the whole $\hat{D}_Z$ and is assigned all the values in D_Z, that is, $R = \{\langle 1, m, 1, n\rangle\}$. Then, it iteratively selects a cluster $\hat{D}_Z^{[\bar{r},\bar{s}]}$, spanning over at least two clusters of $\hat{D}_Z$ (i.e., $\bar{s} > \bar{r}$), along with its associated interval $D_Z^{[\bar{u},\bar{v}]}$, and acts a binary partition on both of them. While the cluster range $[\bar{r},\bar{s}]$ is divided into two equi-numerous sub-ranges (τ ever falls in the middle of $[\bar{r},\bar{s}]$) the partitioning of $[\bar{u}..\bar{v}]$ is performed in a greedy way, based on a suitable cost function Θ, defined as follows: $\Theta(u,v,r,s) = \frac{1}{m}\cdot\sum_{l=r}^{s}\sum_{i=u}^{v}\delta(i,l)$.

Roughly speaking, $\Theta(u, v, r, s)$ gives a pessimistic estimation for the contribution given by $z_u,...,z_v$ to the final mutual information loss, when they are required to be assigned only to some of the clusters $\hat{z}_r,...,\hat{z}_s$. In fact, it is an upper bound for the minimal mutual information loss $\mathcal{G}^*(D_Z^{[u,v]},\hat{D}_Z^{[r,s]})$ arising when $D_Z^{[u,v]}$ is split into the clusters in $\hat{D}_Z^{[r,s]}$, as explained later.

Specifically, the selected interval $[\bar{u}..\bar{v}]$ is split in correspondence of the index τ^* that guarantees the lowest value for $\Theta(\bar{u},\mu^*,\bar{r}\tau) + \Theta(\mu^*+1,\bar{v}\tau+1,\bar{s})$, which estimates the loss in mutual information arising when assigning $z_{\bar{u}},...,z_{\mu^*}$ (resp., $z_{\mu^*+1},...,z_{\bar{v}}$) to some clusters in $\hat{z}_{\bar{r}},...,\hat{z}_\tau$ (resp., $\hat{z}_{\tau+1},...,\hat{z}_{\bar{s}}$).

Moreover, R is updated accordingly, by replacing tuple $\langle\bar{u},\bar{v},\bar{r},\bar{s}\rangle$ with those representing the new clusters originated from it: $\langle\bar{u},\mu^*,\bar{r}\tau^*\rangle$ and $\langle\mu^*+1,\bar{v}\tau^*+1,\bar{s}\rangle$. As a special case, when the tuple extracted from R spots just a single cluster $\hat{z}_{\bar{r}}$ in $\hat{D}_Z$, the algorithm updates the clustering function C_Z by assigning all the values in $D_Z^{[u,v]}$ to $\hat{z}_{\bar{r}}$.

We next formally prove that $\Theta(u, v, r, s)$ us a pessimistic estimation for $\mathcal{G}^*(D_Z^{[u,v]},\hat{D}_Z^{[r,s]})$.

Property 10: $\Theta(u, v, r, s)$ is an upper bound for $\mathcal{G}^*(D_Z^{[u,v]},\hat{D}_Z^{[r,s]})$.

Proof Let $C_Z^*: D_Z^{[u,v]} \mapsto \hat{D}_Z^{[r,s]}$ be the clustering function producing the minimal information loss $\mathcal{G}^*(D_Z^{[u,v]},\hat{D}_Z^{[r,s]})$, and let $\gamma^*(i)$ be the index of the cluster assigned by C_Z^* to each z_i in $D_Z^{[u,v]}$. Then it is:

Figure 4. Algorithm NDP-GR: A greedy implementation for function Partition

```
Input:   A domain D_Z, a set of clusters D̂_Z = {ẑ_1, ..., ẑ_m};
Output:  A clustering function from D_Z to D̂_Z;
-----------------------------------------------------------------
c(u, s) := Σ_{0<i≤u} Σ_{0<l≤r} δ_T(z_u, ẑ_r), ∀u ∈ [1..n] and ∀r ∈ [1..m];
Let Θ(u, v, r, s) = ( c(v, s) − c(u − 1, s) − c(v, r − 1) + c(u − 1, r − 1) ) /(s − r + 1);
R := {⟨1, m, 1, n⟩};
while R ≠ ∅ do
    Let ⟨ū, v̄, r̄, s̄⟩ be a tuple in R, and τ = ⌊(s+r)/2⌋;
    if r̄ = s̄ then
        Set C_Z(z_i) := ẑ_r  for each i in [ū..l*];
    else
        Let l* = argmin_{μ∈[ū..v̄]}( ( Θ(ū, μ, r̄, τ) + Θ(μ + 1, v̄, τ + 1, s̄) );
        R := R ∪ {⟨ū, l*, r̄, τ⟩, ⟨l* + 1, v̄, τ + 1, s̄⟩};
    end if;
end while;
return C_Z;
```

$$\Theta(u,v,r,s)=\frac{1}{m}\cdot\sum_{i=u}^{v}\sum_{l=r}^{s}\delta(i,l)=$$

$$\frac{1}{m}\cdot\left[\mathcal{G}^{*}(D_Z^{[u,v]},\hat{D}_Z^{[r,s]})+\sum_{i=u}^{v}\sum_{l\in[r,s]-\{\gamma^{*}(i)\}}\delta(i,l)\right]\geq$$

$$\frac{1}{m}\cdot\left[\mathcal{G}^{*}(D_Z^{[u,v]},\hat{D}_Z^{[r,s]})+(m-1)\cdot\mathcal{G}^{*}(D_Z^{[u,v]},\hat{D}_Z^{[r,s]})\right]$$

$$=\mathcal{G}^{*}(D_Z^{[u,v]},\hat{D}_Z^{[r,s]})$$

as the term $\sum_{i=u}^{v}\sum_{l\in[r,s]-\{\gamma^{*}(i)\}}\delta(i,l)$ can be seen as the sum of the information losses associated with m–1 clustering functions from $D_Z^{[u,v]}$ to $\hat{D}_Z^{[r,s]}$, all different from the optimal one.

In order to compute Θ efficiently, we rephrase it in terms of a number of prefix sums: $\Theta(u,v,r,s)=\frac{1}{m}\cdot$ [$c(v,s)$ – $c(u\text{-}1,s)$ – $c(v,r\text{-}1)$ +$c(u\text{-}1,r\text{-}1)$], with $c(u,s) = \sum_{0<i\leq u}\sum_{0<l\leq s}\delta(u,s)$. Indeed, all such terms can be computed in constant time in both the number of clusters and the size of D_Z.

Theorem 11: Let $n = |D_Z|$, $m = |\hat{D}_Z|$, *with* $1 < m < n$ *and* $P_{T,Z}$ *be the size of the contingency table relating* D_Z *with* D_T, *i.e., the number of non-zero values in P(T, Z). Then computing algorithm NDP-GR requires* $O(P_{T,Z} + n\times m)$ *steps.*

Proof Pre-computing all partial sums $c(\cdot,\cdot)$ can be done in $O(P_{T,Z} + n\times m)$. The main loop in the algorithm is repeated $O(m)$ times, and for each iteration one of the following major tasks must be carried out: if the selected cluster is a final one, the coclustering C_Z must be updated for all the values currently associated with the cluster; otherwise, every possible binary split of the cluster must be evaluated according to function Θ. As the latter function can be computed in $O(1)$, based on $c(\cdot,\cdot)$, either task can be accomplished in $O(n)$.

The previous result makes algorithm NDP-GR an efficient tool for clustering even large numerical domains. However, as a price for its efficiency, it gives no guarantee about the loss-safe property, instead enjoyed by NDP-OPT. Yet, a number of experiments (discussed in Section ``Experiments") evidence that NDP-GR finds nearly-optimal solutions in many practical cases, hence ensuring reasonable computation time for the co-clustering method.

EXPERIMENTS

The effectiveness of the approach can be assessed according to two different respects. Indeed, a first result is the multidimensional co-clustering discovered, which can be evaluated in terms of the overall information loss—measured according to Equation 3—it produces w.r.t. the case of considering all the original values in the domains.

A second result obtained by our approach are the workflow schemas associated with the different clusters of traces discovered, which constitute, as a whole, an articulated representation for the behavior of the process at hand. To this respect, we define a simple way for comparing the workflow model used for generating the input dataset with the workflow models discovered by algorithm *LogCC*.

Let W be a given workflow schema, $\mathcal{L}$ be a log of traces compliant with W, produced by the generator, and $W_1', W_2', \ldots, W_k'$ be the workflow schemas extracted out of L by means of algorithm *LogCC*, which collectively represent a model for the mined process. The quality of such a process model can be evaluated by resorting the *accuracy* measure presented in Section "Formal Framework," assuming that the soundness is estimated as the percentage of the traces in a log L_{test} (randomly generated from $W_1', \ldots, W_k'$) that are also compliant with the original schema W. In particular, when L_{test} just contains all the possible traces of $W_1', \ldots, W_k'$, the maximal accuracy (i.e., 1) is achieved.

Process Models Discovered for the Example Process HandleOrder

As a first analysis, we discuss the results of *LogCC* on a syntectic log data randomly generated according to the schema in Figure 1. Specifically, following the semantics of the application, we generated 2000 traces in a way that task h is never executed (all the orders are successfully handled), task d does not occur in any trace including r, and that p appears in any trace not containing r.

The co-clustering of the log according to both structure and performance metrics is performed over three correlated dimensions: traces (T), activities (A) and durations (Z).

In order to generate metrics values, we modelled the duration of each single task t as a gaussian random variable with mean $\tau(t)\times 100$ and standard deviation $\tau(t)\times 5$. In particular, we used $\tau(t)=1$ for any task t different from r, while setting $\tau(\mathrm{r})=2$. Moreover, we imposed as a further constraint, that $\tau(\mathrm{x})=1.5$ in all traces containing also r.

We then applied algorithm *LogCC* by requiring 3 clusters for all domains (i.e., $m^T=3$, $m^A=3$ and $m^Z=3$). Moreover, we used the greedy implementation of function Partition (cf. algorithm NDP-GR), and gave the same weight to both auxiliary dimensions (i.e., $\beta_A = \beta_Z = 0.5$).

Figure 5 shows the schemas associated with the discovered clusters, corresponding to three different use cases. Indeed, Figure 5(a) describes

*Figure 5. Workflow schemas for the three trace clusters mined from a log of process **HandleOrder***

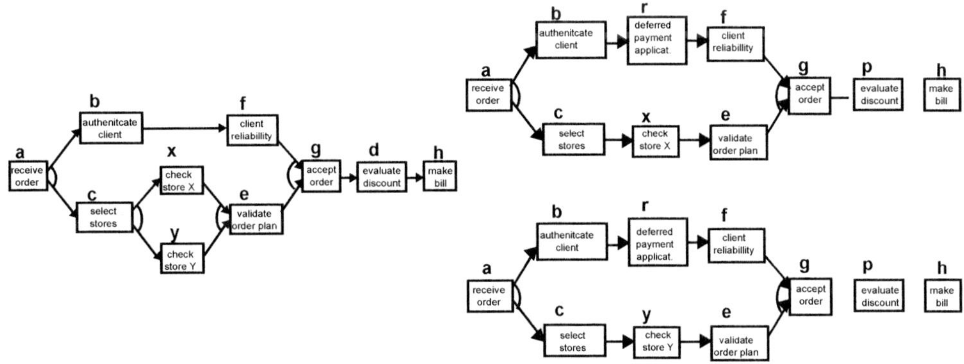

Figure 6. Global information loss (left) and process model accuracy (right) on synthetic data

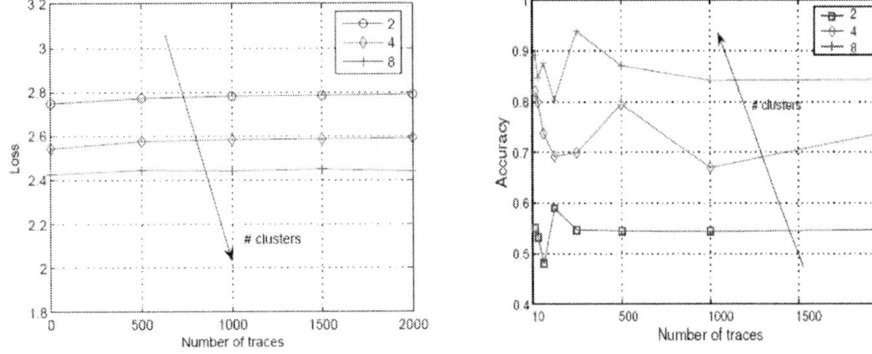

a schema for managing orders without deferred payment (task r does not appear at all), where discount evaluation (task d) is done mandatorily. Conversely, the two other schemas (Figure 5(b) and Figure 5(c)) concern orders with deferred payment, which are, in their turn, split in two different clusters, based on the presence of task x. Note that such a finer grain partition of the traces, which could not have been obtained by considering their structural aspects only, allows to recognize a specific usage case, quite relevant to performance analysis: deferred-payment orders requiring a check over store *X* (cf. Figure 5(b)), which clearly distinguish from the remaining process instances (cf. Figure 5(c)) for higher duration.

All the schemas in Figure 5 collectively constitute a more precise representation for process *HandleOrder*. In fact, the accuracy achieved by this model is maximal (i.e., 1), while the accuracy of the original model, shown in Figure 1, is only 0.5.

Quantitative Results on Synthesized Data

In order to test both the effectiveness and the scalability of the approach, we generated a series of datasets, according to the procedure described next.

We considered a fixed workflow schema consisting of 200 activities, with 100 of them left optional (i.e., a half of activities appear in a certain number of *OR* branches, while the schema does not contain any *XOR*-split). For each dataset, the same number of (true) clusters *m* is chosen for all domains by selecting it from a gaussian distribution with mean 10 and standard deviation 4, while the number *N* of traces is taken from a uniform distribution defined over the integer range 100..2000. The optional activities are then randomly partitioned into the *m* of activities' clusters, and for each of these clusters *N*/*m* traces are randomly generated according to the schema, yet requiring that all the activities in the cluster occur. Conversely, a manually chosen number of metric values are generated according to a uniform distribution defined over the integer range 0..500, and then ordered and split into a *m* intervals. Each such interval is made to correspond to a trace cluster, and all the traces in the latter are randomly associated with some values in the interval. Finally, all correlation tables are altered with noise, by flipping the 20% of their entries, and turned into valid joint probability matrices, by a normalization step.

On each of the datasets obtained as previously specified, algorithm *LogCC* is applied with the same number of clusters (made varying up to 12)

Figure 7. Total computation time for 2 clusters and different numbers of metrics domains (left) and for 2 metrics domains and different numbers of clusters (right)

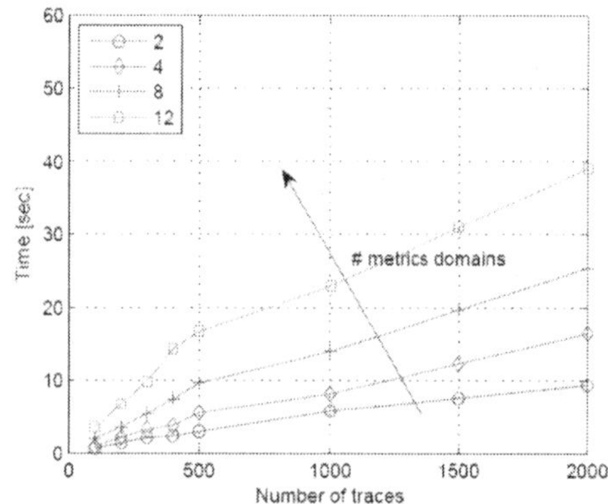

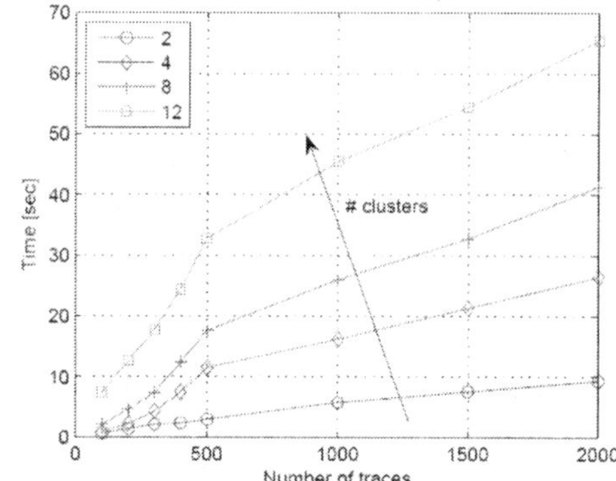

Figure 8. Partitioning time on synthetic data

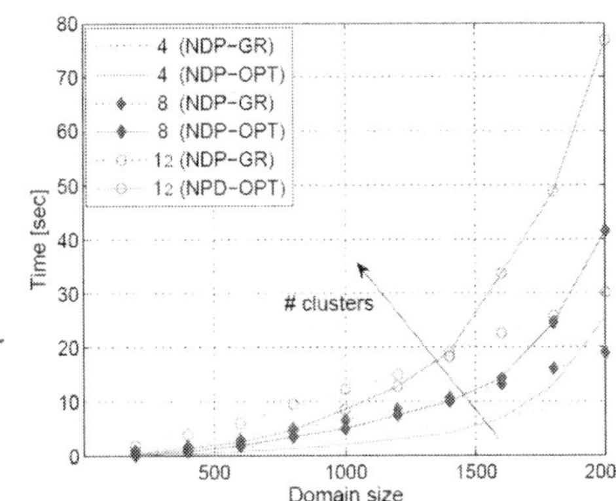

for all the domains (i.e., traces, activities, and all sets of metrics values), while using the same weight for the auxiliary domains (i.e., activities and metrics).

The total computation time of algorithm *LogCC*, equipped with the greedy version of function Partition, is depicted in Figure 6, which, more specifically, reports the trend of the mining time with respect to the log size. In particular, in the left side several curves are plotted which correspond to different numbers of metrics domain considered, with a fixed number of clusters (namely 2) required for all domains; analogously, in the right side the number of clusters has been made varying, while considering only 2 performance metrics. These results, yet confirmed in a wider series of tests, substantiate the scalability of the approach and confirm that the total computation time depends linearly on the number of traces contained in the input log, as well as on the number of desired clusters and on the number of metrics considered.

Figure 7 illustrates the influence of the numbers of clusters and log size on the quality of results obtained, evaluated according to the global information loss produced (in the left side of the figure), and to the accuracy of the discovered process model (in the right side). As expected, no significant role is played by the size of input log, whereas increasing the number of clusters clearly yields lower information loss and notably higher accuracy.

Testing Alternative Partitioning Methods: NDP-OPT vs. NDP-GR

Figure 8 shows results for a comparison between *NDP-OPT* (solid lines) and *NDP-GR* (dashed lines), for different number of clusters and sizes of the partitioned domain. As expected from the theoretical analysis provided in Section "Clustering Numerical Domains (Function Partition)", one may notice that algorithm *NDP-GR* scales linearly w.r.t. the number of elements in the domain, while algorithm *NDP-OPT* suffers from a quadratic dependence, needed to provide an optimal solution for the partitioning problem.

Further insights on this behavior can be achieved from Table 2, where the number of steps (formally, the number of times an element of the domain is considered for being a boundary) of both the algorithms is reported, averaged on four executions.

Note that yet being an approximate solution that does not enjoy the *loss-safe* property, *NDP-GR* allows to effectively split numerical domains, as it appears by contrasting the information loss produced by *NDP-OPT* and *NDP-GR*, still shown in Table 2 for different numbers of elements and of asked clusters

RELATED WORK

We next review some works on process mining, yet noting that a broader overview on this topic can be found in van der Aalst et al. (2003). Maybe this problem was first faced by Cook and Wolf (1995), where it was essentially formulated as a grammar inference problem: the input process traces are considered as sentences in some unknown language (expressing the semantics of the

Table 2. *Loss in mutual information: NDP-OPT vs. NDP-GR*

		Information Loss		Computation Steps	
m	n^2	*NDP-OPT*	*NDP-GR*	*NDP-OPT*	*NDP-GR*
2	400	2,54865	2,54875	193,67	19052,33
2	800	2,57187	2,57186	382	73545
2	1200	2,58146	2,58146	576,67	167161,33
2	1600	2,58550	2,58550	771,67	298920,67
2	2000	2,58951	2,58951	967	469104,33
4	400	2,54267	2,54245	378,67	55205
4	800	2,57483	2,57478	764,67	222167
4	1200	2,58200	2,58200	1146	496971
4	1600	2,58611	2,58610	1555,33	913054
4	2000	2,59054	2,59055	1955,33	1441242
8	400	2,54668	2,54598	579	136089,33
8	800	2,57445	2,57414	1144	520352
8	1200	2,58046	2,58036	1737	1190361,67
8	1600	2,58630	2,58628	2349	2169104
8	2000	2,58918	2,58915	2905	3310708,33

process), and the aim of the discovery procedure is to produce a grammar, in the form of a finite state machine (FSM), which models the language – incidentally, edges in such a model are associated with process activities, and specify state transitions. The discovery of process models is carried out by means of three different strategies: a statistical approach, based on neural networks, a purely algorithmic approach, and a hybrid approach, based on Markov models.

As pointed out in Cook and Wolf (1998), such an approach has a major drawback in the fact that FSMs can only capture the basic structure of sequential processes (sequence, selection and iteration). Conversely, real workflow processes typically exhibit concurrent behaviour, where different threads of control are followed during the enactment of a process instance. In such cases, sequential state machines risk to be an inadequate and overly process model, where a (possibly combinatorial) number of different sequential patterns are likely to be used to represent concurrent activities. As a consequence, the majority of subsequent research works moved towards other kinds of process models, able to explicitly represent concurrency and other control flow constructs.

A special kind of Petri nets, named Workflow nets (or *WF-nets*), is adopted in van der Aalst et al. (2002), van der Aalst and Dongen (2002), van der Aalst et al. (2003), and van der Aalst et al. (2004) for modelling and mining workflow processes: transitions represent tasks, while relationships between tasks are modelled by arcs and places.

A basic algorithm, named α-algorithm, is introduced in van der Aalst et al. (2004), which can derive a WF-net from a workflow log, under the assumption that the log is complete and free of noise. The capability of the algorithm to mine WF-net workflow models and its limitations are analyzed in van der Aalst et al. (2003). In van der Aalst et al. (2002), simple statistics are exploited in the construction of ordering relations for coping with noise.

A further approach to mining a process model from event logs is described in Herbst et al. (2000). The peculiarity of the approach mainly resides in its capability of recognizing duplicate tasks in the control flow graph (i.e., many nodes associated with the same task). Yet another approach is adopted in Schimm, 2003), where a mining tool is presented, which is able to discover hierarchically structured workflow processes. Such a model corresponds to an expression tree, where the leafs represent tasks (operands) while any other node is associated with a control flow operator. In this context, the mining algorithm mainly consists of a suitable set of term rewriting systems.

Recently, an approach has been proposed in Greco et al. (2006), which is capable of discovering a set of workflow schemas out of a workflow log, which separately model different usage scenario of the process at hand. Each of these variants for the process are discovered by partitioning the traces into clusters, through a feature-based clustering scheme.

As pinpointed in the Introduction, all of the process mining approaches previously cited found on an abstraction of logs as a set of task identifiers, and hence disregard all *non-structural* data that are yet kept by many real systems, such as information about activity executors, time-stamps, parameter values, as well as different performance measures. To the best of our knowledge, our work is the first attempt to overcome such a limitation, by making process mining algorithms able to take advantage of performance data.

As far as concerns the co-clustering problem (i.e., the simultaneous clustering of correlated domains), we first note that several approaches have been proposed in the literature for the bi-dimensional case (a.k.a. bi-clustering). These approaches can be grouped in two main families: spectral graph partitioning techniques and information theoretic approaches.

In general, spectral graph partitioning approaches encode the clustering domain as a (weighted and undirected) bipartite graph, and look for a cut, which is *optimal* according some suitable objective function accounting for the strength of association between the resulting groups of nodes. In this setting, the search for an optimal co-clustering can be formulated as a generalized eigenvalue problem, and different approaches (Dhillon, 2001; Ding, He, Zha, Gu, & Simon, 2001; Zha, He, Ding, Simon, & Gu, 2001) were defined for heuristically solving it--via, for example, continuous relaxation or its reformulation as a partial SVD problem).

The seminal work for information theoretic co-clustering approaches is Dhillon et al. (2003), where the notion of bi-clustering based on the minimization of the loss in mutual information has been proposed. Similar information theoretic approaches have been proposed (e.g., in Berkhin et al., 2002; Banerjee, Dhillon, Ghosh, Merugu, & Modha, 2004; Slonim & Tishby, 2000).

Some recent works have generalized the bi-clustering problem to the case of more than two domains (Bekkerman, El-Yaniv, & McCallum, 2005; Gao, Liu, Zheng, Cheng, & Ma, 2005). In particular, (Gao et al., 2005) considered a co-clustering problem for ``star''-structured domains of the form D_X, $D_{Y^1}, \ldots, D_{Y^N}$ (where $N > 1$ and D_X is the central domain), by defining an objective function f_{X,Y^i}, for each "auxiliary" domain D_{Y^i}, whose optimization should yield the best co-clusters over D_X and D_{Y^i}. In order to integrate all such (bi-dimensional) co-clustering sub-problems, a linear combination of all pairwise objective functions is optimized, subject to the constraint that consistent clusters are found for the central (shared) domain. In fact, Gao et al. (2005) addresses this problem through a spectral approach applied in a simplified setting, where each domain can be clustered into two clusters only. Importantly, Gao et al. (2005) asked for both removing this assumption and assessing whether the information-theoretic approach of Dhillon et al. (2003) can be used in this setting. Both issues have been faced in our work (see Section "Multidimensional Co-Clustering of Workflow Logs").

An information-theoretic approach, which can handle general pairwise interactions has been proposed in Bekkerman et al. (2005), where an algorithmic schema is used that interleaves top-down clustering of some domains and bottom-up clustering of the others (where cluster elements are split/merged at random), along with a local correction routine. Notably, this approach produces a hierarchical clustering for each domain, so providing implicit support to the choice of clusters' numbers, which is, in general, a critical issue in clustering applications. However, the algorithm runs in $O(max_W\{log(|\hat{D}_W|),log(|D_W|/|\hat{D}_W|)\}\cdot max\{|D_W|^3\})$, where W denotes any input domain—a quadratic dependence on D_W is achieved only when co-clustering 2 domains. Clearly, this costs is likely to be unfeasible for large data sets.

Finally, an efficient technique is devised in Andritsos, Tsaparas, Miller, and Sevcik (2004) for the clustering of categorical data, which founds on an information theoretic approach. More specifically, a hierarchical clustering is computed for a set of tuples according to a distance measure, defined on the basis of the "information bottleneck" criterion, which captures the correlations between tuples and attribute values. As a further result, the approach in Andritsos et al. (2004) allows for measuring the distance between attribute values as well, which can be exploited to compute a clustering for each of the attributes. This approach, yet exhibiting clear similarities with our work, differs from it in two important respects: it is not a pure co-clustering approach, and it does not support the partitioning of numerical domains into non-overlapping ranges.

CONCLUSION

We devised a novel process mining approach that substantially differs from previous works in taking account performance measures on log traces, beside mere "structural" information on task timing. The approach founds on an information-theoretic framework, where log traces are co-clustered along with their structural elements and metrics values. Each cluster of traces is eventually provided with a workflow schema, hence modeling a specific use case.

Notably, we extended the framework of Dhillon et al. (2003) by considering more than two domains for the clustering, and by introducing numerical domains to be split into non-overlapping ranges. To face the latter issue, we have proposed two alternative implementations of the splitting procedure, based on an exact dynamic-programming approach and on an efficient greedy heuristic, respectively. Encouraging results were obtained by testing the approach on synthesized data, even when using the greedy procedure.

As directions of future work, we intend to empirically analyze the approach in real-world application scenarios, as well as to devise some strategy for supporting the user in properly setting the parameters of the algorithm. Also, given the encouraging results obtained with the greedy heuristic, it would be interesting to investigate whether it yet provides some approximation guarantee for the co-clustering problem addressed in the article.

REFERENCES

Andritsos, P., Tsaparas, P., Miller, R. J., & Sevcik, K. C. (2004). Limbo: Scalable clustering of categorical data. *Proceedings of the 9th International Conference on Extending Database Technology (EDBT'04).*

Banerjee, A., Dhillon, I., Ghosh, J., Merugu, S., & Modha, D. S. (2004). A generalized maximum entropy approach to bregman co-clustering and matrix approximation. *Proceedings of the 10th ACM SIGKDD International Conference on*

Knowledge Discovery and Data Mining (KDD'04) (pp. 509-514).

Bekkerman, R., El-Yaniv, R., & McCallum, A. (2005). Multi-way distributional clustering via pairwise interactions. *ICML '05: Proceedings of the 22nd International Conference on Machine Learning* (pp. 41-48). New York, NY, USA: ACM Press.

Berkhin, P., & Becher, J. D. (2002). Learning simple relations: Theory and applications. *Proceedings of the 2nd SIAM International Conference on data mining (SDM'02).*

Casati, F., Castellanos, M., Dayal, U., & Shan, M.-C. (2005). iBOM: A platform for intelligent business operation management. *Proceedings of the 21st International Conference on Data Engineering, (ICDE'05)* (pp. 1084-1095).

Cook, J. E., & Wolf, A. L. (1998). Event-based detection of concurrency. *Proceedings of the 6th International Symposium on the Foundations of Software Engineering (FSE'98)* (pp. 35-45).

Cook, J. E., & Wolf, A. L. (1995). Automating process discovery through event-data analysis. *Proceedings of the 17th International Conference on Software Engineering (ICSE'95)* (pp. 73-82).

Dhillon, I. S. (2001). Co-clustering documents and words using bipartite spectral graph partitioning. *Proceedings of the 7th ACM SIGKDD International Conference on Knowledge Discovery and Data Mining (KDD'01)* (pp. 269-274).

Dhillon, I. S., Mallela, S., & Modha, D. S. (2003). Information-theoretic co-clustering. *Proceedings of the 9th ACM SIGKDD International Conference on Knowledge Discovery and Data Mining (KDD'03)* (pp. 89-98).

Ding, C. H. Q., He, X., Zha, H., Gu, M., & Simon, H. D. (2001). A min-max cut algorithm for graph partitioning and data clustering. *Proceedings of the 1st IEEE International Conference on Data Mining (ICDM'01)* (pp. 107-114).

Gao, B., Liu, T. Y., Zheng, X., Cheng, Q. S., & Ma, W. Y. (2005). Consistent bipartite graph co-partitioning for star-structured high-order heterogeneous data co-clustering. *Proceedings of the 11th ACM SIGKDD International Conference on Knowledge Discovery and Data Mining (KDD'05))* (pp. 41-50).

Greco, G., Guzzo, A., Pontieri, L., & Sacca, D. (2006). Discovering expressive process models by clustering log traces. *IEEE Trans. on Knowledge and Data Engineering, 18*(8), 1010-1027.

Herbst, J., & Karagiannis, D. (2000). Integrating machine learning and workflow management to support acquisition and adaptation of workflow models. *Journal of Intelligent Systems in Accounting, Finance, and Management, 9*, 67-92.

Parson, L., Haque, E., & Liu, H. (2004). Subspace clustering for high dimensional data: A review. *SIGKDD Explorations, 6*(1), 90-105.

Schimm, G. (2003). Mining most specific workflow models from event-based data. In *Proceedings of the International Conference on Business Process Management* (pp. 25-40).

Slonim, N., & Tishby, N. (2000). Document clustering using word clusters via the information bottleneck method. In *Proceedings of the 23rd International ACM SIGIR Conference on Research and Development in Information Retrieval (SIGIR'00)* (pp. 208-215).

van der Aalst, W. M. P., & van Dongen, B. F. (2002). Discovering workflow performance models from timed logs. *Proceedings of the International Conference on Engineering and Deployment of Cooperative Information Systems (EDCIS'02)* (pp. 45-63).

van der Aalst, W. M. P., & van Hee, K. M. (2002). *Workflow management: Models, methods, and systems.* MIT Press.

van der Aalst, W. M. P., van Dongen, B. F., Herbst, J., Maruster, L., Schimm, G., & Weijters, A. J. M. M. (2003). Workflow mining: A survey of issues and approaches. *Data & Knowledge Engineering, 47*(2), 237-267.

van der Aalst, W. M. P., Weijters, A., & Maruster, L. (2004). Workflow mining: Discovering process models from event logs. *IEEE Transactions on Knowledge and Data Engineering (TKDE), 16*(9), 1128-1142.

Zha, H., He, X., Ding, C., Simon, H., & Gu, M.(2001). Bipartite graph partitioning and data clustering. In *Proceedings of the 10th ACM International Conference on Information and Knowledge Management (CIKM'01)* (pp. 25-32).

This work was previously published in International Journal of Data Warehousing and Mining, Vol. 3, Issue 4, edited by D. Taniar, pp. 99-119, copyright 2007 by IGI Publishing, formerly known as Idea Group Publishing (an imprint of IGI Global).

Chapter 2.20
Domain-Driven Data Mining: A Practical Methodology

Longbing Cao
University of Technology, Sydney, Australia

Chengqi Zhang
University of Technology, Sydney, Australia

ABSTRACT

Extant data mining is based on data-driven methodologies. It either views data mining as an autonomous data-driven, trial-and-error process or only analyzes business issues in an isolated, case-by-case manner. As a result, very often the knowledge discovered generally is not interesting to real business needs. Therefore, this article proposes a practical data mining methodology referred to as domain-driven data mining, which targets actionable knowledge discovery in a constrained environment for satisfying user preference. The domain-driven data mining consists of a DDID-PD framework that considers key components such as constraint-based context, integrating domain knowledge, human-machine cooperation, in-depth mining, actionability enhancement, and iterative refinement process. We also illustrate some examples in mining actionable correlations in Australian Stock Exchange, which show that domain-driven data mining has potential to improve further the actionability of patterns for practical use by industry and business.

INTRODUCTION

Extant data mining is presumed as an automated process that produces automatic algorithms and tools without human involvement and the capability to adapt to external environment constraints. As a result, many patterns are mined, but few are workable in real business.

However, actionable knowledge discovery can afford important grounds to business decision makers for performing appropriate actions. In the panel discussions of SIGKDD 2002 and 2003 (Ankerst, 2002, Fayyad et al 2003), it was highlighted by the panelists as one of the grand challenges for extant and future data mining. This situation partly resulted from the scenario that extant data mining is a data-driven trial-and-error process (Ankerst, 2002) in which data mining algorithms extract patterns from converted data via some predefined models based on experts' hypotheses.

Data mining in the real world (e.g., financial data mining and crime pattern mining) (Bagui, 2006) is highly constraint-based (Boulicaut & Jeudy, 2005; Fayyad & Shapiro, 2003). Constraints involve technical, economic, and social aspects in the process of developing and deploying actionable knowledge. Real-world business problems and requirements often are embedded tightly in domain-specific business processes and business rules in charge of expertise (domain constraint). Patterns that are actionable to business often are hidden in large quantities of data with complex data structures, dynamics, and source distribution (data constraint). Often, mined patterns are not actionable to business, even though they are sensible to research. There may be big interestingness conflicts or gaps between academia and business (interestingness constraint). Furthermore, interesting patterns often cannot be deployed to real life, if they are not integrated with business rules, regulations, and processes (deployment constraint). Some other types of constraints include knowledge type constraint, dimension/level constraint, and rule constraint (Han, 1999).

For actionable knowledge discovery from data embedded with the previous constraints, it is essential to slough off the superficial and capture the essential information from the data mining. However, this is a nontrivial task. While many methodologies have been studied, they either view data mining as an automated process or deal with real-world constraints in a case-by-case manner.

Our experience (Cao & Dai, 2003a, 2003b) and lessons learned in data mining in capital markets (Lin & Cao, 2006) show that the involvement of domain knowledge and experts, the consideration of constraints, and the development of in-depth patterns are essential for filtering subtle concerns while capturing incisive issues. Combining these aspects, a sleek data mining methodology can be developed in order to find the distilled core of a problem. It can advise the process of real-world data analysis and preparation, the selection of features, the design and fine-tuning of algorithms, and the evaluation and refinement of mined results in a manner more effective to business. These are our motivations to develop a practical data mining methodology referred to as domain-driven data mining.

Domain-driven data mining consists of a domain-driven in-depth pattern discovery (DDID-PD) framework. The DDID-PD takes I^3D (i.e., interactive, in-depth, iterative, and domain-specific) as real-world KDD bases. I^3D means that the discovery of actionable knowledge is an iteratively interactive in-depth pattern discovery process in domain-specific context. I^3D is further embodied through (1) mining constraint-based context, (2) incorporating domain knowledge through human-machine-cooperation, (3) mining in-depth patterns, (4) enhancing knowledge actionability, and (5) supporting loop-closed iterative refinement in order to enhance knowledge actionability. Mining constraint-based context requests to effectively extract and transform domain-specific datasets with advice from domain experts and their knowledge.

In the DDID-PD framework, data mining and domain experts complement each other in regard to in-depth granularity through interactive interfaces. The involvement of domain experts and their knowledge can assist in developing highly effective domain-specific data mining techniques and can reduce the complexity of the knowledge-producing process in the real world. In-depth pattern mining discovers more interesting and actionable patterns from a domain-specific perspective. A system following the DDID-PD framework can embed effective supports for domain knowledge and experts' feedback, and refines the life cycle of data mining in an iterative manner.

Taking financial data mining as an example, this article introduces some case studies that deploy the domain-driven data mining methodology.

Deep correlations in stock markets are mined through parallel computing to provide measurable benefits for trading. It shows that domain-driven data mining can benefit the actionable knowledge mining in a more effective and efficient manner than data-driven methodology such as CRISP-DM (CRISP).

The remainder of this article is organized as follows. The second section discusses actionable knowledge discovery. The third section presents the DDIP-DM framework. In the fourth section, key components in domain-driven data mining are stated. The fifth section demonstrates case studies on mining actionable correlations in stock markets. We conclude this article and present future work in the sixth section.

ACTIONABLE KNOWLEDGE DISCOVERY

One of the fundamental objectives of KDD is to discover knowledge of main interest to real business needs and user preference. However, this presents a big challenge to extant and future data mining research and applications. Before talking about actionable knowledge discovery, a prerequisite is about what is knowledge actionability. Then, further research can be on developing methodologies and facilities in order to support the discovery of actionable knowledge.

KDD Challenge: Mining Actionable Knowledge

Discovering actionable knowledge has been viewed as the essence of KDD. However, even up to now, it is still one of the great challenges to extant and future KDD, as pointed out by the panel of SIGKDD 2002 and 2003 (Ankerst, 2002) and retrospective literature (Chen & Liu, 2005). This situation partly results from the limitation of extant data mining methodologies, which view KDD as a data-driven, trial-and-error process targeting automated hidden knowledge discovery (Ankerst, 2002; Cao & Zhang, 2006). The methodologies do not take the constrained and dynamic environment of KDD into much consideration, which naturally excludes human and problem domain from the loop. As a result, very often, data mining research mainly aims at developing, demonstrating, and pushing the use of specific algorithms, while it runs off the rails in producing actionable knowledge of main interest to specific user needs.

To revert to the original objectives of KDD, the following three key points recently have been highlighted: comprehensive constraints around the problem (Boulicaut & Jeudy, 2005), domain knowledge (Pohle, n.d.; Yoon, Henschen, Park, & Makki, 1999), and human role (Ankerst, 2002; Cao & Dai, 2003a; Han, 1999) in the process and environment of real-world KDD. A proper consideration of these aspects in the KDD process has been reported to make KDD promising in digging out actionable knowledge satisfying real life dynamics and requests, even though this is very tough issue. This pushes us to think of what is knowledge actionablility and how to support actionable knowledge discovery.

We further study a practical methodology called domain-driven data mining for actionable knowledge discovery (Cao & Zhang, 2006). On top of the data-driven framework, domain-driven data mining aims to develop proper methodologies and techniques for integrating domain knowledge, human role and interaction, as well as actionability measures into the KDD process in order to discover actionable knowledge in the constrained environment. This research is very important for developing the next-generation data mining methodology and infrastructure (Ankerst, 2002; Cao & Zhang, 2006). It can assist in a paradigm shift from data-driven hidden pattern mining to domain-driven actionable knowledge discovery, and provides supports for KDD to be translated to real business situations, as widely expected.

Knowledge Actionability

Often, mined patterns are nonactionable to real needs due to the interestingness gaps between academia and business (Gur & Wallace, 1997). Therefore, measuring actionability of knowledge is essential in order to recognize interesting links that permit users to react to them to better service business objectives. The measurement of knowledge actionability should be from perspectives of both objective and subjective.

Let $I = \{i_1, i_2, \ldots, i_m\}$ be a set of items, *DB* be a database that consists of a set of transactions, and *x* be an itemsct in *DB*. Let *P* be an interesting pattern discovered in *DB* through utilizing a model *M*. The following concepts are developed for the DDID-PD framework.

Definition 1. Technical Interestingness. The technical interestingness *tech_int*() of a rule or a pattern is highly dependent on certain technical measures of interest specified for a data mining method. Technical interestingness is measured further in terms of technical objective measures *tech_obj*() and technical subjective measures *tech_sub*().

Definition 2. Technical Objective Interestingness. Technical objective measures *tech_obj*() capture the complexities of a link pattern and its statistical significance. It could be a set of criteria. For instance, the following logic formula indicates that an association rule *P* is technically interesting if it satisfies *min_support* and *min_confidence*.

$$\forall x \in I, \exists P : x.\text{min\_support}(P) \wedge x.\text{min\_confidence}(P) \rightarrow x.\text{tech\_obj}(P)$$

Definition 3. Technical Subjective Interestingness. On the other hand, technical subjective measures *tech_subj*(), also focusing and based on technical means, recognize to what extent a pattern is of interest to a particular user's needs. For instance, probability-based belief (Padmanabhan & Tuzhilin, 1998) is developed for measuring the expectedness of a link pattern.

Definition 4. Business Interestingness. The business interestingness *biz_int*() of an itemset or a pattern is determined from domain-oriented social, economic, user preference and/or psychoanalytic aspects. Similar to technical interestingness, business interestingness also is represented by a collection of criteria from both objective *biz_obj*() and subjective *biz_subj*() perspectives.

Definition 5. Business Objective Interestingness. The business objective interestingness *biz_obj*() measures to what extent the findings satisfy the concerns from business needs and user preference based on objective criteria. For instance, in stock trading pattern mining, profit and roi (return on investment) often is used for judging the business potential of a trading pattern objectively. If the profit and roi (return on investment) of a stock price predictor *P* are satisfied, then *P* is interesting to trading.

$$\forall x \in I, \exists P : x.\text{profit}(P) \wedge x.\text{roi}(P) \rightarrow x.\text{biz\_obj}(P)$$

Definition 6. Business Subjective Interestingness. *Biz_subj*() measures business and user concerns from subjective perspectives such as psychoanalytic factors. For instance, in stock trading pattern mining, a kind of psycho-index 90% may be used to indicate that a trader thinks it is very promising for real trading.

A successful discovery of an actionable knowledge is a collaborative work between miners and users, which satisfies both academia-oriented technical interestingness measures *tech_obj*() and *tech_subj*() and domain-specific business interestingness *biz_obj*() and *biz_subj*().

Definition 7. Actionability of a Pattern. Given a pattern *P*, its actionable capability *act*() is described as to what degree it can satisfy both the technical and business interestingness.

Figure 1. DDID-PD process model

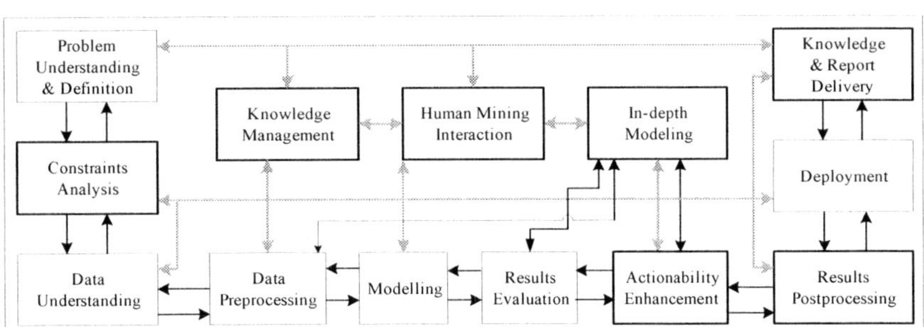

$\forall x \in I, \exists P : act(P) = f(tech\_obj(P) \wedge tech\_subj(P) \wedge biz\_obj(P) \wedge biz\_subj(P))$

If a pattern is discovered automatically by a data mining model while it only satisfies technical interestingness request, it usually is called an (technically) interesting pattern. It is presented as:

$\forall x \in I, \exists P : x.tech\_int(P) \rightarrow x.act(P)$

In a special case, if both technical and business interestingness, or a hybrid interestingness measure integrating both aspects, are satisfied, it is called an actionable pattern. It is not only interesting to data miners but generally interesting to decision makers.

$\forall x \in I, \exists P : x.tech\_int(P) \wedge x.biz\_int(P) \rightarrow x.act(P)$

Therefore, the work of actionable knowledge discovery must focus on knowledge findings that can satisfy not only technical interestingness but also business measures.

DDID-PD FRAMEWORK

The existing data mining methodology (e.g., CRISP) generally supports autonomous pattern discovery from data. The DDID-PD, on the other hand, highlights a process that discovers in-depth patterns from constraint-based context with the involvement of domain experts/knowledge. Its objective is to maximally accommodate both naïve users as well as experienced analysts and satisfy business goals. The patterns discovered are expected to be actionable to solve domain-specific problems and can be taken as grounds for performing effective actions. In order to make domain-driven data mining effective, user guides and intelligent human-machine interaction interfaces are essential through incorporating both human qualitative intelligence and machine quantitative intelligence. In addition, appropriate mechanisms are required to deal with multiform constraints and domain knowledge. This section outlines key ideas and relevant research issues of DDID-PD.

DDID-PD Process Model

The main functional components of the DDID-PD are shown in Figure 1, in which we highlight those processes specific to DDID-PD in thick boxes. The life cycle of DDID-PD is as follows, but be aware that the sequence is not rigid; some phases may be bypassed or moved back and forth in a real problem. Every step of the DDID-PD process may involve domain knowledge and the assistance of domain experts.

P1. Problem understanding
P2. Constraints analysis
P3. Analytical objective definition, feature construction
P4. Data preprocessing
P5. Method selection and modeling or
P5'. In-depth modeling
P6. Initial generic results analysis and evaluation
P7. It is quite possible that each phase from P1 may be iteratively reviewed through analyzing constraints and interaction with domain experts in a back-and-forth manner or
P7': In-depth mining on the initial generic results where applicable
P8. Actionability measurement and enhancement
P9. Back and forth between P7 and P8
P10. Results post-processing
P11. Reviewing phases from P1 may be required
P12. Deployment
P13. Knowledge delivery and report synthesis for smart decision making

The DDID-PD process highlights the following highly correlated ideas that are critical for the success of a data mining process in the real world. They are as follows:

1. **Constraint-Based Context.** Actionable pattern discovery is based on a deep understanding of the constrained environment surrounding the domain problem, data, and its analysis objectives.
2. **Integrating Domain Knowledge.** Real-world data applications inevitably involve domain and background knowledge, which is very significant for actionable knowledge discovery.
3. **Cooperation Between Human and Data Mining System.** The integration of human role and the interaction and cooperation between domain experts and mining systems in the whole process are important for effective mining execution.
4. **In-Depth Mining.** Another round of mining on the first-round results may be necessary for searching patterns really interesting to business.
5. **Enhancing Knowledge Actionability.** Based on the knowledge actionability measures, the actionable capability of findings needs to be further enhanced from modeling and evaluation perspectives.
6. **Loop-Closed Iterative Refinement.** Patterns actionable for smart business decision making in most cases would be discovered through loop-closed iterative refinement.
7. **Interactive and Parallel Mining Supports.** It is necessary and helpful to develop business-friendly system supports for human-mining interaction and parallel mining for complex data mining applications.

The following section outlines each of them respectively.

KEY COMPONENTS SUPPORTING DOMAIN-DRIVEN DATA MINING

In domain-driven data mining, the following seven key components are advocated. They have potential for making KDD different from the existing data-driven data mining if they are appropriately considered and supported from technical, procedural, and business perspectives.

Constraint-Based Context

In human society, everyone is constrained either by social regulations or by personal situations. Similarly, actionable knowledge only can be discovered in a constraint-based context such as environmental reality, expectations, and constraints in the mining process. Specifically, in the first section, we list several types of constraints that play significant roles in a process, effectively discovering knowledge actionable to business. In practice, many other aspects, such as data stream and the scalability and efficiency of algorithms, may be enumerated. They consist of domain-specific, functional, nonfunctional, and environmental constraints. These ubiquitous constraints form a constraint-based context for actionable knowledge discovery. All of the previous constraints to varying degrees must be considered in relevant phases of real-world data mining. In this case, it is even called constraint-based data mining (Boulicaut & Jeudy, 2005; Han, 1999).

Some major aspects of domain constraints include the domain and characteristics of a problem, domain terminology, specific business process, policies and regulations, particular user profiling, and favorite deliverables. Potential matters to satisfy or react on domain constraints could consist of building domain model, domain metadata, semantics, and ontologies (Cao, Zhang, & Liu, 2005); supporting human involvement, human-machine interaction, qualitative and quantitative hypotheses, and conditions; merging with business processes and enterprise information infrastructure; fitting regulatory measures; conducting user profile analysis and modeling; and so forth. Relevant hot research areas include interactive mining, guided mining, knowledge and human involvement, and so forth.

Constraints on particular data may be embodied in terms of aspects such as very large volume, ill structure, multimedia, diversity, high dimensions, high frequency and density, distribution and privacy, and so forth. Data constraints seriously affect the development of and performance requirements on mining algorithms and systems, and constitute some grand challenges to data mining. As a result, some popular researches on data constraints-oriented issues are emerging, such as stream data mining, link mining, multi-relational mining, structure-based mining, privacy mining, multimedia mining, and temporal mining.

What makes this rule, pattern, and finding more interesting than the other? In the real world, simply emphasizing technical interestingness such as objective statistical measures of validity and surprise is not adequate. Social and economic interestingness (we refer to Business Interestingness) such as user preferences and domain knowledge should be considered in assessing whether a pattern is actionable or not. Business interestingness would be instantiated into specific social and economic measures in terms of the problem domain. For instance, profit, return and roi usually are used by traders to judge whether a trading rule is interesting enough or not.

Furthermore, the delivery of an interesting pattern must be integrated with the domain environment such as business rules, process, information flow, presentation, and so forth. In addition, many other realistic issues must be considered. For instance, a software infrastructure may be established to support the full life cycle of data mining; the infrastructure needs to integrate with the existing enterprise information systems and workflow; parallel KDD may be involved with parallel supports on multiple sources, parallel I/O, parallel algorithms, and memory storage; visualization, privacy, and security should receive much deserved attention; and false alarming should be minimized.

In summary, actionable knowledge discovery won't be a trivial task. It should be put into a constraint-based context. On the other hand, tricks not only may include how to find a right pattern with a right algorithm in a right manner, but they also may involve a suitable process-centric support with a suitable deliverable to business.

Integrating Domain Knowledge

It is accepted (Pohle, n.d.; Yoon et al., 1999) gradually that domain knowledge can play significant roles in real-world data mining. For instance, in trading pattern mining, traders often take "beating market" as a personal preference to judge an identified rule's actionability. In this case, a stock mining system needs to embed the formulas calculating market return and rule return, and set an interface in order for traders to specify a favorite threshold and comparison relationship between the two returns in the evaluation process. Therefore, the key is to take advantage of domain knowledge in the KDD process.

The integration of domain knowledge is subject to how it can be represented and filled in to the knowledge discovery process. Ontology-based domain knowledge representation, transformation, and mapping between business and data mining systems is one of the proper approaches (Cao et al., 2005) to model domain knowledge. Further work is to develop agent-based cooperation mechanisms (Cao et al., 2004; Zhang, Zhang, & Cao, 2005) to support ontology-represented domain knowledge in the process.

Domain knowledge in the business field often takes forms of precise knowledge, concepts, beliefs, relations, or vague preference and bias. Ontology-based specifications build a business ontological domain to represent domain knowledge in terms of ontological items and semantic relationships. For instance, in the previous example, return-related items include return, market return, rule return, and so forth. There is *class_of* relationship between return and market return, while market return is associated with rule return in some form of user-specified logic connectors, say beating market if rule return is larger than (>) market return by a threshold φ. We can develop ontological representations to manage these items and relationships.

Further, business ontological items are mapped to a data mining system's internal ontologies. So, we build a mining ontological domain for a KDD system collecting standard domain-specific ontologies and discovered knowledge. To match items and relationships between two domains and to reduce and aggregate synonymous concepts and relationships in each domain, ontological rules, logical connectors, and cardinality constraints are studied in order to support the ontological transformation from one domain to another and the semantic aggregations of semantic relationships and ontological items' intra- or interdomains. For instance, the following rule transforms ontological items from the business domain to the mining domain. Given input item A from users, if it is associated with B by *is_a* relationship, then the output is B from the mining domain: $\forall$ (A AND B), $\exists B ::= is\_a(A, B) \Rightarrow B$, the resulting output is B. For rough and vague knowledge, we can fuzzify and map them to precise terms and relationships. For the aggregation of fuzzy ontologies, fuzzy aggregation and defuzzification mechanisms can be developed in order to sort out proper output ontologies.

Cooperation Between Human and Mining Systems

The real requirements for discovering actionable knowledge in constraint-based context determine that real-world data mining is more likely to be human involved than automated. Human involvement is embodied through the cooperation among humans (including users and business analysts, mainly domain experts) and data mining systems. This is achieved through the complementation between human qualitative intelligence, such as domain knowledge and field supervision, and mining quantitative intelligence like computational capability. Therefore, real-world data mining likely presents as a human-machine-cooperated interactive knowledge discovery process.

The role of humans can be embodied in the full period of data mining from business and data understanding, problem definition, data integration and sampling, feature selection, hypothesis proposal, business modeling, and the evaluation, refinement, and interpretation of algorithms and resulting outcomes. For instance, experience, metaknowledge, and imaginary thinking of domain experts can guide or assist with the selection of features and models, adding business factors into the modeling, creating high-quality hypotheses, designing interestingness measures by injecting business concerns, and quickly evaluating mining results. This assistance largely can improve the effectiveness and efficiency of mining actionable knowledge.

Humans often serve on the feature selection and result evaluation. Humans may play roles in a specific stage or during the full stages of data mining. Humans can be an essential constituent of or the center of a data mining system. The complexity of discovering actionable knowledge in constraint-based context determines to what extent a human must be involved. As a result, the human-mining cooperation could be, to varying degrees, human-centered or guided mining (Ankerst, 2002; Fayyad & Shapiro, 2003), or human-supported or assisted mining, and so forth.

In order to support human involvement, human mining interaction or, in a sense, presented as interactive mining (Aggarwal, 2002; Ankerst, 2002) is absolutely necessary. Interaction often takes explicit forms; for instance, setting up direct interaction interfaces to fine tune parameters. Interaction interfaces may take various forms as well, such as visual interfaces; virtual reality techniques; multi-modal, mobile agents, and so forth. On the other hand, it also could go through implicit mechanisms; for example, accessing a knowledge base or communicating with a user assistant agent. Interaction communication may be message-based, model-based, or event-based. Interaction quality relies on performance such as user-friendliness, flexibility, run-time capability, representability, and even understandability.

Mining In-Depth Patterns

The situation that many mined patterns are interesting more to data miners than to businesspersons has hindered the deployment and adoption of data mining in real applications. Therefore, it is essential to evaluate the actionability of a pattern and to further discover actionable patterns; namely, $\forall P$: $x.tech\_int(P) \wedge x.biz\_int(P) \rightarrow x.act(P)$, to support smarter and more effective decision making. This leads to *in-depth pattern mining.*

Mining in-depth patterns should consider how to improve both technical (*tech_int()*) and business interestingness (*biz_int()*) in the previous constraint-based context. Technically, it could be through enhancing or generating more effective interestingness measures (Omiecinski, 2003); for instance, a series of research has been done on designing right interestingness measures for association rule mining (Tan, Kumar, & Srivastava, 2002). It also could be through developing alternative models for discovering deeper patterns. Some other solutions include further mining actionable patterns on the discovered pattern set. Additionally, techniques can be developed in order to deeply understand, analyze, select, and refine the target data set in order to find in-depth patterns.

More attention should be paid to business requirements, objectives, domain knowledge, and qualitative intelligence of domain experts for their impact on mining deep patterns. This can be through selecting and adding business features, involving domain knowledge into modeling, supporting interaction with users, tuning parameters and data set by domain experts, optimizing models and parameters, adding factors into technical interestingness measures or building business measures, improving result evaluation mechanisms through embedding domain knowledge and human involvement, and so forth.

Enhancing Knowledge Actionability

Patterns that are interesting to data miners may not lead necessarily to business benefits, if deployed. For instance, a large number of association rules often is found, while most of them might not be workable in business situations. These rules are generic patterns or technically interesting rules. Further actionability enhancement is necessary for generating actionable patterns of use to business.

The measurement of actionable patterns is to follow the actionablilty of a pattern. Both technical and business interestingness measures must be satisfied from both objective and subjective perspectives. For those generic patterns identified based on technical measures, business interestingness needs to be checked and emphasized so that the business requirements and user preference can be put into proper consideration.

Actionable patterns in most cases can be created through rule reduction, model refinement, or parameter tuning by optimizing generic patterns. In this case, actionable patterns are a revised optimal version of generic patterns that capture deeper characteristics and understanding of the business and are also called in-depth or optimized patterns. Of course, such patterns also can be directly discovered from a data set with sufficient consideration of business constraints. For instance, the section "Mining Actionable Trading Rules" discusses mining actionable trading rules from a great number of generic rules.

Loop-Closed Iterative Refinement

Actionable knowledge discovery in a constraint-based context is likely to be a closed rather than an open process. It encloses iterative feedback to varying stages such as sampling, hypothesis, feature selection, modeling, evaluation, and interpretation in a human-involved manner. On the other hand, real-world mining process is highly iterative, because the evaluation and refinement of features, models, and outcomes cannot be completed once but, rather, is based on iterative feedback and interaction before reaching the final stage of knowledge and decision-support report delivery.

The previous key points indicate that real-world data mining cannot be dealt with just an algorithm; rather, it is really necessary to build a proper data mining infrastructure in order to discover actionable knowledge from constraint-based scenarios in a loop-closed iterative manner. To this end, agent-based data mining infrastructure (Klusch et al., 2003; Zhang et al., 2005) presents good facilities, since it provides good supports for autonomous problem-solving, user modeling, and user agent interaction.

Interactive and Parallel Mining Supports

To support domain-driven data mining, it is significant to develop interactive mining supports for human-mining interaction and to evaluate the findings. On the other hand, parallel mining supports often are necessary and can greatly upgrade the real-world data mining performance.

For interactive mining supports, intelligent agents and service-oriented computing are some good technologies. They can support flexible, business-friendly, and user-oriented human-mining interaction through building facilities for user modeling; user knowledge acquisition; domain knowledge modeling; personalized user services and recommendation; run-time supports; and mediation and management of user roles, interaction, security, and cooperation.

Based on our experience in building agent service-based stock trading and mining system F-Trade (Cao et al., 2004; F-TRADE), an agent service-based actionable discovery system can be built for domain-driven data mining. User agent, knowledge management agent, ontology services (Cao et al., 2005), and run-time interfaces can be built to support interaction with users, take users' requests, and manage information from users in terms of ontologies. Ontology-represented domain knowledge and user preferences then are mapped to mining domain for mining purposes. Domain experts can help to train, supervise, and evaluate the outcomes.

Parallel (Domingos, 2003; Taniar & Rahayu, 2002) and scalable (Manlatty & Zaki, 2000) KDD supports involve parallel computing and management supports to deal with multiple sources, parallel I/O, parallel algorithms, and memory storage.

For instance, in order to tackle cross-organization transactions, we can design efficient parallel KDD computing and system supports in order to wrap data mining algorithms. This can be through developing parallel genetic algorithms and proper processor-cache memory techniques. Multiple master-client, process-based genetic algorithms and caching techniques can be tested on different CPU and memory configurations in order to find good parallel computing strategies.

The facilities for interactive and parallel mining supports largely can improve the performance of real-world data mining in aspects such as human-mining interaction and cooperation, user modeling, domain knowledge capturing, reducing computation complexity, and so forth. They are some essential parts of next-generation KDD infrastructure.

Reference Model and Questionnaire

Reference models such as those in CRISP-DM are very helpful for guiding and managing the knowledge discovery process. It is recommended that those reference models be respected in domain-oriented, real-world data mining. However, actions and entities for domain-driven data mining, such as considering constraints and integrating domain knowledge, should be paid special attention in the corresponding models and procedures. On the other hand, new reference models are essential for supporting components such as in-depth modeling and actionablility enhancement. For instance, Figure 2 illustrates the reference model for actionability enhancement.

In the field of developing real-world data mining applications, questionnaires are very helpful for capturing business requirements, constraints, requests from organization and management, risk and contingency plans, expected representation of the deliverables, and so forth. It is recommended to design questionnaires for every procedure in the domain-driven actionable knowledge discovery process.

Reports for every procedure must be prepared and recorded in the knowledge management base for organizing well the knowledge and the process of domain-driven data mining applications.

DOMAIN-DRIVEN MINING APPLICATIONS

In this section, we illustrate some of our work in financial data mining (Lin & Cao, 2006) by utilizing domain-driven data mining methodologies.

Figure 2. Actionability enhancement

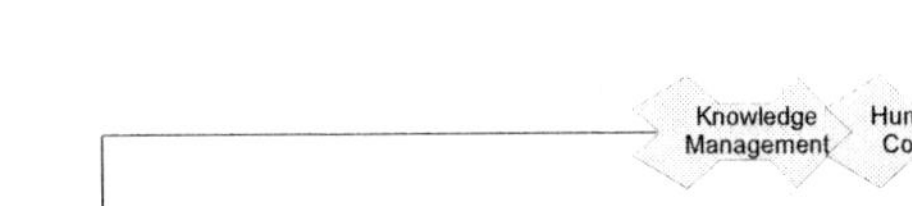

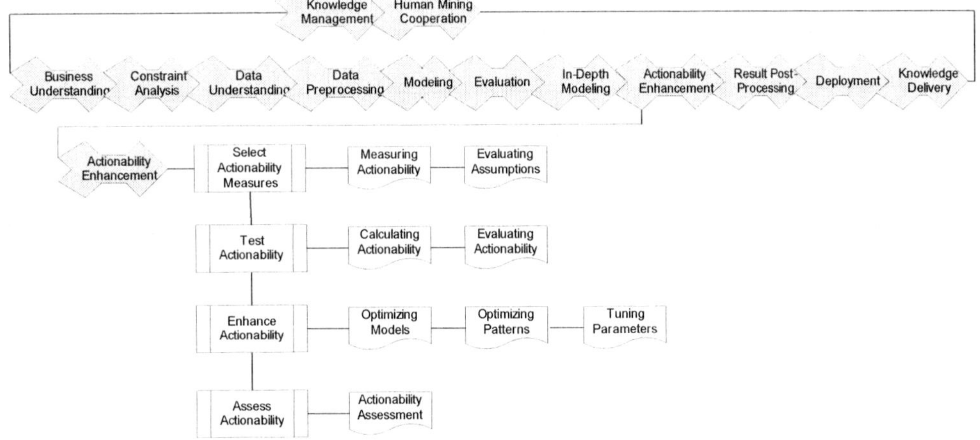

We only highlight some of those key components such as domain knowledge, in-depth rule mining, business interestingness, and parallel mining for pattern pruning in financial trading evidence discovery.

Financial data mining (Kovalerchuk & Vityaev, 2000) is of high interest, since it may benefit trading decision and market surveillance, but it also may be challenging, because financial markets are greatly complex. Taking ASX as an instance, there are more than 1,000 shares listed in this small market. In the Data Mining Program (DMP) of Australian Capital Markets Cooperative Research Center (CMCRC), we deploy the domain-driven data mining methodology to actionable trading evidence discovery, such as mining correlations between stocks, actionable trading rules, and correlations between trading rules and stocks. The following sections illustrate some results of the previous work in ASX data.

Mining Correlated Stocks

In real trading, traders often trade multiple stocks in order to manage risk. Data mining may extract evidences about what stocks are correlated with others. A common hypothesis is that stocks from the same or similar sectors or belonging to a shared production chain to some extent may be correlated. We have developed a set of correlation metrics in order to analyze the relations between stocks in the ASX. The following outlines the basic idea of mining correlated stocks by considering relevant market factors.

Algorithm: Mining Correlated Stocks

C1. Calculating the coefficient ρ of two stocks considering market impact

C2. Determining the scope of ρ interesting to trading through cooperation with traders by considering market aspects, such as market sectors, volatility, liquidity, and index

C3. Evaluation by designing and simulating strategies to trade the correlated stocks

C4. Recommending correlated stocks

In the ASX, we targeted 32 stocks with quality data from January 1997 to June 2002. Thirteen of those stocks were found to be highly correlated. Of all the 78 pairs of combinations, nine pairs were found to be actionable to trading with expectable profits. For instance, we found that stock A (representing some stock) is highly correlated with B. The return on trading the pair A-B was 40.51% on average on historical data from January 1, 1997, to June 19, 2002, without considering the market impact.

In mining correlated stocks benefiting trading, we found the following interesting points: (1) Correlated stocks interesting to trading cannot be determined just by coefficient, but rather, market aspects such as sectors, volatility, liquidity, and index should be considered, as well. (2) Interestingly, all correlated stocks mined in the ASX come from different sectors. This finding means that correlated stocks are not necessary from the same industry, as presumed by financial researchers. (3) The return on trading a correlated pair is affected highly by the liquidity and volatility of a stock.

Mining Actionable Trading Rules

A trading rule actually indicates a possible investment pattern in stock markets. For instance, the trading rule MA (*sr*, *lr*, δ) indicates a correlated trading pattern between features *short-run moving average* (*sr*) and *long-run moving average* (*lr*). The trading strategy is defined as follows (where δ is a fixed difference band between *sr* and *lr*).

IF *sr* *(1-δ) >= *lr* THEN *Buy*
IF *sr* *(1+δ) <= *lr* THEN *Sell*

In market trading, the previous pattern MA actually can be instantiated into millions of individual generic rules such as MA(2, 50, 0.01) and MA(10, 50, 0.01). However, traders do not know which rule is actionable for a specific investment scenario. Therefore, mining actionable trading rules emerge as a worthwhile activity.

Figure 3. Interfaces supporting human-mining system interaction

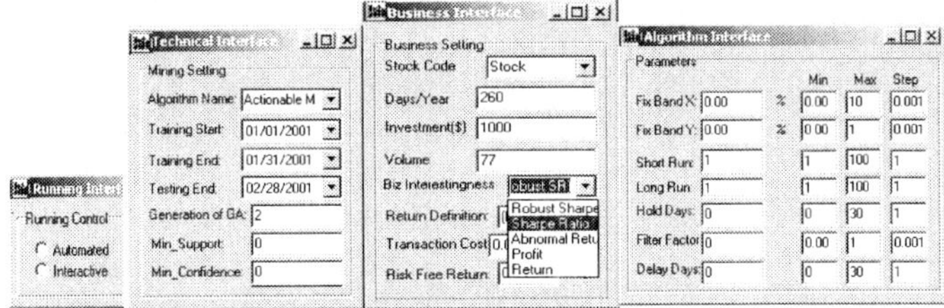

Figure 4. Improved business interestingness by in-depth rules: (a) sharpe ratio with generic MA rules (on left) and (b) sharpe ratio with actionable MA rules (on right)

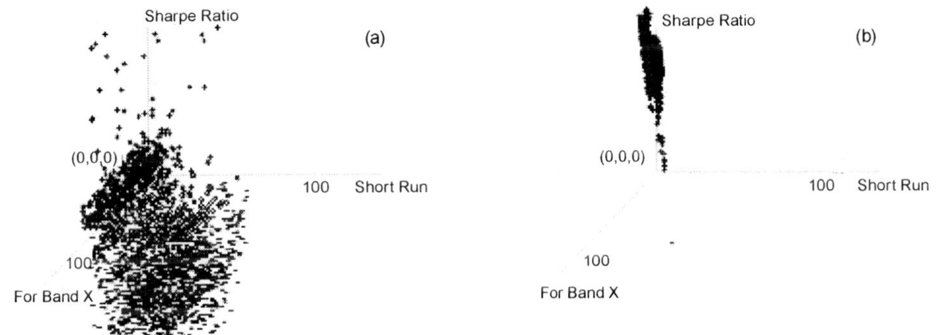

In order to involve domain knowledge in finding actionable rules, we built human mining interaction interfaces. Figure 3 demonstrates some interfaces in which users can trigger the process in terms of automated execution or interactive mode with involvement of users. In interactive mode, technical analysts can advise the previous process as well as refine technical factors for mining setting and algorithm parameter tuning. Business analysts can supervise the construction of features, fine tune the parameters, and set evaluation criteria for business concerns. For instance, measure *sharpe_ratio* is used for evaluating the business actionability of an identified rule. Additionally, the system supports ad-hoc execution, meaning that users can tune the parameters or change interestingness measures to check the results at run time.

$$sharpe\_ratio = (r_P - r_R) / \sigma_P$$

where r_P is the expected portfolio return, r_R is risk-free rate, and σ_P is portfolio-standard deviation. Higher *sharpe_ratio* means more return with lower risk.

We found a collection of actionable rules using our actionable trading rule mining algorithms. For instance, in ASX data, MA(4, 19, 0.033) is a very interesting rule using training data from January 1, 2000, to December 31, 2000, and testing set between January 1, 2001, and December 31, 2001. The number of trading signals generated by this rule is much bigger than other possible rules with good *sharpe_ratio*. Figure 4(b) shows that its *sharpe_ratio* has a greatly improved positive scope compared with (a) the generic results. This demonstrates that the in-depth pattern mining with the involvement of domain knowledge can improve the actionability of trading rules.

Mining Rule-Stock Correlations

In market, some trading rules are tested to be effective to trade a class of stocks, while other rules are more suitable for other stocks. Using data mining, we may evidence that whether there exist correlations between trading rules and stocks. If we do discover some actionable correlations, then it would be helpful for trading. Based on this hypothesis, we developed algorithms to find the correlations between trading rules and stocks in stock market data. The basic ideas of the rule-stock correlation mining algorithms are as follows.

Algorithm: Mining Correlated Trading Rule-Stocks Pairs

C1. Mining actionable rules for an individual stock

C2. Mining highly correlated rule-stock pairs by high dimension reduction

C3. Evaluating and refining the rule-stock pairs by considering traders' concerns

C4. Recommending actionable rule-stock pairs

In discovering actionable rule-stock pairs, traders were invited to give suggestions on designing features, interestingness measure and parameter optimization. They also helped us to design mechanisms for evaluating and refining rule-stock pairs. Taking the ASX as an instance, three types of trading rules (MA, Filter Rule, and Channel Breakout) (Ryan, Allan, & Halbert, 2005) and 26 ASX stocks were chosen for the experiments. For instance, the intraday training data was from January 1, 2001, to January 31, 2001, and the testing set was from February 1, 2001, to February 28, 2001. Five investment plans were conducted on the previous rules and stocks. In organizing pairs, we ranked them based on return and generated 5% pair, 10% pair, and so forth from the whole pair set. The 5% pair means that return for trading these pairs is the top 5% in the whole pair set. Figure 5 illustrates returns for different investment plans on different pairs. These graphs are interesting to traders, allowing them to make smart trading decisions using these mined rule-stock pairs.

Parallel Computing

Mining actionable correlations in a scenario with hundreds of stocks (e.g., more than 1,000) and millions of trading rules on stock data with hundreds

Figure 5. Return on investment with actionable rule-stock pairs

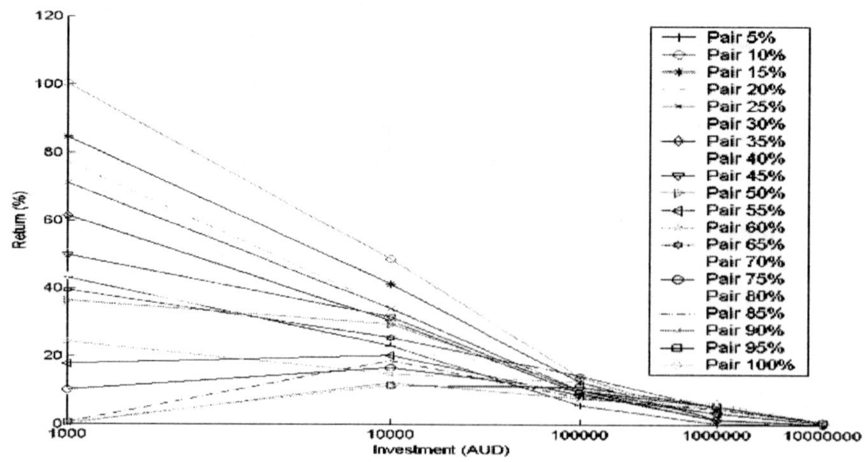

of thousands of intraday stock transactions (e.g., more than 700,000 per day) is very time-consuming. Parallel computing is essential for acceptable response time. Taking the mining actionable trading rules as an example, we designed different parallel algorithms A_l in order to test their performance on ASX stock C (representing a stock in ASX) using intraday data in 2001.

- Alg_1. Loops through all possible combinations of MA (sr, lr, δ).
- Alg_2. Parallelizes Alg_1 by partitioning the search calculations into four processing units.
- Alg_3. Parallelizes Alg_1 by partitioning the search calculations into eight processing units.
- Alg_4. Parallelizes Alg_1 by splitting processes into master and slave subprocesses on four processing units.
- Alg_5. Parallelizes Alg_1 by splitting processes into master and slave subprocesses on eight processing units.

We tested the previous algorithms on a Linux box with eight CPUs (Intel(R) Xeon(TM) MP CPU 2.00GHz) and 4GB memory. The running time for each algorithm is shown in Table 1. The results indicate that parallel computing and efficient implementations can extremely accelerate the computation of data mining. However, in our case, eight CPUs make little difference from four CPUs. This is probably due to the overhead from system and managing master and slave subprocesses.

CONCLUSIONS AND FUTURE WORK

Real-world data mining applications have proposed urgent requests for discovering actionable knowledge of main interest to real-user and business needs. Actionable knowledge discovery is significant and also very challenging. It is nominated as one of great challenges of KDD in the next 10 years. The research on this issue has potential to change the existing situation in which a great number of rules are mined while few of them are interesting to business, and to promote the wide deployment of data mining into business.

This article has developed a new data mining methodology referred to as Domain-Driven Data Mining. It provides a systematic overview of the issues in discovering actionable knowledge and advocates the methodology of mining actionable knowledge in constraint-based context through human mining system cooperation in a loop-closed iterative refinement manner. It is useful for promoting the paradigm shift from data-driven hidden pattern mining to domain-driven actionable knowledge discovery. Further, progress in studying domain-driven data mining methodologies and applications can help the deployment shift from standard or artificial data set-based testing to real data and business environment-based backtesting and development.

On top of data-driven data mining, domain-driven data mining includes almost all phases of the well-known industrial data mining methodol-

Table 1. Running time for mining actionable MA

Algorithms	Running Time (seconds)
Alg_1	860
Alg_2	26
Alg_3	22
Alg_4	13
Alg_5	11

ogy CRISP-DM. However, it also has enclosed some big differences from the data-driven methodologies, such as CRISP-DM. For instance:

- Some new essential components, such as in-depth modeling, the involvement of domain experts and knowledge, and knowledge actionability measurement and enhancement are taken into the life cycle of KDD for consideration.
- In the domain-driven methodology, the phases of CRISP-DM highlighted by thick boxes in Figure 1 are enhanced by dynamic cooperation with domain experts and the consideration of constraints and domain knowledge.
- Knowledge actionability is highlighted in the discovery process. Both technical and business interestingness must be concerned in order to satisfy needs and especially business requests.

These differences actually play key roles in improving the existing knowledge discovery in a more effective way.

In the deployment of the domain-driven data mining methodology, we have demonstrated some of our research results in mining actionable correlations in Australian stock markets. The experiments show that domain-driven data mining has potential for improving the actionable knowledge mining. Our further work is on developing detailed mining process specifications and interfaces for easily deploying domain-driven data mining methodology into real-world mining.

ACKNOWLEDGMENTS

This work was supported in part by the Australian Research Council (ARC) Discovery Projects (DP0449535 and DP0667060), UTS Chancellor and ECRG funds, National Science Foundation of China (60496327), and Overseas Outstanding Talent Research Program of Chinese Academy of Sciences (06S3011S01). We appreciate CMCRC and SIRCA for providing data services. Thanks also go to Dr. Lin Li and Jiarui Ni for implementation supports.

REFERENCES

Aggarwal, C. (2002). Towards effective and interpretable data mining by visual interaction. *ACM SIGKDD Explorations Newsletter, 3*(2), 11–22.

Ankerst, M. (2002). Report on the SIGKDD-2002 panel the perfect data mining tool: Interactive or automated? *ACM SIGKDD Explorations Newsletter, 4*(2), 110–111.

Bagui, S. (2006). An approach to mining crime patterns. *International Journal of Data Warehousing and Mining, 2*(1), 50–80.

Boulicaut, J-F., & Jeudy, B. (2005). Constraint-based data mining. In O. Maimon, & L. Rokach (Eds.), *The data mining and knowledge discovery handbook* (pp. 399–416). New York: Springer.

Cao, L., & Dai, R. (2003a). Human-computer cooperated intelligent information system based on multi-agents. *ACTA Automatica Sinica, 29*(1), 86–94.

Cao, L., & Dai, R. (2003b). Agent-oriented metasynthetic engineering for decision making. *International Journal of Information Technology and Decision Making, 2*(2), 197–215.

Cao, L. et al. (2004). Agent services-based infrastructure for online assessment of trading strategies. In *Proceedings of the 2004 IEEE/WIC/ACM International Conference on Intelligent Agent Technology* (pp. 345–349). IEEE Press.

Cao, L., & Zhang, C. (2006). Domain-driven actionable knowledge discovery in the real world. In *Proceedings of the 10th Pacific-Asia Conference on Knowledge Discovery and Data Mining 2006, LNAI 3918* (pp. 821–830). Singapore: Springer.

Cao, L., Zhang, C., & Liu, J. (2005). *Ontology-based integration of business intelligence* (Technical Report). Sydney, Australia: University of Technology, E-Intelligence Group.

Chen, S.Y., & Liu, X. (2005). Data mining from 1994 to 2004: an application-orientated review. *International Journal of Business Intelligence and Data Mining, 1*(1), 4–21.

CMCRC (Captial Markets Cooperative Research Centre). (n.d.). Retrieved from http://www.cmcrc.com

CRISP. (n.d.). Retrieved from http://www.crisp-dm.org

Domingos, P. (2003). Prospects and challenges for multi-relational data mining. *SIGKDD Explorations, 5*(1), 80–83.

Fayyad, U., Shapiro, G., & Uthurusamy, R. (2003). Summary from the KDD-03 panel—Data mining: The next 10 years. *ACM SIGKDD Explorations Newsletter, 5*(2), 191–196.

F-TRADE. (n.d.). Retrieved from http://www.f-trade.info

Gur Ali, O.F., & Wallace, W.A. (1997). Bridging the gap between business objectives and parameters of data mining algorithms. *Decision Support Systems, 21*, 3–15.

Han, J. (1999). *Towards human-centered, constraint-based, multi-dimensional data mining* [Speech]. Minneapolis, MN: University of Minnesota.

Klusch, M. et al. (2003). The role of agents in distributed data mining: Issues and benefits. In *Proceedings of the 2003 IEEE/WIC International Conference on Intelligent Agent Technology (IAT 2003)* (pp. 211–217). IEEE Computer Society Press.

Kovalerchuk, B., & Vityaev, E. (2000). *Data mining in finance: Advances in relational and hybrid methods*. Massachusetts: Kluwer Academic Publishers.

Lin, L., & Cao, L. (2006). *Mining in-depth patterns in stock market* (Technical Report). Sydney, Australia: University of Technology, E-Intelligence Group.

Manlatty, M., & Zaki, M. (2000). Systems support for scalable data mining. *SIGKDD Explorations, 2*(2), 56–65.

Omiecinski, E. (2003). Alternative interest measures for mining associations. *IEEE Transactions on Knowledge and Data Engineering, 15*, 57–69.

Padmanabhan, B., & Tuzhilin, A. (1998). A belief-driven method for discovering unexpected patterns. In *Proceedings of the 4th International Conference on Knowledge Discovery and Data Mining* (pp. 94–100). Menlo Park, CA: AAAI Press.

Pohle, C. (n.d.). *Integrating and updating domain knowledge with data mining*. Retrieved from http://citeseer.ist.psu.edu/668556.html

Ryan, S., Allan, T., & Halbert, W. (1999). Data-snooping, technical trading rule performance, and the bootstrap. *Journal of Financial, 54*(5), 1647–1692.

Tan, P., Kumar, V., & Srivastava, J. (2002). Selecting the right interestingness measure for association patterns. In *Proceedings of the 8th ACM SIGKDD International Conference on Knowledge Discovery and Data Mining* (pp. 32–41). ACM Press.

Taniar, D., & Rahayu, J.W. (2002). Parallel data mining. In H.A. Abbass, R. Sarker, & C. Newton (Eds.), *Data mining: A heuristic approach* (pp. 261–289). Hershey, PA: Idea Group Publishing.

Yoon, S., Henschen, L., Park, E., & Makki, S. (1999). Using domain knowledge in knowledge discovery. In *Proceedings of the Eighth International Conference on Information and Knowledge Management*. ACM Press.

Zhang, C., Zhang, Z., & Cao, L. (2005). Agents and data mining: Mutual enhancement by integration. In *Proceedings of the International Workshop on Autonomous Intelligent Systems: Agents and Data Mining, LNAI 3505* (pp. 50–61). Berlin: Springer.

This work was previously published in International Journal of Data Warehousing and Mining, Vol. 2, Issue 4, edited by D. D. Taniar, pp. 49-65, copyright 2006 by IGI Publishing, formerly known as Idea Group Publishing (an imprint of IGI Global).

Chapter 2.21
Metric Methods in Data Mining

Dan A. Simovici
University of Massachusetts – Boston, USA

ABSTRACT

This chapter presents data mining techniques that make use of metrics defined on the set of partitions of finite sets. Partitions are naturally associated with object attributes and major data mining problem such as classification, clustering and data preparation which benefit from an algebraic and geometric study of the metric space of partitions. The metrics we find most useful are derived from a generalization of the entropic metric. We discuss techniques that produce smaller classifiers, allow incremental clustering of categorical data and help users to better prepare training data for constructing classifiers. Finally, we discuss open problems and future research directions.

INTRODUCTION

This chapter is dedicated to metric techniques applied to several major data mining problems: classification, feature selection, incremental clustering of categorical data and to other data mining tasks.

These techniques were introduced by R. López de Màntaras (1991) who used a metric between partitions of finite sets to formulate a novel splitting criterion for decision trees that, in many cases, yields better results than the classical entropy gain (or entropy gain ratio) splitting techniques.

Applications of metric methods are based on a simple idea: each attribute of a set of objects induces a partition of this set, where two objects belong to the same class of the partition if they have identical values for that attribute. Thus, any metric defined on the set of partitions of a finite set generates a metric on the set of attributes. Once a metric is defined, we can evaluate how far these attributes are, cluster the attributes, find centrally located attributes and so on. All these possibilities can be exploited for improving existing data mining algorithms and for formulating new ones.

Important contributions in this domain have been made by J. P. Barthélemy (1978), Barthélemy and Leclerc (1995) and B. Monjardet (1981) where a metric on the set of partitions of a finite set is introduced starting from the equivalences defined by partitions.

Our starting point is a generalization of Shannon's entropy that was introduced by Z. Daróczy (1970) and by J. H. Havrda and F. Charvat (1967). We developed a new system of axioms for this type of entropies in Simovici and Jaroszewicz (2002) that has an algebraic character (being formulated for partitions rather than for random distributions). Starting with a notion of generalized conditional entropy we introduced a family of metrics that depends on a single parameter. Depending on the specific data set that is analyzed some of these metrics can be used for identifying the "best" splitting attribute in the process of constructing decision trees (see Simovici & Jaroszewicz, 2003, in press). The general idea is to use as splitting attribute the attribute that best approximates the class attribute on the set of objects to be split. This is made possible by the metric defined on partitions.

The performance, robustness and usefulness of classification algorithms are improved when relatively few features are involved in the classification. Thus, selecting relevant features for the construction of classifiers has received a great deal of attention. A lucid taxonomy of algorithms for feature selection was discussed in Zongker and Jain (1996); a more recent reference is Guyon and Elisseeff (2003). Several approaches to feature selection have been explored, including wrapper techniques in Kohavi and John, (1997) support vector machines in Brown, Grundy, Lin, Cristiani, Sugnet, and Furey (2000), neural networks in Khan, Wei, Ringner, Saal, Ladanyi, and Westerman (2001), and prototype-based feature selection (see Hanczar, Courtine, Benis, Hannegar, Clement, & Zucker, 2003) that is close to our own approach. Following Butterworth, Piatetsky-Shapiro, and Simovici (2005), we shall introduce an algorithm for feature selection that clusters attributes using a special metric and, then uses a hierarchical clustering for feature selection.

Clustering is an unsupervised learning process that partitions data such that similar data items are grouped together in sets referred to as clusters. This activity is important for condensing and identifying patterns in data. Despite the substantial effort invested in researching clustering algorithms by the data mining community, there are still many difficulties to overcome in building clustering algorithms. Indeed, as pointed in Jain, Murthy and Flynn (1999) "there is no clustering technique that is universally applicable in uncovering the variety of structures present in multidimensional data sets." This situation has generated a variety of clustering techniques broadly divided into hierarchical and partitional; also, special clustering algorithms based on a variety of principles, ranging from neural networks and genetic algorithms, to tabu searches.

We present an incremental clustering algorithm that can be applied to nominal data, that is, to data whose attributes have no particular natural ordering. In general, objects processed by clustering algorithms are represented as points in an n-dimensional space R^n and standard distances, such as the Euclidean distance, are used to evaluate similarity between objects. For objects whose attributes are nominal (e.g., color, shape, diagnostic, etc.), no such natural representation of objects as possible, which leaves only the Hamming distance as a dissimilarity measure; a poor choice for discriminating among multivalued attributes of objects. Our approach is to view clustering as a partition of the set of objects and we focus our attention on incremental clustering, that is, on clusterings that build as new objects are added to the data set (see Simovici, Singla, & Kuperberg, 2004; Simovici & Singla, 2005). Incremental clustering has attracted a substantial amount of attention starting with algorithm of Hartigan (1975) implemented in Carpenter and Grossberg (1990). A seminal paper (Fisher, 1987) contains an incremental clustering algorithm that involved restructurings of the clusters in addition to the incremental additions of objects. Incremental clustering related to dynamic aspects of databases were discussed in Can (1993) and Can,

Box 1.

$$\{\{a\},\{b\},\{c\},\{d\}\} \quad \{\{a,b\},\{c\},\{d\}\} \quad \{\{a,c\},\{b\},\{d\}\} \quad \{\{a,d\},\{b\},\{c\}\}$$
$$\{\{a\},\{b,c\},\{d\}\} \quad \{\{a\},\{b,d\},\{c\}\} \quad \{\{a\},\{b\},\{c,d\}\} \quad \{\{a,b,c\},\{d\}\}$$
$$\{\{a,b\},\{c,d\}\} \quad \{\{a,b,d\},\{c\}\} \quad \{\{a,c\},\{b,d\}\} \quad \{\{a,c,d\},\{b\}\}$$
$$\{\{a,d\},\{b,c\}\} \quad \{\{a\},\{b,c,d\}\} \quad \{\{a,b,c,d\}\}$$

Fox, Snavely, and France (1995). It is also notable that incremental clustering has been used in a variety of areas (see Charikar, Chekuri, Feder, & Motwani, 1997; Ester, Kriegel, Sander, Wimmer, & Xu, 1998; Langford, Giraud-Carrier, & Magee, 2001; Lin, Vlachos, Keogh, & Gunopoulos, 2004). Successive clusterings are constructed when adding objects to the data set in such a manner that the clusterings remain equidistant from the partitions generated by the attributes.

Finally, we discuss an application to metric methods to one of the most important pre-processing tasks in data mining, namely data discretization (see Simovici & Butterworth, 2004; Butterworth, Simovici, Santos, & Ohno-Machado, 2004).

PARTITIONS, METRICS, ENTROPIES

Partitions play an important role in data mining. Given a nonempty set S, a *partition of S* is a nonempty collection $\pi = \{B_1, ..., B_n\}$ such that $i \neq j$ implies $B_i \cap B_j = \emptyset$, and:

$$\bigcup_{i=1}^{n} B_i = S.$$

We refer to the sets $B_1, ..., B_n$ as the *blocks* of π. The set of partitions of S is denoted by *PARTS(S)*.

The set of partitions of S is equipped with a partial order by defining $\pi \leq \sigma$ if every block B of π is included in a block C of σ. Equivalently, we have $\pi \leq \sigma$ if every block C of σ is a union of a collection of blocks of π. The smallest element of the partially ordered set ($PART(S)$ $\leq$) is the partition α_S whose blocks are the singletons $\{x\}$ for $x \in S$; the largest element is the one-block partition ω_S whose unique block is S.

Example 1

Let $S = \{a, b, c, d\}$ be a four-element set. The set *PARTS(S)* consists of the 15 partitions shown in Box 1.

Among many chains of partitions we mention that as shown in Box 2.

A partition σ *covers* another partition π (denoted by $\pi \prec \sigma$) if $\pi \leq \sigma$ and there is no partition τ such that $\pi \leq \tau \leq \sigma$. The partially ordered set *PARTS(S)* is actually a lattice. In other words, for every two partitions $\pi, \sigma \in PARTS(S)$ both $\inf\{\pi, \sigma\}$ and $\sup\{\pi, \sigma\}$ exist. Specifically, $\inf\{\pi, \sigma\}$ is easy to describe. It consists of all nonempty intersections of blocks of π and σ:

Box 2.

$$\{\{a\},\{b\},\{c\},\{d\}\} \leq \{\{a\},\{b,c\},\{d\}\} \leq \{\{a,b,c\},\{d\}\} \leq \{\{a,b,c,d\}\}$$

$$\inf\{\pi,\sigma\} = \{B \cap C \mid B \in \pi, C \in \sigma, B \cap C \neq \varnothing\}.$$

We will denote this partition by $\pi \cap \sigma$. The supremum of two partitions $\sup\{\pi,\sigma\}$ is a bit more complicated. It requires that we introduce the graph of the pair π,σ as the bipartite graph $G(\pi,\sigma)$ having the blocks of π and σ as its vertices. An edge (B,C) exists if $B \cap C \neq \varnothing$. The blocks of the partition $\sup\{\pi,\sigma\}$ consist of the union of the blocks that belong to a connected component of the graph $G\{\pi,\sigma\}$. We will denote $\sup\{\pi,\sigma\}$ by $\pi \cup \sigma$.

Example 2

The graph of the partitions $\pi = \{\{a,b\}, \{c\}, \{d\}\}$ and $\sigma = \{\{a\}, \{b,d\}, \{c\}\}$ of the set $S = \{a, b, c, d\}$ is shown in Figure 1. The union of the two connected components of this graph are $\{a,b,d\}$ and $\{c\}$, respectively, which means that $\pi \cup \sigma = \{\{a,b,d\}, \{c\}\}$.

We introduce two new operations on partitions. If S,T are two disjoint sets and $\pi \in PARTS(S)$, $\sigma \in PARTS(T)$, the *sum* of π and σ is the partition: $\pi + \sigma = \{B_1,...,B_n, C_1,...,C_p\}$ of $S \cup T$, where $\pi = \{B_1,...,B_n\}$ and $\sigma = \{C_1,...,C_p\}$.

Whenever the "+" operation is defined, then it is easily seen to be associative. In other words, if S,U,V are pairwise disjoint and nonempty sets, and $\pi \in PARTS(S)$, $\sigma \in PARTS(U)$, and $\tau \in PARTS(V)$, then $(\pi+\sigma)+\tau = \pi+(\sigma+\tau)$. Observe that if S,U are disjoint, then $\alpha_S + \alpha_U = \alpha_{S \cup U}$. Also, $\omega_S + \omega_U$ is the partition $\{S,U\}$ of the set $S \cup U$.

For any two nonempty sets S, T and $\pi \in PARTS(S)$, $\sigma \in PARTS(T)$ we define the *product* of π and σ, as the partition $\pi \times \sigma$ $\{B \times C \mid B \in \pi, C \in \sigma\}$ of the set product $B \times C$.

Example 3

Consider the set $S = \{a_1,a_2,a_3\}$, $T = \{a_4,a_5,a_6,a_7\}$ and the partitions p = $\{\{a_1,a_2\},\{a_3\}\}$, s = $\{\{a_4\}\{a_5,a_6\}\{a_7\}\}$ of S and T, respectively. The sum of these partitions is: $\pi + \sigma = \{\{a_1,a_2\},\{a_3\}, \{a_4\}, \{a_5,a_6\}, \{a_7\}\}$, while their product is:

$\pi \times$ s $= \{\{a_1,a_2\} \times \{a_4\}, \{a_1,a_2\} \times \{a_5,a_6\}, \{a_1,a_2\} \times \{a_7\}, \{a_3\} \times \{a_4\}, \{a_3\} \times \{a_5, a_6\}, \{a_3\} \times \{a_7\}\}$.

A *metric* on a set S is a mapping $d: S \times S \rightarrow R_{\geq 0}$ that satisfies the following conditions:

(M1) $d(x, y) = 0$ if and only if $x = y$
(M2) $d(x,y) = d(y,x)$
(M3) $d(x,y) + d(y,z) \geq d(x,z)$

Figure 1. Graph of two partitions

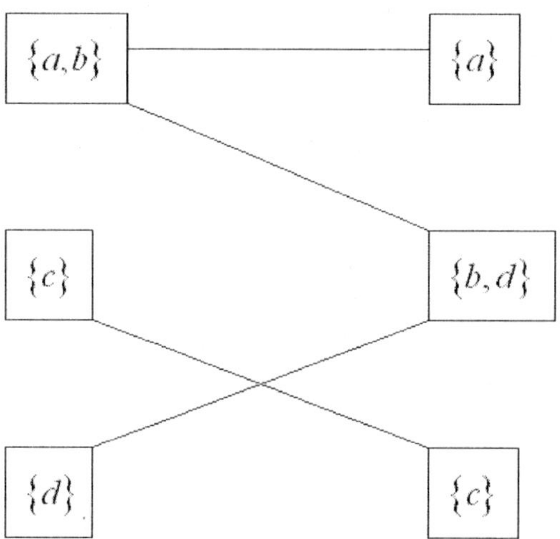

Box 3.

$$H_\beta(\pi + \sigma) = \left(\frac{|S|}{|S|+|T|}\right)^\beta H_\beta(\pi) + \left(\frac{|T|}{|S|+|T|}\right)^\beta H_\beta(\sigma) + H_\beta(\{S,T\})$$

for every $x,y,z \in S$. In equality (M3) is known as the *triangular axiom* of metrics. The pair (S,d) is referred to as a *metric space*.

The *betweeness relation* of the metric space (S,d) is a ternary relation on S defined by [x,y,z] if $d(x,y) + d(y,z) = d(x,z)$. If $[x, y, z]$ we say that *y is between x and z*.

The Shannon entropy of a random variable X having the probability distribution $\boldsymbol{p} = (p_1,...,p_n)$ is given by:

$$H(p_1,...,p_n) = \sum_{i=1}^{n} -p_i \log_2 p_i.$$

For a partition $\pi \in$ *PARTS(S)* one can define a random variable X_π that takes the value *i* whenever a randomly chosen element of the set S belongs to the block B_i of π. Clearly, the distribution of X_π is $(p_1,...,p_n)$, where:

$$p_i = \frac{|B_i|}{|S|}.$$

Thus, the entropy $H(\pi)$ of π can be naturally defined as the entropy of the probability distribution of X and we have:

$$H(\pi) = -\sum_{i=1}^{n} \frac{|B_i|}{|S|} \log_2 \frac{|B_i|}{|S|}.$$

By the well-known properties of Shannon entropy the largest value of $H(\pi)$, $\log_2 |S|$, is obtained for $\pi = \alpha_S$, while the smallest, 0, is obtained for $\pi = \omega_S$.

It is possible to approach the entropy of partitions from a purely algebraic point of view that takes into account the lattice structure of *(PARTS(S)*$\leq$*)* and the operations on partitions that we introduced earlier. To this end, we define the β-entropy, where $\beta>0$, as a function defined on the class of partitions of finite sets that satisfies the following conditions:

(P1) If $\pi_1,\pi_2 \in$ PARTS(S) are such that $\pi_1 \leq \pi_2$, then $H_\beta(\pi_2) \leq H_\beta(\pi_1)$.
(P2) If S,T are two finite sets such that $|S| \leq |T|$, then $H_\beta(\alpha_T) \leq H_\beta(\alpha_S)$.
(P3) For every disjoint sets S,T and partitions p $\in$ *PARTS(S)* and $\sigma \in$ *PARTS(T)* see Box 3.
(P4) If $\pi \in$ *PARTS(S)* and $\sigma \in$ *PARTS(T)*, then $H_\beta(\pi\times\sigma) = \varphi(H_\beta(\pi), H_\beta(\sigma))$, where $\varphi : \mathbb{R}_{\geq 0} \to \mathbb{R}_{\geq 0}$ is a continuous function such that $\varphi(x,y) = \varphi(y,x)$, and $\varphi(x,0) = x$ for $x,y \in \mathbb{R}_{\geq 0}$.

In Simovici and Jaroszewicz (2002) we have shown that if $\pi = \{B_1,...,B_n\}$ is a partition of S, then:

$$H_\beta(\pi) = \frac{1}{2^{1-\beta}-1}\left(\sum_{i=1}^{n}\left(\frac{|B_i|}{|S|}\right)^\beta - 1\right)$$

In the special case, when b $\to$ 1 we have:

$$\lim_{\beta\to 1} H_\beta(\pi) = -\sum_{i=1}^{n} \frac{|B_i|}{|S|} \log_2 \frac{|B_i|}{|S|}$$

This axiomatization also implies a specific form of the function φ. Namely, if $\beta \neq 1$ it follows that $\varphi(x,y) = x+y+(2^{1-\beta}-1)xy$. In the case of Shannon entropy, obtained using $\beta = 1$ we have $\varphi(x,y) = x+y$ for $x,y \in R_{\geq 0}$.

Note that if $|S| = 1$, then *PARTS(S)* consists of a unique partition $\alpha_S = \omega_S$ and $H_\beta(\omega_S) = 0$. Moreover, for an arbitrary finite set S we have $H_\beta(\pi) = 0$ if and only if $\pi = \omega_S$. Indeed, let U,V be two finite disjoint sets that have the same cardinality. Axiom **(P3)** implies Box 4.

Box 4.

$$H_\beta(\omega_U + \omega_V) = \left(\frac{1}{2}\right)^\beta (H_\beta(\omega_U) + H_\beta(\omega_V)) + H_\beta(\{U,V\})$$

Since $\omega_U + \omega_V = \{U,V\}$ it follows that $H_\beta(\omega_U) = H_\beta(\omega_V) = 0$.

Conversely, suppose that $H_\beta(\pi) = 0$. If $\pi \leq \omega_S$ there exists a block B of π such that $\varnothing \subset B \subset S$. Let θ be the partition $\theta = \{B, S - B\}$. It is clear that $\pi \leq \theta$, so we have $0 \leq H_\beta(\theta) \leq H_\beta(\pi)$ which implies $H_\beta(\theta) = 0$. This in turn yields:

$$\left(\frac{|B|}{|S|}\right)^\beta + \left(\frac{|S-B|}{|S|}\right)^\beta - 1 = 0$$

Since the function $f(x) = x^\beta + (1-x)^\beta - 1$ is concave for b > 1 and convex for b < 1 on the interval [0,1], the above equality is possible only if $B = S$ or if $B = \varnothing$, which is a contradiction. Thus, $\pi = \omega_S$.

These facts suggest that for a subset T of S the number $H_\beta(\pi_T)$ can be used as a measure of the purity of the set T with respect to the partition π. If T is π-pure, then $\pi_T = \omega_T$ and, therefore, $H_\beta(\pi_T) = 0$. Thus, the smaller $H_\beta(\pi_T)$, the more pure the set T is.

The largest value of $H_\beta(\pi)$ when p $\in$ *PARTS(S)* is achieved when $\pi = \alpha_S$; in this case we have:

$$H_\beta(\alpha_S) = \frac{1}{2^{1-\beta} - 1}\left(\frac{1}{|S|^{\beta-1}} - 1\right)$$

GEOMETRY OF THE METRIC SPACE OF PARTITIONS OF FINITE SETS

Axiom **(P3)** can be extended as follows:

Theorem 1: Let $S_1, \ldots, S_n$ be n pairwise disjoint finite sets, $S = \prod S_i$ and let $p_1, \ldots, p_n$ be partitions of $S_1, \ldots, S_n$, respectively. We have:

$$H_\beta(\pi_1 + \ldots + \pi_n) = \sum_{i=1}^{n} \left(\frac{|S_i|}{|S|}\right)^\beta H_\beta(\pi_i) + H_\beta(\theta),$$

where θ *is the partition* $\{S_1, \ldots, S_n\}$ *of* S.

The β-entropy defines a naturally conditional entropy of partitions. We note that the definition introduced here is an improvement over our previous definition given in Simovici and Jaroszewicz (2002). Starting from conditional entropies we will be able to define a family of metrics on the set of partitions of a finite set and study the geometry of these finite metric spaces.

Let $\pi, \sigma \in$ *PARTS(S)*, where $\sigma = \{C_1, \ldots, C_n\}$. The β-*conditional entropy of the partitions* $\pi, \sigma \in$ *PARTS(S)* is the function defined by:

$$H_\beta(\pi \mid \sigma) = \sum_{j=1}^{n} \left(\frac{|C_j|}{|S|}\right)^\beta H_\beta(\pi_{C_j}).$$

Box 5.

$$H_\beta(\alpha_S \mid \sigma) = \sum_{j=1}^{n} \left(\frac{|C_j|}{|S|}\right)^\beta H_\beta(\alpha_{C_j}) = \frac{1}{2^{1-\beta} - 1}\left(\frac{1}{|S|^{\beta-1}} - \sum_{j=1}^{n}\left(\frac{|C_j|}{|S|}\right)^\beta\right)$$

where s = $\{C_1, \ldots, C_n\}$

Box 6.

$$H_\beta(\pi \mid \sigma) = \frac{1}{2^{1-\beta}-1} \sum_{i=1}^{m} \sum_{j=1}^{n} \left(\left(\frac{|B_i \cap C_j|}{|S|} \right)^\beta - \left(\frac{|C_j|}{|S|} \right)^\beta \right),$$

where $\pi = \{B_1,...,B_m\}$.

Observe that $H_\beta(\pi|\omega_S) = H_\beta(\pi)$ and that $H_\beta(\omega_S\ \pi) = H_\beta(\pi|\alpha_S) = 0$ for every partition. $\pi \in PARTS(S)$ Also, we can write that which is seen in Box 5. In general, the conditional entropy can be written explicitly as seen in Box 6.

Theorem 2: Let π,σ be two partitions of a finite set *S*. We have H_β (p | s) = 0 if and only if $\sigma \le \pi$.

The next statement is a generalization of a well-known property of Shannon's entropy.

Theorem 3: Let π, σ be two partitions of a finite set *S*. We have:

$$H_\beta(\pi \wedge \sigma) = H_\beta(\pi \mid \sigma) + H_\beta(\sigma) = H_\beta(\sigma \mid \pi) + H_\beta(\pi).$$

The β –conditional entropy is dually monotonic with respect to its first argument and is monotonic with respect to its second argument, as we show in the following statement:

Theorem 4: Let π, σ, σ' be two partitions of a finite set *S*. If $\sigma \le \sigma'$, then $H_\beta(\sigma \mid \pi) \ge H_\beta(\sigma' \mid \pi)$ and $H_\beta(\pi \mid \sigma) \le H_\beta(\beta \mid \sigma')$.

The last statement implies immediately that $H_\beta(\pi) \ge H_\beta(\pi \mid \sigma)$ for every π,σ *PARTS(S)*

The behavior of β -conditional entropies with respect to the sum of partitions is discussed in the next statement.

Theorem 5: Let *S* be a finite set, and let π, $\theta \in PARTS(S)$ where $\theta = \{D_1,...,D_h\}$. If $\sigma_i \in PARTS(D)$ for $1 \le i \le h$, then:

$$H_\beta(\pi \mid \sigma_1 + ... + \sigma_h) = \sum_{i=1}^{h} \left(\frac{|D_i|}{|S|} \right)^\beta H_\beta(\pi_{D_i} \mid \sigma_i).$$

If $t = \{F_1,...,F_k\}$, $\sigma = \{C_1,...,C_n\}$ are two partitions of *S* and $\pi_i \in PARTS(F_i)$ for $1 \le i \le k$ then:

$$H_\beta(\pi_1 + ... + \pi_k \mid \sigma) = \sum_{i=1}^{k} \left(\frac{|F_i|}{|S|} \right)^\beta H_\beta(\pi_i \mid \sigma_{F_i}) + H_\beta(\tau \mid \sigma).$$

López de Màntaras, R. (1991) proved that Shannon's entropy generates a metric $d: S \times S \to R_{\ge 0}$ given by $d(\pi,\sigma) = H(\pi \mid \sigma) + H(\sigma \mid \pi)$, for $\pi,\sigma \in PARTS(S)$. We extended his result to a class of metrics $\{d_\beta \mid \beta \in R_{\ge 0}\}$ that can be defined by β-entropies, thereby improving our earlier results.

The next statement plays a technical role in the proof of the triangular inequality for d_β.

Theorem 6: Let π, σ, τ be three partitions of the finite set *S*. We have:

$$H_\beta(\pi \mid \sigma \wedge \tau) + H_\beta(\sigma \mid \tau) = H_\beta(\pi \wedge \sigma \mid \tau).$$

Corollary 1: Let π, σ, τ be three partitions of the finite set *S*. Then, we have:

$$H_\beta(\pi \mid \sigma) + H_\beta(\sigma \mid \tau) \ge H_\beta(\pi \mid \tau).$$

Proof: By theorem 6, the monotonicity of β-conditional entropy in its second argument and the dual monotonicity of the same in its first argument we can write that which is seen in Box 7, which is the desired inequality. QED.

Box 7.

$$H_\beta(\pi \mid \sigma) + H_\beta(\sigma \mid \tau) \geq H_\beta(\pi \mid \sigma \wedge \tau) + H_\beta(\sigma \mid \tau) = H_\beta(\pi \wedge \sigma \mid \tau) \geq H_\beta(\pi \mid \tau),$$

We can show now a central result:

Theorem 7: The mapping $d_\beta: S \times S \to R_{\geq 0}$ defined by $d_\beta(\pi,\sigma) = H_\beta(\pi \mid \sigma) + H_\beta(\sigma \mid \pi)$ for $\pi, \sigma \in$ *PARTS(S)* is a metric on *PARTS(S)*.

Proof: A double application of Corollary 1 yields $H_\beta(\pi \mid \sigma) + H_\beta(\sigma\ \tau) \geq H_\beta(\pi \mid \tau)$ and $H_\beta(\sigma \mid \pi) + H_\beta(\tau \mid \sigma) \geq H_\beta(\tau \mid \pi)$. Adding these inequality gives: $d_\beta(\pi, \sigma) + d_\beta(\sigma, \tau) \geq d_\beta(\pi, \tau)$, which is the triangular inequality for d_β.

The symmetry of d_β is obvious and it is clear that $d_\beta(\pi, \pi) = 0$ for every $\beta \in$ *PARTS(S)*.

Suppose now that $d_\beta(\pi, \sigma) = 0$. Since the values of β-conditional entropies are non-negative this implies $H_\beta(\pi \mid \sigma) = H_\beta(\sigma \mid \pi) = 0$. By theorem 2, we have both $\sigma \leq \pi$ and $\pi \leq \sigma$, respectively, so $\pi = \sigma$. Thus, d_β is a metric on *PARTS(S)*. QED.

Note that $d_\beta(\pi, \omega_S) = H_\beta(\pi)$ and $d_\beta(\pi, \alpha_S) = H_\beta(\alpha_S \mid \pi)$.

The behavior of the distance d_β with respect to partition sum is discussed in the next statement.

Theorem 8: Let S be a finite set, $\pi, \theta \in$ *PARTS(S)*, where $\theta = \{D_1,\ldots,D_h\}$. If $\sigma_i \in$ *PARTS(D$_i$)* for $1 \leq i \leq h$ then:

$$d_\beta(\pi, \sigma_1 + \ldots + \sigma_h) = \sum_{i=1}^{h} \left(\frac{|D_i|}{|S|}\right)^{\beta} d_\beta(\pi_{D_i}, \sigma_i) + H_\beta(\theta \mid \pi).$$

The distance between two partitions can be expressed using distances relative to the total partition or to the identity partition. Indeed, note that for $\pi, \sigma \in$ *PARTS(S)* where $\pi = \{B_1,\ldots,B_m\}$ and $\sigma = \{C_1,\ldots,C_n\}$ we have:

$$d_\beta(\pi, \sigma) = \frac{1}{(2^{1-\beta} - 1)|S|^\beta} \left(2\sum_{i=1}^{m}\sum_{j=1}^{n} |B_i \cap C_j|^\beta - \sum_{i=1}^{m} |B_i|^\beta - \sum_{j=1}^{n} |C_j|^\beta\right)$$

In the special case, when $\sigma = \omega_S$ we have:

$$d_\beta(\pi, \omega_S) = \frac{1}{(2^{1-\beta} - 1)|S|^\beta} \left(\sum_{i=1}^{m} |B_i|^\beta - |S|^\beta\right).$$

Similarly, we can write:

$$d_\beta(\alpha_S, \sigma) = \frac{1}{(2^{1-\beta} - 1)|S|^\beta} \left(|S| - \sum_{j=1}^{n} |C_j|^\beta\right).$$

These observations yield two metric equalities:

Theorem 9: Let $\pi, \sigma \in$ *PARTS(S)* be two partitions. We have:

$$\begin{aligned} d_\beta(\pi, \sigma) &= 2d_\beta(\pi \wedge \sigma, \omega_S) - d_\beta(\pi, \omega_S) - d_\beta(\sigma, \omega_S) \\ &= d_\beta(\alpha_S, \pi) + d_\beta(\alpha_S, \sigma) - 2d_\beta(\alpha_S, \pi \wedge \sigma). \end{aligned}$$

It follows that for $\theta, \tau \in$ *PARTS(S)*, if $q \leq t$ and we have either $d_\beta(\theta, \omega_S) = d_\beta(\tau, \omega_S)$ or $d_\beta(\theta, \alpha_S) = d_\beta(\tau, \alpha_S)$, then $\theta = \tau$. This allows us to formulate:

Theorem 10: Let $\pi, \sigma \in$ *PARTS(S)*. The following statements are equivalent:

1. $\sigma \leq \pi$
2. We have $[\sigma, \pi, \omega_S]$ in the metric space *(PARTS(S), d_β)*
3. We have $[\alpha_S, \sigma, \pi]$ in the metric space *(PARTS(S), d_β)*

Metrics generated by β-conditional entropies are closely related to lower valuations of the upper semimodular lattices of partitions of finite sets. This connection was established Birkhoff (1973) and studied by Barthèlemy (1978), Barthèlemy and Leclerc (1995) and Monjardet (1981).

A *lower valuation* on a lattice $(L, \wedge, \vee)$ is a mapping $v: L \rightarrow R$ such that:

$$v(\pi \wedge \sigma) + v(\pi \vee \sigma) \geq v(\pi) + v(\sigma)$$

for every $\pi, \sigma \in L$. If the reverse inequality is satisfied, that is, if:

$$v(\pi \wedge \sigma) + v(\pi \vee \sigma) \leq v(\pi) + v(\sigma)$$

for every $\pi, \sigma \in L$ then v is referred to as an *upper valuation*.

If v is both a lower and upper valuation, that is, if:

$$v(\pi \wedge \sigma) + v(\pi \vee \sigma) = v(\pi) + v(\sigma)$$

for every $\pi, \sigma \in L$ then v is a valuation on L. It is known (see Birkhoff (1973) that if there exists a positive valuation v on L, then L must be a modular lattice. Since the lattice of partitions of a set is an upper-semimodular lattice that is not modular it is clear that positive valuations do not exist on partition lattices. However, lower and upper valuations do exist, as shown next.

Theorem 11: Let S be a finite set. Define the mappings v_β:*PARTS(S)* $\rightarrow R$ and w_β:*PARTS(S)* $\rightarrow R$ by $v_\beta(\pi) = d_\beta(\alpha_S, \pi)$ and $w_\beta(\pi) = d_\beta(\pi, \omega_S)$, respectively, for $\sigma \in$ *PARTS(S)*. Then, v_β is a lower valuation and w_β is an upper valuation on the lattice (*PARTS(S)*, $\wedge, \vee$).

METRIC SPLITTING CRITERIA FOR DECISION TREES

The usefulness of studying the metric space of partitions of finite sets stems from the association between partitions defined on a collection of objects and sets of features of these objects. To formalize this idea, define an *object system* as a pair $\mathrm{T} = (T, H)$, where T is a sequence of objects and H is a finite set of functions, $H = \{A_1,...,A_n\}$, where $A_i: T \rightarrow D_i$ for $1 \leq i \leq n$. The functions A_i are referred to as *features* or *attributes* of the system. The set D_i is the domain of the attribute A_i; we assume that each set A_i contains at least to elements. The cardinality of the domain of attribute A will be denoted by m_A. If $X = (A_{i_1},...,A_{i_n})$ is a sequence of attributes and $t \in T$ the projection of t on is the sequence $t[X] = (A_{i_1}(t),...,A_{i_n}(t))$. The partition π^X defined by the sequence of attributes is obtained by grouping together in the same block all objects having the same projection on X. Observe that if X, Y are two sequences of attributes, then $\pi^{XY} = \pi^X \wedge \pi^Y$.

Thus, if U is a subsequence of V (denoted by $U \subseteq V$) we have $\pi^V \leq \pi^U$.

For example, if X is a set of attributes of a table T, any SQL phrase such as:

select count(*) **from** *T* **group by** *X*

computes the number of elements of each of the blocks of the partition π^X of the set of tuples of the table T.

To introduce formally the notion of decision tree we start from the notion of tree domain. A tree domain is a nonempty set of sequences D over the set of natural numbers N that satisfies the following conditions:

1. Every prefix of a sequence $\sigma \in D$ also belongs to D.
2. For every $m \geq 1$, if $(p_1,...,p_{m-1}, p_m) \in D$, then $(p_1,...,p_{m-1}, q) \in D$ for every $q \leq p_m$.

The elements of D are called the *vertices* of D. If u,v are vertices of D and u is a prefix of v, then we refer to v as a *descendant* of u and to u as an *ancestor* of v. If $v = ui$ for some $i \in N$, then we call v an *immediate descendant* of u and u an *immediate ancestor* of v. The *root* of every tree domain is the null sequence λ. A *leaf* of D is a vertex of D with no immediate descendants.

Let S be a finite set and let D be a tree domain. Denote by $\mathsf{P}(S)$ the set of subsets of S. An S-tree is a function $\mathsf{T}: D \rightarrow \mathsf{P}(S)$ such that $\mathsf{T}(\mathsf{l}) = S$, and if $u1,...,um$ are the immediate descendants of a vertex u, then the sets $\mathsf{T}(u1),...,\mathsf{T}(um)$ form a partition of the set $\mathsf{T}(u)$.

A *decision tree* for an object system $\mathsf{T} = (T,H)$ is an S-tree T, such that if the vertex v has the descendants $v1, \ldots, vm$, then there exists an attribute A in H (called the *splitting attribute* in v) such that $\{\mathsf{T}(vi) \mid 1 \le i \le m\}$ is the partition $\pi_A^{\mathsf{T}(v)}$.

Thus, each descendant vi of a vertex v corresponds to a value a of the attribute A that was used as a splitting attribute in v. If $\mathsf{l} = v_1,...,v_k = u$ is the path in T that was used to reach the vertex u, $A_{i},...,A_{k-1}$ are the splitting attributes in $v_1,...,v_{k-1}$ and $a_1,...,a_{k-1}$ are the values that correspond to $v_2,...,v_k$, respectively, then we say that u is reached by the selection $A_{i_1} = a_1 \wedge ... \wedge A_{i_{k-1}} = a_{k-1}$.

It is desirable that the leaves of a decision tree contain C-pure or almost C-pure sets of objects. In other words, the objects assigned to a leaf of the tree should, with few exceptions, have the same value for the class attribute C. This amounts to asking that for each leaf w of T we must have $H_\beta(\pi_{S_w}^C)$ as close to 0 as possible. To take into account the size of the leaves note that the collection of sets of objects assigned to the leafs is a partition κ of S and that we need to minimize:

$$\sum_w \left(\frac{|S_w|}{|S|}\right)^\beta H_\beta(\pi_{S_w}^C),$$

which is the conditional entropy $H_\beta(\pi^C \mid \kappa)$. By theorem 2 we have $H_\beta(\pi^C \mid \kappa) = 0$ if and only if $\kappa \le \pi^C$, which happens when the sets of objects assigned to the leafs are C-pure.

The construction of a decision tree $\mathsf{T}_\beta(\mathsf{T})$ for an object system $\mathsf{T} = (T,H)$ evolves in a top-down manner according to the following high-level description of a general algorithm (see Tan, 2005). The algorithm starts with an object system $\mathsf{T} = (T,H)$, a value of β and with an impurity threshold ε and it consists of the following steps:

1. If $H_\beta(\pi_S^C) \le \varepsilon$, then return T as a one-vertex tree; otherwise go to 2.
2. Assign the set S to a vertex v, choose an attribute A as a splitting attribute of S (using a splitting attribute criterion to be discussed in the sequel) and apply the algorithm to the object systems $\mathsf{T}_1 = (T_1, H_1),...,\mathsf{T}_p = (T_p, H_p)$, where for $1 \le i \le p$. Let $\mathsf{T}_1,...,\mathsf{T}_p$ the decision trees returned for the systems $\mathsf{T}_1,...,\mathsf{T}_p$, respectively. Connect the roots of these trees to v.

Note that if ε is sufficiently small and if $H_\beta(\pi_S^C) \le \varepsilon$, where $S = \mathsf{T}(u)$ is the set of objects at a node u, then there is a block Q_k of the partition π_S^C that is dominant in the set S. We refer to Q_k as the *dominant class* of u.

Once a decision tree T is built it can be used to determine the class of a new object $t \notin S$ such that the attributes of the set H are applicable. If $A_{i_1}(t) = a_1,...,A_{i_{k-1}}(t) = a_{k-1}$, a leaf u was reached through the path $\mathsf{l} = v_1,...,v_k = u$, and $a_1a_2,...,a_{k-1}$ are the values that correspond to $v_2,...,v_k$, respectively, then t is classified in the class Q_k, where Q_k is the dominant class at leaf u.

The description of the algorithm shows that the construction of a decision tree depends essentially on the method for choosing the splitting attribute. We focus next on this issue.

Classical decision tree algorithms make use of the information gain criterion or the gain ratio to choose splitting attribute. These criteria are formulated using Shannon's entropy, as their designations indicate.

Box 8.

$$d_\beta(\pi^C \wedge \kappa, \kappa) = d_\beta(\pi^C \wedge \kappa, \omega_S) - d_\beta(\kappa, \omega_S)$$
$$= H_\beta(\pi^C \wedge \kappa) - H_\beta(\kappa) \geq H_\beta(\pi^C \wedge \kappa') - H_\beta(\kappa'),$$

In our terms, the analogue of the information gain for a vertex w and an attribute A is: $H_\beta(\pi^C_{S_w}) - H_\beta(\pi^C_{S_w} | \pi^A_{S_w})$. The selected attribute is the one that realizes the highest value of this quantity. When $\beta \to 1$ we obtain the information gain linked to Shannon entropy. When $\beta = 2$ one obtains the selection criteria for the Gini index using the CART algorithm described in Breiman, Friedman, Olshen and Stone (1998).

The monotonicity property of conditional entropy shows that if A,B are two attributes such that $\pi^A \leq \pi^B$ (which indicates that the domain of A has more values than the domain of B), then $H_\beta(\pi^C_{S_w} | \pi^A_{S_w}) \leq H_\beta(\pi^C_{S_w} | \pi^B_{S_w})$, so the gain for A is larger than the gain for B. This highlights a well-known problem of choosing attributes based on information gain and related criteria: these criteria favor attributes with large domains, which in turn, generate bushy trees. To alleviate this problem information gain was replaced with the information gain ratio defined as $(H_\beta(\pi^C_{S_w}) - H_\beta(\pi^C_{S_w} | \pi^A_{S_w})) / H_\beta(\pi^A_{S_w})$, which introduces the compensating divisor $H_\beta(\pi^A_{S_w})$.

We propose replacing the information gain and the gain ratio criteria by choosing as splitting attribute for a node w an attribute that minimizes the distance $d_\beta(\pi^C_{S_w}, \pi^A_{S_w}) = H_\beta(\pi^C_{S_w} | \pi^A_{S_w}) + H_\beta(\pi^A_{S_w} | \pi^C_{S_w})$. This idea has been developed by López de Màntaras (1991) for the metric d_1 induced by Shannon's entropy. Since one could obtain better classifiers for various data sets and user needs by using values of β that are different from one, our approach is an improvement of previous results.

Besides being geometrically intuitive, the minimal distance criterion has the advantage of limiting both conditional entropies $H_\beta(\pi^C_{S_w} | \pi^A_{S_w})$ and $H_\beta(\pi^A_{S_w} | \pi^C_{S_w})$. The first limitation insures that the choice of the splitting attribute will provide a high information gain; the second limitation insures that attributes with large domains are not favored over attributes with smaller domains.

Suppose that in the process of building a decision tree for an object system $\mathsf{T} = (T, H)$ we constructed a stump of the tree T that has m leaves and that the sets of objects that correspond to these leaves are $S_1,\dots,S_n$. This means that we created the partition $\kappa = \{S_1,\dots,S_n\} \in PARTS(S)$, so $\kappa = \omega_{S_1} + \dots + \omega_{S_n}$. We choose to split the node v_i using as splitting attribute the attribute A that minimizes the distance $d_\beta(\pi^C_{S_i}, \pi^A_{S_i})$. The new partition κ' that replaces κ is:

$$\kappa' = \omega_{S_1} + \dots + \omega_{S_{i-1}} + \pi^A_{S_i} + \omega_{S_{i+1}} + \dots + \omega_{S_n}.$$

Note that $\kappa \leq \kappa'$. Therefore, we have that which is seen in Box 8 because $[\pi^C \wedge \kappa, \kappa, \omega_S]$. This shows that as the construction of the tree advances the current partition k gets closer to the partition $\pi^C \wedge \kappa$. More significantly, as the stump of the tree grows, κ gets closer to the class partition π^C. Indeed, by theorem 8 we can write:

$$d_\beta(\pi^C, \kappa) = d_\beta(\pi^C, \omega_{S_1} + \dots + \omega_{S_n})$$
$$= \sum_{j=1}^{n} \left(\frac{|S_j|}{|S|} \right)^\beta d_\beta(\pi^C_{S_j}, \omega_{S_j}) + H_\beta(\theta | \pi^C),$$

where $\theta = \{S_1,\dots,S_n\}$. Similarly, we can write that which is seen in Box 9.

These equalities imply that which is seen in Box 10.

We tested our approach on a number of data sets from the University of California Irvine (see Blake & Merz, 1978). The results shown in Table 1 are fairly typical. Decision trees were

Box 9.

$$\kappa') = d_\beta(\pi^C, \omega_{S_1} + \ldots + \omega_{S_{i-1}} + \pi^A_{S_i} + \omega_{S_{i+1}} + \ldots + \omega_{S_n})$$
$$= \sum_{j=1, j\neq i}^{n} \left(\frac{|S_j|}{|S|}\right)^\beta d_\beta(\pi^C_{S_j}, \omega_{S_j}) + \left(\frac{|S_i|}{|S|}\right)^\beta d_\beta(\pi^C_{S_i}, \pi^C_{S_i}) + H_\beta(\theta\,|\pi^C).$$

Box 10.

$$d_\beta(\pi^C, \kappa) - d_\beta(\pi^C, \kappa') = \left(\frac{|S_i|}{|S|}\right)^\beta \left(d_\beta(\pi^C_{S_i}, \omega_{S_i}) - d_\beta(\pi^C_{S_i}, \pi^A_{S_i})\right)$$
$$= \left(\frac{|S_i|}{|S|}\right)^\beta \left(H_\beta(\pi^C_{S_i}) - d_\beta(\pi^C_{S_i}, \pi^A_{S_i})\right)$$

constructed using metrics d_β, where β varied between 0.25 and 2.50. Note that for β =1 the metric algorithm coincides with the approach of Lopez de Màntaras (1991).

If the choices of the node and the splitting attribute are made such that $H_\beta(\pi^C_{S_i}) > d_\beta(\pi^C_{S_i}, \pi^A_{S_i})$, then the distance between π^C and the current partition k of the tree stump will decrease. Since the distance between $\pi^C \wedge$ k and k decreases in any case when the tree is expanded it follows that the "triangle" determined by π^C, $\pi^C \wedge$ k, and k will shrink during the construction of the decision tree.

In all cases, accurracy was assessed through 10-fold crossvalidation. We also built standard decision trees using the J48 technique of the well-known WEKA package (see Witten & Frank, 2005), which yielded the results shown in Table 2.

The experimental evidence shows that β can be adapted such that accuracy is comparable, or better than the standard algorithm. The size of the trees and the number of leaves show that the proposed approach to decision trees results consistently in smaller trees with fewer leaves.

INCREMENTAL CLUSTERING OF CATEGORICAL DATA

Clustering is an unsupervised learning process that partitions data such that similar data items are grouped together in sets referred to as clusters. This activity is important for condensing and identifying patterns in data. Despite the substantial effort invested in researching clustering algorithms by the data mining community, there are still many difficulties to overcome in building clustering algorithms. Indeed, as pointed in Jain (1999) "there is no clustering technique that is universally applicable in uncovering the variety of structures present in multidimensional data sets."

We focus on an incremental clustering algorithm that can be applied to nominal data, that is, to data whose attributes have no particular natural ordering. In general clustering, objects to be clustered are represented as points in an n-dimensional space R^n and standard distances, such as the Euclidean distance is used to evaluate similarity between objects. For objects whose attributes are nominal (e.g., color, shape, diagnostic,

Table 1. Decision trees constructed by using the metric splitting criterion

Audiology			
β	Accuracy	Size	Leaves
2.50	34.81	50	28
2.25	35.99	31	17
2.00	37.76	33	18
1.75	36.28	29	16
1.50	41.89	40	22
1.25	42.18	38	21
1.00	42.48	81	45
0.75	41.30	48	27
0.50	43.36	62	35
0.25	44.25	56	32

Hepatitis			
β	Accuracy	Size	Leaves
2.50	81.94	15	8
2.25	81.94	9	5
2.00	81.94	9	5
1.75	83.23	9	5
1.50	84.52	9	5
1.25	84.52	11	6
1.00	85.16	11	6
0.75	85.81	9	5
0.50	83.23	5	3
0.25	82.58	5	3

Primary Tumor			
β	Accuracy	Size	Leaves
2.50	34.81	50	28
2.25	35.99	31	17
2.00	37.76	33	18
1.75	36.28	29	16
1.50	41.89	40	22
1.25	42.18	38	21
1.00	42.48	81	45
0.75	41.30	48	27
0.50	43.36	62	35
0.25	44.25	56	32

Vote			
β	Accuracy	Size	Leaves
2.50	94.94	7	4
2.25	94.94	7	4
2.00	94.94	7	4
1.75	94.94	7	4
1.50	95.17	7	4
1.25	95.17	7	4
1.00	95.17	7	4
0.75	94.94	7	4
0.50	95.17	9	5
0.25	95.17	9	5

Table 2. Decision trees built by using J48

Data Set	**Accuracy**	**Size**	**Leaves**
Audiology	77.88	54	32
Hepatitis	83.87	21	11
Primary Tumor	39.82	88	47
Vote	94.94	7	4

etc.), no such natural representation of objects is possible, which leaves only the Hamming distance as a dissimilarity measure, a poor choice for discriminating among multivalued attributes of objects.

Incremental clustering has attracted a substantial amount of attention starting with Hartigan (1975). His algorithm was implemented in Carpenter and Grossberg (1990). A seminal paper, Fisher (1987), introduced COBWEB, an incremental clustering algorithm that involved restructurings of the clusters in addition to the incremental additions of objects. Incremental clustering related to dynamic aspects of databases were discussed in Can (1993) and Can et al. (1995). It is also notable that incremental clustering has been used in a variety of applications: Charikar et al. (1997), Ester et al. (1998), Langford et al. (2001) and Lin et al. (2004)). Incremental clustering is interesting because the main memory usage is minimal since there is no need to keep in memory the mutual distances between objects and the algorithms are scalable with respect to the size of the set of objects and the number of attributes.

A clustering of an object system (*T, H*) is defined as a partition k of the set of objects *T* such that similar objects belong to the same blocks of the partition, and objects that belong to distinct blocks are dissimilar. We seek to find clusterings starting from their relationships with partitions induced by attributes. As we shall see, this is a natural approach for nominal data.

Our clustering algorithm was introduced in Simovici, Singla and Kuperberg (2004); a semisupervised extension was discussed in Simovici and Singla (2005). We used the metric space (*PARTS(S), d*), where d is a multiple of the d_2 metric given by that which is seen in Box 11.

This metric has been studied in Barthélemy (1978) and Barthélemy and Leclerc (1978) and in Monjardet (1981), and we will refer to it as the *Barthélemy-Monjardet distance*. A special property of this metric allows the formulation of an incremental clustering algorithm.

The main idea of the algorithm is to seek a clustering $\kappa = \{C_1,...,C_n\} \in PARTS(T)$, where T is the set of objects such that the total distance from k to the partitions of the attributes:

$$D(\kappa) = \sum_{i=1}^{n} d(\kappa, \pi^{A_i})$$

is minimal. The definition of d allows us to write:

$$d(\kappa, \pi^A) = \sum_{i=1}^{n} |C_i|^2 + \sum_{j=1}^{m_A} |B^A_{a_j}|^2 - 2\sum_{i=1}^{n}\sum_{j=1}^{m_A} |C_i \cap B^A_{a_j}|^2 .$$

Suppose now that t is a new object, $t \notin T$, and let $Z = T \cup \{t\}$. The following cases may occur:

1. The object t is added to an existing cluster C_k.
2. A new cluster, C_{n+1} is created that consists only of t.

Also, the partition π^A is modified by adding t to the block $B^A_{t[A]}$, which corresponds to the value t[A] of the A-component of t. In the first case let:

$$\kappa_k = \{C_1,...,C_{k-1}, C_k \cup \{t\}, C_{k+1},...,C_n\},$$
$$\pi^{A\prime} = \{B^A_{a_1},..., B^A_{t[A]} \cup \{t\},..., B^A_{m_A}\}$$

be the partitions of Z. Now, we have what is shown in Box 12.

The minimal increase of $d(\kappa_k, \pi^{A\prime})$ is given by $\min_k \sum_A 2|C_k \oplus B^A_{t[A]}|$.

In the second case we deal with the partitions:

$$\kappa_k = \{C_1,...,C_n, \{t\}\},$$
$$\pi^{A\prime} = \{B^A_{a_1},..., B^A_{t[A]} \cup \{t\},..., B^A_{m_A}\}$$

and we have $d(\kappa', \pi^{A\prime}) - d(\kappa, \pi^A) = 2|B^A_{t[A]}|$. Consequently, we have:

$$D(\kappa') - D(\kappa) = 2\sum_A |C_k \oplus B^A_{t[A]}|$$

Box 11.

$$d(\pi,\sigma) = \sum_{i=1}^{m} |B_i|^2 + \sum_{j=1}^{n} |C_j|^2 - 2\sum_{i=1}^{m}\sum_{j=1}^{n} |B_i \cap C_j|^2 = \frac{|S|^2}{2} d_2(\pi,\sigma),$$

where $\pi = \{B_1,\ldots,B_m\}$ and $\sigma = \{C_1,\ldots,C_n\}$.

Box 12.

$$d(\kappa_k,\pi^{A}{}') - d(\kappa,\pi^A) = (|C_k|+1)^2 - |C_k|^2 + (|B^A_{t[A]}|+1)^2 - |B^A_{t[A]}|^2 - 2(|C_k \cap B^A_{t[A]}|+1)$$
$$= 2|C_k \oplus B^A_{t[A]}|.$$

in the first case and in the second case. Thus, if:

$$D(\kappa') - D(\kappa) = 2\sum_A |B^A_{t[A]}|$$

we add t to a cluster C_k for which:

$$\min_k \sum_A |C_k \oplus B^A_{t[A]}|$$

is minimal; otherwise, we create a new one-object cluster.

Incremental clustering algorithms are affected, in general, by the order in which objects are processed by the clustering algorithm. Moreover, as pointed out in Cornuéjols (1993), each such algorithm proceeds typically in a hill-climbing fashion that yields local minima rather than global ones. For some incremental clustering algorithms certain object orderings may result in rather poor clusterings. To diminish the ordering effect problem we expand the initial algorithm by adopting the "not-yet" technique introduced by Roure and Talavera (1998). The basic idea is that a new cluster is created only when the inequality:

$$r(t) = \frac{\sum_A |B^A_{t[A]}|}{\min_k \sum_A |C_k \oplus B^A_{t[A]}|} < \xi,$$

is satisfied, that is, only when the effect of adding the object t on the total distance is significant enough. Here ξ is a parameter provided by the user, such that $\xi \leq 1$.

Now we formulate a metric incremental clustering algorithm (referred to as AMICA, an acronym of the previous five words) that is using the properties of distance d. The variable nc denotes the current number of clusters.

If $\xi < r(t) \leq 1$ we place the object t in a buffer known as the NOT-YET buffer. If $r(t) \leq \xi$ a new cluster that consists of the object t is created. Otherwise, that is, if $r(t) > 1$, the object t is placed in an existing cluster C_k that minimizes:

$$\min_k \sum_A |C_k \oplus B^A_{t[A]}|;$$

this limits the number of new singleton clusters that would be otherwise created. After all objects of the set T have been examined, the objects contained by the NOT-YET buffer are processed with $\xi = 1$. This prevents new insertions in the buffer and results in either placing these objects in existing clusters or in creating new clusters. The pseudocode of the algorithm is given next:

Input: Data set T and threshold ξ
Output: Clustering $\{C_1,\ldots,C_{nc}\}$
Method: $nc = 0$; $l = 1$

while ($T \neq \emptyset$) do
select an object t;
$T = T - \{t\}$;
if $\sum_A |B^A_{t[A]}| \leq \xi \min_{1 \leq k \leq nc} \sum_A |C_k \oplus B^A_{t[A]}|$
then
nc + +;
create a new single-object cluster $C_{nc} = \{t\}$;
else

$$r(t) = \frac{\sum_A |B^A_{t[A]}|}{\min_k \sum_A |C_k \oplus B^A_{t[A]}|};$$

if ($r(t) \geq 1$)
then
$k = \arg\min_k \sum_A |C_k \oplus B^A_{t[A]}|$;
add t to cluster C_k;
else /* this means that $\xi < r(t) \leq 1$ */
place t in NOT-YET buffer;
end if;
endwhile;
process objects in the NOT-YET buffer as above with $\xi = 1$.

We applied AMICA to synthetic data sets produced by an algorithm that generates clusters of objects having real-numbered components grouped around a specified number of centroids. The resulting tuples were discretized using a specified number of discretization intervals, which allowed us to treat the data as nominal. The experiments were applied to several data sets with an increasing number of tuples and increased dimensionality and using several permutations of the set of objects. All experiments used $\xi = 0.95$.

The stability of the obtained clusterings is quite remarkable. For example, in an experiment applied to a set that consists of 10,000 objects (grouped by the synthetic data algorithm around 6 centroids) a first pass of the algorithm produced 11 clusters; however, most objects (9895) are concentrated in the top 6 clusters, which approximate very well the "natural" clusters produced by the synthetic algorithm.

Table 3 compares the clusters produced by the first run of the algorithm with the cluster produced from a data set obtained by applying a random permutation.

Note that the clusters are stable; they remain almost invariant with the exception of their numbering. Similar results were obtained for other random permutations and collections of objects.

Table 3. Comparison between clusters produced by successive runs

Initial Run		*Random Permutation*		
Cluster	*Size*	*Cluster*	*Size*	*Distribution (Original Cluster)*
1	1548	1	1692	1692 (2)
2	1693	2	1552	1548 (1),3 (3), 1 (2)
3	1655	3	1672	1672 (5)
4	1711	4	1711	1711 (4)
5	1672	5	1652	1652 (3)
6	1616	6	1616	1616(6)
7	1	7	85	85 (8)
8	85	8	10	10 (9)
9	10	9	8	8 (10)
10	8	10	1	1 (11)
11	1	11	1	1 (7)

Table 4. Time for three random permutations

Number of Objects	Time for 3 permutations (ms)	Average time (ms)
2000	131, 140, 154	141.7
5000	410, 381, 432	407.7
10000	782,761, 831	794.7
20000	1103, 1148, 1061	1104

Table 5. Purity of clusters for the mushrooms data set

Class number	Poisonous/Edible	Total	Dominant Group (%)
1	825/2752	3557	76.9%
2	8/1050	1058	99.2%
3	1304/0	1304	100%
4	0/163	163	100%
5	1735/28	1763	98.4%
6	0/7	7	100%
7	1/192	192	100%
8	36/16	52	69%
9	8/0	8	100%

As expected with incremental clustering algorithms, the time requirements scale up very well with the number of tuples. On an IBM T20 system equipped with a 700 MHz Pentium III and with a 256 MB RAM, we obtained the results shown in Table 4 for three randomly chosen permutations of each set of objects.

Another series of experiments involved the application of the algorithm to databases that contain nominal data. We applied AMICA to the mushroom data set from the standard UCI data mining collection (see Blake & Merz, 1998). The data set contains 8124 mushroom records and is typically used as test set for classification algorithms. In classification experiments the task is to construct a classifier that is able to predict the poisonous/edible character of the mushrooms based on the values of the attributes of the mushrooms. We discarded the class attribute (poisonous/edible) and applied AMICA to the remaining data set. Then, we identified the edible/poisonous character of mushrooms that are grouped together in the same cluster. This yields the clusters $C_1,\ldots,C_9$.

Note that in almost all resulting clusters there is a dominant character, and for five out of the total of nine clusters there is complete homogeneity.

A study of the stability of the clusters similar to the one performed for synthetic data shows the same stability relative to input orderings. The clusters remain essentially stable under input data permutations (with the exception of the order in which they are created).

Thus, AMICA provides good quality, stable clusterings for nominal data, an area of clustering that is less explored than the standard clustering algorithms that act on ordinal data. Clusterings produced by the algorithm show a rather low sensitivity to input orderings.

CLUSTERING FEATURES AND FEATURE SELECTION

The performance, robustness and usefulness of classification algorithms are improved when relatively few features are involved in the classification. The main idea of this section, which was developed in Butterworth et al. (2005), is to introduce an algorithm for feature selection that clusters attributes using a special metric, and then use a hierarchical clustering for feature selection.

Hierarchical algorithms generate clusters that are placed in a cluster tree, which is commonly known as a *dendrogram.* Clusterings are obtained by extracting those clusters that are situated at a given height in this tree.

We show that good classifiers can be built by using a small number of attributes located at the centers of the clusters identified in the dendrogram. This type of data compression can be achieved with little or no penalty in terms of the accuracy of the classifier produced. The clustering of attributes helps the user to understand the structure of data, the relative importance of attributes. Alternative feature selection methods, mentioned earlier, are excellent in reducing the data without having a severe impact on the accuracy of classifiers; however, such methods cannot identify how attributes are related to each other.

Let m, $M \in N$ be two natural numbers such that $m \leq M$. Denote by $PARTS(S)_{m,M}$ the set of partitions of S such that for every block $B \in \pi$ we have $m \leq |B| \leq M$. The lower valuation v defined on $PARTS(S)$ is given by:

$$v(\theta) = \sum_{i=1}^{p} |D_i|^2,$$

where $\theta = \{D_1,\ldots,D_p\}$.

Let $\pi = \{B_1,\ldots,B_n\}$, $\sigma = \{C_1,\ldots,C_p\}$ be two partitions of a set S. The *contingency matrix* of π, σ is the matrix $P_{\pi,\sigma}$ whose entries are given by $p_{ij} = |B_i \cap C_j|$ for $1 \leq i \leq n$ and $1 \leq j \leq p$. The Pearson $\chi^2_{\pi,\sigma}$ association index of this contingency matrix can be written in our framework as:

$$\chi^2_{\pi,\sigma} = \sum_i \sum_j \frac{(p_{ij} - |B_i||C_j|)^2}{|B_i||C_j|}.$$

It is well known that the asymptotic distribution of this index is a χ^2 -distribution with $(n-1)(p-1)$ degrees of freedom. The next statement suggests that partitions that are correlated are close in the sense of the *Barthélemy-Monjardet distance*; therefore, if attributes are clustered using the corresponding distance between partitions we could replace clusters with their centroids and, thereby, drastically reduce the number of attributes involved in a classification without significant decreases in accuracy of the resulting classifiers.

Theorem 12: Let S be a finite set and let p,s $\in PARTS(S)_{m,M}$, where $\pi = \{B_1,\ldots,B_n\}$ and $\sigma = \{C_1,\ldots,C_p\}$. We have that which is seen in Box 13.

Thus, the Pearson coefficient decreases with the distance and, thus, the probability that the partitions π and σ and are independent increases with the distance.

We experimented with several data sets from the UCI dataset repository (Blake & Merz, 1998); here we discuss only the results obtained with the *votes* and *zoo* datasets, which have a relative small number of categorical features. In

Box 13.

$$\frac{v(\pi) + v(\sigma) - d(\pi,\sigma)}{2M^2} - 2np + |S|^2 \leq \chi^2_{\pi,\sigma} \leq \frac{v(\pi) + v(\sigma) - d(\pi,\sigma)}{2m^2} - 2np + |S|^2.$$

Table 6.

1	Handicapped infants	9	Mx missile
2	Water project cost sharing	10	Immigration
3	Budget resolution	11	Syn fuels corporation cutback
4	Physician fee freeze	12	Education spending
5	El Salvador aid	13	Superfund right to sue
6	Religious groups in schools	14	Crime
7	Antisatellite test ban	15	Duty-free exports
8	Aid to Nicaraguan contras		

each case, starting from the matrix $(d(\pi^{A_i}, \pi^{A_j}))$ of *Barthélemy-Monjardet distances* between the partitions of the attributes $A_1,\ldots,A_n$, we clustered the attributes using *AGNES*, an agglomerative hierarchical algorithm described in

Kaufman and Rousseeuw (1990) that is implemented as a component of the *cluster* package of system R (see Maindonald & Brown, 2003).

Clusterings were extracted from the tree produced by the algorithm by cutting the tree at various heights starting with the maximum height of the tree created above (corresponding to a single cluster) and working down to a height of 0 (which consists of single-attribute clusters). A "representative" attribute was created for each cluster as the attribute that has the minimum total distance to the other members of the cluster, again using the Barthélemy-Monjardet distance. The J48 and the Naïve Bayes algorithms of the WEKA package from Witten and Frank (2005) were used for constructing classifiers on data sets obtained by projecting the initial data sets on the sets of representative attributes.

The dataset *votes* records the votes of 435 U.S. Congressmen on 15 key questions, where each attribute can have the value "y", "n", or "?" (for abstention), and each Congressman is classified as a Democrat or Republican.

The attributes of this data set are listed in Table 6.

It is interesting to note that by applying the *AGNES* clustering algorithm with the Ward method of computing the intercluster distance the voting issues group naturally into clusters that involve larger issues, as shown in Figure 1.

For example, "El Salvador aid," "Aid to Nicaraguan contras," "Mx missile" and "Antisatellite test ban" are grouped quite early into a cluster that can be described as dealing with defense policies. Similarly, social budgetary legislation issues such as "Budget resolution," "Physician fee freeze" and "Education spending," are grouped together.

Two types of classifiers (J48 and Naïve Bayes) were generated using ten-fold cross validation by extracting centrally located attributes from cluster obtained by cutting the dendrogram at successive levels. The accuracy of these classifiers is shown in Figure 2.

This experiment shows that our method identifies the most influential attribute 5 (in this case "*El_Salvador_aid*"). So, in addition to reducing number of attributes, the proposed methodology allows us to assess the relative importance of attributes.

Figure 2. Dendrogram of votes data set using AGNES and the Ward method

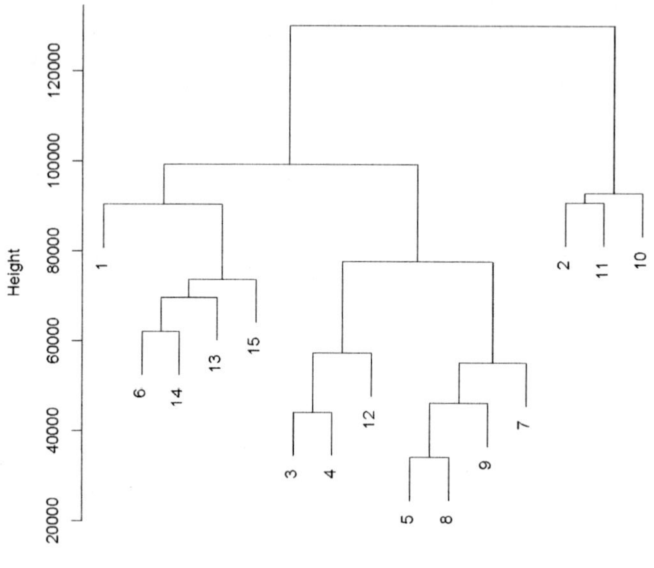

Table 7. Accuracy of classifiers for the votes data set

Attribute Set (class attribute not listed)	J48%	ND%
1,2,3,4,5,6,7,8,9,10,11,12,13,14,15	96.78	90.34
1,2,3,4,5,6,7,9,10,11,12,13,14,15	96.78	91.03
1,2,3,4,5,6,7,10,11,12,13,14,15	96.55	91.26
1,2,4,5,6,7,10,11,12,13,14,15	95.17	92.18
1,2,4,5,6,10,11,12,13,14,15	95.17	92.64
1,2,4,5,6,10,11,13,14,15	95.40	92.18
1,2,6,8,10,11,13,14,15	86.20	85.28
1,2,8,10,11,13,14,15	86.20	85.74
1,2,8,10,11,14,15	84.13	85.74
1,2,8,10,11,14	83.69	85.74
2,8,10,11,14	83.67	84.36
2,5,10,11	88.73	88.50
2,5,10	84.82	84.82
2,5	84.82	84.82
5	84.82	84.82

Table 8.

1	hair	10	breathes
2	feathers	11	venomous
3	eggs	12	fins
4	milk	13	legs
5	airborne	14	tail
6	aquatic	15	domestic
7	predator	16	cat size
8	toothed	17	type
9	backbone		

Figure 3. Dendrogram of zoo dataset using AGNES and the Ward method

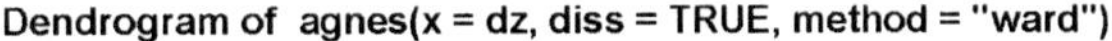

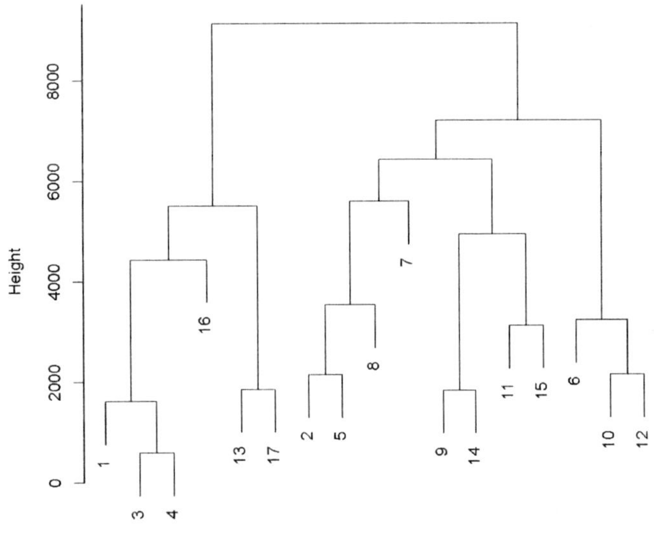

A similar study was undertaken for the *zoo* database, after eliminating the attribute *animal* which determines uniquely the type of the animal. Starting from a dendrogram build by using the Ward method shown in Figure 3 we constructed J48 and Naïve Bayes classifiers for several sets of attributes obtained as successive sections of the cluster tree.

The attributes of this data set are listed in Table 8.

The results are shown in Figure 3. Note that attributes that are biologically correlated (e.g., hair, milk and eggs, or aquatic, breathes and fins) belong to relatively early clusters.

The main interest of the proposed approach to attribute selection is the possibility of the supervi-

Table 9. Accuracy of classifiers for the zoo data set

Attribute Set (class attribute not listed)	J48%	NB%
1,2,3,4,5,6,7,8,9,10,11,12,13,14,15,16	92.07	93.06
1,2,4,5,6,7,8,9,10,11,12,13,14,15,16	92.07	92.07
2,4,5,6,7,8,9,10,11,12,13,14,15,16	87.12	88.11
2,4,5,6,7,8,9,10,11,12,13,15,16	87.12	88.11
2,4,6,7,8,9,10,11,12,13,15,16	88.11	87.12
2,4,6,7,8,9,10,11,13,15,16	91.08	91.08
2,4,6,7,8,9,10,11,13,16	89.10	90.09
2,4,7,8,9,10,11,13,16	86.13	90.09
2,4,7,9,10,11,13,16	84.15	90.09
2,4,7,9,10,11,13	87.12	89.10
4,5,7,9,10,11	88.11	88.11
4,5,7,9,10	88.11	90.09
4,5,9,10	89.10	91.09
4,5,10	73.26	73.26
4,10	73.26	73.26
4	60.39	60.39

sion of the process allowing the user to opt between quasi-equivalent attributes (that is, attributes that are close relatively to the Barthélemy-Monjardet distance) in order to produce more meaningful classifiers.

We compared our approach with two existing attribute set selection techniques: the correlation-based feature (CSF) selection (developed in Hall (1999) and incorporated in the WEKA package) and the wrapper technique, using the "best first" and the greedy method as search methods, and the J48 classifier for the classifier incorporated by the wrapper. The comparative results for the zoo database show that using either the "best first" or the "greedy stepwise" search methods in the case of CSF the accuracy for the J48 classifier is 91.08%, and for the naïve Bayes classifier is 85.04%; the corresponding numbers for the wrapper method with J48 are 96.03% and 92.07%, respectively. These results suggest that this method is not as good for accuracy as the wrapper method or CSF. However, the tree of attributes helps to understand the relationships between attributes and their relative importance.

Attribute clustering helps to build classifiers in a semisupervised manner allowing analysts a certain degree of choice in the selection of the features that may be considered by classifiers, and illuminating relationships between attributes and their relative importance for classification.

A METRIC APPROACH TO DISCRETIZATION

Frequently, data sets have attributes with numerical domains which makes them unsuitable for certain data mining algorithms that deal mainly with nominal attributes, such as decision trees and naïve Bayes classifiers. To use such algorithms we need to replace numerical attributes with nominal attributes that represent intervals of numerical domains with discrete values. This process, known to as *discretization*, has received a great deal of attention in the data mining literature and includes a variety of ideas ranging from fixed k-interval discretization (Dougherty, Kohavi, & Sahami, 1995), fuzzy discretization (see Kononenko, 1992; 1993), Shannon-entropy discretization due to Fayyad and Irani presented in Fayyad (1991) and Fayyad and Irani (1993), proportional κ-interval discretization (see Yang & Webb, 2001; 2003), or techniques that are capable of dealing with highly dependent attributes (cf. Robnik & Kononenko, 1995).

The discretization process can be described generically as follows. Let B be a numerical attribute of a set of objects. The set of values of the components of these objects that correspond to the attribute B is the *active domain* of B and is denoted by *adom(B)*. To discretize B we select a sequence of numbers $t_1 < t_2 <...< t_l$ in *adom(B)*. Next, the attribute B is replaced by the nominal attribute B' that has l+1 distinct values in its active domain $\{k_0, k_1,...k_l\}$. Each B -component b of an object t is replaced by the discretized B'-component k defined by $k=k_0$ if $b \le t_1$, $t = k$ if $t_i < b \le t_{i+1}$ for $1 \le i \le l-1$, and $k=k_l$ if $t_l < b$. The numbers t_1, t_2,...t_l define the discretization process and they will be referred to as *class separators*.

There are two types of discretization (see Witten & Frank, 2005): *unsupervised discretization*, where the discretization takes place without any knowledge of the classes to which objects belong, and *supervised discretization* which takes into account the classes of the objects. Our approach involves supervised discretization. Within our framework, to discretize an attribute B amounts to constructing a partition of the active domain *adom(B)* taking into account the partition π^A determined by the nominal class attribute A.

Partitions of active attribute domains induce partitions on the set of objects. Namely, the partition of the set of objects T that corresponds to a partition θ of *adom(B)*, where B is a numerical attribute, is denoted by θ_* . A block of θ_* consists of all objects whose B -components belong to the same block of θ. For the special case when $\theta=\alpha_{adom(B)}$ observe that $\theta_* = \pi^B$.

Let $P = (t_1,...,t_l)$ a sequence of class separators of the active domain of an attribute B, where $t_1 < t_2 <...< t_l$. This set of cut points creates a partition $\theta_B^P = \{Q_0,...,Q_l\}$ of *adom(B)* where Q $Q_i = \{b \in adom(B) \mid t_i \le b < t_{i+1}\}$ for $0 \le i \le l$, where $t_0 = -\infty$ and $t_{l+1} = \infty$.

It is immediate that for two sets of cut points P, P' we have $\theta_B^{P \cup P'} = \theta_B^P \wedge \theta_B^{P'}$. If the sequence P consists of a single cut point t we shall denote θ_B^P simply by θ_B^t. The discretization process consists of replacing each value that falls in the block Q_i of θ_B^P by i for $0 \le i \le l$.

Suppose that the list of objects *sorted on the values of a numerical attribute* B is $t_1,...,t_n$ and let $t_1[B],...,t_n[B]$ be the sequence of B -components of those objects, where $t_1[B] <...< t_n[B]$. For a nominal attribute A define the partition $\pi_{B,A}$ of *adom(B)* as follows. A block of $\pi_{B,A}$ consists of a maximal subsequence $t_i[B],...,t_l[B]$ of the previous sequence such that every object $t_i,...,t_l$ of this subsequence belongs to the same block K of the partition π^A. If $x \in adom(B)$, we shall denote the block of $\pi_{B,A}$ that contains x by $\langle x \rangle$. The *boundary points* of the partition $\pi_{B,A}$ are the least and the largest elements of each of the blocks of the partition $\pi_{B,A}$. The least and the largest elements of $\langle x \rangle$ are denoted by $x^{\downarrow}$ and $x^{\uparrow}$, respectively. It is clear that $\pi_{B,A} \le \pi^A$ for any attribute B.

Example 4

Let $t_1,...,t_9$ be a collection of nine objects such that the sequence $t_1[B],...,t_9[B]$ is sorted in increasing order of the value of the B -components as seen in Table 10.

The partition π^A has two blocks corresponding to the values "Y" and "N" and is given by

$$\pi^A = \{\{t_1, t_3, t_4, t_7, t_8, t_9\}, \{t_2, t_5, t_6\}\}.$$

Table 10.

	…	B	…	A
t_1	…	95.2	…	Y
t_2	…	110.1	…	N
t_3	…	120.0	…	Y
t_4	…	125.5	…	Y
t_5	…	130.1	…	N
t_6	…	140.0	…	N
t_7	…	140.5	…	Y
t_8	…	168.2	…	Y
t_9	…	190.5	…	Y

The partition $\pi_{B,A}$ is:

$$\pi_{B,A} = \{\{t_1\}, \{t_2\}, \{t_3, t_4\}, \{t_5, t_6\}, \{t_7, t_8, t_9\}\}.$$

The blocks of this partition correspond to the longest subsequences of the sequence $t_1,\ldots,t_9$ that consists of objects that belong to the same π^A-class.

Fayyad (1991) showed that to obtain the least value of the Shannon's conditional entropy $H(\pi^A | \theta^P_{B*})$ the cut points of P must be chosen among the boundary points of the partition $\pi_{B,A}$. This is a powerful result that drastically limits the number of possible cut points and improves the tractability of the discretization.

We present two new basic ideas: a generalization of Fayyad-Irani discretization techniques that relies on a metric on partitions defined by generalized entropy, and a new geometric criterion for halting the discretization process. With an appropriate choice of the parameters of the discretization process the resulting decision trees are smaller, have fewer leaves, and display higher levels of accuracy as verified by stratified cross-validation; similarly, naïve Bayes classifiers applied to data discretized by our algorithm yield smaller error rates.

Our main results show that the same choice of cut points must be made for a broader class of impurity measures, namely the impurity measures related to generalized conditional entropy. Moreover, when the purity of the partition π^A is replaced as a discretization criterion by the minimality of the entropic distance between the partitions π^A and θ^P_{B*} the same method for selecting cut points can be applied. This is a generalization of the approach proposed in Cerquides and López de Màntaras (1997).

We are concerned with supervised discretization, that is, with discretization of attributes that takes into account the classes where the objects belong. Suppose that the class of objects is determined by the nominal attribute A and we need to discretize a numerical attribute B. The discretization of B aims to construct a set P of cut points of *adom(B)* such that the blocks of π^A are as pure as possible relative to the partition θ^P_{B*}, that is, the conditional entropy $H(\pi^A | \theta^P_{B*})$ is minimal.

The following theorem extends a result of Fayyad (1991):

Theorem 13: Let T be a collection of objects, where the class of an object is determined by the attribute A and let $\beta \in (1,2]$. If P is a set of cut points such that the conditional entropy $H(\pi^A | \theta^P_{B*})$ is minimal among the set of cut points with the same number of elements, then P consists of boundary points of the partition $\pi_{B,A}$ of *adom(B)*.

The next theorem is a companion to Fayyad's (1991) result and makes use of the same hypothesis as theorem 13.

Theorem 14: Let β be a number, $\beta \in (1,2]$. If P is a set of cut points of *adom(B)* such that the distance $d_\beta(\pi^A | \theta^P_{B*})$ is minimal among the set of cut points with the same number of elements, then P consists of boundary points of the partition $\pi_{B,A}$ of *adom(B)*.

This result plays a key role. To discretize *adom(B)* we seek a set of cut points P such that:

$$d_\beta(\pi^A, \theta^P_{B*}) = H_\beta(\pi^A | \theta^P_{B*}) + H_\beta(\theta^P_{B*} | \pi^A)$$

is minimal. In other words, we shall seek a set of cut points such that the partition θ^B_{B*} induced on the set of objects T is as close as possible to the target partition π^A.

Initially, before adding cut points, we have $P = \emptyset$, $\theta^B_{B*} = \omega_S = \{S\}$, and therefore $H_\beta(\pi^A | \omega_S) = H_\beta(\pi^A)$. Observe that when the set P grows the entropy $H_\beta(\pi^A | \theta^P_{B*})$ decreases. Note that the use of conditional entropy $H_\beta(\pi^A | \theta^P_{B*})$ tends to favor large cut point sets for which the partition θ^B_{B*} is small in the partial ordered set $(PARTS(P), \leq)$. In the extreme case, every point would be a cut point, a situation that is clearly unacceptable. Fayyad-Irani (Fayyad & Irani, 1993) technique halts the discretization process using the principle of minimum description. We adopt another technique that has the advantage of being geometrically intuitive and produces very good experimental results.

Using the distance $d_\beta(\pi^A | \theta^P_{B*})$ the decrease of $H_\beta(\pi^A | \theta^P_{B*})$ when the set of cut points grows is balanced by the increase in $H_\beta(\theta^P_{B*} | \pi^A)$. Note that initially we have $H_\beta(\omega_T \mid \pi^A) = 0$. The discretization process can thus be halted when the distance $d_\beta(\pi^A | \theta^P_{B*})$ stops decreasing. Thus, we retain as a set of cut points for discretization the set P that determines the closest partition to the class partition π^A. As a result, we obtain good discretizations (as evaluated through the results of various classifiers that use the discretize data) with relatively small cut point sets.

The greedy algorithm shown below is used for discretizing an attribute B. It makes successive passes over the table and, at each pass it adds a new cut point chosen among the boundary points of $\pi_{B,A}$.

Input: An object system (T, H), a class attribute A and a real-valued attribute B
Output: A discretized attribute B
Method: Sort (T, H) on attribute B;
compute the set BP of boundary points of partition $\pi_{B,A}$;
$P = \emptyset$, $d = \infty$;

while BP $\neq \emptyset$ **do**
let $p = \arg\min_{p \in BP} d_\beta(\pi^A, \theta^{P\cup\{p\}}_{B*})$;
if $d \geq d_\beta(\pi^A, \theta^{P\cup\{p\}}_{B*})$ **then**
begin
$P = P \cup \{t\}$;
$BP = BP - \{t\}$;
$d = d_\beta(\pi^A, \theta^P_{B*})$;
end
else
exit while loop;
end while;
for $\theta^P_{B*} = \{Q_0, \ldots, Q_l\}$ replace every value in Q_i by i for $0 \leq i \leq l$.

The while loop runs for as long as candidate boundary points exist, and it is possible to find a new cut point p such that the distance $d_\beta(\pi^A | \theta^P_{B*})$ is less than the previous distance $d_\beta(\pi^A | \theta^P_{B*})$. An experiment performed on a synthetic database shows that a substantial amount of time (about 78% of the total time) is spent on decreasing the distance by the last 1%. Therefore, in practice we run a search for a new cut point only if:

$$| d - d_\beta(\pi^A, \theta^P_{B*}) | > 0.01d.$$

To form an idea on the evolution of the distance between $\pi^A = \{P_1, ..., P_n\}$ and the partition of objects determined by the cut points θ^P_{B*} let $p \in BP$ be a new cut point added to the set P. It is clear that the partition θ^P_{B*} covers the partition $\theta^{P\cup\{p\}}_{B*}$ because $\theta^{P\cup\{p\}}_{B*}$ is obtained by splitting a block of θ^P_{B*}. Without loss of generality we assume that the blocks Q_{m-1} and Q_m of $\theta^{P\cup\{p\}}_{B*}$ result from the split of the block $Q_{m-1} \cup Q_m$ of θ^P_{B*}. Since $\beta > 1$ we have $d_\beta(\pi^A, \theta^{P\cup\{p\}}_{B*}) < d_\beta(\pi^A, \theta^P_{B*})$ if and only if:

$$|\sum_{i=1}^{n}|P_i| + \sum_{j=1}^{m}|Q_j|^\beta - 2\sum_{i=1}^{n}\sum_{j=1}^{m}|P_i \cap Q_j|^\beta < \sum_{i=1}^{n}|P_i|^\beta + \sum_{j=1}^{m-2}|Q_j|^\beta + |Q_{m-1} \cup Q_m|^\beta - 2\sum_{i=1}^{n}\sum_{j=1}^{m-2}|P_i \cap Q_j|^\beta \, \beta$$

which is equivalent to:

$$| Q_{m-1} |^\beta + | Q_m |^\beta - 2\sum_{i=1}^{n} | P_i \cap Q_{m-1} |^\beta - 2\sum_{i=1}^{n} | P_i \cap Q_m |^\beta < | Q_{m-1} \cup Q_m |^\beta - 2\sum_{i=1}^{n} (| P_i \cap Q_{m-1} | + | P_i \cap Q_m |)^\beta .$$

Suppose that $Q_{m-1} \cup Q_m$ is intersected by only by P_1 and P_2 and that $\beta = 2$Then, the previous inequality that describes the condition under which a decrease of $d_\beta(\pi^A | \theta^P_{B*})$ can be obtained becomes:

$$(| P_1 \cap Q_{m-1} | - | P_2 \cap Q_{m-1} |)(| P_1 \cap Q_m | - | P_2 \cap Q_m |) < 0$$

and so, the distance may be decreased by splitting a block $Q_{m-1} \cup Q_m$ into Q_{m-1} and Q_m only when the distribution of the fragments of the blocks P_1 and P_2 in the prospective blocks Q_{m-1} and Q_m satisfies the previous condition. If the block $Q_{m-1} \cup Q_m$ of the partition θ^P_{B*} contains a unique boundary point, then choosing that boundary point as a cut point will decrease the distance.

We tested our discretization algorithm on several machine learning data sets from UCI data sets that have numerical attributes. After discretizations performed with several values of β (typically with $\beta \in \{1.5, 1.8, 1.9, 2\}$) we built the decision trees on the discretized data sets using the WEKA J48 variant of C4.5. The size, number of leaves and accuracy of the trees are described in Table 11, where trees built using the Fayyad-Irani discretization method of J48 are designated as "standard."

It is clear that the discretization technique has a significant impact of the size and accuracy of the decision trees. The experimental results shown in Table 8 suggest that an appropriate choice of β can reduce significantly the size and number of leaves of the decision trees, roughly maintaining the accuracy (measured by stratified five-fold cross validation) or even increasing the accuracy.

Our supervised discretization algorithm that discretizes each attribute B based on the relationship between the partition π^B and π^A (where A is the attribute that specifies the class of the objects). Thus, the discretization process of an attribute is carried out independently of similar processes performed on other attributes. As a result, our algorithm is particularly efficient for naïve Bayes classifiers, which rely on the essential assumption of attribute independence. The error rates of naïve Bayes Classifiers obtained for different discretization methods are shown in Table 12.

The use of the metric space of partitions of the data set in discretization is helpful in preparing the data for classifiers. With an appropriate choice of the parameter β that defines the metric used in discretization, standard classifiers such as C4.5

Table 11. Comparative experimental results for decision trees

Database	Experimental Results			
	Discretization method	Size	Number of Leaves	Accuracy (strat. cross-validation)
Heart-c	*Standard*	*51*	*30*	*79.20*
	$\beta = 1.5$	20	14	77.36
	$\beta = 1.8$	28	18	77.36
	$\beta = 1.9$	35	22	76.01
	$\beta = 2$	54	32	76.01
Glass	*Standard*	*57*	*30*	*57.28*
	$\beta = 1.5$	32	24	71.02
	$\beta = 1.8$	56	50	77.10
	$\beta = 1.9$	64	58	67.57
	$\beta = 2$	92	82	66.35
Ionosphere	*Standard*	*35*	*18*	*90.88*
	$\beta = 1.5$	15	8	95.44
	$\beta = 1.8$	19	12	88.31
	$\beta = 1.9$	15	10	90.02
	$\beta = 2$	15	10	90.02
Iris	*Standard*	*9*	*5*	*95.33*
	$\beta = 1.5$	7	5	96
	$\beta = 1.8$	7	5	96
	$\beta = 1.9$	7	5	96
	$\beta = 2$	7	5	96
Diabetes	*Standard*	*43*	*22*	*74.08*
	$\beta = 1.8$	5	3	75.78
	$\beta = 1.9$	7	4	75.39
	$\beta = 2$	14	10	76.30

Table 12. Error rate for naive Bayes classifiers

Discretization method	Diabetes	Glass	Ionosphere	Iris
$\beta = 1.5$	34.9	25.2	4.8	2.7
$\beta = 1.8$	24.2	22.4	8.3	4
$\beta = 1.9$	24.9	23.4	8.5	4
$\beta = 2$	25.4	24.3	9.1	4.7
Weighted proportional	25.5	38.4	10.3	6.9
Proportional	26.3	33.6	10.4	7.5

or J48 generate smaller decision trees with comparable or better levels of accuracy when applied to data discretized with our technique.

CONCLUSION AND FUTURE RESEARCH

The goal of this chapter is to stress the significance of using metric methods in typical data mining tasks. We introduced a family of metrics on the set of partitions of finite sets that is linked to the notion of generalized entropy and we demonstrated its use in classification, clustering, feature extraction and discretization.

In the realm of classification these metrics are used for a new splitting criterion for building decision trees. In addition to being more intuitive than the classic approach, this criterion results in decision trees that have smaller sizes and fewer leaves than the trees built with standard methods, and have comparable or better accuracy. The value of β that results in the smallest trees seems to depend on the relative distribution of the class attribute and the values of the feature attributes of the objects.

Since clusterings of objects can be regarded as partitions, metrics developed for partitions present an interest for the study of the dynamics of clusters, as clusters are formed during incremental algorithms, or as data sets evolve.

As stated in Guyon and Elisseeff (2003), in early studies of relevance published in the late 1990's (Blum & Langley, 1997; Kohavi & John, 1997), few applications explored data with more than 40 attributes. With the increased interest of data miners in bio-computing in general, and in microarray data in particular, classification problems that involve thousands of features and relatively few examples came to the fore. Applications of metric feature selection techniques should be useful to the analysis of this type of data.

An important open issue is determining characteristics of data sets that will inform the choice of an optimal value for the β parameter. Also, investigating metric discretization for data with missing values seems to present particular challenges that we intend to consider in our future work.

REFERENCES

Barthélemy, J. P. (1978). Remarques sur les propriétés metriques des ensembles ordonnés, *Math. sci. hum.*, 61, 39-60.

Barthélemy, J. P., & Leclerc, B. (1995). The median procedure for partitions. In *Partitioning data sets*, (pp. 3-34), Providence, RI: American Mathematical Society.

Blake, C. L., & Merz, C. J. (1998). *UCI Repository of machine learning databases*. Retrieved February 27, 2007, from University of California Irvine Department of Information and Computer Science Web site: http://www.ics.uci.edu/~mlearn/ML-Repository.html

Blum A., & Langley, P. (1997). Selection of relevant features and examples in machine learning. *Artificial Intelligence*, 245-271.

Breiman, L., Friedman, J. H., Olshen, R. A., & Stone, C. J. (1998). *Classification and regression trees*. Chapman and Hall, Boca Raton.

Brown, M. P. S., Grundy, W. N., Lin, D., Cristiani, N., Sugnet, C. W., Furey, T. S, Ares, M., & Haussler, D. (2000). Knowledge-based analysis of microarray gene expression data by using support vector machines. *PNAS, 97*, 262-267.

Butterworth, R., Piatetsky-Shapiro, G., & Simovici, D. A. (2005). On feature extraction through clustering. In *Proceedings of ICDM*, Houston, Texas.

Butterworth, R., Simovici, D. A., Santos, G. S., & Ohno-Machado, L. (2004). A greedy algorithm for supervised discretization. *Journal of Biomedical Informatics*, 285-292.

Can, F. (1993). Incremental clustering for dynamic information processing. *ACM Transactions for Information Systems, 11*, 143-164.

Can, F., Fox, E. A., Snavely, C. D. & France, R. K. (1995). Incremental clustering for very large document databases: Initial {MARIAN} experience. *Inf. Sci., 84*, 101-114.

Carpenter, G., & Grossberg, S. (1990). Art3: Hierachical search using chemical transmitters in self-organizing pattern recognition architectures. *Neural Networks, 3*, 129-152.

Cerquides, J., & López de Mántaras, R. (1997). Proposal and empirical comparison of a parallelizable distance-based discretization method. In *Proceedings of the 3rd International Conference on Knowledge Discovery and Data Mining (KDD '97).*

Charikar, M., Chekuri, C., Feder, T., & Motwani, R (1997). Incremental clustering and dynamic information retrieval. In *STOC*, (pp. 626-635).

Cornujols, A. (1993). Getting order independence in incremental learning. In *Proceeding of the European Conference on Machine Learning*, pages (pp. 192-212).

Daróczy, Z. (1970). Generalized information functions. *Information and Control, 16*, 36-51.

Dougherty, J., Kohavi, R., & Sahami, M. (1995). Supervised and unsupervised discretization of continuous features. In *Proceedings of the 12th International Conference on Machine Learning*, (pp. 194-202).

Ester, M., Kriegel, H. P., Sander, J., Wimmer, M., & Xu, X. (1998). Incremental clustering for mining in a data warehousing environment. In *VLDB*, (pp. 323-333).

Fayyad, U. M. (1991). *On the induction of decision trees for multiple concept learning.* Unpublished doctoral thesis, University of Michigan.

Fayyad, U. M., & Irani, K. (1993). Multi-interval discretization of continuous-valued attributes for classification learning. In *Proceedings of the 12th International Joint Conference of Artificial intelligence*, (pp. 1022-1027).

Fisher, D. (1987). Knowledge acquisition via incremental conceptual clustering. *Machine Learning, 2*, 139-172.

Guyon, E., & Elisseeff, A. (2003). An introduction to variable and feature selection. *Journal of Machine Learning Research*, 1157-1182.

Hall, M.A. (1999). Correlation-based feature selection for machine learning. Unpublished doctoral thesis, University of Waikato, New Zeland.

Hartigan, J. A. (1975). *Clustering algorithms.* New York: John Wiley.

Hanczar, B., Courtine, M., Benis, A., Hannegar, C., Clement, K., & Zucker, J. D. (2003). Improving classification of microarray data using prototype-based feature selection. *SIGKDD Explorations*, 23-28.

Havrda, J. H., & Charvat, F. (1967). Quantification methods of classification processes: Concepts of structural α-entropy. *Kybernetica, 3*, 30-35.

Jain, A. K., Murty, M. N., & Flynn, P. J. (1999). Data clustering: A review. *ACM Computing Surveys, 31*, 264-323.

Kaufman, L., & Rousseeuw, P. J. (1990). *Finding groups in data—An introduction to cluster analysis.* New York: Wiley Interscience.

Kohavi, R., & John, G. (1997). Wrappers for feature selection. *Artificial Intelligence*, 273-324.

Khan, J., Wei, J. S., Ringner, M., Saal, L. H., Ladanyi, M., Westerman, F., Berthold, F., Schwab, M., Antonescu, C. R., Peterson, C., & Meltzer, P.

S. (2001). Classification and diagnostic prediction of cancers using gene expression profiling and artificial neural networks. *Nature Medicine, 7*, 673-679.

Kononenko, I. (1992). Naïve bayes classifiers and continuous attributes. *Informatica, 16*, 1-8.

Kononenko, I. (1993). Inductive and Bayesian learning in medical diagnosis. *Applied Artificial Intelligence, 7*, 317-337.

Langford, T., Giraud-Carrier, C. G., & Magee, J. (2001). Detection of infectious outbreaks in hospitals through incremental clustering. In *Proceedings of the 8th Conference on AI in Medicine* (AIME).

Lin, J., Vlachos, M., Keogh, E. J., & Gunopulos, D. (2004). Iterative incremental clustering of time series. In *EDBT*, (pp. 106-122).

López de Màntaras, R. (1991). A distance-based attribute selection measure for decision tree induction. *Machine Learning, 6*, 81-92.

Maindonald, J., & Brown, J. (2003). *Data analysis and graphics using R*. Cambridge: Cambridge University Press.

Monjardet, B. (1981). Metrics on partially ordered sets—A survey. *Discrete Mathematics, 35*, 173-184.

Robnik, M. & Kononenko, I. (1995). Discretization of continuous attributes using relieff. In *Proceedings of ERK-95*.

Roure, J., & Talavera, L. (1998). Robust incremental clustering with bad instance ordering: A new strategy. In *IBERAMIA*, 136-147.

Simovici, D. A., & Butterworth, R. (2004). A metric approach to supervised discretization. In *Proceedings of the Extraction et Gestion des Connaisances* (EGC 2004) (pp. 197-202). Toulouse, France.

Simovici, D. A., & Jaroszewicz, S. (2000). On information-theoretical aspects of relational databases. In C. Calude & G. Paun (Eds.), *Finite versus infinite*. London: Springer Verlag.

Simovici, D. A., & Jaroszewicz, S. (2002). An axiomatization of partition entropy. *IEEE Transactions on Information Theory, 48*, 2138-2142.

Simovici, D. A., & Jaroszewicz, S. (2003). Generalized conditional entropy and decision trees. In *Proceedings of the Extraction et gestion des connaissances* - EGC 2003 (pp. 363-380). Paris, Lavoisier.

Simovici, D. A., & Jaroszewicz, S. (in press). A new metric splitting criterion for decision trees. In *Proceedings of PAKDD 2006*, Singapore.

Simovici, D. A., & Singla, N. (2005). Semi-supervised incremental clustering of categorical data. In *Proceedings of EGC* (pp. 189-200).

Simovici, D. A., Singla, N., & Kuperberg, M. (2004). Metric incremental clustering of categorical data. In *Proceedings of ICDM* (pp. 523-527).

Witten, I., & Frank, E. (2005). *Data mining–Practical machine learning tools and techniques (2nd ed)*. Amsterdam: Morgan Kaufmann.

Yang, Y., & Webb, G. I. (2001). Proportional k-interval discretization for naive Bayes classifiers. In *Proceedings of the 12th European Conference on Machine Learning*, (pp. 564-575).

Yang, Y., & Webb, G. I. (2003). Weighted proportional k-interval discretization for naive Bayes classifiers. In *Proceedings of the PAKDD*.

Zongker, D., & Jain, A. (1996). Algorithms for feature selection: An evaluation. In *Proceedings of the International Conference on Pattern Recognition* (pp. 18-22).

NOTE

To Dr. George Simovici,
In memoriam

This work was previously published in Data Mining Patterns: New Methods and Applications, edited by P. Poncelet, F. Masseglia, and M. Teisseire, pp. 1-31, copyright 2008, by Information Science Reference, formerly known as Idea Group Reference (an imprint of IGI Global).

Chapter 2.22
Mining Geo-Referenced Databases:
A Way to Improve Decision-Making

Maribel Yasmina Santos
University of Minho, Portugal

Luís Alfredo Amaral
University of Minho, Portugal

ABSTRACT

Knowledge discovery in databases is a process that aims at the discovery of associations within data sets. The analysis of geo-referenced data demands a particular approach in this process. This chapter presents a new approach to the process of knowledge discovery, in which qualitative geographic identifiers give the positional aspects of geographic data. Those identifiers are manipulated using qualitative reasoning principles, which allows for the inference of new spatial relations required for the data mining step of the knowledge discovery process. The efficacy and usefulness of the implemented system — PADRÃO — has been tested with a bank dataset. The results support that traditional knowledge discovery systems, developed for relational databases and not having semantic knowledge linked to spatial data, can be used in the process of knowledge discovery in geo-referenced databases, since some of this semantic knowledge and the principles of qualitative spatial reasoning are available as spatial domain knowledge.

INTRODUCTION

Knowledge discovery in databases is a process that aims at the discovery of associations within data sets. Data mining is the central step of this process. It corresponds to the application of algorithms for identifying patterns within data. Other steps are related to incorporating prior domain knowledge and interpretation of results.

The analysis of geo-referenced databases constitutes a special case that demands a particular approach within the knowledge discovery

Figure 1. Knowledge Discovery Process

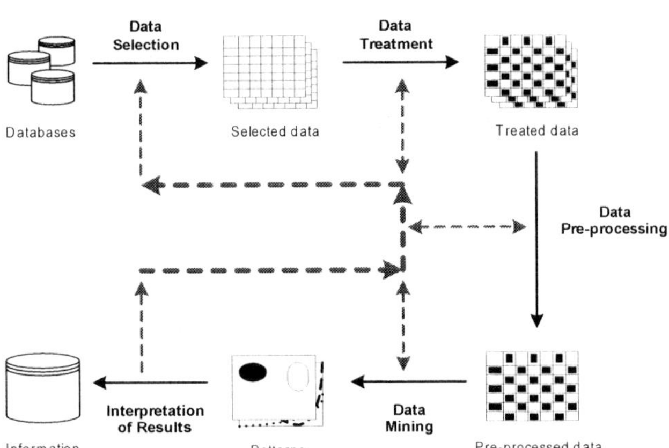

process. Geo-referenced data sets include allusion to geographical objects, locations or administrative sub-divisions of a region. The geographical location and extension of these objects define implicit relationships of spatial neighborhood. The data mining algorithms have to take this spatial neighborhood into account when looking for associations among data. They must evaluate if the geographic component has any influence in the patterns that can be identified.

Data mining algorithms available in traditional knowledge discovery tools, which have been developed for the analysis of relational databases, are not prepared for the analysis of this spatial component. This situation led to: (i) the development of new algorithms capable of dealing with spatial relationships; (ii) the adaptation of existing algorithms in order to enable them to deal with those spatial relationships; (iii) the integration of the capabilities for spatial analysis of spatial database management systems or geographical information systems with the tools normally used in the knowledge discovery process.

Most of the geographical attributes normally found in organizational databases (e.g., addresses) correspond to a type of spatial information, namely qualitative, which can be described using indirect positioning systems. In systems of spatial referencing using geographic identifiers, a position is referenced with respect to a real world location defined by a real world object. This object is termed a *location*, and its identifier is termed a *geographic identifier*. These geographic identifiers are very common in organizational databases, and they allow the integration of the spatial component associated with them in the process of knowledge discovery.

This chapter presents a new approach to the analysis of geo-referenced data. It is based on qualitative spatial reasoning strategies, which enable the integration of the spatial component in the knowledge discovery process. This approach, implemented in the PADRÃO system, allowed the analysis of geo-referenced databases and the identification of implicit relationships existing between the geo-spatial and non-spatial data.

The following sections, in outline, include: (i) an overview of the process of knowledge discovery and its several phases. The approaches usually followed in the analysis of geo-referenced databases are also presented; (ii) a description of qualitative spatial reasoning presenting its principles and the several spatial relations—direction, distance and topology. For the relations, an integrated spatial reasoning system was constructed and made available in the Spatial Knowledge Base of the PADRÃO system. The rules stored enable the inference of new spatial relations needed in the data mining step of the knowledge discovery process; (iii) a presentation of the PADRÃO system describing its architecture and its implementation achieved through the adoption of several technologies. This section continues with the analysis of a geo-referenced database, based on the several steps of the knowledge discovery process considered by the PADRÃO system; and (iv) a conclusion with some comments about the proposed research and its main advantages.

KNOWLEDGE DISCOVERY IN DATABASES

Large amounts of operational data concerning several years of operation are available, mainly from middle-large sized organizations. Knowledge discovery in databases is the key to gaining access to the strategic value of the organizational knowledge stored in databases for use in daily operations, general management and strategic planning.

The Knowledge Discovery Process

Knowledge Discovery in Databases (KDD) is a complex process concerning the discovery of relationships and other descriptions from data. Data mining refers to the application algorithms used to extract patterns from data without the additional steps of the KDD process, e.g., the incorporation of appropriate prior knowledge and the interpretation of results (Fayyad & Uthurusamy, 1996).

Different tasks can be performed in the knowledge discovery process and several techniques can be applied for the execution of a specific task. Among the available tasks are *classification*, *clustering*, *association*, *estimation* and *summarization*. KDD applications integrate a variety of data mining algorithms. The performance of each technique (algorithm) depends upon the task to be carried out, the quality of the available data and the objective of the discovery. The most popular Data Mining algorithms include *neural networks*, *decision trees*, *association rules* and *genetic algorithms* (Han & Kamber, 2001).

The steps of the KDD process (*Figure 1*) include data selection, data treatment, data pre-processing, data mining and interpretation of results. This process is interactive, because it requires user participation, and iterative, because it allows for going back to a previous phase and then proceeding forward with the knowledge discovery process. The steps of the KDD process are briefly described:

- *Data Selection.* This step allows for the selection of relevant data needed for the execution of a defined data mining task. In this phase the minimum sub-set of data to be selected, the size of the sample needed and the period of time to be considered must be evaluated.
- *Data Treatment.* This phase concerns with the cleaning up of selected data, which allows for the treatment of corrupted data and the definition of strategies for dealing with missing data fields.
- *Data Pre-Processing.* This step makes possible the reduction of the sample destined for analysis. Two tasks can be carried out here: (i) the reduction of the number of rows or, (ii) the reduction of the number of columns. In the reduction of the number of rows, data can be generalized according to the defined hierarchies or attributes with continuous values can be transformed into discreet values according to the defined classes. The reduction of the number of columns attempts to verify if any of the selected attributes can now be omitted.
- *Data Mining.* Several algorithms can be used for the execution of a given data mining task. In this step, various available algorithms are

evaluated in order to identify the most appropriate for the execution of the defined task. The selected one is applied to the relevant data in order to find implicit relationships or other interesting patterns that exist in the data.

- *Interpretation of Results.* The interpretation of the discovered patterns aims at evaluating their utility and importance with respect to the application domain. It may be determined that relevant attributes were ignored in the analysis, thus suggesting that the process should be repeated.

Knowledge Discovery in Spatial Databases

The main recognized advances in the area of KDD (Fayyad, Piatetsky-Shapiro, Smyth & Uthurusamy, 1996) are related with the exploration of relational databases. However, in most organizational databases there exists one dimension of data, the *geographic* (associated with addresses or post-codes), the semantic of which is not used by traditional KDD systems.

Knowledge Discovery in Spatial Databases (KDSD) is related with "*the extraction of interesting spatial patterns and features, general relationships that exist between spatial and non-spatial data, and other data characteristics not explicitly stored in spatial databases*" (Koperski & Han, 1995).

Spatial database systems are relational databases with a concept of spatial location and spatial extension (Ester, Kriegel & Sander, 1997). The explicit location and extension of objects define implicit relationships of spatial neighborhood. The major difference between knowledge discovery in relational databases and KDSD is that the neighbor attributes of an object may influence the object itself and, therefore, must be considered in the knowledge discovery process. For example, a new industrial plant may pollute its neighborhood entities depending on the distance between the objects (regions) and the major direction of the wind. Traditionally, knowledge discovery in relational databases does not take into account this spatial reasoning, which motivates the development of new algorithms adapted to the spatial component of spatial data.

The main approaches in KDSD are characterized by the development of new algorithms that treat the position and extension of objects mainly through the manipulation of their coordinates. These algorithms are then implemented, thus extending traditional KDD systems in order to accommodate them. In all, a quantitative approach is used in the spatial reasoning process although the results are presented using qualitative identifiers.

Lu, Han & Ooi (1993) proposed an attribute-oriented induction approach that is applied to spatial and non-spatial attributes using conceptual hierarchies. This allows the discovery of relationships that exist between spatial and non-spatial data. A spatial concept hierarchy represents a successive merge of neighborhood regions into large regions. Two learning algorithms were introduced: (i) non-spatial attribute-oriented induction, which performs generalization on non-spatial data first, and (ii) spatial hierarchy induction, which performs generalization on spatial data first. In both approaches, the classification of the corresponding spatial and non-spatial data is performed based on the classes obtained by the generalization. Another peculiarity of this approach is that the user must provide the system with the relevant data set, the concept hierarchies, the desired rule form and the learning request (specified in a syntax similar to SQL – Structured Query Language).

Koperski & Han (1995) investigated the utilization of interactive data mining for the extraction of spatial association rules. In their approach the spatial and non-spatial attributes are held in different databases, but once the user identifies the attributes or relationships of interest, a selection process takes place and a unified database is created. An algorithm, implemented for the discovery of spatial association rules, analyzes the stored data. The rules obtained represent relationships between objects, described using spatial predicates like *adjacent to* or *close to.*

These approaches are two examples of the efforts made in the area of KDSD. One approach uses two different databases, storing spatial and non-spatial data separately. Once the user identifies the attributes of interest, an interface between the two databases ensures the selection and treatment of data without the creation of a new integrated repository. The other approach also requires two different databases, but the selection phase leads to the creation of a unified database where the analysis of data takes place. In both approaches new algorithms were implemented and the user is asked for the specification of the relevant attributes and the type of results expected.

Two approaches for the analysis of spatial data with the aim of knowledge discovery have been presented. Independently of the adopted approach, several tasks can be performed in this process, among them: *spatial characterization, spatial classification, spatial association* and *spatial trends analysis* (Koperski & Han, 1995; Ester, Frommelt, Kriegel & Sander, 1998; Han & Kamber, 2001).

A *spatial characterization* corresponds to a description of the spatial and non-spatial properties of a selected set of objects. This task is achieved analyzing not only the properties of the target objects, but also the properties of their neighbors. In a characterization, the relative frequency of incidence of a property in the selected objects, and their neighbors, is different from the relative frequency of the same property verified in the remaining of the database (Ester, Frommelt, Kriegel & Sander, 1998). For example, the incidence of a particular disease can be higher in a set of regions closest or holding a specific industrial complex, showing that a possible *cause-effect* relationship exists between the disease and the industry pollution.

Spatial classification aims to classify spatial objects based on the spatial and non-spatial features of these objects in a database. The result of the classification, a set of rules that divides the data into several classes, can be used to get a better understanding of the relationships among the objects in the database and to predict characteristics of new objects (Han, Tung & He, 2001; Han & Kamber, 2001). For example, regions can be classified into *rich* or *poor* according to the average family income or any other relevant attribute present in the database.

Spatial association permits the identification of spatial-related association rules from a set of data. An association rule shows the frequently occurring patterns of a set of data items in a database. A spatial association rule is a rule of the form "$X \rightarrow Y$ (s%, c%)," where X and Y are sets of spatial and non-spatial predicates (Koperski & Han, 1995). In an association rule, s represents the support of the rule, the probability that X and Y exist together in the data items analyzed, while c indicates the confidence of the rule, i.e., the probability that Y is true under the condition of X. For example, the spatial association rule "is_a (x, House) $\wedge$ close_to (x, Beach) $\rightarrow$ is_expensive (x)" states that houses which are close to the beach are expensive.

A *spatial trend* (Ester, Frommelt, Kriegel & Sander, 1998) describes a regular change of one or more non-spatial attributes when moving away from a particular spatial object. Spatial trend analysis allows for the detection of changes and trends along a spatial dimension. Examples of spatial trends are the changes in the economic situation of a population when moving away from the center of a city or the trend of change of the climate with the increasing distance from the ocean (Han & Kamber, 2001).

After the presentation of two approaches and some of the most popular tasks associated with the analysis of spatial data with the aim of knowledge discovery, this chapter posits a new approach to the process of KDSD (more specifically in geo-referenced datasets). This approach integrates qualitative principles in the spatial reasoning system used in the knowledge discovery process. Since the use of coordinates for the identification of a spatial object is not always needed, this work investigates how traditional KDD systems (and their generic data mining algorithms) can be used in KDSD.

QUALITATIVE SPATIAL REASONING

Human beings use qualitative identifiers extensively to simplify reality and to perform spatial reasoning more efficiently. *Spatial reasoning* is the process by which information about objects in space and their relationships are gathered through measurement, observation or inference and used to arrive at valid conclusions regarding the relationships of the objects (Sharma, 1996). *Qualitative spatial reasoning* (Abdelmoty & El-Geresy, 1995) is based on the manipulation of qualitative spatial relations, for which composition[1] tables facilitate reasoning, thereby allowing the inference of new spatial knowledge.

Spatial relations have been classified into several types (Frank, 1996; Papadias & Sellis, 1994), including *direction relations* (Freksa, 1992) (that describe order in space), *distance relations* (Hernández, Clementini & Felice, 1995) (that describe proximity in space) and *topological relations* (Egenhofer, 1994) (that describe neighborhood and incidence). Qualitative spatial relations are specified by using a small set of symbols, like *North, close, etc.*, and are manipulated through a set of inference rules.

The inference of new spatial relations can be achieved using the defined qualitative rules, which are compiled into a composition table. These rules allow for the manipulation of the qualitative identifiers adopted. For example, knowing the facts, A North, very far from B and B Northeast, very close to C, it is possible, by consulting the composition table for integrated direction and distance spatial reasoning (presented later), to infer the relationship that exists between A and C, that is A North, very far from C.

The inference rules can be constructed using quantitative methods (Hong, 1994) or by manipulating qualitatively the set of identifiers adopted (Frank, 1992; Frank, 1996), an approach that requires the definition of axioms and properties for the spatial domain.

Later in this section the construction of the qualitative spatial reasoning system used by PADRÃO is presented. The qualitative system integrates *direction*, *distance* and *topological* spatial relations. Its conception was achieved based on the work developed by Hong (1994) and Sharma (1996). The application domain in which this qualitative reasoning system will be used is characterized by objects that represent administrative subdivisions.

Direction Spatial Relations

Direction relations describe where objects are placed relative to each other. Three elements are needed to establish an orientation: two objects and a fixed point of reference (usually the North Pole) (Frank, 1996; Freksa, 1992). Cardinal directions can be expressed using numerical values specifying degrees (0°, 45°...) or using qualitative values or symbols, such as North or South, which have an associated acceptance region. The regions of acceptance for qualitative directions can be obtained by projections (also known as half-planes) or by cone-shaped regions (*Figure 2*).

A characteristic of the cone-shaped system is that the region of acceptance increases with distance, which makes it suitable for the definition of direction relations between extended objects[2] (Sharma, 1996). It also allows for the definition of finer resolutions, thus permitting the use of eight (*Figure 3*) or 16 different qualitative directions. This model uses triangular acceptance areas that are drawn from the *centroid* of the reference object towards the primary object (in the spatial relation

Figure 2. Direction Relations Definition by Projection and Cone-Shaped Systems

Figure 3. Cone-Shaped System with Eight Regions of Acceptance

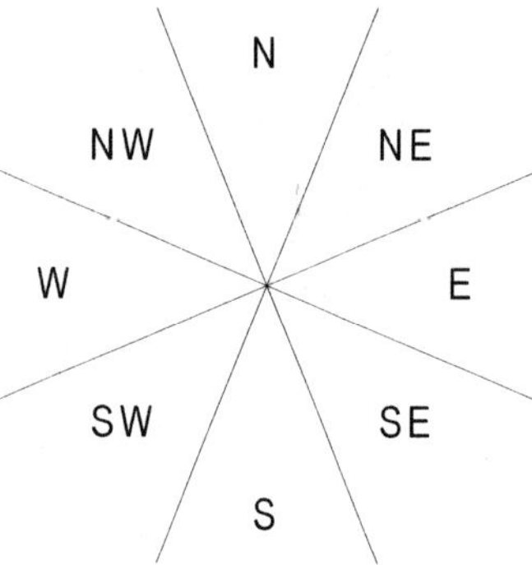

A North B, B represents the reference object, while A constitutes the primary object).

Distance Spatial Relations

Distances are quantitative values determined through measurements or calculated from known coordinates of two objects in some reference system. The frequently used definition of distance can be achieved using the Euclidean geometry and Cartesian coordinates. In a two-dimensional Cartesian system, it corresponds to the length of the shortest possible path (a straight line) between two objects, which is also known as the Euclidean distance (Hong, 1994). Usually a metric quantity is mapped onto some qualitative indicator such as very close or far for human common-sense reasoning (Hernández et al., 1995).

Qualitative distances must correspond to a range of quantitative values specified by an interval and they should be ordered so that comparisons are possible. The adoption of the qualitative distances very close – vc, close – c, far – f and very far – vf, intuitively describe distances from the nearest to the furthest. An order relationship exists among these relations, where a lower order (vc) relates to shorter quantitative distances and a higher order (vf) relates to longer quantitative distances (Hong, 1994). The length of each successive qualitative distance, in terms of quantitative values, should be greater or equal to the length of the previous one (*Figure 4*).

Topological Spatial Relations

Topological relations are those relationships that are invariant under continuous transformations of space such as rotation or scaling. There are eight topological relations that can exist between two planar regions without holes[3]: disjoint, contains, inside, equal, meet, covers, covered by and overlap (*Figure 5*). These relations can be defined considering intersections between the two regions, their boundaries and their complements (Egenhofer,

Figure 4. Qualitative Distances Intervals

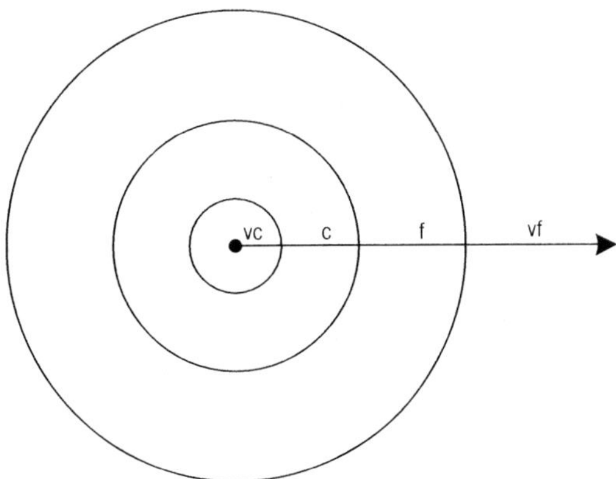

Figure 5. Topological Spatial Relations

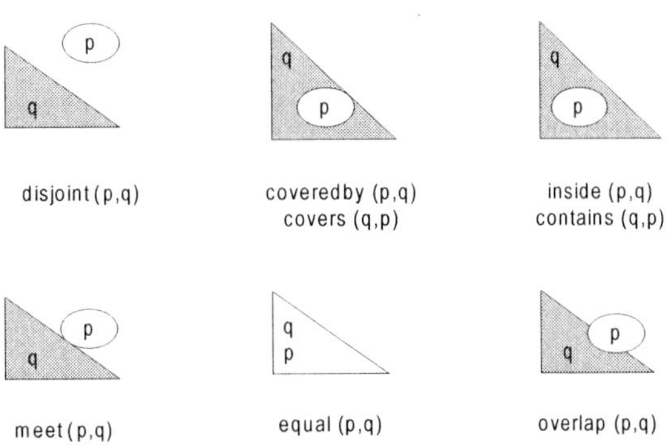

1994). These eight relations, which can exist between two spatial regions without holes, will be the exclusive focus of topological relations in this chapter.

In some exceptional cases, the geographic space cannot be characterized, in topological terms, with reference to the eight topological primitives presented above. One of these cases is related with application domains in which the geographic regions addressed are administrative subdivisions. Administrative subdivisions, represented in this work by full planar graphs[4], can only be related through the topological primitives disjoint, meet and contains (and the corresponding inverse inside), since they cannot have any kind of overlapping. The topological primitives used in this chapter are disjoint and meet, since the implemented qualitative inference process only considers regions at the same geographic hierarchical level.

Integrated Spatial Reasoning

Integrated reasoning about qualitative directions necessarily involves qualitative distances and directions. Particularly in objects with extension, the size and shape of objects and the distance between them influence the directions. One of the ways to determine the direction and distance[5] between regions is to calculate them from the *centroids* of the regions. The extension of the geographic entities is somehow implicit in the topological primitive used to characterize their relationships.

Integration of Direction and Distance

An example of *integrated spatial reasoning* about qualitative distances and directions is as follows. The facts A is very far from B and B is very far from C do not facilitate the inference of the relationship that exists between A and C. A can be very close or close to C, or A may be far or very far from C, depending on the orientation between B and C.

For the integration of qualitative distances and directions the adoption of a set of identifiers is required, which allows for the identification of the considered directions and distances and their respective intervals of validity. Hong (1994) analyzed some possible combinations for the number of identifiers and the geometric patterns that should characterize the distance intervals. The *localization system* (*Figure 6*) suggested by Hong is based on eight symbols for direction relations (North, Northeast, East, Southeast, South, Southwest, West, Northwest) and four symbols for the identification of the distance relations (very close, close, far and very far).

In the case of direction relations, for the cone-shaped system with eight acceptance regions, the quantitative intervals adopted were: [337.5, 22.5), [22.5, 67.5), [67.5, 112.5), [112.5, 157.5), [157.5, 202.5), [202.5, 247.5), [247.5, 292.5), [292.5, 337.5) from North to Northwest respectively.

The definition of the validity interval for each distance identifier must obey some rules (Hong, 1994). In these systems, as can be seen in Table 1, there should exist a constant ratio (ratio = length $(dist_i)$/length $(dist_{i-1})$) relationship between the lengths of two neighboring intervals. The presented simulated intervals allow for the definition of new distance intervals by magnification of the original intervals. For example, the set of values for ratio 4[6] can be increased by a factor of 10 supplying the values $dist_0$ (0, 10], $dist_1$ (10, 50], $dist_2$ (50, 210]

Figure 6. Integration of Direction and Distance Spatial Relations

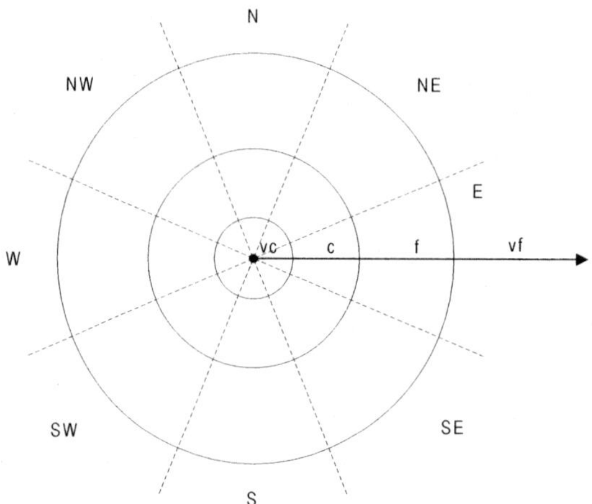

Table 1. Simulated Intervals for Four Symbolic Distance Values

Ratio	$dist_0$	$dist_1$	$dist_2$	$dist_3$
1	(0, 1]	(1, 2]	(2, 3]	(3, 4]
2	(0, 1]	(1, 3]	(3, 7]	(7, 15]
3	(0, 1]	(1, 4]	(4, 13](	13, 40]
4	(0, 1]	(1, 5]	(5, 21](	21, 85]
5	(0, 1]	(1, 6]	(6, 31](	31, 156]
...	...	...	...	...

Figure 7. Graphical Representation of Direction and Distance Integration

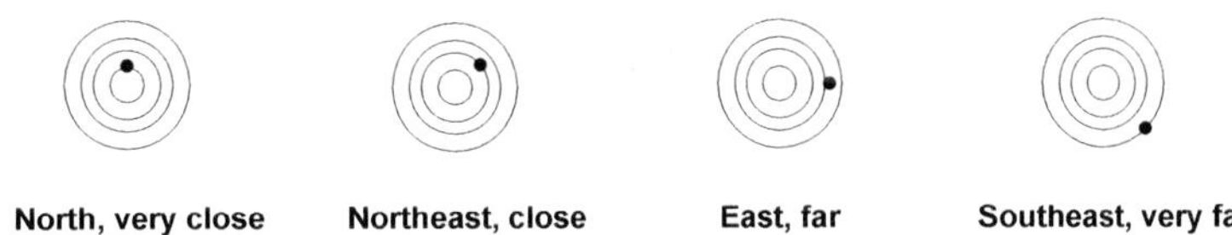

Figure 8. Extract of the Final Composition Table — Integration of Direction and Distance

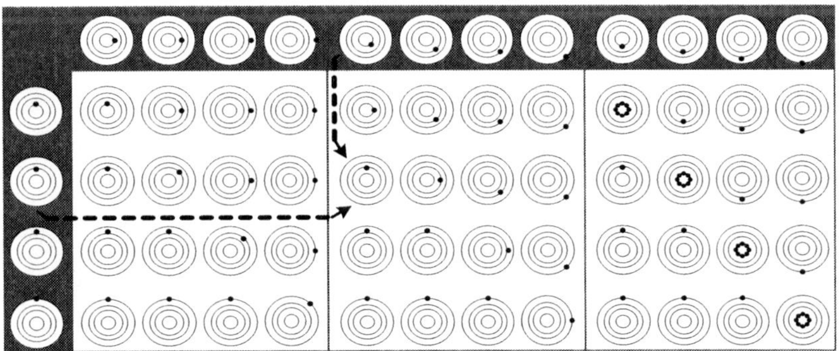

and $dist_3$ (210, 850]. Since the same scale magnifies all intervals and quantitative distance relations, the qualitative compositions will remain the same, regardless of the scaled value.

It is important to know that the number of distance symbols used and the ratio between the quantitative values addressed by each interval play an important role in the robustness of the final system, i.e., in the validity of the composition table for the inference of new spatial relations (Hong, 1994).

The final composition table, a 32x32 matrix for the localization system adopted, was constructed following the suggestions made by Hong (1994) and it is presented in this work through an iconic representation (*Figure 7*). This matrix represents part of the knowledge needed for the inference of new spatial information in the localization system used. Due to its great size, *Figure 8* exhibits an extract of the final matrix. An example of the composition operation: suppose that A North, close B and that B Southeast, very close C. Consulting the composition table (this example is marked in *Figure 8* with two traced arrows) it is possible to identify the relation that exists between A and C: A North, close C. For the particular case of the composition of opposite directions with equal qualitative distances, the system is unable to identify the direction between the objects. For this reason, the composition of these particular cases presents all the qualitative directions as possible results of the inference (*Figure 8*).

Integration of Direction and Topology

The relative position of two objects in the bi-dimensional space can be achieved through the dimension and orientation of the objects. Looking at each of these characteristics separately implies two classes of spatial relations: *topological*, which ignores orientations in space; and *direction* that ignores the extension of the objects.

The integration of these two kinds of spatial relations enables the definition of a system for qualitative spatial reasoning that describes the relative position existing between the objects and how the limits (frontiers) of them are related.

Sharma (1996) integrated direction and topological spatial relations using the principles of qualitative temporal reasoning defined by Allen (1983). The approach undertaken by Sharma (1996) was possible through the adaptation of the temporal principles to the spatial domain. The 13 temporal primitives (Allen, 1983) are: before, after, during, contain, overlap, overlapped by, meet, met by, start, started by, finish, finished by and equal (*Figure 9*).

The temporal primitives (that are one-dimensional) were analyzed by Sharma (1996) along two dimensions (axes *xx* and *yy*) allowing their use in the spatial domain (restricted in this case to a two-dimensional space).

The construction of the composition tables was facilitated by the knowledge representation framework adopted for the integration of direction and topology. Topological relations are independent of the order existing between the objects when analyzed along a given axis. Direction relations depend on the order and are defined by verifying the objects position along a specific axis.

Figure 9. Temporal Primitives

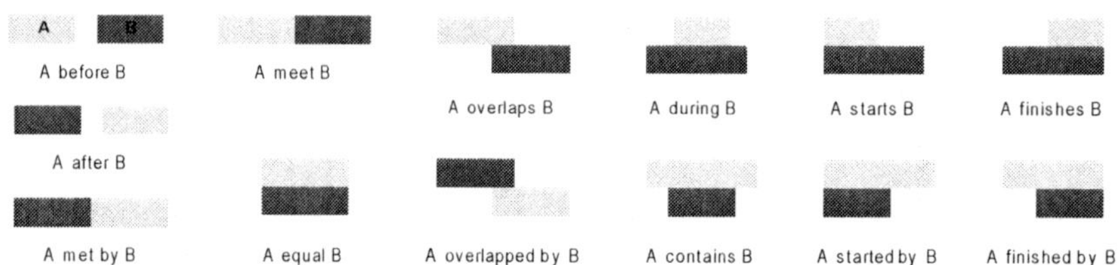

Figure 10. Integration of Direction and Topological Relations

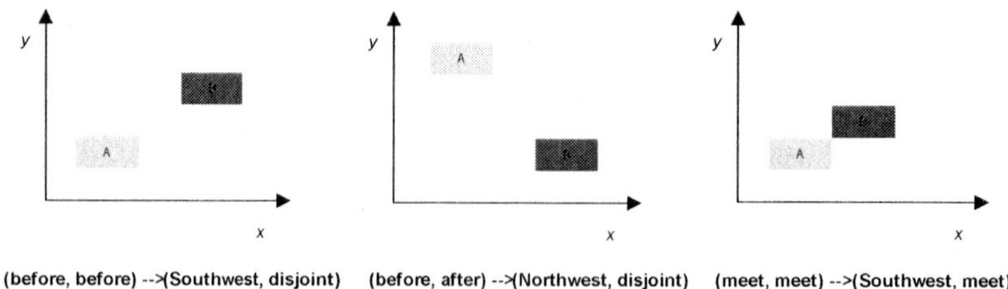

Figure 11. Interval Relations for Direction Relations Representation

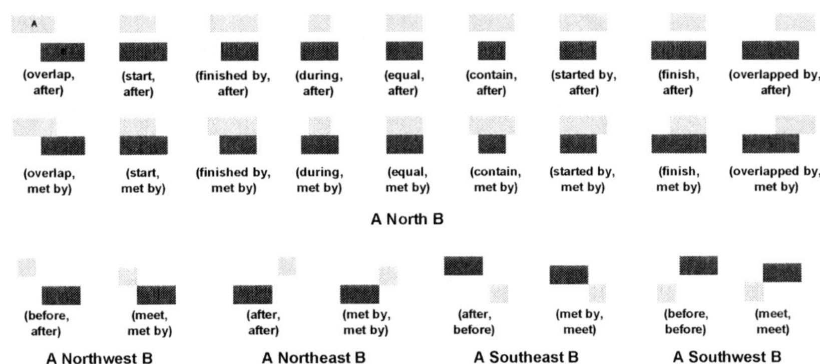

Figure 12. Temporal Relations — Graphical Representation

Notation suggested by Sharma (1996)

The representation of each pair (direction, topology) is accomplished through temporal primitives. The transformation of the one-dimensional characteristics to the two-dimensional space is achieved analyzing the pair of temporal primitives that represent the behavior of the pair (direction, topology) along *x* and *y* (*Figure 10* supplies three examples of selection of the appropriate pair of temporal primitives, verifying the position of A and B along *x* and along *y*, for the characterization of the pair (direction, topology)).

Restricting the integration domain to objects that represent administrative subdivisions without overlap between them, the two topological relations considered were disjoint and meet. These two topological relationships can be represented by the temporal primitives before and meet, and by the corresponding inverses (after and met by). Attending to the direction relations, all the temporal primitives defined by Allen (1983) can be used in their characterization. *Figure 11* shows how the temporal primitives are used in the definition of a particular direction relation.

For the identification of the inference rules it is necessary to identify the temporal primitives that characterize each pair (direction, topology) and then do their composition to achieve the result. *Table 2* presents an extract of the composition table for the temporal domain. This table, graphically presented using the notation showed in *Figure 12*, will be afterwards used for the spatial domain.

The composition of pairs of relations (direction, topology) is performed consulting *Table 2*. An example of the composition[7] operation for the spatial domain is the composition of the pair (Northeast, disjoint) with the pair (Northeast, disjoint). The result of the composition is achieved by the steps:

Table 2. An Extract of the Composition Table for Temporal Intervals

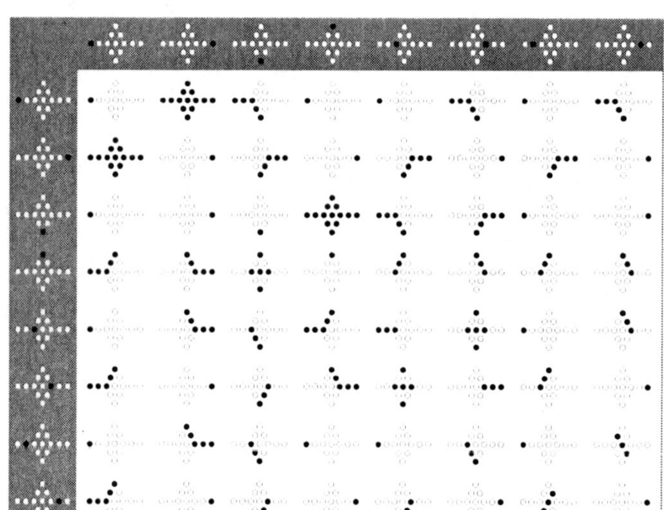

Figure 13. Graphical Representation of Direction and Topological Spatial Relations

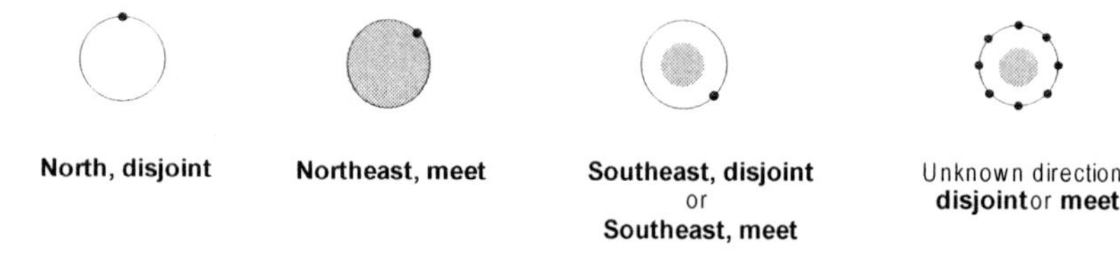

(Northeast, disjoint) ; (Northeast, disjoint) =
(after, after) ; (after, after)
=
(after; after) x (after; after)
=
(after) x (after)
=
(after, after)
=
(Northeast, disjoint)

Following this composition process, Sharma obtained the several composition tables that integrate direction with the several topological pairs disjoint;disjoint, disjoint;meet, meet;disjoint and meet;meet. *Figure 13* presents the graphical symbols used in this chapter to represent the integration of direction and topology. *Table 3* shows one of the composition tables of Sharma, integrating direction with the topological pair disjoint;disjoint.

Integration of Direction, Distance and Topological Spatial Relations

With the integration of direction and distance spatial relations a set of inference rules were obtained. These rules present a unique pair (direction, distance) as outcome, with the exception of the result of the composition of pairs with opposite directions and equal qualitative distances. In the integration of direction and topological spatial relations some improvements can be achieved, since several inference rules present as the result a set of outcomes.

Table 3. Composition Table for the Integration of Direction with the Topological Pair disjoint;disjoint

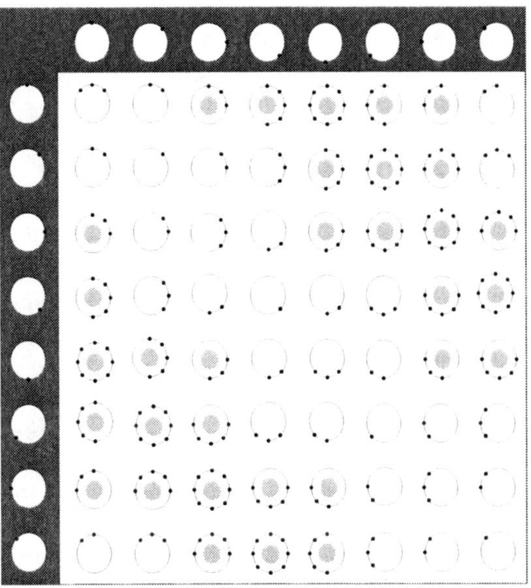

Figure 14. Temporal Intervals for the Characterization of Direction and Topology for Administrative Subdivisions

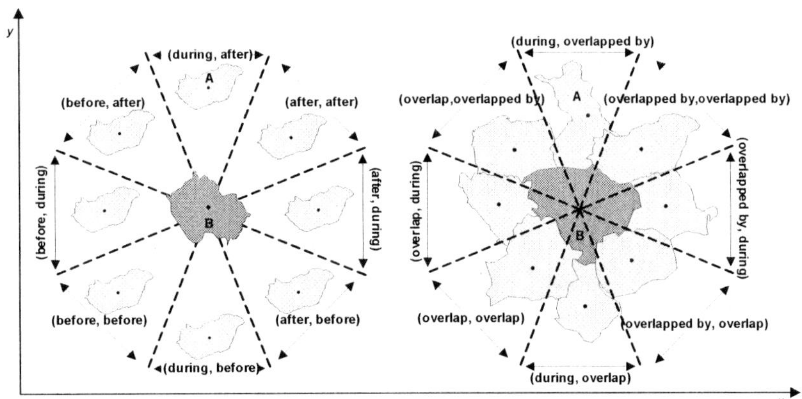

Looking at the work developed by Hong and Sharma it was evident that the integration of the three types of spatial relations, direction, distance and topology, would lead to more accurate composition tables.

Since Hong adopted a cone-shaped system in the definition of the direction relations, and Sharma used a projection-based system for the same task, the integration of the three types of spatial relations was preceded by the adaptation[8] of the principles used by Sharma and the construction of new composition tables for the integration of direction and topology.

In the characterization of the integration of direction and topological relations, for the particular case of administrative subdivisions, new temporal pairs were defined, which allowed for the identification of new inference rules. *Figure 14* shows the several pairs of temporal primitives adopted according to the direction relations and the topological primitives disjoint and meet.

The adoption of the temporal intervals shown in *Figure 14* was motivated by the fact that administrative subdivisions have irregular limits, which impose several difficulties in the identification of the *correct* direction between two regions. Sometimes the *centroid* is positioned in a place that suggests one direction, although the administrative region may have parts of its territory at other acceptance areas in the cone-shaped system. The adoption of the during temporal primitive for the characterization of North, East, South and West directions was motivated by the assumption that the *centroid* of the primary object is located in the zone of acceptance for those directions, as defined by the reference object.

In the case of adjacency it is clear by an analysis of *Figure 14* that some overlapping between the regions can exist, when analyzed in a temporal perspective. This fact influenced the adoption of the overlap and overlapped by primitives instead of the meet and met by primitives adopted by Sharma.

Following the assumptions described above new composition tables were constructed. *Table 4* shows the particular case of integration of direction with the topological pair disjoint;disjoint. The other composition tables, for the topological pairs disjoint;meet, meet;disjoint and meet;meet, are available in Santos (2001).

After the identification of the composition tables that integrate direction and topology under the principles of the cone-shaped system, it was possible to integrate these tables with the composition table proposed by Hong (1994), with respect to direction and distance. This step was preceded by a detailed analysis of the application domain in which the system will be used, in particular the composition of regions that represent administrative subdivisions that cover all the territory considered, without any gap or overlap (Santos, 2001). Concerning the distance spatial relation, it was defined that the qualitative distance very

Table 4. Composition Table for the Integration of Direction with the Topological Pair disjoint;disjoint (particular case of administrative subdivisions)

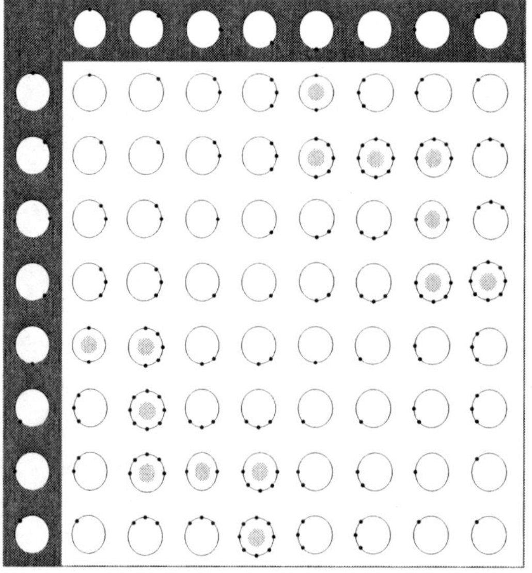

Figure 15. Graphical Representation of Direction, Distance and Topological Spatial Relations

Figure 16. Integration of Direction, Distance and Topological Spatial Relations

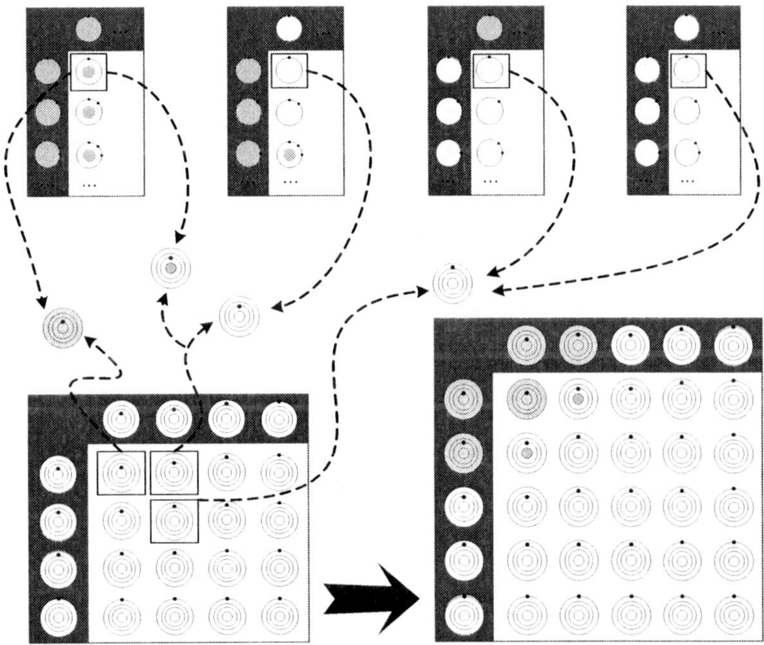

close is restricted to adjacent regions. When the qualitative distance is close the regions may be, or may not be, adjacent. The far and very far qualitative distances can only exist between regions that are disjoint from each other.

The basic assumption for the integration process was that the outcome direction in the integration of direction and distance is the same outcome direction in the integration of direction and topology, or it belongs to the set of possible directions inferred by the last one. The direction that guides the integration process is the direction suggested by the composition table of direction and distance (it is more accurate since it considers the distance existing between the objects).

The final composition table, which is shown with the graphical symbols expressed in *Figure 15*, was obtained through an integration process that is diagrammatically demonstrated in *Figure 16*. For example, the composition of (North, very close) with (North, very close) has as result (North, very close). The composition of (North, meet) with (North, meet) has as the result (North, disjoint or meet). The integration of the three spatial relations leads to (North, very close, disjoint or meet). As the qualitative distance relation very close was restricted to

Figure 17. Influence of the Regions Dimension in the Inference Result

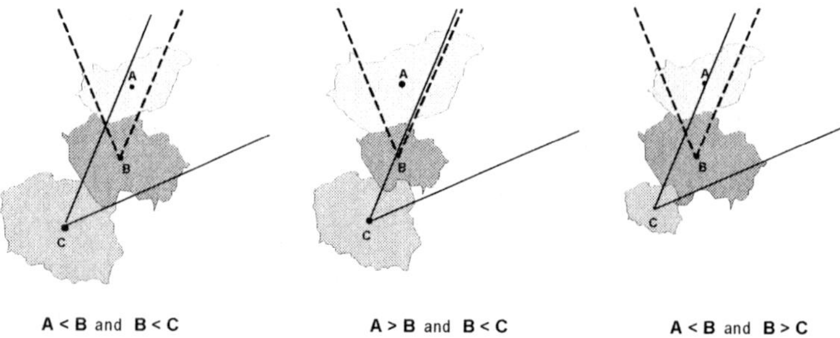

Figure 18. Municipalities of the Braga District

adjacent regions, the result of the integration is (North, very close, meet). Another example explicit in *Figure 16* is the integration of the result of (North, close);(North, close) with (North, disjoint);(North, disjoint). The result of the first composition is (North, far) while the result of the second is (North, disjoint). The integration generates the value (North, far, disjoint), which matches the principles adopted in this work for the distance relation: if the regions are far from each other, then topologically they are disjoint.

In the evaluation of the composition table constructed it was realized that the dimensions of the regions influenced (sometimes negatively) the results achieved. Qualitative reasoning with administrative subdivisions is a difficult task, which is influenced not only by the irregular limits of the regions but also by their size. As can be noted in *Figure 17*, if the dimension of A is less than the dimension of B, and the dimension of B is less than the dimension of C, then the inference result must be A Northeast C. But if the dimension of A is greater than the dimension of B and the dimension of B is less than the dimension of C, then the inference result must be A North C. A detailed analysis of these situations was undertaken, allowing the identification of several rules that integrate the dimensions of the regions in the qualitative reasoning process of the PADRÃO system. Through this process, the reasoning process was improved, and more accurate inferences were obtained.

The performance of the qualitative reasoning system was evaluated (Santos, 2001). The approach followed in this performance test was to compare the spatial relations obtained through the qualitative inference process with the spatial relations obtained by quantitative methods. A Visual Basic module was implemented for the execution of this task. This module calculated quantitatively all the spatial relations existing between the Municipalities of three districts of Portugal, looking at the position of the respective *centroids*. This information was stored in a table and compared with the spatial relations inferred qualitatively. The results achieved were, in the worst scenario, exact[10] for 75% of the inferences obtained in Districts with higher differences between the dimensions of their regions (two of the analyzed Districts). For the Braga District, a District that integrates regions with homogeneous dimensions, the inferences obtained were 88% exact for direction and 81% exact for distance. For topology, the inferences were in all cases 100% exact. The approximate inferences obtained were verified in regions that have parts of their territory in more than one acceptance area for the direction relation. For these cases, the *centroid* of the region is sometimes positioned in one acceptance area, although the region has parts of its territory in other acceptance areas. Another situation, as shown in *Figure 18* for two Municipalities, is verified when the *centroid* is positioned in the line that divides the acceptance areas, which makes even more difficult the identification of the direction between the regions and, as a consequence, the qualitative reasoning process.

After the evaluation of the qualitative reasoning system implemented and the analysis of the inferences obtained, which provided a good approximation to the reality, the system will later be used in the knowledge discovery process.

THE PADRÃO SYSTEM

PADRÃO is a system for knowledge discovery in geo-referenced databases based on qualitative spatial reasoning. This section presents its architecture, gives some technical details about its implementation and tests the system in a geo-referenced data set.

Figure 19. Architecture of Padrão

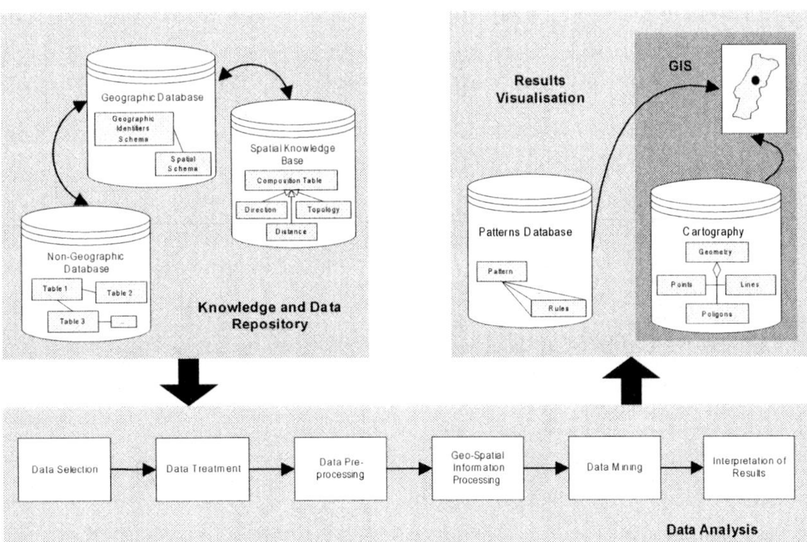

Architecture of Padrão

The architecture of Padrão (*Figure 19*) aggregates three main components: Knowledge and Data Repository, Data Analysis and Results Visualization. The Knowledge and Data Repository component stores the data and knowledge needed in the knowledge discovery process. This process is implemented in the Data Analysis component, which allows for the discovery of patterns or other relationships implicit in the analyzed geo-spatial and non-spatial data. The discovered patterns can be visualized in a map using the Results Visualization component. These components are described below.

The Knowledge and Data Repository component integrates three central databases:

1. A *Geographic Database* (GDB) constructed under the principles established by the European Committee for Normalization in the CEN TC 287 pre-standard for Geographic Information. Following the pre-standard recommendations it was possible to implement a GDB in which the positional aspects of geographic data are provided by a *geographic identifiers system* (CEN/TC-287, 1998). This system characterizes the administrative subdivisions of Portugal at the municipality and district level. Also it includes a geographic gazetteer containing the several geographic identifiers used and the concept hierarchies existing between them. The geographic identifiers system was integrated with a *spatial schema* (CEN/TC-287, 1996) allowing for the definition of the *direction*, *distance* and *topological* spatial relations that exist between adjacent regions at the Municipality level.
2. A *Spatial Knowledge Base* (SKB) that stores the qualitative rules needed in the inference of new spatial relations. The knowledge available in this database aggregates the constructed composition table (integrating direction, distance and topological spatial relations), the set of identifiers used, and the several rules that incorporate the dimension of the regions in the reasoning process. This knowledge base is used in conjunction with the GDB in the inference of unknown spatial relations.
3. A *non-Geographic Database* (nGDB) that is integrated with the GDB and analyzed in the Data Analysis component. This procedure enables the discovery of implicit relationships that exist between the geo-spatial and non-spatial data analyzed.

The Data Analysis component is characterized by six main steps. The five steps presented above for the knowledge discovery process plus the Geo-Spatial Information Processing step. This step verifies if the geo-spatial information needed is available in the GDB. In many situations the spatial relations are implicit due to the properties of the spatial schema implemented. In those cases, and to ensure that all geo-spatial knowledge is available for the data mining algorithms, the implicit relations are transformed into explicit relations through the inference rules stored in the SKB.

The Results Visualization component is responsible for the management of the discovered patterns and their visualization in a map (if required by the user and when the geometry[11] of the analyzed region is available). For that Padrão uses a Geographic Information System (GIS), which integrates the discovered patterns with the geometry of the region. This component aggregates two main databases:

1. The *Patterns Database* (PDB) that stores all relevant discoveries. In this database each discovery is catalogued and associated with the set of rules that represents the discoveries made in a given data mining task.
2. A *Cartographic Database* (CDB) containing the cartography of the region. It aggregates a set of points, lines and polygons with the geometry of the geographical objects.

Implementation of Padrão

Padrão was implemented using the relational database system Microsoft Access, the knowledge discovery tool Clementine (SPSS, 1999), and Geomedia Professional (Intergraph, 1999), the GIS used for the graphical representation of results.

The databases that integrate the Knowledge and Data Repository and the Results Visualization components were implemented in Access. The data stored in them are available to the Data Analysis component or from it, through ODBC (Open Database Connectivity) connections.

Clementine is a data mining toolkit based on visual programming[12], which includes machine learning technologies like rule induction, neural networks, association rules discovery and clustering. The knowledge discovery process is defined in Clementine through the construction of a *stream* in which each operation on data is represented by a *node*.

The workspace of Clementine comprises three main areas. The main work area, the Stream Pane, constitutes the area for the streams construction. The palettes area in which the several available icons are grouped according to their functions: links to sources of information, operations on data (rows or columns), visual facilities and modeling techniques (data mining algorithms). The models area stores the several models generated in a specific stream. These models can be directly re-used in other streams or they can be saved providing for their later use. *Figure 20* shows the work environment of Clementine and presents some of the several nodes available according to their functionality. Circular nodes represent links to data sources and constitute the first node of any stream. Nodes with a hexagonal shape are for data manipulation, including operations on records (lines of a table) or operations on fields (columns of a table). Triangular nodes allow for data exploration and visualization, providing a set

Figure 20. Workspace of Clementine

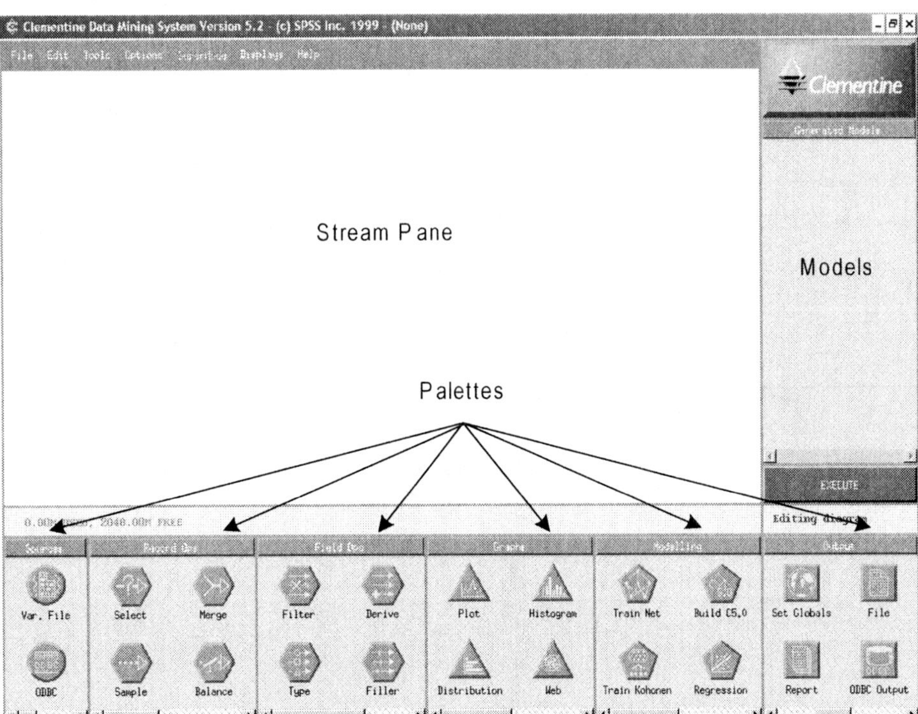

of graphs that can be used to get a better understanding of data. Nodes with a pentagonal shape are modeling nodes, i.e., data mining algorithms that can be used to identify patterns in data. The last group of square-shaped nodes is related to the output functions, which make available a set of nodes for reporting, storing or exporting data.

The Data Analysis component of PADRÃO is based on the construction of several streams that implement the knowledge discovery process. The several models obtained in the data mining phase represent knowledge about the analyzed data and can be saved or reused in other streams. In PADRÃO, these models can be exported through an ODBC connection to the PDB. The integration of the PDB with the CDB allows the visualization of the rules explicit in the models in a map. The visualization is achieved through the VisualPadrão application, a module implemented in Visual Basic. VisualPadrão manipulates the library of objects available in Geomedia. This application was integrated in the Clementine workspace using a *specification file*, i.e., a mechanism provided by the Clementine system that allows for the integration of new capabilities in its environment. This approach provides an integrated workspace in which all tasks associated with the knowledge discovery process can be executed.

Analysis of a Geo-Referenced Database

Several datasets have been analyzed by the PADRÃO system. Among them are demographic databases storing the Parish records of several Municipalities of Portugal (Santos & Amaral, 2000a; Santos & Amaral, 2000b; Santos & Amaral, 2000c). Another dataset analyzed was a component of the Portuguese Army Database (Santos & Ramos, 2003). The several data mining objectives defined allowed for the identification of the implicit relationships existing between the geo-spatial data and non-spatial data.

The dataset selected for description in this chapter integrates data from a financial institution, which supplies credit for the acquisition of several types of goods. To overcome confidentiality issues with the data and the several identifiable patterns, the data was manipulated in order to create a random data set. Through this process the confidentially is ensured and the knowledge discovery process in the PADRÃO system can be described.

Figure 21. Data Exploration

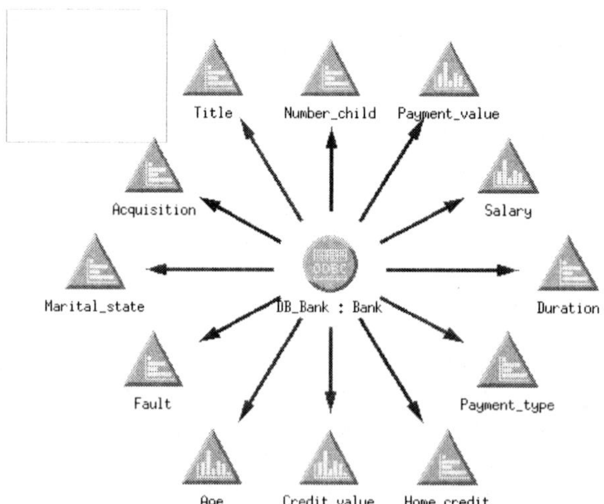

Figure 22. Distribution of Categorical Data

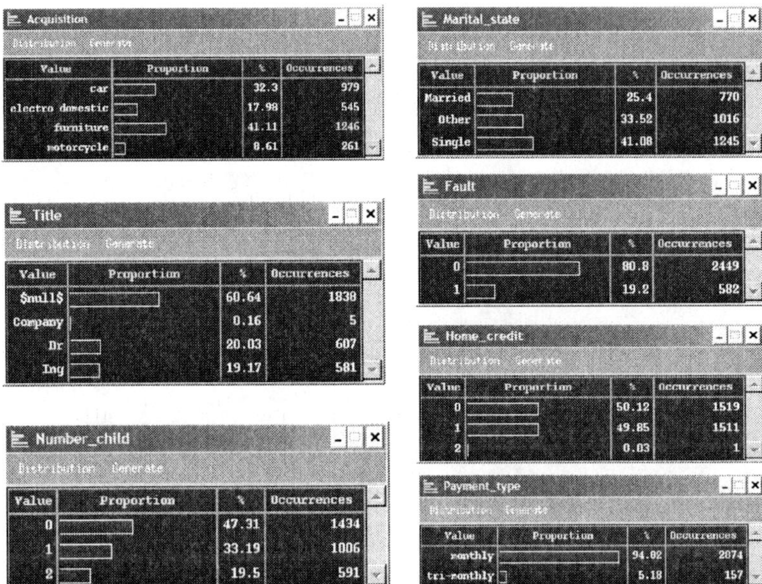

The bank database aggregates a set of 3,031 records that characterize the behavior of the bank clients. The following data mining objective was defined: "*identify the profile of the clients in order to minimize the institutional risk of investment.*" This profile will be identified for the Braga District of Portugal.

The knowledge discovery process is preceded by the business understanding phase in which the meaning and importance of each attribute for this process is evaluated. The attributes integrated in the database are: identification number (ID), VAT number (VAT_number), client title (Title), name (Name), good purchased (Acquisition), contract duration (Duration), income (Salary), overall value of credit (Credit_value), payment type (Payment_type), credit for home acquisition (Home_credit), lending value (Payment_value), marital state (Marital_state), number of children (Number_child), age (Age) and the accomplishment or not of the credit (Fault).

At this phase Distribution and Histogram nodes of Clementine were used to explore the several attributes, identify their values, and distribution, and determine if any of them present anomalies. *Figure 21* shows the stream constructed for this exploration phase.

The results obtained by each Distribution[13] graph are showed in *Figure 22*. It can be seen that the majority of attributes present a distribution of values that are the normal operation of the organization. However, exceptions were verified for the Home_credit and Title attributes. Namely:

- The attribute Home_credit, which shows if the client has or does not have a credit for home acquisition (values 1 and 0 respectively), also includes a record with the 2 value. As this value constitute an error, the respective record must be removed from the dataset;
- The Title attribute integrates five cases of credit for organizations (value Company). As a result, these records must be removed from the database since they represent a minority class[14] in the overall set.

Figure 23 shows the Histograms with the distribution of attributes with continuous values. The analysis of the distributions allows for the verifica-

Figure 23. Distribution of Continuous Values

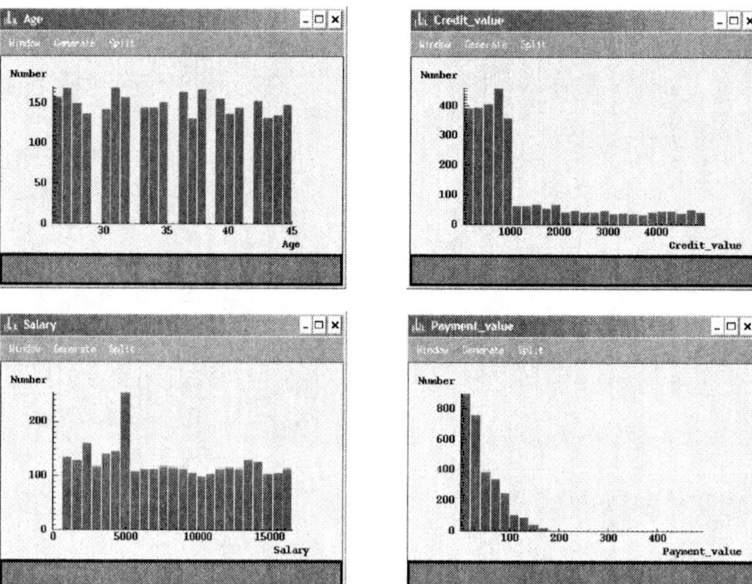

Figure 24. Data Selection and Data Treatment Steps

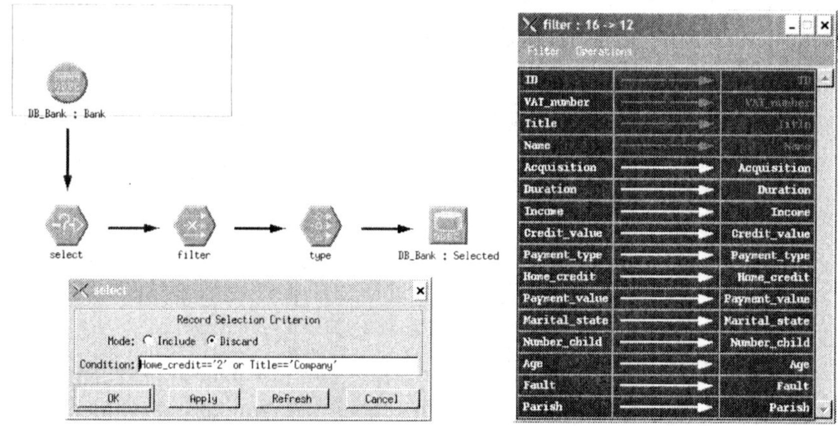

tion of the several classes that will be created in order to transform continuous values into discreet values. The defined classes are presented in *Table 5*. Their definition is based on the assumption that the data available for analysis must be distributed homogeneously across the several classes.

This exercise of exploration and comprehension of the available data allowed the identification of the attributes for analysis and the definition of the several classes that will be used in the pre-processing step, i.e., to transform continuous values into discreet values. Next, the six steps considered in the PADRÃO system for the knowledge discovery process (Data Selection, Data Treatment, Data Pre-processing, Geo-spatial Information Processing, Data Mining and Interpretation of Results) are described.

Table 5. Classes for Attributes with Continuous Values

Attributes	Classes
Age	(25..31] → '**26-31**', (31..38] → '**32-38**', (38..45] → '**39-45**'
Credit_value	(0..350] → '**0-350**', (350..650] → '**351-650**', (650..900] → '**651-900**', (900..2500] → '**901-2500**', (2500..5000] → '**2501-5000**'
Salary	(0..4500] → '**0-4500**', (4500..8000] → '**4501-8000**', (8000..12500] → '**8001-12500**', (12500..17000] → '**12501-17000**'
Payment_value	(0..17] → '**0-17**', (17..30] → '**18-30**', (30..50] → '**31-50**', (50..80] → '**51-80**', (80..500] → '**81-500**'

Data Selection and Data Treatment

The data selection step allows for the exclusion of attributes that have no influence in the knowledge discovery process. Among them are ID, VAT_number, Title and Name, since they only have an informative role. The other attributes will be considered in order to evaluate the contribution of each one to the definition of the profile of the clients.

Figure 24 shows the stream constructed for the data selection and data treatment steps. The stream integrates a source node (DB_Bank:Bank) that makes the data available to the knowledge discovery process through an ODBC connection. The select node discards records with anomalies. As previously mentioned, the record with the value 2 in the Home_credit attribute must be deleted. All records associated with the value Company in the Title attribute also need to be removed. The filter node is used to select the attributes that will be excluded from the process. The type node allows for the specification of the data type (numeric, character...) of the attributes that will be exported to the database. As result of the several tasks undertaken, a new table (DB_Bank:SelectedData) is created in the bank database.

Data Pre-Processing

The data pre-processing step (*Figure 25*) allows for the transformation of the attributes with continuous values into attributes with discreet values (nodes SalaryClass, CreditClass, PaymentClass and AgeClass), according to the classes presented in *Table 5*. In this step, web nodes, exploration graphs available in Clementine, are also used for the identification of associations[15] among the analyzed attributes (nodes Acquisition x Fault, SalaryClass x AgeClass x Fault, Marital_state x Number_child x Fault and PaymentClass x Fault). The last task undertaken is associated with the creation of the two datasets (nodes DB_Bank:Training and DB_Bank:Test) that will be used from now on. They are the Training and the Test datasets, and in which the original data is randomly distributed. The Training file is used in the model construction (data mining step) while the Test dataset evaluates the model confidence when applied to unknown data.

The web nodes constructed are shown in *Figure 26*. They combine several attributes and through the analysis of them it is possible to identify associations between attributes. Strong associations between attributes are represented by bold lines, while weak associations are symbolized by dotted lines. For the several acquisitions that

Figure 25. Data Pre-Processing Step

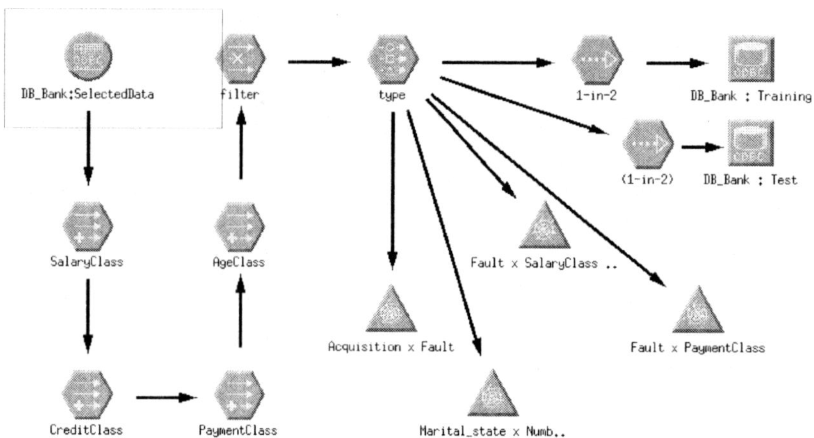

Figure 26. Data Exploration with Web Nodes

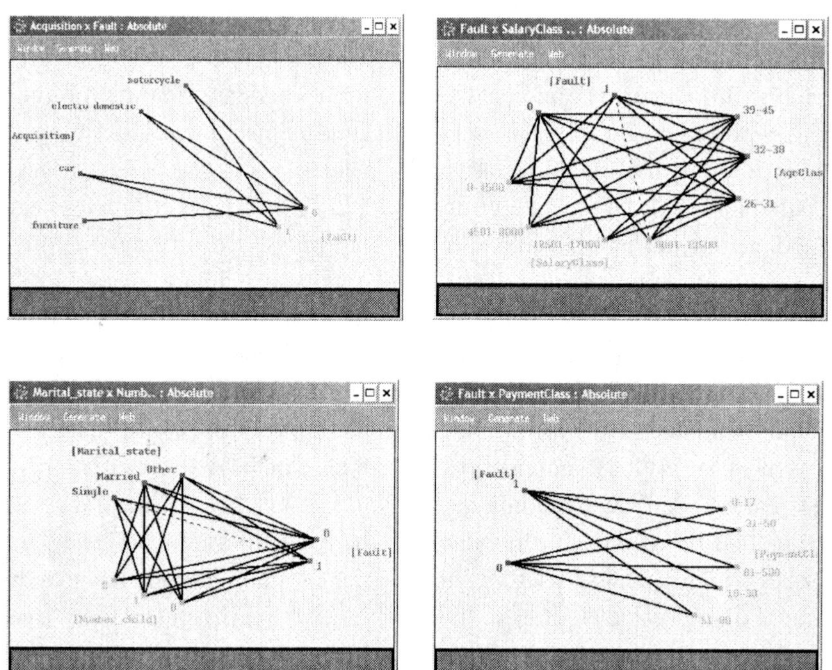

can be effected, *Figure 26a* points out that no association exists between the good furniture and the value 1 of the Fault attribute, indicating that faults were not usual with the credit supplied for this specific acquisition. Analyzing the income and age attributes with Fault in *Figure 26b*, it is evident that individuals with a higher income honor their payments, since the value 12501-17000 of the SalaryClass attribute presents no association with value 1 of Fault. Between value 8001-12500 and value 1 of Fault there exists a weak association, which indicates that this specific group may or may not be able to honor its credit payments. Similarly, a weak association is verified between the marital state Single and value 1 of Fault, *Figure 26c.* PaymentClass and Fault present strong connections between all attribute values as seen in *Figure 26d*, thus indicating that all type of payment values are associated with *good* and *bad* clients.

Geo-Spatial Information Processing

As the GDB only stores spatial relations for adjacent regions and, as it is necessary to analyze if the geographical component has any influence in the identification of the profile of the clients, all the other relationships that exist between non-adjacent regions and needed in the data mining step will be inferred. In Clementine, a rule induction[16] algorithm is able to learn the inference rules available in the composition table stored in the SKB. That enables the inference of new spatial relations.

The models created, nodes infDir, infDis and infTop, can now be used in the inference process. With these models and as shown in *Figure 27* it is possible to infer the unknown spatial relationships existing in the Municipalities of the Braga District. The spatial relations for adjacent regions stored in the GBD are gathered through the source node (GDB:geoBraga) of the stream and combined (node Inflection) in order to obtain new associations between regions. The spatial relations existing among these new associations are identified by the models infDir, infDis and infTop. After the inferential process, the knowledge obtained is recorded in the GDB (output node GDB:geoBraga). In the stream of *Figure 27*, the super nodes SuperNodeDir1 and SuperNodeDir3 are responsible for the integration of the dimension of the regions in the reasoning process. In this process, there is validation if the several inferences obtained for a particular region agree independently of the composed regions. Several paths can be followed in order to infer a specific spatial relation. For example, knowing the facts A North B, A East D, B East C and D North C, the direction relation existing between A and C may be obtained composing A North B with B East C or combining A East D with D North C. If several compositions can be effected and if the results obtained from each one do not match, then the super node VerInferences excludes those results from the set of accepted ones.

Data Mining

In the data mining step (*Figure 28*) an appropriate algorithm is selected to carry out a specific data mining task. Three different tasks were undertaken (see *Figure 28*). First, a decision

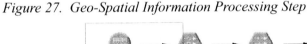

Figure 27. Geo-Spatial Information Processing Step

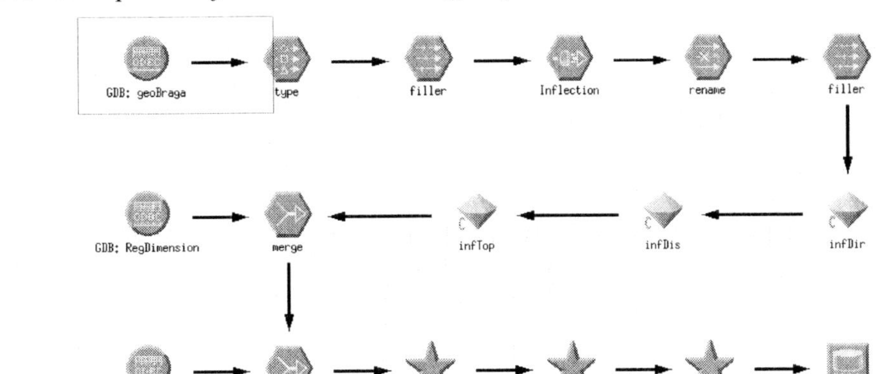

Figure 28. Data Mining Step

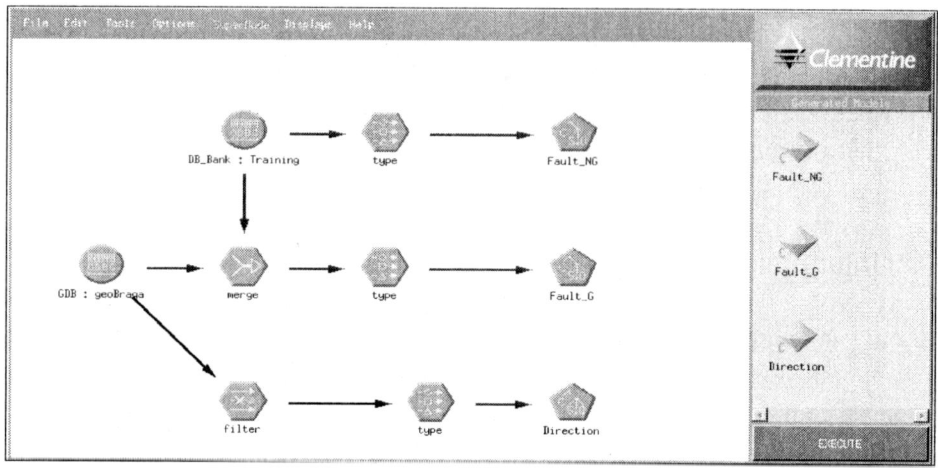

Figure 29. Generated Models for the Profile of the Clients

```
workbuff Rule browser 1 for fault_ng
File  Folding  Select  Generate  View

Acquisition [car furniture] -> 0
Acquisition ['electro domestic' motorcycle]
    Marital_state Single -> 0
    Marital_state [Married Other]
        SalaryClass [0-4500 4501-8000] -> 1
        SalaryClass [12501-17000 8001-12500]
            Marital_state Married
                Acquisition electro domestic -> 0
                Acquisition motorcycle
                    CreditClass 351-650 -> 1
                    CreditClass [0-350 2501-5000 651-900] -> 0
                    CreditClass 901-2500
                        Home_credit 0 -> 1
                        Home_credit 1 -> 0
            Marital_state Other
                Acquisition motorcycle -> 1
                Acquisition electro domestic
                    SalaryClass 12501-17000 -> 0
                    SalaryClass 8001-12500
                        PaymentClass [0-17 31-50 81-500] -> 1
                        PaymentClass [18-30 51-80]
                            Number_child [0 2] -> 0
                            Number_child 1 -> 1
```

```
workbuff Rule browser 2 for fault_g
File  Folding  Select  Generate  View

Acquisition car -> 0
Acquisition furniture -> 0
Acquisition motorcycle
    Marital_state Married -> 1
    Marital_state Other -> 1
    Marital_state Single -> 0
Acquisition electro domestic
    Marital_state Other -> 1
    Marital_state Single -> 0
    Marital_state Married
        direction E -> 0
        direction N -> 1
        direction NE -> 0
        direction NW -> 1
        direction W -> 1
        direction S -> 0
        direction SE -> 1
        direction SW -> 1
```

tree (node Fault_NG) that characterizes the profile of the clients without considering the location of the clients was generated. Second, the training set (DB_Bank:Training) was integrated with the spatial relations for the District in analysis (GDB:GeoBraga) in order to include the geographical component in the analysis of the profile of the clients (node Fault_G). Third, the geographical model of the District was created. This latter model (Direction) indicates the direction of each Municipality in the District and was obtained by analyzing the spatial relations inferred in the geo-spatial information processing step. All models were obtained with the C5.0 algorithm that allows for the induction of decision trees. *Figure 28* highlights the stream constructed for the generation of the three models. These models are available in the Generated Models palette and have the shape of a diamond (right hand side of *Figure 28*).

The Fault_NG model (*Figure 29*, left side) integrates a set of rules that are represented in a decision tree, which characterizes the profile of the clients. Through the analysis of the model it is possible to verify that the acquisition of car and furniture is traditionally associated with clients that honor their payments, while the acquisition of electro domestic and motorcycle have other attributes (Marital_state, Salary_class ...) that influence the profile of the clients. One explicit rule in the model for clients that the institution has no interest in supplying with credit, is: IF SalaryClass = '12501-17000' and Marital_state= 'Married' and Acquisition = 'motorcycle' and CreditClass = '351-600' THEN 1. The Fault_G model (*Figure 29*, right hand side) allows for the verification of geographic zones that have associated clients with a higher incidence of faults in credit payments. These zones are represented in directions, which partition the District into eight areas. The analysis of the model points out that Northeast (NE), East (E) and South (S) are associated with clients that pay the credit assumed.

Interpretation of Results

The Test set (DB_Bank:Test) is used in the interpretation of results step to verify the confidence of the models built in the Data Mining step. With respect to the Fault_NG model, *Figure 30* shows a percentage of confidence of 94.18%. The Fault_G model presents a percentage of confidence of 93.26%. This decrease in the model confidence, when considering the geographical component, may be caused by the aggregation of Municipalities into eight regions (the Cardinal directions), which represents a loss of specificity in favor of generality. Although the Direction model was obtained through the analysis of spatial relations inferred by qualitative rules, the results obtained in the Fault_G model maintain a high level of confidence.

The PADRÃO system permits the visualization of the results of the knowledge discovery process on a map. In this system the several rules that integrate a model are recorded in the PDB (Santos & Amaral, 2000b). At the same time the user has the option to run the VisualPadrão tool and visualize the desired model (*Figure 31*). As can be noted in the figure, Municipalities located at Northeast, East and South of the District contain clients mainly associated with no faults in their credit payments (information explicit in the Fault_G model obtained in the data mining step). This geographic characterization enabled the identification of regions where the relative incidence of clients with faults is higher than elsewhere in the District.

Risk zones were identified, aggregating together regions that have clients with similar behavior. The geographic segments can be cataloged by the bank, looking at similarities like proximity with other regions, population density, population qualification and other relevant issues.

Figure 30. Percentage of Confidence of the Generated Models

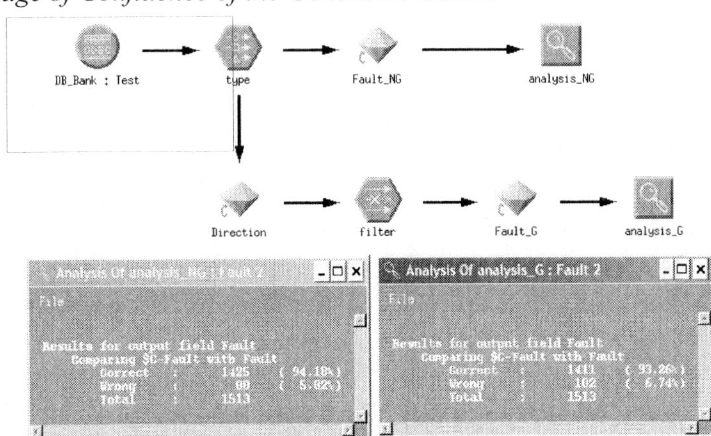

Figure 31. Visualization of Results

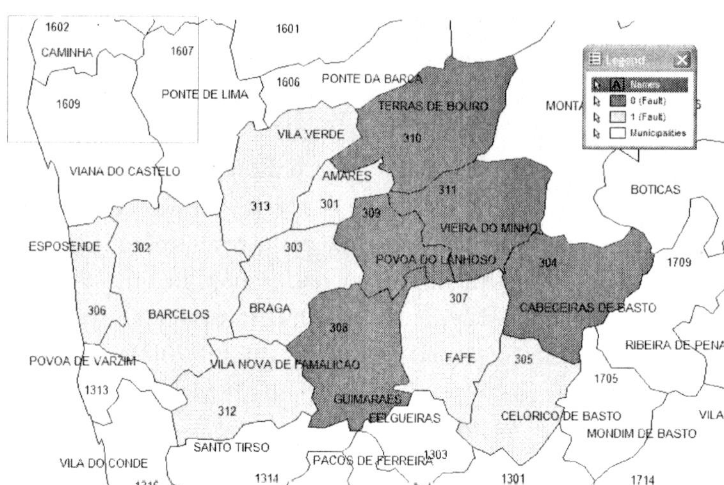

Figure 32. Classification of New Clients by the Model

table (10 records)

Acquisition	Home_credit	Marital_state	Number_child	SalaryClass	CreditClass	PaymentClass	$C-Fault	$CC-Fault
furniture	1	Single	0	8001-12500	351-650	51-80	0	0.94
motorcycle	1	Other	1	4501-8000	901-2500	81-500	1	0.981
car	0	Single	1	12501-17000	2501-5000	51-80	0	0.94
motorcycle	1	Married	0	8001-12500	651-900	18-30	0	1.0
furniture	1	Other	1	4501-8000	0-350	0-17	0	0.94
car	0	Other	2	4501-8000	2501-5000	51-80	0	0.94
car	0	Other	1	8001-12500	2501-5000	81-500	0	0.94
electro domestic	1	Other	1	0-4500	651-900	18-30	1	0.981
electro domestic	0	Other	1	4501-8000	351-650	31-50	1	0.981
furniture	0	Married	0	8001-12500	351-650	18-30	0	0.94

The models obtained in the data mining step define the profile of the bank clients. They integrate the attributes and the corresponding values related to the classification of the clients bearing in mind the risk of investment in specific classes of clients. For the available segments, the several rules identified can support managers in the decision-making process. In the granting of new credit, the organization is now supported by models that track the previous behavior of its clients, indicating groups of clients in which the organization has to pay more attention in the granting of credit and those groups without difficulties in the assignment of credit.

Suppose that 10 new potential clients request credit to the organization. *Figure 32* shows the relevant data on each client and the classification (column $C-Fault) of the model according to the rules explicit in it. The column $CC-Fault indicates the confidence of the classification, which is equal or superior to 94%. Looking at the classification achieved, for seven clients the decision of the model is 0 in the $C-Fault attribute, which means that, based on the past experience of the organization, these are *good* clients. For 3 clients the result was 1 in the $C-Fault attribute, labeling these clients as *risk* clients and suggesting that a more detailed analysis must be undertaken in order to identify the appropriate decision (grant credit or not).

The use of predictive models assumes that the past is a good predictor of the future. However, there are situations where the past may not be a good predictor, if the facts occurred were influenced by external events not present in the analyzed data (Berry & Linoff, 2000). For this reason, in making predictions, the organization cannot only be supported by the models obtained from the knowledge discovery process, in order to avoid penalizing new potential *good* clients as a result of the behavior of past clients. The models obtained should be seen as tools that support the decision-making process, not as the decision-maker.

The knowledge discovery process should support the creation of organizational knowledge through the incorporation of the information expressed in the several models in its daily activities. This procedure will contribute to fulfill the information requirements of the bank and help in the accomplishment and improvement of its mission.

CONCLUSION

This chapter presented an approach for knowledge discovery in geo-referenced databases based on qualitative spatial reasoning, where the position of geographical data was provided by qualitative identifiers.

Direction, distance and topological spatial relations were defined for a set of Municipalities of Portugal. This knowledge and the composition table constructed for integrated spatial reasoning enabled the inference of new spatial relations analyzed in the data mining step of the knowledge discovery process.

The integration of a bank database with the GDB (storing the administrative subdivisions of Portugal) made possible the discovery of general descriptions that exploit the relationships that exist between the geo-spatial and non-spatial data analyzed. The models obtained in the data mining step define the profile of the clients, bearing in mind the risk of investment of the organization for specific segments of clients. For the available classes, the several rules identified support the managers of the organization in the decision-making process. The latter represents one of the organizational processes that can benefit from data mining technology through the incorporation of its results in the evaluation of critical and uncertain situations.

The results obtained with the PADRÃO system support that traditional KDD systems, which were developed for the analysis of relational databases and that do not have semantic knowledge linked to spatial data, can be used in the process of knowledge discovery in geo-referenced databases, since some of this semantic knowledge and the principles of qualitative spatial reasoning are available as domain knowledge. Clementine, a KDD system, was used in the assimilation of the geographic domain knowledge such as composition tables, in the inference of new spatial relations, and in the discovery of spatial patterns.

The main advantages of the proposed approach, for mining geo-referenced databases, include the use of already existing data mining algorithms developed for the analysis of non-spatial data; an avoidance of the geometric characterization of spatial objects for the knowledge discovery process; and the ability of data mining algorithms to deal with geo-spatial and non-spatial data simultaneously, thus imposing no limits and constraints on the results achieved.

ACKNOWLEDGMENTS

We thank NTech – Sistemas de Informação, Lda. for making the database available for analysis. We thank Tony Lavender for his help in improving the English writing of this chapter.

REFERENCES

Abdelmoty, A. I., & El-Geresy, B. A. (1995). A general method for spatial reasoning in spatial databases. *Proceedings of the Fourth International Conference on Information and Knowledge Management* (pp. 312-317). Baltimore, Maryland.

Allen, J. F. (1983). Maintaining knowledge about temporal intervals. *Communications of the ACM, 26* (11), 832-843.

Berry, M., & Linoff, G. (2000). *Mastering data mining: The art and science of customer relationship management.* New York: John Wiley & Sons.

CEN/TC-287. (1996). *Geographic information: Data description, spatial schema* (prENV 12160). European Committee for Standardization.

CEN/TC-287. (1998). *Geographic information: Referencing, geographic identifiers* (prENV 12661). European Committee for Standardization.

Egenhofer, M. J. (1994). Deriving the composition of binary topological relations. *Journal of Visual Languages and Computing, 5* (2), 133-149.

Ester, M., Frommelt, A., Kriegel, H.-P., & Sander, J. (1998). Algorithms for characterization and trend detection in spatial databases. *Proceedings of the Fourth International Conference on Knowledge Discovery and Data Mining.* AAAI Press.

Ester, M., Kriegel, H.-P., & Sander, J. (1997). Spatial data mining: A database approach. *Proceedings of the Fifth International Symposium on Large Spatial Databases* (pp. 47-68). Germany.

Fayyad, U., & Uthurusamy, R. (1996). Data mining and knowledge discovery in databases. *Communications of the ACM, 39* (11), 24-26.

Fayyad, U. M., Piatetsky-Shapiro, G., Smyth, P., & Uthurusamy, R. (eds.). (1996). *Advances in knowledge discovery and data mining.* MA: The MIT Press.

Frank, A. U. (1992). Qualitative spatial reasoning about distances and directions in geographic space. *Journal of Visual Languages and Computing, 3*, 343-371.

Frank, A. U. (1996). Qualitative spatial reasoning: Cardinal directions as an example. *International Journal of Geographical Information Systems, 10* (3), 269-290.

Freksa, C. (1992). Using orientation information for qualitative spatial reasoning. In A. U. Frank, I. Campari, & U. Formentini (Eds.), *Theories and methods of spatio-temporal reasoning in geographic space* (Lectures Notes in Computer Science 639). Berlin: Springer-Verlag.

Han, J., & Kamber, M. (2001). *Data mining: Concepts and techniques.* CA: Morgan Kaufmann Publishers.

Han, J., Tung, A., & He, J. (2001). SPARC: Spatial association rule-based classification. In R. Grossman, C. Kamath, P. Kegelmeyer, V. Kumar & R. Namburu (Eds.), *Data mining for scientific and engineering applications* (pp. 461-485). Kluwer Academic Publishers.

Hernández, D., Clementini, E., & Felice, P. D. (1995). Qualitative distances. *Proceedings of the International Conference COSIT'95* (pp. 45-57). Austria.

Hong, J.-H. (1994). *Qualitative distance and direction reasoning in geographic space.* Unpublished doctoral dissertation, University of Maine, Maine.

Intergraph. (1999). *Geomedia professional v3* (Reference Manual). Intergraph Corporation.

Koperski, K., & Han, J. (1995). Discovery of spatial association rules in geographic information systems. *Proceedings of the 4th International Symposium on Large Spatial Databases* (pp. 47-66). Maine.

Lu, W., Han, J., & Ooi, B. (1993). Discovery of general knowledge in large spatial databases. *Proceedings of the 1993 Far East Workshop on Geographic Information Systems* (pp. 275-289). Singapore.

Papadias, D., & Sellis, T. (1994). On the qualitative representation of spatial knowledge in 2D space. *Very Large Databases Journal, Special Issue on Spatial Databases, 3* (4), 479-516.

Santos, M., & Amaral, L. (2000a). Knowledge discovery in spatial databases through qualitative spatial reasoning. *Proceedings of the 4th International Conference and Exhibition on Practical Applications of Knowledge Discovery and Data Mining* (pp. 73-88). Manchester.

Santos, M., & Amaral, L. (2000b, November). Knowledge discovery in spatial databases: The PADRÃO's qualitative approach. *Cities and Regions, GIS special issue,* 33-49.

Santos, M., & Amaral, L. (2000c). A qualitative spatial reasoning approach in knowledge discovery in spatial databases. *Proceedings of Data Mining 2000: Data Mining Methods and Databases for Engineering, Finance and Others Fields* (pp. 249-258). Cambridge.

Santos, M. Y. (2001). *PADRÃO: Um sistema de descoberta de conhecimento em bases de dados geo-referenciadas (in Portuguese)*. Unpublished doctoral dissertation, Universidade do Minho, Portugal.

Santos, M. Y., & Ramos, I. (2003). Knowledge construction: The role of data mining tools. *Proceedings of the UKAIS 2003 Conference "Co-ordination and Co-operation: the IS role".* Warwick.

Sharma, J. (1996). *Integrated spatial reasoning in geographic information systems: Combining topology and direction.* Unpublished doctoral dissertation, University of Maine, USA.

SPSS. (1999). *Clementine* (user guide, version 5.2). SPSS Inc.

ENDNOTES

1. GISs allow for the storage of geographic information and enable users to request information about geographic phenomena. If the requested spatial relation is not explicitly stored in databases, it must be inferred from the information available. The inference process requires searching relations that can form an *inference path* between the two objects where the relation is requested (Hong, 1994). The composition operation combines two contiguous paths in order to infer a third spatial relation. A composition table integrates a set of inference rules used to identify the result of a specific composition operation.
2. Extended objects are not point-like, so represent objects for which their dimension is relevant (Frank, 1996). In this work, extended objects are geometrically represented by a polygon, indicating that their position and extension in space are relevant.
3. In IR^2, there are eight topological relations between two planar regions without holes (two-dimensional, connected objects with connected boundaries); 18 topological relations between spatial regions with holes; 33 between two simple lines and 19 between a spatial region without holes and a simple line (Egenhofer, 1994).
4. The topology of a full planar graph refers to a planar graph that integrates regions completely covering the plane without any gap or overlap. Regions are topologically

represented by faces, which are defined without holes (CEN/TC-287, 1996).

5 Defining distances between regions is a complex task, since the size of each object plays an important role in determining the possible distances. Sharma (1996) gives the following ways to define distances between regions: (i) taking the distance between the *centroids* of the two regions; (ii) determining the shortest distance between the two regions; or (iii) determining the furthest distance between the two regions.

6 Other validity intervals, for different ratios, can by found in Hong (1994).

7 The symbol used to represent the composition operation is ";".

8 Since the system will be used with administrative subdivisions, the orientation between the several regions is calculated according to the position of the respective *centroids*.

9 The dotted lines define the acceptance area defined for the North direction (designed from the *centroid* of B), while the whole lines represent the acceptance area defined for the Northeast direction (designed from the *centroid* of C).

10 In this work, an inference is considered exact if the result achieved with the correspondent qualitative rule is the same as if the data was translated to quantitative information and manipulated through analytical functions. Otherwise, it is considered approximate.

11 The geometry is not required in the knowledge discovery process, since the manipulation of the geographic information is undertaken by a qualitative approach.

12 Visual programming involves placing and manipulating icons representing processing nodes.

13 Distribution nodes are used for the analysis of categorical data.

14 The data mining algorithms may be negatively influenced by classes with a great number of values.

15 The several associations identified anticipate the importance of each attribute in the definition of the profile of the clients.

16 A rule induction algorithm creates a decision tree aggregating a set of rules for classifying the data into different outcomes. This technique only includes in the rules the attributes that really matter in the decision-making process.

This work was previously published in Geographic Information Systems in Business, edited by J. Pick, pp. 113-150, copyright 2005 by IGI Publishing, formerly known as Idea Group Publishing (an imprint of IGI Global).

Chapter 2.23
Ontology-Based Construction of Grid Data Mining Workflows

Peter Brezany
University of Vienna, Austria

Ivan Janciak
University of Vienna, Austria

A.Min Tjoa
Vienna University of Technology, Austria

ABSTRACT

This chapter introduces an ontology-based framework for automated construction of complex interactive data mining workflows as a means of improving productivity of Grid-enabled data exploration systems. The authors first characterize existing manual and automated workflow composition approaches and then present their solution called GridMiner Assistant (GMA), which addresses the whole life cycle of the knowledge discovery process. GMA is specified in the OWL language and is being developed around a novel data mining ontology, which is based on concepts of industry standards like the predictive model markup language, cross industry standard process for data mining, and Java data mining API. The ontology introduces basic data mining concepts like data mining elements, tasks, services, and so forth. In addition, conceptual and implementation architectures of the framework are presented and its application to an example taken from the medical domain is illustrated. The authors hope that the further research and development of this framework can lead to productivity improvements, which can have significant impact on many real-life spheres. For example, it can be a crucial factor in achievement of scientific discoveries, optimal treatment of patients, productive decision making, cutting costs, and so forth.

INTRODUCTION

Grid computing is emerging as a key enabling infrastructure for a wide range of disciplines in science and engineering. Some of the hot topics in current Grid research include the issues associated with data mining and other analytical processes performed on large-scale data repositories integrated into the Grid. These processes are not implemented as monolithic codes. Instead, the standalone processing phases, implemented as Grid services, are combined to process data and extract knowledge patterns in various ways. They can now be viewed as complex workflows, which are highly interactive and may involve several subprocesses, such as data cleaning, data integration, data selection, modeling (applying a data mining algorithm), and postprocessing the mining results (e.g., visualization). The targeted workflows are often large, both in terms of the number of tasks in a given workflow and in terms of the total execution time. There are many possible choices concerning each process's functionality and parameters as well as the ways a process is combined into the workflow but only some combinations are valid. Moreover, users need to discover Grid resources and analytical services manually and schedule these services directly on the Grid resources essentially composing detailed workflow descriptions by hand. At present, only such a "low-productivity" working model is available to the users of the first generation data mining Grids, like **GridMiner** (Brezany et al., 2004) (a system developed by our research group), DiscoveryNet (Sairafi et al., 2003), and so forth. Productivity improvements can have significant impact on many real-life spheres, for example, it can be a crucial factor in achievement of scientific discoveries, optimal treatment of patients, productive decision making, cutting costs, and so forth. There is a stringent need for automatic or semiautomatic support for constructing valid and efficient data mining workflows on the Grid, and this (long-term) goal is associated with many research challenges.

The objective of this chapter is to present an ontology-based workflow construction framework reflecting the whole life cycle of the knowledge discovery process and explain the scientific rationale behind its design. We first introduce possible workflow composition approaches—we consider two main classes: (1) manual composition used by the current Grid data mining systems, for example, the GridMiner system, and (2) automated composition, which is addressed by our research and presented in this chapter. Then we relate these approaches to the work of others. The kernel part presents the whole framework built-up around a data mining ontology developed by us. This ontology is based on concepts reflecting the terms of several standards, namely, the predictive model markup language, cross industry process for data mining, and Java data mining API. The ontology is specified by means of OWL-S, a Web ontology language for services, and uses some concepts from Weka, a popular open source data mining toolkit. Further, conceptual and implementation architectures of the framework are discussed and illustrated by an application example taken from a medical domain. Based on the analysis of future and emerging trends and associated challenges, we discuss some future research directions followed by brief conclusions.

BACKGROUND

In the context of modern service-oriented Grid architectures, the data mining workflow can be seen as a collection of Grid services that are processed on distributed resources in a well-defined order to accomplish a larger and sophisticated data exploration goal. At the highest level, functions of Grid workflow management systems could be characterized into build-time functions and run-time functions. The build-time functions are concerned with defining and modeling workflow tasks and their dependencies while the run-time functions are concerned with managing the workflow execution and interactions with Grid resources for processing workflow applications. Users interact with workflow modeling tools to generate a workflow specification, which is submitted for execution to a run-time service called workflow enactment service, or *workflow engine*. Many languages, mostly based on XML, were defined for workflow description, like XLANG (Thatte, 2001), WSFL (Leymann, 2001), DSCL (Kickinger et al., 2003) and BPML (Arkin, 2002). Eventually the WSBPEL (Arkin et al., 2005) and BPEL4WS (BEA et al., 2003) specifications emerged as the de facto standard.

In our research, we consider two main workflow composition models: *manual* (implemented in the fully functional GridMiner prototype (Kickinger et al., 2003)) and *automated* (addressed in this chapter), as illustrated in Figure 1. Within manual composition, the user constructs the target workflow specification graphically in the workflow editor by means of the advanced graphical user interface. The graphical form is converted into a *workflow description* document, which is passed to the workflow engine. Based on the workflow description, the engine sequentially or in parallel calls the appropriate analytical services (database access, preprocessing, OLAP, classification, clustering, etc.). During the workflow execution, the user only has the ability to stop, inspect, resume, or cancel the execution. As a result, the user has limited abilities to interact with the workflow and influence the execution process. A similar approach was implemented in the DiscoveryNet (Sairafi et al., 2003) workflow management system.

The automated composition is based on an intensive support of five involved components: *workflow composer*, *resources monitoring*, *workflow engine*, *knowledge base*, and *reasoner*.

Workflow composer: Is a specialized tool, which interacts with a user during the workflow composition process. This chapter describes its functionality in detail.

Resources monitoring: Its main purpose is obtaining information concerning the utilization of system resources. Varieties of different systems exist for monitoring and managing distributed Grid-based resources and applications. For exam-

Figure 1. Workflow composition approaches

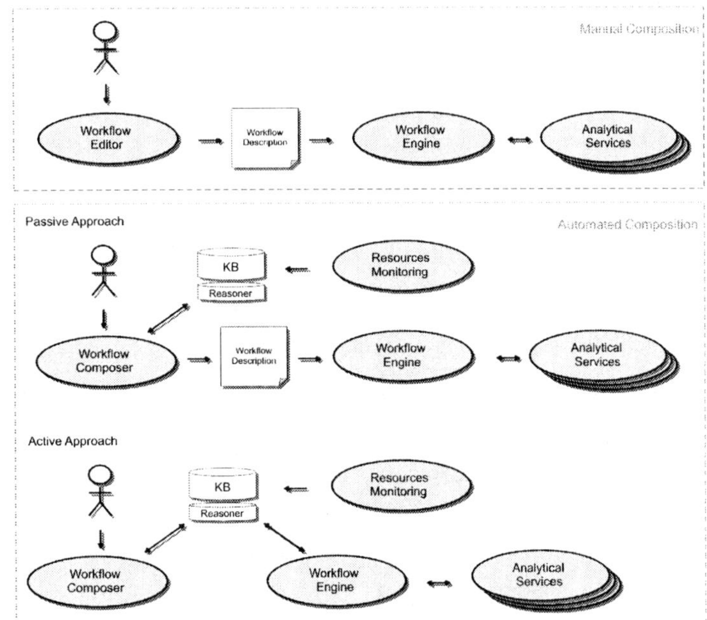

ple, the monitoring and discovery system (MDS) is the information services component of the Globus Toolkit (Globus Alliance, 2005), which provides information about available resources on the Grid and their status. Moreover, MDS facilitates the discovery and characterization of resources and monitors services and computations. The information provided by resource monitoring can be continuously updated in the knowledge base (KB) to reflect the current status of the Grid resources.

Workflow engine: Is a runtime execution environment that performs the coordination of services as specified in the workflow description expressed in terms of a workflow language. The workflow engine is able to invoke and orchestrate the services and acts as their client, that is, listen to the notification messages, deliver outputs, and so forth.

Knowledge base (KB) and **reasoner:** A set of ontologies can be used for the specification of the KB structure, which is built-up using a set of instances of ontology classes and rules. The reasoner applies deductive reasoning about the stored knowledge in a logically consistent manner; it assures consistency of the ontology and answers given queries.

Due to different roles and behaviors of the presented components, we distinguish two modes of automated workflow composition: *passive* and *active*.

Passive Workflow Construction

The passive approach is based on the assumption that the *workflow composer* is able to compose a reasoning-based complete workflow description involving all possible scenarios of the workflow engine behavior and reflecting the status of the involved Grid resources and task parameters provided by the user at the workflow composition time. Although the KB is continuously modified by the user's entries and by information retrieved from the resource monitoring services, the composition of involved services is not updated during the workflow execution. Therefore, the composition does not reflect the 'state of the world,' which can be dynamically changed during the execution. It means that the workflow engine does not interact with the inference engine to reason about knowledge in the KB. Thus, the behavior of the engine (the decisions it takes) is steered by fixed condition statements as specified in the workflow document.

The essential tasks leading to a final outcome of the passive workflow composition approach can be summarized as follows:

1. The workflow composer constructs a complete workflow description based on the information collected in KB and presents it to the workflow engine in an appropriate workflow language.
2. The workflow engine executes each subsequent composition step as presented in the workflow description, which includes all possible scenarios of the engine behavior.

Active Workflow Construction

The active approach assumes a kind of intelligent behavior by the workflow engine supported by an inference engine and the related KB. Workflow composition is done in the same way as in the passive approach, but its usability is more efficient because it reflects a 'state of the world.' It means that the outputs and effects of the executed services are propagated to the KB together with changes of the involved Grid resources. Considering these changes, the workflow engine dynamically makes decisions about next execution steps. In this approach, no workflow document is needed because the workflow engine instructs itself using an inference engine which queries and updates the KB. The KB is queried each time the workflow engine needs information to invoke a consequent service, for example, it decides which concrete service should be executed, discovers the values of its input parameters in the KB, and so forth. The workflow engine also updates the KB when there is a new result returned from an analytical service that can be reused as input for the other services.

The essential tasks leading to a final outcome in active workflow composition approach can be summarized as follows:

1. The workflow composer constructs an abstract workflow description based on the information collected in the KB and propagates the workflow description back into the KB. The abstract workflow is not a detailed description of the particular steps in the workflow execution but instead a kind of path that leads to the demanded outcome.
2. The workflow engine executes each subsequent composition step as a result of its interaction with the KB reflecting its actual state. The workflow engine autonomously constructs directives for each service execution and adapts its behavior during the execution.

Related Work

A main focus of our work presented in this chapter is on the above mentioned passive approach of the automated workflow composition. This research was partially motivated by (Bernstein et al., 2001). They developed an intelligent discovery assistant (IDA), which provides users (data miners) with (1) systematic enumerations of valid data mining processes according to the constraints imposed by the users' inputs, the data, and/or the data mining ontology in order that important and potentially fruitful options are not overlooked, and (2) effective rankings of these valid processes by different criteria (e.g., speed and accuracy) to facilitate the choice of data mining processes to execute. The IDA performs a search of the space of processes defined by the ontology. Hence, no standard language for ontology specification and appropriate reasoning mechanisms are used in their approach. Further, they do not consider any state-of-the-art workflow management framework and language.

Substantial work has already been done on automated composition of Web services using Semantic Web technologies. For example, Majithia et al., (2004) present a framework to facilitate automated service composition in service-oriented architectures (Tsalgatidou & Pilioura, 2002) using Semantic Web technologies. The main objective of the framework is to support the discovery, selection, and composition of semantically-described heterogeneous Web services. The framework supports mechanisms to allow users to elaborate workflows of two levels of granularity: abstract and concrete workflows. Abstract workflows specify the workflow without referring to any specific service implementation. Hence, services (and data sources) are referred to by their logical names. A concrete workflow specifies the actual names and network locations of the services participating in the workflow. These two level workflow granularities are also considered in our approach, as shown in an application example.

Challenges associated with Grid workflow planning based on artificial intelligence concepts and with generation of abstract and concrete workflows are addressed by (Deelman et al., 2003). However, they do not consider any service-oriented architecture. Workflow representation and enactment are also investigated by the NextGrid Project (NextGrid Project, 2006). They proposed the OWL-WS (OWL for workflow and services) (Beco et al., 2006) ontology definition language. The myGrid project has developed the Taverna Workbench (Oinn et al., 2004) for the composition and execution of workflows for the life sciences community. The assisted composition approach of

Sirin (Sirin et al., 2004) uses the richness of Semantic Web service descriptions and information from the compositional context to filter matching services and help select appropriate services.

UNDERLYING STANDARDS AND TECHNOLOGIES

CRoss Industry Standard Process for Data Mining

Cross industry standard process for data mining (CRISP-DM) (Chapman et al., 1999) is a data mining process model that describes commonly used approaches that expert data miners use to tackle problems of organizing phases in data mining projects. CRISP-DM does not describe a particular data mining technique; rather it focuses on the process of a data mining projects' life cycle. The CRISP-DM data mining methodology is described in terms of a hierarchical process model consisting of sets of tasks organized at four levels of abstraction: phase, generic task, specialized task, and process instance. At the top level, the life cycle of a data mining project is organized into six phases as depicted in Figure 2.

The sequence of the phases is not strict. Moving back and forth between different phases is always required. It depends on the outcome of each phase, which one, or which particular task of a phase has to be performed next. In this chapter, we focus our attention on the three phases of data mining projects' life cycle, namely: data understanding, data preparation, and modeling.

Data understanding: This phase starts with an initial data collection and proceeds with analytic activities in order to get familiar with the data, to identify data quality problems, to discover first insights into the data, or to detect interesting subsets to form hypotheses for hidden information.

Figure 2. Phases of CRISP-DM reference model (Chapman et al., 1999)

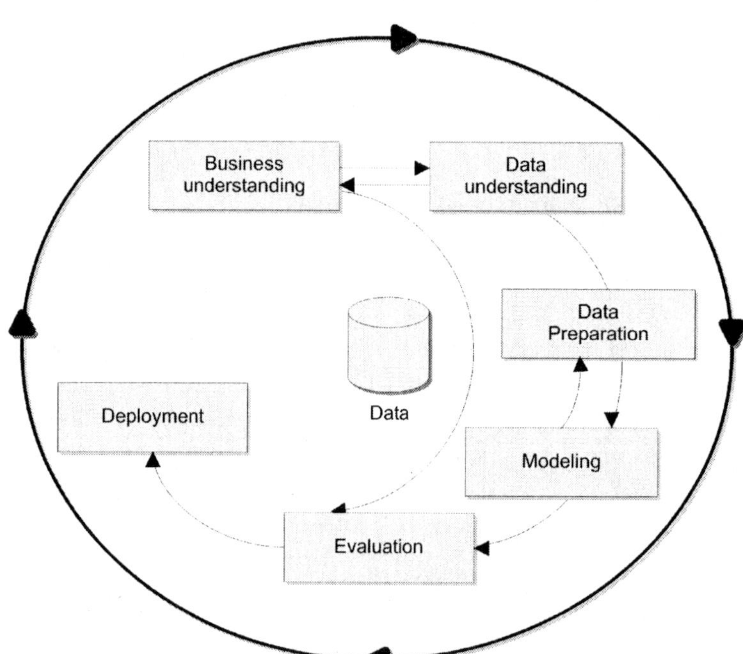

Table 1. Generic tasks and outputs of the CRISP-DM reference model

Data Understanding	Data Preparation	Modeling
Collect Initial Data • Initial Data Collection Report **Describe Data** • Data Description Report **Explore Data** • Data Exploration Report **Verify Data Quality** • Data Quality Report	**Data Set** • Data Set Description **Select Data** • Rationale for Inclusion/ Exclusion **Clean Data** • Data Cleaning Report **Construct Data** • Derived Attributes • Generated Records **Integrate Data** • Merged Data **Format Data** • Reformatted Data	**Select Modeling Techniques** • Modeling Techniques • Modeling Assumption **Generate Text Design** • Text Design **Build Model** • Parameter Settings • Models • Model Description **Assess Model** • Model Assessment • Revised Parameter Settings

Data preparation: This phase covers all activities to construct the final data set from the initial raw data. Data preparation tasks are likely to be performed multiple times and not in any prescribed order. The tasks include table, record, and attribute selection as well as transforming and cleaning data for the modeling phase.

Modeling: In this phase, various modeling techniques are selected and applied, and their parameters are calibrated to optimal values. Typically, there are several techniques for the same data mining problem type. Some techniques have specific requirements on the form of the data. Therefore, stepping back to the data preparation phase is often required.

The presented phases can be delimitated into a set of tasks defined by their outputs as presented in Table 1.

Predictive Model Markup Language

Predictive model markup language (PMML) (Data Mining Group, 2004) is an XML-based language that provides a way for applications to define statistical and data mining models and to share these models between PMML compliant applications. More precisely, the language's goal is to encapsulate a model in application and in a system independent fashion so that its producer and consumer can easily use it. Furthermore, the language can describe some of the operations required for cleaning and transforming input data prior to modeling. Since PMML version 3.1 is an XML based standard, its specification comes in the form of an XML schema that defines language primitives as follows:

- **Data Dictionary:** It defines fields that are the inputs for models and specifies their types and value ranges. These definitions are assumed to be independent of specific data mining models. The values of a categorical field can be organized in a hierarchy as defined by the taxonomy element, and numeric fields can be specified by their intervals.

- **Mining schema:** The mining schema is a subset of fields as defined in the data dictionary. Each model contains one mining schema that lists fields as used in that model. The main purpose of the mining schema is to list fields, which a user has to provide in order to apply the model.
- **Transformations:** It contains descriptions of derived mining fields using the following transformations: normalization—mapping continuous or discrete values to numbers; discretization—mapping continuous values to discrete values; value mapping—mapping discrete values to discrete values; aggregation—summarizing or collecting groups of values, for example, compute averages; and functions—derive a value by applying a function to one or more parameters.
- **Model statistics:** It stores basic uni-variate statistics about the numerical attributes used in the model such as minimum, maximum, mean, standard deviation, median, and so forth.
- **Data mining model:** It contains specification of the actual parameters defining the statistical and data mining models. The latest PMML version addresses the following classes of models: association rules, decision trees, center-based clustering, distribution-based clustering, regression, general regression, neural networks, naive bayes, sequences, text, ruleset, and support vector machine.

The models presented in PMML can be additionally defined by a set of extensions that can increase the overall complexity of a mining model as follows:

- **Built-in functions:** PMML supports functions that can be used to perform preprocessing steps on the input data. A number of predefined built-in functions for simple arithmetic operations like sum, difference, product, division, square root, logarithm, and so forth, for numeric input fields, as well as functions for string handling such as trimming blanks or choosing substrings are provided.
- **Model composition:** Using simple models as transformations offers the possibility to combine multiple conventional models into a single new one by using individual models as building blocks. This can result in models being used in sequence, where the result of each model is the input to the next one. This approach, called 'model sequencing,' is not only useful for building more complex models but can also be applied to data preparation. Another approach, 'model selection,' is used when the result of a model can be used to select which model should be applied next.
- **Output:** It describes a set of result values that can be computed by the model. In particular, the output fields specify names, types and rules for selecting specific result features. The output section in the model specifies default names for columns in an output table that might be different from names used locally in the model. Furthermore, they describe how to compute the corresponding values.
- **Model verification:** A verification model provides a mechanism for attaching a sample data set with sample results so that a PMML consumer can verify that a model has been implemented correctly. This will make model exchange much more transparent for users and inform them in advance in case compatibility problems arise.

Figure 3. Taxonomy of algorithms as presented in Weka API

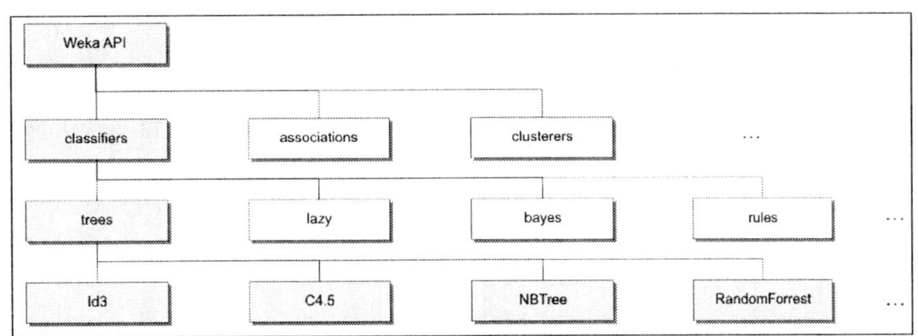

Weka Toolkit

Weka (Witten & Eibe, 2005) is a collection of machine learning algorithms, especially classifications, for data mining tasks. Moreover, Weka contains tools for data preprocessing, regression, clustering, association rules, and visualization. It is also well-suited for developing new machine learning schemes. The Weka's API is organized in a hierarchical structure, and the algorithms are delimitated by their relevancy to the classes of data mining tasks as presented in Figure 3.

Java Data Mining Application Programming Interface

The Java data mining API (JDM) (Hornick et al., 2003) proposes a pure Java API for developing data mining applications. The idea is to have a common API for data mining that can be used by clients without users being aware or affected by the actual vendor implementations for data mining. A key JDM API benefit is that it abstracts out the physical components, tasks, and even algorithms of a data mining system into Java classes. It gives a very good basis for defining concrete data mining algorithms and describing their parameters and results. JDM does not define a large number of algorithms, but provides mechanisms to add new ones, which helps in fine tuning the existing algorithms. Various data mining functions and techniques like statistical classification and association, regression analysis, data clustering, and attribute importance are covered by this standard.

Web Ontology Language for Services

Web ontology language for services (OWL-S) (Martin et al., 2004) consists of several interrelated OWL ontologies that provide a set of well defined terms for use in service applications. OWL-S leverages the rich expressive power of OWL together with its well-defined semantics to provide richer descriptions of Web services that include process preconditions and effects. This enables the encoding of service side-effects that are often important for automated selection and composition of Web services. OWL-S also provides means for the description of nonfunctional service constraints that are useful for automated Web service discovery or partnership bindings. OWL-S uses OWL to define a set of classes and their properties specific to the description of Web services. The class *Service* is at the top of this ontology (see Figure 4), which provides three essential types of knowledge about a service represented as classes: *ServiceProfile*, *ServiceGrounding* and *ServiceModel*.

- The *ServiceProfile* describes "what the service does." The profile provides information about a service that can be used in the

Figure 4. Selected classes and their relations in OWL-S ontology (Martin et al., 2004)

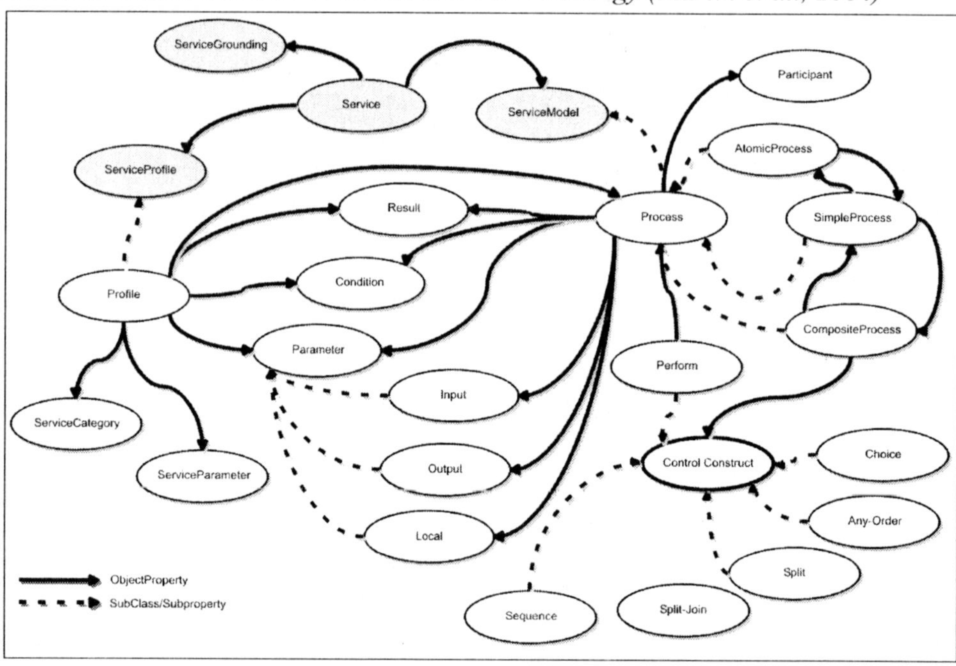

process of service discovery to determine whether the service meets one's needs.

- The *ServiceModel* informs "how to use the service." In more detail, the model gives information about the service itself and describes how to perform a specific task composed by subtasks involving certain conditions.
- The *ServiceGrounding* specifies the service-specific details of how to access the service, for example communication protocols, message formats, port numbers, and so forth. It is a kind of mapping from abstract activity description to its concrete implementation.

As we deal with the services composition, the aspects of *ServiceModel* and its main class *process*, including subclasses *AtomicProcess*, *Simple Process*, and *Composite Process* and their properties are discussed here in more detail.

Atomic process: The atomic process specifies an action provided by the Web service that expects one message as an input and returns one message in response. It means that the atomic processes are directly invokable and have no other subprocesses to be executed in order to produce a result. By definition, for each atomic process there must be grounding provided, which is associated with a concrete service implementation.

Simple process: The simple process gives a higher abstraction level of the activity execution. It is not associated with groundings and is not directly invokable, but like the atomic process, it is conceived of having a single step execution.

Composite process: Web services composition is a task of combining and linking Web services to create new processes in order to add value to the collection of services. In other words, it means that composition of several services can be viewed as one composite process with its defined inputs and outputs.

Moreover, OWL-S enables inclusion of some expressions to represent logical formulas in Semantic Web rule language (SWRL) (Horrocks et al., 2004). SWRL is a rule language that combines OWL with the rule markup language providing a rule language compatible with OWL. SWRL includes a high-level abstract syntax for Horn-like rules in OWL-DL and OWL-Lite, which are sublanguages of OWL. SWRL expressions may be used in OWL-S preconditions, process control conditions (such as if-then-else), and in effects expressions.

GRIDMINER ASSISTANT

Design Concepts

To achieve the goals presented in the Introduction section, we have designed a specialized tool—**GridMiner Assistant** (GMA)—that fulfils the role of the workflow composer shown in Figure 1. It is implemented as a Web application able to navigate a user in the phases of the knowledge discovery process (KDD) and construct a workflow consisting of a set of cooperating services aiming to realize concrete data mining objectives. The main goal of the GMA is to assist the user in the workflow composition process. The GMA provides support in choosing particular objectives of the knowledge discovery process and manage the entire process by which properties of data mining tasks are specified and results are presented. It can accurately select appropriate tasks and provide a detailed combination of services that can work together to create a complex workflow based on the selected outcome and its preferences. The GMA dynamically modifies the tasks composition depending on the entered values, defined process preconditions and effects, and existing description of services available in the KB. For this purpose we have designed a **data mining ontology** (DMO), which takes advantage of an explicit ontology of data mining techniques

Figure 5. Concept overview of the abstract data mining service

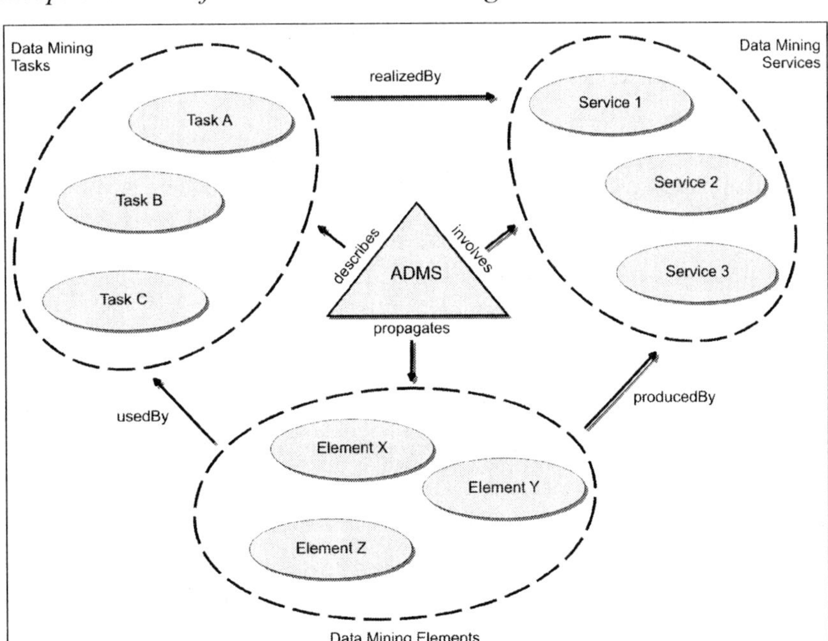

Figure 6. Basic setting classes used to describe input parameters

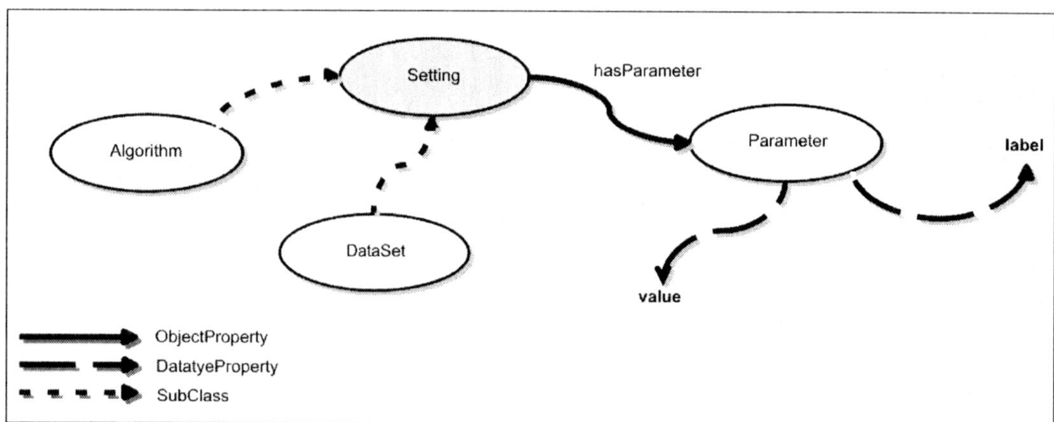

and standards (as presented in the above sections) using the OWL-S concepts to describe an abstract Semantic Web service for data mining and its main operations.

The service named **abstract data mining service** (ADMS) simplifies the architecture of the DMO as the realization of the OWL-S service with a detailed description of its *profile* and *model*. To clearly present the process of workflow composition using operations of the ADMS, we define three essential types of data mining components involved in the assisted workflow composition: DM-*elements, DM-tasks* and DM-*services,* as depicted in Figure 5. In order to design the ADMS, we consider a set of transactions representing its functionality described by DM-tasks. The DM-tasks can be seen as operations of the ADMS realized by concrete operations of involved DM-services using DM-elements as their inputs and outputs.

The following paragraphs introduce the data mining ontology, which is built through the description of the DM-tasks, DM-elements and involved DM-services. The ontology covers all phases of the knowledge discovery process and describes available data mining tasks, methods, algorithms, their inputs and results they produce. All these concepts are not strictly separated but are rather used in conjunction forming a consistent ontology.

Data Mining Elements

The DM-elements are represented by OWL classes together with variations of their representations in XML. It means that a concept described by an OWL class can have one or more related XML schemas that define its concrete representation in XML. The elements are propagated by the ADMS into the KB and can be used in any phase of data mining process. The instances of OWL classes and related XML elements are created and updated by the ADMS service operations as results of concrete services or user inputs. The elements can also determine the behavior of a workflow execution if used in SWRL rules and have an influence on preconditions or effects in the OWL-S processes. In the DMO, we distinguish two types of DM-elements: settings and results. The settings represent inputs for the DM-tasks, and on the other hand, the results represent outputs produced by these tasks. From the workflow execution point of view, there is no difference between inputs and outputs because it is obvious that an output from one process can be used, at the same time, as an input for another process.

The main reason why we distinguish inputs and outputs as settings and results is to simplify the workflow composition process, to ease searching in the KB, and to exactly identify and select requested classes and their properties.

The **settings** are built through enumeration of properties of the data mining algorithms and characterization of their input parameters. Based on the concrete Java interfaces, as presented in the Weka API and JDM API, we constructed a set of OWL classes and their instances that handle input parameters of the algorithms and their default values (see Figure 6). The settings are also used to define different types of data sets that can be involved in the KDD process. Class *DataSet* and its derived subclasses collect all necessary information about the data set (file location, user name, SQL etc.) that can be represented by different data repositories such as a relational database, CSV, WebRowSet file, and so forth. Properties of the *DataSet* are usually specified by a user at the very beginning of the KDD process composition.

The following example shows a concrete instance of the OWL class *algorithm* keeping input parameters of an Apriory-type algorithm (Agrawal et al., 1994), which produces an association model. The example is presented in OWL abstract syntax (World Wide Web Consortium, 2004).

Class (Setting partial Element)
Class (Algorithm partial Element)
Class (Parameter partial Element)

ObjectProperty(hasParameter domain(Setting)
range(Parameter))

Individual(_algorithm_AprioryType_Setting
annotation(rdfs:label "Apriori-type algorithm")
type(Algorithm)

Figure 7. Basic classes used to describe Results in DMO

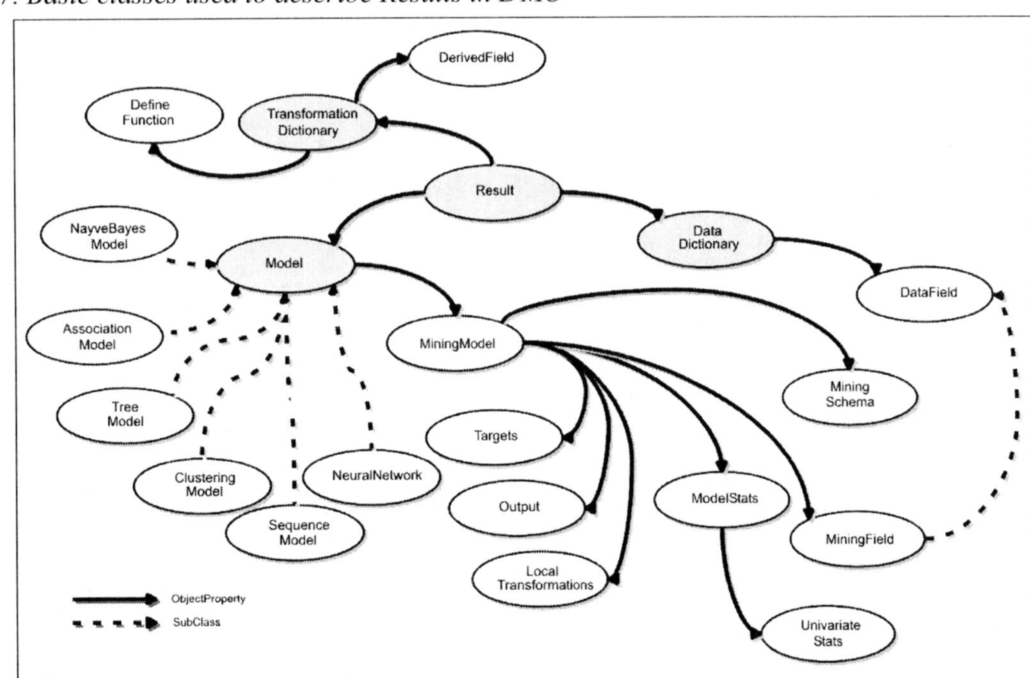

```
value(hasParameter _number_of_rules)

value(hasParameter _minimum_support)
value(hasParameter _minimun_rule_confidence))

Individual(_number_of_rules
  annotation(rdfs:label "The required number of rules")
  type(Parameter)
  value(value "10"))

Individual(_minimum_support
  annotation(rdfs:label "The delta for minimum support")
  type(Parameter)
  value(value "0.05"))

Individual(_minimun_rule_confidence
  annotation(rdfs:label "The minimum confidence of a rule")
  type(Parameter)
  value(value "0.9"))
```

The **results** are built on taxonomy of data mining models and characterization of their main components as presented in the PMML specification, therefore, the terminology used for naming the result elements is tightly linked with the names of the elements in PMML. As a result, it is easy to map its concepts to the concrete XML representations as done in the PMML schema. Figure 7 depicts the basic classes and their relations used to describe the Result DM-elements in the DMO.

From the perspective of a Web service, the DM-elements can be seen as messages exchanged between service and client (XML elements), and from the abstract workflow point of view, as items exchanged between activities of simple or atomic process inside a composite process (instances of OWL classes). The following example shows how the PMML element DataDictionary, having subelements DataField and taxonomy, can be represented as *DataDictionary* class in the OWL.

DataDictionary — XML Schema:

```
<element name="DataDictionary">
  <complexType>
   <sequence>
        <element ref="DataField" maxOccurs="unbounded" />
     <element ref="Taxonomy" minOccurs="0" maxOccurs="unbounded" />
     </sequence>
       <attribute name="numberOfFields" type="nonNegativeInteger" />
   </complexType>
 </element>
```

DataDictionary — OWL Class:

```
<Class rdf:ID="DataDictionary">
  <Restriction>
    <onProperty rdf:resource="#hasDataField"/>
  </Restriction>
  <Restriction>
    <onProperty rdf:resource="#hasTaxonomy"/>
   <minCardinality rdf:datatype="#nonNegativeInteger">0</minCardinality>
  </Restriction>
  <Restriction>
    <onProperty rdf:resource="#numberOfFields"/>
  </Restriction>
 </rdfs:subClassOf>
</Class>
```

Data Mining Tasks

The tasks are specialized operations of the ADMS organized in the phases of the KDD process as presented in the CRISP-DM reference model. The GMA composes these tasks into consistent and valid workflows to fulfill selected data mining

Table 2. DM-tasks and their DM-elements

	crisp-dm task	dmo dm-task	input	output
data understanding	collect initial data	setdataset	datasetsettings	dataset
	describe data	getdatadictionary	dataset	datadictionary
		settaxonomy	taxonomysettings	taxonomy
	explore data	getmodelstats	dataset	modelstats
	verify data quality			
data preparation	select data	setminingschema	miningschemasettings	miningschema
	clean data	gettransformation	definefunction	dataset
	construct data		derivedfield	dataset
	integrate data		mediationschema	dataset
	format data		miningschema	dataset
modeling	select modeling technique	setminingmodel	miningmodelsettings	miningmodel
	generate test design	settestset	datasetsettings	dataset
	build model	getclassificationmodel getassociationmodel getclusteringmodel getsequentialmodel getneuralnetworksmodel	mininingmodelsettings	model
	assess model	getmodelverification	mininingmodelsettings	model

objectives. The tasks are workflow's building blocks and are realized by concrete operations of involved DM-Services using DM-elements as their settings and results. Furthermore, GMA can automatically select and insert additional tasks into the workflow to assure validity and logical consistency of the data mining processes. We distinguish two types of DM-tasks that are forming the OWL-S *ServiceModel* of the ADMS—*setters* and *getters*.

Setters and getters give a functional description of the ADMS expressed in terms of the transformation produced by the abstract service. Furthermore, the setters are used to specify the input parameters for data mining tasks, and the getters are designed to present results of concrete service operations. The setters interact with a user who specifies values of the input parameters represented as properties of the *settings* class, for example, location of data source, selection of target attributes, the number of clusters, and so forth. The setters do not return any results but usually have an effect on creating and updating the DM-elements. The setters are not realized by concrete operations of involved services but are used to compose compact workflows and assure interaction with the user. The getters are designed to describe actual data mining tasks at different levels of abstraction. Thus a getter can be represented by an instance of the *CompositeProcess* class as, for example, a sequence of several subtasks, or a getter can be directly defined as an instance of the *AtomicProcess* class realized by a concrete operation of a DM-service.

Table 2 presents some of the setters and getters on the highest level of abstraction organized according to the phases of the CRISP-DM reference model and lists their input and output DM-elements.

The setters are designed to interact with the user, therefore, each *setter* has a related HTML input form used by the user to insert or select the input parameters' values of the examined DM-element. The GMA presents the form implemented as a dynamic Web page to the user, and based on his/her inputs, the GMA updates parameters of the DM-elements.

Data Mining Services

Realization of a particular DM-task is done by invoking concrete operations of involved DM-services described in OWL-S as an atomic, simple or composite process related to its *ServiceGrounding* (operators that can be executed) as defined in the appropriate WSDL document. The operations produce DM-elements that can be reused by other operations in further steps of the workflow execution. Within our project, several data mining services were developed including decision tree, clustering, associations, sequences, and neural networks with detailed descriptions of their functionality in OWL-S.

Data Mining Ontology

Based on the concepts and principal classes in the preceding sections, we have constructed the final DMO as depicted in Figure 8. The DMO incorporates the presented OWL-S ontology and its classes describing DM-tasks and DM-services as well as *Result* and *Setting* classes, which describe the DM-elements. The ontology is also supplemented by a set of semantic rules that determine in detail particular relations between involved classes, but its presentation is out of the scope of this chapter.

WORKFLOW CONSTRUCTION

In order to create the final workflow, the GMA follows a combination of the backward and forward chaining approaches. It means that the process begins with a user-based selection of

Figure 8. Basic classes and their relations in DMO

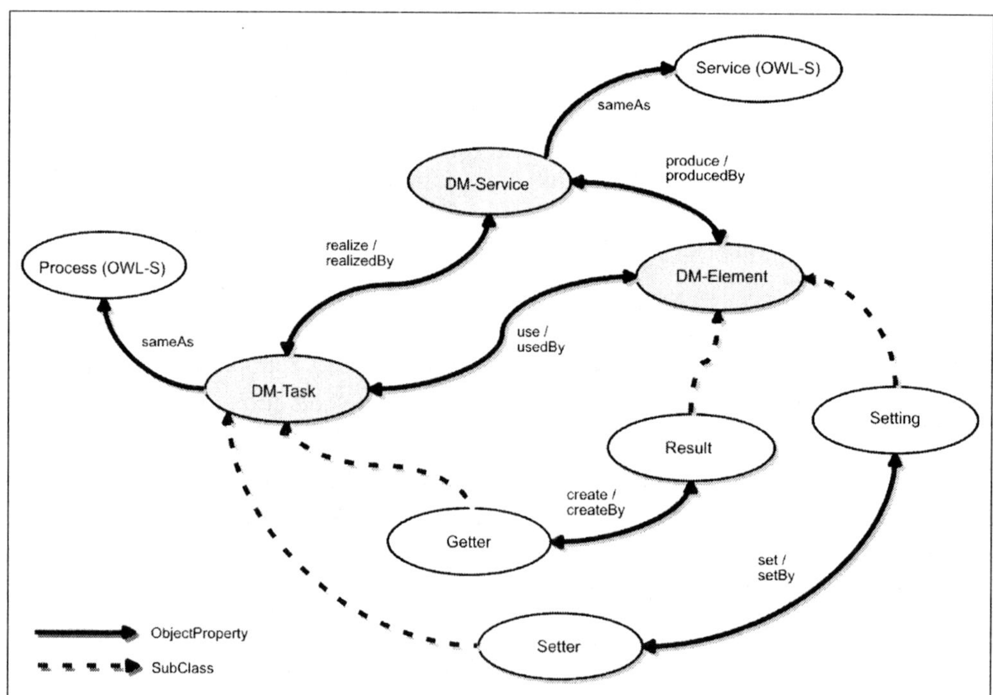

a target task, which produces the desired data exploration output. Additional tasks are automatically inserted into a chain before the target task until a task without any or already satisfied preconditions is encountered (backward phase). Next, by insertion of additional tasks, this chain is automatically extended into a form in which all matching preconditions and inputs parameters are satisfied (forward chain). According to this, our approach to workflow construction is based on two phases as follows.

Tasks Composition

The aim of this phase is to create an abstract workflow consisting of a sequence of DM-tasks. Figure 9 presents an example of the abstract workflow composed of DM-tasks. 'Task D' is the initial task inserted into the workflow in the sense of the previously mentioned backward phase of the workflow composition, and the task's result, represented by a DM-element, is the final goal of the abstract workflow. The DM-element can be, for example, a decision tree model in the data mining phase, a list of all available statistics in the data understanding phase, or the data preparation phase can result in a new transformed data set. Selection of the final result is the only interaction with the user in this phase; the other steps are hidden. The composition then continues with an examination of preconditions and inputs of the target task 'Task D.' If the task has an input which does not exist (KB does not contain an instance of the required DM-element) or condition that has to be satisfied, then the KB is queried for such a task that can supply the required DM-elements or can satisfy these preconditions by its effects; the missing task can be 'Task C' in our case. The design of the ontology ensures that there is only one such task that can be selected and inserted into the workflow prior to the examined task. For example, if we want to obtain a list of statistics (getModelStats task) then there must be an existing DM-element DataSet. It means that a task which creates the DataSet element must anticipate the getModelStats task in the workflow composition (it can be the setDataSet task in our case). The newly added tasks are treated in the same way until a task without any preconditions or already satisfied preconditions is encountered, or a task without any input that is produced as result of another task is reached, which is 'Task A' in our example.

Values Acquisition

Figure 10 presents the same workflow but viewed from another perspective: now 'Task A' is the initial task and 'Task D' is the final one. In this phase of the workflow construction, the task parameters are set up. Their values can be obtained in the following ways: (a) as effects of DM-tasks (getters) or (b) entered directly by a user (setters). In other words, not all values of input parameters can be obtained automatically as results of previous operations and therefore must be supplied by a user. This phase of the values acquisition

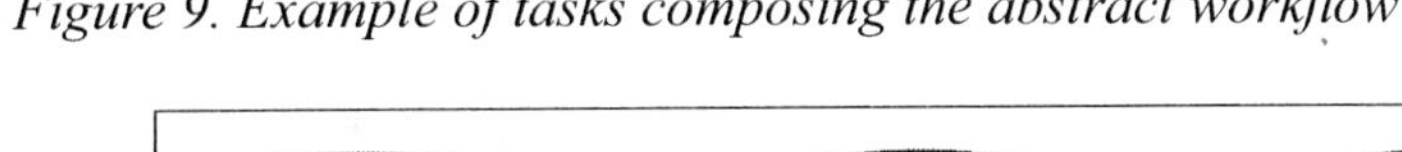

Figure 9. Example of tasks composing the abstract workflow

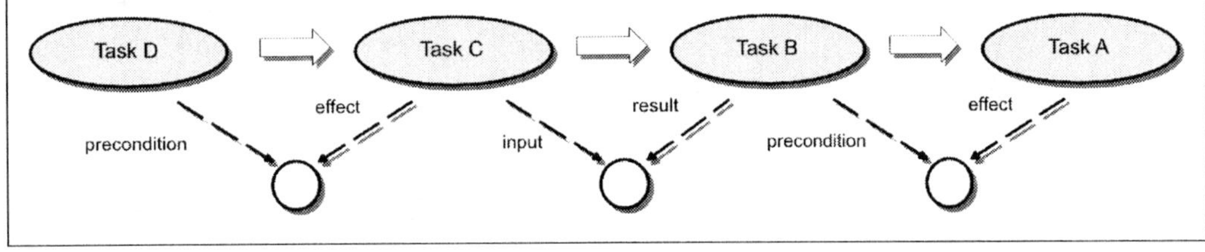

Figure 10. Example of values acquisition phase

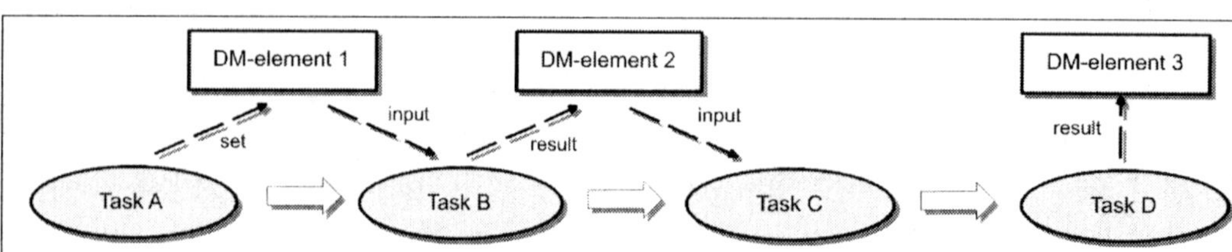

starts by tracing the abstract workflow from its beginning, 'Task A', and supplying the values by abstract interpretation of the partial workflow or providing them from a user. The user can enter the values directly by filling input fields offered by an appropriate graphical user interface or by selecting them from a list created as a result of a KB query, e.g., a list of data mining algorithms for a specific method is determined by available implementations of services able to perform the task. If the user selects a list item value that has influence on the precondition or effect that has to be satisfied in the next steps, then the KB is searched for such a task that can satisfy this request. The newly discovered tasks are inserted automatically into the workflow. It can be, for example, a case when the user wants to increase the quality of used data adding some transformation tasks, presenting the resulting model in different form, and so forth.

To illustrate the main features of the GMA and explain the phases of the tasks composition and values acquisition, we present a practical scenario addressing step-by-step construction of a simple workflow aiming at discovering of classification model for a given data set. This scenario is taken from a medical application dealing with patients suffering from serious traumatic brain injuries (TBI).

Workflow Construction Example

At the first clinical examination of a TBI patient (Brezany et al., 2003), it is very common to assign the patient into a category, which allows to define his/her next treatment and helps to predict the final outcome of the treatment. There are five categories of the final outcome defined by the Glasgow outcome scale (GOS): dead, vegetative, severely disabled, moderately disabled, and good recovery.

It is obvious that the outcome is influenced by several factors that are usually known and are often monitored and stored in a hospital data warehouse. For TBI patients, these factors are for example: injury severity score (ISS), abbreviated injury scale (AIS), Glasgow coma score (GCS), age, and so forth. It is evident that if we want to categorize the patient, then there must be a prior knowledge based on cases of other patients with the same type of injury. This knowledge can be mined from the historical data and represented as a classification model. The mined model is then used to assign the patient to the one of the outcome categories. In particular, the model can assign one of the values from the GOS to a concrete patient.

As we mentioned in the previous section, in the first phase, the composition of the abstract workflow proceeds by using the backward chaining approach starting with the task and then producing the demanded result. In our case, the classification model is represented by a decision tree. Moreover, in this example, we assume that the data understanding phase of the KDD process was successfully finished, and we have all the necessary information about the data set to be mined. It means that appropriate records corresponding

Box 1.

Query:

```
PREFIX dmo: <http://dmo.gridminer.org/v1#>
PREFIX rdfs: <http://www.w3.org/2000/01/rdf-schema#>
PREFIX rdf: <http://www.w3.org/1999/02/22-rdf-syntax-ns#>
SELECT ?ModelName ?Task
FROM    <http://www.gridminer.org/dmo/v1/dmo.owl>
WHERE {
            ?model rdf:type <#Model> .
            ?model rdfs:label ?ModelName .
            ?model dmo:createdBy ?Task
      }
ORDER BY ?ModelName
```

Result:

ModelName T	ask
Association Model	getAssociationModel
Classification Model	getClassificationModel
Clustering Model	getClusteringModel
...	...

Box 2.

Query:

```
PREFIX dmo: <http://dmo.gridminer.org/v1#>
SELECT ?Setting ?Task
FROM    <http://www.gridminer.org/dmo/v1/dmo.owl>
WHERE {
            dmo:getClassificationModel dmo:hasSettings ?Setting .
            ?Task dmo:create ?Setting
      }
```

Result:

Setting	Task
MiningModel	setMiningModel

to the *DataSet* and *DataDictionary* DM-elements are already available in the KB, and the workflow can start with the data preprocessing task.

Phase 1: Tasks Composition

As we presented previously, the first step of the task composition is the interactive selection of the final model from a list of all available models. The list can be obtained as a result of the following SPARQL (SPARQL, 2006) query returning a list of DM-tasks and models they produce. This query is issued by the GMA automatically. (See Box 1.)

Selection of the classification model gives us a direct link to the *getClassificationModel* DM-task that can be realized by a concrete service operation. Information about its input DM-elements and the corresponding DM-task producing them can be retrieved from the KB by submitting the following SPARQL query, which is also issued by the GMA automatically (see Box 2).

The discovered DM-task *setMiningModel* is inserted into the workflow prior to the *getClassificationModel* task, and its preconditions and inputs are examined. The only precondition of

Figure 11. Abstract workflow after the phase of tasks composition

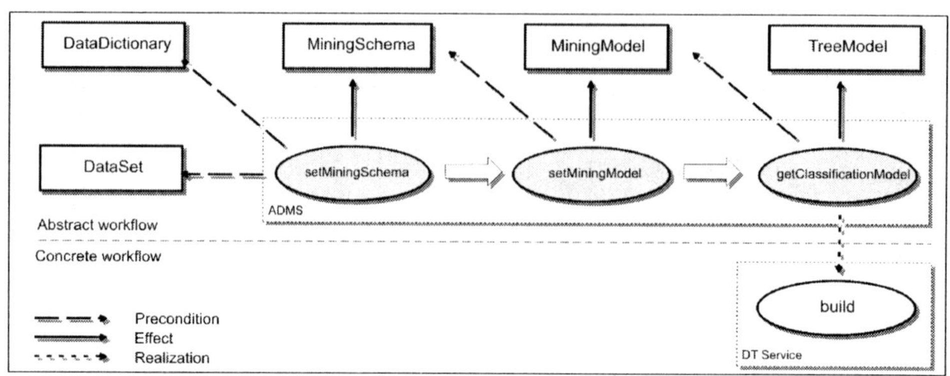

the *setMiningModel* task is the existence of the *MiningSchema* DM-element. This requirement can be satisfied by inserting the *setMiningSchema* task into the workflow, whose effect is the creation of the *MiningSchema* DM-element. The *setMiningSchema* task has two preconditions: the existence of the *DataSet* and *DataDictionary* DM-elements. Their corresponding records are already available in the KB, so no additional tasks are inserted into the abstract workflow. As the result, an abstract workflow consisting of three DM-tasks (see Figure 11) is created and is instanced as a new composite process of the ADMS in the KB. The figure also presents the DM-elements identified during the composition phase as preconditions of the involved tasks and a fragment of the concrete workflow.

Phase 2: Values Acquisition

The second phase of the workflow construction starts with the examination of the first DM-task in the abstract workflow (*setMiningSchema*). In this phase, the values of the DM-elements' properties, identified in the previous phase, are supplied by the user and additional DM-tasks are inserted as needed. The following paragraphs describe in more detail the steps of setting the DM-elements produced and used by the involved tasks.

setMiningSchema: This task can be seen as a simple data preprocessing step where data fields (attributes) used in the modeling phase can be selected and their usage types can be specified. The primary effect of this task is a new *MiningSchema* element instanced in the KB, keeping all the schema's parameters specified by the user. Moreover, the user can specify whether some preprocessing methods should be used to treat missing values and outliers of the numerical attributes. Selection of a preprocessing method requires an additional DM-task, which is able to perform the data transformations and produce a new data set that can be used in the next steps. If one of the transformation methods is selected then the KB is queried again for a task able to transform the data set. The *getTransformation* task has the ability to transform the selected data set, therefore, can be inserted into the abstract workflow in the next step.

As we presented in previous paragraphs, the *setters* are designed to interact with the user, therefore, each *setter* has a related HTML input form used by the user to insert or select the values of the examined DM-element input parameters. The GMA presents the form implemented as a dynamic Web page to the user, and based on its inputs, the GMA updates values of the DM-elements' parameters.

Figure 12. Input HTML form for the MiningSchema

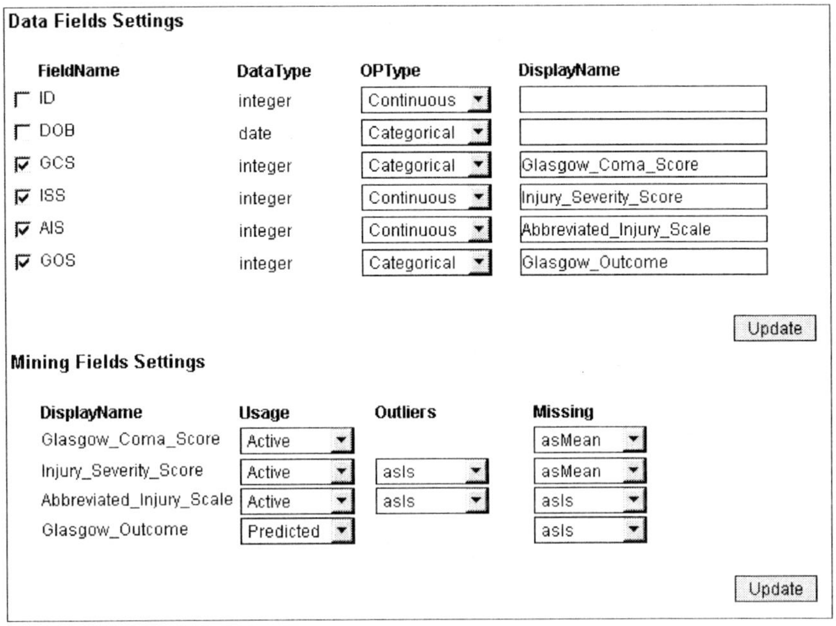

Data Fields Settings

	FieldName	DataType	OPType	DisplayName
☐	ID	integer	Continuous	
☐	DOB	date	Categorical	
☑	GCS	integer	Categorical	Glasgow_Coma_Score
☑	ISS	integer	Continuous	Injury_Severity_Score
☑	AIS	integer	Continuous	Abbreviated_Injury_Scale
☑	GOS	integer	Categorical	Glasgow_Outcome

Update

Mining Fields Settings

DisplayName	Usage	Outliers	Missing
Glasgow_Coma_Score	Active		asMean
Injury_Severity_Score	Active	asIs	asMean
Abbreviated_Injury_Scale	Active	asIs	asIs
Glasgow_Outcome	Predicted		asIs

Update

Figure 13. Input HTML form for the MiningModel

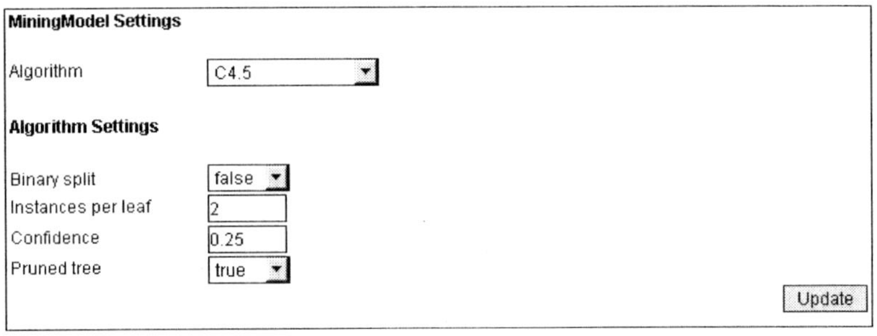

MiningModel Settings

Algorithm: C4.5

Algorithm Settings

Binary split	false
Instances per leaf	2
Confidence	0.25
Pruned tree	true

Update

Figure 12 presents the input form used by the GMA to construct the *MiningSchema* DM-element. In this form, there is one mandatory property for the classification task — 'target attribute.' It is one of the categorical *DataFields* from the *DataDictionary* element, which is the GOS in our case. Therefore, the 'target attribute' must be marked as 'predicted' in the *MiningSchema* DM-element. The effect of the *setMiningSchema* task is a newly created DM-element *MiningSchema,* which describes mined fields and their transformations.

getTransformation: This task is inserted into the workflow right after the *setMiningSchema* task. It does not require interaction with the user because its input parameters are already specified in the *MiningSchema* created as the effect of the previous task. The task just examines the *MiningSchema* element and selects a concrete operation from

Figure 14. Abstract and concrete workflow after the phase of values acquisition

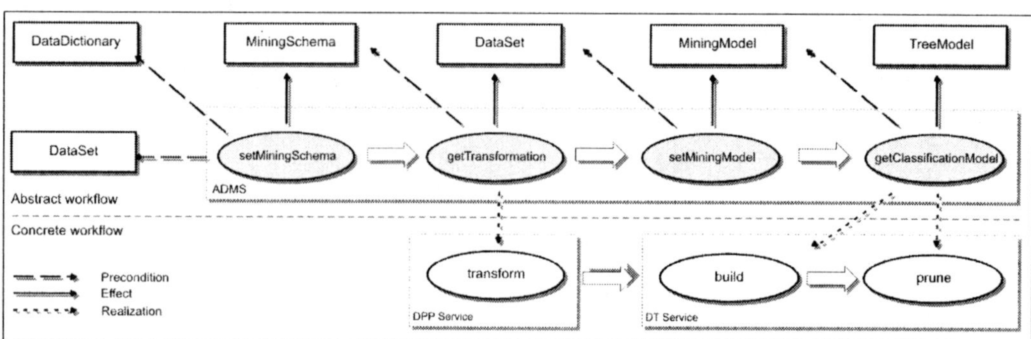

Box 3.

Query:

```
PREFIX dmo: <http://dmo.gridminer.org/v1#>
PREFIX rdfs: <http://www.w3.org/2000/01/rdf-schema#>
SELECT ?ParameterName ?DefaultValue
FROM <http://www.gridminer.org/dmo/v1/dmo.owl>
WHERE {
        dmo:_algorithm_c4.5_Settings dmo:hasParameter ?Parameter .
        ?Parameter rdfs:label ?ParameterName .
        ?Parameter dmo:value ?DefaultValue
      }
ORDER BY ?ParameterName
```

DM-Services available in the KB, which can satisfy the chosen data preprocessing objectives. The task can select operation 'transform' of the specialized *DataPreprocessing* service (DPP service) and insert it into the concrete workflow (see Figure 14).

setMiningModel: Specification of the properties of the selected model is the main purpose of this task. The GMA presents a list of all available data mining algorithms producing classification models and selects its input parameters. Based on the selected parameters, a new DM-element *MiningModel* describing model properties is created as an effect of this task. The following SPARQL query retrieves all parameters for the C4.5 classification algorithm (Quinlan, 1993) that is used to setup the *MiningModel* element in our example. (See Box 3.)

The GMA presents the results to the user in the HTML form presented in Figure 13, where the user specifies values of the input parameters needed to build the classification model using the C4.5 algorithm.

getClassificationModel: This task examines the *MiningModel* element created in the previous task and identifies the appropriate operation that can build the classification model using *MiningModel* parameters. The task can be the operation 'build' implemented by the DecisionTree Service (DT Service), which returns the classification model represented by the PMML element TreeModel. Moreover, if parameter 'pruned tree' is marked as true (false by default) then the additional operation of the DT Service 'prune' is inserted into to the concrete workflow to assure that the discovered decision tree is modified using a pruning mechanism.

Box 4.

```
Variables:

<variable name="DataSet" element="wrs:webRowSet"/>
<variable name="TransformedDataset" element="wrs:webRowSet "/>
<variable name="TreeModel" element="pmml:TreeModel"/>
<variable name="PrunedTreeModel" element="pmml:TreeModel"/>
<variable name="TreeSettings" element="dmo:Setting"/>

Sequence:

<sequence>
    < flow>
      <invoke partnerLink="DPPService" operation="transform"   inputVariable="DataSet"
      outputVariable="TransformedDataset" />

      <invoke partnerLink="DTService" operation="build"
      inputVariable="TreeSettings" outputVariable="TreeModel"/>

      <invoke partnerLink="DTService" operation="prune"
      inputVariable="TreeSettings"  outputVariable="PrunedTreeModel"/>
    </ flow>
</ sequence>
```

If all required parameters and preconditions of the tasks involved in the abstract workflow are satisfied then the GMA constructs a concrete workflow specification in the BPEL language and presents it to the workflow engine. The concrete workflow is a sequence of the real services and is related to the abstract DM-tasks as presented in Figure 14.

The final output returned from the workflow engine is a PMML document containing a TreeModel element that represents the demanded model that can be used to classify a particular patient into the GOS category.

The following BPEL document created in our scenario contains five variables representing the DM-elements used as inputs and outputs of the invoked operations. The variable DataSet is an XML in WebRowSet format (`RowSet` Java object in XML format) storing all the initial data. TransformedDataset is a new WebRowSet created by the 'transform' operation, and TreeSettings is used as input for the 'build' and 'prune' operations. The variable TreeModel stores the PMML document with the full decision tree, and the PrunedTreeModel stores its pruned version. The BPEL flow reflects the composition as done in the concrete workflow consisting of three operations invoked in sequence. (See Box 4.)

SYSTEM PROTOTYPE

An overview of the first system prototype is shown in Figure 15. We use the OWL editor Protégé (Noy et al., 2001) to create and maintain the DMO, which is stored in the KB. To reason about knowledge in the KB, we use the Pellet reasoner (Sirin & Parsia, 2004), which is an open-source Java based OWL DL reasoner and provides a description logic interface (DIG) (Bechhofer et al., 2003). The GMA is implemented as a standalone Web application supported by the Jena Toolkit (McBride, 2003) and is able to interact with a user to assemble the required information. The GMA communicates over the DIG interface with the reasoner, which is able to answer a subset of RDQL queries (Seaborn, 2004). The GMA queries KB every time it needs to enumerate some

Figure 15. Overview of the prototype system

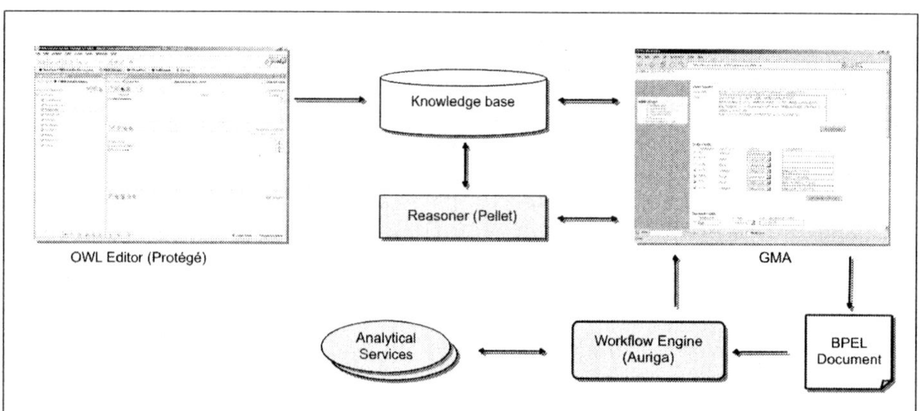

parameters or find a data mining task, algorithm, service, and so forth. Moreover, the GMA also updates the KB with instances of DMO classes and values of their properties. The final outcome of the GMA is a workflow document presented to the workflow engine Auriga (Brezany et al., 2006) in the BPEL4WS language. The GMA also acts as a client of the workflow engine, which executes appropriate services as described in the BPEL document and returns their outputs back to the GMA. A more detailed characterization of these major components follows.

Auriga WEEP workflow engine is an easy to execute and manage workflow enactment service for Grid and Web services. The core of the engine is implemented as a standalone application referred to as the Auriga WEEP Core, which orchestrates the services as specified in a BPEL. Auriga WEEP has also a specialized version, which is wrapped by a Grid service implementation focused on using the Globus 4 container as the running environment. The engine has a pluggable architecture, which allows additional Grid specific functionality to be used in the Auriga Core extensions.

- **Jena** is a Java framework for building Semantic Web applications. It provides a programming environment for RDF, RDFS, OWL and SPARQL and includes a rule-based inference engine.
- **Pellet** provides functionalities to see the species validation, check consistency of ontologies, classify the taxonomy, check entailments and answer a subset of RDQL queries. Pellet is based on the tableaux algorithms developed for expressive Description Logics and supports the full expressivity of OWL DL.
- **Protégé** is an ontology editor and knowledge acquisition system. It implements a rich set of knowledge-modeling structures and actions that support the creation, visualization, and manipulation of ontologies in various representation formats including OWL.

FUTURE WORK

We envision the following key directions for future extension of the research presented in this chapter:

- **Active workflow engine:** This approach was already briefly discussed in the background section and sketched in Figure 1. In this case, the interaction mode between the user, workflow composer and the functionality of the composer basically remain the same as in the described passive approach. The functionality of the existing GridMiner workflow engine will be extended to be able to make dynamic decisions about the next workflow execution step based on the actual context of the knowledge base and the results of the reasoning. Moreover, the workflow composer can listen to the changes in the knowledge base and automatically interact with the user when some additional information or hints have to be supplied.

- **Workflow ranking:** The data mining ontology will be extended by estimations of each operation's effects on workflow attributes such as speed, model accuracy, etc. Due to the user's preferences (e.g., speed vs. accuracy) the composer can then better optimize individual selection steps, derive a set of workflows with the corresponding ranking and supply the best option to the workflow engine. In this process, information about the current Grid resource utilization provided by standard Grid information services can also be included into this optimization process.
- **Workflow planning:** We consider upgrading the intelligence of the workflow composer with the development of a supporting planning system which will be able to propose an abstract workflow from the specification of the goals and the initial state. We will exploit and adapt AI planning optimizations.
- **Support by autonomic computing:** We will investigate how the presented framework should be extended to be able to include some functionality of autonomic computing into the workflows composed. This involves investigating workflow patterns, categorizing requirements and objectives, and designing corresponding rule templates.

CONCLUSION

The characteristics of data exploration in scientific environments impose unique requirements for workflow composition and execution systems. In this chapter, we addressed the issues of composing workflows with automated support developed on top of Semantic Web technologies and the workflow management framework elaborated in our Grid data mining project. The kernel part of that support is a tool called the GridMiner workflow assistant (GMA), which helps the user interactively construct workflow description expressed in a standard workflow specification language. The specification is then passed to the workflow engine for execution. The GMA operations are controlled by the data mining ontology based on the concepts of PMML, JDM, WEKA and CRISP-DM. A practical example taken from a medical application addressing management of patients with traumatic brain injuries illustrates the use of the GMA. The results achieved will be extended in our future research whose key issues were outlined in the chapter. Although this research is conducted in the context of the GridMiner project, its results can be used in any system involving workflow construction activities.

FUTURE RESEARCH DIRECTIONS

In this section, we identify three future challenges and research problems in the ontology-based workflow construction and execution.

1. Extended Data Mining Ontology

Data mining as a scientific discipline is a huge domain which is still expanding. New approaches to data analyses, visualization techniques, or even new algorithms are continuously being developed.

There are also plenty of real applications tailored to the application domain specifically for data mining tasks. Therefore, it is nearly impossible to completely describe this dynamic field of data mining with a static ontology. The ontology proposed in our chapter can only be used for a subset of the high number of data mining tasks. Hence we see new opportunities in extending the proposed data mining ontology with different, application domain specific, tasks that would better express the functionality of the constructed workflows.

2. Quality of Services and Workflows
Another issue that is not fully covered in the proposed ontology is the description of the quality of the involved data mining services. Especially in the Grid infrastructures, the properties of the involved resources (e.g., performance, price, bandwidth, etc.) play the crucial role in their discovery and right selection. So we see another opportunity in the detailed description of the data mining services' properties which can be done as a direct extension of the OLW-S language. Moreover, there can also be a detailed description of the composed workflows' quality which can be used for effective ranging of the entire workflows.

3. Autonomic Behavior of the Workflow Enactment Engine
Autonomic computing is one of the hottest topics in information technologies. Different areas in computer science, ranging from hardware to software implementation on the application level, try to apply some autonomic features (like, e.g., self-tuning, self-configuration, self-healing, etc.) to assure stability and availability of the system. The autonomic behavior of the Workflow Engine can ensure that the execution of the data mining workflows results in a required goal even in such a dynamic environment as the Grid where the Workflow Engine must react to the changes of the involved resources and adopt its behavior to new conditions and reflect the actual 'State of the Grid'.

REFERENCES

Agrawal, R., & Srikant, R. (1994). Fast algorithms for mining association rules in large databases. In *Proceedings of the International Conference on Very Large Databases* (pp. 478-499). Santiage, Chile: Morgan Kaufmann.

Antonioletti, M., Krause, A., Paton, N. W., Eisenberg, A., Laws, S., Malaika, S., et al. (2006). The WS-DAI family of specifications for web service data access and integration. *ACM SIGMOD Record, 35*(1), 48-55.

Arkin, A. (2002). *Business process modeling language* (BPML). Specification. BPMI.org.

Arkin, A., Askary, S., Bloch, B., Curbera, F., Goland, Y., Kartha, N., et al. (2005). *Web services business process execution language version* 2.0. wsbpel-specificationdraft-01, OASIS.

BEA, IBM, Microsoft, SAP, & Siebel. (2003). *Business process execution language for Web services.* Version 1.1. Specification. Retrieved May 15, 2006, from ftp://www6.software.ibm.com/software/developer/library/ws-bpel.pdf

Bechhofer, S., Moller, R., & Crowther, P. (2003). *The DIG description logic interface.* International Workshop on Description Logics, Rome, Italy.

Beco, S., Cantalupo, B., Matskanis, N., & Surridge M. (2006). *Putting semantics in Grid workflow management: The OWL-WS approach.* GGF16 Semantic Grid Workshop, Athens, Greece.

Bernstein, A., Hill, S., & Provost, F. (2001). An intelligent assistant for the knowledge discovery process. *In Proceedings of the IJCAI-01 Workshop on Wrappers for Performance Enhancement in KDD.* Seattle, WA: Morgan Kaufmann.

Brezany, P., Tjoa, A.M., Rusnak, M., & Janciak, I. (2003). Knowledge Grid support for treatment of traumatic brain injury victims. *International Conference on Computational Science and its Applications.* Montreal, Canada.

Brezany, P., Janciak, I., Woehrer, A., & Tjoa, A.M. (2004). *GridMiner: A Framework for knowledge discovery on the Grid - from a vision to design and implementation.* Cracow Grid Workshop, Cracow, Poland: Springer.

Brezany, P., Janciak, I., Kloner, C., & Petz, G. (2006). *Auriga — workflow engine for WS-I/WS-RF services.* Retrieved September 15, 2006, from http://www.Gridminer.org/auriga/

Bussler ,C., Davies, J., Dieter, F., & Studer , R. (2004). The Semantic Web: Research and applications. In *Proceedings of the 1st European Semantic Web Symposium, ESWS. Lecture Notes in Computer Science, 3053.* Springer.

Chapman, P., Clinton, J., Khabaza, T., Reinartz, T., & Wirth. R. (1999). *The CRISP-DM process model.* Technical report, CRISM-DM consortium. Retrieved May 15, 2006, from http://www.crisp-dm.org/CRISPWP-0800.pdf

Christensen, E., Curbera, F., Meredith, G., & Weerawarana, S. (2001). *Web Services Description Language* (WSDL) 1.1. Retrieved May 10, 2006, from http://www.w3.org/TR/wsdl

Data Mining Group. (2004). *Predictive model markup language.* Retrieved May 10, 2006, from http://www.dmg.org/

Deelman, E., Blythe, J., Gil, Y., & Kesselman, C. (2003). Workflow management in GriPhyN. *The Grid Resource Management.* The Netherlands: Kluwer.

Globus Alliance (2005). *Globus Toolkit 4.* http://www.globus.org

Globus Alliance, IBM, & HP (2004). *The WS-Resource framework.* Retrieved May 10, 2006, from http://www.globus.org/wsrf/

Hornick, F. M., et al. (2005). *Java data mining 2.0.* Retrieved June 20, 2006, from http://jcp.org/aboutJava/communityprocess/edr/jsr247/

Horrocks, I., Patel-Schneider, P. F., Boley, H., Tabet, S., Grosof, B., & Dean, M. (2004). *SWRL: A Semantic Web rule language combining OWL and RuleML.* W3C Member Submission. Retrieved May 10, 2006, from http://www.w3.org/Submission/2004/SUBM-SWRL-20040521

Kickinger, G., Hofer, J., Tjoa, A.M., & Brezany, P. (2003). Workflow mManagement in GridMiner. *The 3rd Cracow Grid Workshop.* Cracow, Poland: Springer.

Leymann, F. (2001). *Web services flow language (WSFL 1.0).* Retrieved September 23, 2002, from www4.ibm.com/software/solutions/webservices/pdf/WSFL.pdf

Majithia, S., Walker, D. W., & Gray, W.A. (2004). *A framework for automated service composition in service-oriented architectures* (pp. 269-283). ESWS.

Martin, D., Paolucci, M., McIlraith, S., Burstein, M., McDermott, D., McGuinness, D., et al.(2004). Bringing semantics to Web services: The OWL-S approach. In *Proceedings of the 1st International Workshop on Semantic Web Services and Web Process Composition.* San Diego, California.

McBride, B. (2002). Jena: A Semantic Web toolkit. *IEEE Internet Computing,* November /December, 55-59.

Oinn, T. M., Addis, M., Ferris, J., Marvin, D., Senger, M., Greenwood, R. M., et al. (2004). Taverna: A tool for the composition and enactment of bioinformatics workflows. *Bioinformatics, 20*(17), 3045-3054.

Noy, N. F. , Sintek, M., Decker, S., Crubezy, M., Fergerson, R. W., & Musen, M.A. (2001). Creating Semantic Web contents with Protege-2000. *IEEE Intelligent Systems, 16*(2), 60-71.

Quinlan, R. (1993). *C4.5: Programs for machine learning.* San Mateo, CA: Morgan Kaufmann Publishers.

Sairafi, S., A., Emmanouil, F. S., Ghanem, M., Giannadakis, N., Guo, Y., Kalaitzopolous, D., et al. (2003). The design of discovery net: Towards open Grid services for knowledge discovery. *International Journal of High Performance Computing Applications, 17*(3).

Seaborne, A. (2004). *RDQL: A query language for RDF.* Retrieved May 10, 2006, from http://www.w3.org/Submission/RDQL/

Sirin, E., & Parsia, B. (2004). *Pellet: An OWL DL Reasoner, 3rd International Semantic Web Conference*, Hiroshima, Japan. Springer.

Sirin, E.B. Parsia, B., & Hendler, J. (2004). Filtering and selecting Semantic Web services with interactive composition techniques. *IEEE Intelligent Systems, 19*(4), 42-49.

SPARQL. Query Language for RDF, W3C Working Draft 4 October 2006. Retrieved October 8, 2006, from http://128.30.52.31/TR/rdf-sparql-query/

Thatte, S. (2001). *XLANG: Web services for business process design.* Microsoft Corporation, Initial Public Draft.

Tsalgatidou, A., & Pilioura, T. (2002). An overview of standards and related technology in web services. *Distributed and Parallel Databases. 12*(3).

Witten, I.H., & Eibe, F. (2005). *Data mining: Practical machine learning tools and techniques.* (2nd ed.). San Francisco: Morgan Kaufmann.

World Wide Web Consortium. (2004). *OWL Web ontology language semantics and abstract syntax.* W3C Recommendation 10 Feb, 2004.

ADDITIONAL READING

For more information on the topics covered in this chapter, see http://www.Gridminer.org and also the following references:

Alesso, P. H., & Smith, F. C. (2005). *Developing Semantic Web services.* A.K. Peterson Ltd.

Antoniou, G., & Harmelen, F. (2004). *A Semantic Web primer.* MIT Press.

Davies, J., Studer, R., & Warren P. (2006). *Semantic Web technologies: Trends and research in ontology-based systems.* John Wiley & Sons.

Davies, N. J., Fensel, D., & Harmelen, F. (2003). *Towards the Semantic Web: Ontology-driven knowledge management.* John Wiley & Sons.

Foster, I., & Kesselman, C. (1999). *The Grid: Blueprint for a new computing infrastructure.* Morgan Kaufmann.

Fox, G.C., Berman, F., & Hey, A.J.G. (2003). *Grid computing: Making the global infrastructure a reality.* John Wiley & Sons.

Han, J., & Kamber, M. (2000) *Data mining: Concepts and techniques.* Morgan Kaufmann.

Lacy, L.W. (2005). *Owl: Representing information using the Web ontology language.* Trafford Publishing.

Li, M., & Baker, M. (2005). *The Grid: Core technologies.* John Wiley & Sons.

Marinescu, D.C. (2002) *Internet-based workflow management: Toward a Semantic Web.* John Wiley & Sons.

Matjaz, B.J., Sarang, P.G., & Mathew, B. (2006). *Business process execution language for Web services* (2nd ed.). Packt Publishing.

Murch, R. (2004). *Autonomic computing.* Published by IBM Press.

Oberle, D. (2005). *The semantic management of middleware.* Springer.

Singh, M.P., & Huhns, M.N. (2006). *Service-oriented computing: Semantics, processes, agents.* John Wiley & Sons.

Sotomayor, B., & Childers, L. (2006). *Globus Toolkit 4: Programming Java services.* Morgan Kaufmann.

Stojanovic, Z., & Dahanayake. A. (2005). *Service oriented software system engineering: Challenges and practices.* Idea Group Inc.

Taylor, I. J., Deelman E., Gannon, D. B., & Shields, M. (2007). *Workflows for e-science.* Springer.

Zhong, N., Liu, J., & Yao, Y.(2003). *Web intelligence.* Springer.

Zhu, X., & Davidson, I. (2007). *Knowledge discovery and data mining: Challenges and realities.* Idea Group Inc.

Zhuge, H. (2004). *The knowledge Grid.* World Scientific.

This work was previously published in Data Mining with Ontologies: Implementations, Findings, and Frameworks, edited by H. O. Nigro, S. Gonzalez Cisaro, and D. Xodo, pp. 182-210, copyright 2008 by Information Science Reference, formerly known as Idea Group Reference (an imprint of IGI Global).

Chapter 2.24
Exploratory Time Series Data Mining by Genetic Clustering

T. Warren Liao
Louisiana State University, USA

ABSTRACT

In this chapter, we present genetic-algorithm(GA)-based methods developed for clustering univariate time series with equal or unequal length as an exploratory step of data mining. These methods basically implement the k-medoids algorithm. Each chromosome encodes, in binary, the data objects serving as the k-medoids. To compare their performance, both fixed-parameter and adaptive GAs were used. We first employed the synthetic control-chart data set to investigate the performance of three fitness functions, two distance measures, and other GA parameters such as population size, crossover rate, and mutation rate. Two more sets of time series with or without a known number of clusters were also experimented: one is the cylinder-bell-funnel data and the other is the novel battle simulation data. The clustering results are presented and discussed.

INTRODUCTION

Before prediction models can be built in data mining or knowledge discovery, it is often advisable to first explore the data. Clustering is known to be a good exploratory data-mining tool. The goal of clustering is to create structure for unlabeled data by objectively forming data into homogeneous groups, where the within-group object similarity and the between-group object dissimilarity are optimized. The bulk of clustering analyses has been performed on data associated with static features, that is, feature values that do not change with time, or the changes are negligible.

Two major classes of clustering methods are partitioning and hierarchical clustering. Well-known partitioning-based clustering methods include *k*-means (MacQueen, 1967), *k*-medoids (Kaufman & Rousseeuw, 1990), and the corresponding fuzzy versions: fuzzy *c*-means (FCM) (Bezdek, 1987) and fuzzy *c*-medoids (Krishnapuram, Joshi, Nasraoui, & Yi, 2001). Hierarchical clustering methods are either of the agglomerative type or the divisive type. Lately, soft computing technologies, including neural networks and genetic algorithms, have emerged as another class of clustering techniques. Two prominent methods of the neural network approach to clustering are competitive learning and self-organizing feature maps. Most genetic clustering methods implement the spirit of partitioning methods, especially the *k*-means algorithm (Krishna & Murty, 1999; Maulik & Bandyopadhyay, 2000), and the fuzzy *c*-means algorithm (Hall, Özyurt, & Bezdek, 1999).

Just like static feature data, forming groups of similar time series given a set of unlabeled time series is often desirable. These unlabeled time series could be monitoring data collected during different periods from a particular process, or from several different processes. These processes could be natural, engineered, business, economical, or medical related. Unlike static feature values, the time series of a feature consists of dynamic values, that is, values changed with time. This greatly increases the dimensionality of the problem and calls for somewhat different, and often more complicated, clustering methods. This study will focus only on time-series data.

In surveying work related to time-series clustering, Liao (2005) distinguished three different time-series clustering approaches: those working with full data either in the time or frequency domain, those working with extracted features, and model-based approaches with models built from the raw data. An example of the first approach is Golay *et al.* (Golay, Kollias, Stoll, Meier, Valavanis, & Boesiger, 1998). They applied the fuzzy *c*-means algorithm to provide the functional maps of human brain activity on the application of a stimulus. In their study, three different distances (the Euclidean distance and two cross-correlation-based distances) were alternately used for comparison purposes. Goutte *et al.* (Goutte, Toft, & Rostrup, 1999) and Fu *et al.* (Fu, Chung, Ng, & Luk, 2001) took the feature-based approach. Goutte *et al.* clustered functional magnetic resonance imaging (fMRI) time series in groups of voxels with similar activations using two algorithms: *k*-means and Ward's hierarchical clustering. The cross-correlation function, instead of the raw fMRI time series, was used as the feature space. Fu *et al.* described the use of self-organizing maps for grouping similar temporal patterns dispersed along the time series. Two enhancements were made: consolidating the discovered clusters by a redundancy removal step, and introducing the perceptually important point-identification method to reduce the dimension of the input data sequences.

Three model-based time-series clustering methods are described next. Li and Biswas (1999) described a clustering methodology for temporal data using hidden Markov model representation with a sequence-to-model likelihood distance measure. The temporal data was assumed to have the Markov property. Time series were considered similar when the models characterizing individual series were similar. Policker and Geva (2000) presented a model for nonstationary time series with time varying mixture of stationary sources, comparable to the continuous hidden Markov model. Fuzzy clustering methods were applied to estimate the continuous drift in the time-series distribution, and the resultant membership matrix was given an interpretation as weights in a time varying, mixture probability distribution function. Kalpakis *et al.* (Kalpakis, Gada, & Puttagunta, 2001) studied the clustering of ARIMA time-series by using the Euclidean distance between the Linear Predictive Coding Cepstra of two time-series as their dissimilarity measure and the Partition around Medoids (PAM) method as the clustering algorithm.

To the best of our knowledge, the only study that applied genetic algorithms to cluster time series is Baragona (2001). He evaluated three metaheuristic methods to partition a set of time series into clusters in such a way that (1) the maximum absolute cross-correlation value between each pair of time series that belong to the same cluster is greater than some given threshold, and (2) the *k*-min cluster criterion is minimized with a specified number of clusters. The cross-correlations were computed from the residuals of the models of the original time series. Among all methods evaluated, Tabu search was found to perform better than single linkage, pure random search, simulation annealing, and genetic algorithms, based on a simulation experiment on 10 sets of artificial time series generated from low-order univariate and vector ARMA models. For each time series, 300 observations were generated.

This study proposes *k*-medoids-based genetic algorithms for clustering time-series data of equal or unequal length as an exploratory step of data mining. Though few selected time series were tested in this study due to space constraint, the methods proposed can definitely be generalized to time-series data in other domains. We chose to directly process raw data to avoid the need either to extract features or to fit some appropriate models for the data at hand. Our GA differs from that of Baragona in three major aspects: (1) working with the original data rather than the residuals, (2) capable of handling unequal length of time series by using the dynamic time warping distance rather than the cross-correlation function, and (3) implementing different GAs. In the next two sections, the proposed genetic clustering method and the distance measures used are presented. The subsequent three sections present the test results of the synthetic control-chart data, cylinder-bell-funnel data, and battle simulation data, respectively. A discussion then follows and finally, the chapter is concluded.

GENETIC CLUSTERING OF TIME-SERIES DATA

Genetic algorithms have the following elements: population of chromosomes, selection according to fitness, crossover to produce new offspring, and random mutation of new offspring (Mitchell, 1996). In this section, the proposed genetic algorithms for clustering time-series data are detailed element by element.

In summary, four different chromosome representations have been employed by the genetic clustering techniques that implemented either the *k*-means or fuzzy *c*-means algorithm. They include integer-coded cluster for each datum with length equal to the number of data points, as in Krishna and Murty (1999); real-coded cluster centers, as in Maulik and Bandyopadhyay (2000) and Bandyopadhyay and Maulik (2002); binary-coded cluster centers (in gray coding), as in Hall *et al.* (1999); and binary-coded representation of *p*-median problems with length equaling to the number of data points, in which a digit value of "1" denotes a median and "0" not a median (Lorena & Furtado, 2000). These chromosome representation methods can be extended to time-series data, though they were initially designed for static feature data.

We implemented the *k*-medoids algorithm rather than *k*-means (or fuzzy *c*-means) because of the difficulty involved in defining the cluster centers for time series with unequal length. Since the data dimension is relatively large in our application, either binary-coded or real-coded cluster centers are inappropriate. Therefore, we elected to use a binary-coded representation of data objects serving as the cluster medoids. (The other alternative is integer-coded representation, which we will investigate, and hope to share the results once they become available). The prespecified number of cluster medoids and the number of digits used to represent each medoid together determine the chromosome length.

Each chromosome in the population is evaluated in two steps: first distributing each data point to the closest medoid according to some distance measure, and then computing the fitness value. The dynamic time warping distance, d_{DTW}, is chosen because of its ability to handle time series with varying lengths. More details are presented in the next section. The cluster of each datum is determined to be the one medoid closest to it, based on the nearest neighbor concept. Three fitness functions, as given next, are evaluated in this study.

1. 10^6 / (TWCV × (1 + nbr0 × a large integer)), where TWCV and nbr0 denote the total within-cluster distance and number of clusters with zero members, respectively. Krishna and Murty (1999) used the total within-cluster distance for static feature values but not time series. Our implementation differs from theirs also in the distance measure (dynamic time warping instead of Euclidean), the clustering algorithm (*k*-medoids instead of *k*-means), the chromosome representation, and other GA details. Let v_i, i =1, …, c be the cluster medoids representing cluster C_i, i=1, …, c and x_j, j = 1,…, n be the data vectors. If $x_j \in C_i$, then w_{ij} = 1; else, w_{ij} = 0. In this study, the TWCV is computed as:

$$TWCV = \sum_{i=1}^{c} WCV_i = \sum_{i=1}^{c}\sum_{j=1}^{n} w_{ij} d_{DTW}(v_i, x_j). \quad (1)$$

2. 10^6 / (DB × (1 + nbr0 × a large integer)), where DB is the Davies-Bouldin index that was initially proposed as a measure of the validity of the clusters by Davies and Boulden (1979) and later used by Bandyopadhyay and Maulik (2002) in their study of genetic clustering of satellite images. Let n_i denote the number of data points in cluster *i*. For this study, the DB index is modified as:

$$DB = \frac{1}{c}\sum_{i=1}^{c} \max_{j=1, j\neq i}^{c} \left\{ \frac{WCV_i / n_i + WCV_j / n_j}{d_{DTW}(v_i, v_j)} \right\}. \quad (2)$$

3. 10^6 / (R_m(V) × (1 + nbr0 × a large integer)), where R_m(V) is the reformulated FCM functional (Hall *et al.*, 1999). In the following equation, we replace the original Euclidean distance with the DTW distance and *m* is the fuzzy weight (*m*>1).

$$Rm(V) = \sum_{j=1}^{n} \left(\sum_{i=1}^{c} d_{DTW}(v_i, x_j)^{\frac{1}{(1-m)}} \right)^{1-m}. \quad (3)$$

Note that each fitness function includes a penalty term to discourage the formation of clusters with zero members ("a large integer" was consistently set at 10 in this study). For comparing time-series data with equal length, Euclidean distance, d_E, was also implemented. In this case, d_{DTW}, in the previous equations, is replaced by d_E.

Standard roulette-wheel selection is used to reproduce offspring for the next generation. Each current chromosome in the population has a roulette-wheel slot, sized in proportion to its fitness. Chromosomes with a higher fitness value thus have a higher probability of contributing more offspring. Pairs of chromosomes are randomly chosen to perform the one-point crossover operation according to the specified crossover rate. The mutation is then performed to flip some of the bits in a chromosome from "0" to "1" or "1" to "0," according to the specified rate. If the resultant chromosome contains an invalid cluster medoid, its fitness is then set to zero to prevent it surviving to the next generation (a simple repair procedure to take care of the infeasible solutions).

The GA process has the following steps:

- Set the generation value to zero.
- Initialize the population.
- Evaluate the population.
- While the maximum number of generations is not reached,
- Select chromosomes.
- Perform the crossover operation.
- Perform the mutation operation.
- Evaluate the new pool of chromosomes.
- Increment the generation value.

Several GA parameters need to be set for the GA process to run. They include the population size, s, the crossover rate, p_c, the mutation rate, p_m, and the maximum number of generations, g_{max}. In addition, a different selection strategy could be used. Proper selection of these parameters greatly affects the GA behavior that is strongly determined by the balance between exploration (to explore new and unknown areas in the search space) and exploitation (to make use of knowledge acquired by exploration to reach better positions in the search space). Most GA studies choose these parameters by trial and error, without a systematic investigation.

Attempts to find the optimal and general set of parameters have been made by testing a wide range of problems (Grefenstette, 1986). To determine the relevancy and relative importance of these parameters, Rojas *et al.* (Rojas, González, Pomares, Merelo, Castillo, & Romero, 2002) applied the analysis of the variance (ANOVA) technique. The response variables used to perform the statistical analysis were the maximum fitness in the last generation that measures the capacity to find a local/global optimum, and the average fitness in the last generation that measures the diversity in the population. In terms of the best solution, all variables were found significant, with the first three most significant ones being the selection operator, the population size, and the type of mutation. Regarding the diversity, the significant variables in descending order were the type of selection operators, the mutation rate, the mutation type, and the number of generations. These results were obtained based on their tests on a 0/1 knapsack problem, the Riolo function, the prisoner's dilemma problem, and three Michalewicz's functions.

Another school of approaches for setting GA parameters is using some mechanism to adapt them depending upon the state of the GA learning process, instead of fixing them from the outset. Srinivas and Patnaik (1994) proposed the adaptive genetic algorithm (AGA) for multimodal function optimization to realize the dual goals of maintaining diversity in the population and sustaining the convergence capacity of the GA. The AGA varies the probabilities of crossover and mutation depending upon the fitness values of the solutions. Let f_{max} and f_{avg} be the maximum fitness and the average fitness of the entire population, f' be the larger of the fitness values of the solutions to be crossed, and f be the fitness values of the solutions to be mutated. The expressions for p_c and p_m are given as,

$$p_c = \begin{cases} k_1(f_{\max} - f')/(f_{\max} - f_{avg}), & if\ f' \geq f_{avg} \\ k_3, & if\ f' < f_{avg} \end{cases} \quad (4)$$

and

$$p_m = \begin{cases} k_2(f_{\max} - f)/(f_{\max} - f_{avg}), & if\ f \geq f_{avg} \\ k_4, & if\ f < f_{avg} \end{cases} \quad (5)$$

where k_1, k_2, k_3, $k_4 \leq 1.0$. They set k_2 and k_4 to 0.5 to ensure the disruption of those solutions with average or subaverage fitness. To force all solutions with a fitness value less than or equal to the average fitness to undergo crossover, they assigned k_1 and k_3 a value of 1.

Eiben *et al.* (Eiben, Hinterding, & Mizhalewicz, 1999) classified parameter control (or parameter adaptation) studies based on two aspects: how the mechanism works and what component of the evolutionary algorithms (that includes GA) is affected by the mechanism. They classified parameter control mechanisms into three categories: deterministic in the sense that parameter values are altered by some deterministic rule, adaptive by using feedback from the search to determine the direction and/or magnitude of the change, and self-adaptive by encoding the parameters into the chromosomes that undergo mutation and recombination. They identified six components being adapted: representation, evaluation function, mutation operators and their probabilities, crossover operators and their probabilities, parent selection, and replacement operator. Most previous parameter adaptation studies used one mechanism to adapt one or two components. For example, the work of Srinivas and Patnaik (1994) employed an adaptive mechanism to vary two components: mutation rate and crossover rate. To date, there are insufficient research studies and results to conclude how much parameter control is most useful. Herrera and Lozano (2003) reviewed different aspects of fuzzy adaptive genetic algorithms (FAGA) from three points of view: design, taxonomy, and future directions. The steps for designing the fuzzy logic controller used by FAGAs were shown with an example. They categorized FAGAs based on two criteria: how the rule base is obtained, and the level where the adaptation takes place. They also discussed future directions and some challenges for FAGA research.

The parameter adaptation approaches have become more popular than the selection approaches as one gradually realizes the difficulty in coming up with a general rule, and that different problems really require different GA parameters for satisfactory performance. Therefore, this study will employ a parameter adaptation approach. Specifically, we employ the AGA proposed by Srinivas and Patnaik (1994) to adapt both the crossover rate and the mutation rate (or only the mutation rate) in order to evaluate its performance in clustering time-series data. Proposing a new parameter adaptation method is beyond the scope of this study. Nevertheless, both the results with adapted parameters and those with fixed parameters are obtained and compared.

SIMILARITY/DISTANCE MEASURES

One key issue in clustering is how to measure the similarity between two data objects being compared. For static feature values, Euclidean distance or generalized Mikowski distance is often used. The Euclidean distance has been used to measure the distance between two time series of the same length, for example, by Pham and Chan (1998). The Euclidean distance or generalized Mikowski distance is applicable only when the lengths of time-series data are equal, which is not the case for the battle simulation data to be studied in the sequel. Therefore, we resort to the dynamic-time-warping distance that is known capable of coping with unequal time series.

Dynamic time warping (DTW) is a generalization of classical algorithms for comparing discrete sequences to sequences of continuous values. Given two time series, $Q = q_1, q_2, \ldots, q_i, \ldots, q_n$ and $C = c_1, c_2, \ldots, c_j, \ldots, c_m$, DTW aligns the two series so that their difference is minimized. To this end, an n by m distance matrix was used in which the (i, j) element contains the distance $d(q_i, c_j)$ between two points q_i and c_j that is often measured by the Euclidean distance. A warping path, $W = w_1, w_2, \ldots, w_k, \ldots, w_K$ where $\max(n, m) \leq K \leq m+n-1$, is a set of matrix elements that satisfies three constraints: boundary condition, continuity, and monotonicity. The boundary condition constraint requires the warping path to start and to finish in diagonally opposite corner cells of the matrix. That is $w_1 = (1, 1)$ and $w_K = (n, m)$. The continuity constraint restricts the allowable steps to adjacent cells. The monotonicity constraint forces the points in the warping path to be monotonically spaced in time. Of interest is the warping path that has the minimum distance between the two series. Mathematically,

$$d_{DTW} = \min \frac{\sum_{k=1}^{K} w_k}{K}. \quad (6)$$

Dynamic programming can be used to effectively find this path by evaluating the recurrence function given as equation (7), which defines the cumulative distance as the sum of the distance of the current element and the minimum of the cumulative distances of the adjacent elements:

$$d_{cum}(i,j) = d(q_i, c_j) + \min\{d_{cum}(i-1, j-1), d_{cum}(i-1, j), d_{cum}(i, j-1)\} \quad (7)$$

CLUSTERING RESULTS OF CONTROL-CHART DATA

The genetic clustering methods were first applied to clustering 30 synthetic control-chart data, taken from the UCI Data Mining Archive (http://kdd.ics.uci.edu/), that were initially generated by Pham and Chan (1998). Figure 1 shows the 30 synthetic control-chart data used in this study. There are six known patterns (or clusters): normal, cyclic, increasing trend, decreasing trend, upward shift, and downward shift. For each pattern, there are five time series. It is not difficult to see that there is some overlapping between similar patterns (adjacent clusters), for example, between increasing trend and upward shift. We were able to compute the clustering accuracy rates for the control-chart data set because the ground truth is known. Just like any clustering algorithm, the proposed genetic clustering algorithm arbitrarily labels each cluster. This inconsistent labeling makes the accuracy-checking task somewhat tedious.

Table 1 summarizes the clustering results of using fixed parameter GAs. The fixed GA parameters include population size of 20, maximum generation of 30, crossover rate of 0.8, and three mutation rates: 0.01, 0.05, and 0.1.

For each combination of GA parameters, five repetitions were made with different random seeds. Among all the fixed parameter GAs tested, the highest average clustering accuracy of 0.733 was obtained when using the TWCV-based fitness function, DTW distance measure, and mutation rate of 0.01. This particular combination also produced the highest clustering accuracy of 86.7% among all runs executed for this dataset. Figure 2 shows the cluster medoids of one of the five runs for this particular fixed parameter GA. Note that the six patterns can be clearly seen in the figure.

Table 2 summarizes the clustering results of using AGAs that adapt both crossover and mutation rates. For the AGAs, the GA parameter fixed is maximum generation of 30. Factors varied are fitness function, distance measure, and population size. For each combination of GA parameters, five repetitions were made with different random seeds. Among all the AGAs tested, the highest average clustering accuracy of 0.7 was attained when using the TWCV-based fitness function, Euclidean distance, and population size of 60. Note that the highest clustering accuracy produced by AGA is 83.3%, which was produced by using the *Rm*-based fitness function, DTW distance, and population size of 20. Figure 3 shows the cluster

Figure 1. 30 synthetic control-chart data

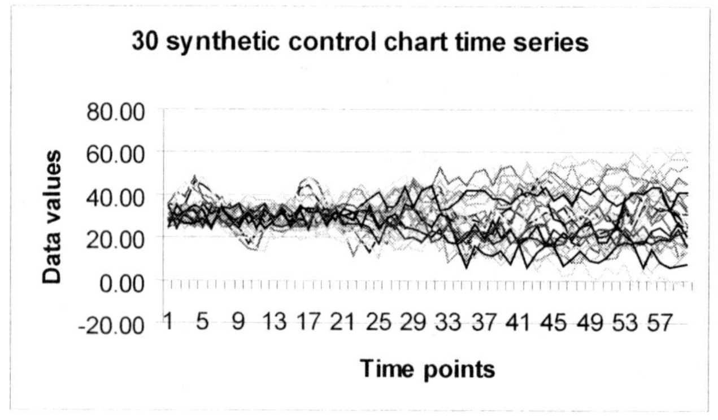

Table 1. Clustering results of control-chart data by fixed parameter GAs

Fitness Function	Distance	Mutation Rate	Run Accuracy	Avg. Accuracy
TWCV	Euclidean	0.01	.667, .667, .8, .567, .667	.673
TWCV	Euclidean	0.05	.633, .667, .7, .7, .633	.667
TWCV	Euclidean	0.1	.667, .733, .633, .7, .7	.687
TWCV	DTW	0.01	.667, **.867**, .767, .7, .667	**.733**
TWCV	DTW	0.05	.667, .7, .6, .7, .667	.667
TWCV	DTW	0.1	.6, .733, .633, .8, .667	.687
DB	Euclidean	0.01	.4, .6, .467, .533, .467	.493
DB	Euclidean	0.05	.433, .6, .433, .467, .433	.473
DB	Euclidean	0.1	.433, .4, .467, .5, .467	.453
DB	DTW	0.01	.567, .633, .5, .7, .667	.601
DB	DTW	0.05	.6, .6, .467, .633, .567	.573
DB	DTW	0.1	.533, .567, .667, .567, .6	.587
Rm	Euclidean	0.01	.467, .533, .533, .567, .7	.560
Rm	Euclidean	0.05	.8, .633, .733, .8, .533	.700
Rm	Euclidean	0.1	.5, .667, .533, .633, .633	.593
Rm	DTW	0.01	.533, .6, .567, .6, .567	.573
Rm	DTW	0.05	.533, .667, .567, .567, .5	.567
Rm	DTW	0.1	.567, .533, .533, .533, .633	.560

Figure 2. Cluster means of a selected fixed parameter GA run

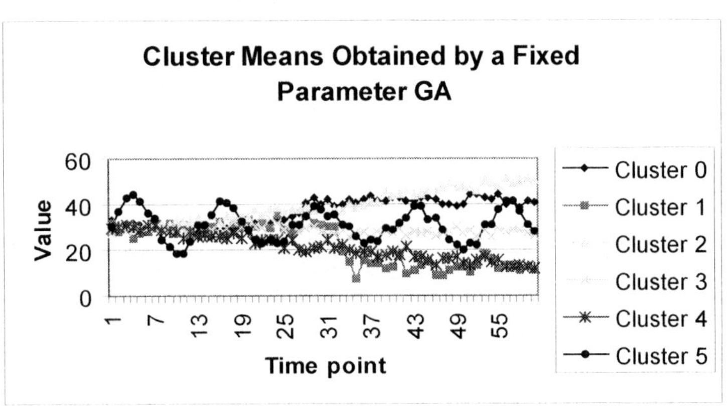

medoids of one of the five runs for this particular AGA. Note that the six patterns can also be clearly seen as in Figure 2.

To determine the effect of fitness function, distance measure, and GA parameter, we performed ANOVA tests on the clustering results given in Tables 1 and 2. The results indicate that for both fixed parameter GAs and AGAs, the fitness function, and the interaction between the fitness function and the distance measure are highly significant (with p value < 0.005).

Table 2. Clustering results of control-chart data by AGA

Fitness Function	Distance	Population Size	Run Accuracy	Avg. Accuracy
TWCV	Euclidean	20	.633, .7, .667, .733, .633	.673
TWCV	Euclidean	40	.667, .767, .633, .767, .633	.693
TWCV	Euclidean	60	.667, .733, .7, .7, .7	**.700**
TWCV	DTW	20	.633, .7, .7, .6, .6	.647
TWCV	DTW	40	.7, .6, .667, .6, .667	.647
TWCV	DTW	60	.667, .633, .6, .667, .733	.660
DB	Euclidean	20	.433, .633, .367, .467, .567	.493
DB	Euclidean	40	.5, .4, .433, .5, .467	.460
DB	Euclidean	60	.433, .4, .5, .4, .467	.440
DB	DTW	20	.6, .667, .667, .533, .5	.593
DB	DTW	40	.533, .667, .633, .633, .6	.601
DB	DTW	60	.5, .633, .633, .633, .6	.600
Rm	Euclidean	20	.533, .633, .767, .667, .533	.627
Rm	Euclidean	40	.533, .633, .767, .667, .533	.627
Rm	Euclidean	60	.533, .533, .533, .533, .633	.553
Rm	DTW	20	.7, .7, .6, .533, **.833**	.673
Rm	DTW	40	.533, .633, .633, .7, .467	.593
Rm	DTW	60	.567, .567, .5, .567, .5	.540

Figure 3. Cluster means of a selected AGA run

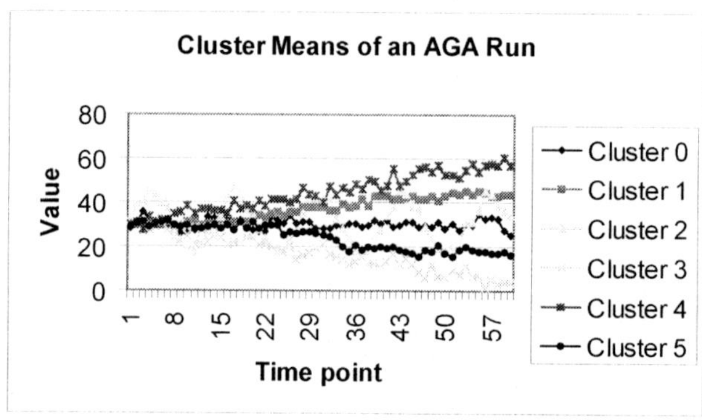

To evaluate the performance of adapting only one parameter, we also tried the adaptive GA that adapts only the mutation rate. Three levels of crossover rates were experimented: 0.7, 0.8, and 0.9. The two fixed parameters are the population size at 20 and the maximum number of generation at 30. Table 3 summarizes the clustering accuracies. Among all the AGAs tested, the highest average clustering accuracy was attained at 70.7% when using the TWCV-based fitness

Table 3. Clustering results of control-chart data by AGA that adapts mutation rate

Fitness Function	Distance	Crossover Rate	Run Accuracy	Avg. Accuracy
TWCV	Euclidean	0.7	.6, .7, .633, .633, .733	.660
TWCV	Euclidean	0.8	.7, .7, .7, .633, .633	.673
TWCV	Euclidean	0.9	.7, .733, .7, .733, .677	**.707**
TWCV	DTW	0.7	.7, .633, .7, .633, .7	.673
TWCV	DTW	0.8	.8, .567, .633, .8, .677	.693
TWCV	DTW	0.9	.7, .733, **.833**, .6, .677	**.707**
DB	Euclidean	0.7	.6, .467, .467, .433, .467	.487
DB	Euclidean	0.8	.467, .467, .367, .467, .467	.447
DB	Euclidean	0.9	.6, .467, .567, .433, .467	.507
DB	DTW	0.7	.633, .567, .533, .733, .667	.627
DB	DTW	0.8	.5, .6, .633, .567, .633	.587
DB	DTW	0.9	.6, .667, .667, .6, .633	.633
Rm	Euclidean	0.7	.633, .667, .533, .767, .6	.640
Rm	Euclidean	0.8	.567, .567, .533, .633, .8	.620
Rm	Euclidean	0.9	.633, .567, .533, .567, .567	.573
Rm	DTW	0.7	.733, .7, .6, .533, .5	.613
Rm	DTW	0.8	.667, .567, .6, .567, .5	.580
Rm	DTW	0.9	.733, .7, .6, .533, .533	.620

function and crossover rate of 0.9, regardless the distance measure. The highest clustering accuracy produced by AGA is 83.3%, which was produced by using the TWCV-based fitness function, DTW distance, and crossover rate of 0.9. The ANOVA test indicates that the fitness function, the distance measure, and their interaction are significant in affecting the clustering accuracy (with p value < 0.005). From Tables 2 and 3, it can be observed that adapting both crossover rate and mutation rates does not have any advantage over adapting only the mutation rate in terms of finding the highest clustering accuracy.

Comparing Table 2 (only those results based on population size of 20) and Table 3 (only those results based on crossover rate of 0.8) with Table 1, one can easily see that for the control-chart data, AGAs do not always perform better than fixed parameter GAs. Depending upon the combination of fitness-function and distance measure, AGA could be better than all, none, or some fixed parameter GAs tested. Therefore, how to devise an AGA that always performs better than all fixed-parameter GAs does require further investigation.

CLUSTERING RESULTS OF CYLINDER BELL FUNNEL DATA

This section presents results obtained from one relatively larger data set of univariate time series with known number of clusters. The data set contains 300 series generated by implementing the cylinder, bell, and funnel equations given in the UCR Time Series Data Mining Archive (http://www.cs.ucr.edu/~eamonn/TSDMA/). One hundred series were generated for each pattern, with each series having a length of 80 data points.

Table 4. Clustering results of cylinder-bell-funnel data by AGA that adapts mutation rate

Fitness Function	Distance	Crossover Rate	Run Accuracy	Avg. Accuracy
TWCV	Euclidean	0.7	.823, .843, .823, .873, .853	.843
TWCV	Euclidean	0.8	.877, .820, .847, **.883**, .807	.847
TWCV	Euclidean	0.9	.867, .857, .840, .847, .857	.854
TWCV	DTW	0.7	.607, .537, .647, .460, .710	.592
TWCV	DTW	0.8	.643, .570, .543, .647, .760	.633
TWCV	DTW	0.9	.757, .763, .640, .517, .660	.667
DB	Euclidean	0.7	.800, .817, .827, .817, .847	.822
DB	Euclidean	0.8	.830, .820, .713, .817, .693	.775
DB	Euclidean	0.9	.863, .880, .823, .833, .880	**.856**
DB	DTW	0.7	.517, .607, .527, .563, .540	.551
DB	DTW	0.8	.533, .617, .553, .557, .567	.565
DB	DTW	0.9	.593, .503, .487, .547, .543	.535
Rm	Euclidean	0.7	.623, .610, .653, .557, .527	.594
Rm	Euclidean	0.8	.777, .610, .710, .697, .773	.713
Rm	Euclidean	0.9	.703, .617, .670, .760, .627	.675
Rm	DTW	0.7	.520, .550, .520, .603, .650	.569
Rm	DTW	0.8	.777, .570, .543, .613, .493	.599
Rm	DTW	0.9	.443, .607, .483, .560, .493	.515

We run the adaptive GA (that adapts only the mutation rate) 18 times by varying the fitness function, the distance measure, and the crossover rate. In each run, the GA was repeated five times. Three levels of crossover rates were experimented: 0.7, 0.8, and 0.9. The two fixed parameters are the population size at 60 and the maximum number of generation at 30. Table 4 summarizes the clustering results. The ANOVA test on these results reveals that the fitness function, the distance measure, and their interaction are significant in affecting the clustering accuracy (with p value < 0.005). Obviously, the Euclidean distance outperforms the dynamic time warping for this dataset. As far as the fitness function is concerned, the TWCV-based fitness function is the best and the R_m-based index is the worst. The best average clustering accuracy is 85.6%, obtained by the combination using the DB-based fitness, Euclidean distance, and crossover rate of 0.9. The highest clustering accuracy among all runs is 88.3%, which was produced in one of the five replicates by the combination using the TWCV-based fitness, Euclidean distance, and crossover rate of 0.8.

CLUSTERING RESULTS OF BATTLESIMULATION DATA

The OneSAF combat simulation software was used to create a battle scenario for our experiments (Heilman *et al.*, 2002). Time-series data were collected using a modified version of the Killer-Victim Scoreboard (KVS) method originally developed by O'May and Heilman (2002) to collect static-feature data. The KVS method modifies OneSAF to provide critical battlespace

Figure 4. Five time series of a sample battle simulation run.

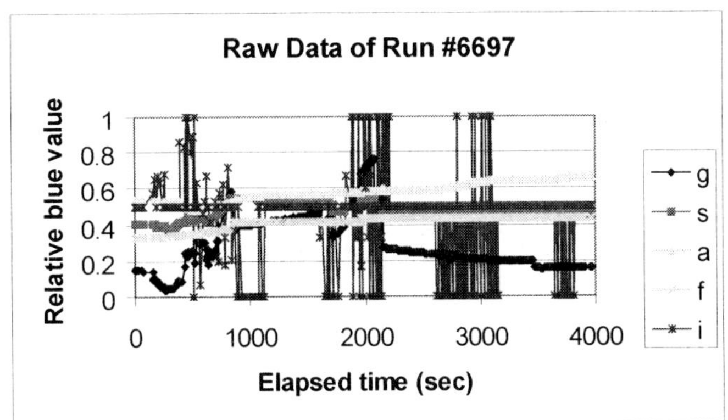

Figure 5. Interpolated results of time series shown in Figure 4

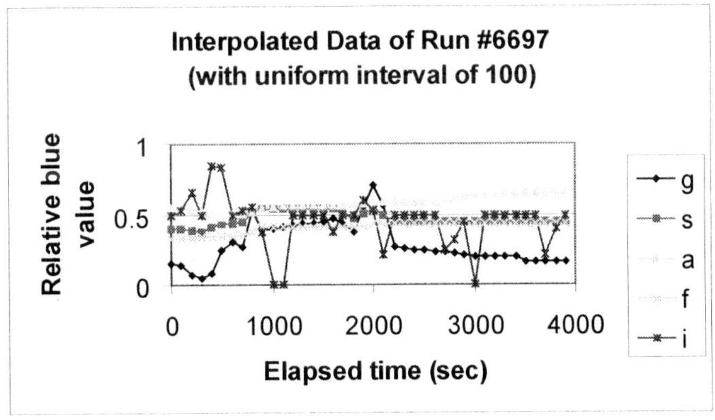

data. A set of three files was generated during each simulation execution. These files include entity identification data, firer-target interaction data, and logistics and appearance data. Each run of the raw experimental data collected was then processed into five time series (by arranging them in the order of timestamps) reflecting the state of ongoing battle from the viewpoint of the "blue" force, which includes:

- Relative territory ownership (denoted by *g* in figures)
- Relative firepower strength (*s*)
- Relative ammunition support (*a*)
- Relative fuel support (*f*)
- Relative firing intensity (*i*)

Time-series data of 15 battle simulation runs were prepared for this study according to the outlined procedure. Note that these data were nonuniformly sampled as a result of the event-triggered data-collection mechanism. The result of a sample run is shown in Figure 4.

We intentionally use only a few runs in this study because one often can afford only a limited number of simulation runs in actual applications, due to time constraint. To enable real-time response, which is desirable in order to provide

Figure 6. Fifteen series of relative territory ownership of the blue forces

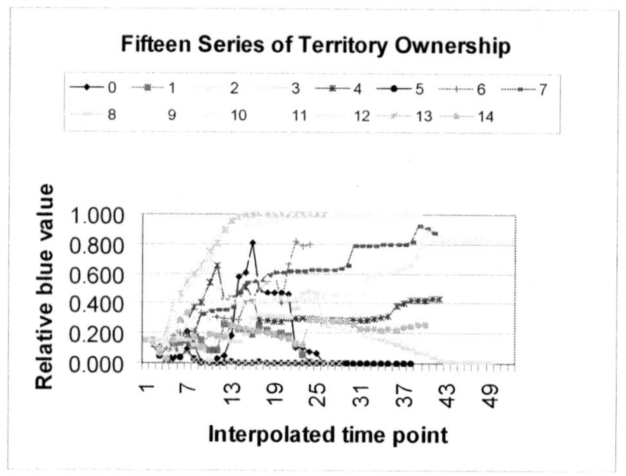

Figure 7. Fifteen series of relative ammunition values of the blue forces

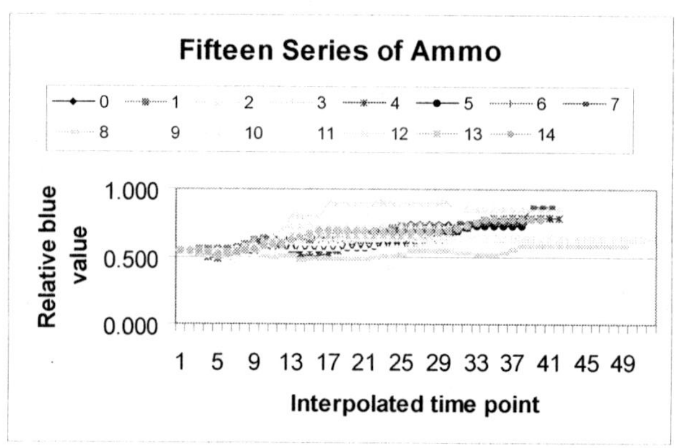

Figure 8. Fifteen series of relative fire intensity values of the blue forces

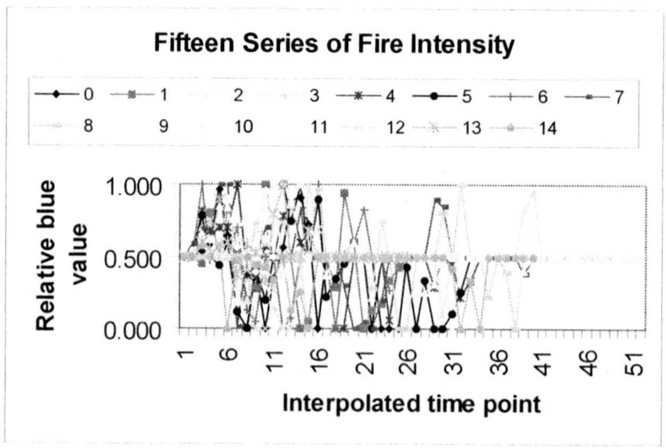

fast decision support, our analyses also attempt to use as few attributes and data points as possible. First, we applied the linear interpolation method to convert the original nonuniform series into uniform ones. Figure 5 shows the interpolated results of a battle simulation run. The uniform interval is consistently set at 100 seconds, which is relatively large.

Comparing Figure 5 with Figure 4, it is observed that the overall trends are retained for all series, except that some high-frequency activities of the intensity series are lost. The empirical results indicate that this relatively large down sampled size is sufficient for the clustering task. Nevertheless, it might be desirable to determine the optimal down sampled size in the future by investigating the tradeoff between improved clustering (an unknown) and increased computational cost. Naturally, a better interpolation method might also exist for our data. To limit the scope of this study, we elect to address these issues in the future.

Correlation analyses were performed on the five interpolated time series shown in Figure 5. It was discovered that series *g* and *s* are highly correlated (0.858); the same is true for *f* and *a* (0.771). These correlations generally apply to all simulation runs. Therefore, in the following, we will only discuss three indicators: *g*, *a*, and *i*.

Figures 6, 7, and 8 show the 15 interpolated time series for each one of the three indicators: *g*, *a*, and *i*, respectively. In these figures, the series numbered 0, 1, …, and 14 come from simulation run x0521, x4672, x0996, x5703, x7553, x3017, x0620, x8646, x4757, x0250, x2739, x6687, x5414, x7554, and x2514, respectively. Based on the clustering results of control-chart data and the cylinder-bell-funnel data, we chose the TWCV-based fitness function for both the fixed parameter GAs and AGAs. The DTW distance measure was chosen

Table 5. Clustering results of territory ownership series

Method	Number of clusters	Repl.	Best fitness	Clusters
FPGA	2	1-4	3308602	{0, 1, 4, 5, 8, 11, 12, 14}{2, 3, 6, 7, 9, 10, 13}
		5	3224082	{0, 1, 4, 5, 8, 11, 12, 14}{2, 3, 6, 7, 9, 10, 13}
	3	1-2	3907545	{1, 5, 12}{0, 4, 6, 8, 9, 11, 14}{2, 3, 7, 10, 13}
		3	3948684	{1, 5, 12}{0, 4, 8, 11, 14}{2, 3, 6, 7, 9, 10, 13}
		4	3842933	{1, 5, 12}{0, 4, 8, 11, 14}{2, 3, 6, 7, 9, 10, 13}
		5	3884901	{0, 1, 5, 8, 12}{2, 4, 6, 9, 11, 14}{3, 7, 10, 13}
	4	1-3	4365496	{1, 5, 12}{0, 8, 11, 14}{2, 4, 6, 9}{3, 7, 10, 13}
		4	4424450	{1, 5, 12}{0, 8, 11, 14}{2, 4, 6, 9}{3, 7, 10, 13}
		5	**4462072**	{1, 5, 12}{0, 8, 11, 14}{2, 4, 6, 9}{3, 7, 10, 13}
AGA	2	1-4	3308602	{0, 1, 4, 5, 8, 11, 12, 14}{2, 3, 6, 7, 9, 10, 13}
		5	3229174	{0, 1, 4, 5, 8, 11, 12, 14}{2, 3, 6, 7, 9, 10, 13}
	3	1-3	3978621	{1, 5, 12}{0, 4, 8, 11, 14}{2, 3, 6, 7, 9, 10, 13}
		4-5	3948684	{1, 5, 12}{0, 4, 8, 11, 14}{2, 3, 6, 7, 9, 10, 13}
	4	1	4386211	{1, 5, 12}{0, 8, 11, 14}{2, 4, 6, 9}{3, 7, 10, 13}
		2	4259005	{0}{1, 5, 8, 12}{2, 4, 6, 9, 11, 14}{3, 7, 10, 13}
		3	4294529	{1, 5, 8, 12}{0, 11, 14}{2, 4, 6, 9}{3, 7, 10, 13}
		4	4248009	{0}{1, 5, 12}{4, 8, 11, 14}{2, 3, 6, 7, 9, 10, 13}
		5	**4462072**	{1, 5, 12}{0, 8, 11, 14}{2, 4, 6, 9}{3, 7, 10, 13}

Table 6. Clustering results of ammunition series

Method	Number of clusters	Repl.	Best fitness	Clusters
FPGA	2	1-5	**4505133**	{8}{0, 1, 2, 3, 4, 5, 6, 7, 9, 10, 11, 12, 13, 14}
	3	1-3	4302312	{8}{2}{0, 1, 3, 4, 5, 6, 7, 9, 10, 11, 12, 13, 14}
		4	4297057	{8}{2, 3, 4}{0, 1, 5, 6, 7, 9, 10, 11, 12, 13, 14}
		5	4178824	{5}{11}{0, 1, 2, 3, 4, 6, 7, 8, 9, 10, 12, 13, 14}
	4	1	4086145	{8}{2}{14}{0, 1, 3, 4, 5, 6, 7, 9, 10, 11, 12, 13}
		2	4081404	{8}{2, 3, 4}{14}{0, 1, 5, 6, 7, 9, 10, 11, 12, 13}
		3	4070924	{8}{11}{5}{0, 1, 2, 3, 4, 6, 7, 9, 10, 12, 13, 14}
		4	4029205	{2}{3, 4}{11}{0, 1, 5, 6, 7, 8, 9, 10, 12, 13, 14}
		5	4060230	{8}{2, 3, 4}{7}{0, 1, 5, 6, 9, 10, 11, 12, 13, 14}
AGA	2	1-5	**4505133**	{8}{0, 1, 2, 3, 4, 5, 6, 7, 9, 10, 11, 12, 13, 14}
	3	1-4	4302312	{8}{2}{0, 1, 3, 4, 5, 6, 7, 9, 10, 11, 12, 13, 14}
		5	4297057	{8}{2, 3, 4}{0, 1, 5, 6, 7, 9, 10, 11, 12, 13, 14}
	4	1-3	4107672	{8}{2}{3, 4}{0, 1, 5, 6, 7, 9, 10, 11, 12, 13, 14}
		4	4101704	{8}{2, 3, 4}{11}{0, 1, 5, 6, 7, 9, 10, 12, 13, 14}
		5	4086145	{8}{2}{14}{0, 1, 2, 3, 5, 6, 7, 9, 10, 11, 12, 13}

Table 7. Clustering results of fire intensity series

Method	Number of clusters	Repl.	Best fitness	Clusters
FPGA	2	1	3794554	{0, 8, 11}{1, 2, 3, 4, 5, 7, 9, 10, 12, 13, 14}
		2-5	3820013	{0, 11, 14}{1, 2, 3, 4, 5, 6, 7, 8, 9, 10, 12, 13}
	3	1	3919665	{6}{0, 11, 14}{1, 2, 3, 4, 5, 7, 8, 9, 10, 12, 13}
		2	3892864	{6}{0, 8, 11}{1, 2, 3, 4, 5, 7, 9, 10, 12, 13, 14}
		3	3811803	{6}{4, 14}{0, 1, 2, 3, 5, 7, 8, 9, 10, 11, 12, 13}
		4	3870329	{8}{0, 11, 14}{1, 2, 3, 4, 5, 6, 7, 9, 10, 12, 13}
		5	3864623	{6}{11, 14}{0, 1, 2, 3, 4, 5, 7, 8, 9, 10, 12, 13}
	4	1	3956146	{6}{11, 14}{0}{1, 2, 3, 4, 5, 7, 8, 9, 10, 12, 13}
		2-3	3969378	{0, 8}{6}{11, 14}{1, 2, 3, 4, 5, 7, 9, 10, 12, 13}
		4	3900812	{6}{4, 14}{0}{1, 2, 3, 5, 7, 8, 9, 10, 11, 12, 13}
		5	3938876	{2}{0, 11, 14}{6}{1, 3, 4, 5, 7, 8, 9, 10, 12, 13}
AGA	2	1-5	3820013	{0, 11, 14}{1, 2, 3, 4, 5, 6, 7, 8, 9, 10, 12, 13}
	3	1-5	3919665	{6}{0, 11, 14}{1, 2, 3, 4, 5, 7, 8, 9, 10, 12, 13}
	4	1-4	**3972659**	{8}{6}{0, 11, 14}{1, 2, 3, 4, 5, 7, 9, 10, 12, 13}
		5	3956145	{0}{6}{11, 14}{1, 2, 3, 4, 5, 7, 8, 9, 10, 12, 13}

Figure 9. Four-cluster of territory ownership series generated by AGA

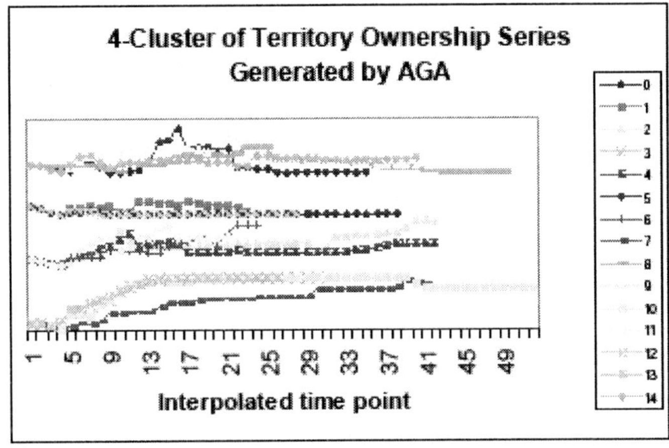

Figure 10. Four-cluster of ammunition series generated by AGA

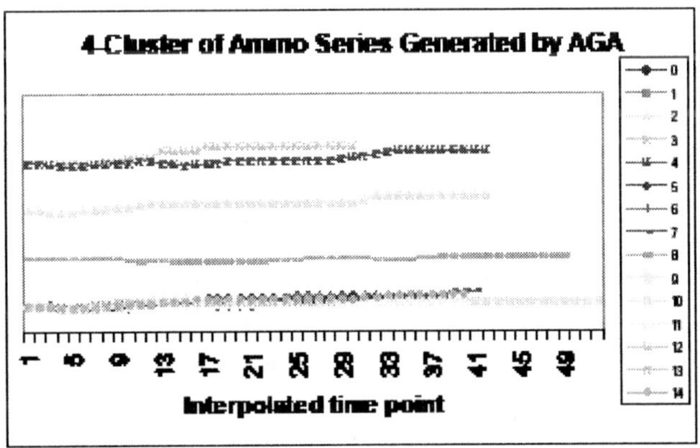

Figure 11. Four-cluster of fire intensity series generated by AGA

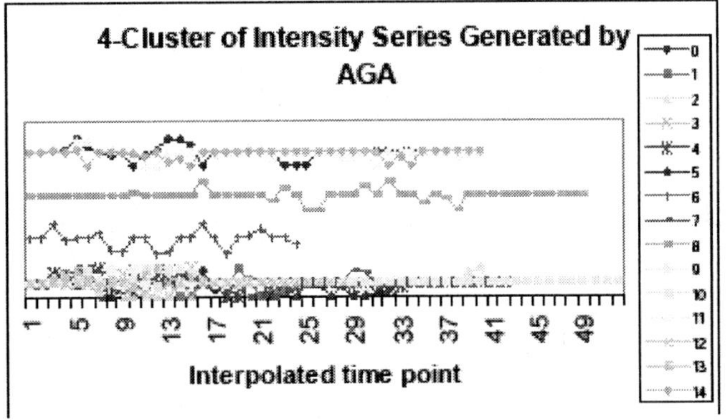

Figure 12. An example of dynamic time warping

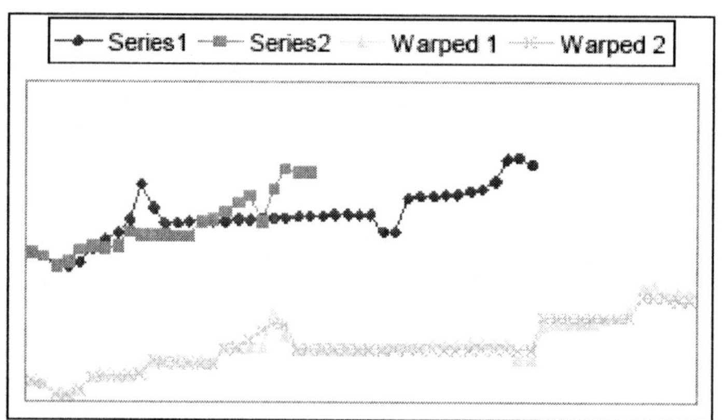

to handle these time series of unequal length. The maximum number of generations was kept at 30 throughout. For better results, the population size of 20 and mutation rate of 0.01 were selected for the fixed parameter GAs, whereas the population size of 60 was chosen for the AGAs. For each GA, five repetitions were made.

Tables 5-7 summarize the clustering results of territory ownership series, ammunition series, and firing intensity series, respectively, generated by both fixed parameter GAs and AGAs. In all tables, the medoids used by GA to form clusters are underlined. The best fitness value is shown in boldface for each indicator series. Based on these tables, the following observations can be made:

1. The AGAs, in general, generate more consistent results (with the only exception in generating four clusters of territory ownership series) and are able to find solutions with equivalent or higher best fitness than the fixed parameter GAs. Therefore, the results generated by AGAs are more trustworthy.
2. As the number of clusters increases, the consistency decreases for both AGAs and fixed parameter GAs.
3. A different set of medoids could lead to the same clustering results. For instance, the fixed-parameter GA forms the same set of four clusters of territory ownership series based on three different sets of medoids (for replications 1-3, 4, and 5, respectively).

Since there is no ground truth for this data set, it is not easy to conclude which clustering result is better because the clustering accuracy cannot be computed for this particular data set. As a standard practice in this situation, we resort to visual inspection for qualitative evaluation. Figures 9-11 show the four-cluster results generated by the AGA for each indicator series, respectively.

Note that in the previous three figures, the values along the vertical axis were changed from the original (for the series in three clusters) in order to separate one cluster from the others. Also note that the time-warping distance warps the time axis in comparing two series of unequal length. The No. 2 and No. 6 series of territory ownership were taken as examples to show how the time axis is warped to compare the two in Figure 12. Since the warping is done on a pair-by-pair basis, it would take a lot of space to show all the warped pairs. To better visualize the similarity between any pair of two series, one must also exercise this warping ability, as demonstrated in Figure 12.

Based on these results, we argue that the proposed genetic clustering method is effective for grouping time-series data produced in battle-simulation experiments. The notable feature of the proposed method is that it takes the entire battle sequence into consideration. Similar battle sequences, grouped in the same cluster, indicate that they are affected by the same set of mechanisms in play during the battle. Therefore, a model should be built for each cluster, to explain the mechanisms. Without using clustering as an exploratory step, one might build just one global model for all runs by assuming that there is only one mechanism in play. The predictions thus derived from this single global model are expected to be less accurate than from a set of local models developed for each cluster of data. The existence of more than one cluster for the same scenario indicates the nondeterministic nature of battles with several different mechanisms in play at the same time. This is exactly the knowledge discovered by exploratory data mining. The intriguing question is what tips one set of mechanisms from the other. Further investigation is necessary to answer this question. The study by Bodt *et al.* (Bodt, Forester, Hansen, Heilman, E., Kaste, & O'May, 2002) is a step in this direction, except that they did not consider the entire battle sequence.

DISCUSSION

To explain why one combination of fitness function and distance measure performs better than another, the correlation between the best fitness value and the clustering accuracy was computed for the synthetic control-chart data, as given in Table 8. It is interesting to see that for both fixed-parameter GAs and AGAs, the best clustering results were generated by the combination having the highest positive correlation between the best fitness value and the clustering accuracy. The negative correlations are undesirable because such fitness functions are counterproductive in the sense that maximizing the fitness by the GA actually leads to lower clustering accuracy. The low positive correlation indicates that the best fitness function used in this study is marginal, and a better one should be developed in order to achieve even higher clustering accuracy. This will be a topic for future studies.

For comparison, the clustering results of *k*-means were also obtained for both the synthetic control-chart data and the cylinder-bell-funnel data. For each data set, five runs were made. The clustering accuracies obtained are 63.3%, 70%, 60%, 56.7%, and 60% for the synthetic control-chart data, and are 87%, 87.3%, 87%, 87.3%, and 87% for the cylinder-bell-funnel data. Therefore, genetic clustering methods could obtain higher accuracy than *k*-means if an appropriate combination of fitness function and distance measure is used.

CONCLUSION

This chapter presented an exploratory data-mining study of time-series data. To this end, genetic algorithms were developed for clustering the data. First, the effects of three fitness functions, two distance measures, and some selected GA parameters were investigated using the synthetic control-chart data that is available in the public domain. The results indicate that the two most significant factors are the fitness function and the interaction between the fitness function and the distance measure. It was also discovered that the best fitness value has a low positive correlation with the clustering accuracy. To increase the clustering accuracy, a better fitness function must be developed, which poses both an opportunity and a challenge for researchers interested in this area of research. In addition, adaptive GAs did not always outperform fixed-parameter GAs. Better adaptive schemes are thus needed. Further testing on the cylinder-bell-funnel data confirms once again the significance of fitness function, distance measure, and their interaction. In addition, it was found that dynamic time warping did not work as well as Euclidean for this dataset. The

Table 8. Correlation between best fitness and accuracy

Fitness function	Distance measure	Fixed parameter GAs	AGAs
TWCA	Euclidean	0.327	0.264
TWCA	DTW	-0.100	0.267
DB	Euclidean	-0.783	-0.605
DB	DTW	-0.458	0.182
Rm	Euclidean	-0.129	-0.556
Rm	DTW	-0.676	-0.241

GA parameters such as population size, crossover rate, and mutation rate are relatively insignificant compared with fitness function and distance measure in the context of clustering.

The proposed genetic clustering methods were also applied to time-series data acquired in battle simulation experiments. The potential use of the proposed methods in grouping similar battle profiles and separating dissimilar battle profiles are shown and discussed. As far as future study is concerned, this study can be further improved by investigating the following general topics:

1. Develop a better fitness function.
2. Develop a better GA parameter adaptation method.
3. Try other chromosome coding schemes, such as integer coding rather than binary coding.
4. Develop a viable validation index, either application dependent or application independent, to determine how many clusters are optimal.
5. Extend the current method, which is currently applicable only to univariate time-series data, to multivariate time-series data.

Specific topics for the battle simulation study in the future include:

1. Complement this study with an in-depth analysis of the driving mechanisms/events that shape each group of battle sequences (or time-series profiles).
2. Investigate the effect of the interpolation method and sampling interval in converting nonuniform data into uniform data.

Finally, we shall point out that each time series was clustered individually in this study. For the battle simulation data, some multivariate time-series clustering methods such as hidden Markov models should be applicable. Alternatively, one can make use of a two-stage clustering procedure that was recently developed by Liao (2007), which involves converting multivariate time series into a univariate discrete time series in the first step and then applying the proposed genetic clustering algorithm in the second step. The results of the first stage have been presented in Liao *et al.* (Liao, Bodt, Forester, Hansen, Heilman, Kaste, & O'May, 2002). The complete two-stage procedure and results will be presented in our future work.

ACKNOWLEDGMENT

Dr. Liao acknowledges the support of ASEE-ARL Postdoctoral Fellowship (contract number DAAL01-96-C-0038) that helped make this study possible.

REFERENCES

Bandyopadhyay, S., & Maulik, U. (2002). Genetic clustering for automatic evolution of clusters and application to image classification. *Pattern Recognition, 35*, 1197-1208.

Baragona, R. (2001). A simulation study on clustering time series with metaheuristic methods. *Quaderni di Statistica, 3*, 1-26.

Bezdek, J. C. (1987). *Pattern recognition with fuzzy objective function algorithms.* New York: Plenum Press.

Bodt, B., Forester, J., Hansen, C., Heilman, E., Kaste, R., & O'May, J. (2002). Data mining combat simulations: a new approach for battlefield parameterization. *Army Research Conference*, Orlando FL, December 2002.

Davies, D. L., & Bouldin, D. W. (1979). A cluster separation measure. *IEEE Trans. Pattern Analysis and Machine Intelligence, 1*, 224-227.

Eiben, A. E., Hinterding, R., & Mizhalewicz, Z. (1999). Parameter control in evolutionary algorithms. *IEEE Transactions on Evolutionary Computation, 3*(2), 124-141.

Fu, T.-C., Chung F.-L., Ng, V., & Luk, R. (2001). Pattern discovery from stock time series using self-organizing maps. Paper presented at the *KDD 2001 Workshop on Temporal Data Mining*, August 26-29, San Francisco.

Golay, X., Kollias, S., Stoll, G., Meier, D., Valavanis, A., & Boesiger, P. (1998). A new correlation-based fuzzy logic clustering algorithm for fMRI. *Magnetic Resonance in Medicine, 40*, 249-260.

Goutte, C., Toft, P., & Rostrup, E. (1999). On clustering fMRI time series. *Neuroimage, 9*(3), 298-310.

Grefenstette, J. J. (1986). Optimization of control parameters for genetic algorithms. *IEEE Transactions on Systems, Man, and Cybernetics*, SMC-16, 122-128.

Hall, L. O., Özyurt, B., & Bezdek, J. C. (1999). Clustering with a genetically optimized approach. *IEEE Transactions on Evolutionary Computation, 3*(2), 103-112.

Heilman, E., *et al.* (2002) Identifying battlefield metrics through experimentation. *Proceedings of the 7th International Command & Control Research & Technology Symposium.*

Herrera, F., & Lozano, M. (2003). Fuzzy adaptive genetic algorithms: design, taxonomy, and future directions. *Soft Computing, 7*, 545-562.

Kalpakis, K., Gada, D., & Puttagunta, V. (2001). Distance measures for effective clustering of ARIMA time-series. *Proceedings of the 2001 IEEE International Conference on Data Mining* (pp. 273-280)., San Jose, CA, Nov. 29 – Dec. 2, 2001.

Kaufman, L., & Rousseeuw, P. J. (1990). *Finding groups in data: An introduction to cluster analysis.* New York: John Wiley & Sons.

Krishna, K., & Murty, M. N. (1999). Genetic k-means algorithms. *IEEE Transactions on Systems Man and Cybernetics - Part B: Cybernetics, 29*(3), 433-439.

Krishnapuram, R., Joshi, A., Nasraoui, O., & Yi, L. (2001). Low-complexity fuzzy relational clustering algorithms for web mining. *IEEE Transactions on Fuzzy Systems, 9*(4), 595-607.

Li, C., & Biswas, G. (1999). Temporal pattern generation using hidden Markov model based unsupervised classification. In D. J. Hand, J. N. Kok, M. R. Berthold (Eds.), IDA '99, *LNCS 164*, Springer-Verlag: Berlin, 245-256.

Liao, T. W. (2005). Clustering of time series data - a survey. *Pattern Recognition, 38*(11), 1857-1874.

Liao, T. W. (2007). A clustering procedure for exploratory mining of vector time series. *Pattern Recognition*, doi:10.1016/j.patcog.2007.01.005.

Liao, T. W., Bodt, B., Forester, J., Hansen, C., Heilman, E., Kaste, R., & O'May, J. (2002). *Army Research Conference*, Orlando FL, December 2002.

Lorena, L. A. N., & Furtado, J. C. (2001). Constructive genetic algorithms for clustering problems. *Evolutionary Computation*, *9*(3), 309-327.

MacQueen, J. (1967). Some methods for classification and analysis of multivariate observations. In L. M. LeCam & J. Neyman (Eds.), *Proceedings of the* 5th *Berkeley Symposium on Mathematical Statistics and Probability*, Vol. 1 (pp. 281-297), University of California Press, Berkeley.

Maulik, U., & Bandyopadhyay, S. (2000). Genetic algorithm-based clustering technique. *Pattern Recognition*, *33*, 1455-1465.

Mitchell, M. (1996). *An introduction to genetic algorithms.* Cambridge, MA: The MIT Press.

O'May, J., & Hailman, E. (2002). OneSAF killer/victim scoreboard capability. ARL-TR-2829.

Pham, D. T., & Chan, A. B. (1998). Control chart pattern recognition using a new type of self-organizing neural network. *Proc. Instn. Mech. Engrs.*, *212*(1), 115-127.

Policker, S., & Geva, A. B. (2000). Nonstationary time series analysis by temporal clustering. *IEEE Transactions on Systems, Man, and Cybernetics—Part B: Cybernetics*, *30*(2), 339-343.

Rojas, Ignacio, González, J., Pomares, H., Merelo, J. J., Castillo, P. A., & Romero, G. (2002). Statistical analysis of the main parameters involved in the design of a genetic algorithm. *IEEE Transactions on Systems, Man, and Cybernetics-Part C: Applications and Reviews*, *32*(1), 31-37.

Srinivas, M., & Patnaik, L. M. (1994). Adaptive probabilities of crossover and mutation in genetic algorithms. *IEEE Transactions on Systems, Man, and Cybernetics*, *24*(4), 656-666.

This work was previously published in Mathematical Methods for Knowledge Discovery and Data Mining, edited by G. Felici and C. Vercellis, pp. 157-178, copyright 2008 by Information Science Reference, formerly known as Idea Group Reference (an imprint of IGI Global).

Chapter 2.25
Two Rough Set Approaches to Mining Hop Extraction Data

Jerzy W. Grzymala-Busse
University of Kansas, USA and Institute of Computer Science, PAS, Poland

Zdzislaw S. Hippe
University of Information Technology and Management, Poland

Teresa Mroczek
University of Information Technology and Management, Poland

Edward Roj
Fertilizer Research Institute, Poland

Boleslaw Skowronski
Fertilizer Research Institute, Poland

ABSTRACT

Results of our research on using two approaches, both based on rough sets, to mining three data sets describing bed caking during the hop extraction process, are presented. For data mining we used two methods: direct rule induction by the and generation of belief networks associated with conversion belief networks into rule sets by the BeliefSEEKER system. Statistics for rule sets are presented, including an error rate. Finally, six rule sets were ranked by an expert. Our results show that both our approaches to data mining are of approximately the same quality.

INTRODUCTION

In the past, addition of hop to beer was quite simple: hop was added to the boiling wort, while spent hop formed a filter bed for the wort. Bitterness was controlled by varying the used hop. A

Table 1. Composition of hop extract

Components		Contents
Total resins		15 %
	Soft resins - alpha acids	8 %
	Soft resins - beta acids	4 %
	Hard and uncharacterized soft resins	3 %
Essential oils		1 %
Tannins		4 %
Proteins		15 %
Water		9.5 %
Monosacharides		2 %
Lipids and waxes		3 %
Amino acids		0.1 %
Pectin		2 %
Ash		8 %
Cellulose, lignin, and so forth.		40.4 %

few decades ago the process was changed, and the technology of beer production increased in complexity and sophistication.

The following factors are essential for beer production:

- Reduction of hop extraction cost
- Unification of production process
- Environmental protection regulations (pesticide residues, nitrate residues, heavy metals, etc.)
- Requirements of particular consumer groups for special beers

Currently, fresh hop cones are dried in a dryer and then compressed. Hop cones are usually traded in bales that have been compressed to reduce shipping volumes. Some of the main types of the product on the market are either hop pellets, made from unmilled hop cones, or hop powder pellets, made from milled hop cones and presented at different levels of enrichment.

Hop pellets are prepared by freeing hops from foreign matter (stems, rhizomes, metal rods, wires, etc.) and then pelleting them without milling. Hop pellets are not significant in modern beer production, but still can be seen in use in some countries.

Hop powder pellets in various forms are dominant in the brewing industry around the world. Basic preparation consists of removing foreign matter, milling in a hammer mill, blending batches of several hop bales together for product consistency, pelleting through a standardized pellet die, cooling and packing in packs.

Principles of hop extraction have been known for 150 years, when extraction in water and ethanol was, for the first time, successfully applied. An ethanol extraction is still in use, but the predominant extracts are either organic or the CO_2 based extracts (in liquid or supercritical phase). Recently, due to the properties of CO_2, the supercritical extraction is getting more popular among commercial producers.

The soft resins and oils comprise usually about 15% of the hop mass (new varieties sometimes exceed that value) and strongly depend on hop variety (Stevens, 1987). An exemplary composition of hop extract is presented in Table 1 (the example is based on a mid-alpha acid content variety).

There are other commercial methods to extract oil or resins from hop or other commercial plants, such as herbs, flax, hemp, and so forth. Organic solvent-based methods are commonly applied worldwide. But during the last two decades, a supercritical CO_2 extraction has been used intensively in various applications replacing the classic organic solvent methods (Chassagnes-Mendez, Machado, Araujo, Maia, & Meireles, 2000; Skowronski, 2000; Skowronski & Mordecka, 2001). The extraction process under supercritical conditions is usually carried out in a high-pressure reactor vessel filled with a preprocessed material in the form of tablets, pellets, granules, and so forth., thus, forming a granular bed through which flows a compressed gas. The gas under a supercritical condition is able to dissolve oil products contained in the bed. A mixture of oil products and solvent (CO_2) flows together through a low-pressure separator where liquid and gas fractions can be efficiently separated. The oil mixture is ecologically clean and does not require any additional treatment. The spent gas expands to the atmosphere or is recirculated after compression. Produced liquid resin and oil fractions are directed to the storage tanks.

As we have already mentioned, the extraction process uses pressed organic material (cones or leafs) mainly in the form of tablets or extrusions in order to load more material for each batch. After the bed is loaded, the pressure and temperature are increased to reach a supercritical condition. Once a supercritical condition is reached, the gas flows through the bed at a constant pressure and temperature. After a few hours of running the process, the tank is decompressed to ambient pressure and the batch process is finished. Sometimes during the extraction process, high-pressure drops occur and, as a result, the bed is caked. This event has a negative influence on the extraction process. A pressure drop reduces gas circulation through the circulation loop (pressure tank and the rest of the equipment) and therefore extends the extraction time. This also strongly affects process efficiency by increasing production costs.

The objective of our work was to investigate system variables that would contribute to the bed-caking tendency during the hop extraction process. Commercial experience indicated two parameters determine the bed's state, namely, content of extract and moisture of the granular bed. The content of the extract concentration varies significantly, between 14% and 32%. For that reason, three easy–to-identify input parameters have been identified, namely:

- Variety
- Content of extract (α_{kg})
- Moisture of the bed (m_{wg})

In one of the data sets, two additional variables were used:

- Content of alpha acid in spent hop, that is, remaining concentration (α_{kw}), and
- Moisture of the postprocessed bed (m_{ww}).

Certainly, there are other unknown factors that have not been considered here due to essential measurement problems. Output variables were:

- The caking rate of the bed and
- Extraction time

In fact, there is a strong relationship between the caking rate and time of extraction. If a bed is caked, then extraction time increases. The caking rate description is based on process operator reports. The following four grades of bed caking rates have been used:

1. loose material,
2. loose material with dust,
3. caked and slightly raised bed, and
4. strongly caked and raised bed.

The extraction period has been also quantified according to the following principles:

1. A standard time of extraction over a loose bed.
2. Extension of time by 25% with reference to the standard time.
3. Extension of extraction time by 50% with reference to the standard time.
4. Extension of extraction time by 75% with reference to the standard time.

The experimental data have been processed using two data-mining approaches: rule induction, represented by the MLEM2 algorithm, an option of the LERS (learning from examples based on rough sets) system and generation of Bayesian networks by the BeliefSEEKER system.

A preliminary version of this paper was presented at the ISDA'2005 Conference (Grzymala-Busse, Hippe, Mroczek, Roj, & Skowronski, 2005). In this chapter, we present different results, since we changed our methodology, running MLEM2 directly from raw data instead of running LEM2 on discretized data as in Grzymala-Busse *et al.* (2005). Here we consider both certain and possible rule sets induced by LEM2. Additionally, we present new results on ranking six rule sets by an expert.

DISCRETIZATION

In our data sets, all attributes, except *Variety*, were numerical. The numerical attributes should be discretized before rule induction. The data-mining system LERS uses for discretization a number of discretization algorithms (Chmielewski & Grzymala-Busse, 1996). To compare both approaches, LERS and BeliefSEEKER, used for data mining, all data sets were discretized using the same LERS discretization method, namely, a polythetic agglomerative method of cluster analysis (Everitt, 1980). *Polythetic* methods use all attributes, while *agglomerative* methods begin with all cases being singleton clusters. Our method was also *hierarchical*, that is, the final structure of all formed clusters was a tree.

More specifically, we selected to use the median cluster analysis method (Everitt, 1980) as a basic clustering method. Different numerical attributes were standarized by dividing the attribute values by the corresponding attribute's standard deviation (Everitt, 1980). Cluster formation was started by computing a distance matrix between every cluster. New clusters were formed by merging two existing clusters that were the closest to each other. When such a pair was founded (clusters *b* and *c*), they were fused to form a new cluster *d*. The formation of the cluster *d* introduces a new cluster to the space, and hence its distance to all the other remaining clusters must be recomputed. For this purpose, we used the Lance and Williams Flexible Method (Everitt, 1980). Given a cluster a and a new cluster *d* to be formed from clusters *b* and *c*, the distance from *d* to *a* was computed as:

$$d_{ad} = d_{da} = 0.5 * d_{ab} + 0.5 * d_{ac} - 0.25 * d_{bc}.$$

At any point during the clustering process, the formed clusters induce a partition on the set of all cases. Cases that belong to the same cluster are indiscernible by the set of numerical attributes. Therefore, we should continue cluster formation until the level of consistency of the partition formed by clusters is equal to or greater than the original data's level of consistency.

Once clusters are formed, the postprocessing starts. First, all clusters are projected on all attributes. Then the resulting intervals are merged to reduce the number of intervals and, at the same time, preserving consistency. Merging of intervals begins from safe merging where, for each attribute, neighboring intervals labeled by the same decision value are replaced by their union. The next step of merging intervals is based on checking every pair of neighboring intervals, whether their merging will result in preserving consistency. If so, intervals are merged permanently. If not, they are marked as unmergeable. Obviously, the order in which pairs of intervals are selected affects the final outcome. In our experiments, we started from attributes with the largest conditional entropy of the decision-given attribute.

LERS

The data system LERS (Grzymala-Busse, 1992, 1997, 2002) induces rules from inconsistent data, that is, data with conflicting cases. Two cases are conflicting when they are characterized by the same values of all attributes, but they belong to different concepts. LERS handles inconsistencies using rough-set theory, introduced by Z. Pawlak in 1982 (1982, 1991). In rough-set theory, lower and upper approximations are computed for concepts involved in conflicts with other concepts.

Rules induced from the lower approximation of the concept *certainly* describe the concept; hence, such rules are called *certain* (Grzymala-Busse, 1988). On the other hand, rules induced from the upper approximation of the concept describe the concept *possibly*, so these rules are called *possible* (Grzymala-Busse, 1988). In general, LERS uses two different approaches to rule induction: one is used in machine learning, the other in knowledge acquisition. In machine learning, the usual task is to learn the smallest set of minimal rules, describing the concept. To accomplish this goal, LERS uses three algorithms: LEM1, LEM2, and MLEM2 (LEM1, LEM2, and MLEM2 stand for learning from examples module, version 1, 2, and modified, respectively). In our experiments, rules were induced using the algorithm MLEM2.

MLEM2

The LEM2 option of LERS is most frequently used for rule induction since, in most cases, it gives better results. LEM2 explores the search space of attribute-value pairs. Its input data set is a lower or upper approximation of a concept, so its input data set is always consistent. In general, LEM2 computes a local covering and then converts it into a rule set. We will quote a few definitions to describe the LEM2 algorithm (Chan & Grzymala-Busse, 1991; Grzymala-Busse, 1997).

The LEM2 algorithm is based on an idea of an attribute-value pair block. Let U be the set of all cases (examples) of the data set. For an attribute-value pair $(a, v) = t$, a *block* of t, denoted by $[t]$, is a set of all cases from U, such that for attribute a have value v. Let B be a nonempty lower or upper approximation of a concept represented by a decision-value pair (d, w). Set B *depends* on a set T of attribute-value pairs $t = (a, v)$ if and only if:

$$\emptyset \neq [T] = \bigcap_{t \in T} [t] \subseteq B.$$

Set T is a *minimal complex* of B if and only if B depends on T and no proper subset T' of T exists such that B depends on T'. Let **T** be a nonempty collection of nonempty sets of attribute-value pairs. Then **T** is a *local covering* of B if and only if the following conditions are satisfied:

1. Each member T of **T** is a minimal complex of B.
2. $\bigcup_{T \in \mathbf{T}} [T] = B$
3. **T** is minimal, that is, **T** has the smallest possible number of members.

The procedure LEM2, based on rule induction from local coverings, is presented below.

Procedure LEM2
(input: a set B;
output: a single local covering **T** of set B);
begin
$G := B$;
T$:= \emptyset$;
while $G \neq \emptyset$ **do**
begin

$T := \emptyset$
$T(G) := \{t \mid [t] \cap G \neq \emptyset\}$;
while $T = \emptyset$ **or not** $([T] \subseteq B)$ **do**
begin

select a pair $t \in T(G)$ with the highest attribute priority, if a tie occurs, select a pair $t \in T(G)$ such that $|[t] \cap G|$ is maximum; if another tie occurs, select a pair $t \in T(G)$ with the smallest cardinality of $[t]$; if a further tie occurs, select first pair:

$T := T \cup \{t\}$;
$G := [t] \cap G$;
$T(G) := \{t \mid [t] \cap G \neq \emptyset\}$;
$T(G) := T(G) - T$;

end; {while}
for each t in T **do**
if $[T-\{t\}] \subseteq B$ **then** $T := T - \{t\}$;
$\mathbf{T} := \mathbf{T} \cup \{T\}$;
$G := B - \cup_{T \in \mathbf{T}} [T]$;
end {while};
for each $T \in \mathbf{T}$ **do**
if $\cup_{S \in \mathbf{T} - \{T\}} [S] = B$ **then** $\mathbf{T} := \mathbf{T} - \{T\}$;
end {procedure}.

For a set X, $|X|$ denotes the cardinality of X.

MLEM2, a modified version of LEM2, processes numerical attributes differently than symbolic attributes. For numerical attributes, MLEM2 sorts all values of a numerical attribute. Then it computes cutpoints as averages for any two consecutive values of the sorted list. For each cutpoint q, MLEM2 creates two blocks: the first block contains all cases for which values of the numerical attribute are smaller than q, the second block contains remaining cases, that is, all cases for which values of the numerical attribute are larger than q. The search space of MLEM2 is the set of all blocks computed this way, together with blocks defined by symbolic attributes. Starting from that point, rule induction in MLEM2 is conducted the same way as in LEM2. At the very end, MLEM2 simplifies rules by merging appropriate intervals for numerical attributes.

Classification System

Rule sets, induced from data sets, are used mostly to classify new, unseen cases. Such rule sets may be used in rule-based expert systems.

There are a few existing classification systems, for example, associated with rule induction systems LERS or AQ. A classification system used in LERS is a modification of the well-known bucket brigade algorithm (Grzymala-Busse, 1997; Stefanowski, 2001). In the rule induction system AQ, the classification system is based on a rule estimate of probability. Some classification systems use a decision list in which rules are ordered; the first rule that matches the case classifies it. In this section, we will concentrate on a classification system used in LERS.

The decision to which concept a case belongs to is made on the basis of three factors: *strength*, *specificity*, and *support*. These factors are defined as follows: *strength* is the total number of cases correctly classified by the rule during training. *Specificity* is the total number of attribute-value pairs on the left-hand side of the rule. The matching rules with a larger number of attribute-value pairs are considered more specific. The third factor, *support*, is defined as the sum of products of strength and specificity for all matching rules indicating the same concept. The concept C for which the support, that is, the following expression:

$$\sum_{R \in Rul} Strength(R) * Specifity(R),$$

is the largest is the winner, and the case is classified as being a member of C, where Rul denotes the set of all matching rules R describing the concept C.

In the classification system of LERS, if complete matching is impossible, all partially matching rules are identified. These are rules with at least one attribute-value pair matching the corresponding attribute-value pair of a case.

For any partially matching rule R, the additional factor, called *matching_factor* (R), is computed. Matching_factor (R) is defined as the ratio of the number of matched attribute-value pairs with a case to the total number of attribute-value pairs. In partial matching, the concept C for which the following expression is the largest:

$$\sum_{R \in Rul'} Matching\_factor(R) * Strength(R)$$
$$*Specifity(R),$$

is the winner, and the case is classified as being a member of C, where *Rul'* denotes the set of all matching rules R describing the concept C.

VALIDATION

The most important performance criterion of rule induction methods is the error rate. If the number of cases is less than 100, the *leaving-one-out* method is used to estimate the error rate of the rule set. In leaving-one-out, the number of learn-and-test experiments is equal to the number of cases in the data set. During the i-th experiment, the i-th case is removed from the data set, a rule set is induced by the rule induction system from the remaining cases, and the classification of the omitted case by rules produced is recorded. The error rate is computed as the ratio of the total number of misclassifications to the number of cases.

On the other hand, if the number of cases in the data set is greater than or equal to 100, the ten-fold cross-validation should be used. This technique is similar to leaving-one-out in that it follows the learn-and-test paradigm. In this case, however, all cases are randomly reordered, and then a set of all cases is divided into 10 mutually disjoint subsets of approximately equal size. For each subset, all remaining cases are used for training, that is, for rule induction, while the subset is used for testing. This method is used primarily to save time at the negligible expense of accuracy.

Ten-fold cross validation is commonly accepted as a standard way of validating rule sets. However, using this method twice, with different preliminary random reordering of all cases yields, in general, two different estimates for the error rate (Grzymala-Busse, 1997).

For large data sets (at least 1,000 cases), a single application of the train-and-test paradigm may be used. This technique is also known as *holdout*. Two thirds of cases should be used for training, one third for testing.

In yet another way of validation, *resubstitution*, it is assumed that the training data set is identical with the testing data set. In general, an estimate for the error rate is here too optimistic. However, this technique is used in many applications. For some applications, it is debatable whether the ten-fold cross-validation is better or not, see, for example, Braga-Neto, Hashimoto, Dougherty, Nguyen, & Carroll, 2004).

BeliefSEEKER

BeliefSEEKER is a computer program generating belief networks for any type of decision tables prepared in the format described in Pawlak (1995). Various algorithms may be applied for learning such networks (Mroczek, Grzymala-Busse, & Hippe, 2004). The development of belief networks is controlled by a specific parameter, representing the maximum dependence between variables, known as marginal likelihood and defined as follows:

$$ML = \prod_{i=1}^{v} \prod_{j=1}^{q_i} \frac{\Gamma(\alpha_{ij})}{\Gamma(\alpha_{ij} + n_{ij})} \prod_{k=1}^{c_i} \frac{\Gamma(\alpha_{ijk} + n_{ijk})}{\Gamma(\alpha_{ijk})},$$

where v is the number of nodes in the network, $i = 1,\ldots,v$, q_i is the number of possible combinations of parents of the node X_i (if a given attribute does not contain nodes of the parent type, then $q_i = 1$), c_i is the number of classes within the attribute X_i, $k = 1,\ldots, c_i$, n_{ijk} is the number of rows in the database, for which parents of the attribute X_i have value *j*, and this attribute has the value of *k*, and α_{ijk} and n_{ij} are parameters of the initial Dirichlet's distribution (Heckerman, 1995).

It is worth to mention that the developed learning models (belief networks) can be converted into some sets of belief rules, characterized by a specific parameter called a *certainty factor*, *CF*, that reveals indirectly the influence of the most significant descriptive attributes on the dependent variable. Also, to facilitate the preliminary evaluation of generated rules, an additional mechanism supports the calculation of their specificity, strength, generality, and accuracy (Hippe, 1996). Characteristic features of the system are:

- Capability to generate various exhaustive learning models (Bayesian networks) for different values of Dirichlet's parameter α and the certainty factor *CF* (Heckerman, 1995).
- Capability to convert generated belief networks into possible rule sets.
- Built-in classification mechanism for unseen cases.

RESULTS OF EXPERIMENTS

Tables 2 and 3 show basic statistics about our three data sets representing granular bed caking during hop extraction. In Tables 4–6, results of our experiments are presented. Note that for every data set, the MLEM2 rule induction algorithm was run once, while BeliefSEEKER generated many rule sets for selected values of α and *CF*. Table 4 shows error rates, computed by ten-fold cross validation for MLEM2 and resubstitution for BeliefSEEKER. Resubstitution as a validation technique for BeliefSEEKER has been selected because of time concerns. Furthermore, Table 4 shows the best results of BeliefSEEKER in terms of error rates. These rule sets were generated with $\alpha = 50$ and $CF = 0.4$ for the first data set, with $\alpha = 10$ and $CF = 0.5$ for the second data set, and with $\alpha = 1$ and $CF = 0.5$ for the third data set.

Table 2. Data sets

Data set	Number of cases	Number of attributes	Number of concepts
Data-set-1	288	5	4
Data-set-2	288	3	4
Data-set-3	179	3	4

Table 3. Number of conflicting cases

Data set	Before discretization	After discretization
Data-set-1	2	6
Data-set-2	56	137
Data-set-3	23	53

Table 4. Error rates

Data set	MLEM2		Belief-SEEKER
	Certain rule set	Possible rule set	
Data-set-1	9.38%	9.72%	6.60%
Data-set-2	11.11%	9.72%	10.76%
Data-set-3	9.50%	8.94%	8.94%

Table 5 presents the cardinalities of all nine rule sets from Table 4. Finally, Table 6 shows ranking of all possible rule sets induced by MLEM2 and BeliefSEEKER. We restricted our attention only to possible rule sets, since the procedure was conducted manually by an expert who ranked rule after rule, judging rules on scale from 1 (unacceptable) to 5 (excellent). Table 6 presents averages of all rule rankings for all six rule sets.

CONCLUSION

This section is divided into three subsections, describing results of MLEM2, BeliefSEEKER, and then general conclusions.

1. *Discussion of results obtained by running the MLEM2 system.*

The basic results obtained by running the MLEM2 system against our three data sets are:

- *Marynka* variety and its modification contains a lesser amount of extract than *Magnum* variety and its modification.
- The bed state is desirable, 1 or 2, if:
    - Content of extract in a bed is ranged below 22%, and moisture of the bed is ranged below 7%, or a content of extract is below 14% and moisture of the bed is ranged below 10.95%.
    - Content of extract in *Marynka* variety does not exceed 14% and moisture 8.35%,
    - content of extract in *Magnum* variety does nor exceed 20% and moisture 8.15%,
- Both varieties, *Marynka* and *Magnum*, can be caked sometimes, but the absolute content of extract is not a main contributor of caking.
- Moisture of granulated hops plays a stabilizing role at the determined ranges; below the determined value, the extracted bed is loose, and above this value it tends to cake.

2. *Discussion of results obtained by using the Bayesian approach.*

The basic results obtained by using the Bayesian networks against our three data sets are:

- *Marynka* variety and its modification contains less extract compared to the *Magnum* variety and its modification.

Table 5. Number of rules

Data set	MLEM2		Belief-SEEKER
	Certain rule set	Possible rule set	
Data-set-1	34	34	37
Data-set-2	28	22	9
Data-set-3	12	22	3

Table 6. Ranking of rule sets

Data set	MLEM2	BeliefSEEKER
Data-set-1	4.48	4.62
Data-set-2	4.39	4.22
Data-set-3	3.42	1.00

- Caking ability is a common feature for both varieties and their modifications. In general, the less amount of extract in the granulated hops of the particular variety, the less tendency to caking.
- Higher moisture content in the bed is a favorable caking factor.
- Cake ability usually extends extraction time.

GENERAL CONCLUSIONS

Experts found that the results of both data-mining systems: MLEM2 and BeliefSEEKER are consistent with the current state of the art in the area. In general, possible rule sets were highly ranked by the expert for the first two data sets. The possible rule sets induced from the third data set were ranked lower.

Moreover, stating knowledge in the form of rule sets and Bayesian networks makes it possible to utilize knowledge in a form of an expert system. Thus, results of our research may be considered as a successful first step towards building a system of monitoring the process of hop extraction.

Results of MLEM2 and BeliefSEEKER, in terms of an error rate, indicate that both systems are approximately of the same quality.

REFERENCES

Braga-Neto, U., Hashimoto, R., Dougherty, R. E, Nguyen, D.V., & Carroll, R.J. (2004). Is cross-validation better than resubstitution for ranking genes? *Bioinformatics*, *20*, 253-258.

Chan, C.-C., & Grzymala-Busse, J.W. (1991). *On the attribute redundancy and the learning programs ID3, PRISM, and LEM2* (TR-91-14). Lawrence, KS: Department of Computer Science, University of Kansas.

Chassagnez-Mendez, A.L., Machado, N.T., Araujo, M.E., Maia, J.G., & Meireles, M.A. (2000). Supercritical CO_2 extraction of eurcumis and essential eil from the rhizomes of turmeric (Curcuma longa L.), *Ind. Eng. Chem. Res.*, *39*, 4729-4733.

Chmielewski, M. R., & Grzymala-Busse, J. W. (1996). Global discretization of continuous attributes as preprocessing for machine learning. *International Journal of Approximate Reasoning*, *15*, 319-331.

Everitt, B. (1980). *Cluster analysis* (2nd ed.). London: Heinemann Educational Books.

Grzymala-Busse, J.W. (1988). Knowledge acquisition under uncertainty—A rough set approach. *Journal of Intelligent & Robotic Systems*, *1*, 3-16.

Grzymala-Busse, J.W. (1992). LERS—A system for learning from examples based on rough sets. In R. Slowinski (Ed.), *Intelligent decision support. Handbook of applications and advances of the rough set theory* (pp. 3-18). Dordrecht, Boston, London: Kluwer Academic Publishers.

Grzymala-Busse, J.W. (1997). A new version of the rule induction system LERS. *Fundamenta Informaticae*, *31*, 27-39.

Grzymala-Busse, J.W. (2002). MLEM2: A new algorithm for rule induction from imperfect data. In *Proceedings of the 9th International Conference on Information Processing and Management of Uncertainty in Knowledge-Based Systems* (pp. 243-250). Annecy, France.

Grzymala-Busse, J.W., Hippe, Z.S., Mroczek, T., Roj, E., & Skowronski, B. (2005). Data mining analysis of granular bed caking during hop extraction. In *Proceedings of the ISDA'2005, Fifth International Conference on Intelligent System Design and Applications* (pp. 426-431). IEEE Computer Society, Wroclaw, Poland.

Heckerman, D. (1995). *A tutorial on learning Bayesian networks* (MSR-TR-95-06). Retrieved from heckerman@microsoft.com

Hippe, Z.S. (1996). Design and application of new knowledge engineering tool for solving real world problems, *Knowledge-Based Systems, 9*, 509-515.

Mroczek, T., Grzymala-Busse, J.W., & Hippe, Z.S. (2004). Rules from belief networks: A rough set approach. In S. Tsumoto, R. Slowinski, J. Komorowski, and J.W. Grzymala-Busse (Eds.), *Rough sets and current trends in computing* (pp. 483-487). Berlin, Heidelberg, New York: Springer-Verlag.

Pawlak, Z. (1982). Rough sets. *International Journal of Computer and Information Sciences, 11*, 341-356.

Pawlak, Z. (1991). *Rough sets. Theoretical aspects of reasoning about data.* Dordrecht, Boston, London: Kluwer Academic Publishers.

Pawlak, Z. (1995). Knowledge and rough sets (in Polish). In W. Traczyk (Ed.), *Problems of artificial intelligence* (pp. 9-21). Warsaw, Poland: Wiedza i Zycie.

Skowronski, B. (2000). Interview by L. Dubiel. Hop extract—Polish at last (in Polish), *Przem. Ferment. i Owocowo -Warzywny, 9*, 30-31.

Skowronski B., & Mordecka, Z. (2001). Polish plant for supercritical extraction of hop (in Polish), *Przem. Chem., 80*, 521-523.

Stefanowski, J. (2001). *Algorithms of decision rule induction in data mining.* Poznan, Poland: Poznan, University of Technology Press.

Stevens, R. (1987). *Hops, An introduction to brewing science and technology.* Series II, Vol. I, p. 23. London: Institute of Brewing.

This work was previously published in Rough Computing: Theories, Technologies and Applications, edited by A. E. Hassanien, Z. Suraj, D. Slezak, and P. Lingras, pp. 227-238, copyright 2008 by Information Science Reference, formerly known as Idea Group Reference (an imprint of IGI Global).

Chapter 2.26
Semantics-Aware Advanced OLAP Visualization of Multidimensional Data Cubes

Alfredo Cuzzocrea
University of Calabria, Italy

Domenico Saccà
University of Calabria, Italy

Paolo Serafino
University of Calabria, Italy

ABSTRACT

Efficiently supporting advanced OLAP visualization of multidimensional data cubes is a novel and challenging research topic, which results to be of interest for a large family of data warehouse applications relying on the management of spatio-temporal (e.g., mobile) data, scientific and statistical data, sensor network data, biological data, etc. On the other hand, the issue of visualizing multidimensional data domains has been quite neglected from the research community, since it does not belong to the well-founded conceptual-logical-physical design hierarchy inherited from relational database methodologies. Inspired from these considerations, in this article we propose an innovative advanced OLAP visualization technique that meaningfully combines (i) the so-called OLAP dimension flattening process, which allows us to extract two-dimensional OLAP views from multidimensional data cubes, and (ii) very efficient data compression techniques for such views, which allow us to generate "semantics-aware" compressed representations where data are grouped along OLAP hierarchies.

INTRODUCTION

OLAP systems (Chaudhuri & Dayal, 1997; Codd, Codd, & Salley, 1993; Inmon, 1996; Kimball, 1996) have rapidly gained momentum in both the academic and research communities, mainly due to their capability of exploring and querying huge amounts of data sets according to a multidimensional and multi-resolution vision. Research-wise, three relevant challenges of OLAP have captured the attention of researchers during the last years: (*i*) the *data querying* problem, which concerns with how data are accessed and queried to support summarized knowledge extraction from massive data cubes; (*ii*) the *data modeling* problem, which concerns with how data are represented and, thus, processed inside OLAP servers (e.g., during query evaluation); and (*iii*) the *data visualization* problem, which concerns with how data are presented to OLAP users and decision makers in data warehouse environments. Indeed, research communities have mainly studied and investigated the first two problems, whereas the last one, even if important-with-practical-applications, has been very often neglected.

Approximate query answering (AQA) techniques address the first challenge, and can be reasonably considered as one of the most important topics in OLAP research. The main proposal of AQA techniques consists in providing approximate answers to resource-consuming OLAP queries (e.g., *range-* (Ho, Agrawal, Megiddo, & Srikant, 1997), *top-k* (Fang, Shivakumar, Garcia-Molina, Motwani, & Ullman, 1998), and *iceberg* (Xin, Han, Cheng, & Li, 2006) queries) instead of computing exact answers, as decimal precision is usually negligible in OLAP query and report activities (e.g., see Cuzzocrea, 2005). Due to a relevant interest from the data warehouse research community, AQA techniques have been intensively investigated during the last years with the achievement of important results. Among the others, *histograms* (e.g., Acharya, Poosala, & Ramaswamy, 1999; Bruno, Chaudhuri, & Gravano, 2001; Gunopulos, Kollios, Tsotras, & Domeniconi, 2000; Muralikrishna & DeWitt, 1998; Poosala & Ioannidis, 1997), *wavelets* (Vitter, Wang, & Iyer, 1998), and *sampling* (e.g., Babcock, Chaudhuri, & Das, 2003; Chaudhuri, Das, Datar, Motwani, & Rastogi, 2001; Cuzzocrea & Wang, 2007; Gibbons & Matias 1998) are the most successful techniques, and they have also inducted several applications in contexts even different from OLAP, like P2P data management (e.g., Gupta, Agrawal, & El Abbadi, 2003). Summarizing, with respect to the OLAP context, AQA techniques propose (*i*) computing compressed representations of multidimensional data cubes, and (*ii*) evaluating (approximate) answers against such representations via ad-hoc query algorithms that, usually, meaningfully take advantages from their hierarchical nature, which, in turn, is inherited from the one of input data cubes.

Conceptual data models for OLAP are widely recognized as based on data cube concepts like *dimension*, *hierarchy*, *level*, *member*, and *measure*, first introduced by Gray et al. (1997), which inspired various models for multidimensional databases and data cubes (e.g., Agrawal et al., 1997; Hacid & Sattler, 1998; Thanh Binh, Min Tjoa, & Wagner, 2000; Tsois, Karayannidis, & Sellis, 2001; Vassiliadis, 1998; Vassiliadis & Sellis, 1999)). Nevertheless, despite this effort, several papers have recently put in evidence some formal limitations of accepted conceptual models for OLAP (e.g., Cabibbo & Torlone, 1998), or theoretical failures of popular data cube operations, like aggregation functions (e.g., Lehner, Albrecht, & Wedekind, 1998; Lenz & Shoshani, 1997; Lenz & Thalheim, 2001).

Contrarily to data querying and modeling issues, since *data presentation models* do not properly belong to the well-founded conceptual-logical-physical design hierarchy for relational databases (which has also been inherited from multidimensional models (Vassiliadis et al., 1999)), the problem of OLAP data visualization has been studied and investigated so far only (Gebhardt, Jarke, & Jacobs, 1997; Inselberg, 2001; Keim, 1997; Maniatis, Vassiliadis, Skiadopoulos, & Vassiliou, 2003a, 2003b). On the other hand, being OLAP a technology focused at supporting decision making, thus based on (sensitive) information exploration and browsing, it is easy to understand that, in future years, tools for advanced visualization of multidimensional data cubes will quickly conquest the OLAP research scene.

Starting from fundamentals of data cube compression techniques and OLAP data visualization research issues, in this article we argue to meaningfully exploit the main results coming from the former and the goals of the latter in a combined manner, and propose *a novel technique for supporting advanced OLAP visualization of multidimensional data cubes*. The basic motivation of such an approach is realizing that (*i*) compressing data is an efficient way of visualizing data, and (*ii*) this intuition is well-founded at large (i.e., for any data-intensive system relying on massive data repositories), and, more specifically, it is particularly targeted to the OLAP context where accessing multidimensional data cubes can become a realistic bottleneck for data warehouse systems and applications. For instance, as we better motivate in Section 2, this is the case of *mobile OLAP*, which, recently, has attracted considerable attention from the data warehouse research community.

Another contribution of our work is represented by the wide experimental analysis we conducted in order to test the effectiveness of our proposed technique. To this end, we performed various kinds of experiments with respect to several metrics, and against different classes of data cubes; specifically, we have taken into consideration synthetic, benchmark, and real data cubes. Results of these experiments confirm that our proposed technique outperforms similar state-of-the-art initiatives with respect to both the accuracy and visualization goals.

Technique Overview

Briefly, our proposed technique relies on two steps. The first one consists of generating a two-dimensional OLAP view *D* from the input multidimensional data cube *A* by means of an innovative approach that allows us to *flatten OLAP dimensions* (of *A*), and, as a consequence, effectively support exploration and browsing activities against *A* (via *D*), by overcoming the natural disorientation and refractoriness of human beings in dealing with hyper-spaces. Specifically, the (two) OLAP dimensions on which *D* is defined are built from the dimensions of *A* according to the analysis goals of the target OLAP user/application. The idea of using views to tame computational overheads due to data management and query processing tasks against massive data warehouses is not novel in literature, and it has been extensively investigated across the last decade (e.g., (Ezeife, 2001; Harinarayan, Rajaraman, & Ullman, 1996), with relevant results. The second step consists of generating a bucket-based compressed representation of *D*, namely *hierarchy-driven indexed quad-tree summary* (H-IQTS), denoted by *H-IQTS*(*D*), which meaningfully extends the compression technique for two-dimensional summary data domains presented by us in Buccafurri, Furfaro, Saccà, and Sirangelo (2003), via introducing the amenity of *generating semantics-aware buckets* (i.e., buckets that "follow" groups of the OLAP hierarchies of *D*). In other words, *we use the OLAP hierarchies defined on the dimensions of D to drive the compression process*. This allows us to achieve space efficiency, while, at the same time, support approximate query answering and advanced OLAP visualization features against multidimensional data cubes.

Article Outline

The remaining part of this article is organized as follows. In the second section, we describe mobile OLAP application scenarios where the advanced visualization technique for multidimensional data cubes we propose assumes a critical role in the vest of *enabling technology*. In the third section, we outline the background of our proposal, which is represented by the compression techniques presented in Buccafurri et al. (2003). In the fourth section, we provide a motivating example stating the goodness of our idea of making use of semantics-aware compressed representations of two-dimensional OLAP views. In the fifth section, we provide fundamentals and basic definitions used throughout the article. The sixth section is devoted to the description of our innovative OLAP dimension flattening process. In the seventh section, we illustrate the hierarchy-driven two-dimensional OLAP view compression algorithm we propose. The eighth section focuses on a comprehensive experimental evaluation of our technique against different classes of data cubes. Finally, the ninth section we derive conclusions of our work, and draw future directions for further research in this field.

APPLICATION SCENARIOS

The technique we propose in this article can be successfully applied to all those scenarios in which accessing and exploring massive multidimensional data cubes is a critical requirement. For instance, this is the case of *mobile OLAP systems and applications*, where users access corporate OLAP servers via handheld devices. In fact, mobile devices are usually characterized by specific properties (e.g., small storage space, small size of the display screen, discontinuance of the connection to the WLAN, etc) that are often incompatible with the need of browsing and querying summarized information extracted from massive multidimensional data cubes made accessible through wireless networks.

In such application scenarios, flattening multidimensional data cubes into two-dimensional OLAP views represents an effective solution yet an enabling technology for mobile OLAP environments, as, contrarily to what happens for hyper-spaces, handheld devices can easily visualize two-dimensional spaces on conventional (e.g., 2D) screens. This property, along with the realistic need of compressing data to be transmitted and processed by handheld devices, makes perfect sense to our idea of using data compression techniques as a way of visualizing OLAP data. Moreover, the amenity of driving the compression process by means of OLAP hierarchies, thus meaningfully generating semantics-aware buckets, further corroborates the application of our proposed technique to mobile OLAP environments, as the limited computational capabilities of handheld devices impose us to definitively process *useful knowledge*, by discarding the useless one, being resource-consuming transactions infeasible to be processed by such kind of devices.

Figure 1. The system Hand-OLAP: Overview

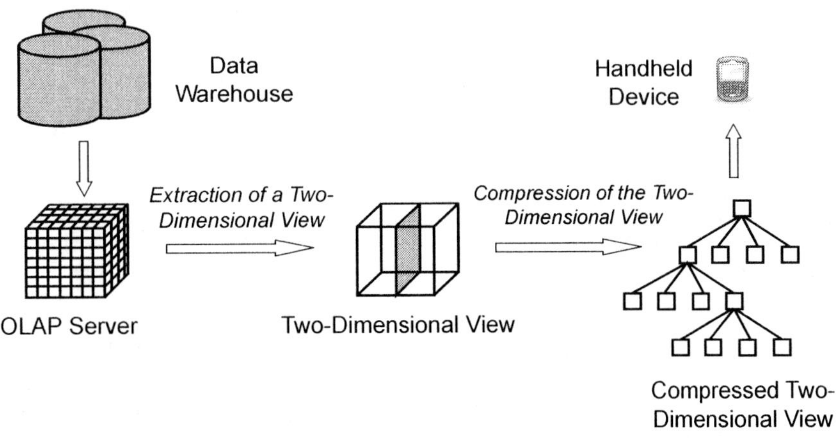

Figure 2. The system Hand-OLAP: Logical architecture

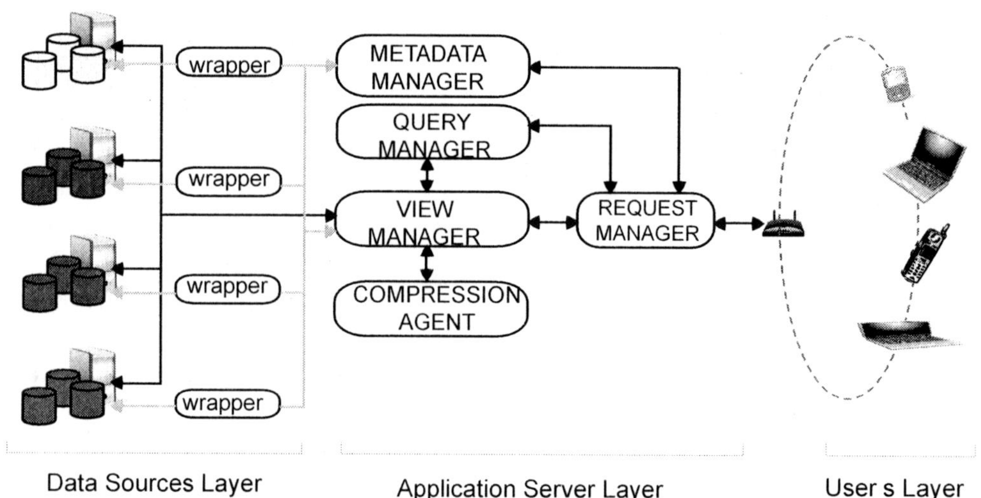

As a motivating application scenario, these results can be successfully applied to the system *Hand-OLAP*, proposed by us in Cuzzocrea et al. (2003), which allows OLAP users to extract, browse, and query compressed two-dimensional views (which are computed via the technique (Buccafurri et al., 2003)) coming from a remote OLAP server (see Figure 1). Specifically, according to the guidelines of algorithms proposed in Buccafurri et al. (2003), Hand-OLAP is targeted at supporting range-queries, a very popular class of queries useful to extract summarized knowledge from data cubes in the vest of aggregate information (Ho et al., 1997). The basic idea that Hand-OLAP is based on is: rather than querying the original multidimensional data, it may be more convenient to generate a compressed view of them, store the view into the handheld device, and query it locally, even if the WLAN is off, thus obtaining approximate answers that, as discussed in the first section, are perfectly suitable for OLAP goals (e.g., see Cuzzocrea, 2005).

According to well-known design patterns, Hand-OLAP is a multi-tier system, and every software layer corresponds to a specific application logic (see Figure 2). Specifically, the following software layers can be identified in the Hand-OLAP logical architecture:

- **Data Sources Layer:** It is the collection of (*i*) OLAP servers from which the desired information can be retrieved, and (*ii*) wrappers that extract meta-information about the available data cubes as well as the actual data;
- **Application Server Layer:** It is the layer that (*i*) elaborates OLAP-users' requests, (*ii*) interacts with OLAP servers, (*iii*) computes the compressed representation of the extracted OLAP view, and (*iv*) sends it to the handheld device;
- **User's Layer:** It includes the client-side tool that allows a handheld device to acquire and elaborate the desired information, by enabling useful functionalities such as connectivity services, metadata querying and browsing, range-query managing (e.g., editing, executing, browsing, refreshing etc).

The application server layer, which is the most interesting component of Hand-OLAP with respect to the Data Engineering point of view, consists of three components which cooperate to fulfill OLAP-users' requests:

- **Request Manager:** It is the component that receives the request of OLAP users, and translates it either into a request to the *Metadata Manager* for retrieving meta-information about the content of the target data cube, or into a request to the *View Manager* for retrieving a compressed representation of the two-dimensional OLAP view defined by OLAP users:
  - **Metadata Manager:** It is the component that extracts meta-information about the OLAP server it is connected to, and returns them in a XML format;
- **View Manager:** It is the component that (*i*) extracts from the selected data cube the two-dimensional view defined by OLAP users, (*ii*) uses the *compression agent* for summarizing it, and (*iii*) returns the compressed representation to the handheld device;
- **Compression Agent:** It is the component that receives a two-dimensional view from the *view manager* and returns its compressed representation to it--in particular, the *view manager* sends the extracted two-dimensional view to the *compression agent* together with the value of the desired *compression ratio*, which depends on both the amount of storage space available at the handheld device and the size of the view;
- **Query Manager:** It is the component that is in charge of supporting range-query evaluation on the compressed two-dimensional view, and visualizing the results on the handheld device according to a partitioned hierarchical representation.

Indeed, as we discuss next, the actual capabilities of Hand-OLAP can be further improved by integrating inside its core layer advanced OLAP visualization features developed on top of the technique we propose in this article.

BACKGROUND

Given a two-dimensional summary data domain *D*, the technique proposed in Buccafurri et al. (2003) allows us to obtain a compact data structure called *quad-tree summary* (QTS), which founds on a quad-tree-based partitioned representation of *D*, denoted by *QTS*(*D*), where, at each iteration of the generating partition process, (*i*) the current bucket *b* in *QTS*(*D*) to be split is greedily chosen by selecting the one having maximum *sum of the squared errors* (SSE), and (*ii*) *b* is split in four equal-size square sub-buckets that are added to the current partition. Specifically, given a bucket *b*, the SSE of *b*, denoted by SSE(*b*), is defined as follows:

$$SSE(b) = \sum_{k \in b} (D[k] - AVG(b))^2 \qquad (1)$$

such that: (*i*) *k* denotes a position inside *b*, (*ii*) *D*[*k*] is the value of *D* at position *k*, and (*iii*) *AVG*(*b*) is the average of values inside *b*.

This task is iterated until the storage space *B* available for housing *QTS*(*D*) is consumed. The "natural" representation of *QTS*(*D*) is like a quad-tree, such that nodes are corresponding to buckets of the partition and store the sum of all the items contained within such buckets. As shown in Buccafurri et al. (2003), due to its hierarchical nature, QTS is particularly suitable for evaluating range-queries. To further improve query capabilities, the leaf buckets of QTS having non-uniform data distribution, on which traditional interpolation techniques would fail, are equipped with very compact data structures called *indexes*. Indexes can be efficiently represented in few bytes, and provide *succinct descriptions* of data distributions of buckets they summarize. This approach leads to the definition of an extended version of QTS called *indexed quad-tree summary* (IQTS). Indexes allow us to definitively augment the quality of intra-bucket query estimation, thus overcoming general-purpose state-of-the-art compression techniques like histograms and wavelets (Buccafurri et al., 2003).

Figure 3. Building a 2/3LT-index and its compressed representation on memory

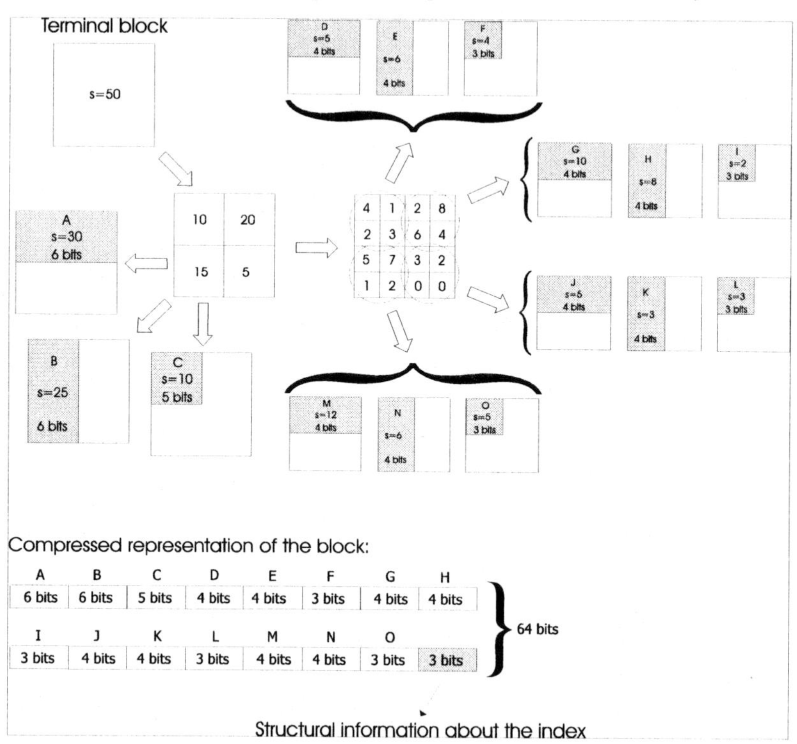

In Buccafurri et al. (2003), we define three index types with different organization of sub-buckets, so that we select the index, which better approximates the data distribution inside a given bucket: (*i*) the *2/3LT-Index*, which is suitable for distributions with no strong asymmetry; (*ii*) the *2/4LT-Index*, which is oriented to biased distributions; (*iii*) the *2/p(eak)LT-Index*, which is designed for capturing distributions having a few high density peaks. As an example, here we focus on an instance of Index, the 2/3LT-index (see Figure 3), which is built for a leaf bucket of a compressed two-dimensional OLAP view. The sum of all the items contained in such bucket is equal to 50. The index is obtained as follows: the bucket is partitioned into four equal-size sub-buckets and, in turn, each of the four sub-buckets into other four equal-size sub-sub-buckets. The index stores approximate aggregate data about both the generated sub-buckets and sub-sub-buckets. Such aggregate data consist of the sums of items contained in the regions, which are colored in grey (see Figure 3). The values of the sums are stored using less than 32 bits, introducing some approximation. The number of bits used for each stored value depends on the size of the corresponding sub-bucket. That is, referring to Figure 3, we use 6 bits for both the regions *A* and *B* (which have the same size), and 5 bits for *C*, whose size is an half of *A* and *B*. Analogously, we use 4 bits for *D* and *E* (whose size is an half of *C*), and so on. We point out that saving one bit for storing the sum of *C* with respect to *A* can be justified by considering that, on average, the value of the sum of items inside *C* is an half of the sum corresponding to *A*, since the size of *C* is an half of the size of *A*. Thus, on the average, the accuracy of representing *A* using 6 bits is the same as the accuracy of representing *C* using 5 bits.

The previously described index is based on a balanced quad-tree partition of a bucket. Different types of index can be used, based on different partitions. For instance, we can build an index based on an unbalanced quad-tree partition. Such an Index is more suitable for a bucket where items are distributed *heterogeneously* (i.e., buckets consisting of some regions containing very skewed data distributions and other regions with rather uniform distributions). The detailed description of these kinds of index, along with their experimental evaluation on both synthetic and real data sets, can be found in Buccafurri et al. (2003).

As regards the issue of selecting the most suitable Index for a bucket on the basis of the actual distribution of data inside the bucket. That is, we measure the approximation error carried out by the index, and select the index, which provides the best accuracy. In order to measure the approximation error of an Index *I*(*b*) on a bucket *b*, we use the following error metrics:

$$\varepsilon(I(b)) = \sum_{i=1}^{64} (sum(q_i) - sum_I(q_i))^2 \qquad (2)$$

where q_i represents the *i*-th (among 64 ones) sub-bucket of *b* obtained by dividing its sides into eight equal-size ranges, and $sum_I(q_i)$ represents the estimation of the sum of items occurring in q_i which can be done by using *I*(*b*) and the knowledge of *sum*(*b*). Given a bucket *b*, we choose the 2/nLT-Index which "originates" the minimum value of ε(*I*(*b*)) (see Buccafurri et al., 2003) for further details).

With respect to the results achieved in Buccafurri et al. (2003), in this article we investigate the problem of providing a compressed representation of a given two-dimensional OLAP view *D*, being *D* extracted from the target data cube *A* by means of the previously mentioned OLAP dimension flattening process instead of processing two-dimensional summary data domains (like in Buccafurri et al., 2003). This imposes us to *handle OLAP hierarchies* defined on the dimensions of *D*, thus achieving an innovative contribution with respect to goals of Buccafurri et al. (2003). Indeed, summary data considered in Buccafurri et al. (2003) resemble OLAP data in the application scenario we address in this article, but summary data do not expose hierarchies and do not impose us to handle and deal with the semantics of hierarchies.

Due to the need of handling OLAP hierarchies, *H-IQTS*(*D*) adds to *IQTS*(*D*) the amenity of generating a quad-tree based partitioned representation of *D according to the semantics provided by hierarchies* defined on the dimensions of *D* (i.e., as highlighted in the first section, using the hierarchies to drive the compression process). In fact, in the presence of hierarchies on the dimensions, neglecting such information (as would happen by adopting the quad-tree based partitioning scheme (Buccafurri et al., 2003)) could involve in the wrong condition of obtaining buckets storing aggregate values computed over OLAP data related to items belonging to *different* groups within a *same* hierarchy. To go in further details, due to its generating process and contrarily to *IQTS*(*D*), *H-IQTS*(*D*) can also house *rectangular buckets* instead of square buckets only, since an arbitrary data cube exposes, without any loss of generality, arbitrary groups in the hierarchies. Similarly to *IQTS*(*D*), *H-IQTS*(*D*) is shaped as a quad-tree, and the information stored in its buckets is still the sum of items contained within them. Just like *IQTS*(*D*), leaf nodes of *H-IQTS*(*D*) are equipped with Indexes in order to improve query capabilities.

MOTIVATING EXAMPLE

To become convinced of the benefits coming from the idea of using OLAP hierarchies to drive the compression process, consider the following example. Let *A* be a two-dimensional data cube defined on top of relational data sources storing sale data, and having as measure the total amount of *Sales* (e.g., 15,000 €) of a given product (e.g., t-shirt), belonging to the dimension *product*, in a given city (e.g., Lisbon), belonging to the dimension *zone*. Consider the quad-tree-based partitioning scheme for *A* depicted in the left side of Figure 4. This scheme presents "wrong" buckets as items related to Chicago (belonging to the dimension *zone*) are aggregated in the left-down bucket along with items related to cities located in Europe (i.e., Prague, Berlin, Munich, etc), instead of being aggregated in the left-up bucket along with items related to cities located in America (i.e., New York, Vancouver, Toronto, etc). The same happens with Raincoat (belonging to the dimension *product*), whose items are aggregated in the right-down bucket along with items related to summer clothes (i.e., sunglasses, bikini, t-shirt etc), instead of being aggregated in the left-down bucket along with items related to winter clothes (i.e., gloves, hat, scarf, etc). On the contrary, consider the hierarchy-driven partitioning scheme for *A* depicted in the right side of Figure 4. As an alternative to the previous one, this scheme follows the hierarchies defined on the dimensions, and, as a consequence, buckets are computed on top of measures *related to the same semantic domain*. It should be note that this condition is desirable at large, but it assumes a more relevant role for the context we address, as, typically, the compression process causes the loss of the structure (in terms of OLAP schemas) of data cubes.

Now, consider the benefits due to the described approach in a mobile OLAP setting like the one drawn by the system Hand-OLAP. In Hand-OLAP, compressed views extracted from remote OLAP servers are mainly explored and browsed via popular DRILL-DOWN OLAP operations (i.e., increasing the level of detail of OLAP data) implemented via splits over buckets of the view. Nevertheless, since each split partitions the current bucket into four equal-size sub-buckets, OLAP users could be required to perform many splits before to access the summarized knowledge he/she is interested in, as "wrong" buckets could be accessed during the exploration task. On the contrary, by admitting semantics-aware buckets, since OLAP analysis is subject-oriented (Han & Kamber, 2000), OLAP users access the summarized knowledge of interest in a faster manner rather than the previous case, as each split partitions the current bucket into four sub-buckets computed over semantically-related OLAP data.

FOUNDAMENTALS AND BASIC DEFINITIONS

In order to better understand our proposal, it is needed to introduce some fundamentals and basic definitions regarding the constructs of OLAP conceptual data model we adopt, along with the notation we use in the rest of the article. These definitions are compatible with main results of previous popular models (e.g., Gray et al., 1997)—a complete survey can be found in Vassiliadis et al. (1999).

Hierarchy, Member, Level, and OLAP Metadata

Given an OLAP dimension d_i and its domain of *members* $\Psi(d_i)$, each of them denoted by ρ_j, a *hierarchy* defined on d_i, denoted by $H(d_i)$ can be represented as a general tree (i.e., such that each node of the tree has a number $n \geq 0$ of child nodes) built on top of $\Psi(d_i)$. $H(d_i)$ is usually obtained according to a bottom-up strategy by (*i*) setting as leaf nodes of $H(d_i)$ members in $\Psi(d_i)$, and (*ii*) iteratively aggregating sets of members in $\Psi(d_i)$ to obtain other (internal) members, each of them denoted by σ_j, which correspond to internal nodes in $H(d_i)$. In turn, internal members in $\Psi(d_i)$ (equally, nodes in $H(d_i)$) can be further aggregated to form other super-members until a unique aggregation of members is obtained; the latter corresponds to the root node of $H(d_i)$, and it is known in literature as the *aggregation ALL*. More precisely, ALL is only an *artificial aggregation* introduced to obtain a tree (i.e., $H(d_i)$) instead of a list of trees, each of them rooted in the second-level-internal-nodes σ_j, which should be the "effective" highest-level

Figure 4. Equal-size quad-tree based partition (left) and hierarchy-driven quad-tree based partition (right) of the product-zone data cube

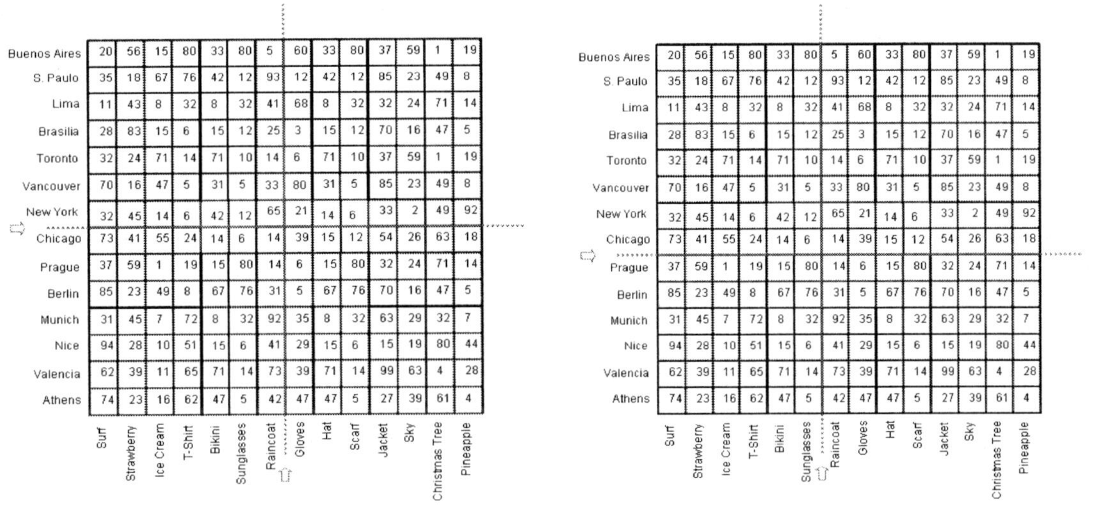

Left:

	Surf	Strawberry	Ice Cream	T-Shirt	Bikini	Sunglasses	Raincoat	Gloves	Hat	Scarf	Jacket	Sky	Christmas Tree	Pineapple
Buenos Aires	20	56	15	80	33	80	5	60	33	80	37	59	1	19
S. Paulo	35	18	67	76	42	12	93	12	42	12	85	23	49	8
Lima	11	43	8	32	8	32	41	68	8	32	32	24	71	14
Brasilia	28	83	15	6	15	12	25	3	15	12	70	16	47	5
Toronto	32	24	71	14	71	10	14	6	71	10	37	59	1	19
Vancouver	70	16	47	5	31	5	33	80	31	5	85	23	49	8
New York	32	45	14	6	42	12	65	21	14	6	33	2	49	92
Chicago	73	41	55	24	14	6	14	39	15	12	54	26	63	18
Prague	37	59	1	19	15	80	14	6	15	80	32	24	71	14
Berlin	85	23	49	8	67	76	31	5	67	76	70	16	47	5
Munich	31	45	7	72	8	32	92	35	8	32	63	29	32	7
Nice	94	28	10	51	15	6	41	29	15	6	15	19	80	44
Valencia	62	39	11	65	71	14	73	39	71	14	99	63	4	28
Athens	74	23	16	62	47	5	42	47	47	5	27	39	61	4

Right:

	Surf	Strawberry	Ice Cream	T-Shirt	Bikini	Sunglasses	Raincoat	Gloves	Hat	Scarf	Jacket	Sky	Christmas Tree	Pineapple
Buenos Aires	20	56	15	80	33	80	5	60	33	80	37	59	1	19
S. Paulo	35	18	67	76	42	12	93	12	42	12	85	23	49	8
Lima	11	43	8	32	8	32	41	68	8	32	32	24	71	14
Brasilia	28	83	15	6	15	12	25	3	15	12	70	16	47	5
Toronto	32	24	71	14	71	10	14	6	71	10	37	59	1	19
Vancouver	70	16	47	5	31	5	33	80	31	5	85	23	49	8
New York	32	45	14	6	42	12	65	21	14	6	33	2	49	92
Chicago	73	41	55	24	14	6	14	39	15	12	54	26	63	18
Prague	37	59	1	19	15	80	14	6	15	80	32	24	71	14
Berlin	85	23	49	8	67	76	31	5	67	76	70	16	47	5
Munich	31	45	7	72	8	32	92	35	8	32	63	29	32	7
Nice	94	28	10	51	15	6	41	29	15	6	15	19	80	44
Valencia	62	39	11	65	71	14	73	39	71	14	99	63	4	28
Athens	74	23	16	62	47	5	42	47	47	5	27	39	61	4

partition of members in $\Psi(d_i)$. Each member in $H(d_i)$ is characterized by a *level* (of the hierarchy), denoted by L_j, such that $L_j \geq 0$ (note that, when $L_j = 0$, $\sigma_j \equiv \rho_j$); as a consequence, we can define a level L_j in $H(d_i)$ as a *collection* of members. For each level L_j, the ordering of L_j, denoted by $O(L_j)$, is the one exposed by the OLAP server platform for the target data cube. Note that such ordering depends on how knowledge held in (OLAP) data is produced, processed, and delivered.

Given a multidimensional data cube A such that $Dim(A) = \{d_0, d_1, \ldots, d_{n-1}\}$ is the set of dimensions of A, and $Hie(A) = \{H(d_0), H(d_1), \ldots, H(d_{n-1})\}$ the set of hierarchies defined on the latter dimensions, the collection of members σ_j at level L_j of each hierarchy $H(d_i)$ in $Hie(A)$ univocally refers, in a multidimensional fashion, a certain (OLAP) data cell $C_{j,h}$ in A *at level* L_j. In other words, $C_{j,h}$ is the OLAP aggregation of data cells in A *at level* L_j. We name such collection as *j-level OLAP Metadata* (for $C_{j,h}$), and denote them as $\mathcal{M}(C_{j,h})$. Given a level L_j, the data cell $C_{j,h}$ at L_j and the corresponding collection of OLAP metadata $\mathcal{M}(C_{j,h})$, if we move up towards the level L_{j+1} (i.e., by performing a *roll-up* (OLAP) operation on the hierarchy $H(d_i)$), we increase the *level of abstraction* and decrease the *level of detail* of both $C_{j,h}$ and $\mathcal{M}(C_{j,h})$, thus accessing the upper-level components $C_{j+1,m}$ and $\mathcal{M}(C_{j+1,m})$, with $m \neq h$. Contrarily to this, if we move down towards the level L_{j-1} (i.e., by performing a *drill-down* (OLAP) operation on the hierarchy $H(d_i)$), we decrease the level of abstraction and increase the level of detail of both $C_{j,h}$ and $\mathcal{M}(C_{j,h})$, thus accessing the lower-level components $C_{j-1,m}$ and $\mathcal{M}(C_{j-1,m})$, with $m \neq h$.

For instance, consider a four-dimensional data cube A defined on the top of relational data sources containing insurance data and having as measure the average value of the refunds (e.g., 12,000 €) allocated in a given region (e.g., Seattle), during a given time interval (e.g., 2005), for a given employment class (e.g., bank clerk), and for a given kind of accident (e.g., car crash), such that *city*, *year*, *employment*, and *kindofaccident* are the members of the third level of the hierarchies defined on the dimensions *zone*, *time*, *userclass*, and *accidentclass* respectively. Under the described OLAP schema, given the data cell $C_{3,h}$ in A with value $Val(C_{3,h}) = 2{,}000$ € at third level L_3, the metadata set $\mathcal{M}(C_{3,h})$ could be defined as follows: $\mathcal{M}(C_{3,h}) = \{IL, 1{,}500, Accountant, AccidentsInIL\}$; by rolling-up, we could access the upper-level components $C_{2,m}$ with value $Val(C_{2,m})$ = 3,500 € and $\mathcal{M}(C_{2,m})$ = {*USA, 2,000, AdministrativeManager, AccidentsInUSA*}; by drilling-down, we could access the lower-level components $C_{4,m}$ with value $Val(C_{4,m})$ = 1,000 € and $\mathcal{M}(C_{4,m})$ = {*Chicago, 900, AdministrativeOfficer, WorkAccidents*}.

Left Boundary Member (LBM) and Right Boundary Member (RBM)

Given a member σ_j at level L_j of the hierarchy $H(d_i)$ defined on an OLAP dimension d_i and the set of its child nodes $Child(\sigma_j)$, which are members at level L_{j+1}, we define as the *Left Boundary Member* (LBM) of σ_j the child node of σ_j in $Child(\sigma_j)$ that is the *first* in the ordering $O(L_{j+1})$. Analogously, we define as the *Right Boundary Member* (RBM) of σ_j the child node of σ_j in $Child(\sigma_j)$ that is the *last* in the ordering $O(L_{j+1})$. As an example, consider the hierarchy $H(d_i)$ depicted in Figure 5; here, (*i*) *e* is the LBM of *b*, (*ii*) *f* is the RBM of *b*, (*iii*) *b* is the LBM of *a*, (*iv*) *d* is the RBM of *a* etc.

OLAP DIMENSION FLATTENING

OLAP dimension flattening process is the first step of our technique for supporting advanced OLAP visualization of multidimensional data cubes. In more detail, we flatten dimensions of the input multidimensional data cube A into two specialized dimensions called *visualization dimensions* (VD) that support advanced OLAP visualization of A via constructing an ad-hoc two-dimensional OLAP view D defined on the VDs.

Figure 5. An OLAP hierarchy

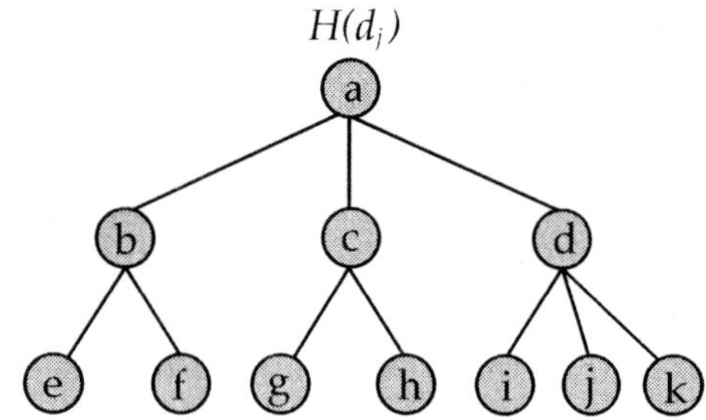

The process that allows us to obtain the two VDs from the dimensions of A works as follows. Let $Dim(A)$ and $Hie(A)$ be the set of dimensions and the set of hierarchies of A, respectively. Each VD is a tuple $v_i = \langle d_i, H^*(d_i)\rangle$ such that (*i*) d_i is the dimension selected by the target OLAP user/application, (*ii*) $H^*(d_i)$ is a hierarchy built from meaningfully merging the "original" hierarchy $H(d_i)$ of d_i with the hierarchies of other dimensions in A according to an ordered definition set $\mathcal{D}(v_i)$, defined as follows $\mathcal{D}(v_i) = \{\langle HL_i, d_j, P_j\rangle, \langle HL_j, d_{j+1}, P_{j+1}\rangle, \ldots, \langle HL_{j+K-1}, d_{j+K}, P_{j+K}\rangle\}$, where $K = |\mathcal{D}(v_i)| - 1$. In more detail, for each pair of consecutive tuples $\langle\langle HL_j, d_{j+1}, P_{j+1}\rangle, \langle HL_{j+1}, d_{j+2}, P_{j+2}\rangle\rangle$ in $\mathcal{D}(v_i)$, the sub-tree of $H(d_{j+2})$ rooted in the root node of $H(d_{j+2})$ and having depth equal to P_{j+2}, denoted by $H_S^{P_{j+2}}(d_{j+2})$, is merged to $H(d_{j+1})$ by appending a *clone* of it to *each* member σ_{j+1} of level HL_{j+1}, named as *hooking level*, in $H(d_{j+1})$. From the described approach, it follows that: (*i*) the ordering of items in $\mathcal{D}(v_i)$ defines the way of building $H^*(d_i)$; (*ii*) the first hierarchy to be processed is just $H(d_i)$. As an example of the flattening process of two OLAP dimensions into a new one, consider Figure 6, where the hierarchy $H^*(d_j)$ is obtained by merging $H(d_{j+1})$ to $H(d_j)$ via setting $P_{j+1} = 1$ and $HL_j = 1$.

As regards data processing issues, it should be noted that, in order to finally compute D, due to the OLAP dimension flattening task above, it is needed to re-aggregate multidimensional data in A according to the new VDs.

Algorithm build2DOLAPViewViaFlattening (see Figure 7) implements the OLAP dimension flattening process. It takes as input the following parameters: (*i*) the multidimensional data cube A; (*ii*) the dimension d_i of A selected as first VD; (*iii*) the dimension d_j of A selected as second VD; (*iv*) the definition set for the first output VD v_i, $\mathcal{D}(v_i)$; (*v*) the definition set for the second output VD v_j,

Figure 6. Merging OLAP hierarchies

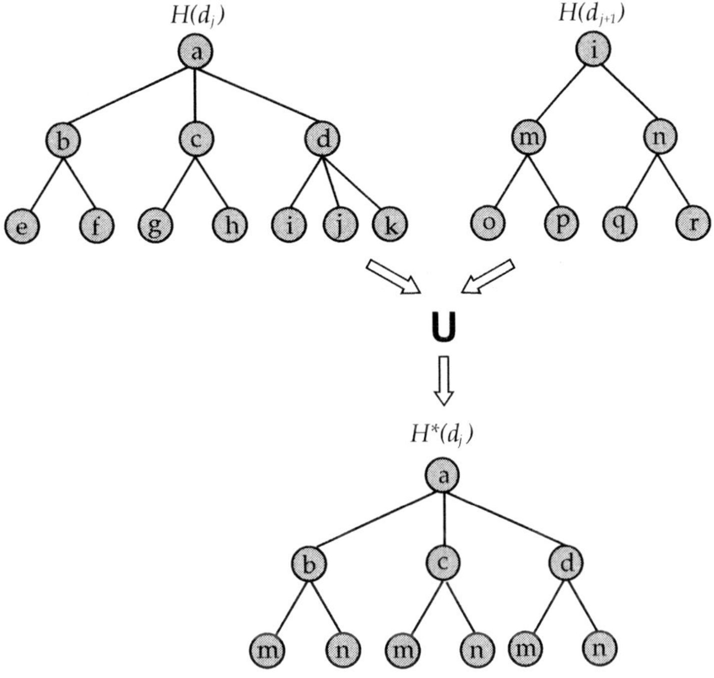

$\mathcal{D}(v_j)$. It returns as output the two-dimensional OLAP view D extracted from A via flattening dimensions of A into the VDs according to the input definition sets. Specifically, build2DOLAPViewViaFlattening makes use of the following procedures: (*i*) buildVisualizationDimension, which takes as input the data cube A, a dimension d of A and the definition set $\mathcal{D}(v)$, and returns as output the VD v built on top of d according to the guidelines previous given above; (*ii*) getHierarchy, belonging to the utility package OLAPTools, which, applied to a VisualizationDimension object v, returns the modified hierarchy H^* of v; (*iii*) aggregate2DOLAPView, belonging to the utility package OLAPTools, which takes as input a data cube A and two OLAP hierarchies H_i and H_j of A, and returns as output the two-dimensional OLAP view D extracted from A by re-aggregating multidimensional data in A according to H_i and H_j.

HIERARCHY-DRIVEN COMPRESSION OF TWO-DIMENSIONAL OLAP VIEWS

Compressing the two-dimensional OLAP view D (extracted from A according to the OLAP dimension flattening process described in the sixth section) is the second step of our proposed technique. Given D, for each step j of our compression algorithm, we need to (*i*) greedily select the leaf bucket b of *H-IQTS*(D) having maximum SSE (see the third Section), and (*ii*) split b in four sub-buckets through investigating, for each dimension d_k of D, levels of the hierarchy $H(d_k)$. The first task is similar to what proposed in Buccafurri et al. (2003) for two-dimensional summary data domains, whereas the novelty proposed in this article consists in the second task properly.

Formally, given the current bucket $b_j = D[l_{j,0}:u_{j,0}][l_{j,1}:u_{j,1}]$ to be split at step j of our compression algorithm, such that $[l_{j,k}:u_{j,k}]$ is the range of b_j on the dimension d_k of D, the problem is finding, for each dimension d_k of D, a *splitting position* $S_{j,k}$ belonging to $[l_{j,k}:u_{j,k}]$, i.e. $l_{j,k} \leq S_{j,k} \leq u_{j,k}$. To this end, for each dimension d_k of D, our splitting strategy aims at (*i*) grouping items into buckets related to the same semantic domain, and (*ii*) maintaining the hierarchy $H(d_k)$ balanced as more as possible. Particularly, the first aspect allows us to achieve the benefits highlighted in the fourth section; the second aspect allows us to sensitively improve query estimation capabilities as, on the basis of this approach, we finally obtain buckets with balanced "numerousness" (of items) that introduce a smaller approximation error in the evaluation of (OLAP) queries involving several buckets rather than the contrary case (see Buccafurri et al., 2003) for further investigations). On the other hand, this evidence has been already recognized in the context of *Equi-Width* histograms (Piatetsky-Shapiro & Connell, 1984).

A Hierarchy-Driven Algorithm for Compressing Two-Dimensional OLAP Views

For the sake of simplicity, we will present our hierarchy-driven compression algorithm for two-dimensional OLAP views through showing how to handle the hierarchy of an OLAP dimension d_k (i.e., how to determine a splitting position $S_{j,k}$ on d_k). Obviously, this technique must be performed for both the dimensions of the target (two-dimensional) OLAP view D, thus obtaining, for each *pair* of splits at step j of our algorithm (i.e., $S_{j,0}$ and $S_{j,1}$), four two-dimensional buckets to be added to the current partition of D (i.e., *H-IQTS*(D)).

Let D be a two-dimensional data cube, and $D[0:|d_k| - 1]$ be a one-dimensional OLAP view of D obtained by projecting D with respect to d_k (see Figure 8). Let $b_j = D[l_{j,k}:u_{j,k}]$ be the current (one-dimensional) bucket of $D[0:|d_k| - 1]$ to be split at step j. To determine $S_{j,k}$ on $[l_{j,k}:u_{j,k}]$, we denote as $T_{j,k}(l_{j,k}:u_{j,k})$ the sub-tree of $H(d_k)$ whose (*i*) leaf nodes are the members of the sets $\mathcal{M}(C_{0,h})$ defined

Figure 7. Algorithm build2DOLAPViewViaFlattening

```
ALGORITHM build2DOLAPViewViaFlattening
Input: The multidimensional data cube A; the dimension of A selected as first VD, d_i;
   the dimension of A selected as second VD, d_j; the definition set for the first output VD v_i, D(v_i);
   the definition set for the second output VD v_j, D(v_j).
Output: The two-dimensional OLAP view D.
import OLAPTools.*;
begin
      OLAPTools.OLAPView D ← null;
      OLAPTools.VisualizationDimension v_i ← null;
      OLAPTools.VisualizationDimension v_j ← null;
      OLAPTools.Hierarchy H*_i ← null;
      OLAPTools.Hierarchy H*_j ← null;
      v_i ← buildVisualizationDimension(A,d_i, D(v_i));
      v_j ← buildVisualizationDimension(A,d_j,D(v_j));
      H*_i ← v_i.getHierarchy();
      H*_j ← v_j.getHierarchy();
      D ← OLAPTools.aggregate2DOLAPView(A,H*_i,H*_j);
      return D;
end;
```

Figure 8. Modeling the splitting strategy

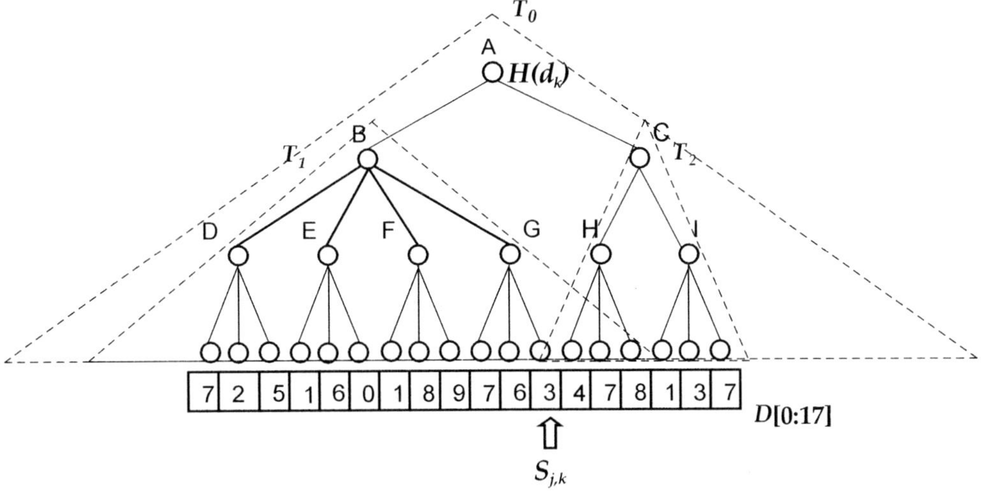

on data cells $C_{0,h}$ in $D[l_{j,k}:u_{j,k}]$ with $l_{j,k} \leq h \leq u_{j,k}$, and (*ii*) the root node is the (singleton) member of the set $\mathcal{M}(C_{p,r})$ defined on the data cell $C_{p,r}$ that is the aggregation of $D[l_{j,k}:u_{j,k}]$ at level L_p of $H(d_k)$, being p the depth of $T_{j,k}(l_{j,k}:u_{j,k})$. To give an example, consider Figure 8. Here, tree T_0, properly denoted by $T_{j,k}(0{:}17)$, is related to the whole OLAP view $D[0{:}17]$, and corresponds to the whole $H(d_k)$. At step j, d_k is split in the position $S_{j,k} = 11$, thus generating the buckets $D[0{:}11]$ and $D[12{:}17]$. In consequence of this, tree T_1, properly denoted by $T_{j+1,k}(0{:}11)$, is related to $D[0{:}11]$, whereas tree T_2, properly denoted by $T_{j+1,k}(12{:}17)$, is related to $D[12{:}17]$.

Let (*i*) d_k be the dimension of D to be processed, (*ii*) $H(d_k)$ the hierarchy defined on d_k, (*iii*) $b_j = D[l_{j,k}:u_{j,k}]$ the current (one-dimensional) bucket to be split at step j of our algorithm, (*iv*) $T_{j,k}(l_{j,k}:u_{j,k})$ the tree related to b_j, (v) $T^1_{j,k}(l_{j,k}:u_{j,k})$ be the *second* level of $T_{j,k}(l_{j,k}:u_{j,k})$. In order to select the splitting position $S_{j,k}$ on $[l_{j,k}:u_{j,k}]$, we initially consider the data cell $C_{0,k}$ in $D[l_{j,k}:u_{j,k}]$ whose indexer is in the middle of $D[l_{j,k}:u_{j,k}]$, denoted by $X_{j,D}$, which is defined as follows:

$$X_{j,D} = \left\lfloor \frac{1}{2} \cdot \left| D[l_{j,k} : u_{j,k}] \right| \right\rfloor \tag{3}$$

It should be noted that processing the second level of $T_{j,k}(l_{j,k}:u_{j,k})$ (i.e., $T^1_{j,k}(l_{j,k}:u_{j,k})$) derives from the usage of the aggregation ALL in OLAP conceptual models, which, in total, introduces an *additional* level in the general tree modeling an OLAP hierarchy (see the fifth section).

Then, starting from ρ_k, being ρ_k the (*singleton-*-see the fifth section) member in the set $M(C_{0,k})$, we go up on $H(d_k)$ until the parent of ρ_k at level $T^1_{j,k}(l_{j,k}:u_{j,k})$, denoted by σ_k, is reached, and we decide how to determine $S_{j,k}$ on the basis of the nature of σ_k. If σ_k is the LBM of the root node of $T_{j,k}(l_{j,k}:u_{j,k})$, denoted by $R_{j,k}$, then we have:

$$S_{j,k} = \left\lfloor \frac{1}{2} \cdot \left| D[l_{j,k} : u_{j,k}] \right| \right\rfloor - 1 \tag{4}$$

and, as a consequence, we obtain the following two (one-dimensional) buckets as child buckets of b_j:

$$b'_{j+1} = D\left[l_{j,k} : \left\lfloor \frac{1}{2} \cdot \left| D[l_{j,k} : u_{j,k}] \right| \right\rfloor - 1 \right] \tag{5}$$

and:

$$b''_{j+1} = D\left[\left\lfloor \frac{1}{2} \cdot \left| D[l_{j,k} : u_{j,k}] \right| \right\rfloor : u_{j,k} \right] \tag{6}$$

Otherwise, if σ_k is the RBM of $R_{j,k}$, then we have:

$$S_{j,k} = \left\lfloor \frac{1}{2} \cdot |D[l_{j,k} : u_{j,k}]| \right\rfloor \quad (7)$$

and, as a consequence, we obtain the following buckets:

$$b'_{j+1} = D\left[l_{j,k} : \left\lfloor \frac{1}{2} \cdot |D[l_{j,k} : u_{j,k}]| \right\rfloor \right] \quad (8)$$

and:

$$b''_{j+1} = D\left[\left\lfloor \frac{1}{2} \cdot |D[l_{j,k} : u_{j,k}]| \right\rfloor + 1 : u_{j,k} \right] \quad (9)$$

Finally, if σ_k is different from both the LBM and the RBM of $R_{j,k}$ (i.e. it *follows* the LBM of $R_{j,k}$ in the ordering $O(T^1_{j,k}(l_{j,k}:u_{j,k}))$ and *precedes* the RBM of $R_{j,k}$ in the ordering $O(T^1_{j,k}(l_{j,k}:u_{j,k}))$, we perform a finite number of shift operations on the indexers of $D[l_{j,k}:u_{j,k}]$ starting from the middle indexer $X_{j,D}$ and within the range:

$$\Gamma_{j,k} = \left[\gamma^{lo}_{j,k}, \gamma^{up}_{j,k} \right] \quad (10)$$

such that:

$$\gamma^{lo}_{j,k} = \left\lfloor \frac{1}{2} \cdot |D[l_{j,k} : u_{j,k}]| \right\rfloor - \left\lfloor \frac{1}{3} \cdot |D[l_{j,k} : u_{j,k}]| \right\rfloor \quad (11)$$

and:

$$\gamma^{lo}_{j,k} = \left\lfloor \frac{1}{2} \cdot |D[l_{j,k} : u_{j,k}]| \right\rfloor + \left\lfloor \frac{1}{3} \cdot |D[l_{j,k} : u_{j,k}]| \right\rfloor \quad (12)$$

These shift operations are repeated until a data cell $V_{j,k}$ in $D[l_{j,k}:u_{j,k}]$ such that the corresponding member σ_k at level $T^1_{j,k}(l_{j,k}:u_{j,k})$ is the LBM or the RBM of $R_{j,k}$. It should be noted that admitting a maximum offset of $\pm\left\lfloor \frac{1}{3} \cdot |D^k[l_{j,k} : u_{j,k}]| \right\rfloor$ with respect to the middle of the current bucket is coherent with the aim of maintaining the hierarchy $H(d_k)$ balanced as more as possible, which allows us to take advantages from the previously highlighted benefits (see the fourth section).

To this end, starting from the middle of $\Gamma_{j,k}$ (which is equal to the one of $D[l_{j,k}:u_{j,k}]$, $X_{j,D}$), we search for the data cell $V_{j,k}$ by iteratively considering indexers $I_{j,q}$ within $\Gamma_{j,k}$ defined by the following function:

$$I_{j,q} = \begin{cases} X_{j,D} & q = 0 \\ I_{j,q-1} + (-1)^q \cdot q & q > 1 \end{cases} \quad (13)$$

If such data cell $V_{j,k}$ exists, then $S_{j,k}$ is set as equal to the so-determined indexer $I^*_{j,q}$, and, as a consequence, we obtain the pairs of buckets:

$$b'_{j+1} = D\left[l_{j,k} : I^*_{j,k} - 1 \right] \quad (14)$$

and:

$$b''_{j+1} = D\left[I^*_{j,q} : u_{j,k} \right] \quad (15)$$

if $I^*_{j,q}$ is the LBM of $R_{j,k}$, or, alternatively, the pairs of buckets:

$$b'_{j+1} = D\left[l_{j,k} : I^*_{j,q} \right] \quad (16)$$

and:

$$b''_{j+1} = D\left[I^*_{j,q} + 1 : u_{j,k} \right] \quad (17)$$

if $I^*_{j,q}$ is the RBM of $R_{j,k}$. On the contrary, if such data cell $V_{j,k}$ does not exist, then we do not perform any split on $D[l_{j,k}:u_{j,k}]$, and we "remand" the splitting at the next step of the algorithm (i.e., $j+1$) where the splitting position $S_{j+1,k}$ is determined by processing the *third* level $T^2_{j+1,k}(l_{j+1,k}:u_{j+1,k})$ of the tree $T_{j+1,k}(l_{j+1,k}:u_{j+1,k})$ (i.e., by decreasing the aggregation level of OLAP data with respect to the previous step). The latter approach is iteratively repeated until a data cell $V_{j,k}$ verifying the condition above is found; otherwise, if the leaf level of $T_{j,k}(l_{j,k}:u_{j,k})$ is reached without finding any admissible splitting point, then $D[l_{j,k}:u_{j,k}]$ is added to the current partition of the OLAP view without being split. We point out that this way to do still pursues the aim of obtaining balanced partitions of the input OLAP view.

Figure 9. Algorithm compress2DOLAPView

```
ALGORITHM compress2DOLAPView

Input: The two-dimensional OLAP view D; the storage space available for housing
H-IQTS(D), B.
Output: The compressed representation of D, H-IQTS(D).

import Sets.*;
import CompressionToolkit.*;
import OLAPTools.*;
begin
        Sets.Set H-IQTS(D) ← new Sets.Set();
        CompressionToolkit.Bucket b_j ← null;
        CompressionToolkit.Bucket b_{j+1} ← null;
        Sets.Set bucketsToBeProcessed ← null;
        int SUM ← 0;
        int ℓ_{j,0} ← 0;
        int ℓ_{j,1} ← 0;
        int S_{j,0} ← 0;
        int S_{j,1} ← 0;
        int k ← 0;
        b_j ← new CompressionToolkit.Bucket(0,|d_0| - 1,0,|d_1| - 1);
        SUM ← OLAPTools.computeBucketSum(D,b_j);
        CompressionToolkit.setSum(b_j,SUM);
        H-QTS(D).add(b_j);
        bucketsToBeProcessed ← new Sets.Set();
        bucketsToBeProcessed.add(b_j);
        B ← B - CompressionToolkit.computeOccupancy(H-IQTS(D));
        while (B > 0 && bucketsToBeProcessed.size() > 0) do
          b_j ← CompressionToolkit.findLeafBucketWithMaxSSE(bucketsToBeProcessed);
          ℓ_{j,0} ← OLAPTools.getLevel(b_j,H(d_0));
          ℓ_{j,1} ← OLAPTools.getLevel(b_j,H(d_1));
          S_{j,0} ← computeSplittingPosition(b_j,H(d_0),ℓ_{j,0});
          S_{j,1} ← computeSplittingPosition(b_j,H(d_1),ℓ_{j,1});
          while (S_{j,0} = -1 && S_{j,1} = -1 &&
                    ℓ_{j,0} < OLAPTools.getDepth(H(d_0)) &&
                    ℓ_{j,1} < OLAPTools.getDepth(H(d_1)) do
            ℓ_{j,0} ← ℓ_{j,0} + 1;
            ℓ_{j,1} ← ℓ_{j,1} + 1;
            S_{j,0} ← computeSplittingPosition(b_j,H(d_0),ℓ_{j,0});
            S_{j,1} ← computeSplittingPosition(b_j,H(d_1),ℓ_{j,1});
          endwhile
          if (S_{j,0} <> -1 || S_{j,1} <> -1) then
            if (CompressionToolkit.hasIndex(b_j) = true) then
              B ← B + CompressionToolkit.computeOccupancy(b_j.getIndex());
              CompressionToolkit.removeIndex(b_j);
            endif
          endif
          while (k < 4) do
            b_{j+1} ← CompressionToolkit.getSubBucket(b_j,S_{j,0},S_{j,1},ℓ_{j,0},ℓ_{j,1},k);
            if (b_{j+1} <> null) then
              I_{j+1} ← CompressionToolkit.computeIndex(b_{j+1});
              if (I_{j+1} <> null) then
                CompressionToolkit.equipeWith(b_{j+1},I_{j+1});
              endif
              H-IQTS(D).add(b_{j+1});
              B ← B – CompressionToolkit.computeOccupancy(H-IQTS(D));
              bucketsToBeProcessed.add(b_{j+1});
            endif
          endwhile
          bucketsToBeProcessed.remove(b_j);
        endwhile
        return H-IQTS(D);
end;
```

The described approach is implemented by algorithm compress2DOLAPView (see Figure 9), which takes as input the two-dimensional OLAP view D and the amount of storage space B available for housing the compressed representation of D, and returns as output the data structure *H-IQTS*(*D*).

compress2DOLAPView makes use of the following procedures: (*i*) computeBucketSum, belonging to the utility package OLAPTools, which takes as input an OLAP view D and a Bucket object b (which implements a bucket of the partition of D), and returns as output the sum of the items contained in b; (*ii*) setSum, belonging to the utility package CompressionToolkit, which takes as input a Bucket object b and an integer *Sum*, and sets the sum stored in b to the value *Sum*; (*iii*) add, belonging to the utility package Sets, which takes as input an item a and, applied to a Set object s, adds a to s; (*iv*) computeOccupancy, belonging to the utility package CompressionToolkit, which takes as input the (current) compressed data structure *H-IQTS*(*D*), and returns as output its occupancy in KB; (*v*) findLeafBucketWithMaxSSE, belonging to the utility package CompressionToolkit, which takes as input an array of (current) leaf buckets V, and returns as output the bucket having maximum SSE among them; (*vi*) getLevel, belonging to the utility package OLAPTools, which takes as input a bucket b and a hierarchy H, and returns as output the level L of b in H; (*vii*) computeSplittingPosition, belonging to the utility package CompressionToolkit, which takes as input a bucket b, a hierarchy H and a level L, and returns as output the splitting position S of b at level L, according to the guidelines given above; (*viii*) getDepth, belonging to the utility package OLAPTools, which takes as input a hierarchy H, and returns as output the depth p of H; (*ix*) hasIndex, belonging to the utility package CompressionToolkit, which, applied to a Bucket object b, returns TRUE if b is equipped with an index $I(b)$, otherwise FALSE; (*x*) removeIndex, belonging to the utility package CompressionToolkit, which takes as input a Bucket object b, and removes the Index $I(b)$ which b is equipped with; (*xi*) getSubBucket, belonging to the utility package CompressionToolkit, which takes as input a Bucket object b_j, two splitting positions S_i and S_j defined on b_j along the two dimensions d_i and d_j at depths ℓ_i and ℓ_j of levels L_i and L_j of b_j in $H(d_i)$ and $H(d_j)$, respectively, and an integer k ranging in [0:3], and returns as output the sub-bucket b_{j+1} of b_j by (*xi.i*) splitting b_j on S_i and S_j, thus obtaining four sub-buckets, and (*xi.ii*) selecting among the latter the bucket b_{j+1} on the basis of the value of k – (i.e., if $k = 0$, then the left up sub-bucket of b_j is selected) and so on; (*xii*) computeIndex, belonging to the utility package CompressionToolkit, which takes as input a Bucket object b, and returns as output the "best" index $I(b)$ built on it if the accuracy provided by linear interpolation is not higher than that provided by $I(b)$ – otherwise, it returns the Null object; (*xiii*) equipWith, belonging to the utility package CompressionToolkit, which takes as input a Bucket object b and an Index $I(b)$, and equips b with $I(b)$; (*xiv*) remove, belonging to the utility package Sets, which takes as input an item a and, applied to a Set object s, removes a from s.

Example

Consider a three-dimensional data cube A defined on top of relational data sources storing sale data, and having (*i*) as measure the total amount of *sales* (i.e., the SQL aggregation operator SUM is exploited), and (*ii*) as dimensions the set: *Dim*(*A*) = {*Product*, *Zone*, *Time*}. The hierarchies *H*(*Product*), *H*(*Zone*), and *H*(*Time*) are depicted in Figure 10, Figure 11, and Figure 12, respectively. Here, we provide a compression process example, where the VDs are (*i*) *Product*/*Time*, whose hierarchy *H*(*Product*/*Time*) is obtained by merging the hierarchies *H*(*Product*) and *H*(*Time*) using $P_{Product} = 2$ and $HL_{Time} = 1$ (see Figure 13), and (*ii*) *Zone*, which is the same as the one defined on the target data cube A. Figure 14 shows the two-dimensional OLAP view D extracted from A by re-aggregating multidimensional data in A

according to the (new) VDs *Product/Time* and *Zone*. Finally, Figure 15 shows the steps of the compression process of *D*.

EXPERIMENTAL STUDY

In order to test the effectiveness of our proposed technique, we defined two kinds of experiments. The first one is oriented to probe the data cube *compression performance* (or, equally, the *accuracy*) of our technique, whereas the second one is instead oriented to probe the *visualization capabilities* of our technique in meaningfully supporting advanced OLAP visualization of multidimensional data cubes.

Data Layer

Ins regards the data layer of our experimental framework, we engineered three kinds of data cubes and we extracted from them two-dimensional OLAP views by means of a *random* flattening process on the data cube dimensions. The usage of different classes of data cubes allowed us to submit our proposed technique to a comprehensive and "rich" experimental analysis, and, as a consequence, carefully test its performance. Data cube classes we considered are the following: (*i*) *synthetic data cubes*, which allow us to completely control the variation of input parameters determining the nature of OLAP data distributions as well as the one of the OLAP hierarchies (e.g., acting on the topology of the hierarchies etc); (*ii*) *benchmark data cubes*, which allow us

Figure 10. Hierarchy H(Product)

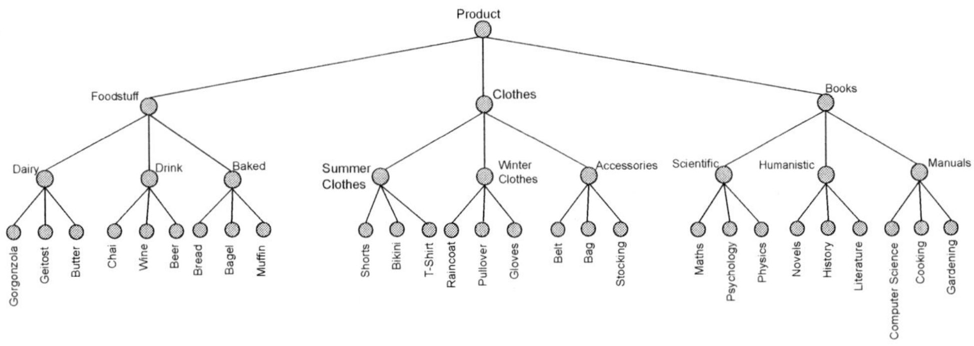

Figure 11. Hierarchy H(Zone)

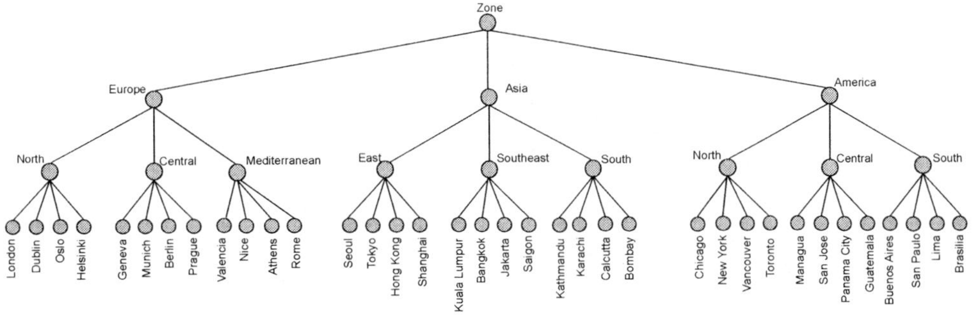

Figure 12. Hierarchy H(Time)

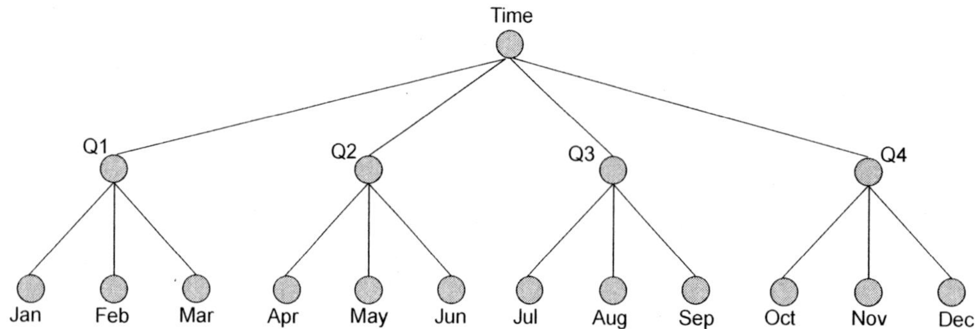

Figure 13. Hierarchy H(Product/Time)

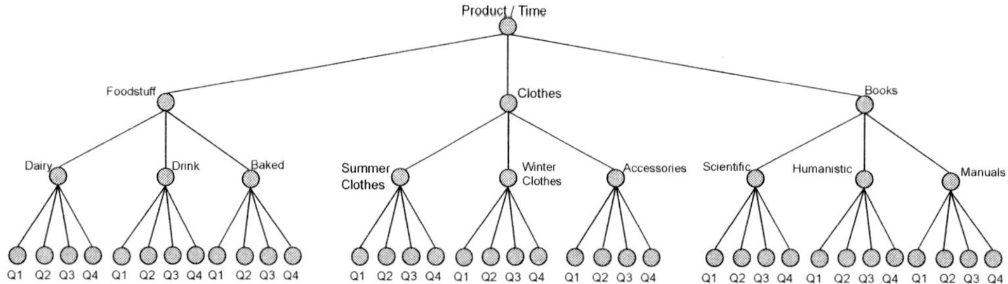

Figure 14. Product/time-zone OLAP view D

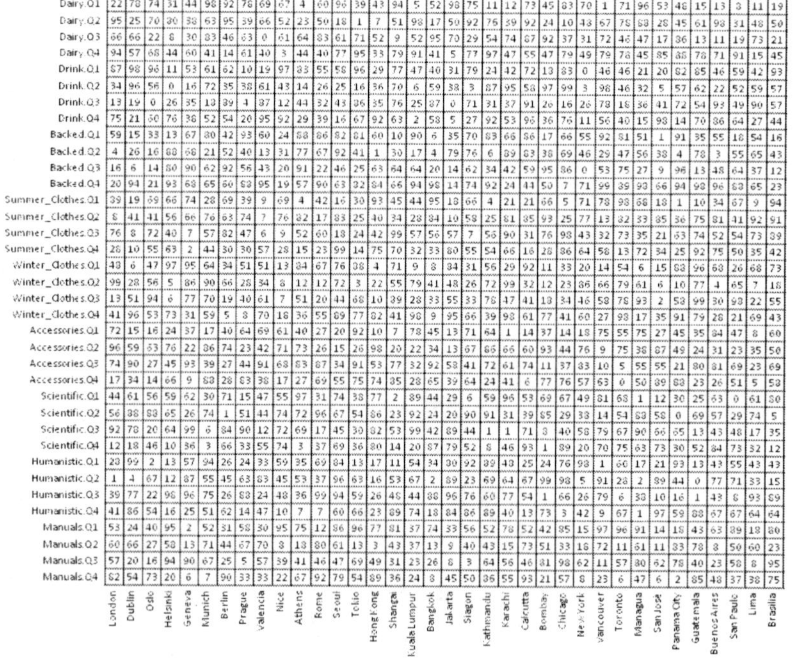

	London	Dublin	Oslo	Helsinki	Geneva	Munich	Berlin	Prague	Valencia	Nice	Athens	Rome	Seoul	Tokio	Hong Kong	Shangai	Kuala Lumpur	Bangkok	Jakarta	Siagon	Kathmandu	Karachi	Calcutta	Bombay	Chicago	New York	Vancouver	Toronto	Managua	San Jose	Panama City	Guatemala	Buenos Aires	San Paulo	Lima	Brasilia
Dairy.Q1	22	78	74	31	44	98	92	78	69	67	4	60	96	39	43	94	5	52	98	75	11	12	73	45	83	70	1	71	96	53	48	15	13	3	11	19
Dairy.Q2	95	25	70	30	38	63	95	39	66	52	23	50	18	1	7	51	98	17	50	92	76	39	92	24	10	48	67	78	68	28	45	61	98	31	48	50
Dairy.Q3	66	66	22	8	30	83	46	63	0	61	64	83	61	71	52	9	52	95	70	29	54	74	87	92	37	31	72	46	47	17	86	13	11	19	73	21
Dairy.Q4	94	57	68	44	60	41	14	61	40	3	44	40	77	95	33	79	91	41	5	77	97	47	55	47	79	49	79	78	45	85	88	78	71	91	15	45
Drink.Q1	87	98	96	11	53	61	62	10	19	97	83	55	58	96	29	77	47	40	31	79	24	42	72	13	83	0	46	46	21	20	82	85	46	59	42	93
Drink.Q2	34	96	56	0	16	72	35	38	61	43	14	26	25	16	36	70	6	59	38	3	87	95	58	97	99	3	98	46	32	5	57	62	22	52	59	57
Drink.Q3	13	19	0	26	35	13	89	4	87	12	44	32	43	86	35	76	25	87	0	71	31	37	91	26	16	26	78	18	36	41	72	54	93	49	90	57
Drink.Q4	75	21	60	76	36	52	54	20	95	92	29	39	16	67	92	63	2	58	5	27	92	53	96	36	76	11	56	40	15	98	14	70	86	64	27	44
Backed.Q1	59	15	33	13	67	80	42	93	60	24	88	86	82	81	60	10	90	6	35	70	83	66	86	17	66	55	92	81	51	1	91	35	55	18	54	16
Backed.Q2	4	26	16	88	68	21	52	40	13	31	77	67	92	41	1	30	17	4	79	76	6	89	83	38	69	46	29	47	56	38	4	78	3	55	65	43
Backed.Q3	16	6	14	80	90	62	92	56	43	20	91	22	46	25	63	64	64	20	14	62	34	42	59	95	86	0	53	75	27	9	96	13	48	64	37	12
Backed.Q4	20	94	21	93	68	65	60	88	95	19	57	90	63	32	84	66	94	98	14	74	92	24	44	50	7	71	99	39	93	66	94	98	96	88	65	23
Summer_Clothes.Q1	39	19	69	66	74	28	69	39	9	69	4	42	16	30	93	45	44	95	18	66	4	21	21	66	5	71	78	98	68	18	1	10	34	67	9	94
Summer_Clothes.Q2	8	41	41	56	66	76	63	74	7	76	82	17	83	25	40	34	28	84	10	58	25	81	85	93	25	77	13	82	33	85	36	75	81	41	92	91
Summer_Clothes.Q3	76	8	72	40	7	57	82	47	6	9	52	60	18	24	42	99	57	56	57	7	56	90	31	76	98	43	32	73	35	21	63	74	52	54	73	89
Summer_Clothes.Q4	28	10	55	63	2	44	30	30	57	28	15	23	99	14	75	70	32	33	80	55	54	66	16	28	86	64	58	13	72	34	25	92	75	50	35	42
Winter_Clothes.Q1	48	6	47	97	95	64	34	51	51	13	84	67	76	38	4	71	9	8	84	31	56	29	92	11	33	20	14	54	6	15	88	96	68	26	68	73
Winter_clothes.Q2	99	28	56	5	86	90	66	28	34	8	12	12	72	3	22	55	79	41	48	26	72	99	32	12	23	86	66	79	61	6	10	77	4	65	7	18
Winter_Clothes.Q3	13	51	94	6	77	70	19	40	61	7	51	20	44	68	10	39	28	33	55	33	78	47	41	18	34	46	58	78	93	2	58	99	30	98	22	55
Winter_Clothes.Q4	41	96	53	73	31	59	5	8	70	18	36	55	89	77	82	41	98	9	95	66	39	98	61	77	41	60	27	98	17	35	91	79	28	21	69	43
Accessories.Q1	72	15	16	24	37	17	40	64	69	61	40	27	20	92	10	7	78	45	13	71	64	1	14	37	14	18	75	55	75	27	45	35	84	47	8	60
Accessories.Q2	96	59	63	76	22	86	74	23	42	71	73	26	15	26	98	20	22	34	13	67	86	66	60	93	44	76	9	75	38	87	49	24	31	23	35	50
Accessories.Q3	74	90	27	45	93	39	27	44	91	68	83	87	34	91	53	77	32	92	58	41	72	61	74	11	37	33	10	5	55	55	21	80	81	69	23	69
Accessories.Q4	17	34	14	66	9	88	28	83	38	17	27	69	55	75	74	35	28	65	39	64	24	41	6	77	76	57	63	0	50	89	88	23	26	51	5	58
Scientific.Q1	44	61	56	59	62	30	71	15	47	55	97	31	74	38	77	2	89	44	29	6	59	96	53	69	67	49	81	68	1	12	30	25	63	0	61	80
Scientific.Q2	56	88	88	65	26	74	1	51	44	74	72	96	67	54	86	23	92	24	20	90	91	31	39	85	29	38	14	54	88	58	0	69	57	29	74	5
Scientific.Q3	92	78	20	64	99	6	84	90	12	72	69	17	45	30	82	53	99	42	89	44	1	1	71	3	40	58	79	67	90	66	65	13	43	48	17	35
Scientific.Q4	12	18	46	10	36	3	66	33	55	74	3	37	69	36	80	14	20	87	79	52	8	46	93	1	89	20	70	75	63	73	30	52	84	73	32	12
Humanistic.Q1	28	99	2	13	57	94	26	24	33	59	35	69	84	13	17	11	54	34	30	92	89	48	25	24	76	98	1	60	17	21	93	13	43	55	43	43
Humanistic.Q2	1	4	67	12	87	55	45	63	83	45	53	37	96	63	16	53	67	2	89	23	69	64	67	99	98	5	91	28	2	89	44	0	77	71	33	15
Humanistic.Q3	39	77	22	98	96	75	26	88	24	48	36	99	94	59	26	48	44	88	96	76	60	77	54	1	66	26	79	6	38	10	16	1	43	8	93	89
Humanistic.Q4	41	86	54	16	25	51	62	14	47	10	7	7	60	66	23	89	74	18	84	86	89	40	13	73	3	42	9	67	1	97	59	88	67	67	64	64
Manuals.Q1	53	24	40	95	2	52	31	58	30	95	75	12	86	96	77	81	37	74	33	56	52	78	52	42	85	15	97	96	91	14	18	43	63	89	18	80
Manuals.Q2	60	66	27	58	13	71	44	67	70	8	18	80	61	13	3	43	37	13	9	40	43	15	73	51	33	18	72	11	61	11	83	78	8	50	60	23
Manuals.Q3	57	20	16	94	90	67	25	5	57	39	41	46	47	69	49	31	23	26	8	3	64	56	46	81	98	62	11	57	80	62	78	40	23	58	8	95
Manuals.Q4	82	54	73	20	6	7	90	33	33	22	67	92	79	54	89	36	24	8	45	50	36	55	93	21	57	8	23	6	47	6	2	85	48	37	38	75

Figure 15. Compression process of the product/time-zone OLAP view D

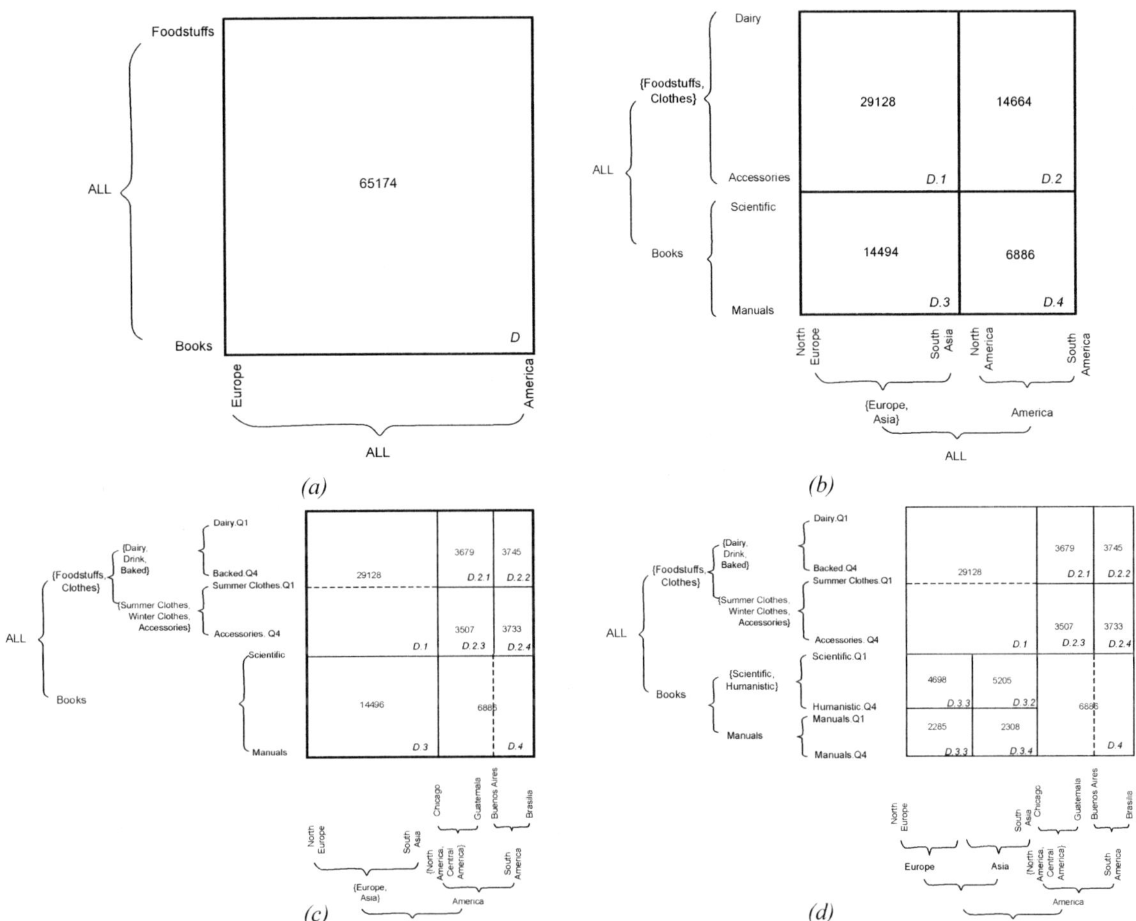

to test the effectiveness of our technique under the stressing of an in-laboratory-built input, and to evaluate our technique against competitor ones on "well-referred" data sets that have been widely used in similar research experiences; (*iii*) *real data cubes*, which allow us to probe the efficiency of our technique against real-life data sets.

For what regards synthetic data sets, we finally obtained two kinds of two-dimensional OLAP views: (*i*) the view $D_C(L_1,L_2)$, for which data are uniformly distributed on a given range $[L_1,L_2]$, with $L_1 < L_2$, (i.e., the well-known *Continuous Values Assumption* (CVA) (Colliat, 1996) holds), and (*ii*) the view $D_Z(z_{min},z_{max})$, for which data are distributed according to a Zipf (Zipf, 1949) distribution whose parameter z is randomly chosen on a given range $[z_{min},z_{max}]$, with $z_{min} < z_{max}$. Uniform and Zipf-based views allow us to probe the benefits of our technique under two "opposite" cases of (OLAP) data distributions, being the latter, due to its generating process, closer to real-life instances. In both views, we generated, for each dimension, an *artificial hierarchy* having depth equal to 15, which is a reasonable value to be considered with respect to the goals of our experimental analysis. In more detail, each artificial hierarchy has been generated by means of a bottom-up approach that, starting from the lowest-level members of the view, progressively aggregates members in internal members until

Figure 16. Two-dimensional OLAP view extracted from the benchmark data set TPC-H

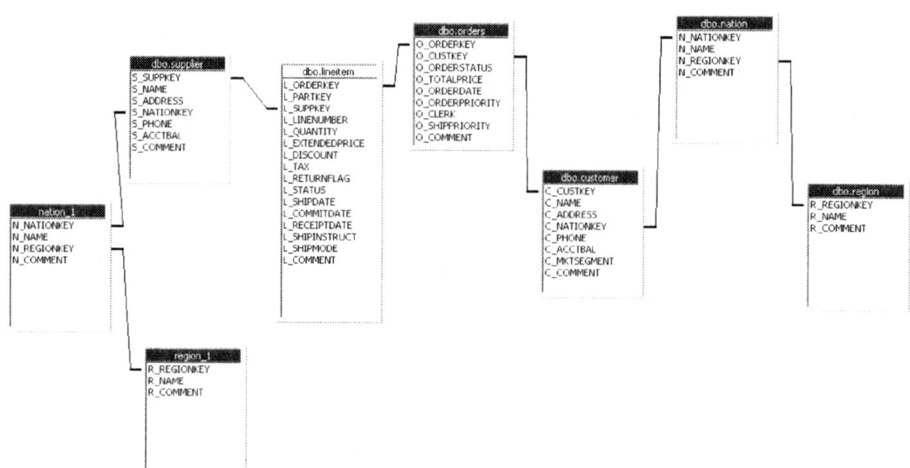

Figure 17. Two-dimensional OLAP view extracted from the benchmark data set APB-1

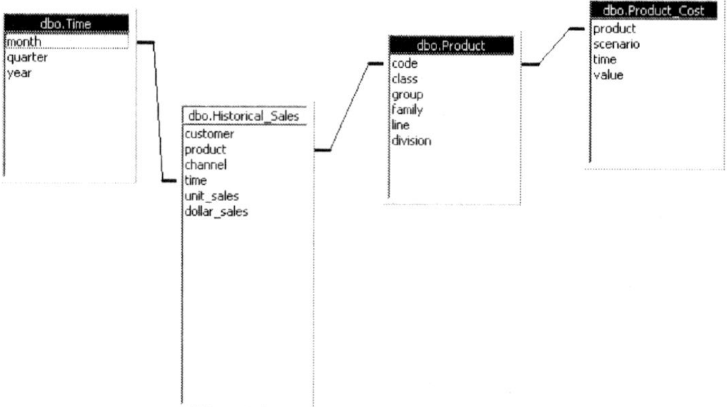

the desired depth is obtained. It should be noted that these artificial hierarchies *implicitly* define the semantics of the (synthetic) view.

For what regards benchmark data sets, we considered two popular benchmarks: *TPC-H* (Transaction Processing Council, 2006) and *APB-1* (OLAP Council, 1998). By exploiting data generation routines made available at the respective benchmark Web sites, we built benchmark databases and, based on the latter, multidimensional (benchmark) data cubes from which we extracted two-dimensional OLAP views. In more detail, from the benchmark data set TPC-H, we extracted a two-dimensional OLAP view (see Figure 16) having as dimensions the attributes (*i*) *C_Address*, belonging to the dimensional table *dbo.Customer* and linked to the fact table *dbo.Lineitem* through the dimensional table *dbo.Orders*, and (*ii*) *S_Address*, belonging to the dimensional table *dbo.Supplier*. Since the original hierarchies in TPC-H have limited depth, thus being inappropriate to the scope of our experimental analysis, we equipped the dimensions *C_Address* and *S_Address* with artificial hierarchies having depth equal to 15, similarly to what done with synthetic data sets. From the benchmark data set APB-1, we extracted

Figure 18. Two-dimensional OLAP view built on top of the real-life data set USCensus1990

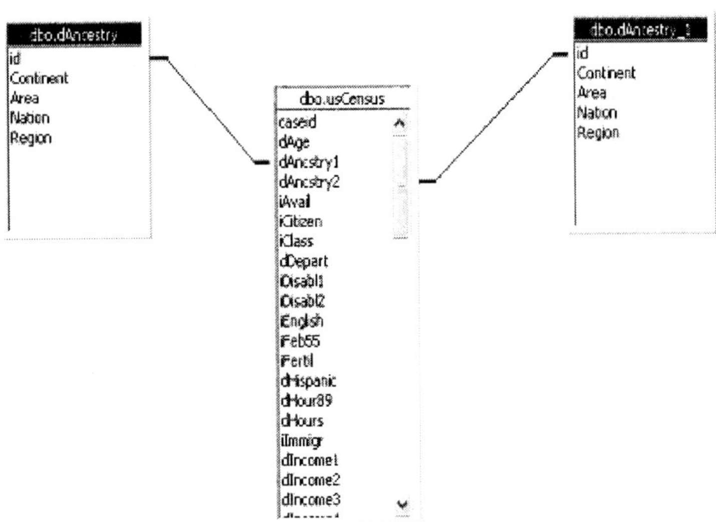

a two-dimensional OLAP view (see Figure 17) having as dimensions the attributes (*i*) *Class*, belonging to the dimensional table *dbo.Product*, and (*ii*) *Month*, belonging to the dimensional table *dbo.Time*. Just like the TPC-H case, we equipped the dimensions *Class* and *Month* with 15-depth artificial hierarchies.

Finally, for what regards real-life data sets, we considered the popular data set *USCensus1990* (University of California, Irvine, 2001) made available from *UCI KDD Archive* (University of California, Irvine, 2005), and we built a two-dimensional OLAP view (see Figure 18) by defining two new dimensional tables *dbo.dAncestry* and *dbo.dAncestry_1*, which both store data on the origin regions of parents of each person whose data are stored in the fact table *dbo.usCensus*. *dbo.dAncestry* and *dbo.dAncestry_1* have been built starting from the definition of *USCensus1990* attributes available at (University of California, Irvine, 2001), and populated with tuples coming from *dbo.usCensus*. Then, they have been linked to *dbo.usCensus* via simple ID-based relationships (i.e., *dbo.usCensus.dAncstry1* ↔ *dbo.dAncestry.Id* and *dbo.usCensus.dAncstry2* ↔ *dbo.dAncestry_1.Id*). The resulting two-dimensional OLAP view has as dimensions the attributes (*i*) *Id*, belonging to the dimensional table *dbo.dAncestry_1*, and (*ii*) *Id*, belonging to the dimensional table *dbo.dAncestry*. Finally, 15-depth artificial hierarchies have been embedded to the latter dimensions.

Metrics

As regards the outcomes of our experimental study, we defined the following metrics. For the first kind of experiments (i.e., that focused on the accuracy), given a population of synthetic range-SUM queries Q_S, we measure the *average relative error* (ARE) between exact and approximate answers to queries in Q_S, defined as follows:

$$\overline{E}_{rel} = \frac{1}{|Q_S|} \cdot \sum_{k=0}^{|Q_S|-1} E_{rel}(Q_k) \tag{18}$$

such that, for each query Q_k in Q_S, we have:

$$E_{rel}(Q_k) = \frac{|A(Q_k) - \tilde{A}(Q_k)|}{A(Q_k)} \tag{19}$$

Figure 19. A HRB (left) and its implementation on the data cube sales (right)

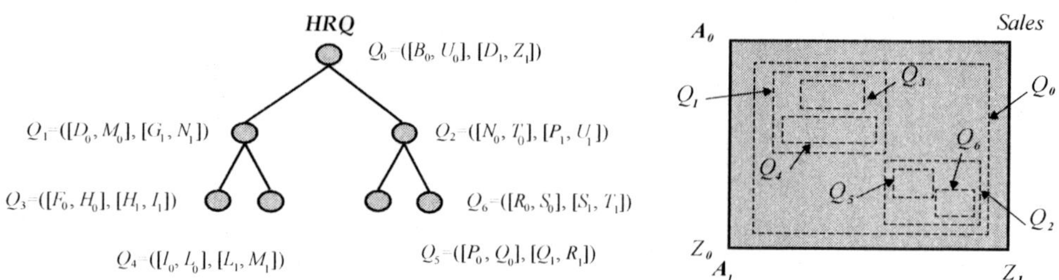

Figure 20. Experimental results for the accuracy metrics with respect to the query selectivity $||Q||$ on the 1,000 × 1,000 two-dimensional synthetic OLAP views $D_C(25,70)$ (left) and $D_Z(0.5,1.5)$ (right) with $r = 10\,\%$

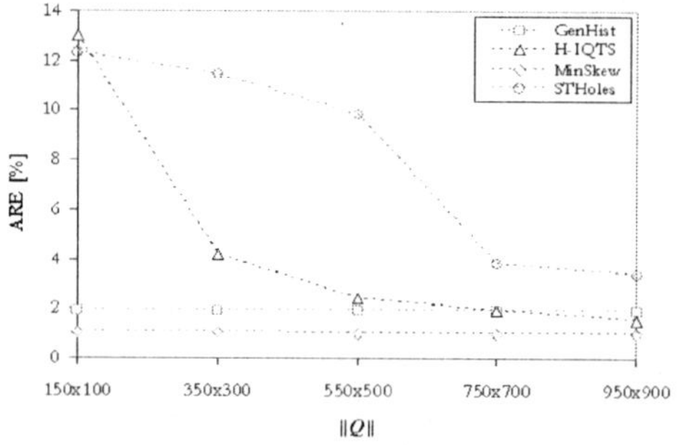

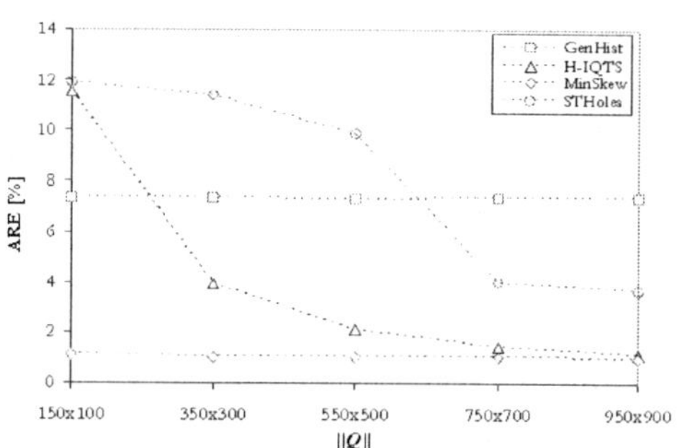

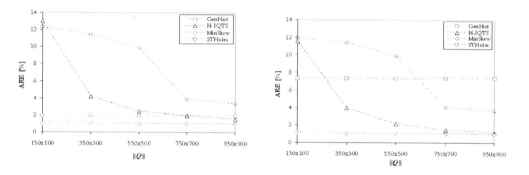

where (*i*) $A(Q_k)$ is the exact answer to Q_k, and (*ii*) $\tilde{A}(Q_k)$ is the approximate answer to Q_k. Specifically, having fixed a range size Δ_k for each dimension d_k of the target synthetic OLAP view *D*, we generated queries in Q_S through spanning *D* by means of the "seed" $\Delta_0 \times \Delta_1$ query Q^s whose left-up corner moves across two-dimensional references $\langle i, j \rangle$ of *D*.

For the second kind of experiments, we have been inspired from *hierarchical range queries* (HRQ) introduced by Koudas, Muthukrishnan and & Srivastava (2000). In our implementation, a HRQ $Q_H(W_H, P_H)$ is a full tree such that: (*i*) the depth of such tree is equal to P_H; (*ii*) each internal node N_i has a fan-out degree equal to W_H; (*iii*) each node N_i stores the definition of a ("traditional") range-SUM query Q_i; (*iv*) for each node N_i in $Q_H(W_H, P_H)$, there not exists any sibling node N_j of N_i such that $Q_i \cap Q_j <> \emptyset$.

As an example, consider Figure 19, where a HRQ having depth equal to 2 is depicted. This query could model a typical *business intelligence* (BI) scenario where two local companies which are joined to a common main company pose queries to specialized sub-domains of the data cube *sales* according to their business goals. Also following the previous simple yet effective example, it should be noted that HRQs have a wide range of applications in OLAP systems (as also highlighted in Koudas et al. (2000)), since they allow us to extract "hierarchically-shaped" summarized knowledge from massive data cubes.

Similarly to the previous kind of experiments, for each node N_i in $Q_H(W_H, P_H)$, the population of queries $Q_{S,i}$ to be used as input query set has been generated by means of the above-described spanning technique (i.e., based on the seed query Q_i^s). In more detail, since, due the nature of HRQs, the selectivity of seed queries $Q^s_{i,k}$ of nodes N_i at level k of $Q_H(W_H,P_H)$ *must* decreases as the depth P_k of $Q_H(W_H,P_H)$ increases, we first imposed that the selectivity of the seed query of the root node N_0 in $Q_H(W_H,P_H)$, denoted by $\| Q^s_{0,0} \|$, is equal to the γ % of $\|D\|$, being γ an input parameter and $\|D\|$ the selectivity of the target OLAP view D, respectively. Then, for each internal node N_i in $Q_H(W_H,P_H)$ at level k, we randomly determined the seed queries of the child nodes of N_i by checking the following constraint:

$$\sum_{i=0}^{|(W_H)^{k+1}|-1} \| Q^s_{i,k+1} \| \leq \| Q^s_{i,k} \| \qquad (20)$$

with:

$$Q^s_{i,k+1} \cap Q^s_{j,k+1} = \varnothing \qquad (21)$$

for each i and j in $[0, |(W_H)^{k+1}|-1]$, with $i <> j$, and adopting the criterion of *maximizing* each $\|Q^s_{i,k}\|$.

Given a HRB $Q_H(W_H,P_H)$, we measure the *average accessed bucket number* (AABN), which models the average number of buckets accessed during the evaluation of $Q_H(W_H,P_H)$, and it is defined as follows:

$$AABN(Q_H(W_H,P_H)) = \sum_{k=0}^{P_H} \frac{1}{(W_H)^k} \cdot \sum_{\ell=0}^{|(W_H)^k|-1} AABN(N_\ell) \qquad (22)$$

where, in turn, AABN(N_ℓ) is the average number of buckets accessed during the evaluation of the population of queries $Q_{S,\ell}$ of the node N_ℓ in $Q_H(W_H,P_H)$, defined as follows:

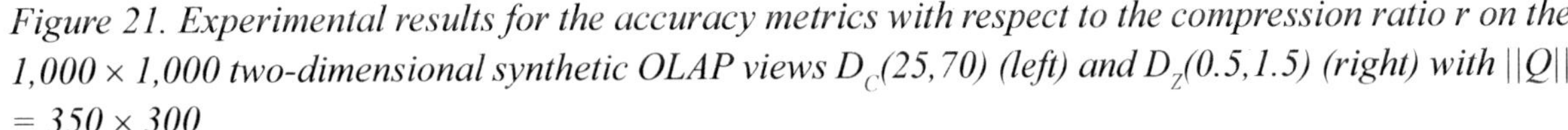

Figure 21. Experimental results for the accuracy metrics with respect to the compression ratio r on the 1,000 × 1,000 two-dimensional synthetic OLAP views $D_C(25,70)$ (left) and $D_Z(0.5,1.5)$ (right) with $||Q||$ = 350 × 300

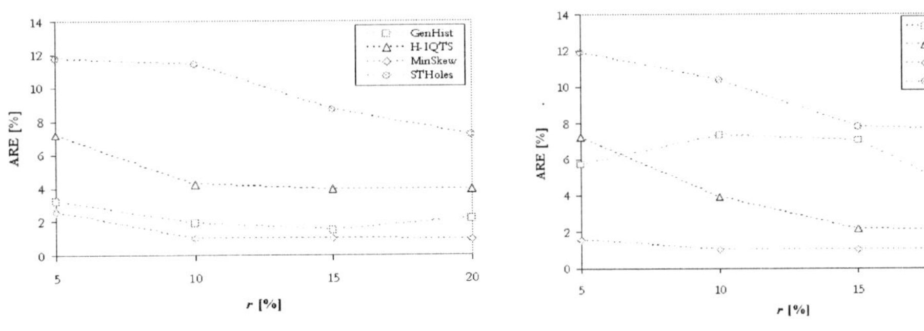

Figure 22. Experimental results for the visualization metrics with respect to the depth of HRQs P on the 1,000 × 1,000 two-dimensional synthetic OLAP views $D_C(25,70)$ (left) and $D_Z(0.5,1.5)$ (right) with W_H = 5, r = 10 %, and γ = 70 %

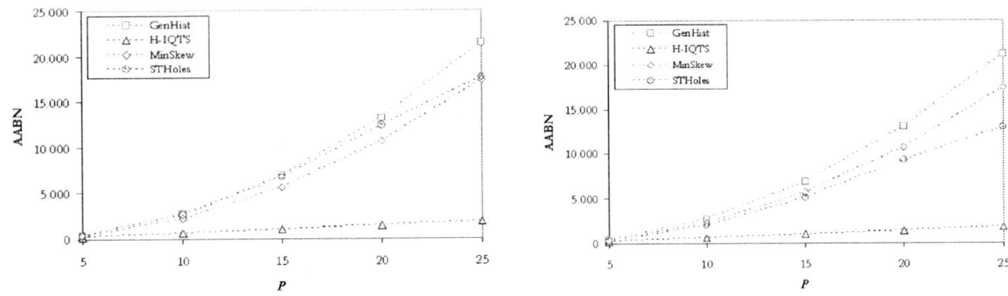

$$AABN(N_t) = \frac{1}{|Q_{S,t}|} \cdot \sum_{k=0}^{|Q_{S,t}|-1} ABN(Q_k) \quad (23)$$

such that, for each query Q_k in $Q_{S,t}$, ABN(Q_k) is the number of buckets accessed during the evaluation of Q_k.

Summarizing, given a compression technique *T*, AABN allows us to measure the capabilities of *T* in supporting advanced OLAP visualization of multidimensional data cubes as the number of buckets accessed can be reasonably considered as a measure of the computational cost needed to extract summarized knowledge. This resembles a sort of measure of the *entropy* of the overall knowledge extraction process. As stated in the fourth Section, this aspect assumes a leading role in mobile OLAP settings (e.g., Hand-OLAP).

COMPARISON TECHNIQUES

In our experimental study, we compared the performance of our proposed technique (under the two metrics previously defined) against the following well-known histogram-based techniques for compressing data cubes: *MinSkew* by Acharya et al. (1999), *GenHist* by Gunopulos et al. (2000), and *STHoles* by Bruno et al. (2001).

In more detail, having fixed the space budget *B* (i.e., the storage space available for housing the compressed representation of the input OLAP view), we derived, for each comparison technique, the configuration of the input parameters that respective authors consider the best in their papers. This ensures a *fair* experimental analysis (i.e., an analysis such that each comparison technique provides its *best* performance). Furthermore, for all the comparison techniques, we set the space budget *B* as equal to the *r* % of *size*(*D*), being *r* the *compression ratio* and *size*(*D*) the total occupancy of the input OLAP view *D*. As an example, *r* = 10 % (i.e., *B* is equal to the 10 % of *size*(*D*)) is widely recognized as a reasonable setting (e.g., see Bruno et al., 2001)).

Experimental Results

Figure 20 shows our experimental results for what regards the accuracy of the compression techniques with respect to the selectivity of queries in Q_S on the 1,000 × 1,000 two-dimensional synthetic OLAP views $D_C(25,70)$ (left side) and $D_Z(0.5,1.5)$ (right side), respectively. Figure 21 shows the results of the same experiment when ranging *r* on the interval [5, 20] (i.e., *B* on the interval [5, 20] % of *size*(*D*)), and fixing the selec-

Figure 23. Experimental results for the accuracy metrics with respect to the query selectivity ||Q|| on the 1,000 × 1,000 two-dimensional benchmark OLAP views extracted from the data sets TPC-H (left) and APB-1 (right) with r = 10 %

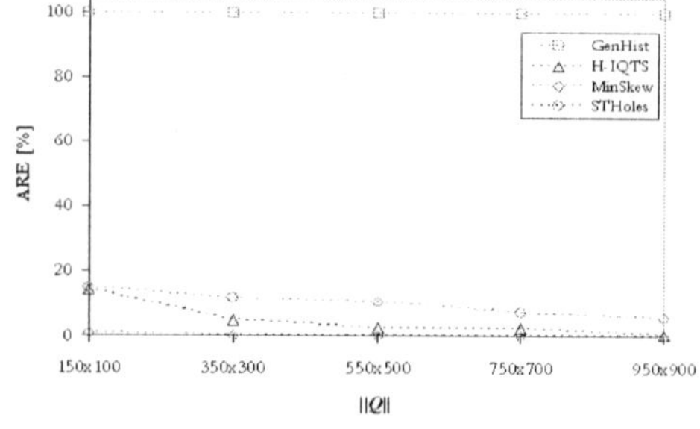

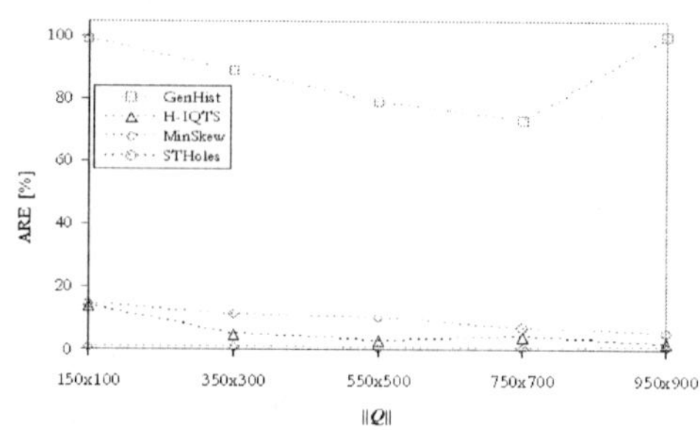

Figure 24. Experimental results for the accuracy metrics with respect to the compression ratio r on the 1,000 × 1,000 two-dimensional benchmark OLAP views extracted from the data sets TPC-H (left) and APB-1 (right) with ||Q|| = 350 × 300

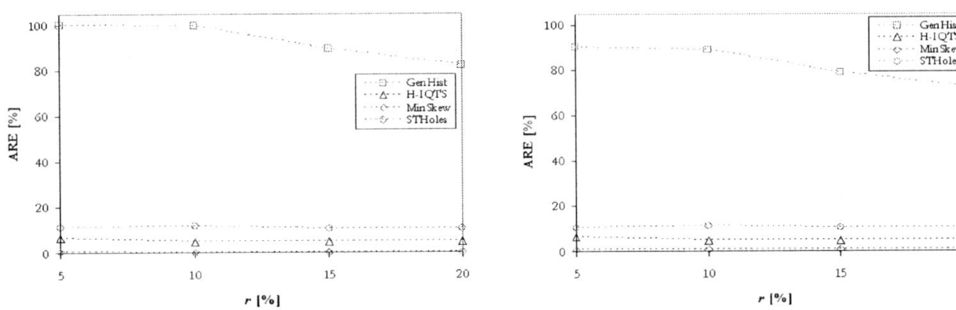

Figure 25. Experimental results for the visualization metrics with respect to the depth of HRQs P on the 1,000 × 1,000 two-dimensional benchmark OLAP views extracted from the data sets TPC-H (left) and APB-1 (right) with W_H = 5, r = 10 %, and γ = 70 %

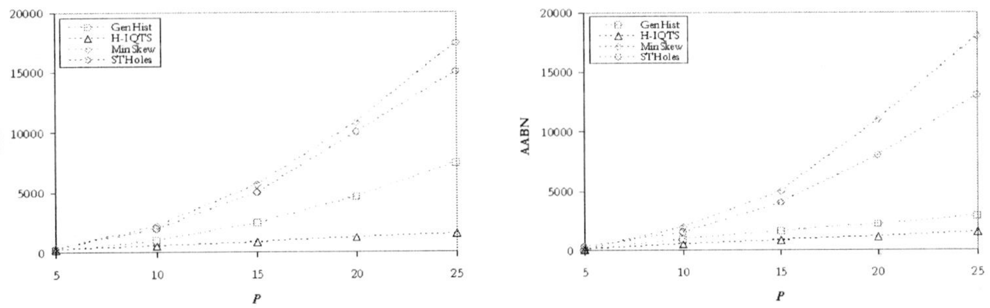

Figure 26. Experimental results for the accuracy metrics with respect to the query selectivity ||Q|| on the 1,000 × 1,000 two-dimensional real-life OLAP view built on top of the data set USCensus1990 with r = 10 %

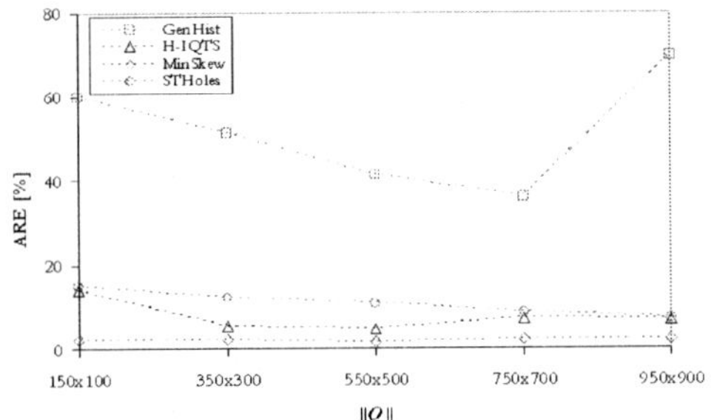

Figure 27. Experimental results for the accuracy metrics with respect to the compression ratio r on the 1,000 × 1,000 two-dimensional real-life OLAP view built on top of the data set USCensus1990 with ||Q|| = 350 × 300

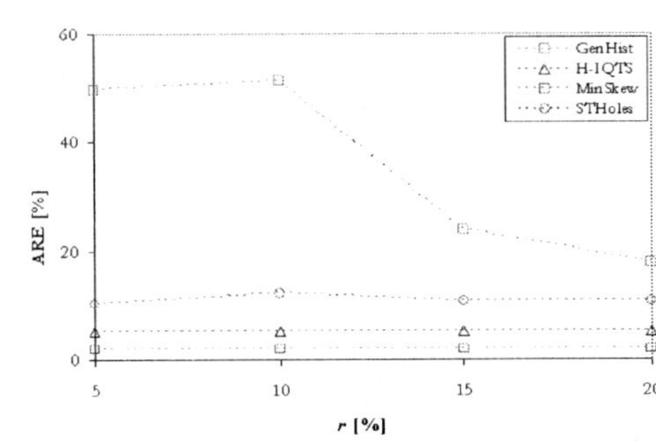

tivity of queries $||Q||$. This allows us to measure the *scalability* of the compression techniques, which is a critical aspect in OLAP systems (e.g., see Cuzzocrea, 2005). Finally, Figure 22 shows our experimental results for what regards the "visualization capabilities" of the comparison techniques (according to the guidelines drawn through the article) with respect to the depth of HRQs (i.e., P_H) having fan-out degree W_H equal to 5 and the parameter γ equal to 70 %. The input two-dimensional OLAP views and the value of the parameter r are the same of the previous experiments.

Figure 23, 24, and 25 show the results of the same experiment set described above when 1,000 × 1,000 two-dimensional benchmark OLAP views extracted from the data sets TPC-H (left sides of figures) and APB-1 (right sides of figures), respectively, are considered as input. Finally, Figure 26, 27, and 28 show the experimental results when a 1,000 × 1,000 two-dimensional real-life OLAP view built on top of the data set *USCensus1990* is considered as input.

From the analysis of the set of experimental results on two-dimensional synthetic, benchmark and real-life OLAP views, it follows that, with respect to the accuracy metrics, our proposed technique is comparable with *MinSkew*, which represents the best on two-dimensional views (indeed, as well-recognized-in-literature, *MinSkew* presents severe limitations on multidimensional domains); instead, with respect to the visualization metrics, our proposed technique overcomes the comparison techniques, thus confirming its suitability in efficiently supporting advanced OLAP visualization of multidimensional data cubes.

CONCLUSION AND FUTURE WORK

In this article, we have presented an innovative technique for supporting advanced OLAP visualization of multidimensional data cubes, which is particularly suitable for mobile OLAP scenarios (like, for instance, those addressed by the system Hand-OLAP). Founding on very efficient two-dimensional summary data domain compression solutions (Buccafurri et al., 2003), our technique meaningfully exploits the data compression paradigm that, in this article, has been proposed as a way of visualizing multidimensional OLAP domains to overcome the natural disorientation and refractoriness of human beings in dealing with hyper-spaces. In this direction, the OLAP dimension flattening process and the amenity of computing semantics-aware buckets are, to the best of our knowledge, innovative contributions to the state-of-the-art OLAP research. Finally, various experimental results performed on different kinds of two-dimensional OLAP views extracted from synthetic, benchmark, and real-life multidimensional data cubes have clearly confirmed the benefits of our proposed technique in the OLAP visualization context, also in comparison with well-known data cube compression techniques.

Future work is mainly focused on making the proposed technique capable of building m-dimensional OLAP views over massive n-dimensional data cubes, with $m << n$ and $m > 2$, by extending the algorithms presented in this article. A possible solution could be found in the results coming

Figure 28. Experimental results for the visualization metrics with respect to the depth of HRQs P on the 1,000 × 1,000 two-dimensional real-life OLAP view built on top of the data set USCensus1990 with W_H = 5, r = 10 %, and γ = 70 %

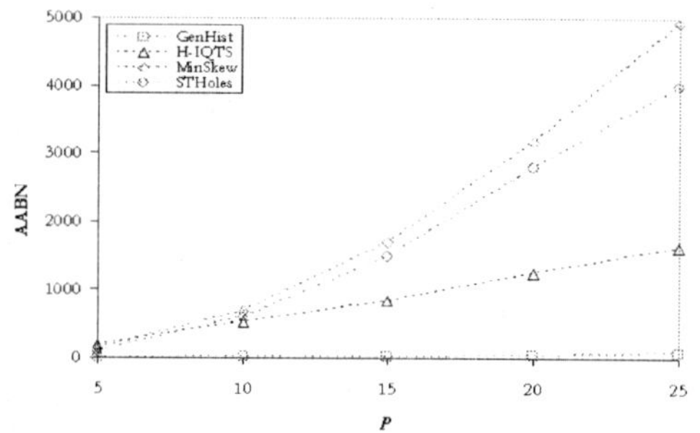

from the *high-dimensional data and information visualization* research area (e.g., see University of Hannover, Germany, 2005), which are already suitable to be applied to the problem of visualizing multidimensional databases and data cubes.

ACKNOWLEDGMENT

Authors are very grateful to Francesco Cristofaro for having developed and performed the experimental section on benchmark and real-life data sets.

REFERENCES

Agrawal, R., Gupta, A., & Sarawagi, S. (1997). Modeling multidimensional databases. *Proceedings of 13th IEEE ICDE International Conference* (pp. 232-243), Bangalore, India.

Acharya, S., Poosala, V., & Ramaswamy, S. (1999). Selectivity estimation in spatial databases. *Proceedings of 1999 ACM SIGMOD International Conference* (pp. 13-24), Philadelphia, PA, USA.

Babcock, B., Chaudhuri, S., & Das, G. (2003). Dynamic sample selection for approximate query answers. *Proceedings of 2003 ACM SIGMOD International Conference* (pp. 539-550), San Diego, CA, USA.

Buccafurri, F., Furfaro, F., Saccà, D., & Sirangelo, C. (2003). A quad-tree based multiresolution approach for two-dimensional summary data. *Proceedings of 15th IEEE SSDBM International Conference* (pp. 127-140), Cambridge, MA, USA.

Bruno, N., Chaudhuri, S., & Gravano, L. (2001). STHoles: A multidimensional workload-aware histogram. *Proceedings of 2001 ACM SIGMOD International Conference* (pp. 211-222), Santa Barbara, CA, USA.

Cabibbo, L., & Torlone, R. (1998). From a procedural to a visual query language for OLAP. *Proceedings of 10th IEEE SSDBM International Conference* (pp. 74-83), Capri, Italy.

Chaudhuri, S., & Dayal, U. (1997). An overview of data warehousing and OLAP technology. *ACM SIGMOD Record, 26*(1), 65-74.

Chaudhuri, S., Das, G., Datar, M., Motwani, R., & Rastogi, R. (2001). Overcoming limitations of sampling for aggregation queries. *Proceedings of 17th IEEE ICDE International Conference* (pp. 534-542), Heidelberg, Germany.

Codd, E. F., Codd, S. B., & Salley, C. T. (1993). *Providing OLAP to user-analysts: An IT mandate*. E. F. Codd and Associates Technical Report.

Colliat, G. (1996). OLAP, relational, and multidimensional database systems. *SIGMOD Record, 25*(3), 64-69.

Cuzzocrea, A. (2005). Overcoming limitations of approximate query answering in OLAP. *Proceedings of 9th IEEE IDEAS International Conference* (pp. 200-209), Montreal, Canada.

Cuzzocrea, A., & Wang, W. (2007). Approximate range-sum query answering on data cubes with probabilistic guarantees. *Journal of Intelligent Information Systems, 28*(2), 161-197.

Cuzzocrea, A., Furfaro, F., & Saccà, D. (2003). Hand-OLAP: A system for delivering OLAP services on handheld devices. *Proceedings of 6th IEEE ISADS International Conference* (pp. 80-87), Pisa, Italy.

Ezeife, C. I. (2001). Selecting and materializing horizontally partitioned warehouse views. *Data & Knowledge Engineering, 36*(2), 185-210.

Fang, M., Shivakumar, N., Garcia-Molina, H., Motwani, R., & Ullman, J.D., (1998). Computing iceberg queries efficiently. *Proceedings of 24th VLDB International Conference* (pp. 299-310), New York City, NY, USA.

Gebhardt, M., Jarke, M., & Jacobs, S. (1997). A toolkit for negotiation support interfaces to multi-dimensional data. *Proceedings of 1997 ACM SIGMOD International Conference* (pp. 348-356), Tucson, AZ, USA.

Gibbons, P. B., & Matias, Y. (1998). New sampling-based summary statistics for improving approximate query answers. *Proceedings of 1998 ACM SIGMOD International Conference* (pp. 331-342), Seattle, WA, USA.

Gray, J., Chaudhuri, S., Bosworth, A., Layman, A., Reichart, D., & Venkatrao, M. (1997). Data cube: A relational aggregation operator generalizing group-by, cross-tab, and sub-totals. *Data Mining and Knowledge Discovery, 1*(1), 29-53.

Gunopulos, D., Kollios, G., Tsotras, V. J., & Domeniconi, C. (2000). Approximating multi-dimensional aggregate range queries over real attributes. *Proceedings of 2000 ACM SIGMOD International Conference* (pp. 463-474), Dallas, TX, USA.

Gupta, A., Agrawal, D., & El Abbadi, A. (2003). Approximate range selection queries in peer-to-peer systems. *Proceedings of 1st CIDIR International Conference*, Asilomar, CA, USA. Retrieved from http://www-db.cs.wisc.edu/cidr/cidr2003/program/p13.pdf

Hacid, M. S., & Sattler, U. (1998). Modeling multidimensional databases: A formal object-centered approach. *Proceedings of 6th ECIS International Conference* (pp. 1-15), Aix-en-provence, France.

Han, J., & Kamber, M. (2000). *Data mining: Concepts and techniques*. Morgan Kauffmann Publishers.

Harinarayan, V., Rajaraman, A., & Ullman, J. (1996). Implementing data cubes efficiently. *Proceedings of 1996 ACM SIGMOD International Conference* (pp. 205-216), Montreal, Canada.

Ho, C. T., Agrawal, R., Megiddo, N., & Srikant, R. (1997). Range queries in OLAP data cubes. *Proceedings of 1997 ACM SIGMOD International Conference* (pp. 73-88), Tucson, AZ, USA.

Inmon, W. H. (1996). *Building the data warehouse*. John Wiley & Sons.

Inselberg, A. (2001). Visualization and knowledge discovery for high dimensional data. *Proceedings of 2nd IEEE UIDIS International Workshop* (pp. 5-24), Zurich, Switzerland.

Keim, D. A. (1997). Visual data mining. *Tutorial at 23rd VLDB International Conference*, Athens, Greece. Retrieved from http://www.dbs.informatik.uni-muenchen.de/daniel/VLDBTutorial.ps

Kimball, R. (1996). *The data warehouse toolkit*. John Wiley & Sons.

Koudas N., Muthukrishnan S., & Srivastava D. (2000). Optimal histograms for hierarchical range queries. *Proceedings of 19th ACM PODS International Symposium* (pp. 196-204), Dallas, TX, USA.

Lehner, W., Albrecht, J., & Wedekind, H. (1998). Normal forms for multivariate databases. *Proceedings of 10th IEEE SSDBM International Conference* (pp. 63-72), Capri, Italy.

Lenz, H. J., & Shoshani, A. (1997). Summarizability in OLAP and statistical databases. *Proceedings of 9th IEEE SSDBM International Conference* (pp. 132-143), Olympia, WA, USA.

Lenz, H. J., & Thalheim, B. (2001). OLAP databases and aggregation functions. *Proceedings of 13th IEEE SSDBM International Conference* (pp. 91-100), Fairfax, VA, USA.

Maniatis, A., Vassiliadis, P., Skiadopoulos, S., & Vassiliou, Y. (2003a). CPM: A cube presentation model for OLAP. *Proceedings of 5th DaWaK International Conference* (pp. 4-13), Prague, Czech Republic.

Maniatis, A., Vassiliadis, P., Skiadopoulos, & S., Vassiliou, Y. (2003b). Advanced visualization for OLAP. *Proceedings of 6th ACM DOLAP International Workshop* (pp. 9-16), New Orleans, LO, USA.

Muralikrishna, M., & DeWitt, D.J. (1998). Equi-depth histograms for estimating selectivity factors for multi-dimensional queries. *Proceedings of 1998 ACM SIGMOD International Conference* (pp. 28-36), Seattle, WA, USA.

OLAP Council. (1998). *Analytical processing benchmark 1, Release II*. Retrieved from http://www.symcorp.com/downloads/OLAP_Council-WhitePaper.pdf

Piatetsky-Shapiro, G., & Connell, C. (1984). Accurate estimation of the number of tuples satisfying a condition. *Proceedings of 1984 ACM SIGMOD International Conference* (pp. 265-275), Boston, MA, USA.

Poosala, V., & Ioannidis, Y. (1997). Selectivity estimation without the attribute value independence assumption. *Proceedings of 23rd VLDB International Conference* (pp. 486-495), Athens, Greece.

Thanh Binh, N., Min Tjoa, A., & Wagner, R. (2000). An object oriented multidimensional data model for OLAP. *Proceedings of 1st WAIM International Conference* (pp. 69-82), Shanghai, China.

Transaction Processing Council. (2006). *TPC benchmark H*. Retrieved from http://www.tpc.org/tpch/

Tsois, A., Karayannidis, N., & Sellis, T. (2001). MAC: Conceptual data modeling for OLAP. *Proceedings of 3rd DMDW International Workshop*, Interlaken, Switzerland. Retrieved from http://sunsite.informatik.rwth-aachen.de/Publications/CEUR-WS/Vol-39/paper5.pdf

University of California, Irvine. (2001). *1990 US Census Data*. Retrieved from http://kdd.ics.uci.edu/databases/census1990/USCensus1990.html

University of California, Irvine. (2005). *Knowledge discovery in databases archive*. Retrieved from http://kdd.ics.uci.edu/

University of Hannover, Germany. (2005). *2D, 3D, and high-dimensional data and information visualization research group*. Retrieved from http://www.iwi.uni-hannover.de/lv/seminar_ss05/bartke/home.htm

Vassiliadis, P. (1998). Modeling multidimensional databases, cubes, and cube operations. *Proceedings of 10th IEEE SSDBM International Conference* (pp. 53-62), Capri, Italy.

Vassiliadis, P., & Sellis, T. (1999). A survey of logical models for OLAP databases. *SIGMOD Record, 28*(4), 64-69.

Vitter, J. S., Wang, M. & Iyer, B. (1998). Data cube approximation and histograms via wavelets. *Proceedings of 7th ACM CIKM International Conference* (pp. 96-104), Bethesda, MD, USA.

Xin, D., Han, J., Cheng, H., & Li, X. (2006). Answering top-k queries with multi-dimensional selections: The ranking cube approach. *Proceedings of 32nd VLDB International Conference* (pp. 463-475), Cairo, Egypt.

Zipf, G. K. (1949). *Human behaviour and the principle of least effort*. Addison-Wesley.

This work was previously published in International Journal of Data Warehousing and Mining, Vol. 3, Issue 4, edited by David Taniar, pp. 1-30, copyright 2007 by IGI Publishing, formerly known as Idea Group Publishing (an imprint of IGI Global).

Chapter 2.27
A Presentation Model and Non-Traditional Visualization for OLAP

Andreas Maniatis
National Technical University of Athens, Greece

Panos Vassiliadis
University of Ioannina, Greece

Spiros Skiadopoulos
National Technical University of Athens, Greece

Yannis Vassiliou
National Technical University of Athens, Greece

George Mavrogonatos
National Technical University of Athens, Greece

Ilias Michalarias
National Technical University of Athens, Greece

ABSTRACT

Data visualization is one of the major issues of database research. OLAP a decision support technology, is clearly in the center of this effort. Thus far, visualization has not been incorporated in the abstraction levels of DBMS architecture (conceptual, logical, physical); neither has it been formally treated in this context. In this paper we start by reconsidering the separation of the aforementioned abstraction levels to take visualization into consideration. Then, we present the Cube Presentation Model (CPM), a novel presentational model for OLAP screens. The proposal lies on the fundamental idea of separating the logical part of a data cube computation from the presentational part of the client tool. Then, CPM can be naturally mapped on the Table Lens, which is an advanced visualization technique from the Human-Computer Interaction area, particularly tailored for cross-tab reports. Based on the particularities of Table Lens, we propose automated proactive support to the user for the interaction with an OLAP screen. Finally, we discuss implementation and usage issues in the context of an academic prototype system (CubeView) that we have implemented.

INTRODUCTION

In the last years, Online Analytical Processing (OLAP) and data warehousing (DW) have become major research areas in the database community (Abiteboul et al., 2003; Inmon, 1996). Although the *modeling* of data (Tsois et al., 2001; Vassiliadis & Sellis, 1999) has been extensively dealt with, an equally important issue in the OLAP domain, the *presentation* of data, has not been adequately investigated.

As the Lowell report (Abiteboul et al., 2003) mentions, visualization is one of the big issues of database research for the next years. To cite the Lowell report:

The original Laguna-Beach report lamented that there was little research on user interfaces to DBMSs. ... There have not been comparable advances in the last 15 years. There is a crying need for better ideas in this area.

It is easy to understand that of all fields of database research, decision support, and OLAP are the ones to be affected most out of this phenomenon.

In the context of OLAP, data visualization deals with the techniques and tools used for presenting OLAP-specific information to end users and decision makers. During the next years, the database community expects visualization to be of

Figure 1. General Framework for CPM

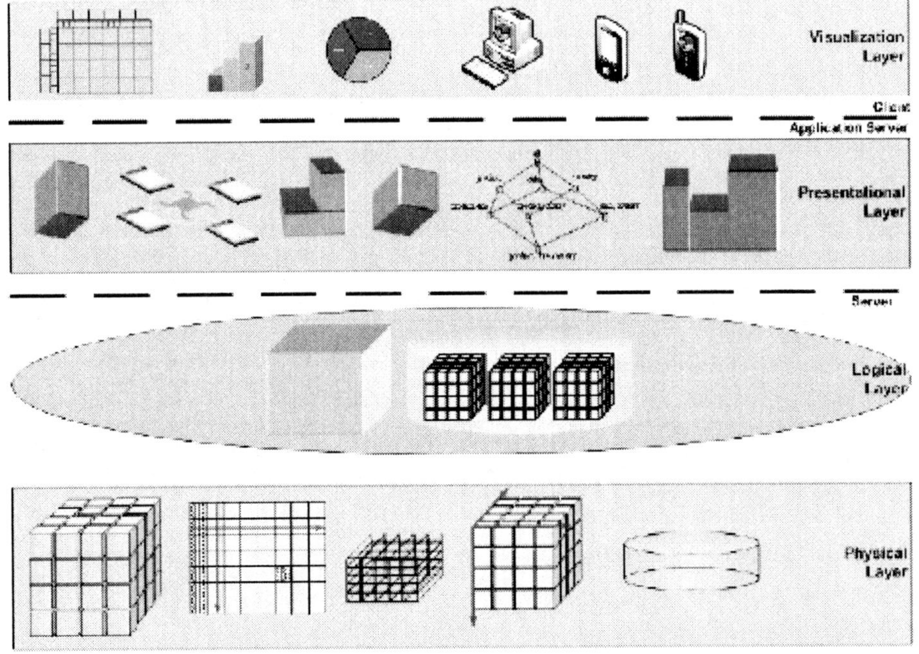

significant importance in the area (Abiteboul et al., 2003), and although research has provided results dealing with the presentation of vast amounts of data (Gebhardt et al., 1997; Inselberg, 2001; Keim, 1997), to our knowledge, OLAP has not been part of advanced visualization techniques so far.

For us, it is clear that one of the main reasons for the research community not dealing with visualization issues so far is the heritage of the computing paradigm of the past three decades. This paradigm silently made the assumption that the user sitting in front of a console makes *one* query and retrieves *one* answer (as would have happened in a UNIX terminal 30 years ago). Still, this is not the case with modern user interfaces for datasets, especially in the context of OLAP. A single front-end screen typically involves the combination of more than one back-end query. Still, to the best of our knowledge, there are no modeling techniques and languages (from the relational model to SQL and the OLAP modeling efforts proposed in the academia) that build upon this fact. Our effort tries to formalize the simultaneous presence of more than one query, which is done in two layers. In the presentational layer, we provide a uniform and generic model for the user interface, which hides the complexity of answer retrieval, detached in the logical layer. As a second interesting difference, note that the users work in *sessions* of queries, as opposed to *sequences* of unrelated queries. OLAP is a typical, but not the only, case for this behavior.

In this paper, we try to approach the problem from a clean sheet of paper. Although we do not claim to provide a generic answer for all kinds of database visualization problems, we focus on the specifics of the OLAP field. Having observed that presentational models are not really part of the classical conceptual-logical-physical hierarchy of database models (depicted in Figure 1), we propose a new separation of layers. In the sequel, we will refer to the different layers of abstraction (models) that help us design, manage, and operate an OLAP environment through the term layers.

In the middle, there is a *logical layer* that abstracts from the particularities of data storage and describes cubes and dimensions. This layer is naturally mapped to physical storage entities, like relational tables (ROLAP), or proprietary structures like multidimensional matrices (MOLAP). These kinds of physical entities form the *physical layer*. Having these structures covers well enough the part pertaining to the query formulation. Still, although the logical layer deals with the representation of data in an abstract form, as well as the formulation of queries and operations over them, we need a way to model how the answer to a query is represented in the client part. The role of the logical layer for the server is played by the *presentational layer* for the client, which involves a simple and generic model to abstract from the particularities of data retrieval. The ultimate representation is performed by the specific visualization techniques (pies, bar charts, etc.) handled by the *visualization layer*. The presentational layer is generic enough to discuss the broad strategy of how data are to be visualized (e.g., by 2D vs. 3D means, focused for tabular vs. multimedia data, and so on), whereas the visualization layer deals with the particularities of the final visualization means (e.g., palmtop, printer, virtual reality environment, and so on).

As one can see in Figure 1, there is a part of the functionality that pertains to the server and part of it that pertains to the application server and the client. Naturally, the distinguishing lines between this three-tier architecture can be rearranged easily for a two-tier or a four-tier architecture. Note also that conceptual modeling is orthogonal to this classification for two reasons: (1) conceptual models are not really part of the OLAP engines or environments, and (2) in the OLAP field, the logical and conceptual levels are quite often indistinguishable (Kimbal et al., 1998).

Someone could possibly question the need for new models. Is it really necessary to depart from the well-known classification of models? Our answer is positive, and we base our proposal on the following reasons:

- First, we need to allow the formal definition of the presentation of the result of a database query — in our case, a cube. This is currently done in an ad hoc manner from the client tools; no formal foundations are given for this kind of representation.
- Second, we need to decouple the definition of the logical underpinnings of user operations from the way the result is presented. Even if we create a model for the presentation of the results, we need to keep it loosely coupled with the model of the underlying data. One could claim that a generic, complete model could cover all cases; still, to our understanding, such a model is too hard to achieve. Therefore, the natural antidote to the lack of completeness should be applied, and this is genericity: by decoupling logical and presentational models, we can decide how we match a particular logical model to a presentational one, among many choices.
- Third, we need to be able to depart from the traditional thinking of treating visualization as appropriate only for computer screens. On the contrary, we live in an age where a computer screen is just one of the possible choices for the presentation of data. Bundling the choices for different presentation devices with the logical models and languages would probably create hard-to-use constructs.

In the context of all the aforementioned issues, we move on in this paper to make the following contributions.

First, we introduce CPM (Cube Presentation Model), a presentation model for OLAP, and we combine it with non-traditional visualization techniques. The main idea behind CPM lies in the separation of *logical data retrieval* (which we encapsulate in the logical layer of CPM) and *data presentation* (captured from the presentational layer of CPM). The logical layer that we propose is based on an extension of a previous proposal (Vassiliadis & Skiadopoulos, 2000) to incorporate more complex cubes. At the same time, the presentational layer provides a formal model for OLAP screens. To our knowledge, there is no such result in the related literature.

Once CPM has been introduced, we move on to give a mapping of the generic presentational scheme of CPM to the particularities of an advanced visualization technique coming from the field of Human Computer Interaction. The Table Lens technique (Pirollo & Rao, 1996; Rao & Card, 1994) is particularly tailored for cross-tab reports, which are most commonly used for OLAP purposes and accompanied by a set of handy features for the exploration of data sets that are presented in this way. In the sequel, we provide algorithms for the automated proactive support of the user during his or her interaction with an OLAP screen, based on the particularities of Table Lens. Specifically, Table Lens employs a particular distortion of the presentation to highlight areas of increasing interest to the user. We provide a generic algorithm to support this task proactively and customize it to a particular instance to show how it could actually work. By exploiting Table Lens, along with suitable coloring schemes, we provide a new presentation technique, which we call OLAP Lens.

Moreover, an academic software platform specifically designed and implemented to support both CPM and OLAP Lens is introduced. The architecture of the platform, which is called CubeView, is particularly tailored to support Mobile OLAP, a term used to denote the porting of OLAP Visualization and Analysis applications onto portable, mobile, and wireless devices.

To motivate the discussion, we will use throughout the paper a running example, where we customize the example presented in Microsoft (1998) to an international publishing company, with traveling salesmen selling books and CDs to other bookstores all over the world (Figure 2). In this example, we assume that a cube — SalesCube — is defined over the dimensions — Products, Salesman, Time, and Geography — each involving

Figure 2: Motivating Example for the Cube Model (taken from Microsoft, 1998)

```
SELECT  CROSSJOIN ({Venk,Netz}, {USA_N.Children, USA_S,Japan}) ON COLUMNS
        {Qtr1.CHILDREN, Qtr2, Qtr3, Qtr4.CHILDREN} ON ROWS
FROM    SalesCube
WHERE   Sales, [1991], Products.ALL)
```

Year = 1991			C1		C2	C3	C4		C5	C6
Product = ALL			Venk				Netz			
			USA			Japan	USA			Japan
			USA_N		USA_S		USA_N		USA_S	
		Size (City)	Seattle	Boston			Seattle	Boston		
R1	**Qtr1**	Jan	20	32	62	97	23	40	75	12
		Feb	25	40	74	121	18	32	51	20
		Mar	18	12	36	110	42	48	65	3
R2	**Qtr2**		56	63	150	253	50	70	280	50
R3	**Qtr3**		52	65	147	200	53	64	270	50
R4	**Qtr4**	Oct	25	24	64	98	32	12	64	76
		Nov	28	28	76	102	40	21	83	69
		Dec	23	30	68	150	42	29	99	77

several levels of aggregation. In this query, we restrict the Time dimension to the sales of Year 1991. We ignore the Products dimension (Products. ALL) in the subsequent aggregation of detailed data. Whenever we need to present a 2D screen and more than two dimensions are involved, we need to merge (CROSSJOIN in [Microsoft, 1998] terminology) as many dimensions as necessary in a single axis. In this case, we combine the dimensions Salesman (restricted by the query author to two particular salesmen — Salesman in ['Venk','Netz'] — and Geography on the COLUMNS axis and leave the dimension Time on the ROWS axis. Note that the Geography dimension involves more than one level of aggregation (both City and Region). The same applies for the Time dimension, where both Quarters and Months are employed.

The remainder of this paper is structured as follows: In Section 2, we summarize the logical layer of CPM. In Section 3, we present in detail the presentation layer of CPM. In Section 4, we show how CPM can be naturally combined with Table Lens and how we can automate the task of proactively supporting the user with highlighted areas of interest. In Section 5, we describe a prototype platform — CubeView — where we have implemented the proposed visualization schemes. In Section 6, we present work closely related to our research, and finally, in Section 7, we conclude our results and present topics for future work.

LOGICAL FOUNDATIONS

In this section, we present the logical layer of CPM; to this end, we extend a logical model (Vassiliadis & Skiadopoulos, 2000) in order to compute more complex cubes. In a nutshell, the logical layer involves (a) *dimensions* defined as lattices of dimension *levels*, (b) *ancestor functions* (in the form of) mapping values between related levels of a dimension, (c) *detailed data sets*, practically modeling fact tables at the lowest granule of information for all their dimensions, and (d) *cubes*, defined as aggregations over detailed data sets.

Formally, the constructs of the model (Vassiliadis & Skiadopoulos, 2000) are:

- Four countable pairwise disjoint infinite sets exist: a set of *level names* (or simply *levels*) U_L; a set of *measure* names (or simply *measures*) $U_{\mathcal{M}}$; a set of *dimension* names (or simply *dimensions*) $U_{\mathcal{D}}$; and a set of *cube* names (or simply *cubes*) $U_{\mathcal{C}}$. The set of *attributes* U is defined as $U=U_{\mathcal{L}} \cup U_{\mathcal{M}}$. For each $A \in U_{\mathcal{L}}$, we define a countable totally ordered set dom(A), the domain of A, which is isomorphic to the integers. Similarly, for each $A \in U_{\mathcal{M}}$, we define an infinite set dom(A), the domain of A, which is isomorphic to the real numbers. We can impose the usual comparison operators to all the values participating to totally ordered domains $\{<, >, \leq, \geq\}$. We also assume the existence of two attributes — ALL and RANK. The role of the special attribute ALL will be analyzed in the sequel. Level ALL has a single value in its domain, namely "all". RANK is a special purpose measure, which will be used for the ordering of a cube. The domain of RANK is the set of integers.

- A *dimension* D is a lattice $(\mathbf{L}, \prec)$ such that:
  - $\mathbf{L}=(L_1,\ldots,L_n)$ is a finite subset of $U_{\mathcal{L}}$.
  - $dom(L_i) \cap dom(L_j) = \varnothing$ for every $i \neq j$.
  - $\prec$ is a partial order defined among the levels of $\mathbf{L}$.

  Each path in the dimension lattice, beginning from its upper bound and ending in its lower bound, is called a *dimension path.*
- A family of functions is defined, satisfying the following conditions:
  - For each pair of levels L_1 and L_2 such that $L_1 \prec L_2$ the function maps each element of $dom(L_1)$ to an element of $dom(L_2)$.
  - Given levels L_1, L_2 and L_3 such that $L_1 \prec L_2 \prec L_3$, the function equals to the composition $_\circ$. This implies that:

  1. (x)=x.
  2. if y=(x) and z=(y), then z=a(x).

  - for each pair of levels L_1 and L_2 such that $L_1 \prec L_2$ the function is monotone (preserves the ordering of values). In other words:

    $\forall x,y \in dom(L_1)$:
    $x<y \Rightarrow (x) \leq (y), L_1 \prec L_2$
- A *schema* S is a finite subset of U. Normally, we will represent a schema as divided in two parts: $S=[D_1.L_1, \ldots, D_n.L_n, A_1, \ldots, A_m]$, where:
  - $(L_1,\ldots,L_n)$ are levels from a dimension set $\mathbf{D}=(D_1,\ldots,D_n)$ and level L_i comes from dimension D_i, for $1 \leq i \leq n$.
  - $(A_1,\ldots,A_m)$ are attributes (i.e., measures and levels).
- A *detailed schema* S^0 is a schema whose levels are the lowest in the respective dimensions. When we refer to a level L as the *lowest* in the dimension, it means that there does not exist any other level L', such that $L' \prec L$.
- A *tuple* t over a schema $S=[L_1, \ldots, L_n, A_1, \ldots, A_m]$ is a total and injective mapping from S to $dom(L_1) \times \ldots \times dom(L_n) \times dom(A_1) \times \ldots \times dom(A_m)$, such that $t[X] \in dom(X)$ for each $X \in S$.
- A *data set* DS over a schema $S=[L_1, \ldots, L_n, A_1, \ldots, A_m]$ is a finite set of tuples over S such that:
  - $\forall t_1,t_2 \in DS, t_1[L_1,\ldots,L_n]=t_2[L_1,\ldots,L_n] \Rightarrow t_1=t_2$.
  - for no strict subset $X \subset \{L_1,\ldots,L_n\}$, the previous also holds.

  In other words, $A_1,\ldots,A_m$ are functionally dependent (in the relational sense) on levels $(L_1,\ldots,L_n)$ of schema S. A *detailed data set* DS^0 is a data set over a detailed schema S.
- An atom is true, false (with obvious semantics) or an expression of the form $x \partial y$, where x and y can be one of the following: (a) a level L_1 (i.e., not a measure); (b) a value l; (c) an expression of the form (L_1) where $L_1 \prec L_2$; (d) an expression of the form (l) where $L_1 \prec L_2$ and $l \in dom(L_1)$. If x and y are levels then they should belong to isomorphic dimensions. ∂ is an operator from the set $(>, <, =, \geq, \leq, \neq)$.
- A *selection condition* φ is a formula involving atoms and the logical connectives $\wedge$, $\vee$, and $\neg$. A selection condition is always applied to a data set such that all the level

names occurring in the selection condition—either\in the form (1) or (3)—belong to the schema of the data set. Let DS be a data set over schema S. The expression φ(DS) is a set of tuples **X** belonging to DS such that when, for all the occurrences of level names in φ, we substitute the respective level values of every $x \in \mathbf{X}$, the formula φ becomes true. A *detailed selection condition* φ^0 is a selection condition where all participating levels are the detailed levels of their dimensions.

- A *primary cube* c (over the schema $[L_1, \ldots, L_n, M_1, \ldots, M_m]$) is an expression of the form:

$c=(DS^0, \varphi, [L_1, \ldots, L_n, M_1, \ldots, M_m], [agg_1(), \ldots, agg_m()])$, where:

- DS^0 is a detailed data set over the schema $S=[, \ldots, , , , \ldots,]$, $m \leq k$.
- φ is a detailed selection condition.
- $M_1, \ldots, M_m$ are measures.
- and L_i are levels such that $\prec L_i$, $1 \leq i \leq n$.
- $agg_i \in \{sum, min, max, count\}$, $1 \leq i \leq m$.

The expression characterising a cube has the following semantics:

$c=\{x \in Tup(L_1, \ldots, L_n, M_1, \ldots, M_m) \mid \exists y \in \varphi(DS^0)$,
$x[L_i]=(y[]), 1 \leq i \leq n$,
$x[M_j]=agg_j(\{q \mid \leq z \in \varphi(DS^0)$,
$x[L_i]= \quad (z[]), 1 \leq i \leq n$,
$q=z[]\}), 1 \leq j \leq m\}$

In other words, a cube c is a set of tuples. To compute it, first we apply the selection condition to the detailed data set. Then, we replace the values of the levels for the tuples of the result,

Figure 3. Dimensions arrival date, departure date, location, product and salesman

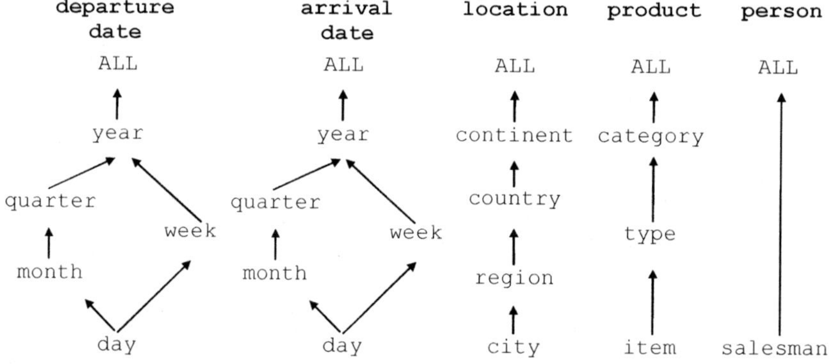

Figure 4. Dimension person

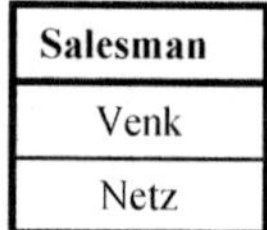

Salesman
Venk
Netz

Figure 5. Dimension Product

Category	Type	Item
Books	Literature "	Report to El Greco"
		"Karamazof brothers"
	Philosophy	"Zarathustra"
		"Symposium"
Music	Heavy Metal "	Piece of Mind"
		"Ace of Spades"

Figure 6. Dimension location

Continent	Country	Region	City
Europe	Greece	Greece-North	Salonica
		Greece-South	Athens
			Rhodes
North America	USA	USA-East	New York
			Boston
		USA-West	Los Angeles
			San Francisco
		USA-North	Seattle
Asia	Japan	Kiusiu	Nagasaki
		Hondo	Tokyo
			Yokohama
			Kioto

with their respective ancestor values (at the levels of the schema of c) and group them into a single value for each measure, through the application of the appropriate aggregate function. Coming back to our motivating example, we can detect the following dimensions:

- arrival date and departure date (when the salesman arrives/leaves the store).
- salesman (instantiated in Figure 4).
- product (instantiated in Figure 5).
- location (instantiated in Figure 6).

The functionally dependent measures are Sales, PercentChange. Our data set DS_0 is depicted in Figure 7. Based on this data set we can define a basic primary cube as:

c^0=(DS^0,true,[arrival day.day, departure day.day, product.item, person.salesman, location.city,sales],sum (sales))

For brevity we can write c^0 as:

c^0=(DS^0,true,[arrival day,departure day,item, salesman,city,sales],sum(sales)).

A primary cube can be defined as follows:

c=(DS^0,='1997', [arrival.month, departure. month,category, salesman.ALL,continent,sum_ sales],sum(sales))

with the data values shown in Figure 8.

The limitations of primary cubes is that although they model accurately SELECT-FROM-WHERE-GROUPBY queries, they fail to model (a) ordering, (b) computation of values through functions, and (c) selection over computed or aggregate values (i.e., the HAVING clause of a SQL query). To compensate this shortcoming, we extend the aforementioned model with the following entities:

- Let **F** be a set of *functions* mapping sets of attributes to attributes. We distinguish the following major categories of functions: property functions, arithmetic functions, and control functions. For example, for the level Day, we can have the property function holiday(Day) indicating whether a day is a holiday or not. An arithmetic function is for example Profit =(Price-Cost)*Sold_Items. Control functions simulate the control state-

Figure 7. Data set

Arrival Day	Departure day	Item	Salesman	City	%Change	Sales
1-Jan-97	3-Jan-97	"Report to El Greco"	Netz	Rhodes	10	10
1-Jan-97	3-Jan-97	"Symposium"	Netz	Rhodes	20	5
6-Feb-97	17-Feb-97	"Symposium"	Netz	Athens	-30	7
6-Feb-97	17-Feb-97	"Karamazof brothers"	Netz	Athens	-50	10
6-Feb-97	17-Feb-97	"Piece of Mind"	Netz	Athens	+35	13
18-Feb-97	10-May-97	"Karamazof brothers"	Netz	Seattle	-50	5
11-May-97	7-Jun-97	"Report to El Greco"	Netz	Los Angeles	100	2
11-May-97	7-Jun-97	"Ace of Spades"	Netz	Los Angeles	100	20
3-Sep-97	5-Sep-97	"Zarathustra"	Netz	Nagasaki	0	50
3-Sep-97	5-Sep-97	"Report to El Greco"	Netz	Nagasaki	0	30
6-Sep-97	16-Dec-97	"Piece of Mind"	Netz	Tokyo	10	10
1-Jul-97	4-Aug-97	"Ace of Spades"	Venk	Salonica	30	13
1-Jul-97	4-Aug-97	"Piece of Mind"	Venk	Salonica	50	34
6-Sep-97	12-Oct-97	"Symposium"	Venk	Boston	-30	7
6-Sep-97	12-Oct-97	"Zarathustra"	Venk	Boston	0	10
1-Feb-98	10-Apr-98	"Ace of Spades"	Venk	Seattle	50	15
1-Feb-98	10-Apr-98	"Piece of Mind"	Venk	Seattle	6	53
4-May-98	7-Jun-98	"Report to El Greco"	Venk	Kyoto	-30	14
13-Jun-98	15-Jul-98	"Zarathustra"	Venk	Nagasaki	0	50
13-Jun-98	15-Jul-98	"Report to El Greco"	Venk	Nagasaki	0	30

ments of the programming languages.

- A *secondary selection condition* ψ is a formula in disjunctive normal form. An atom of the secondary selection condition is true, false or an expression of the form $x\,\theta\,y$, where x and y can be one of the following: (a) an attribute A_i (including RANK), (b) a value l, an expression of the form $f_i(\mathbf{A}_i)$, where $\mathbf{A}_i$ is a set of attributes (levels and measures), and (c) θ is an operator from the set ($>, <, =, \geq, \leq, \neq$). With this kind of formulae, we can compute relationships between measures (Cost>Price), ranking and range selections (ORDER BY...;STOP after 200, RANK[20:30]), measure selections (sales>3000), property-based selection (Color(Product)='Green').
- Suppose a data set DS over the schema $[A_1,A_2,\ldots,A_z]$. Without loss of generality, suppose a non-empty subset of the schema $S=A_1,\ldots,A_k, k\leq z$. Then, there is a set of *ordering operations* used to sort the values of the data set, with respect to the set of attributes participating to S. θ belongs to the set $\{<,>,\emptyset\}$ in order to denote ascending, descending, and no order, respectively. An ordering operation is applied over a data set and returns another data set, which obligatorily encompasses the measure RANK.
- A *secondary cube* over the schema $S=[L_1,\ldots,L_n,M_1,\ldots,M_m,A_{m+1},\ldots,A_{m+p}, RANK]$ is an expression of the form:
$s=[c,[A_{m+1}:f_{m+1}(\mathbf{A}_{m+1}),\ldots,A_{m+p}:f_{m+p}(\mathbf{A}_{m+p})],\psi]$

Figure 8. A primary cube

Arrival month	Departure month	Category	Salesman.ALL	Continent	Sales
Jan-97	Jan-97	Books	all	Europe 1	5
Feb-97	Feb-97	Books	all	Europe 1	7
Feb-97	Feb-97	Music	all	Europe 1	3
Feb-97	May-97	Books	all	North-America	5
May-97	Jun-97	Books	all	North-America	2
May-97	Jun-97	Music	all	North-America	20
Jul-97	Aug-97	Music	all	Europe 4	7
Sep-97	Sep-97	Books	all	Asia 8	0
Sep-97	Oct-97	Books	all	North-America	17
Sep-97	Dec-97	Music	all	Asia 1	0

where

$c=(DS^0,\varphi, [L_1,\ldots,L_n,M_1,\ldots,M_m],[agg_1(),\ldots, agg_m()])$

is a primary cube,

$[\mathbf{A}_{m+1},\ldots,\mathbf{A}_{m+p}]\subseteq[L_1,\ldots,L_n,M_1,\ldots,M_m]$, $\mathbf{A}\subseteq S\text{-}\{RANK\}$,

$f_{m+1},\ldots,f_{m+p}$ are functions belonging to **F** and ψ is a secondary selection condition.

A secondary cube has the following formal semantics:

$s=\{x\in Tup(L_1,\ldots,L_n,M_1,\ldots,M_m,A_{m+1},\ldots,A_{m+p},RANK)|\exists$

data set DS^s defined over the schema:

$[L_1,\ldots,L_n,M_1,\ldots,M_m,A_{m+1},\ldots,A_{m+p}]$,
$y\in DS^s, y_1\in c^1: y[L_i]=y_1[L_i]$,
$1\leq i\leq n$,
$y[M_i]=y_1[M_i], 1\leq i\leq m$,
$y[A_{m+i}]=f_{m+i}[\mathbf{A}_{m+i}], 1\leq i\leq p, x\in\psi$ $(;A(DS^s))\}$

- A *star schema* $(\mathbf{D},S^0)$ is a couple comprising a finite set of dimensions **D** and a detailed schema S^0 defined over (a subset of) these dimensions.
- A *star databases instance* over a star schema $(\mathbf{D},S^0)$ is a triplet $[DS^0,\mathbf{C},\mathbf{C}^S]$, where
  - DS^0 is a detailed data set defined over S^0;
  - **C** is a finite set of cubes, defined over DS^0;
  - $\mathbf{C}^S$ is a finite set of secondary cubes, defined over **C**.

With these additions, primary cubes are extended to secondary cubes that incorporate: (a) computation of new attributes ($\mathbf{A}_{m+i}$) through the respective functions ($\mathbf{f}_{m+i}$); (b) ordering (); and (c) the HAVING clause, through the secondary selection condition ψ.

PRESENTATIONAL LAYER

In this section, we present the *presentational layer* of CPM. As already mentioned, the reason for introducing a presentational layer is twofold: (a) decouple the definition of the logical underpinnings of user operations or queries from the way the result is presented; and (b) allow the formal definition of the presentation of the result. Our approach is based on the introduction of points, axes, and layers with a particular focus on two-dimensional spaces that can be represented easily in the regular devices used for data representation (e.g., screens, paper, etc.). Areas of this space can

Figure 9. Mapping CPM objects to 3D and 2D cross tabular layouts

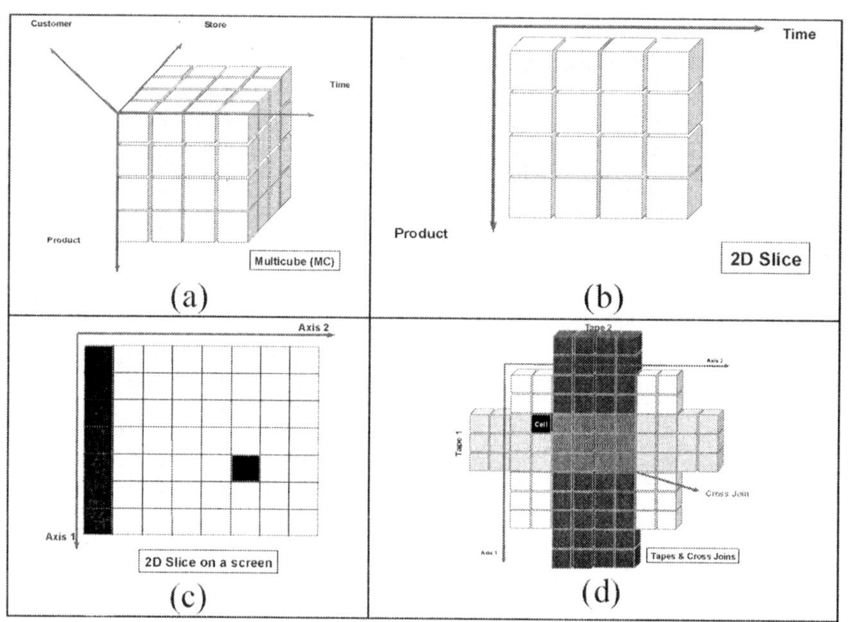

then be mapped to underlying database constructs, such as cubes. In the rest of this section, we will first give an intuitive, informal description of the presentation layer. Then, we will present its formal definition. Throughout this section, we will use the example of Figure 2 as our reference example.

The most important entities of the presentational layer of CPM (depicted in Figure 9) include:

- **Axis:** OLAP is based on the *multidimensional* representation of information. Each dimension of an OLAP environment conceptually represents a different categorization of measures, in the mind of the knowledge worker. In principle, we would prefer each dimension to be graphically represented as an axis (e.g., an axis for Salesman, another axis for Product, and so on). In this case, an axis can be viewed as a set of points, and combinations of axes define a multi-dimensional space, just like in geometry. Nevertheless, in CPM we focus on two-dimensional devices for the representation of information. Therefore, for presentation reasons, we will see how we can handle the reduction of spaces with high dimensionality to two-dimensional representations in order to depict data on 2D devices.
- **Points:** A *point over an axis* resembles the classical notion of points over axes in mathematics. In the simple case, a point is characterized by an equality selection condition over a level (e.g., City=Seattle) and represents a dimension value in any of the levels of the dimension. Still, since we need to multiplex several logical dimensions to one presentational axis, a point will be formally defined to handle this kind of situations, too.
- **Multicubes:** A combination of axes forms a multidimensional space. The data values of the points of the multidimensional spaces

Figure 10. The 2D-Slice SL for the example of Figure 2

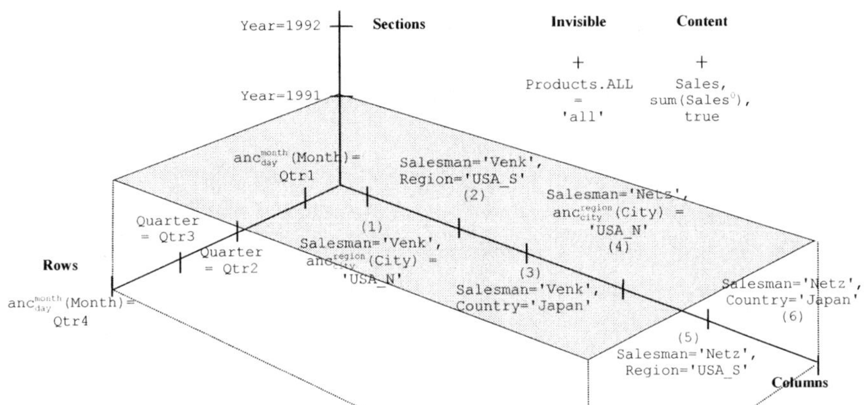

are the (aggregate) measure values that have to be computed over the detailed data of the logical model. In this section, we show how constructs in the logical model will be mapped to constructs of the multidimensional space.

- **2D-slice:** As already mentioned, the focus of CPM is set on two-dimensional devices. A 2D-slice is a two-dimensional layer of data over the multidimensional space that can be presented on the screen. Consider a multicube MC, composed of K axes. A *2D-slice over MC* can be sufficiently defined by a set of (K-2) points, each from a separate axis. Intuitively, a 2D-slice pins the axes of the multicube to specific points, except for 2 axes, which will be presented on the screen (or a printout). In both Figure 2 and Figure 10, we depict such a 2D-slice over a multicube.
- **Tape:** Intuitively, a tape is a column or a row over a 2D-slice (i.e., a construct parallel to one of the axis of the 2D — slice). Again, if we consider a 2D-slice SL over a multicube MC, composed of K axes, a tape is sufficiently defined by a set of (K-1) points, where the (K-2) points are the points of SL. A tape is always parallel to a specific axis: out of the two "free" axes of the 2D-slice, we pin one of them to a specific point that distinguishes the tape from the 2D-slice.
- **Cross-join:** Intuitively, if we take one tape parallel to the horizontal axis and another parallel to the vertical axis, their intersection is a cell. In the most general case, as we shall see, it can be a set of cells. In both cases, the intersection of two non-parallel tapes is called a cross-join. Cross-joins are directly mapped to secondary cubes of the logical model and form the basis of mapping an OLAP screen to a *set* of queries over the underlying database.

In terms of *CPM* terminology, the query of Figure 2 is a 2D-Slice, say SL (see also Figure 3). In SL one can identify four horizontal tapes denoted as R1, R2, R3 and R4 in Figures 2 and 6 vertical tapes (numbered from C1 to C6). The meaning of the horizontal tapes is straightforward; they represent the Quarter dimension, expressed either as quarters or as months. The meaning of the vertical tapes is somewhat more complex; they represent the combination of the dimensions Salesman and Geography, with the latter expressed in City, Region, and Country level. Moreover, two constraints are superimposed over these tapes

— the Year dimension is pinned to a specific value and the Product dimension is ignored. In this multidimensional world of five axes, the tapes C1 and R1 are defined as:

C1 = [(Salesman='Venk'∧(city)= 'USA_N'),(Year='1991'),
(Products)='all'),(Sales, sum(Sales))]

R1 = [((Month)='Qtr1'∧Year ='1991'),(Year='1991'),
(Products)='all'),(Sales, sum(Sales))]

One can also consider the cross-join t1 defined by the common cells of the tapes R1 and C1. Remember that City defines an attribute group along with [Size(City)].

t1=([SalesCube,(Salesman='Venk'∧(city)= 'USA_N ∧
(Month)='Qtr1'∧Year='1991' ∧(Products)='all'),
[Salesman,City,Month,Year,Products.ALL, Sales],sum],[Size(City)],true)

In the rest of this section, we will describe the presentational layer of CPM in its formality. First, we will introduce the multidimensional space, and then we will show how the contents of the space are computed.

The Multidimensional Space

In this subsection, we will introduce the multidimensional space of the presentational layer. The main entities of the multidimensional space are *axes* and *points*. Before giving their formal definition, however, we will introduce auxiliary entities necessary to cover the multiplexing of more than one logical dimension to a single axis.

Preliminaries

We assume the existence of the following pairwise disjoint infinitely countable sets: a set of *point names* (or simply *points*) $U_{\wp}$, a set of *axes* names (or simply *axes*) U_{L}, and a set of *multicube* names (or simply *multicubes*) U_{MC}.

We will also assume their finite subsets, **P** for points, **A** for axes and **MC** for multicubes, each time that we deal with a particular instance.

Before proceeding, we need to extend the definition of dimensions and to deal with multiplexing of dimensions. First, we extend the notion of dimension to incorporate any kind of *attributes* (i.e., results of functions, measures, etc.). Consequently, we consider every attribute not already belonging to some dimension to belong to a single-level dimension (with the same name as the attribute), with no ancestor functions or properties defined over it. We will distinguish between the dimensions comprising levels and functionally dependent attributes through the terms *level dimensions* and *attribute dimensions*, wherever necessary. The dimensions involving arithmetic measures will be called *measure dimensions*.

Now we are ready to deal with multiplexing dimensions in a single axis. This is necessary due to the fact that typically data are presented by 2D means (e.g., a screen), meaning that the multidimensional space has to be folded in a 2D projection. For example, observe Figure 11: the logical dimensions Salesman and Geography have been multiplexed in order to be presented on the same axis. This practically means that for every value of Salesman, all the values of Geography are repeated. Therefore, in order to be able to represent these kinds of structures we need to define groups of attributes to be multiplexed in the same axis.

An *attribute group* AG is a pair **[A,DA]**, where **A** is a list of attributes (called the *key* of the group) and **DA** is a list of attributes *dependent* on the attributes of **A**. With the term *dependent* we mean (a) measures dependent over the respective levels of the data set and (b) function results depending on the arguments of the function. One can consider examples of the attribute groups such as:

ag_1= ([City],[Size(City)]),ag_2=([Sales,Expenses],[Profit]).

A *dimension group* DG is a pair **[D,DD]**, where **D** is a list of dimensions (called the *key* of the dimension group) and **DD** is a list of dimensions *dependent* on the dimensions of **D**. With the term *dependent* we simply extend the respective definition of attribute groups to cover also the respective dimensions. For reasons of brevity, wherever possible we will denote an attribute/dimension group comprising only of its key simply by the respective attribute/dimension.

Axes & Points

An *axis schema* is a pair **[DG,AG]**, where **DG** is a list of K dimension groups and **AG** is an ordered list of K finite ordered lists of attribute groups, where the keys of each (inner) list belong to the same dimension, found in the same position in **DG**, where K>0. The members of each ordered list are not necessarily different. We denote an axis schema as a pair:

AS_K= ([DG_1×DG_2×...×DG_K],[[,,...,]×[,,...,]×...×[, ,...,]]}.

In other words, one can consider an axis schema as the Cartesian product of the respective dimension groups, instantiated at a finite number of attribute groups. For instance, in the example of Figure 1, we can observe two axes schemata having the following definitions:

Row_S = {[Quarter],[Month,Quarter, Quarter,Month]}
Column_S = {[Salesman×Geography], [Salesman]×[[City,Size(City)], Region, Country]}

A *point* is a member of the set U_P.

A *point over an axis schema* AS, is a point tagged with a set of equality selection conditions, one for the key of each attribute group of the axis schema.

For example, given the axis schema [Salesman ,[City,Size(City)]], a point can be defined as:

p_1=([Salesman='Venk',(City)= 'USA_N'])

or, if we wish to incorporate the axis schema in the definition,

p_1=([Salesman,[City,Size(City)]], [Salesman='Venk',(City) ='USA_N'])

An *axis over an axis schema* AS, is a finite list of points, all defined over the axis schema AS.

Practically, an axis is a restriction of an axis schema to specific values through the introduction of specific constraints for each occurrence of a level.

a = (AS_K,[φ_1,φ_2,...,φ_K]),K≤N or
a={[DG_1×DG_2×...×DG_K],[[,,...,]× [,,...,]×...×[, ,...,]],[[,,...,]× [,,...,] ×...×[,,...,]]}

We will denote the set of dimension groups of each axis a by dim(a).

In our motivating example, we can observe the following two axes:

Rows = {Row_S,[(Month)= Qtr1,Quarter=Qtr2,Quarter =Qtr3,(Month)=Qtr4]}
Columns = {Column_S,{[Salesman= 'Venk',Salesman='Netz'],
[(City)='USA_N', Region='USA_S', Country='Japan']}

- **Lemma.** An axis can be reduced to a finite set of points if one calculates the Cartesian products of the attribute groups and their respective selection conditions. In other words:

a=([DG_1×DG_2×...×DG_K],[[p_1,p_2,...,p_l]), l=k_1×k_2×...×k_{kk}.

Proof. Obvious.

In the sequel, we will mostly treat an axis as a finite set of pairwise disjoint points; therefore, we impose the constraint that the selection con-

Figure 11: The Points for the Axes Rows & Columns

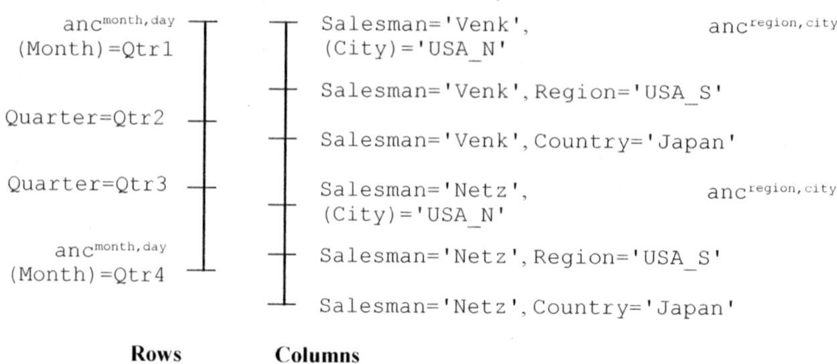

ditions characterizing each point are pairwise disjoint, too.

We will differentiate between two types of points: atomic and hierarchically decomposable. The former constitute points defined over single level or measure values, whereas the latter are defined over sets of values.

Atomic points are characterized by the fact that all the equality selection conditions for their attribute groups involve an attribute (level or measure) and a constant. In other words, atomic points are of the form Level=constant or Measure=constant.

Hierarchically decomposable points are characterized by the fact that the selection condition of one (or possibly more) of their attribute groups involves the usage of an ancestor function.

For example, p_1 is a hierarchically decomposable point:

p_1=([Salesman,[City,Size(City)]], [Salesman='Venk',$anc^{region;city}$(City)= 'USA_N'])

whereas p_2 is an atomic point:

p_2=([Time,[Quarter], [Quarter=Qtr2])

Naturally, a hierarchically decomposable point corresponds to a finite set of atomic points (directly stemming from the finiteness of the domain of ancestor functions). Therefore, the aforementioned point p_1 corresponds to the points $p_{1.1}$ and $p_{1.2}$, defined at the City level.

$p_{1.1}$=([Salesman,[City,Size(City)]], [Salesman='Venk',City= 'Seattle'])
$p_{1.2}$=([Salesman,[City,Size(City)]], [Salesman='Venk',City= 'Boston'])

An axis that comprises only atomic points is an *atomic-level* axis. An atomic-level axis X_a which comprises the atomic points produced from the hierarchical decomposition of the points of an axis X, is the *atomic-level equivalent* of X.

In the sequel, we will refer to points indiscriminately of their type; in the case where we will need to make a distinction, this will be shown clearly.

An *axis tag* is a characterization of an axis with respect to (a) its natural properties and (b) the fact that it can be visualized in a 2D screen or not. Therefore, an axis is characterized as:

- coordinate vs. measure, depending on whether it represents values that determine the coordinates or the internal points of the multidimensional space (see section 3.2)
- visible vs. invisible, depending on whether we allow its representation on a 2D screen or we use the axis simply for pinpointing values of its points without involving them in the

Figure 12. Multidimensional Space for the Variant of the Motivating Example, Extended with Sections

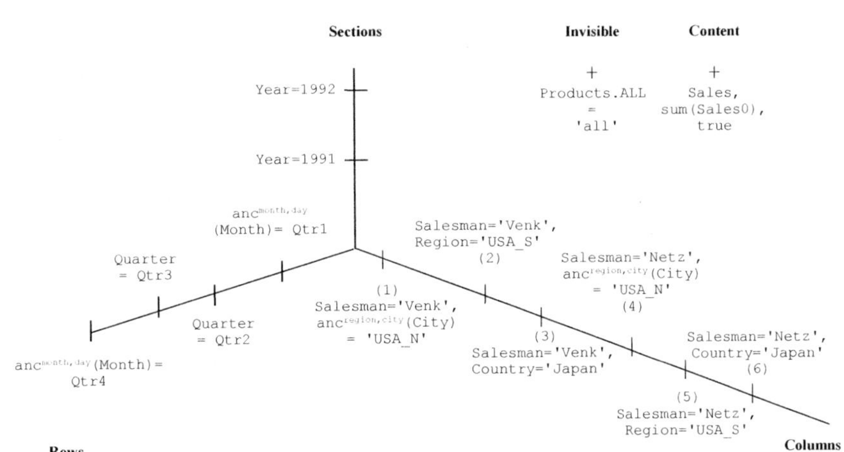

visualization of the result. For example, in Figure 2, there is an invisible axis, pinpointing Year to 1991 and Products to ALL.

Finally, we say that two axes schemata are *joinable over a data set* if their key dimensions (a) belong to the set of dimensions of the data set and (b) their points are disjoint. For instance, Rows_S and Columns_S are joinable.

The Contents of Multidimensional Space

In this subsection, we will introduce the contents of the multidimensional space of the presentational layer. The main entities of the multidimensional space are *multicubes* that more or less correspond to n-dimensional structures. Prevalent entities in their context are *2D-slices*, standing for two-dimensional structures that can be presented on a 2D screen—*tapes*—which are one-dimensional entities and *cross-joins* which are areas of a 2D-slice where two tapes meet.

Multicubes

A *multicube schema* $MC^S=[\mathcal{A}^S]$ is a finite set of axis schemata $\mathcal{A}^S$.

A *multicube* $MC=[\mathcal{A},\mathbf{f}]$, where $\mathcal{A}$ is a finite set of axes and **f** is a contents function, mapping coordinates to measures. We require that $\mathcal{A} = \mathcal{C}$ ⊠ $\{\mathbf{M}\}$, where $\mathcal{C}$ are the *coordinate* axes and **M** is the *measure* axis of the multicube.

Let also **M** be a measure axis. The points of **M** will be computed through queries to the underlying database. Still, it is a regular axis with the only difference that the same point (e.g., Sales = 40) can be repeated more than once (since measures can have identical values). Remember that axes are finite *lists* of points; for measure axes we assume bag semantics underlying this list.

In the simple case, a point is characterized by a single equation of the form [measure=constant]. Still, we can multiplex more than one logical measure in a measure axis and each point of the measure axis is characterized by a set of equations of the form $[measure_1=constant_1,..., measure_k=constant_k]$, depending on the attribute/dimension group that regulates the axis.

In the case of a measure axis, we can tag the schema of the axis with an aggregate function for each of the dimensions participating in the schema. Also, a secondary selection condition can be attached to the schema, acting as a filter for retrieved data.

In our motivating example of Figure 2, we have a measure axis, named Content, comprising 64 points. Observe that the measure axis is defined in terms of the atomic-level equivalents of the involved coordinate axes. The Time axis is hierarchically decomposed in eight values and the Geography×Salesman axis, also in another eight points. The measure axis schema is also tagged with the aggregation Sales=sum($Sales^0$) and the secondary selection condition true (i.e., no selection is performed).

We assume the existence of a *contents* function for **M**. The contents function practically instantiates the points of the measure axis by computing them as queries over the underlying data set. Each point in the measure axes is then dependent on the points of the atomic-level equivalents of the coordinate axes responsible for its identification. Formally, let $contents_M(C)$: $C \to U_p$.

In other words, supposing that there are K-1 axes in C, $contents_M(C)$ is defined as $[\mathbf{A}_1,\ldots,\mathbf{A}_{K-1}]$, therefore, for every combination of points $[p_1,\ldots,p_{K-1}]$ (each point p_i coming from axis $\mathbf{A}_i$) there exists a point μ in U_p, as the result of the $contents_M(C)$ function. Based on the fact that C comprises a finite number of points, then $contents_M(C)$ returns also a finite number of points; nevertheless, as already mentioned, more than one coordinates can map to the same measure value. This fact disqualifies the existence of an inverse function; to compensate for this shortcoming, we assume the mapping coordinates(μ), such that coordinates(μ)=$[p_1,\ldots,p_{K-1}]$.

In our motivating example, we can observe the following axes schemata and axes:

Row_S = {[Quarter],[Month,Quarter, Quarter,Month]}
Column_S = {[Salesman×Geography], [Salesman]×[[City,Size(City)], Region, Country]}
Invisible_S = {[Product×Time],[[Product.ALL]×[Year]]}
Content_S = {[Sales],[Sales=sum($Sales^0$), true]}

and their respective axes:

Rows={[(Month)=Qtr1,Quarter= Qtr2,Quarter=Qtr3, (Month)= Qtr4]}
Columns = {[Salesman='Venk',Salesman ='Netz'] × [(City)='USA_N', Region ='USA_S', Country='Japan']}
Invisible = {[Year=1991] × [ALL='all']}
Content = {64 points}

Then, a multicube MC can be defined as:

MC = {Rows, Columns, Invisible, Content}

Assume now that we want to present data in multiple spreadsheets, and each sheet comprises a certain year (e.g., the first sheet involves 1991 and the second involves 1992). We can resolve this by adding an extra axis, Sections (Figure 12). The changes are as follows:

Axes schemata:
Section_S = {[Time],[Year]}
Invisible_S = {[Product],[Product.ALL]}

and axes:
Sections = {Section_S,[Year=1991,Year=1992]}
Invisible = {Invisible_S,[ALL='all']}
Content = {128 points}

Then, the multicube MC can be defined as:

MC = {Rows, Columns, Sections, Invisible, Content}

2D-Slices

In the beginning of this section, we have informally introduced 2D-Slices. Intuitively, a 2D-slice represents a bounded two-dimensional plane. To achieve this, it is only necessary to pin the axes of the multicube to specific points, except for two axes, which are left free. Then, these two axes define a two-dimensional plane that can be presented on a screen (or a printout).

Formally, consider a multicube MC composed of K axes. A *2D-slice over MC* is a set of (K-2) points, each from a separate axis, where the points of the Invisible and the Content axis are comprised within the points of the 2D-slice.

In our motivating example, Figure 2 and Figure 11 represent the same 2D slice.

Tapes

Tapes represent "one-dimensional" parts of a 2D slice. In fact, out of the two free axes of the 2D slice, we have only one left free and the other pinpointed to a particular point, say p. In this case, a tape is parallel to this particular axis. Tapes are not considered "lines" due to hierarchically decomposable points; if the pinpointed point p is hierarchically decomposable, the tape will be visualized as a set of parallel lines.

Formally, consider a 2D-slice SL over a multicube MC composed of K axes. A *tape over* SL is a set of (K-1) points, where the (K-2) points are the points of SL. A tape is always parallel to a specific axis; out of the two "free" axes of the 2D-slice, we pin one of them to a specific point that distinguishes the tape from the 2D-slice. A tape is more restrictively defined with respect to the 2D-slice by a single point. We will call this point the *key of the tape with respect to its 2D-slice*. Moreover, if a 2D-slice has two axes a_1,a_2 with size(a_1) and size(a_2) points each, then one can define size(a_1)*size(a_2) tapes over this 2D-slice.

Observe Figure 2. All C1, C2, C3, C4, C5, C6 and R1, R2, R3, R4 are tapes. Observe C1 or R1; due to the fact that they are pinpointed to hierarchically decomposable points, they involve more than one "line." The different colors correspond to different vertical tapes (C1-C6).

Cross-Joins

Intuitively, a cross join is a set of cells produced by the intersection of two tapes. If the two tapes are defined over atomic points, the cross-join involves a single cell (e.g., the case of tapes R2 and C3); otherwise, a set of cells is produced, as in the case of tapes R2 and C1. Note that a "cell" corresponds to a point of the measure axis.

Formally, consider a 2D-slice SL over a multicube MC, composed of K axes. Consider also two tapes t_1 and t_2 which are not parallel to the same axis. A *cross-join over t_1 and t_2* is a set of K points, where the (K-2) points are the points of SL and each of the two remaining points is a point on a different axis of the remaining axes of the slice.

Two tapes are *joinable* if they can produce a cross-join.

The only difference between a tape and a cross-join is that the cross-join restricts all of its dimensions with equality constraints, whereas the tape constraints only a subset of them.

Bridging the Presentation & Logical Layers of CPM

Cross-joins form the bridge between the logical and the presentational layers. In this section, we provide a theorem proving that a cross-join is a secondary cube. Then, we show how common OLAP operations can be performed on the basis of our model.

Theorem 1. Assume a star schema database [DS^0,**C**,$\mathbf{C}^S$], over a star schema [**D**,S^0]. Assume also a cross-join, say c, defined over a subset of the dimensions **D**. Then, c can be mapped to a secondary cube over the star schema database.

Proof. We will constructively obtain the definition of the secondary cube. Remember that the cross-join is practically defined by a set of K points over the axes of a multicube.

1. The detailed data set is naturally DS^0.
2. Each of the dimensions of the cross-join is a subset of **D**, and we can assume that the levels referring to each point are [L_1,...,L_n,M_1,...,M_m] with the first being coordinate axes and the latter being measure axes.

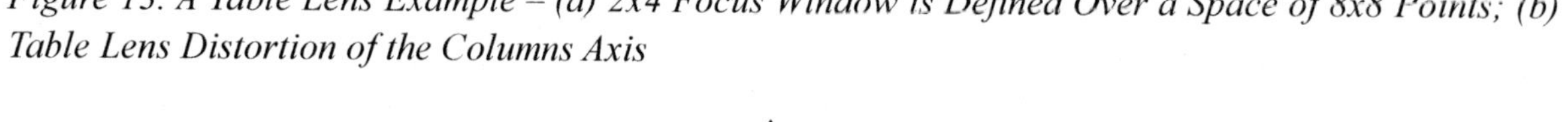

Figure 13. A Table Lens Example – (a) 2x4 Focus Window is Defined Over a Space of 8x8 Points; (b) Table Lens Distortion of the Columns Axis

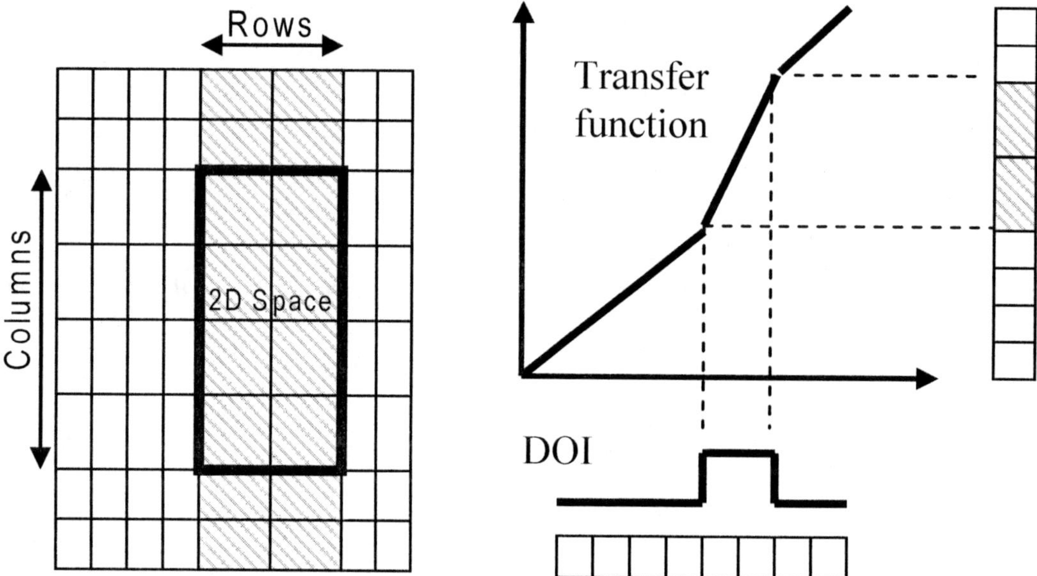

3. A selection condition φ can be derived from the points of the coordinate axes.
4. If there are any functions applied, they are also defined over the attributes of the data set; suppose that we have $A_{m+1}:f_{m+1}(\mathbf{A}_{m+1}),\ldots,A_{m+p}:f_{m+p}(\mathbf{A}_{m+p})$ attributes in the definitions of the attribute groups of the multicube. The measure axis also has a set of aggregate expressions over measures, say $[agg_1(),\ldots,agg_m()]$, and a secondary selection condition.
5. The function **f** of the cross-join's multicube is defined as a mapping of $[L_1,\ldots,L_n]$ to $[M_1,\ldots,M_m]$, possibly exploiting the use of the functions f_{m+i}.
6. Consequently, one can produce the following secondary cube out of the cross-join c:

$$c=[DS^0,\varphi,[L_1,\ldots,L_n,M_1,\ldots,M_m],[agg_1(),\ldots,agg_m()],[A_{m+1}:f_{m+1}(\mathbf{A}_{m+1}),\ldots,A_{m+p}:f_{m+p}(\mathbf{A}_{m+p})],\psi]$$

where φ is the conjunction of the primary selection conditions of the levels and ψ is the conjunction of the secondary selection conditions of the rest of the attributes.

The only difference between a tape and a cross-join is that the cross-join restricts all of its dimensions with equality constraints, whereas the tape constraints only a subset of them. Moreover, from the definition of the *joinable* tapes it follows that a 2D-slice contains as many cross-joins as the number of pairs of joinable tapes belonging to this particular slice. This observation also helps us to understand why a tape can also be viewed as a collection of cross-joins (or cubes). Thus, we have the following lemma.

- **Lemma.** A tape is a finite set of secondary cubes.
- ***Proof.*** Each of the cross-joins is defined from the k-1 points of the tape and one point from all its joinable tapes. This point belongs to the points of the axis the tape is parallel to. Consequently, we are allowed to treat a tape as a set of cross-joins, or cubes: $t=[c_1,\ldots,c_k]$.

MAPPING TO TABLE LENS

In the previous section, we have shown how a generic presentational model—CPM— can represent multi-query screens that can be mapped to constructs of an underlying logical model. At the same time, there is more that we can do over the presentation model, which involves advanced visualization techniques. It would be straightforward to visualize the CPM constructs simply as tabular data. Nevertheless, we can do better than that and apply advanced visualization techniques over the CPM constructs. In this section, we will demonstrate how CPM can be combined with Table Lens (TL) (Pirollo & Rao, 1996; Rao & Card, 1994), a non-traditional cross-tabular presentational model from the Human Computer Interaction area. This model, based on the "focus plus context" technique, is used in applications and platforms for the visualization of tabular data and appears to be quite appropriate for OLAP purposes. Using Table Lens, we can easily examine patterns and correlations in large tables and effectively zoom in without losing the global picture of our data. We have chosen Table Lens as a visualization technique due to the fact that it is based on a cross-tabular paradigm for the presentation of information; a paradigm quite popular in OLAP screens, too.

Mapping CPM to Table Lens

In this subsection, we will present the main features of Table Lens, and then we will link it to the CPM model. The main constructs of the Table Lens technique involve:

- **Axes:** The Table Lens model assumes two axes. For clarity, we will use Rows and Columns to denote these two axes, as shown in Figure13.
- **2D Space:** The 2D Space is constructed from the Cartesian product of the two Table Lens axes. It is a (finite) matrix of cells. One of the basic ideas behind the Table Lens technique is that not all cells are considered equal in terms of presentation. In fact, certain cells comprising a concrete region of the 2D Space are assigned to occupy more surface of the screen than the rest of the cells. This resembles zooming into the particular region of the 2D Space.
- **Degree of Interest Function (DOI):** DOI is a function that maps each axis point to a value that indicates the level of interest for that point. For each axis, a different DOI function is prescribed; thus, a 2D Space is characterized by 2D windows of focus. In the simplest setting of Table Lens, each DOI function is a simple "pulse" function, meaning that it has a standard value for all points, except for the points of a certain interval that are mapped to a higher value. In Figure 13a, we depict an 8x8 space with a 2x4 focus window. In Figure 13b, we show how the originally equally important cells of the Columns axis are assigned importance values by the DOI function (notice the pulse on two particular cells that assigns them greater importance than the rest of the cells).
- **Transfer Function**: A transfer function maps each cell to its physical location, indicating the level of zoom for each cell. Practically, the transfer function is the translation of the respective DOI function (operating at the "interest" space) to the "pixel" space. In Figure 13b, we show how the Transfer function, defined as a weighted integral of the DOI function, maps the points to pixel areas. For reasons of efficient representation (Rao & Card, 1994) (Figure 13b), the produced axis is rotated by 90. Finally, another interesting feature of Table Lens is the ability to define more than one window of focus. This is quite helpful in situations where two areas can be contrasted and compared. As we shall see in the next section, this feature is particularly useful in the case of OLAP.

There is an easy way to map the underlying constructs of the CPM to the ones of the Table Lens. The axis points of CPM are mapped to axis points of Table Lens and a 2D Slice in CPM is implemented as a 2D Space in Table Lens. The contents function provides the values of the cells of the 2D Space. Naturally, CPM is generic enough to lack the particularities of the axis distortion due to the DOI function. The naïve way to overcome the limitation is simply to ask the user to define a certain window of focus over the presented 2D Space, specifying both its size and position. Still, we can automate the process on the basis of the structure and the contents of a 2D Space.

Which Window of Interest to Choose?

In this subsection, we will deal with the problem of providing the user with proactive automated support for the exploration of an OLAP report. Our main tool towards this end is the window of interest as determined by the DOI functions, and the basic idea is to provide an algorithm *to proactively determine the window of interest over a 2D Space*. We want to define an algorithm that automatically determines this window whenever a user invokes an OLAP report. It appears that we can come up with a generic algorithm where the controlling parameters (e.g., stopping conditions, error range, etc.) can be tuned by the user. Actually, we can even treat as a parameter the choice of whether the user is simply interested in having a window of a certain surface or if he or she is actually interested in seeing a focus on a range of cells satisfying certain statistical properties (e.g., minimum/maximum/closest to average set of values). Having determined algorithmically the window of interest, the two involved DOI functions, which are independent of each other, are directly derived.

Motivation & Assumptions

Before providing the generic algorithm, let us clarify our contribution through a specific example. We instantiate the example of Figure 2 with the values in Figure14. Let us assume that when the user activates this OLAP screen, he would like to be informed on three particular cross-joins:

Figure 14. Instantiation of the Motivating Example with Values (Different coloring determines different cross-joins and thick borders highlight the cross-joins with the highest, lowest and closest to average values.)

			C1		C2	C3	C4		C5	C6
			Venk				Netz			
			USA			Japan	USA			Japan
			USA_N		USA_S		USA_N		USA_S	
			Seattle	Boston			Seattle	Boston		
R1	Qtr1	Jan	20	32	62	97	23	40	76	12
		Feb	25	40	74	121	18	32	51	20
		Mar	18	12	36	110	42	48	65	3
R2	Qtr2		56	63	150	253	50	70	280	50
R3	Qtr3		52	65	147	200	53	64	270	50
R4	Qtr4	Oct	25	24	64	98	32	12	64	76
		Nov	28	28	76	102	40	21	83	69
		Dec	23	30	68	150	42	29	99	77

(1) one involving the maximum sales (max); (2) another involving the lowest (min); and (3) a third involving the cross-join with behavior closest to the average (closest-to-avg) of the whole 2D Space. Practically, this involves three windows of focus, which we depict through a thick border around the involved cross-joins. In this particular case, the cross-join R1/C6 is the one with the lowest summary of values, the cross-join R4/C3, the one with the highest sum, and the cross-join R2/C3, the one closest to the average sales per cross-join (which amounts to 240.5 sales per cross-join).

A simple algorithm to compute the aforementioned quantities proceeds as follows: (a) summarizes all cells per cross-join; (b) sorts cross-joins and computes the average cross-join value; and (c) pinpoints the three regions of interest. This algorithm has linear (precisely, one-pass) complexity on the number of cells and *nlogn* (due to sorting) complexity on the number of cross-joins. Actually, if we are simply to keep the max, min or closest-to-avg cross-join, a linear single pass from all the cells is sufficient without any sorting. In the case of avg, each time that we summarize the cells from a cross-join, we can compute the average of the individual cross-join summaries and compute the closest cross-join to the current value of this average.

Assumptions: Underlying this proactive notification to the user, we have made the following assumptions:

- *Cross-joins constitute homogeneous pieces of information.* This means that we can assume a certain level of semantic cohesion among the cells of a certain cross-join. Moreover, we can assume that each cross-join can be considered as a distinct semantic unit and that cross-joins are comparable to each other. For example, we assume that it makes sense to compare sales from Japan to the sales of Southern USA. Naturally, the user choices for the axes points (and the produced cross-joins) may severely affect this assumption.
- *We are allowed to perform certain aggregate operations over our data.* Specifically, we assume that the underlying detailed data set has been summarized by a distributive aggregate function.

In Lenz and Thalheim (2001), aggregation functions are categorized as (a) *distributive functions*, like max, min, sum or count, meaning that there is a way to compute the result of the application of the aggregation function to the overall data set by composing the individual results of its application to subsets of the dataset; (b) *algebraic functions* that are expressed as finite algebraic expressions over distributive functions, like avg; and (c) *holistic functions* for all other functions.

To forestall any possible criticism, we want to point out that the *exact* result of aggregation operations over a 2D Slice is handled by the logical layer. In the case of the logical OLAP model presented in Vassiliadis and Skiadopoulos (2000), all operations are formally defined as operations over the detailed data set; optimization results for the obvious cases are also provided. Nevertheless, in the case of this paper, we want a *quick approximation* of the statistical measures under consideration to be used for the determination of the focus window and not of the values of the report. Thus, problems like the *Simpson's paradox* or the *non-invariance* property (Lenz & Thalheim, 2001) are not considered in the scope of this paper. Finally, as a general comment, since it is quite cumbersome to ask the user each time to characterize the statistical nature of the underlying data, we employ the idea that one can have an *indication* of the statistical nature of the information of screen by observing the aggregate function that has been applied to compute them. Thus, since in our case we are starting with a sum aggregate function, we conclude that we can apply further distributive operations to the measure Sales in order to obtain our indicative approximations.

Figure 15. (a) Algorithm GenericFocusWindow

```
Algorithm GenericFocusWindow
Input:
 A set of cross-joins GJ and a display grid of cells Grid related to GJ.
 Each cell belonging to Grid is characterized by coordinates (x,y) and each CJ belonging to GJ is
 characterized by the coordinates of its upper left and lower right cell. Each cross-join has a surface,
 determined by its coordinates.
Parameters:
 OriginalPick(GJ): a routine to determine the starting cross-join of the algorithm
 GuardCondition: a routine to determine whether the algorithm should stop
 ε: a tolerance or error range for the acceptance of a solution or not
 Qualifies: a Boolean function that determines whether a solution satisfies a set of constraints
 DeterminingQuality: a property of a cross-join like surface, sum of values, ...
 Pick(GJ,Q): a routine picking a cross-join to enlarge the produced solution

Output:

 A set of cross-joins, Q that satisfies the conditions set by the user.

Begin
1.1.   Q={}
1.2.   C = OriginalPick(GJ)
 Add C to Q.
 While (GuardCondition) {
     CJ = Pick(GJ,Q);
     If CJ≠NULL Then add CJ to Q  Else exit the loop
 }
 Return Q

End.
```

Figure 15. (b) Instantiation of the Algorithm (cont.)

```
OriginalPick(GJ){
 L   et the cross-join Cr s.t., |sum(Cr)| is the minimum;
 A   mong equals pick the upper and left-wise;
 R   eturn (Cr); }
DeterminingQuality(Q) {
 R   eturn surface(Q)-surface(3x3); }
GuardCondition (Q,1){
 I   f surface(Q)-surface(3x3) <1 Then Return true;
 E   lse Return false }
Pick(CJ,Q){
 L   et V be the subset of the cross-joins of CJ, s.t., for each v∈V: Qualifies(v,CJ,Q)
 L   et vP∈V be a cross-join s.t., |DeterminingQuality(Q)| is minimum, if vP is added to Q.
 R   eturn vP; }
Qualifies(v,CJ,Q){
 I   f (v is adjacent to a cross-join CJ∈CJ) &&
        (v ∪ Q forms a rectangle)
 T   hen Return true;
 E   lse Return false}
```

Figure 16. The System Architecture of CubeView

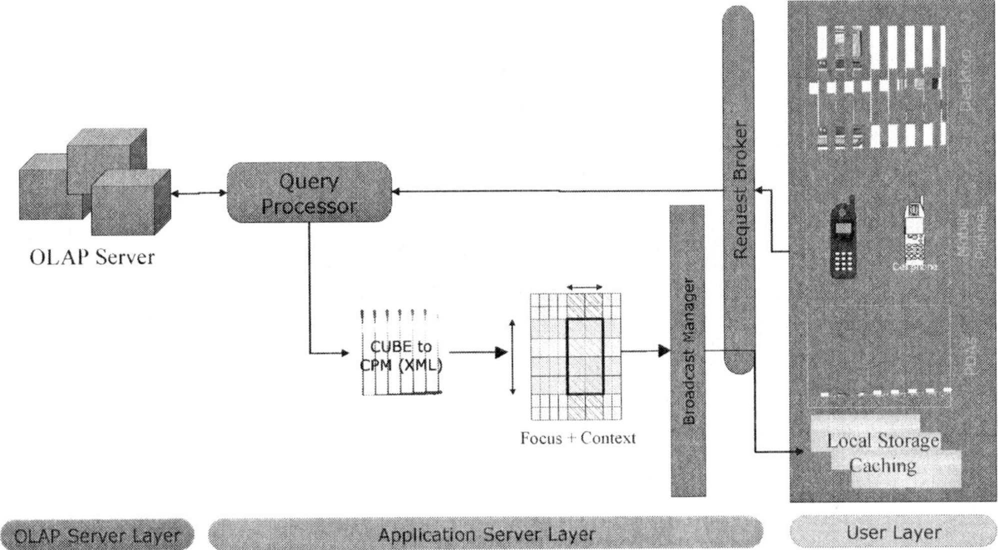

A Generic Algorithm for Determining the Window of Focus

Naturally, we can do better than the aforementioned algorithm by adding extra criteria to the proactive selection of the starting window of focus. We propose a guided greedy generic algorithm, *GenericFocusWindow* (Figure15), to deal with the issue. The simple idea underlying the algorithm is that there are certain conditions to be met for the focus window. For example, one could require that the focus window occupy at most/least a certain percentage of the screen size, or of a certain size of cells. Moreover, the selected window optimizes an objective function. The property *Determining Quality* of the algorithms captures exactly this requirement in the form of a certain function. Since our algorithm is greedy, we need an *Original Pick* routine to start the processing; in general this is closely related to the *Determining Quality* function and we require that it start with a smallest value. Moreover, a *Guard Condition* checks for the satisfaction of the desired property, meaning that we can possibly allow a certain approximation error ε to out obtained solution. Finally, a function *Pick* provides the necessary details for working from the original small-in-value solution towards the final result, practically picking the next cross-join to enlarge the current window of focus.

One implicit assumption that our algorithm makes is that the *Original Pick* fits inside the allowed window. This constraint can easily be relaxed by an extension of the algorithm picking subparts of a cross-join in a similar fashion with the proposed algorithm, if we consider that we pick subparts of a 2D-slice.

We present an example for the instantiation of the aforementioned generic algorithm in Figure 15b, where we are interested in a focus window which (a) includes the window with the minimum summary of values and (b) is not bigger than 3x3 (with a tolerance of the surface ε=1). To accomplish this, we initialize accordingly the parameters of the algorithm (GuardCondition, ε) and define accordingly the functions of the algorithms (OriginalPick, DeterminingQuality and Pick). The greedy algorithm is guided to pick the window of minimum value as its starting point. The first constraint is met by the original pick and the second by the stop condition of the algorithm. During the expansion phase, each time we choose a cross-join such that (a) is neighbouring with the current solution; (b) if merged with the current solution, it comprises a rectangle (easily determined by comparing the lengths of the opposite sides of the new solution; and (c) has the smallest surface.

If we execute the algorithm on the data of Figure14, the result will be **Q**={R1/C6,R2/C6,R3/C6,R4/C6}, which is practically the tape C6. If, instead of the minimum value, in function Pick we had chosen the maximum, then the result would be **Q**={R1/C6,R1/C5}. Another obvious extension would be to employ a 2-greedy algorithm: in this case the small cross-joins R2/C5 and R2/C6, each comprising a single cell, could have been incorporated in the solution, too.

In Maniatis et al. (2003a), we present more examples for the instantiation of the algorithm.

IMPLEMENTATION

In this section we will present a framework in support of *Mobile OLAP*, a term used to express the porting of requirements and specifications for OLAP applications into the wireless and mobile computing world, as introduced in Maniatis (2004). In this context, we present *CubeView*, a pilot academic platform enabling OLAP visualization both for contemporary desktops as well as for mobile devices. We present details about the adopted system and software architecture of the system, along with explanations on the usage of the system.

The Architecture of CubeView

In this section we present the architecture of *CubeView*, organized as (a) system architecture, (b) software architecture, and (c) implementations specifics.

Figure 17: The Software Architecture of CubeView

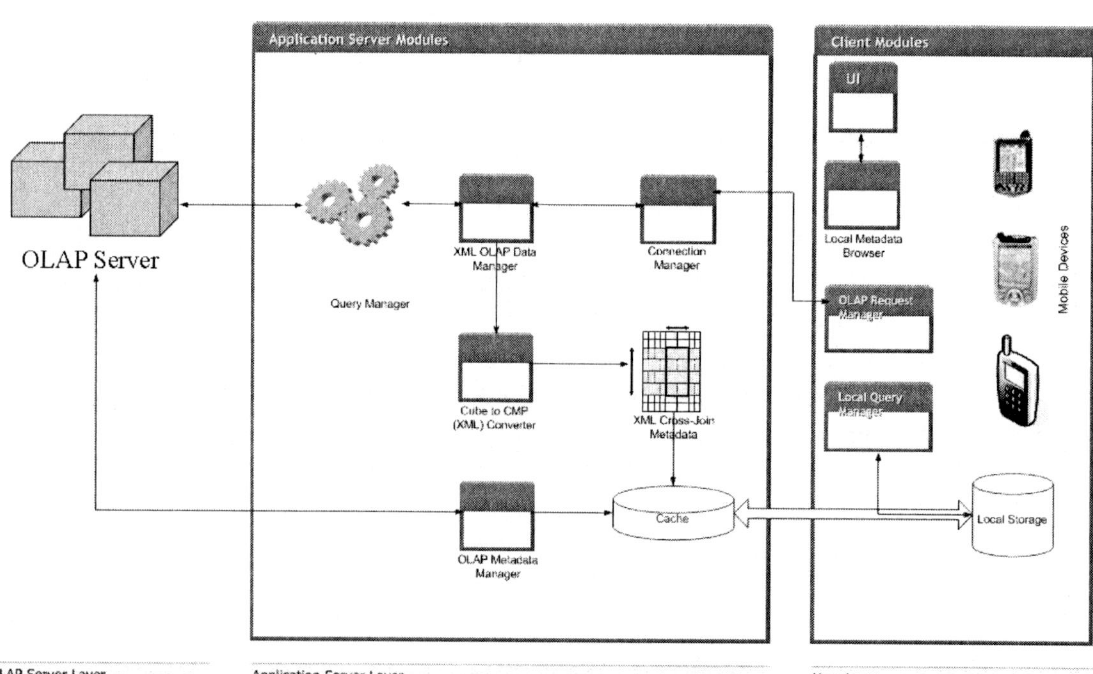

Figure 18. A Prototype Front-End for CubeView on Pocket PC Employing OLAP Lens

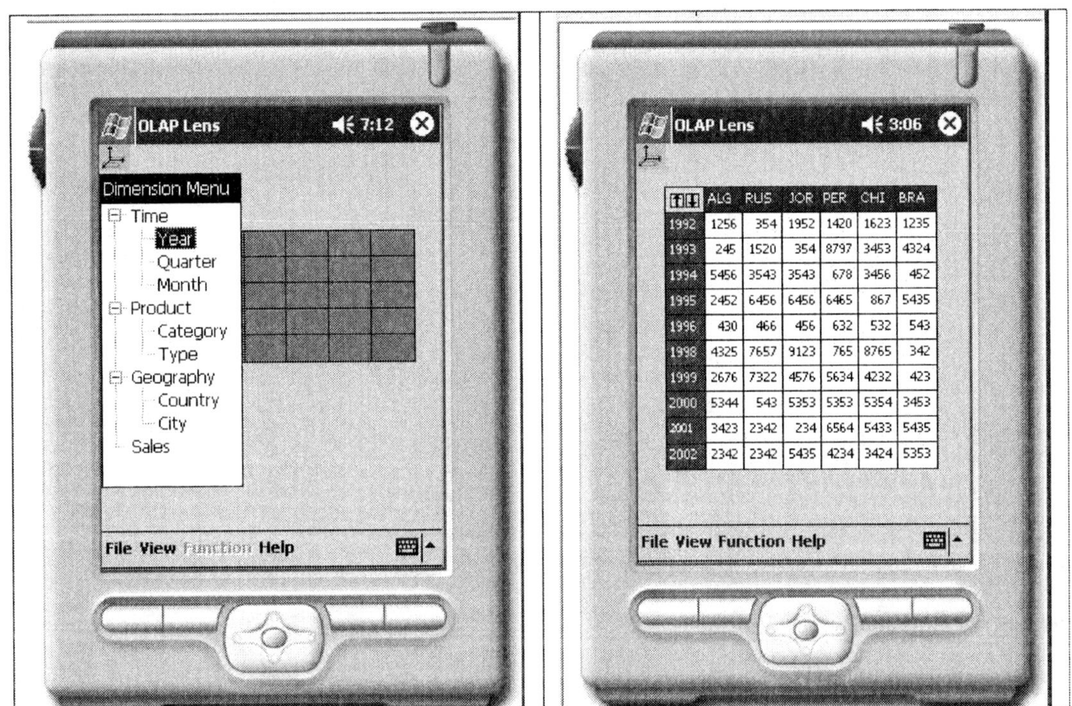

The system architecture for CubeView is depicted in Figure16. The general idea is that the user on the mobile device (PDA, mobile phone, or even remote desktop PC) uses a specific user interface on the user's device to navigate on the screen between OLAP data and perform OLAP analysis in general, based on data stored locally on the device in highly aggregated and summarized format.

The system is composed of three discrete and autonomous modules: A traditional *OLAP Server Module*, used as a black box in the process since it can be any of the existing commercial, open source, or academic ones; a *Middleware Application Server*, serving as the mediator between the data stored in the OLAP Server and the mobile device and the *mobile Front-End Applications*, incorporating the local storage; and the user interface, navigation, and presentation options.

The *Software Architecture of CubeView* is depicted in Figure17. Each distinct system layer holds a number of software modules, each in turn performing specific tasks in the context of the whole system. To be more specific, the software layers of CubeView are:

- The *OLAP Server Layer*, which, as mentioned before, is "transparent" to the user and used as a "*black box*" from the system, meaning that only a specific API is used to query the server for OLAP data. Typically, any available OLAP server (MOLAP or ROLAP), commercial, or academic platform could be used for storing and processing the actual OLAP queries.
- The *Application Server Layer*, which is the software layer holding the Java server-side application logic. It incorporates the following software modules: (*i*) The *Query Manager*, responsible for directly querying the OLAP Server according to the query posed by the mobile user; (*ii*) the *XML OLAP*

Figure 19. OLAP Lens User Interface Details

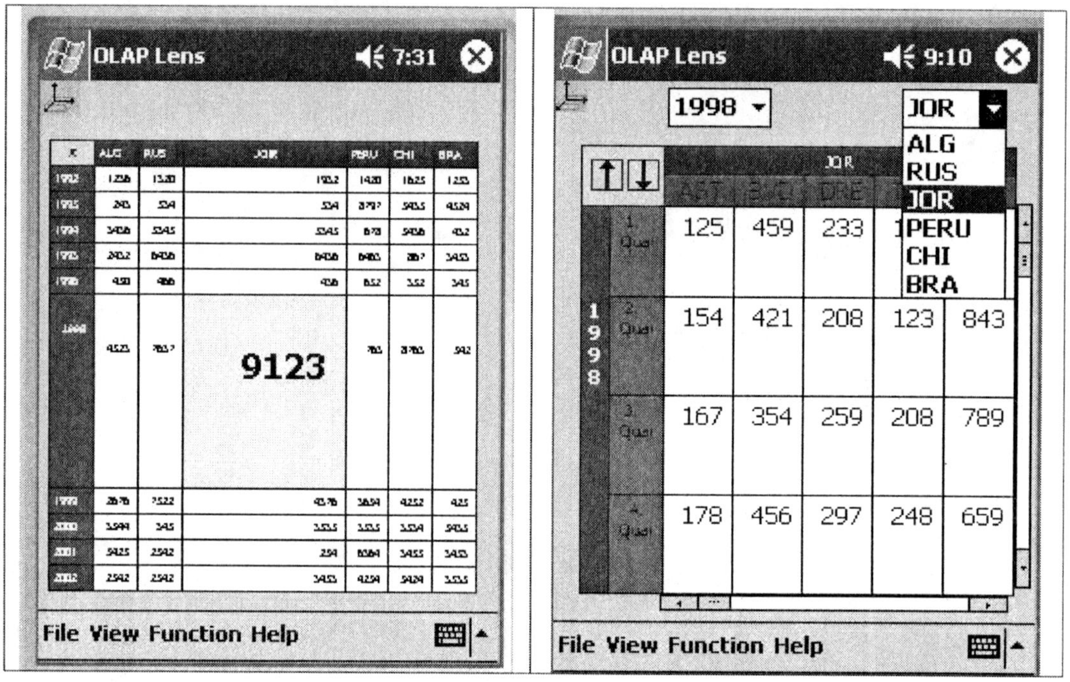

Data Manager, a component responsible for formatting the requested data in XML format and interacts with (*iii*) the *Cube to CMP (XML) Converter*; and (*iv*) the *XML Cross-join Metadata*, which formats the OLAP data queried in a format suitable for presentation using CPM entities. A simple (*v*) *Caching Mechanism* is employed for performance purposes, which holds both the actual OLAP data and the necessary OLAP metadata, retrieved and managed through (*vi*) the *OLAP Metadata Manager*. Finally, (*vii*) the *Connection Manager* is the synchronization mediator between the Application Server and the User Layer, controlling the connection process when data from the OLAP Server are needed to refresh the OLAP data in CPM entities format, stored locally on the mobile device.

- The *User Layer* incorporates a number of software modules and client tools that supplies the mobile user with instruments for (*i*) storing OLAP data locally on the *Local Storage RDBMS*, which holds the metadata of the system as well as highly aggregated OLAP data; (*ii*) executing simple ad hoc local queries, meaning on the OLAP data stored locally through (*iii*) the *OLAP Query Manager*; (*iv*) the *OLAP Request Manager* interacts with the Application Server Layer — actually with the Connection Manager — to retrieve data from the OLAP Server if such a case is enforced due to the query posed by the user; and finally, (*v*) a suitable *User Interface* and *Local Metadata Browser* are the front end tools on the mobile device through which the user can pose queries, browse the local data, and display the results of his queries on OLAP screens, employing the Table Lens visualization technique.

Using CubeView

In this section we give an illustrative example of how CubeView is designed to work and respond to queries posed by mobile users.

An implementation on a Pocket PC environment of the example described in Figure 2 is displayed in Figure 18, where a prototype for the front end tool used is developed and displayed on the simulator provided by Microsoft (http://www.microsoft. com/windowsmobile/information/devprograms/default.mspx). In terms of usability of the system and user interface characteristics, special features are employed. A step-by-step usability of the user interface controls is as follows:

- The user exploits the user interface facilities of the front end application to specify the user's range query, using drag and drop to select from the pop up dimensions window the actual dimension levels to form the initial screen. The first level of the highly aggregated data appears on the screen (Figure 18).
- In a next specialization manipulation, the user selects from a pop up menu a function or formula (system or user defined) to focus on specific characteristics and attributes of the displayed data. For example, it may be that the user wants to locate the cell with the max value of the displayed data and drill into them to locate further details. The appearance of the display screen focuses to the desired cell, which is distorted (in accordance to the Table Lens characteristics) to guide the user to the exact results of the user's query (Figure 19).
- By double clicking on the focused cell, the user drills in to the details of the cell. The cross tab displayed is replaced by the next level of the dimension hierarchies, both for rows and columns. The previous level of the dimension hierarchies is displayed as pull down menus on the top of the screen.
- The user may continue performing the previous steps to perform further analysis. Special controls (such as ↓↑ for drill in and drill out) and menu options assist the user in performing OLAP analysis on the user's mobile device.

Finally, we would also like to point out that the architecture implemented in CubeView is flexible enough to provide a framework suitable for both desktop and mobile OLAP visualization applications.

RELATED WORK

In this section, we will present related work on the topics covered by our research. This includes existing presentation models for databases and multidimensional data, implementations of visualization tools, and a discussion of related efforts in the field of OLAP for mobile environments.

Furthermore, we should mention that preliminary results of the work analysed herein can be found in Maniatis et al. (2003), Maniatis et al. (2003a), Maniatis (2003), and Maniatis (2004). Still, in this paper, (a) we provide the big picture for these works, (b) we have further decoupled the logical and presentational models from the work of Maniatis et al. (2003), and (c) we provide more examples and details on the proofs.

Presentational Models

Although OLAP has been an active research area for the past few years, the efforts devoted to the visualization of OLAP screens are very scarce. To our knowledge, only two such efforts exist.

The first effort is from the industrial field, where Microsoft has issued a commercial standard for multidimensional databases and where the presentational issues form a major part (Microsoft, 1998). In this approach, a powerful query language is used to provide the user with complex reports created from several cubes (or actually subsets of existing cubes). However, this standard suffers from several problems, with two of them being the most prominent: First, the logical and presentational models are mixed, resulting in a complex language that we personally found hard to use (although powerful enough). Secondly, the model is formalized but not thoroughly; for instance, we did not really see a definition for the schema of a multicube. Also, there are specific axes that are predefined, namely "rows," "columns," "pages," "sections," and "chapters"; no other axes are supposed.

The second proposal is an academic approach —the Tape Model (Gebhardt et al., 1997)—based on the notion of "Tapes," called thus due to their look and feel. Tapes are infinite and can overlap (if they contain shared data dimensions) or intersect with each other. A two-dimensional intersection is called a matrix and represents a kind of cross-tab between the corresponding dimensions. Each tape comprises a variable number of *tracks*. The most important operations on tapes include: (a) insertion and deletion of tracks; (b) changing the sequence of tracks (e.g., sorting); and (c) scrolling on tracks. The model offers the possibility of defining *hierarchical structures* within a tape. Tapes are infinite and can overlap (if they contain shared data dimensions) or intersect with each other. A two-dimensional intersection is called a matrix and represents a kind of cross-tab between the corresponding dimensions. Each tape comprises of a variable number of *tracks*.

Compared to CPM, the aforementioned models can be characterized as follows: CPM is a formal approach, with a rigorous formal background. The tape model seems to be limited in its expressive power (with respect to the Microsoft proposal), and, to our knowledge, its formal aspects are not yet publicly available. The Microsoft proposal, on the other hand, appears to be too complicated, without a clean-cut separation of the underlying concepts. Also, its coupling to the underlying logical structures is not clear.

Visualization in Contemporary OLAP Tools

Most vendors offering data warehousing and OLAP tools and platforms have included in their products special modules running on mobile devices and offering OLAP analysis possibilities to mobile decision makers. Vendors such as MicroStrategy Inc. (http://www.micro strategy.com) and Business Objects (http://www.businessobjects.com) have done a great deal towards implementing dedicated broadcast servers that provide OLAP specific information to users in numerous typical formats such as e-mails, beeps on pagers, or specifically designed Web pages, using WAP and WML and employing a specific but typical server-based architecture to offer this functionality.

With respect to academic pilot visualization tools and platforms, numerous have been developed, mainly in the area of the general area of information visualization. Many proposals focused on more specific areas such as statistical and scientific databases, data mining, and multi-dimensional data visualization. In the last area, we can mention VisDB (Keim & Kriegel, 1994), HD-Eye (Hinneburg et al., 2002) and Polaris (Stolte & Hanrahan, 2000).

VisDB (Keim & Kriegel, 1994) and its more recent sibling *HD-Eye* (Hinneburg et al., 2002) originated from the area of general database exploration techniques, with specialization in multi-dimensional visualization. It cannot be considered as an OLAP visualization platform; instead, it is a platform for the exploration of large multi-dimensional data sets using techniques such as the mapping of two dimensions to axes, parallel coordinates, etc., all integrated into an interactive graphical environment. In a sense, ViSDB can be viewed as being closer to data mining than OLAP. *HD-Eye* is a more recent version oriented towards visual clustering of large data sets containing high-dimensional data.

Finally, *Polaris* (Stolte & Hanrahan, 2000) is one of the most recently designed and implemented visual interfaces, designed to explore large multi-dimensional databases that extends the well-known Pivot Table interface. The features of Polaris include an interface for constructing visual specifications of table-based graphical displays and the ability to generate a precise set of relational queries from the visual specification. The visual specifications can be incrementally developed, giving the analyst visual feedback as they construct complex queries and visualizations.

Applications for Mobile OLAP

OLAP-specific functionality for mobile devices provided by the vendors does very little towards exploiting the specific characteristics and power of the mobile devices on which these applications run. Rather, they base their solutions on migrating the desktop OLAP interface of their tools to the mobile device employing WAP and the WML, but fail to take into account recent improved facts about mobile devices such as increased system memory; clearer color screens; increased processing power or their limitations, such as small screen size, different usability, and user interface requirements; common off line work, etc.

To fill this gap, numerous approaches coming mainly from the academia and from various research areas have been proposing solutions and frameworks that address this problem. Many of them propose novel approaches to cope with the case of mobile OLAP and, more importantly, many of them have been actually implemented and used in real case scenarios. We will briefly present some of the most notable (in our judgment) approaches.

MOCHA (Rodriguez-Martinez & Rossopoulos, 2000) was an early, more generic approach to a database middleware for distributed data sources, which, although not specifically addressing the case of mobile devices and OLAP, incorporates many of the notions present in "*Mobile OLAP*," such as the distributed nature of the system, the scaling to a larger environment, and the novel approach of deploying application-specific functionality from one point of the system to all the others through the middleware itself. The system was implemented in Java, which allowed for the shipping of Java code to implement either advanced data types or tailored remote operators to remote data sources and have it executed remotely, and was actually put to work effectively on a large aerospace organization.

A more specific approach is presented in Cuzzocrea et al. (2003), namely Hand-OLAP, a system specifically designed for bringing OLAP functionality to users of mobile devices. This proposal focuses mainly on a number of the drawbacks of handhelds devices, with emphasis on the small storage space and the usual discontinuance of the connection to the Wireless LAN, as opposed to the user needs for querying and browsing information extracted from enormous amounts of data accessible through the network. To cope with this issue, this approach focuses on presenting a solution for storing locally in the mobile device a compressed and highly summarized view of the data that can be more efficiently transmitted from the OLAP server than the original ones. Hand-OLAP is a prototype system with a suitable architecture that seamlessly supports the interaction between mobile device and OLAP server, stores data locally on the mobile device in a compressed format (based on Quadtree representation), and always provides the user with a specific bi-dimensional (tabular) view of the data, even when the connection to the WLAN is off.

Finally, the work of Sharaf and Chrysanthis (2002, 2002a) focused more on matters of wireless network and power consumption on the mobile devices, proposing a suitable mobile OLAP model, along with an on-demand scheduling algorithm to minimize access time and energy consumption on mobile agents. This approach is based also on summary tables, along with the functionality of simple OLAP front-end tools to execute simple SQL queries. What is more, this proposal maximizes the aggregated data sharing between clients and reduces the broadcast length. Finally, the proposed on-demand scheduling algorithm employs user parameters to fine tune the degree of aggregation of data so as to control the tradeoff between access time and energy consumption, and adapts to different request rates, access patterns, and data distributions.

As a whole, as we compare CubeView with all the previously described tools — those used for visualization and those reviewed for Mobile OLAP — we stress the fact that our prototype is the only one that supports the full cycle, starting from a formal and rigorous theory background depicted in CPM itself, and reaching a full fledged implementation covering both worlds, the traditional desktop environments, and the mobile devices. All the other paradigms are departmental in the sense that the they tamper only portions of the big picture, this being either the information visualization area (VisDB [Keim & Kriegel, 1994]; HD-Eye [Hinneburg et al., 2002]; Polaris [Stolte & Hanrahan, 2000]), or specific approaches and implementations for mobile devices (Hand-OLAP [Cuzzocrea et al., 2003]) or simply middleware, like MOCHA (Rodriguez-Martinez & Rossopoulos, 2000), or, finally, a framework for a wireless OLAP model (Sharaf & Chrysanthis, 2002, Sharaf & Chrysanthis, 2002a).

CONCLUSIONS & FUTURE WORK

So far, visualization has not been fully incorporated in the abstraction levels of DBMS architecture (conceptual, logical, or physical). In this paper, we have discussed the separation of the aforementioned abstraction levels to take visualization into consideration. In this context, we have presented the Cube Presentation Model (CPM), a formal presentation model for OLAP data. Our contributions can be listed as follows: (a) we have presented an extension of a previous logical model for cubes to handle more complex cases; (b) we have introduced a novel presentational model for OLAP screens, intuitively based on the geometrical representation of a cube and its human perception in the space; and (c) we have discussed how these two models can be smoothly integrated. Moreover, we have demonstrated how CPM can be naturally mapped into an advanced visualization technique (Table Lens), and we have discussed suitable algorithms for proactive automated support of the user towards the highlighting of interesting areas of a report. Finally, we have discussed implementation and usage issues in the context of an academic prototype system (CubeView) that we have implemented.

Obviously, we do not claim that this is the ultimate solution to the problem of OLAP data visualization, but rather we wish to indicate that there is quite an interesting research field in this area and a supportive body of knowledge from other disciplines such as Human-Computer Interaction and Information Visualization.

An obvious particularity of our approach is that it is crafted mostly for tabular data in the context of OLAP. Should we wish to differentiate the context of data utilization or the data themselves (e.g., perform OLAP over spatial or biological data), the presentation and visualization techniques would have been different. In general, it is an interesting research challenge to discuss the integration of different models in the presence of different contexts, either in terms of the data or their usage.

At the same time, new hardware developments pose new requirements for our visualization techniques. One of our goals has been to implement OLAP visualization techniques for particularly small devices such mobile phones and palmtops. Although the processing power of these gadgets is no more negligible (actually, the buzzword "thin client" seems to disappear from the standard vocabulary of the area), their screen sizes shrink over time. To make OLAP screens presentable to such devices, one can follow several paths such as: (a) showing only high level summaries which involve small 2D slices or (b) showing simply pie or bar charts. We have chosen an alternative approach where (a) the contents of the screen do not have to be squeezed in size in order to fit in the screen, and, most importantly, (b) the report does not have to be rewritten and neither do we have to check for the aggregation level of the presented data. On the contrary, a certain part of the report is presented, depending on the particularities of the device. Here, we make the reasonable assumption that either the device has the computational power to determine the amount of cells that can be presented to the user, or, if this is not an option, the device can at least piggy-back its characteristics to the OLAP server and let the server decide on the focus window. Naturally, as part of future research, different implementation issues (e.g., caching schemes or visualization techniques) can be applied in this context.

Finally, coming back to the visualization issue, we have brought up Table Lens to highlight the possibility of facilitating proactive user decision support in the presence of large datasets (in our case, the value axis is quite larger than the size that someone can handle efficiently). Clearly, as report screens are limited, not only due to hardware constraints but also due to the particularities of human nature (e.g., the classical discussion on the limited capacity of persons in processing information (Miller, 1956), it comes quite natural that automated proactive support to the users is thus one of the new requirements that decision support tools have to provide. Thus, our contribution is related to a broader line of research (Han, 1998, Sarawagi et al., 1998), which is obviously open to a wide range of different possibilities.

REFERENCES

Abiteboul, S. et al. (2003). *The Lowell database research self assessment*. Retrieved from: http://research.microsoft. com/~Gray/lowell/

Cuzzocrea, A., Furfaro, F., & Sacca, D. (2003, April 9-11). Hand-OLAP: A system for delivering OLAP services on handheld devices. In *Proceedings of ISADS 2003*. Pisa, Italy, (pp. 213-224).

Gebhardt, M., Jarke, M., & Jacobs, S. (1997). A toolkit for negotiation support interfaces to multi-dimensional data. In *Proceedings of ACM SIGMOD 1997* (pp. 348-356).

Han, J. (1998). Towards on-line analytical mining in large databases. *SIGMOD Record, 27*(1), 97-107.

Hinneburg, A., Keim, D.A., & Wawryniuk, M. (2002). HD-eye: Visual clustering of high-dimensional data. In *Proceedings of the 2002 ACM SIGMOD 2002*. Madison, Wisconsin, p. 629.

Inmon, W.H. (1996). *Building the Data Warehouse*. John Wiley & Sons.

Inselberg, A. (2001). Visualization and knowledge discovery for high dimensional data. *The Second Workshop Proceedings of UIDIS*. IEEE Press.

Keim, D.A. (1997). *Visual data mining*. Tutorials of the 23rd International Conference on Very Large Data Bases. Athens, Greece.

Keim, D.A. & Kriegel, H.P. (1994): VisDB: Database exploration using multidimensional visualization. *IEEE Computer Graphics and Applications,* September.

Kimbal, R., Reeves, L., Ross, M., & Thornthwaite, W. (1998). *The Data Warehouse Lifecycle Toolkit: Expert Methods for Designing, Developing, and Deploying Data Warehouses*. John Wiley & Sons.

Lenz, H.J. & Thalheim, B. (2001). OLAP databases and aggregation functions. In *Proceedings of the SSDBM 2001*. Fairfax, Virginia, (pp. 91-100).

Maniatis, A. (2003). OLAP presentation modeling with UML and XML. In *Proceedings of BCI 2003*. Thessaloniki, Greece, (pp. 232-241).

Maniatis, A. (2004). The case for mobile OLAP. *Proceedings of the First International Workshop on Pervasive Information Management* (in conjunction with EDBT '04). Heraklion, Greece, (pp. 103-114).

Maniatis, A., Vassiliadis, P., Skiadopoulos, S., & Vassiliou, Y. (2003). CPM: A cube presentation model for OLAP. In *Proceedings of DaWaK 2003*. Prague, Czech Republic (pp. 4-13).

Maniatis, A., Vassiliadis, P., Skiadopoulos, S., & Vassiliou, Y. (2003a). Advanced visualization for OLAP. In *Proceedings of DOLAP 2003*. New Orleans, Louisiana (pp. 9-16).

Microsoft (1998). *OLEDB for OLAP*. Retrieved from: http://www.microsoft. com/data/oledb/olap/

Miller, G.A. (1956). The magical number seven, plus or minus two: Some limits on our capacity for processing information. *Psychological Review*, *63*, 81-97.

Pirollo, P. & Rao, R. (1996). Table lens as a tool for making sense of data. In *Proceedings of the AVI 1996 Workshop*. Gubbio, Italy.

Rao, R. & Card, S.K. (1994). The table lens: Merging graphical and symbolic representations in an effective focus + context visualization for tabular information. In *Proceedings of ACM SIGCHI 1994*. Boston, Massachusetts.

Rodriguez-Martinez, M. & Rossopoulos, N. (2000). MOCHA: A self-extensible database middleware system for distributed data sources. In *Proceedings of ACM SIGMOD 2000*. Dallas, Texas, (pp. 213-224).

Sarawagi, S., Agrawal, R., & Megiddo, N. (1998). Discovery-driven exploration of OLAP data cubes. In *Proceedings of EDBT 1998*. Valencia, Spain, (pp. 168-182).

Sharaf, M.A. & Chrysanthis, P.K. (2002). On-demand broadcast: New challenges and algorithms. In *Proceedings of HDMS 2002*. Athens, Greece.

Sharaf, M.A. & Chrysanthis, P.K. (2002a). Semantic-based delivery of OLAP summary tables in wireless environments. In *Proceedings of CIKM 2002*. McLean (pp. 84-92).

Stolte, C. & Hanrahan, P. (2000). Polaris: A system for query, analysis and visualization of multidimensional relational databases. In *Proceedings of InfoVis 2000*. Salt Lake City, Utah, (pp. 5-14).

Tsois, A., Karayannidis, N., & Sellis, T. (2001). MAC: conceptual data modeling for OLAP. In *Proceedings of DMDW 2001*. Interlaken, Switzerland, (pp. 5.1-5.11).

Vassiliadis, P. & Sellis, T. (1999). A survey on logical models for OLAP databases. *SIGMOD Record, 28*(4), 64-69.

Vassiliadis, P. & Skiadopoulos, S. (2000). Modeling and optimization issues for multidimensional databases. In *Proceedings of CaiSE 2000*. Stockholm, Sweden, (pp. 482-497).

This work was previously published in International Journal of Data Warehousing and Mining, Vol. 1, No. 1, edited by D. Taniar, pp. 1-36, copyright 2005 by IGI Publishing, formerly known as Idea Group Publishing (an imprint of IGI Global).

Chapter 2.28
An Ontology-Based Data Mediation Framework for Semantic Environments

Adrian Mocan
University of Innsbruck, Austria

Emilia Cimpian
University of Innsbruck, Austria

ABSTRACT

In a semantic environment data is described by ontologies and ontology mapping has become a crucial aspect in solving the heterogeneity problems of semantically described data. This means that alignments between ontologies have to be created, most probably during design-time, and used in various run-time processes. Such alignments describe a set of mappings between the source and target ontologies, where the mappings show how instance data from one ontology can be expressed in terms of another ontology. In this article we propose a formal model for creation of mappings and we explore how such a model maps onto a design-time graphical tool that can be used in creating alignments between ontologies. In the other direction, we investigate how such a model helps in expressing the mappings in a logical language, based on the semantic relationships identified using the graphical tool.

INTRODUCTION

Semantic Web technologies have become more and more popular during the last decade. Based on simple and appealing ideas (common understanding of data, common formats for representing these data), Semantic Web aims at providing a framework that would allow information sharing across the Web in a manner understandable not only by humans, but also by the machines.

The agreed upon format for representing this information are the ontologies, but the representation of data using ontologies cannot guarantee the homogeneity and consistency of information. Even if the ontologies are supposed to be a formal explicit specification of a shared conceptualization (Gruber, 1993), they are usually developed in isolation, and shared between well-defined boundaries. This leads to the development of different conceptualizations of the same domain, different ontologies modeling the same aspects but in different manners.

In this context, ontology mapping has become a crucial aspect in solving heterogeneity problems between semantically described data. The benefits of using ontologies, especially in heterogeneous environments where more than one ontology is used, can only be realized if this process is both correct and efficient. A trend is to provide graphical tools capable of creating alignments during design-time in a (semi-)automatic manner (Ehrig, Staab & Sure, 2005; Mocan & Cimpian, 2005; Noy & Munsen, 2003; Silva & Rocha, 2003). These alignments consist of mapping rules, frequently described as *statements* in a logical language, and can be executed for performing the actual mediation when needed. One of the main challenges is to fully isolate the domain expert (who is indispensable if 100% accuracy is required) from the burdens of logics using a graphical tool, and in the same time to be able to create complex, complete and correct mappings between the ontologies.

We consider that it is absolutely necessary to formally describe the mappings creation process and to link it with the instruments available in the graphical tool and with a mapping representation formalism that can be used later during run-time. This allows the capturing of the actions performed by the human user in a meaningful way with respect to the visualized ontology structure and then to associate the results of these actions (mappings) with concrete statements in a mapping language (mapping rules).

The article proposes a set of strategies and enhancements to the classical approach towards *ontology mapping* and *run-time mediators*. Additionally it proposes a formal model that unifies the conceptual models of the design-time and run-time tools, improving and making more explicit the process of translating the domain expert inputs placed in graphical interface in logical formalism that is to be executed by the run-time tool.

The article structure is as follows: the next section presents the context and motivation for the work. The third section introduces the model we propose expressed using First-Order Logic (Genesereth & Nilson, 1987). The fourth section describes how this model can be applied to WSMO (Feier, Polleres, Roman, Domingue, Stollberg & Fensel, 2005) ontologies, while the fifth section presents the creation of mapping rules; the prototype that implements and applies the proposed formal model is described in the sixth section. Following this related work and conclusions are presented.

CONTEXT AND MOTIVATION

The work described in this article has been carried out in the Web Service Execution Environment (WSMX) working group, whose scope is to build a framework that enables discovery, selection, mediation, invocation and interoperation of Semantic Web Services (Mocan, Moran, Cimpian & Zaremba, 2006). Web Services are semantically described using ontologies, but as they are generally developed in isolation, heterogeneity problems appear between the underlying ontologies. Without resolving these problems the communication (data exchange) between different Semantic Web Services cannot take place.

The data mediation process in WSMX includes two phases: a *design-time* and a *run-time* phase. The mismatches between the ontologies are identified at design-time, while the found semantic relationships are expressed as mapping rules; these mapping rules are used at run-time to transform the data passing through the system. The run-time phase can be completely automated, while the design-time phase remains semi-automatic, requiring the inputs of a domain expert.

For the design-time phase a semi-automatic ontology mapping tool was developed that allows the user to create alignments between ontologies and to make these alignments available for the run-time process. There has been much research in the area of graphical mapping tools, for example, Noy and Munsen (2003) and Silva and Rocha (2003), however we believe there are many challenges still to be addressed. In particular, our focus has been on providing the user with proper support (e.g., suggestions and guidance), and in defining strategies that hide the burden of logical languages that are generally used to express ontology alignments, from the domain expert.

The suggestion of correct mappings is accomplished by using a set of algorithms for both lexical and structural analysis of the concepts. A brief description of these algorithms is provided in the sixth section. Additionally, the guidance is offered by decomposition and context updates (as described in the third section). The second aspect, which is to better isolate the domain expert from the burden of logical languages, is achieved by the use of several perspectives, or views[1], on the ontology that help in identifying and capturing various mismatches only by graphical means (more details are offered in the fourth section). All these features are formalized in a model that creates a bridge between the graphical mapping tool and the result of the mapping process (the ontology alignment). In the following section we will describe this model, together with the main principles that support the graphical instruments and how they fit with the underlying logical mechanism.

As described in Mocan and Cimpian (2005), the graphical point of view adopted to visualize the source and target ontologies makes it easier to identify certain types of mappings. We call this viewpoints *perspectives* and argue that only by switching between combinations of these perspectives on the source and target ontologies, can certain types of mappings be created using only one simple operation, *map*, combined with mechanisms for ontology traversal and contextualized visualization strategies. In the following sections, various types of perspectives are provided along with examples of equivalences (mappings) that can be easily identified and described in a certain perspective but difficult or impossible in another ones.

A formal model that describes the general principles of the perspectives allows for a better understanding of the human user actions in the graphical tool and of the effects of these actions on the ontology alignment (i.e., the set of mapping rules) that is being created. This model defines the main principles that support the graphical instruments (e.g., perspectives) and how they fit with the underlying logical mechanisms (e.g., decomposition, context updates). The same model is also used to describe how the inputs placed through these graphical instruments by the domain expert affect the generated mappings. Having this formal model as a link between the graphical elements and the mappings, defines precisely the process of hiding from the domain expert the complexity of the underlying logical languages; it also allows some of the mapping properties such as (in)completeness or (in)consistency to be reflected back into the graphical tool. Additionally, such a model allows experts to become more familiar with the tools and to create extensions that are more suited for capturing certain types of mismatches.

A MODEL FOR MAPPINGS CREATION

This section defines a model and all its elements to be used in the creation of mappings between ontologies. The roles that appear in the graphical user interface, and which will be later associated with the ontological entities, are defined here.

First-Order Logic (Genesereth & Nilson, 1987) is used as a formalism to represent this model.

Running Example

This section introduces an example that will be further used to illustrate the concepts introduced in the rest of the article. Consider the following fragment from an ontology O_1 (see Table 1).

Different elements from this ontology fragment need to be mapped to elements from another ontology, O_2 (see Table 2).

Table 1 presents an example of concepts and their attributes, and some instances of these concepts. The concept *person* is modeled as having five attributes, each of them having a type (i.e., a range) that is either another concept or a data type. For the concept *gender* there are two instances defined (i.e., *male* and *female*) that have attributes pointing to values of the corresponding types.

The ontology fragment in Table 2 defines the concept of *human* having three attributes, *name*, *age*, and *noOfChildren*, of type *string*, *integer*, and *integer* respectively. The concepts of *man* and *woman* are subclasses of the concept *human*. Additionally, a third concept called *marriage* is defined having two attributes *hasParticipant* and *date* of type *human* and *date* respectively.

In WSML all data types' names are prefixed with "_" - in this example *_integer*, *_string*, and *_date* are data types. Actually, WSML defines a number of built-in functions (de Bruijn et al., 2005) for the use of XML Schema data types as they are described in (Biron & Malhotra, 2004). As such, it is important to note that in WSML, attributes can have as types both data types and concepts.

Assuming that we want to map the concept *person* from O_1 to the concept *human* of O_2 a straight forward way to visualize the ontologies is by using a frame based approach where concepts are considered to be the main elements of the visualization, and the attributes and their range (value type) can only be displayed as part of the concept; the concepts' instances are not displayed in this perspective. We call this visualization mode the *PartOf* perspective. The two fragments of ontologies presented using this perspective, are shown in Figure 1.

Using these perspectives it will be very easy for a domain expert to identify, for example, all the elements that need to be mapped for mapping

Table 1. O_1 Fragment[2]

```
concept person                     concept gender
  name ofType _string                value ofType _string
  age ofType _integer              instance male memberOf gender
  hasGender ofType gender            value hasValue "male"
  hasChild ofType person           instance female memberOf gender
  marriedTo ofType person            value hasValue "female"
```

Table 2. O_2 fragment

```
concept human                      concept woman subConceptOf human
  name ofType _string              concept marriage
  age ofType _integer                hasParticipant ofType human
  noOfChildren ofType _integer       date ofType _date
concept man subConceptOf human
```

Figure 1. PartOf perspective for O_1 and O_2

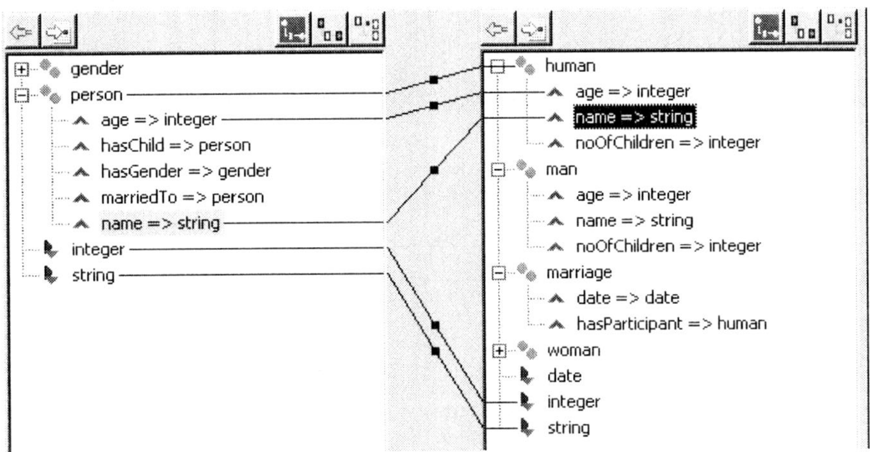

Figure 2. RelatedBy perspective for O_1 and PartOf perspective for O_2

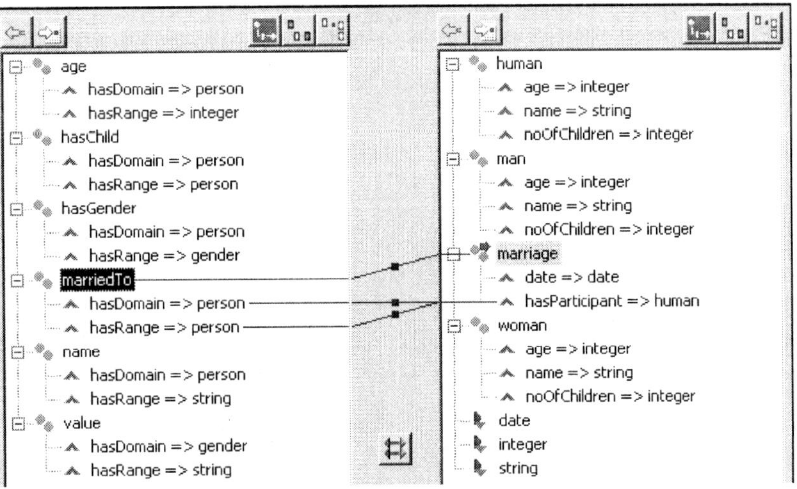

the *person* concept from O_1 to the *human* concept from O_2. However, several of *person*'s attributes cannot be directly mapped (at least not in a *natural* way[3]) using this perspective, but changing the perspectives the mappings are possible. For example, the *marriedTo* attribute from O_1 should be mapped to the *marriage* concept from O_2. For this we need to change the perspective in the first ontology, so that *marriedTo* is displayed as a main entity, and the concept it belongs to and as well as its range (value type) are displayed as part of the attribute. We call this perspective *RelatedBy* (see Figure 2).

Another problem occurs when trying to map the gender attribute from O_1. By simple looking at this short fragment of ontologies, a human user can easily determine that if the attribute has the value "male" it should be mapped to the *man* concept, and if its value is "female" it should be mapped to the *woman* concept. The graphical

Figure 3. InstanceOf Perspective for O_1, and PartOf Perspective for O_2

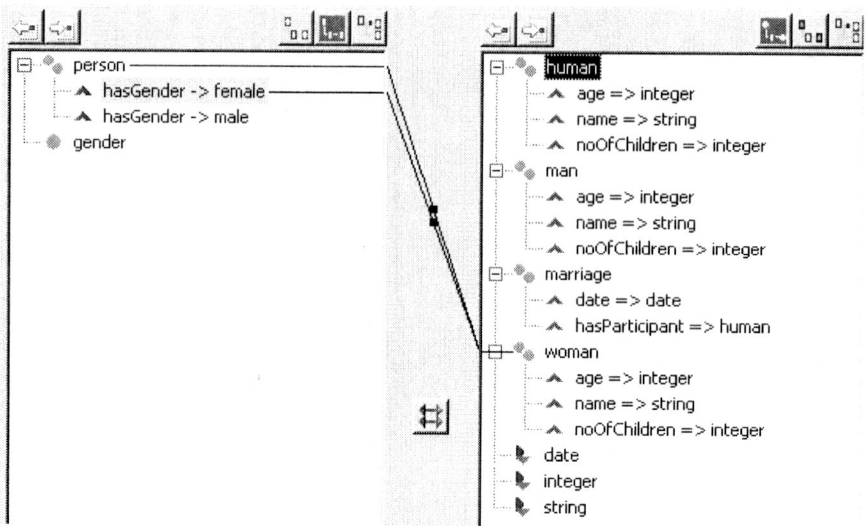

user interface should offer support for modeling this kind of mappings conditioned by the values of the attributes (Figure 3 – the mappings to *man* concept have been omitted to reduce the number of lines for clarity). For this we introduce a third perspective *InstanceOf*; in this perspective the focus is on the actual instances of the ontology.

It is interesting to note that in the examples only one nesting level is expanded. As it is discussed in sequel, a decomposition mechanism provides access to subsequent nesting levels while graphically only one nesting level is displayed at all time. By this the design-time phase can benefit of several advantages, like higher efficiency of the suggestion algorithms and more effective guidance through the mapping process.

Perspectives and Their Elements—Formal Definitions

As mentioned earlier, in our approach the ontologies are presented to the user using *perspectives*. A perspective can be seen as a vertical projection of the ontology and it will be used by the domain expert to visualize and browse the ontologies and to define mappings. We can define several perspectives on an ontology as will be presented in Section 4, all of them being characterized by a set of common elements. These elements are formally defined in this section, referring to the running example for clarity.

We identify four types of such elements (items): *compound*, *primitive*, *description*, and *successor*. We use the following unary relations to denote each of them:

- ci(x) where x is a compound item. An item is a compound item if it has at least another item that describes it.
- pi(x) where x is a primitive item. A primitive item is an item that has no descriptions.
- di(x) where x is a description item. Description items have the role of providing descriptions for compound items.
- si(x) where x is a successor item. A successor item is either a compound or primitive item and encapsulates the descriptive information the description item points to.

Table 3. Types of Items for the O_1's Perspectives in Figure 1, Figure 2, and Figure 3

	PartOf	***InstanceOf***	***RelatedBy***
ci	person, gender	person	name, hasGender, marriedTo, value
pi	string, integer	string, integer, gender	-
di	name, age, hasGender, hasChild, marriedTo	(hasGender,male), (hasGender,female)	hasDomain, hasRange
si	string, integer, gender, person	gender	person, string, gender

Both primitive and compound items represent first-class citizens of a perspective while description and successor items link the compound and the primitive items in a graph-based structure. Table 3 shows examples of such items as identified in different perspectives of ontology O_1. It is important to note that the same item can have different roles across the perspectives.

In addition we define a set of general relationships between these items that hold for all perspectives:

- Each compound item is described by at least one description item:

$$\forall x.(ci(x) \Leftrightarrow \exists y.(di(y) \wedge describes(y, x))) \quad (1)$$

where *describes* is a binary relation that holds between a compound item and one of its description items. The participants in this relation are always a compound item and a description item:

$$\forall x.\forall y.(describes(y, x) \Rightarrow ci(x) \wedge di(y)) \quad (2)$$

- Each description item points to at least one successor item:

$$\forall y.(di(y) \Leftrightarrow \exists z.\ (si(z) \wedge succesor(z, y))) \quad (3)$$

where *successor* is a binary relation that holds between a description item and one of its successor items. The participants in this relation are always a description item and a successor item:

$$\forall y.\forall z.(succesor(z, y) \Rightarrow di(y) \wedge si(z)) \quad (4)$$

The successor items are either primitive or compound items:

$$\forall x.(si(x) \Rightarrow pi(x) \wedge ci(x)) \quad (5)$$

The compound, primitive and description items are mutually exclusive for the same perspective:

$$\forall x.(\neg((ci(x) \wedge pi(x)) \vee (ci(x) \wedge di(x)) \vee (di(x) \wedge pi(x)))) \quad (6)$$

To summarize, a description item can be seen as a link that points to another item, which from a certain point of view, describes the original item. As such, for the concept *person* a description item is *hasGender,* which is a link to the definition of the concept *gender* (where *gender* is a successor item). When defining a perspective, one can choose as description items a class of ontology entities (e.g., the attributes), a general ontology relation (e.g., the *is-a* relationship) or

Table 4. Relations between items in the perspectives illustrated in Figure 1, Figure 2, and Figure 3

	successor(x, y)	***describes(x, y)***
PartOf	*successor*(string, name), *successor*(integer, age), *successor*(gender, hasGender), *successor*(person, hasChild), *successor*(person, marriedTo)	*describes*(name, person), *describes*(hasGender, person), *describes*(age, person), *describes*(marriedTo, person), *describes*(hasChild, person)
InstanceOf	*successor*(gender,(hasGender,male)), *successor*(gender,(hasGender,female))	*describes*((hasGender,male),person) *describes*((hasGender,female),person)
RelatedBy	*successor*(person, hasDomain), *successor*(string, hasRange), *successor*(gender, hasRange), *successor*(person, hasRange), *successor*(gender, hasDomain)	*describes*(hasDomain, name), *describes*(hasRange, name), *describes*(hasRange, hasGender), *describes*(hasDomain, hasGender), *describes*(hasDomain, marriedTo), *describes*(hasRange, marriedTo), *describes*(hasDomain, value), *describes*(hasRange, value)

even a custom defined relation (e.g., *hasChild*). The only requirement is to be able to define for the chosen description item the *describes* and *successor* binary relationships.

This is a set of minimal descriptions for our model, but by inference some other useful consequences can be determined. For example, note that sentences *(1)* and *(6)* imply that primitive items have no description items. Table 4 shows examples of relationships for the perspectives in Figure 1, Figure 2 and Figure 3.

As a consequence we can define a perspective as a set $\phi=\{x_1, x_2, \dots, x_n\}$ for which we have:

$$\forall x.\forall\phi.(member(x, \phi) \Rightarrow pi(x) \vee ci(x) \vee di(x)) \quad (7)$$

In addition, for any perspective the following sentences hold:

$$\forall x.\forall y.\forall\phi.(describes(y, x) \Rightarrow (member(x, \phi) \Leftrightarrow member(y, \phi)) \quad (8a)$$

$$\forall x.\forall y.\forall\phi.(successor(y, x) \Rightarrow (member(x, \phi) \Leftrightarrow member(y, \phi)) \quad (8b)$$

Sentences *(8a)* and *(8b)* together with *(2)* and *(4)* state that the description of a compound item appears in the perspective if and only if (*iff*) the compound item appears in the perspective as well. Similarly, a successor of a description item appears in a perspective *iff* the description item appears in the perspective too.

Contexts

Not all of the information modeled in the ontology is useful in all stages of the mapping process. The previous section shows that a perspective represents only a subset of an ontology, but we can go further and define the notion of *context*. A context is a subset of a perspective that contains only those ontological entities, from that perspective, relevant to a concrete operation. For example when mapping the concept *person* from ontology O_1 (Table 1) with the corresponding concept(s) from ontology O_2, not all the other information contained in ontology O_1 (which for example might also contain job descriptions and qualifications hierarchies) is of interest for this particular mapping case.

We can say that γ_ϕ is a context of the perspective ϕ if:

$$\forall x.(member(x, \gamma_\phi) \Rightarrow member(x, \phi)) \tag{9}$$

For a context from formulas *8a* and *8b* only *8a* holds, such that:

$$\forall x.\forall y.\forall \phi.(describes(y, x) \Rightarrow (member(x, \gamma_\phi) \Leftrightarrow member(y, \gamma_\phi)) \tag{10}$$

This means that if compound item is part of a context all its description items are included as well, On the other hand if a description item is included in a context its successors are not necessarily part of the same context (but they can be reached by decomposition). As a consequence we can say that all perspectives are contexts but not all contexts are perspectives.

It is important to note that our notion of *context* matches the one used in the community. As stated in Giunchiglia (1993), a context is a *theory of the world* (it encodes an individual's subjective perspective), it is a *partial theory* (the individual's complete description of the world is given by the set of all the contexts), and it is an *approximate theory* (the world is never described in full details). Each of these properties holds for our context in the following way:

- **Theory of the World:** A context is a subset of a perspective which in its turn is a projection of the ontology in respect with certain aspects and an ontology is a conceptualization of domain of discourse.
- **Partial Theory:** A context is a subset of a perspective and the result of the union of all the possible contexts is that particular perspectives.
- **Approximate Theory:** The perspectives themselves are approximations over the ontology. Depending on the type of perspective used, different aspects of the ontology are shown or hidden.

A notion tightly related with contexts is the process of *decomposition*. A context can be created from another context (this operation is called *context update*) by applying decomposition on an item from a perspective or a context. Decomposition allows navigating between contexts and links consecutive nested levels; the way the contexts are navigated when creating mappings influences the creation types of mappings that are created.

Let *decomposition*(x, ϕ) be a binary function which has as value a new context obtained by decomposing x in respect with the context γ_ϕ. We can define the following axioms:

$$\forall x.\forall y.\forall \gamma_\phi.(member(x, \gamma_\phi) \wedge pi(x) \Rightarrow (member(y, decomposition(x, \gamma_\phi)) \Leftrightarrow member(y, \gamma_\phi))) \tag{11}$$

$$\forall x.\forall y.\forall \gamma_\phi.(member(x, \gamma_\phi) \wedge ci(x) \Rightarrow (member(y, decomposition(x, \gamma_\phi)) \Leftrightarrow y = x \vee describes(y, x))) \tag{12}$$

$$\forall x.\forall y.\forall z.\forall \gamma_\phi.(member(x, \gamma_\phi) \wedge di(x) \wedge successor(z, x) \wedge (pi(z) \vee ci(z) \wedge member(z, \gamma_\phi))) \Rightarrow (member(y, decomposition(x, \gamma_\phi)) \Leftrightarrow member(y, \gamma_\phi))) \tag{13}$$

Intuitively, formula *(11)* specifies that the decomposition of a primitive item does not update the current context (the context remains unchanged). Decomposition of a compound item *(12)* generates a new context that contains the item that was just decomposed and all its description items. Also, decomposition applied on a description item that has a primitive successor (formula *(13)*) leaves the current context unchanged. The same formula also does not allow the decomposition of those description items that have as successor a compound item already contained by the current context (recursive structures). Otherwise, for the

Table 5. Decomposition and context updates

Original Context		
• string • integer • person ├ name → string ├ age → integer ├ hasGender → integer ├ hasChild → integer └ marriedTo → person • gender └ value → person	• string • integer • person ├ name → string ├ age → integer ├ hasGender → integer ├ hasChild → integer └ marriedTo → person • gender └ value → person	• string • integer • person ├ name → string ├ age → integer ├ hasGender → integer ├ hasChild → integer └ marriedTo → person • gender └ value → person
New Context		
• string • integer • person ├ name → string ├ age → integer ├ hasGender → integer ├ hasChild → integer └ marriedTo → person • gender └ value → person	• person ├ name → string ├ age → integer ├ hasGender → integer ├ hasChild → integer └ marriedTo → person	• gender └ value → person

rest of the description items (formula *(14)*) the new context includes the successors of that particular description items together with all successor's description items.

Table 5 presents some examples of decomposition and context updates: each column shows how the context changes by decomposing any of the marked items in the top row. The decomposition can be applied simultaneously on multiple items, and the result of decomposing each item is contributing to the new context. Note that as described above, for column *(1)* no change occurs as none of the marked items can trigger decomposition conforming to the formulae *(11)* and *(13)*.

Mappings

To create mappings between ontologies, the source and the target ontologies are each represented using one of the perspectives. We refer to this approach as interactive mapping creation. It means that the mapping creation process relies upon the domain expert, who has the role of choosing an item from the source perspective and one from the target perspective (or contexts) and explicitly marking them as mapped items. We call this action *map* and using this, the domain expert states that there is a semantic relationship between the selected items. Choosing the right pair of items to be mapped is not necessarily a manual task: a semi-automatic solution can offer suggestions that are eventually validated by the domain expert (Mocan & Cimpian, 2005).

We define a *mapping context* as a quadruple $Mc = \langle \phi_S, \gamma_{\phi_S}, \phi_T, \gamma_{\phi_T} \rangle$ where ϕ_S and ϕ_T are the source and target perspectives associated to the source and target ontologies. γ_{ϕ_S} and γ_{ϕ_T} are the current contexts derived out of the two perspectives ϕ_S and ϕ_T. Initially, $\gamma_{\phi_S} \equiv \phi_S$ and $\gamma_{\phi_T} \equiv \phi_T$. Additionally we define $map_{Mc}(x, y)$ the action of marking the two items x and y as being semantically related with respect to the mapping context Mc. Thus, we have the following axiom:

$$\forall x.\forall y.\forall \phi_S.\forall \phi_T.\forall \gamma_{\phi_S}.\forall \gamma_{\phi_T}.map_{Mc}(x, y) \wedge Mc = \langle \phi_S, \gamma_{\phi_S}, \phi_T, \gamma_{\phi_T} \rangle \Rightarrow ((ci(x) \vee pi(x)) \wedge (ci(y) \vee pi(y)) \vee (di(x) \wedge di(y)) \wedge member(x, \gamma_{\phi_S}) \wedge member(y, \gamma_{\phi_T}) \quad (15)$$

Formula 15 defines the allowed types of mappings. Thus we can have mappings between primitive and/or compound items and between description items. It is not allowed to map between a description item from one perspective and a compound or primitive item from the other perspective, since the easiest and more natural way to do it is to first change the perspective in one of the ontologies, and only then create the actual mapping. As described in Mocan and Cimpian (2005) the set of the allowed mappings can be extended or restricted by a particular, concrete perspective.

Each time a *map* action occurs the mapping context is updated; we denote the updates using: $Mc \mapsto Mc'$ meaning that at least one element of the quadruple defining Mc has changed and the new mapping context is Mc'. The mapping context updates occur as defined in axiom 16:

$$\forall x.\forall y.\forall \phi_S.\forall \phi_T.\forall \gamma_{\phi_S}.\forall \gamma_{\phi_T}.map_{Mc}(x, y) \wedge Mc = \langle \phi_S, \gamma_{\phi_S}, \phi_T, \gamma_{\phi_T} \rangle \Rightarrow Mc' = \langle \phi_S, decomposition(x, \gamma_{\phi_S}), \phi_T, decomposition(y, \gamma_{\phi_T}) \rangle \wedge Mc \mapsto Mc' \quad (16)$$

There are cases when Mc and Mc' are identical; such situations occur when the source and target context remain unchanged, for example, when creating mappings between primitive items.

GROUNDING THE MODEL TO ONTOLOGIES

This section explores the way in which the model presented earlier can be applied to a real ontological model and how we can use it to define concrete perspectives that could be used to create meaningful mappings between ontologies. WSMO ontologies are used for this purpose since the tools implementing these conceptual ideas are part of Web service execution environment (WSMX) which is a references implementation of WSMO; however this model can be potentially grounded to any ontology representation language. We first introduce the main aspects of WSMO ontologies and a mechanism to link these ontologies with our model and then we will present the three types of concrete perspectives (already informally introduced in the third section) identified as being useful in our mediation scenario. It is important to note that the concrete perspectives are not defined inside of the model. This is because a perspective's definition depends on the ontology language used, while the model itself is an abstract model, ontology language neutral.

The first sub-section gives a short overview of WSML, followed by the definition of the three perspectives we have identified as useful for our scenario.

Web Service Modeling Language (WSML)

The Web service modeling ontology (WSMO) defines the main aspects related to Semantic Web services: ontologies, web services, goals and mediators (Feier et al., 2005). From these four, only ontologies are of interest for this work. We will focus on concepts, attributes, and instances in this article; however we intend to address other ontological elements in the future. WSMO ontologies are expressed using the Web service modeling language WSML (Polleres, Lausen, de Bruijn & Fensel, 2005), which is a language for the specification of different aspects of SWS; it takes into account all aspects identified by WSMO.

WSML comprises different formalisms, most notably description logics and logic programming, in order to investigate their applicability in the context of ontologies and Web services. Three main areas can benefit from the use of formal methods in service descriptions: ontology description, declarative functional description of goals and Web services, and description of dynamics. So far, WSML defines a syntax and semantics for ontology descriptions. The underlying formalisms which were mentioned earlier are used to give a formal meaning to ontology descriptions in WSML, resulting in different variants of the language, which differ in logical expressiveness and in the underlying language paradigms, and allow users to make the trade-off between provided expressiveness and the implied complexity for ontology modeling on a per-application basis. We shortly emphasize these variants in the following:

- **WSML-Core** is based on by the intersection of the description logic SHIQ (**D**) (Horrocks, Sattler & Tobies, 1999) and horn logic, based on description logic programs. It has the least expressive power of all the WSML variants. The main features of the language are concepts, attributes, binary relations and instances, as well as concept and relation hierarchies and support for datatypes.
- **WSML-DL** captures the description logic SHIQ (**D**), which is a major part of the (DL species of) Web ontology language OWL (Dean & Schreiber, 2004).
- **WSML-Flight** is an extension of WSML-Core which provides a powerful rule language. It adds features such as metamodeling, constraints and nonmonotonic negation. WSML-Flight is based on a logic programming variant of F-Logic (Kifer, Lausen & Wu, 1995) and is semantically equivalent to Datalog with inequality and (locally) stratified negation. WSML-Flight is a direct syntactic extension of WSML-Core and it is a semantic extension in the sense that the WSML-Core subset of WSML-Flight agrees with WSML-Core on ground entailments).

- **WSML-Rule** extends WSML-Flight with further features from Logic Programming, namely the use of function symbols, unsafe rules, and unstratified negation under the Well-Founded semantics.
- **WSML-Full** unifies WSML-DL and WSML-Rule under a First-Order umbrella with extensions to support the nonmonotonic negation of WSML-Rule.

Several features make WSML unique from other language proposals for the SW and SWS, amongst them the most important are: one syntactic framework for a set of layered languages (no single language paradigm will be sufficient for all SWS use cases, thus different language variants of different expressiveness are needed); normative, human readable syntax (allows for easier adoption of the language by the users); separation of conceptual and logical modeling (the conceptual syntax allows for easy modeling of ontologies, Web services, goals, and mediators, and the logical expression syntax allows expert users to refine definitions on the conceptual syntax), semantics based on well known formalisms (WSML captures well known logical formalisms in a unifying syntactical framework, while maintaining the established computational properties of the original formalisms); and a frame-based syntax (it allows the user to work directly on the level of concepts, attributes, instances, and attribute values, instead of at the level of predicates).

PartOf Perspective

The *PartOf* perspective is the most common perspective that can be used to display an ontology, focusing on the concepts, attributes and attributes' types hierarchies. To link this perspective with our model we define the unary relations $ci_{PartOf}(x)$, $pi_{PartOf}(x)$ and $di_{PartOf}(x)$ such that:

$ci(x)$ **iff** $ci_{PartOf}(x)$
$pi(y)$ **iff** $pi_{PartOf}(y)$
$di(z)$ **iff** $di_{PartOf}(z)$

(17)

It basically means that on ontology entity x has the role of a compound item in the formal model if and only if in the WSML ontology the $ci_{PartOf}(x)$ holds. The same applies for primitive items and descriptions items relative to $pi_{PartOf}(x)$ and $di_{PartOf}(x)$.

$ci_{PartOf}(x)$, $pi_{PartOf}(x)$ and $di_{PartOf}(x)$ have to be defined in the logical language used to represent the ontologies to be aligned, in our case WSML[5] as can be seen in *(17)*. In the *PartOf* perspective, the role of compound items is taken by those concepts that have at least one attribute—we call them *compound concepts*. Naturally, the description items are in this case attributes, as stated in (19). Primitive items are data types or those concepts that have no attributes, as expressed by axiom *(20)* where *x* ***subconceptOf*** *true* holds ***iff*** *x* is a concept and *naf* stands for negation as failure. Finally we link the *describes* and *successor* relations with the WSML ontologies in (21).

axiom ci_{PartOf} **definedBy**
$ci_{PartOf}(x)$ **equivalent exists**?y, ?z(?x[?y **ofType**? z])

(18)

axiom di_{PartOf} **definedBy**
$di_{PartOf}(y)$ **equivalent exists**?y, ?z(?x[?y **ofType**? z])

(19)

axiom pi_{PartOf} **definedBy**
$pi_{PartOf}(x)$: −?x **subconceptOf** *true* **and naf** $ci_{PartOf}(x)$

(20)

$describes(y,x) \wedge successor(z,y)$ **iff** ?x[? y **ofType** ?z]

(21)

The fragment of ontology presented in Table 1 can be visualized using the *PartOf* perspective as shown in Figure 1.

InstanceOf Perspective

The *InstanceOf* perspective can be used to create conditional mappings based on predefined values and instances. Conditional mappings are normal mappings that hold only when certain conditions are fulfilled. Such conditions could for example restrict the allowed values for the source or/and for the target attributes. To link this perspective with our model we define $ci_{InstanceOf}(x)$, $pi_{InstanceOf}(x)$ and $di_{InstanceOf}(y, w)$ such that:

$ci(x)$ **iff** $ci_{InstanceOf}(x)$
$pi(x)$ **iff** $pi_{InstanceOf}(x)$
$di(\langle y, w \rangle)$ **iff** $di_{InstanceOf}(y, w)$

(22)

In the same way as above, $ci_{InstanceOf}(x)$, $pi_{InstanceOf}(x)$ and $di_{InstanceOf}(<y, w>)$ are defined using WSML:

axiom $ci_{InstanceOf}$ **definedBy**
$ci_{InstanceOf}(x)$ **equivalent exists** ?*y*, ?*z*(?*x* [?*y* **ofType** ?*z*])

(23)

In the *InstanceOf* a compound item is a concept that has at least one attribute and that attribute has as type either another compound item or a concept for which there is at least one instance defined in the ontology.

The description items are tuples $< y, w >$ where:

- **y** is an attribute matching the above conditions;
- **w** is an instance member of *y*'s type explicitly defined in the ontology or an anonymous id representing a potential instance of the *y*'s type;

as defined in (24).

axiom $di_{InstanceOf}$ **definedBy**
$di_{InstanceOf}(y, w)$ **equivalent exists** ?*w*, ?*x*(?*x* [?*y* **ofType** ?*z*]) **and**
(?*w* **memberOf** ?*z* or ($ci_{Instanceof}$ (?*z*) **and** ?*w*=_#)))

(24)

Primitive items are those concepts that have no attributes that follow the above condition.

axiom $pi_{Instanceof}$ **definedBy**
$pi_{Instanceof}(x)$: −?*x* **subconceptOf** *true* **and naf** $ci_{Instanceof}(x)$

(25)

Finally we link the *describes* and *successor* relations with the WSML ontologies:

$describes(\langle y, w \rangle, x) \wedge successor(z, y, w\rangle)$ **iff** ?*x*[?*y* **ofType** ?*z*]

(26)

The fragment of ontology presented in Table 1 can be visualized using the *InstanceOf* perspective as shown in Figure 3.

RelatedBy Perspective

The *RelatedBy* perspective focuses on the attributes of the ontology, and describes them from their domain and type point of view.

$ci(x)$ **iff** $ci_{RelatedBy}(x)$
$pi(x)$ **iff** $pi_{RelatedBy}(x)$
$di(x)$ **iff** $di_{RelatedBy}(x)$

(27)

The WSML definition for $ci_{RelatedBy}(x)$, $pi_{RelatedBy}(x)$ and $di_{RelatedBy}(x)$ are described in *(28)*, *(29)* and *(30)*.

axiom $ci_{RelatedBy}$ **definedBy**

$ci_{RelatedBy}(y)$ **equivalent exists** ?x, ?z(?y [?y **ofType** ?z])

(28)

In the *RelatedBy* the attributes are considered compound items and all of them have only two description items: *hasDomain* and *hasRange*.

$$\forall x.(di_{RelatedBy}(x) \Leftrightarrow x = hasDomain \vee x = hasRange) \quad (29)$$

As there are no attributes in the ontology without a domain or without a range, *RelatedBy* perspective does not have any primitive concepts:

$$\forall x.(\neg(pi_{RelatedBy}(x))) \quad (30)$$

Finally we link the *describes* and *successor* relations with the WSML ontologies:

describes(*hasDomain*, *y*) ∧ *successor*(*x*, *hasDomain*) **iff**
exists ?z(?x[?y **ofType** ?z])

(31)

describes(*hasRange*, *y*) ∧ *successor*(*z*, *hasRange*) **iff**
exists ?z(?x[?x **ofType** ?z])

(32)

The fragment of ontology presented in Table 1 can be visualized using the *RelatedBy* perspective as shown in Figure 2.

LINKING THE MODEL TO A MAPPING LANGUAGE

In this section we specify the allowed mappings for each of the perspectives described in Section 4. We start from the following premise $map_{Mc}(x_S, x_T) \wedge Mc = \langle \phi_S, \gamma_{\phi_S}, \phi_T, \gamma_{\phi_T} \rangle$, which means that the elements x_S and y_T from the source and target ontology, respectively, are to be mapped in the mapping context *Mc*. In the following subsection we will discuss the situations that can occur for a pair of perspectives (due to space reasons we address only those cases when the source and target perspectives are of the same type). The types of mappings that can be created will be analyzed with respect to the abstract mapping language proposed in Scharffe and de Bruijn (2005), briefly described in the fifth section.

Abstract Mapping Language

We chose to express the mappings in the Abstract Mapping Language proposed in de Bruijn, Foxvog, and Zimmerman(2004) and Scharffe and de Bruijn (2005) because it does not commit to any existing ontology representation language. Later, a formal semantic has to be associated with it and to ground the mappings to a concrete language.

A mapping in the Abstract Mapping Language has the following form[6]:

mapping ::== 'Mapping('**mappingId [{annotation}] {statement}**')'

The *mappingId* is a unique identifier of the mapping, while *annotation* is a free form explanatory text providing a textual description of the mappings. The mapping language statements are briefly described next:

- *classMapping* - By using this statement, mappings between classes in the source and the target ontologies are specified. Such a statement can be conditioned by class conditions (*attributeValueConditions*, *attributeTypeConditions*, *attributeOccurenceConditions*) and it has the form seen in Box 1.

The *attributeValueCondition* specifies for what values of a certain attribute the given mapping holds. The *attributeTypeCondition* indicates the

Box 1.

statement ::= 'classMapping(' 'one-way'|'two-way' [**{annotation}**]
classExpr classExpr {innerCM_AttributeMapping}
{classCondition} ['{'**logicalExpression**'}'] ')'
classCondition ::= 'attributeValueCondition('**attributeId**
(individualId | dataLiteral) ')'
classCondition ::= 'attributeTypeCondition('**attributeId classExpr** ')'
classCondition ::= 'attributeOccurenceCondition('**attributeId** ')'

Box 2.

innerCM_AttributeMapping ::= 'attributeMapping(' 'one-way'|'two-way' [**{annotation}**]
attributeExpr attributeExpr {attributeCondition}
statement ::= 'attributeMapping(' 'one-way'|'two-way' [**{annotation}**]
attributeExpr attributeExpr
{attributeCondition} ['{'**logicalExpression**'}'] ')'
attributeCondition ::= 'valueCondition('**(individualId | dataLiteral)** ')'
attributeCondition ::= 'typeCondition(' **classExpr** ')'

type (range) a given attribute should have, while *attributeOccurenceCondition* only imposes that a given attribute has to be present in the in the source or the target concept definitions for that mapping to hold.

- *attributeMapping* - Specifies mappings between attributes. Such statements can appear together with or outside *classMappings* and can be conditioned by attribute conditions (*valueConditions*, *typeConditions*) (see Box 2).

The *attributeConditions* are similar with the *classConditions*, with the difference that the first are applied on attributes, while the second are applied on classes. They specify the value or the type a certain attribute should have in order for the mapping to hold.

- *classAttributeMapping* - It specifies mappings between a class and an attribute (or the other way around) and it can be conditioned by both class conditions and attribute conditions (see Box 3).

Since this type of mappings involves classes as well as attributes, both *classConditions* and *attributeConditions* are allowed.

- *instanceMapping* - It states a mapping between two individuals, one from the source and the other from the target (see Box 4).

The *classExpr* and *attributeExpr* can be class identifiers or attribute identifiers, respectively, but also more complex expressions on these identifiers, for example, conjunctions, disjunction, negation, and so forth. The class and attribute identifiers take the form of the identifiers used in the given ontology languages the abstract mapping language is grounded to. For example, if there are WSML ontologies to be mapped these identifiers are IRIs (Polleres et al., 2005). The

Box 3.

statement ::= 'classAttributeMapping(' 'one-way'|'two-way' [**{annotation}**] **(classExpr attributeExpr)|(attributeExpr classExpr)** **{classCondition}** **{atttibuteCondition}** ['{'**logicalExpression**'}'| ')'

Box 4.

statement ::= 'instanceMapping('[**{annotation}**] **individualId individualId** ')'

logicalExpressions represent extra refinements that can be applied to particular mappings. An example of such a refinement is the usage of built-in functions to specify transformations of the data values affected by that mapping (e.g., a currency transformation function).

In the next sections we illustrate how these mapping language statements are generated during design time by using different pairs of perspectives. Even if some of the generated mappings are *two-way*, for brevity in this work all the mappings are considered to be *one-way*.

Partof to *Partof* Mappings

When using the *PartOf* perspective for both the source and the target ontologies to create mappings, we have the following allowed cases (derived from axiom *(15)*):

- $pi_{PartOf}(x_S) \wedge pi_{PartOf}(x_T)$: In this case, the mapping generates a *classMapping* statement in the mapping language and leaves the mapping context unchanged (axioms *(11)* and *(16)*).
- $ci_{PartOf}(x_S) \wedge ci_{PartOf}(x_T)$: Generates a *classMapping* statement and updates the context for the source and target perspectives (axioms *(12)* and *(16)*).
- $di_{PartOf}(x_S) \wedge di_{PartOf}(x_T)$: In this case $successor(y_X, x_S) \wedge successor(y_T, x_T)$ holds and we can distinguish the following situations:
  - $pi_{PartOf}(y_S) \wedge pi_{PartOf}(y_T)$: An *attributeMapping* is generated between x_S and x_T followed by a *classMapping* between y_S and y_T. Conforming to the axioms *(13)* and *(16)*, the mapping context remains unchanged.
  - $ci_{PartOf}(y_S) \wedge ci_{PartOf}(y_T)$: An *attributeMapping* is generated having x_S and x_T as participants. The mapping context is updated conform to the axioms *(13)*, *(14)* and *(16)*.
  - $pi_{PartOf}(y_S) \wedge ci_{PartOf}(y_T)$: Generates a *classAttributeMapping* between z_S and the x_T, where *describes*(x_S, z_S). The new mapping context keeps the source context unchanged while decomposing the target context over y_T.
  - $ci_{PartOf}(y_S) \wedge pi_{PartOf}(y_T)$: This case is symmetric with the one presented previously, and it generates a *classAttributeMapping* between x_S and the z_T where *describes*(x_T, z_T).
- $ci_{PartOf}(x_S) \wedge pi_{PartOf}(x_T)$: It is not allowed for this combination of perspectives. To take an example, such a case would involve a mapping between $ci_{PartOf}(person)$ and

$pi_{PartOf}(string)$ where $describes(hasName, person) \wedge successor(string, hasName)$, which does not have any semantic meaning. A correct solution would be a mapping between $ci_{PartOf}(person)$ and $ci_{PartOf}(u_T)$ such as $\exists v_T(describes(v_T, u_T) \wedge successor(string, v_T)$.

- $pi_{PartOf}(x_S) \wedge ci_{PartOf}(x_T)$: The same explanation applies as earlier.

InstanceOf to *InstanceOf* Mappings

When using the *InstanceOf* perspectives we can create similar mappings to those created with the *PartOf* perspectives, the difference being that conditions are added to the mappings, and by this, the mappings hold only if the conditions are fulfilled. The mappings between two primitive items or between two compound items in the *InstanceOf* perspective are identical with the ones from the *PartOf* perspective. For the remaining cases we have:

- $di_{InstanceOf}(x_T, w_T) \wedge di_{InstanceOf}(x_T, w_T)$: In this case, we have:

$$successor(\langle x_S, w_S\rangle, y_S) \wedge successor(\langle x_T, w_T\rangle, y_T)$$

and we can distinguish the following situations:

  - $pi_{InstanceOf}(y_S) \wedge pi_{InstanceOf}(y_T)$: An *attributeMapping* is generated between x_S and x_T conditioned by two *attributeValueConditions* imposing the presence of w_S and w_T in the mediated data. Also a *classMapping* between y_S and y_T is generated. Conforming to the axioms (13) and (16) the mapping context remains unchanged.
  - $ci_{InstanceOf}(y_S) \wedge ci_{InstanceOf}(y_T)$: An *attributeMapping* is generated having x_S and x_T as participants conditioned by two *typeConditions*. The mapping context is updated conforming to the axioms (13), (14), and (16).
  - $pi_{InstanceOf}(y_S) \wedge ci_{InstanceOf}(y_T)$: This case generates a *classAttributeMapping* between z_S and the x_T, where $describes(x_S, z_S)$. A *typeCondition* is added for x_T attribute. The new mapping context keeps the source context unchanged while decomposing the target context over y_T.
  - $ci_{InstanceOf}(y_S) \wedge pi_{InstanceOf}(y_T)$: This case is symmetric with the one presented previously and it generates a *classAttributeMapping* between x_S and the z_T where $describes(x_T, z_T)$. A *typeCondition* is added for x_S.
- $di_{InstanceOf}(x_S, w_S) \wedge pi_{InstanceOf}(x_T)$: *InstanceOf* extends the set of allowed mappings as defined in *(15)*. For z_S such that $describes(<x_S, w_S>, z_S)$, a *classMapping* between z_S and x_T is generated, conditioned by an *attributeValueCondition* on the attribute x_S and value w_S.
- $pi_{InstanceOf}(x_S) \wedge di_{InstanceOf}(x_T, w_T)$: Similar with the above case.
- $ci_{InstanceOf}(x_S) \wedge pi_{InstanceOf}(x_T)$: It is not directly allowed for this combination of perspectives, but the intended mapping can be created as described by the previous case.
- $pi_{InstanceOf}(x_S) \wedge ci_{InstanceOf}(x_T)$: The same explanation applies as earlier.

RelatedBy to *RelatedBy* Mappings

In the *RelatedBy* perspective attributes are seen as root elements, having only two descriptions: their domain and their type. We identify the following cases:

- $pi_{\mathbf{RelatedBy}}(x_S) \wedge pi_{\mathbf{RelatedBy}}(x_T)$: This case does not appear as we do not have primitive items in the *RelatedBy* perspective.
- $ci_{\mathbf{RelatedBy}}(x_S) \wedge ci_{\mathbf{RelatedBy}}(x_T)$: The mapping will generate an *attributeMapping* state-

ment in the mapping language having as participants x_S and x_T.

- $di_{\textbf{RelatedBy}}(x_S) \wedge di_{\textbf{RelatedBy}}(x_T)$: The source and the target perspectives are changed from *RelatedBy* to *PartOf* and the context is obtained by decomposing the perspectives over z_S and z_T, where $sucessor(z_S, x_S) \wedge successor(z_T, x_T)$.

Grounding the Abstract Mapping Language

As presented in the fourth section, in order to use our model (i.e., an abstract model) it had to be "grounded" to concrete ontologies. In the same fashion, in order to use the abstract mapping language in real mediation scenarios, it has to be grounded to a concrete ontology representation language: the same language the ontologies to be mapped are expressed in.

When creating a grounding for the abstract mappings, a meaning (i.e., a formal semantics) is associated with these mappings. We can say that the grounding prescribes how the mappings are interpreted and how are they going to be used in the context of that particular ontology language. As a consequence, it is very likely for different mediation scenarios (e.g., instance transformation vs. query rewriting) to need different groundings while using the same set of abstract mappings.

The grounding mechanism described in this section is a grounding to WSML-Rule (Polleres et al., 2005) (the interested reader can find groundings to OWL or RDF in (de Bruijn et al., 2004; Scharffe, 2005) designed for instance transformation. The output of this mechanism is a set of WSML axioms representing rules; when evaluated by a WSML reasoner that already contains in its knowledge base the source instances, these rules will generate a set of target instances, that is the mediation results. The new instances can be retrieved by simple queries and used in further computations. For example such a query would look like "*?x* ***memberOf*** *O2#man*" which translates into "what are the instances of concept *man* in the target ontology ?". The WSML mapping rules fire and new instances of *man* are created based on the information in the source instances.

It is important to note that we assume that the mappings represented in the abstract mapping language are correct and consistent. By correctness in this context it is understood that the given mappings conform to the domain expert's view on each particular heterogeneity issue. Unfortunately the correctness as defined above cannot be checked or insured by tools since this is the very role of the domain expert: to validate and to approve the tool's suggestions.

Part of this validation also involves the consistency checking. A set of mappings would be inconsistent if the mediation results (in our case the mediated instances) do not conform to the target ontology constraints and its entities formal semantics. Since the abstract mapping language does not have a formal semantic associated, the consistency of mappings cannot be checked at this level. However, after the grounding is applied and the mapping rules are loaded into the reasoner together with the required source and target schema information and constraints, such checking is possible. If the mappings are grounded to a rule-based ontology representation language (e.g., WSML-Rule) it is necessary to also include samples of instances to be mediated in order to detect the potential inconsistencies. If the mappings are grounded to a DL-based ontology language several mappings' properties (including consistency) can be checked as in Stuckenschmidt, Serafini, and Wache (2006).

Mapping Examples

Table 6 contains examples of mappings expressed in the abstract mapping language (left column) between the fragments in Table 1 and Table 2. The right column shows what these mappings look like in WSML when mapping the concept *person* in the source ontology with *human* (and *man*) in the target ontology.

Table 6. Mapping examples

Abstract Mapping Language	Mapping Rules in WSML
Mapping(o1#persono2#man **classMapping**(*one-way* person man))	**axiom** mapping001 **definedBy** mediated(X_1, o2#man) **memberOf** o2#man:- X_1 **memberOf** o1#person.
Mapping(o1#ageo2#age **attributeMapping**(*one-way* [(person)age=>integer] [(human)age=>integer]))	**axiom** mapping001 **definedBy** mediated(X_2, o2#human) **memberOf** o2#human:- X_2 **memberOf** o1#person.
Mapping(o1#nameo2#name **attributeMapping**(*one-way* [(person)name => string] [(human)name => string]))	**axiom** mapping005 **definedBy** mediated(X_5, o2#human)[o2#age **hasValue** Y_6]:- X_5[o1#age **hasValue** Y_6]:o1#person. **axiom** mapping006 **definedBy** mediated(X_7, o2#human)[o2#name **hasValue** Y_8]:- X_7[o1#name **hasValue** Y_8]:o1#person.
Mapping(o1#hasGendero2#man **attributeClassMapping**(*one-way* [(person)hasGender => gender] man)) **valueCondition**([(person)hasGender => gender] male)	**axiom** mapping007 **definedBy** mediated(Y_11, o2#man)[A_9 **hasValue** AR_10]:- mediated(Y_11, o2#human)[A_9 **hasValue** AR_10], Y_11[o1#hasGender **hasValue** o1#male].

When evaluated, the WSML mapping rules will generate instances of *man* if the *gender* condition is met or of *human* otherwise. The construct *mediated(X, C)* represents the identifier of the newly created target instance, where *X* is the source instance that is transformed, and *C* is the target concept we map to.

As mentioned above, the building of the new mediated instances is performed by the reasoner (through the mapping rules) following the very same principles of decomposition described in the third section (a decomposition algorithm is presented in the sixth section as well). In layman terms, a rule will indicate that an instance of *man* has to be created if an instance of person is encountered. Several rules are attaching attributes (description items in our model) to this newly created instance of *man* and they specify the values for this attributes. These values are either literals (when the successor of the description item is a data-type) or other mediated instances (when the successor of the description item is an ontology concept).

IMPLEMENTATION AND PROTOTYPE

The ideas and methods presented in this article are used in the mediation component of the WSMX architecture (Mocan et al., 2006). The WSMX Data Mediation component is designed to support data transformation, that is it transforms the source ontology instances entering the system into instances expressed in terms of the target ontology. As described previously, in order to make this possible the data mediation process consists of design-time and run-time phases. Each of these two phases has its own implementations: the *Ontology Mapping Tool* and the *Run-time Data Mediator*. Figure 4 gives an overview if the two tools and of the relationship between them.

Figure 4. Overview on the design-time and run-time components

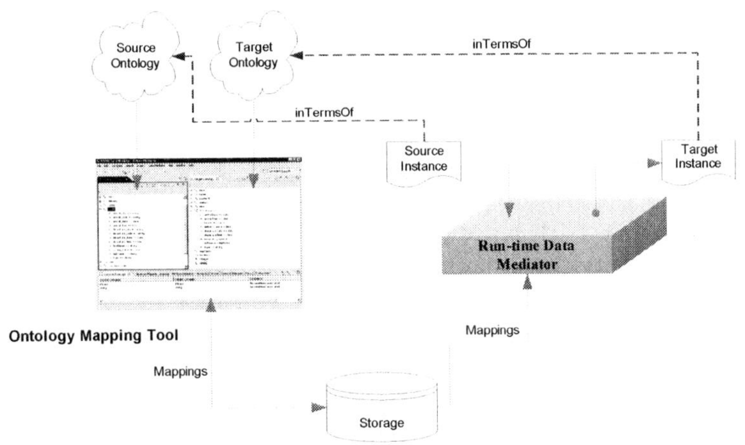

The Ontology Mapping Tools offers a graphical interface for creating the mappings and supports the storage of the mappings in an external storage for further usages or refinements. The Run-time Data Mediator is able to load the mappings created during design-time and use them to transform, during run-time, the incoming data from terms of one ontology (the source ontology) in terms of another ontology (target ontology).

The mapping creation (during design-time) is a semi-automatic, computer-assisted process that keeps the domain expert in control of the process. The run-time phase on the other hand, is a completely automatic process that runs from end-to-end with no human intervention as long as mappings between the source and target ontology exist in the mapping repository.

In the following subsections the main features of these tools are presented, emphasizing their relation with the formal model presented above.

Ontology Mapping Tool

The ontology mapping tool is implemented as an Eclipse plug-in, part of the Web service modeling toolkit (WSMT)[7] (Kerrigan, 2006), an integrated environment for ontology creation, visualization and mapping. The ontology mapping tool is currently compatible with WSMO ontologies (but by providing the appropriate wrappers different ontology languages could be supported); it offers different modalities of browsing the ontologies using perspectives and allows the domain expert to create mappings between two ontologies (source and target) and to store them in a persistent mapping storage.

Perspectives

Currently only the *PartOf* and *InstanceOf* perspectives are fully implemented. They follow the principles of the formal model and support the features described in the next subsections.

Figure 5 shows how the perspectives are represented in the ontology mapping tool. From the technical point of view the notion of perspective corresponds in the implementation to the notion of eclipse view.

Decomposition Algorithm

The decomposition algorithm is one of the most important algorithms in our approach and it is used to offer guidance to the domain expert during the

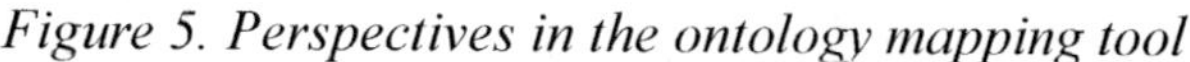

Figure 5. Perspectives in the ontology mapping tool

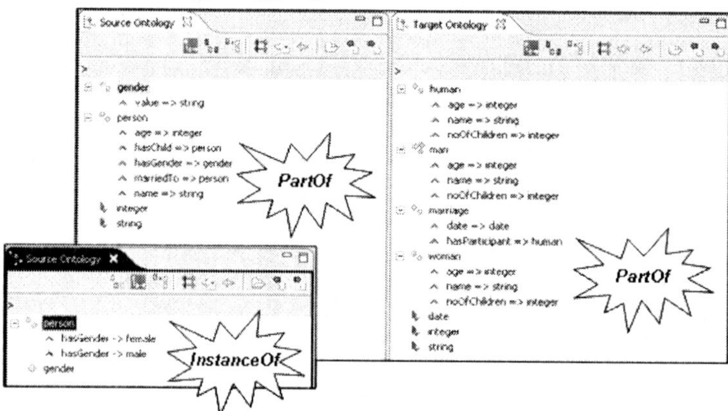

Table 7. The decomposition algorithm

```
decompose(Collection collectionOfItems){
  Collection result;
  for each item in collectionOfItems do{
    result = result + item;
    if isCompound(item){
      Collection itemsDescriptions = getDescriptions(item);
      result = result + itemDescriptions;
    }
    if isDescriptionItem(item){
      Item successorItem = getSuccessor(item);
      if (not createsLoop(succesorItem)){
        result = result + successorItem;
        Collection itemsDescriptions = getDescriptions(item);
        result = result + itemDescriptions;
      }
    }
  }
  return result;
}
```

mapping process and to compute the structural factor as part of the suggestions algorithms (described later in this section). By decomposition we expose the descriptions of a compound item and make them available to the mapping process. That is, the decomposition algorithm can be applied on description items and it returns the description items (if any) for the successors of that particular description items. An overview of this algorithm is presented in Table 7.

The implementation of *isCompound, isDescriptionItem, getDescriptions, getSuccessor,* and *createsLoop* differ from one view to another—for example, the cases when loops are encountered (i.e., the algorithm will not terminate) have to be addressed for each view in particular.

Suggestion Algorithms

Suggestion algorithms do not represent the focus of the work presented in here. However, in order to deliver a truly semi-automatic mapping tool, suggestion algorithms are a necessity.

The suggestion algorithms are used for helping the domain expert in taking decisions during the mapping process, regarding the possible semantic relationships between source and target items in the current mapping context. We propose a combination of two types of such algorithms: lexical based algorithms, where the focus is on syntactic similarities, and structural algorithms that consider description items in their computations. Brief descriptions of these algorithms are provided below

As a result, for each pair of items we compute a so-called *eligibility factor* (EF), which indicates the degree of similarity between the two items: the smallest value (0) means that the two items are completely different, while the greatest value (1) indicates that the two items are similar. For dealing with the values between 0 and 1 a threshold value is used: the values lower than this value, indicate different items and values greater than this value indicate similar items. Setting a lower threshold assures a greater number of suggestions, while a higher value for the threshold restricts the number of suggestion to a smaller subset.

The EF is computed as a weighted average between a *structural factor* (SF), referring to the structural properties and a *lexical factor* (LF), referring to the lexical relationships determined for a given pair of items. The weights can be chosen based on the characteristics of the ontologies to be mapped. For example when mapping between ontologies developed in different spoken languages the weight of LF should be close to 0 in contrast with the case when mapping between ontology developed in the same working group or institution (the usage of similar names for related terms is more likely to happen).

Even if the structural factor is computed using the decomposition algorithm, the actual heuristics used are dependent on the specific perspectives where it is applied. In a similar manner the current perspectives determine the weight for the structural and lexical factors as well as the exact features of the items to be used in computations.

Lexical Factor

The lexical factor is computed based on the syntactic similarities between the names of a given pair of items. There are two main aspects used in these computations: first, the lexical relationships between the terms as given by WordNet[8] and second, the results returned by string analysis algorithms. WordNet is an on-line lexical reference, inspired by current psycholinguistic theories of human lexical memory (Miller, Beckwith, Felbaum, Gross & Miller, 1990). English nouns, verbs, adjectives, and adverbs are organized into synonym sets, each representing one underlying lexical concept. Different relations such as hyponymy and hypernymy link the synonym sets.

The first step in computing the lexical factor involves the splitting of both source and target term in tokens. Tokens are separated in a term by one or more symbols like '_', '-', or by the usage of capital letters inside the term's name. For example the term *PurchaseOrder_Request* is formed of the following tokens: *purchase*, *order*, and *request*. For each of the source tokens, a set of lexical related terms are retrieved by using WordNet (generically referred further as synonyms for brevity even if hyponims and hypernims are considered as well). The Monge and Elkan's (1996) recursive matching algorithm is used to compute the matching score (*ME*) between each of the source tokens' synonyms and the target term. As such, the lexical factor is computed by considering the number of tokens in a term and the sum of the maximum matching score between each source token's collection of synonyms and the target term. The following formula is used for determining the LF:

$$LF = M\frac{1}{S}\sum_{i=1}^{S}\underset{k=1}{\overset{SS_i}{Max}}(ME(A_{i,k}, B))$$

where:

- S represents the number of tokens in the source term. T is used in later formulas to denote the number of tokens in the target term.
- A represents the initial source term. A_i represents the set of synonyms for the i-th token. $A_{i,k}$ is the k-th synonym term from the i-th set of synonyms. SS_i gives the number of synonyms for the i-th token.
- B represents the target term.
- ME returns the matching score conform to Monge and Elkan's algorithm.
- M is a multiplicity factor (*multiplier*) used to compensate for the difference in the number of tokens from the source and the target:

$$M = \begin{cases} \frac{\alpha}{S} & if \quad S > \alpha \\ \frac{S}{\alpha} & if \quad S \leq \alpha \end{cases}$$

where α is the average of number of tokens in source and target: $\alpha = \frac{S+T}{2}$. As such M can be calculated as:

$$M = \begin{cases} \frac{S+T}{2S} & if \quad S > T \\ \frac{2S}{S+T} & if \quad S \leq T \end{cases}$$

A graphical representation of the values M takes when S and T vary from 1 to 10 is presented in Figure 6.

It is important to note that M does not affect the score given by the string matching algorithms when the numbers of tokens in the source and the target are the same (i.e., $M=1$). In the rest of the cases, M decreases as the difference between the numbers of tokens increases.

Structural Factor

The structural factor is computed based on the structural similarities between the two terms. The following formula is used:

$$SF = \begin{cases} \frac{2D}{S+T}\frac{S}{T} & if \quad S < T \\ \frac{2D}{S+T}\frac{T}{S} & if \quad S \geq T \end{cases}$$

Figure 6. Values taken by the multiplicity factor when the number of tokens in the source and target terms varies from 1 to 10

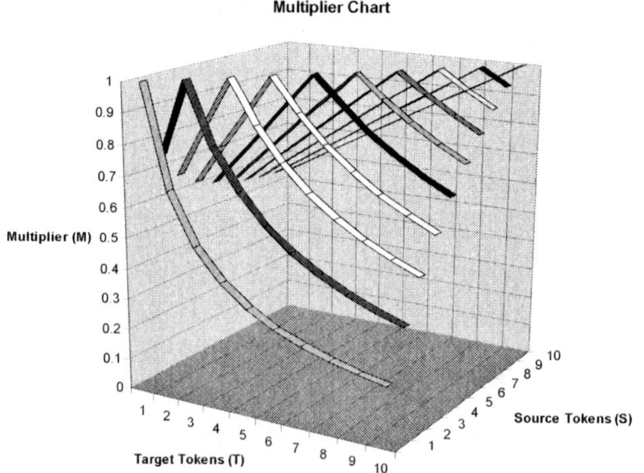

where:

- D is determined by the number of mappings between primitive description items that can be reached by decomposition (see Table 7) starting from the given source and target items. A primitive description item is a description item that has as successor a primitive item. For example in the PartOf perspective, a typical primitive description item would be an attribute that has as range a data-type.
- S represents the number of primitive description items that can be reached by decomposition from the initial source item.
- T represents the number of primitive description items that can be reached by decomposition from the initial source item.

The eligibility factor might seem very expensive to compute between a selected source item and all the items from the target, especially when mapping large ontologies. The performance is drastically improved by the use of contexts since the set of item pairs for which the eligibility factor has to be computed is significantly cut down.

Bottom-Up vs. Top-Down Approach

Considering the algorithms and methods described, two possible approaches regarding ontology mapping are supported by the Ontology Mapping Tool: bottom-up and top-down approaches.

The *bottom-up approach* means that the mapping process starts with the mappings of the primitive items (if possible) and then continues with items having more and more complex descriptions. By this the pairs of primitive items act like a minimal, agreed upon set of mappings between the two ontologies, and starting from this minimal set more complex relationships could be gradually discovered. This approach is useful when a complete alignment of the two ontologies is desired.

The *top-down approach* implies that the mapping process starts with mappings of compound items and it is usually adopted when a concrete heterogeneity problem has to be resolved. That means that the domain expert is interested only in resolving particular item mismatches and not in fully aligning the input ontologies. The decomposition algorithm and the mapping contexts it updates will help the user to identify all the relationships that can be captured by using a specific type of view and that are relevant to the problems to be solved.

In the same way as for the other algorithms, the applicability and advantages/disadvantages of each of these approaches depends on the type of view used.

Run-Time Data Mediator

The *Run-time Data Mediator* plays the role of the data mediation component in WSMX (available together with the WSMX system[9]). It uses the abstract mappings created during design-time, grounds them to WSML and uses a reasoner to evaluate them against the incoming source instances. The mapping rules, the source instances and if necessary, source and target schema information are loaded into the reasoning space in what could be seen as a "pseudo-merged" ontology (i.e., the entities from the source and the target and the rules are strictly separated by namespaces). By querying the reasoning space for instances of target concepts, if semantically related source instances exist, the rules fire and produce as results the target instances.

The storage used is a relational database but the mappings could be stored using the mapping documents in the file system as well. **Figure 7** gives a more detailed overview of the mediation tools and their relationships, together with the several possible usage scenarios identified for the run-time data mediator.

The mappings grounding module is part of the run-time component. In this way, the same set of mappings (i.e., abstract, ontology language independent mappings) can be grounded to different languages depending on the scenario the run-time mediator is used in. A second reason is that by grounding, a formal semantics is associated with the mappings, meaning that only at this stage it is stated what exactly it means to have two items mapped to each other. These formal semantics differ from one mediation scenario to another, that is different grounding has to be applied when using the abstract mappings in instance transformation then when using them in query rewriting. The third reason is an easier management of the mappings that form the ontology alignment. If the ontologies evolve, the mappings have to be updated accordingly and it becomes more efficient to perform these updates free of any language related peculiarities.

The main scope of the run-time data mediator is to be deployed as a component in the WSMX environment. Additionally it can be made available as a Web service (i.e., a Semantic Web service) that can be invoked directly via SOAP from specialized tools (one of such Invokers is deployed as a plug-in in WSMT) or through a Web interface using a Web browser. And finally, the run-time mediation can be offered as a stand-alone application that helps in testing and evaluating the mappings during the mapping process from the design-time. The stand-alone version can be also integrated and delivered together with WSMT as a helper tool for the ontology mapping.

RELATED WORK

The resource decoupling and isolation principles implemented by Web services could become from an advantage a disadvantage when considering the heterogeneity problems introduced by this approach. As such, this area have been intensely investigated in the last years, in order to offer robust solutions towards solving such problems that may appear to the level of all operations the Web services are involved in.

For example, Wang and Stroulia (2003) propose a mechanism to solve the differences between the requester's needs and provider's offer in the discovery process. This approach applies syntactic and structure algorithms to asses the similarity between the desired WSDL[10] specifications and the set of WSDL specifications advertised in UDDI[11]. The syntactical algorithm uses a *WordNet-Powered Vector-space model* to retrieve published WSDL services that are most similar to the input description. The structural algorithm involves the comparison of the operation set offered by the services, which is based on the comparison of the structures of the operations input and output messages, which, in turn, is based on the comparison of the data types of the objects communicated by these messages.

Even if such a method might present clear advantages over the "traditional" manual browsing of UDDI repository, we think that it is not enough to a truly dynamic and (semi-)automatic web service discovery. We consider that comprehensive semantic descriptions based on ontologies have to be associated with Web services and used in solving the heterogeneity problems. Ontology mapping-based mediation is the key to dynamic and (semi-)automatic solutions due to its generality and its great potential to reuse. Our approach proposes methodologies and solutions that could be applied in most of the Web services' operations (including discovery, selection, invocation, etc.) where heterogeneity problems arise. However, some of the lexical algorithms main principles presented in Wang and Stroulia (2003) (including the use of WordNet) can be also found in our approach in the suggestion algorithms (see the sixth section for more details).

Another important reference point for our work is represented by the work done in the area of database integration in general and database schemas matching in particular. A survey of approaches towards schema matching is presented

Figure 7. Run-time data mediator usage scenario

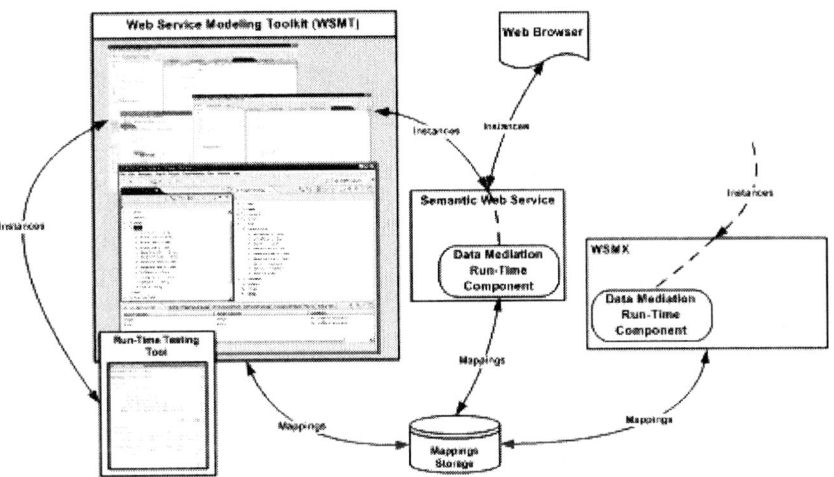

in (Rahm & Bernstein, 2001) and a classification of the Match operator is provided.

In our approach, we use *schema-only*-based[12] mechanisms to semi-automatically determine the matches between elements of the source and target schema. The mappings created by using the perspectives presented above are 1:1 element level mappings but by layering support for transformation functions on top of perspectives 1:n and n:1 *element level* can be created. These mappings represented in the Abstract Mapping Language are grounded to a set of rules in an ontology language to be evaluated into a reasoner. By using reasoning and appropriate queries, the existing rules behave like virtual *structure-level* mappings allowing the creation of data structures expressed in terms of the target ontology based on data expressed in terms of the source ontology. Furthermore, by not explicitly creating mappings between whole structures, enables the reuse of the mapping information from one particular heterogeneity case to another. Regarding the suggestion mechanisms, we use a set of lexical algorithms for *name matching* as described in Section 6.1.3. Additionally, the usage of the already existing mappings in the structural algorithms can be categorized as a *constraint-based* approach, conform to Rahm and Bernstein (2001).

Kalfoglou and Schorlemmer (2003) provide a survey of ontology mapping approaches and classifies them in several categories based on the type of work behind them. We have selected three of them which we consider relevant for our work: *Frameworks*, *Methods and Tools*, and *Translators*. For each of these categories one approach has been identified and analyzed below in comparison to our work: MAFRA (Silva & Rocha, 2003) for Frameworks, PROMPT (Noy & Munsen, 2003) for Methods and Tools and Abiteboul and colleagues' approach (Abiteboul, Cluet & Milo, 1997) for Translators.

MAFRA and PROMPT have been chosen because of the similarities they share with our work: they both take a semi-automatic approach towards ontology mapping having the human user involved in an interactive mapping process where the user has to choose between a set of available actions to solve the existing mismatches between the ontologies. The work regarding "correspondence and translation for heterogeneous data" presented in Abiteboul et al. (1997), even if it tackles integration of heterogeneous data in the databases world, have been chosen due to the relation with our approach to instance transformation based on ontology mappings created before hand.

MAFRA proposes a Semantic Bridge Ontology to represent the mappings. This ontology has as central concept, the so-called "Semantic bridge", which is the equivalent of our mapping language statements. The main difference to our approach is that MAFRA does not define any explicit relation between the graphical representation of the ontologies in their tool and the generation of these Semantic Bridges or between the user's actions and the particular bridges to be used. The formal abstract model we propose links the graphical elements of the user interface with the mapping representation language, ensuring a clear correspondence between the user actions and the generated mappings.

PROMPT is an interactive and semi-automatic algorithm for ontology merging. The user is asked to apply a set of given operations to a set of possible matches, based on which the algorithm recomputes the set of suggestions and signals the potential inconsistencies. The fundamental difference in our approach is that instead of defining several operations we have only one operation (*map*) which will take two ontology elements as arguments, and multiple *perspectives* to graphically represent the ontologies in the user interface. Based on the particular types of perspectives used and on the roles of the *map* action arguments in that perspective the tool is able to determine the type of mapping to be created. Such that, by switching between perspectives different ontology mismatches can be addressed using a single *map* action. An interesting aspect is that PROMPT defines the term *local context,* which perfectly matches our *context* definition: the set of descriptions attached to an item together with the items these descriptions point to. While PROMPT uses the local context in decision-making when computing the suggestions, we also use the context when displaying the ontology.

Instead of allowing browsing on multiple hierarchical layers, as PROMPT and MAFRA do, we adopt a context based browsing that allows the identification of the domain expert's intentions and generates mappings.

Abiteboul et al. (1997) propose a middleware data model as basis for the integration task, and declarative rules to specify the integration. The model is a minimal one and the data structure consists of ordered label trees. The authors claim that "even though a mapping from a richer data model to this model may loose some of the original semantics, the data itself is preserved and the integration with other models is facilitated". Since the main scenario they target is data translation while in our approach we apply our framework to instance transformation, the relation is obvious. The tree model they use, resemble our items model for the perspectives with the difference that our model can be grounded to several "view points" on the two conceptualizations to be mapped. The authors also consider two types of rules: *correspondence rules* used to express relationships between tree nodes (similar with our mappings in the Abstract Mapping Language but expressed in Datalog) and *translation rules*, a decidable subcase for the actual data translation (resembling our mapping rules in a concrete ontology language, e.g. WSML). Additionally, in our approach, we extend our model to the design-time phase as well in order to directly capture the relationships between the ontologies in terms of the "middleware" model.

CONCLUSION AND FURTHER WORK

In this article we define a formal model for mapping creation. This model sits between the graphical elements used to represent the ontologies and the result of the mapping process that is the ontology alignment. By defining both the graphical instruments and the mapping creation strategies in terms of this model we assure a direct and complete correspondence between the human user actions and the effect on the generated ontology alignment. It is formally defined what kind of effect a user's action has in terms of the generated alignment (for example that a mapping between two compound items in the *PartOf* view always generates only a *classMapping* in the alignment). Additionally, it is also formally defined what are the allowed user's actions in the graphical interface and what each of them means in terms of the model and in terms of the alignment that is generated.

In addition we propose a set of different graphical perspectives that can be linked with the same model, each of them offering a different viewpoint on the displayed ontology. By combining these types of perspectives different types of mismatches can be addressed in an identical way from one pair of perspectives to the other. On top of these perspectives a lexical and a structural algorithm offer suggestions to the domain expert; a mechanism of context updates and decomposition offers guidance through the whole mapping process.

As future work, we plan to focus on identifying more relevant perspectives and to investigate the possible combination of these perspectives from the types of mappings that can be generated point of view. These would lead in the end to defining a set of mapping patterns in terms of our model, which will significantly improve the mapping finding mechanism. Another point to be investigated is the mapping with multiple participants from the source and from the target ontology. In this article we investigated only the cases when exactly one element from the source and exactly one element from the target can be selected at a time to be mapped. We plan to also address transformation functions from the perspective of our model. Such transformation functions (e.g., string concatenation), would allow the creation of new target data, based on a combination of given source data.

ACKNOWLEDGMENT

This work is funded by the European Commission under the projects ASG, DIP, enIRaF, InfraWebs, Knowledge Web, Musing, Salero, SEKT, Seemp, SemanticGOV, Super, SWING and TripCom; by Science Foundation Ireland under the DERI-Líon Grant No.SFI/02/CE1/I13; by the FFG (Österreichische Forschungsförderungsgeselleschaft mbH) under the projects Grisino, RW[2], SemNetMan, SeNSE, TSC, OnTourism.

REFERENCES

Abiteboul, S., Cluet, S., & Milo, T. (1997). Correspondence and translation for heterogeneous data. In *Proceedings of the International Conference on Database Theory.*

Biron, P.V., & Malhotra, A. (2004). XML schema part 2: Datatypes (2nd ed). *W3C Recommendation.*

Dean, M. ,& Schreiber, G. (Eds.). (2004). OWL web ontology language reference. *W3C Recommendation.* Retrieved April 3, 2007, from http://www.w3.org/TR/2004/REC-owl-ref-20040210/

de Bruijn, J., Foxvog, D., & Zimmerman, K. (2004). Ontology mediation patterns library. *SEKT Project Deliverable D4.3.2.*

de Bruijn, J., Lausen, H., Krummenacher, R., Polleres, A., Predoiu, L., Kifer, M., & Fensel, D. (2005): The web service modeling language (WSML). *WSML Working Draft.* Retrieved April 3, 2007, from http://www.wsmo.org/TR/d16/d16.1/v0.2/

Ehrig, M., Staab, S., & Sure, Y. (2005). Bootstrapping ontology alignment methods with APFEL. In *Proceedings of the Fourth International Semantic Web Conference (ISWC-2005)* (pp. 186-200).

Feier, C., Polleres, A., Roman, D., Domingue, J., Stollberg, M., & Fensel, D. (2005). Towards intelligent web services: The web service modeling ontology (WSMO). In *Proceedings of the International Conference on Intelligent Computing (ICIC2005).*

Genesereth, M. R., & Nilson, N. J. (1987). *Logical foundations of artificial intelligence.* San Francisco: Morgan-Kaufmann Publishers Inc.

Giunchiglia, F. (1993). Contextual reasoning. *Epistemologia, special issue on I Linguaggi e le Macchine, XVI,* 345-364.

Gruber, T. R. (1993). A translation approach to portable ontology specifications. *Knowledge Acquisition, 5,* 199-220.

Horrocks, I., Sattler, U., & Tobies, S. (1999). Practical reasoning for expressive description logics. In *Proceedings of the 6th International Conference on Logic for Programming and Automated Reasoning (LPAR99)* (pp. 161-180).

Kalfoglou, Y., & Schorlemmer, M. (2003). Ontology mapping: The state of the art. *The Knowledge Engineering Review Journal (KER), 18*(1), 1-31.

Kerrigan, M. (2006). WSMOViz: An ontology visualization approach for WSMO. In *Proceedings of the 10th International Conference on Information Visualization.*

Kifer, M., Lausen, G., & Wu, J. (1995). Logical foundations of object-oriented and frame-based languages. *Journal of the ACM, 42,* 741-843.

Miller, G.A., Beckwith, R., Felbaum, C., Gross, D., & Miller, K. (1990). Introduction to WordNet: An on-line lexical database. *International Journal of Lexicography, 3*(4), 235-244.

Mocan, A., & Cimpian, E. (2005). Mapping creation using a view based approach. In *Proceedings of the 1st International Workshop on Mediation in Semantic Web Services (Mediate 2005).*

Mocan, A., Moran, M., Cimpian, E., & Zaremba, M. (2006). Filling the gap: Extending service oriented architectures with semantics. In *Proceedings of the Second IEEE International Symposium on Service-Oriented Applications, Integration and Collaboration (SOAIC-2006).*

Monge, A., & Elkan, C. (1996). The field-matching problem: Algorithm and applications. In *Proceedings of the Second International Conference on Knowledge Discovery and Data Mining* (pp. 267-270).

Noy, N. F., & Munsen, M. A. (2003). The PROMPT suite: Interactive tools for ontology merging and mapping. *International Journal of Human-Computer Studies, 6*(59).

Polleres, A., Lausen, H., de Bruijn, J., & Fensel, D. (2005). WSML - A language framework for semantic web services. In *Proceedings of the W3C Workshop on Rule Languages for Interoperability.*

Rahm, E., & Bernstein, P. A. (2001). A survey of approaches to automatic schema matching. *International Journal on Very Large Data Bases, 10*(4), 334-350.

Scharffe, F. (2005). Mapping and merging tool design. *Ontology Management Working Group (OMWG).* Unpublished manuscript.

Scharffe, F., & de Bruijn, J. (2005). A language to specify mappings between ontologies. In *Proceedings of the IEEE Conference on Internet-Based Systems.*

Silva, N., & Rocha, J. (2003). Semantic web complex ontology mapping. In *Proceedings of the IEEE/WIC International Conference on Web Intelligence (WI'03),* 82.

Stuckenschmidt, H., Serafini, L., & Wache, H. (2006). Reasoning about ontology mappings. In *Proceedings of the 17th European Conference on Artificial Intelligence (ECAI 2006), Workshop on Context Representation and Reasoning.*

Wang, Y., & Stroulia, E. (2003). Semantic structure matching for assessing web-service similarity. In *Proccedigns of the First International Conference on Service Oriented Computing.*

ENDNOTES

1. In Mocan and Cimpian (2005) the perspectives are called views. In order to avoid suggesting a straight forward connection with the data base views (even if they share indeed some similarities) to the readers from the data base community, the term perspective is used from now on in this article to denote this concept.
2. The fragments of ontologies analyzed in this section are represented using the Human Readable Syntax of Web Service Modeling Language (WSML) (Polleres et al., 2005).
3. As further described in this article what we informally call here "a natural way" of creating mappings is actually reflected in the formal model as a consistent and uniform methodology for creating ontology mappings.
4. *member* is a relationship expressing the membership of an element to a list. It can be defined as:
   $\forall x.\forall l.(member(x, x.l)$
   $\forall x.\ \forall y.\ \forall l.\ (member(x,\ l) \Rightarrow member(x, y.l)$
5. In WSML α[β **ofType** γ] is an atomic formulas called *molecule*; in here both α and γ identifies concepts while β identifies an attribute and '?' is used to denote variables. An example of a molecule for the ontology fragment in Table 1 is *person*[*name* **ofType** *string*].
6. An abstract syntax expressed in EBNF is used. The content between ' and ' literally appears in the mapping language, elements between { and } can have multiple occurrences and elements between [and] are optional.
7. Open Source Project available at http://sourceforge.net/projects/wsmt
8. More details available at http://www.cogsci.princeton.edu/~wn/
9. Open Source Project available at http://sourceforge.net/projects/wsmx
10. Web services description language (WSDL), http://www.w3.org/TR/wsdl
11. UDDI technical paper, http://www.uddi.org/pubs/Iru_UDDI_Technical_White_Paper.pdf
12. The *InstanceOf* perspective does make use of instances but they are not application specific instances part of an instance store, but instances that are normally shared together with the schema. They are later used as "constants" when creating an instance base based on that particular schema.

This work was previously published in International Journal on Semantic Web & Information Systems, Vol. 3, Issue 2, edited by A. P. Sheth and M.D. Lytras, pp. 69-98, copyright 2007 by IGI Publishing, formerly known as Idea Group Publishing (an imprint of IGI Global).

Chapter 2.29
Engineering Conceptual Data Models from Domain Ontologies:
A Critical Evaluation

Haya El-Ghalayini
University of the West of England (UWE), UK

Mohammed Odeh
University of the West of England (UWE), UK

Richard McClatchey
University of the West of England (UWE), UK

ABSTRACT

This article studies the differences and similarities between domain ontologies and conceptual data models and the role that ontologies can play in establishing conceptual data models during the process of developing information systems. A mapping algorithm has been proposed and embedded in a special purpose transformation engine to generate a conceptual data model from a given domain ontology. Both quantitative and qualitative methods have been adopted to critically evaluate this new approach. In addition, this article focuses on evaluating the quality of the generated conceptual data model elements using Bunge-Wand-Weber and OntoClean ontologies. The results of this evaluation indicate that the generated conceptual data model provides a high degree of accuracy in identifying the substantial domain entities, along with their relationships being derived from the consensual semantics of domain knowledge. The results are encouraging and support the potential role that this approach can take part in the process of information system development.

INTRODUCTION

In the last decade, ontologies have been considered as essential components in most knowledge-based application development. As these models are increasingly becoming common, their applicability has ranged from the artificial intelligence domain, such as knowledge engineering/representation and natural language processing, to different fields like information integration and retrieval systems, the semantic Web, and the requirements analysis phase of the software development process. Therefore, the importance of using ontologies in building conceptual data models (CDMS) (CDMs) has already been recognized by different researchers. In our approach, we claim that the differences and similarities between ontologies and CDMs play an important role in the development of CDMs during the information system development process. We indicate that CDMs can be enriched by modeling the consensual knowledge of a certain domain, which, in turn, minimizes the semantic heterogeneities between the different data models (El-Ghalayini, Odeh, McClatchey, & Solomonides, 2005). We chose to study ontologies represented by the Web ontology language (OWL), since it is the most recent Web ontology language released by the World Wide Web Consortium in February 2004 (W3C-World Wide Web Consortium, 2005), and its formal semantics are based on description logics (DL).

The remainder of this article is structured as follows. The next section provides relevant information related to ontologies, CDMs, and the so-called transformation engine (TE). Then the following section discusses the process of evaluating the TE and its parameters, in general, and the qualitative dimension in evaluating the quality of the generated CDM elements, using ontological rules. This evaluation is demonstrated by a real-life case study related to the transparent access

Figure 1. General architecture of the proposed approach

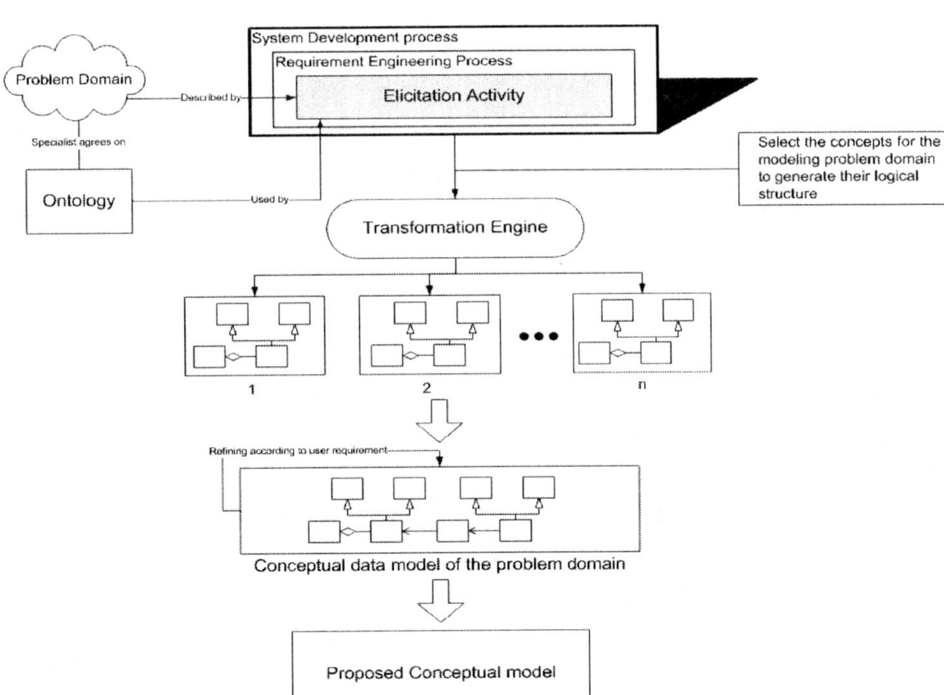

to multiple bioinformatics information sources (TAMBIS) ontology; finally the conclusion and future work are presented.

ONTOLOGY VERSUS CONCEPTUAL DATA MODEL

This section informally explores ontologies and CDMs, including their similarities and differences. The literature shows many definitions of ontologies with the most popular definition proposed by Gruber (1995) as "a formal, explicit specification of a shared conceptualization" (p. 907). In general terms, an ontology may be defined as expressing knowledge in a machine-readable form to permit a common understanding of domain knowledge, so knowledge can be exchanged between heterogeneous environments.

On the other hand, conceptual data models capture the meaning of information for modelling an application and offer means for organizing information for the purposes of understanding and communication(Mylopoulos, 1998). The major role of the CDM is to model the so-called universe of discourse (UoD), entities and relationships in relation to particular user requirements independent of implementation issues. Hirschheim, Klein, and Lyytinen (1995) define the Universe of Discourse in the information systems (IS) world as "the slice of reality to be modelled" (p. 58). Therefore, there are some similarities and differences between ontologies and CDMs. Both are represented by a modeling grammar with similar constructs, such as classes in ontologies that correspond to entity types in CDMs. Thus, the methodologies of developing both models have common activities (Fonseca & Martin, 2005). While ontologies and CDMs share common features, they have some differences. According to Guarino's (1998) proposal of ontology-driven information systems, an ontology can be used at the development or run time of IS, whereas a CDM is a building block of the analysis and design process of an IS.

Moreover Fonseca and Martin (2005) define two criteria that differentiate ontologies from CDMs; the first is the *objectives of modeling* and the second *is objects to model*. Using the first criterion, an ontology focuses on the description of the "invariant features that define the domain of interest," whereas a CDM links the domain invariant features with a set of observations to be defined within an IS. Regarding the second criterion, objects to model, an ontology describes real or factual structures of a domain that enables information integration. Conversely, a CDM object represents a general category of a certain domain linked to its individual events, for example, linking the general category of gene with the size of its DNA sequence. The central question addressed in this research is: *"To what extent can domain ontologies participate in developing CDMs?"*

Having surveyed the literature, the differences between ontology and CDMs have been mainly explored using descriptive studies. Thus, in order to address the main research question, a two-phase approach has been devised to integrate both theoretical and empirical studies. In the first phase, the ontological model provided by Wand and Weber (1993), which is known as the Bunge-Wand-Weber ontology (BWW), has been utilized in interpreting the OWL ontology language. We note that ontology language constructs are related to the structural components of the problem domain.

Other constructs related to time dependency have not been represented in OWL. This result is in line with the observation of Bera and Wand (2004) that OWL concepts can be used to represent multiple BWW concepts. However, Bera and Wand focus on interpreting the basic concepts of OWL (i.e., classes, properties, and individuals), whereas our study is related to OWL constructs such as *owl:class or owl:objectTypeProperty.*

The second phase implements a new algorithm (implemented as a TE component) that maps a domain ontology expressed in OWL to a generated

Figure 2. Excerpt of the GCDM for protein concept

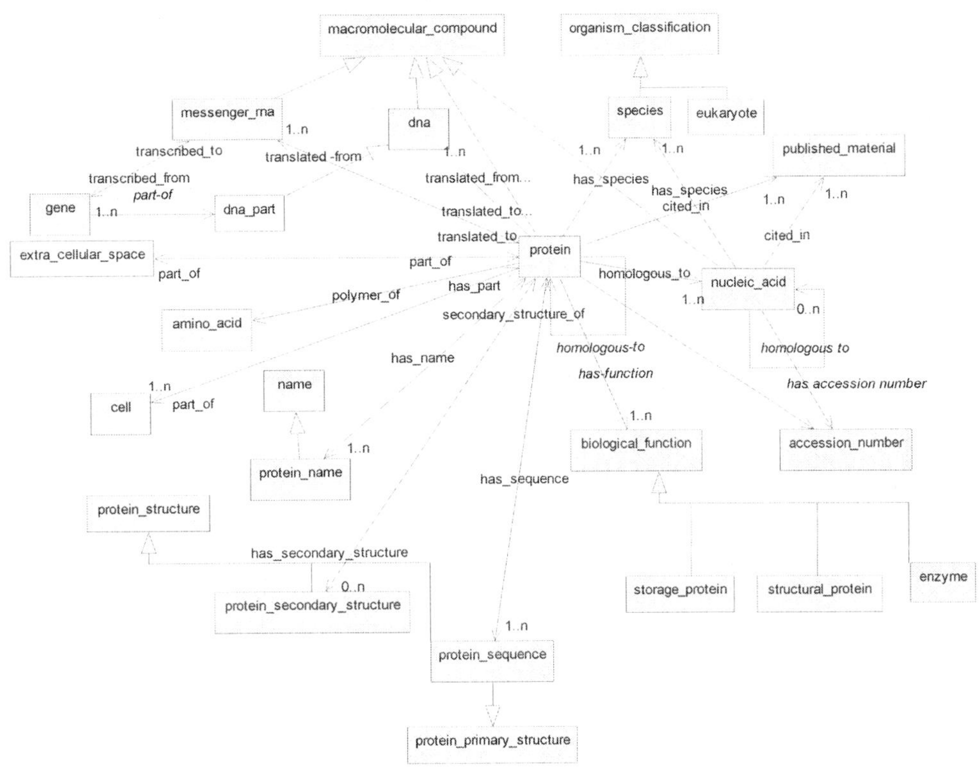

conceptual data model (GCDM) represented as a unified modeling language (UML) class model. The process of developing the CDM begins by selecting an OWL ontology of the domain of interest. Then, the TE applies the mapping rules onto the ontology concepts, thereby generating submodels that are integrated to construct the proposed CDM, as shown in Figure 1.

Briefly the TE mapping rules follow.

Rule 1: The ontology concept or class is mapped into the entity-type construct in the GCDM.

Rule 2: The ontology property is mapped to the relationship construct in the GCDM. In particular, property features such as *owl: inverseOf, owl:functional, owl:domain,* and *owl:range* determine the semantic constraints of the relationships.

Rule 3: The ontology restriction is decom posed to develop a relationship between two entity types, if the related property is a mutual type property. If the filler type of the restriction is a data type, then this relation should be refined to become an attribute of the source entity type.

Rule 4: Using an intrinsic type property, the restriction class is mapped to an attribute of the entity type with a proper data type range.

Rule 5: The subsumption relationship *(rdfs:subClassOf)* between ontology classes is mapped to generalization/specialization relationship between entity types in the GCDM.

Rule 6: The logical expression concept in the ontology language is decomposed into a generalization relationship between the entity types in the GCDM. For example, the *owl:intersectionOf* expression is translated to a multiple inheritance relationship between the operands of the logical expression (as superentity types) and the concept being studied as a subentity type, whereas the *owl:unionOf* expression partitions the concept being studied (i.e., the superentity type) into its operands as subentity types.

Rule 7: A translation of the selected concepts from the domain ontology by the TE is followed by a refinement process of the GCDM (a) by searching for redundant concepts or relationships and removing them, and (b) by merging the same relationships having different cardinalities.

Therefore, to validate the significance of the above adopted approach, we propose a set of measures to evaluate the quality of the GCDM from a given domain ontology using the two prominent works of BWW (Wand & Weber, 1993) and OntoClean (Guarino & Welty, 2002).

THE EVALUATION PROCESS

The evaluation of the TE embodies two components, both qualitative and quantitative methods.

The quantitative dimension proposes a set of measures to evaluate the TE behavior and parameters when applied to different domain ontologies, and these are listed in the numbered sections below.

1. *TE performance* measures the effectiveness of a set of ontological constructs that have been used within the TE mapping algorithm on the GCDM elements.

   To have a quantitative measure of the TE performance in mapping and decomposing the ontology constructs to CDM elements, a straight line regression analysis was used to develop the correlation between ontology constructs (classes, subsumption relation, mutual properties, and intrinsic properties) used in the TE and the GCDMs constructs (entity types, generalization/specialization, relationships, and attributes). The relationships (using R^2) are: 0.999, 0.9981, and 0.9645 for classes versus entity types, subsumption versus generalization/specialization, and mutual properties versus relationships, respectively. This means that on the one hand the TE performance was consistent for the different case studies; therefore, a best fit line can be produced for these constructs. On the other hand, the relationship is poor (0.0762 R^2) for intrinsic properties and attributes. This means that the proportion of the mapping of the attributes in the TE cannot be explained only with the intrinsic properties of the domain ontologies, and there must be some other variable participating in the mapping process. This is because in some domain ontologies intrinsic proprieties are expressed as mutual properties, so the TE refines the mapping of these properties to attribute constructs in the GCDM.
2. *GCDM accuracy* measures the "correct" answers in the GCDM compared to the models developed by human analysts. However, since there is no "gold standard" model for any given application requirements, we have selected a collection of data models, either available in databases texts or provided by the researcher working on different projects to be the gold models (GMs).

   The results of comparing the GCDM by GM show that general knowledge about the domain has been extracted with an overall accuracy of 69% for entity types, 82% for generalization/specialization, and 35% for the relationships. The missing elements in the GCDM can be mainly attributed to modeling the application requirements in the GM that are not expressed in the domain ontologies.

3. *GCDM lexical correctness* measures the "correct" number of lexical names for elements of an ontology and the GCDM, using WordNet (Fellbaum, 1998), a lexical database for English developed at Princeton University. Since most of the terms in ontologies are phrases, we modified phrases, such as "AdministrativeStaff" before searching WordNet. The results of comparing the ontology and GCDM lexical correctness show that there is an overlap in the approaches used in developing a CDM and ontology.

Next, we present the qualitative dimension in evaluating the quality of the GCDM. This criterion addresses the question as to whether the GCDM components conform to the ontological-based rules, provided by philosophical ontologies of conceptual modeling. Consequently, it validates whether the domain ontology provides a proper ontological representation of the respective CDM elements. This will be investigated using a set of ontological rules that merges the BWW ontology and the OntoClean methodology (Guarino & Welty, 2002). The BWW ontology rules are used to validate the accuracy of the ontological meaning of the GCDM elements, whereas the OntoClean axioms are used to evaluate the correctness of the generalization/specialization relationships.

We agree with others that an ontological theory is essential for conceptual modeling, since ontological theories provide conceptual modeling constructs with the semantics of real-world phenomena (Weber, 2003). This impacts the quality of the CDM by reducing the maintenance cost if errors are discovered in the later stages of the software development process (Walrad & Moss, 1993) To describe our proposal, we introduce the main concepts in the BWW ontology followed by an overview of the OntoClean methodology in addition to introducing ontological rules to validate the ontological structure of the GCDM.

Overview of BWW Ontology

Wand and Weber (1993) are among the first researchers who initiated the use of ontology theories in information system analysis and design activity (ISAD). Based on their adaptation of Bunge's ontology, their ontology, the Bunge-Wand-Weber model or BWW, has led to fruitful research areas in ISAD, in general, and in evaluating modeling grammars in particular (Wand, Storey, & Weber, 1999; Guarino & Welty, 2004). For this reason, this ontology is considered as a benchmark ontology for evaluating the expressiveness of modeling languages, since it assists the modeler in constructing ontological CDMs with the maximum semantics about real-world phenomena (Weber, 2003).

In the BWW model, the world is made up of things. A thing can be either simple or composite, where the latter is made up of other things. Composite things possess emergent properties. Things are described by their properties. A property is either intrinsic, depending on only one thing, or mutual, depending on two or more things. A class is a set of things that possesses a common set of characteristic properties. A subclass is a set of things that possess their class properties in addition to other common properties. A natural kind describes a set of things via their common properties and laws connecting them. Properties are restricted by natural or human laws.

The aim of using these concepts is to validate whether the constructs used in the GCDM conform to their ontological meaning or not. For example, what is the proper representation for accession-number? and is it an entity type or an intrinsic property of protein entity type?

Overview of OntoClean

OntoClean is a methodology proposed by Guarino and Welty (2002, 2004) that is based on the philosophical notions for evaluating taxonomical structures. OntoClean mainly constitutes two major building blocks: (1) a set of constraints that formalizes the correctness of the subsumption relationship, and (2) an assignment of the top level unary predicates (concepts) of the taxonomical structure to a number of metaproperties. The four fundamental ontological notions of *rigidity*, *unity*, *identity*, and *dependence* are attached as metaproperties to concepts or classes in a taxonomy structure describing the behavior of the concepts, i.e., these metaproperties clarify the way subsumption is used to model a domain by imposing some constraints (Evermann & Wand, 2001). We briefly and informally introduce these ontological/ philosophical notions.

1. *Rigidity* is based on the idea of an essential property that must hold for all instances of a concept or a class. Thus, a class or concept is rigid (+R) if it holds the essential property for all its instances. The nonrigid concept (-R) holds a property that is not essential to the entire concept instances, however it is necessary for some of the instances. The antirigid (~R) concept holds a property that is optional for all concept instances.
2. The notion of *identity* is concerned with recognizing a common property that identifies the individuals of a concept as being the same or different; and it is known as an identity condition or characteristics property in the philosophical literature. The identity metaproperty (+I) supplies or carries this property. If the class supplies this property then all subclasses carry it as an inherited property. On the contrary, if the concept does not provide the identity condition, then it will be marked with (-I).
3. *Unity* is defined if there is a common unity condition such that all the individuals are intrinsic wholes (+U). A class carries anti-unity (-U) if all its individuals can possibly be nonwholes.
4. *Dependence* (+D) is based on whether the existence of an individual is externally dependent on the existence of another individual with **(-D),** otherwise.

OntoClean classifies concepts into categories based on three metaproperties: identity, rigidity, and dependence. The basic categories are: type category, which describes (+R, ±D, +I); phased-sortal category, which describes (~R, -D, +I); role category, which describes (~R, +D, +I); and attribution category, which describes (-R, ±D, -I).

Also, the OntoClean methodology restricts the correctness of a given taxonomical structure by a set of axioms. The axioms related to identity, rigidity, and dependence metaproperties are:

1. An anti-rigid class cannot subsume rigid class;
2. A class that supplies or carries an identity property cannot subsume a class that does not hold this property;
3. A dependent class cannot subsume an independent class.

Merging OntoClean and BWW to Evaluate the GCDM

As a result of utilizing the BWW ontology for evaluating the expressiveness of different conceptual modeling languages, a set of rules are proposed as a theory of conceptual modeling practice. For example, Wand et al. (1999) derive a set of rules as a theory of constructing the relationships in conceptual modeling practice. Moreover, Evermann and Wand (2001) investigated the mapping between ontological constructs and UML elements; and this led them to suggest modeling rules, in general, and guidelines on how to use UML elements to model real-world systems, in particular.

In our approach, we utilize a set of these general rules in evaluating the quality of the GCDM. However, we suggest that the integration of these rules with the OntoClean methodology would improve the quality of the GCDM, especially in the generalization/specialization relationships. Therefore, the evaluation process has to prove the ontological appropriateness in representing the GCDM elements.

We have to mention here that the integration between different ontologies has been used recently by different researchers but for different purposes. Their purpose is to evaluate and develop an ontological UML and conceptual modeling language. For example, Guizzardi, Wagner, Guarino, & Sinderen (2004) use the general ontology language (GOL) and its underling upper-level ontology in evaluating the ontological correctness of the UML class model. Their approach is influenced by the OntoClean methodology, in addition to the psychological claims proposed by the cognitive psychologist John McNamara (1994). Also, Li (2005) studies the use of the Bunge ontology with the OntoClean methodology for the same objective. In our research, we integrate these prominent ontologies to evaluate the quality of the GCDM by studying the appropriateness of its ontological meanings. In the following sections, we propose a set of ontological rules inspired by BWW and OntoClean (Wand et al., 1999; Evermann & Wand, 2001; Guarino & Welty, 2002, 2004) in order to check the quality of the GCDM elements.

Rule 1: The BWW ontology models only substantial things in the world as entities, that is properties (attributes or relationships) or events cannot be modeled using entity type constructs. According to OntoClean, substantial things are recognized by their identity condition or characteristics property; therefore, substantial entities belong to type, phased-sortal, or role categories.

Rule 2: BWW's intrinsic properties are represented as attributes of an entity type that describe a property of one thing independent of any other entities. Therefore, the BWW property cannot be represented using an entity type construct. According to OntoClean, an attribute of an entity type is assigned (-R, -I, ±D) metaproperties.

Rule 3: Any BWW mutual property is represented as a relationship between two or more substantial things; therefore, it prescribes representing entity types as a mutual property.

Rule 4: A BWW aggregate or composite entity type must have emergent properties in addition to those of its components types; therefore, a composite thing should be recognized with an identity characteristic. Whilst a simple thing is composed of one thing, a composite or aggregate thing is made up of two or more things.

Rule 5: In the BWW ontology, a specialized entity type must define more properties than the general entity type. According to OntoClean, entities are recognized by their identity characteristics. In addition, the generalization/specialization relationship must conform to the OntoClean taxonomical structure axioms.

Applying the Evaluation Methodology Using the TAMBIS Ontology

The TAMBIS ontology contains knowledge about bioinformatics and molecular biology concepts and their relationships. It describes proteins and enzymes, as well as their motifs, and secondary and tertiary structure functions and processes (Goble et al., 2001). We use the TAMBIS ontology (TAO) to demonstrate our approach. TAO has 393 concepts and 94 properties, whereas the GCDM has 392 entity types, 259 relationships, 49 attributes, and 402 generalization/specialization relationships. In this case study, we have selected the concepts that are relevant to protein in order to generate the CDM using the TE. The GCDM has been translated to a set of Java files and reverse engineered to a class diagram by using a UML graphical tool.

In what follows, we present our observations of the GCDM (shown in Figure 2) with respect to the proposed ontological rules.

1. According to ontological Rule 1, protein structure and biological function are not substantial entities, since they do not have any identity property. Therefore, these concepts should not be represented as entity types. Protein structure is an intrinsic property that can be used in classifying protein type according to its internal structure, whereas a biological function can be used in classifying protein types according to their role with other existing entities.
2. According to OntoClean, protein name and accession number are assigned (-I-R+D) metasemantics, which means that these elements belong to the attribution category. By using ontological Rule 2, protein name and accession number are intrinsic properties that describe protein independent from any other entities; therefore, they cannot be represented as entity types, according to ontological Rule 1.
3. We consider an individual protein as a macromolecule of amino acid sequences linked by a peptide bond. We assume that these large molecules have their own essential properties, and their existence is independent of any other concepts. Therefore, +R+I-D seems to be an obvious assignment that classifies them as type category. The structure of a protein is considered as an intrinsic property that classifies proteins according to their internal structure. The primary structure or primary sequence is a linear sequence of amino acids; secondary structure involves the hydrogen bond that forms the alpha helix, beta sheet, and others; the tertiary structure is the three-dimensional structure of the molecule that consists of the secondary structure linked by covalent disulfide bonds and noncovalent bonds; and the quaternary structure is the association of separate polypeptide chains into the functional protein. Hence, each structure of a protein belongs to the phased-sortal category, since this classifier type allows an instance to change certain intrinsic properties while remaining the same entity. Also, according to the OntoClean axioms, the generalization/specialization relation between protein and its different structures is correct.
4. The function of the protein can be used as a classification property that classifies proteins according to their role with other existing entities.
   Therefore, proteins can be classified according to their functions into, for example, enzyme, storage protein, and structural protein. Here we have to mention that these subentity types belong to the role category, since their existence depends on other entities, for example, each enzyme is catalyzed by one reaction.
5. DNA and RNA are polymers of many nucleic acids (adenine, guanine, thymine, or cytosine in DNA; adenine, guanine, uracil, or cytosine in RNA), whereas a protein is a large complex molecule made up of one or more chains of amino acids. According to this, we propose that the macromolecular-compound type specifies two types of compounds: a compound based on nucleic-acid blocks and a compound based on amino-acid blocks. In this case, nucleic-acids and amino-acid compounds belong to the type category with (+R+I-D). Also, DNA and m-RNA are subtypes of nucleic-acid compound, whereas protein is a subtype of amino-acid compound.
6. We propose to replace the species type with the prokaryote type, since species type also can be classified into prokaryote and eukaryote.

Figure 3. The refined model of the protein concepts using the ontological rules

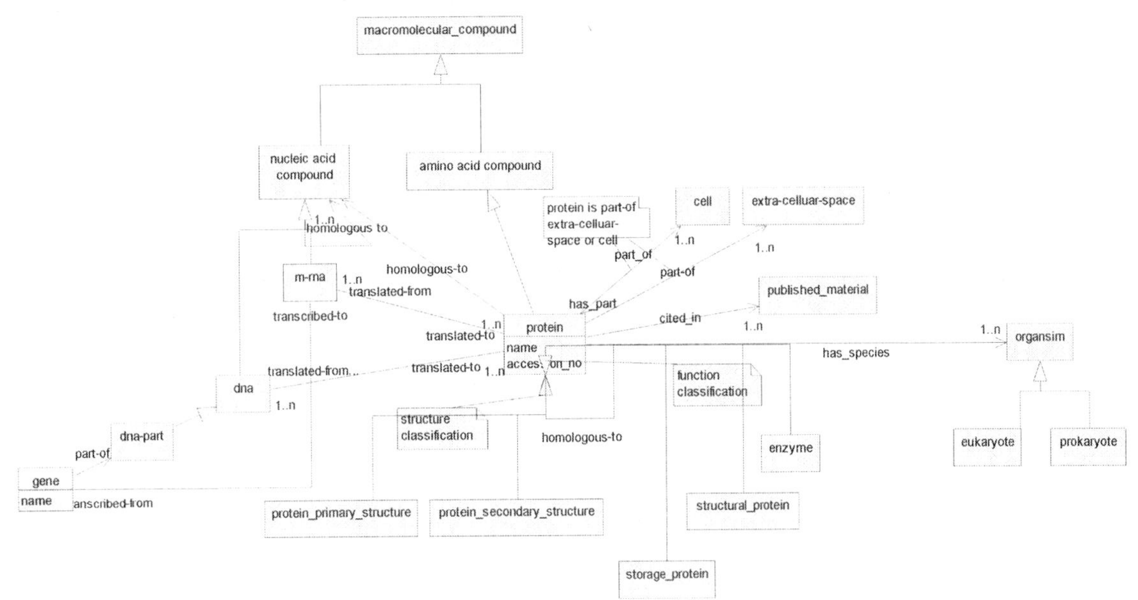

Figure 3 presents the refined GCDM resulting from the application of the ontological rules. In addition it has been approved by a domain expert. However, evaluating the GCDM elements using the ontological rules leads to the following observations.

Firstly the TE achieves good agreement in automating the CDM development activity. This means that the invariant information about the domain can be extracted to a certain extent from the domain ontologies. In other words, the GCDM provides a high degree of accuracy in identifying the substantial entities along with their attributes and relationships. Therefore, the semantics of the CDM elements conform to the consensual knowledge about the interested domain.

Secondly there are some ontological issues that could not be built into the TE. For example, applying Rule 1 in the TE onto all named classes in a domain ontology results in generating some entity types in the GCDM that lack the existence of the identity criteria that is considered an essential property for representing substantial things. For example, accession number and protein name are expressed as named classes in the ontology; thereby, they are mapped onto entity types using Rule 1, where these concepts are better mapped to attributes of protein type. Furthermore, the misinterpretation of Rule 1 in the TE for some ontology classes reflects on the rest of the rules in the TE. For example, Rule 5 in the TE is used to generate the generalization/specialization relation between biological function as a general entity type and enzyme, structural protein, and storage protein as subentity types. And according to the ontological rules, biological function is a mutual property that describes the role of protein kinds (enzyme, structural protein, and storage protein), depending on the existence of other entity types (i.e., the existence of an enzyme depends on the existence of a reaction). Therefore, the misinterpretation of biological function as an entity type leads to misinterpreting the generalization/specialization relationship.

This observation stems from the fact that, OWL class constructs are overloaded to represent all real-world characteristics (i.e., dynamic and static characteristics). This is the same construct used to represent a domain concept, event, process, or transformation. To overcome this problem, we propose extending the ontology language by adding more semantics to the class construct (metaconcept), in order to describe the nature of the represented real-world phenomena. Therefore, the "static" metaconcept represents domain concepts that identify and support the identity property of an entity type (i.e., substantial entities), whereas the "dynamic" metaconcept represents an event or transformation concept that captures the behavior of a given real-world phenomena.

CONCLUSION AND FUTURE WORK

The similarities and differences between ontologies and CDMs led us to study the possibility of engineering CDMs from domain ontologies. In this regard, a new approach has been developed to automate the derivation of CDMs from domain ontologies. The theoretical ontology of BWW has been used to interpret the OWL constructs, which have contributed to the development of a mapping algorithm to generate a CDM from the given domain ontology. An important aspect of this approach is that it accelerates the development of the CDM from an explicit and consensual knowledge model. In addition, a set of measures have been established to evaluate the capabilities and the effectiveness of this approach. The proposed measures in the quantitative dimension reveal that: (1) there is a strong correlation between the ontology and CDM constructs; (2) the domain ontology describes the invariant knowledge about the domain; and hence the development of the GCDM elements, such as entity types, and that their relationships are independent of any application requirements; and (3) the development process of ontologies and CDMs conforms to the same lexical rules for naming their elements. In this work, a set of ontological rules, derived from the BWW and OntoClean ontologies, have been applied to serve the qualitative evaluation of the GCDM. The TAMBIS ontology has been used as the test case, and results have shown that the GCDM provides a high degree of accuracy in identifying the substantial entities along with their attributes and relationships.

However, to improve the quality of the GCDM, we suggest extending the definition of the class construct to incorporate a metaconcept element to distinguish between concepts related to the identity property and concepts representing the events and transformations of a given domain. Furthermore, as the functionality of the TE is restricted to decomposing and mapping domain ontology constructs to CDM constructs, the metaproperties of the OntoClean ontology can be used to validate the ontological correctness of the subsumption relations in the given domain ontology. This must be implemented as a part of the TE mapping algorithm, in order to improve the ontological appropriateness of the GCDM elements.

While this research has been focused on using one domain ontology to generate a possible and relatively appropriate CDM, further work needs to consider the possibility of using more than one related domain ontology to enable the development of a hybrid CDM, such as enterprise data models. This may even suggest enriching the process of ontology development with theoretical ontologies for the improved engineering of domain ontologies and, hence, CDMs.

REFERENCES

Bera, P., & Wand, Y. (2004). Analyzing OWL using a philosophy-based ontology. In *Proceedings of the 2004 Conference on Formal Ontologies in Information Systems FOIS* (pp. 353-362) .

El-Ghalayini, H., Odeh, M., McClatchey, R., & Solomonides, T. (2005). Reverse engineering domain ontology to conceptual data models. In *Proceedings of the 23rd IASTED International Conference on Databases and Applications (DBA), Innsbruck, Austria* (pp. 222-227).

Evermann, J., & Wand, Y. (2001). Towards ontologically-based semantics for UML constructs. In H. S. Kunii, S. Jajodia, & A. Solvberg (Eds.), *Conceptual modeling—ER* (LNCS no. 2224) (pp. 341-354). Springer.

Fellbaum, C. (1998). *WordNet: An electronic lexical database*. MIT Press.

Fonseca, F. and Martin, J. (2005). Learning the differences between ontologies and conceptual schemas through ontology-driven information Systems To appear in *a Journal of the Association of Information Systems Special Issue on Ontologies in the Context of Information Systems*. (fredfonseca@ist.psu.edu).

Goble, C. A., Stevens, R., Ng, G., Bechhofer, S., Paton, N. W., Baker, P. G., Peim, M., & Brass, A. (2001). Transparent access to multiple bioinformatics information sources. *IBM System Journal, 40*(2), 532-551.

Gruber, T. (1995). Toward principles for the design of ontologies used for knowledge sharing. *International Journal of Human and Computer Studies, 43*(5/6), 907-928.

Guarino, N. (1998). *Formal ontology and information systems.* Amsterdam, Netherlands: IOS Press.

Guarino, N., & Welty, C. (2002). Evaluating ontological decisions with OntoClean. *Communications of the ACM, 45*(2), 61-65.

Guarino, N., & Welty, C. (2004). An overview of OntoClean. In S. Staab & R. Studer (Eds.), *Handbook on ontologies* (pp. 51-159). Springer Verlag.

Guizzardi, G., Wagner, G., Guarino, N., & Sinderen, M. (2004). An ontologically well-founded profile for UML conceptual models. In *Proceeding of the 16th Conference on Advanced Information Systems Engineering- CAiSE04* (pp. 112-126).

Hischheim, R., Klein, H., & Lyytinen, K. (1995). *Information systems development and data modelling: Conceptual and philosophical foundations.* Cambridge: Cambridge University Press.

Li, X. (2005). Using UML in conceptual modelling: Towards an ontological xore. In *Proceeding of the 17th Conference on Advanced Information Systems Engineering-CAiSE05* (pp. 13-17).

McNamara, J. (1994). Logic and cognition. In J. McNamara & G. Reyes (Eds.), *The logical foundations of cognition* (Vo. 4) (pp.). Vancouver Studies in Cognitive Science.

Mylopoulos, J. (1998). Information modeling in the time of the revolution. *Information Systems, 23*(3/4), 127-155.

W3C-World Wide Web Consortium. (2005). OWL 1.1 Web Ontology Language Syntax. In P. F. Patel-Schneider (ed.). Retrieved May 2005 from from http://www-db.research.bell-labs.com/user/pfps/owl/syntax.html.

Walrad, C., & Moss, E. (1993). Measurement: The key to application development quality. *IBM Systems Journal, 32*(3), 445-460.

Wand, Y, Storey, V., & Weber, R. (1999). An ontological analysis of the relationship construct in conceptual modeling. *ACM Transactions on Database Systems, 24*(2), 494-528.

Wand, Y., & Weber, R. (1993). On the ontological expressiveness of information systems analysis and design grammars. *Journal of Information Systems, 3*(4), 217-237.

Weber, R. (2003). Conceptual modelling and ontology: Possibilities and pitfalls. *Database Management, 14*(3), 1-2

This work was previously published in International Journal of Information Technology and Web Engineering, Vol. 2, Issue 1, edited by G.I. Alkhatib and D.C. Rine, pp. 57-70, copyright 2007 by IGI Publishing, formerly known as Idea Group Publishing (an imprint of IGI Global).

Chapter 2.30
Data Mining of Bayesian Network Structure Using a Semantic Genetic Algorithm-Based Approach

Sachin Shetty
Old Dominion University, USA

Min Song
Old Dominion University, USA

Mansoor Alam
University of Toledo, USA

ABSTRACT

A Bayesian network model is a popular formalism for data mining due to its intuitive interpretation. This chapter presents a semantic genetic algorithm (SGA) to learn the best Bayesian network structure from a database. SGA builds on recent advances in the field and focuses on the generation of initial population, crossover, and mutation operators. In SGA, we introduce semantic crossover and mutation operators to aid in obtaining accurate solutions. The crossover and mutation operators incorporate the semantic of Bayesian network structures to learn the structure with very minimal errors. SGA has been proven to discover Bayesian networks with greater accuracy than existing classical genetic algorithms. We present empirical results to prove the accuracy of SGA in predicting the Bayesian network structures.

INTRODUCTION

One of the most important steps in data mining is building a descriptive model of the database being mined. To do so, probability-based approaches have been considered an effective tool because of the uncertain nature of descriptive models. Unfortunately, high computational requirements and the lack of proper representation have hindered the building of probabilistic models. To alleviate the above twin problems, probabilistic graphical models have been proposed. In the past decade, many variants of probabilistic graphical models have been developed, with the simplest variant being Bayesian networks (BN) (Pearl, 1988). BN is a popular descriptive modeling technique for available data by giving an easily understandable way to see relationships between attributes of a set of records. It has been employed to reason under uncertainty, with wide varying applications in the field of medicine, finance, and military planning (Pearl, 1988; Jensen, 1996). Computationally, BN provides an efficient way to represent relationships between attributes and allow reasonably fast inference of probabilities. Learning BN from raw data can be viewed as an optimization problem where a BN has to be found that best represents the probability distribution that has generated the data in a given database (Heckerman, Geiger, & Chickering, 1995). This has lately been the subject of considerable research because the traditional designer of a BN may not be able to see all of the relationships between the attributes. In this chapter, we focus on the structure learning of a BN from a complete database. The database stores the statistical values of the variables as well as the conditional dependence relationship among the variables. We employ a genetic algorithm technique to learn the structure of BN.

A typical genetic algorithm works with populations of individuals, each of which needs to be coded using a *representative function* and be evaluated using a *fitness function* to measure the adaptiveness of each individual. These two functions are the basic building blocks of a genetic algorithm. To actually perform the algorithm, three genetic operators are used to explore the set of solutions: *reproduction*, *mutation*, and *crossover*. The reproduction operator promotes the best individual structures to the next generation. That is, the individual with the highest fitness in a population will reproduce with a highest probability than the one with the lowest fitness. The mutation operator toggles a position in the symbolic representation of the potential solutions. Mutation avoids local optima by exploring new solutions by introducing a variation in the population. The crossover operator exchanges genetic material to generate new individuals by selecting a point where pieces of parents are swapped. The main parameters, which influence the genetic algorithm search process, are initial population, population size, mutation, and crossover operators.

In this chapter we first introduce the related work in BN structure learning and present the details of our approach for structure learning in a BN structure using a modified genetic algorithm. Then we experiment with two different genetic algorithms. The first one is the genetic algorithm with classical genetic operatiors. In the second algorithm, we extend the standard mutation and crossover operators to incorporate the semantic of the BN structures. Finally, we conclude the chapter and proposes some thoughts for futher resarch.

RELATED WORK

Larranaga, Kuijpers, Murga, and Yurramendi (1996) proposed a genetic algorithm based on the score-based greedy algorithm. In their algorithm, a directed acyclic graph (DAG) is represented by a connectivity matrix that is stored as a string. The recombination is implemented as one-point crossover on these strings, while mutation is implemented as random bit flipping. In a related work, Larranaga, Poza, Yurramendi, Murga, and Kuijpers (1996) employed a wrapper approach by implementing a genetic algorithm that searches for an ordering that is passed on to K2 (Cooper & Herskovits, 1992), a score-based greedy learning algorithm. The results of the wrapper approach were comparable to those of their previous genetic algorithms. Different crossover operators have been implemented in a genetic algorithm to increase the adaptiveness of the learning problem, with good results (Cotta & Muruzabal, 2002). Lam and Bacchus (1994) proposed a hybrid evolutionary programming (HEP) algorithm that combines the use of independence tests with a quality-based search. In the HEP algorithm, the search space of DAG is constrained in the sense that each possible DAG only connects two nodes if they show a strong dependence in the available data. The HEP algorithm evolves a population of DAG to find a solution that minimizes the minimal description length (MDL) score. A common feature of the aforementioned algorithms is that the mutation and crossover operators were classical in nature. These operators do not help the evolution process reach the best solution.

Wong, Lam, and Leung (1999) developed an approach based on MDL score and evolutionary programming. They have integrated a knowledge-guided genetic operator for optimization in the search process. However, the fitness function is not taken into account to guide the search process. Myers and Levitt (1999) have proposed an adaptive mutation operator for learning structure of BN from incomplete data. It is a generalized approach to influence the current recombination process based on previous population. It does not take into account the fitness of a population either. Blanco, Inza, and Larranaga (2003) have adopted the estimation of distribution algorithms method for learning BN without the use of crossover and mutation operators. This is not in accordance with the classical genetic algorithm due to the lack of recombination operators. Recently, Dijk et al. (2003) built another generalized genetic algorithm to improve the search process without taking into account the specific characteristics of the population. As we see, most of the genetic algorithm-based approaches mentioned above adopt a generalized approach to improve the search process. The mutation and crossover operators proposed in this chapter are semantically oriented and thus they aid in a better convergence to the solution. Hence, our BN structure learning algorithm differs from the above algorithms in the design of mutation and crossover operators.

SEMANTIC GENETIC ALGORITHM-BASED APPROACH

Structure Learning of Bayesian Networks

Formally, a BN consists of a set of nodes that represent variables, and a set of directed edges between the nodes. Each node is featured by a finite set of mutually exclusive states. The directed edges between nodes represent the dependence between the linked variables. The strengths of the relationships between the variables are expressed as conditional probability tables (CPT). Thus, a BN efficiently encodes the joint probability distribution of its variables. For n-dimensional random variable $(X_1, \ldots, X_n)$, the joint probability distribution is determined as follows:

$$P(x_1, \ldots, x_n) = \prod_{i=1}^{n} P(x_i \mid pa(x_i)) \qquad (1)$$

where x_i represents the value of the random variable X_i and $pa(x_i)$ represents the value of the parents of X_i. Thus, the structure learning problem of a BN is equivalent to the problem of searching the optimum in the space of all DAG. During the search process, a trade-off between the structural network complexity and the network accuracy has to be made. The trade-off is necessary as complex networks suffer from over fitting, making the run time of inference very long. A popular measure to balance complexity and accuracy is based on the principle of MDL from information theory (Lam & Bacchus, 1994). In this chapter, the BN structure learning problem is solved by searching for a DAG that minimizes the MDL score.

Representative Function and Fitness Function

The first task in a genetic algorithm is the representation of initial population. To represent a BN as a genetic algorithm individual, an edge matrix or adjacency matrix is needed. The set of network structures for a specific database characterized by n variables can be represented by an $n \times n$ connectivity matrix C. Each bit represents the edge between two nodes where

$$C_{ij} = \begin{cases} 1, & \text{if } j \text{ is a parent of } i \\ 0, & \text{otherwise} \end{cases}.$$

The two-dimensional array of bits can be represented as an individual of the population by the following string $C_{11}\, C_{12} \ldots C_{1n} C_{21} C_{22} \ldots C_{2n} \ldots C_{n1}\, C_{n2} \ldots C_{nn}$, where the first n bits represent the edges to the first node of the network, and so on. It can be easily found that C_{kk} are the irrelevant bits which represent an edge from node k to itself, which can be ignored by the search process.

With the representative function decided, we need to devise the generation of the initial population. There are several approaches to generate initial population. We implemented the Box-Muller random number generator to select how many parents would be chosen for each individual node. The parameters for the Box-Muller algorithm are the desired average and standard deviation. Based on these two input parameters, the algorithm generates a number that fits the distribution. For our implementation, the average corresponds to the average number of parents for each node in the resultant BN. After considerable experimentation with databases whose Bayesian structure is similar to the ASIA network (Lauritzen & Spiegelhalter, 1988), we found that the best average was 1.0 with a standard deviation of 0.5. Although this approach is simple, it creates numerous illegal DAG due to cyclic subnetworks. An algorithm to remove or fix these cyclic structures has to be designed. The basic operation of the algorithm is to remove a random edge of a cycle until cycles are not found in a DAG individual.

Now that the representative function and the population generation have been decided, we need to find a good fitness function. Most of the *state-of-the-art* implementations use the fitness function proposed in the algorithm K2 (Cooper & Herskovits, 1992). The K2 algorithm assumes an ordered list of variables as its input. It maximizes the following function by searching for every node from the ordered list of a set of parent nodes:

$$g(x_i, pa(x_i)) = \prod_{j=1}^{q_i} \frac{(r_i - 1)!}{(N_{ij} + r_i - 1)!} \prod_{k=1}^{r_i} N_{ijk}! \quad (2)$$

where r_i represents the possible value assignments $(v_{i1}, \ldots, v_{ir_i})$ for the variable with index i, N_{ijk} representing the number of instances in a database in which a variable X_i has value v_{ik}, and q_i represents the number of unique instantiations of $pa(x_i)$.

Mutation and Crossover Operators

We introduce two new operators, semantic mutation (SM) and single point semantic crossover (SPSC), to the existing standard mutation and crossover operators. The SM operator is a heuristic operator that toggles the bit value of a position in the edge matrix to ensure that the fitness function $g(x_i, pa(x_i))$ is maximized. The SPSC operator is specific to our representation function. As the function is a two-dimensional edge matrix consisting of columns and rows, our new crossover operator operates on either columns or rows. Thus, the crossover operator generates two offspring by either manipulating columns or rows. The SPSC crosses two parents by manipulating columns or parents and maximizing the function $g(x_i, pa(x_i))$, and b) manipulating rows or children and maximizing the function $\prod_i g(x_i, pa(x_i))$.

By combining SM and SPSC, we implement our new genetic algorithm called semantic genetic algorithm (SGA). Following is the pseudo code for the semantic crossover operation. The algorithm expects an individual as input and returns the modified individual after applying semantic crossover operations. (see Box 1)

Box 1. Pseudo code for semantic crossover

```
Step 1. Initialization
        Read the input individual and populate a parent table
        for each node
Step 2  Generate new individual
        For each node in the individual do the following n
        times
        2.1    Execute the Box Mueller algorithm to find how
               many parents need to be altered.
        2.2    Ensure that the nodes selected as parents do not
               form cycles. If cycles are formed repeat step 2.1
        2.3    Evaluate the network score of the resultant
               structure.
        2.4    If current score is higher than previous score,
               then the chosen parents are the new parents of
               the selected node
        Repeat steps 2.1 through 2.4.
Step 3. Return the final modified individual.
```

SIMULATIONS

SGA Implementation

The SGA algorithm has been implemented and incorporated into the Bayesian network tools in Java (BNJ) (*http://bnj.sourceforge.net*). BNJ is an open-source suite of software tool for research and development using graphical models of probability. Specifically, SGA is implemented as a separate module using the BNJ API. To depict the Bayesian network, BNJ visually provides a visualization tool to create and edit Bayesian networks.

Simulation Methodology

Figure 1 shows the overall simulation setup to evaluate our genetic algorithm. Following are the main steps of the algorithm:

1. Determine a BN and simulate it using a probabilistic logic sampling technique (Henrion, 1988) to obtain a database *D*, which reflects the conditional relations between the variables;
2. Apply our SGA approach to obtain the BN structure B_s, which maximizes the probability $P(D \mid B_s)$; and
3. Evaluate the fitness of the solutions.

Simulations and Analysis

The BN sizes used in our simulations are 8, 12, 18, 24, 30, and 36. The 8-node BN used in the simulations is from the ASIA networks (Lauritzen & Spiegelhalter, 1988) as shown in Figure 2. The ASIA network illustrates their method of propagation of evidence, and considers a small amount of fictitious qualitative medical knowledge. The remaining networks were created by adding extra nodes to the basic ASIA network.

There are several techniques for simulating BN. For our experiments we have adopted the probabilistic logic sampling technique. In this

Figure 1. Simulation setup for learning Bayesian network structure

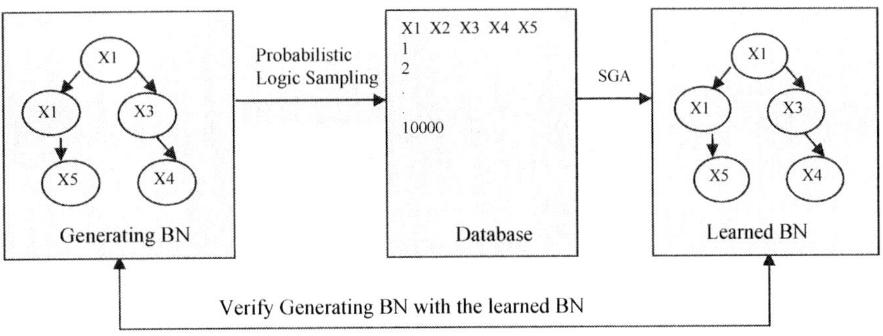

Figure 2. The structure of the ASIA network

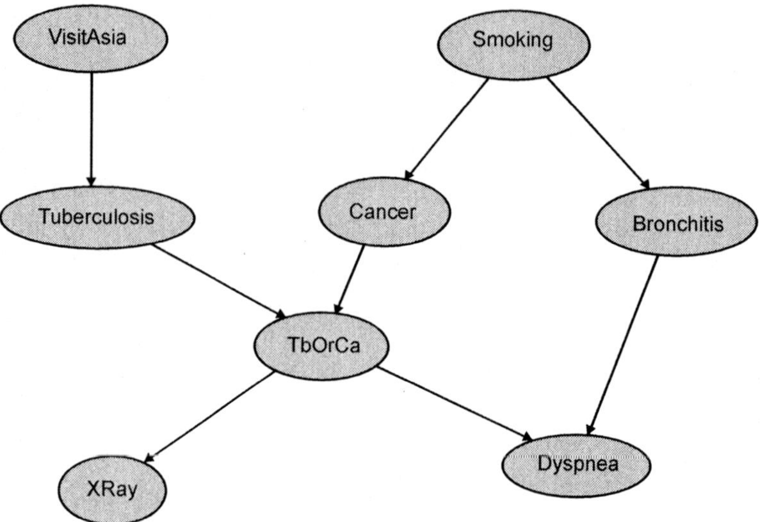

technique, the data generator generates random samples based on the ASIA network's joint probability distribution table. The data generator sorts nodes topologically and picks a value for each root node using the probability distribution, and then generates values for each child node according to its parent's values in the joint probability table. The root mean square error (RMSE) of the data generated compared to the ASIA network is approximately zero. This indicates that the data was generated correctly. We have populated the database with 2000, 3000, 5000, and 10,000 records. This was done to measure the effectiveness of the learning algorithm for a broad range of information sizes. The following input is used in the simulations:

- Population size λ. The experiments have been carried out with λ = 100.
- Crossover probability p_c we chose $p_c = 0.9$.
- Mutation rate p_m we considered $p_m = 0.1$.

The fitness function used by our algorithm is based on the formula proposed by Cooper and Herskovits (1992). For each of the samples (2000, 3000, 5000, 10000), we executed 10 runs with each of the above parameter combinations. We considered the following four metrics to evaluate the behavior of our algorithm.

- **Average fitness value:** This is an average of fitness function values over 10 runs.
- **Best fitness value:** This value corresponds to the best fitness value throughout the evolution of the genetic algorithm.
- **Average graph errors:** This represents the average of the graph errors between the best

Figure 3. Plot of generations vs. average fitness values (10000 Records)

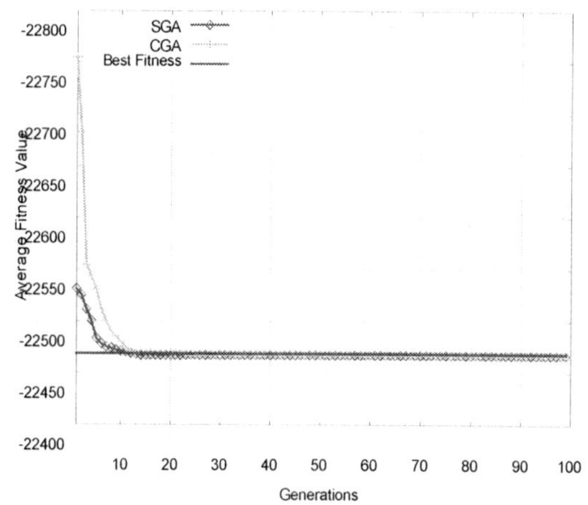

BN structure found in each search, and the initial BN structure. Graph errors are defined to be an addition, a deletion, or a reversal of an edge.
- **Average number of generations:** This represents the number of generations taken to find the best fitness function.

For comparison purposes, we also implemented the classical genetic algorithm (CGA) with classical mutation (CM) and single point cyclic crossover (SPCC) operators. Figure 3 plots the average fitness values for the following parameter combination. The average and best fitness values are expressed in terms of log $P(D \mid B_s)$. The numbers of records are 10,000. The figure also shows the best fitness value for the whole evolution process. One can see that SGA performs better than CGA in the initial 15-20 generations. After 15-20 generations, the genetic algorithm using both operators stabilizes to a common fitness value. The final fitness value is

Figure 4. Learned BN after 100 generations for 5,000 records - graph errors = 3

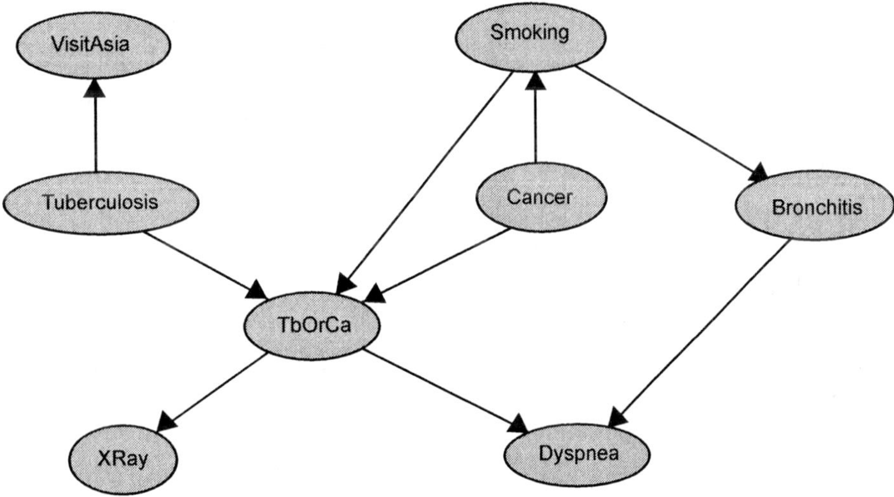

Figure 5. Learned BN after 100 generations for 10,000 records - graph errors = 2

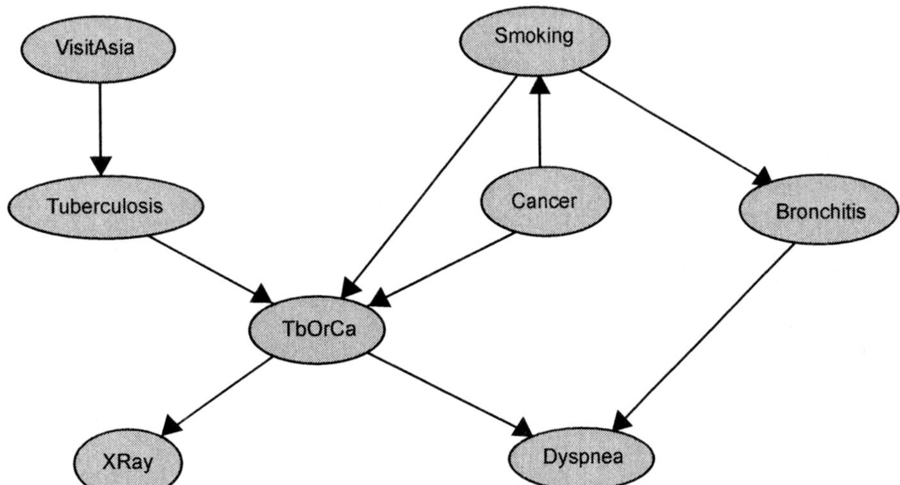

very close to the best fitness value. An important observation is that the average fitness value does not deviate by any significant amount even after 100 generations. The best fitness value is carried over to every generation and is not affected.

The final learned BN was constructed from the final individual generated after 100 generations. Figures 4 and 5 plot the final learned BN for 5,000 records and 10,000 records, respectively. It can be observed that for both the scenarios, the learned BN differs from the actual generating BN shown in Figure 2 by a small number of graph errors. It is also worth noting that the numbers of graph errors reduce when the total numbers of records increase. This could mean that to reduce the total number of graph errors, a large number of records need to be provided.

Tables 1 and 2 provide the average number of generations and the average graph errors for a different number of records. It is obvious that for 2000 records, the total number of generations taken to achieve the stabilized fitness value is very high. Also, the average number of graph errors is too high. For the 3,000, 5,000, and 10,000 records, the values for these metrics are reasonable and acceptable.

To compare the performance of SGA with CGA in the presence of larger BN structures, we modified the 8-node ASIA network and generated five additional BN with node sizes 12, 18, 24, 30, and 36. Tables 3-7 show results for simulations carried out on these additional BNs. The tables compare the average graph errors in both approaches. The accuracy of SGA does not deteriorate under increased network sizes.

CONCLUSION AND FUTURE WORK

In this chapter, we have presented a new semantic genetic algorithm (SGA) for BN structure learning. This algorithm is another effective contribution to the list of structure learning algorithms. Our results show that SGA discovers BN structures

Table 1. Average number of generations

Records	SGA	CGA
3000	25	30
5000	20	15
10000	20	15

Table 2. Average graph errors for 8-node

Records	SGA	CGA
3000	3	4
5000	2	3
10000	2	3

Table 3. Average graph errors for 12-node

Records	SGA	CGA
3000	21	28
5000	24	29
10000	20	25

Table 4. Average graph errors for 18-node

Records	SGA	CGA
3000	19	24
5000	19.5	25.5
10000	20.7	26

Table 5. Average graph errors for 24-node

Records	SGA	CGA
3000	14.8	22.2
5000	15.3	23.1
10000	10.9	13.3

Table 6. Average graph errors for 30-node

Records	SGA	CGA
3000	15.7	21.5
5000	14.9	20.3
10000	14	18.9

Table 7. Average graph errors for 36-node

Records	SGA	CGA
3000	15.1	19.4
5000	15.6	20.5
10000	13.6	22.5

with a greater accuracy than existing classical genetic algorithms. Moreover, for large network sizes, the accuracy of SGA does not degrade and this accuracy improvement does not come with an increase of search space. In all our simulations, 100 to 150 individuals are used in each of the 100 generations. Thus 10,000 to 15,000 networks are completely searched to learn the BN structure. Considering that the exhaustive search space is of 2^{n^2} networks, only a small percentage of the entire search space is needed by our algorithm to learn the BN structure.

One aspect for future work is to change the current random generation of adjacency matrices for the initial population generation. The second is to improve scalability by implementing our genetic algorithm on a distributed platform. We plan to adopt the island model (Tanese, 1989) of computation for implementing the distributed genetic algorithm. The novel aspect of our future work would be to propose a distributed genetic algorithm for a peer-to-peer networking environment. The proposed algorithm would combine the island model of computation with epidemic communications (Birman, Hayden, Ozkasap, Xiao, Budiu, & Minsky, 1999). The epidemic communication paradigm would be adapted to allow the algorithm to be implemented on scalable platforms that would also include fault-tolerance at different levels.

REFERENCES

Birman, P., Hayden, M., Ozkasap, O., Xiao, Z., Budiu, M., & Minsky, Y. (1999). Bimodal multicast. *ACM Transactions on Computer Systems*, *17*(2), 41-88.

Blanco, R., Inza, I., & Larrañaga, P. (2003). Learning Bayesian networks in the space of structures by estimation of distribution algorithms. *International Journal of Intelligent Systems*, *18*(2), 205-220.

Cooper, G., & Herskovits, E.A. (1992). A Bayesian method for the induction of probabilistic networks from data. *Machine Learning*, *9*(4), 309-347.

Cotta, C., & Muruzabal, J. (2002). Towards more efficient evolutionary induction of Bayesian networks. In *Proceedings of the Parallel Problem Solving from Nature VII* (pp. 730-739). Springer-Verlag.

Dijk, S.V., Thierens, D., & Gaag, L.C. (2003). Building a GA from design principles for learning Bayesian Networks. In *Proceedings of the Genetic and Evolutionary Computation Conference* (pp. 886-897).

Heckerman, D., Geiger, D., & Chickering, D.M. (1995). Learning Bayesian networks: The combination of knowledge and statistical data. *Machine Learning*, *20*(3), 197-243.

Henrion, M. (1988). Propagating uncertainty in Bayesian networks by probabilistic logic sampling. *Uncertainty in Artificial Intelligence, 2*, 149-163.

Jensen, F.V. (1996). *Introduction to Bayesian networks*. New York, Inc.; Secaucus, NJ: Springer-Verlag.

Lam, W., & Bacchus, F. (1994). Learning Bayesian Belief Networks: An approach based on the MDL principle. *Computational Intelligence*, *10*(4), 269-293.

Larrañaga, P., Kuijpers, C.M.H., Murga, R.H., & Yurramendi, Y. (1996). Learning Bayesian Network structures by searching for best ordering with genetic algorithm. *IEEE Transactions on Systems, Man, and Cybernetics*, *26*(4), 487-493.

Larrañaga, P., Poza, M., Yurramendi, Y., Murga, R.H., & Kuijpers, C.M.H. (1996). Structure learning of Bayesian Networks by genetic algorithms: A performance analysis of control parameters. *IEEE Transactions on Pattern Analysis and Machine Intelligence*, *18*(9), 912-926.

Lauritzen, S. L., & Spiegelhalter, D.J. (1988). Local computations with probabilities on graphical structures and their application on expert systems. *Journal of the Royal Statistical Society Series, 50*(2), 157-224.

Myers, J., & Levitt, T. (1999). Learning Bayesian Networks from incomplete data with stochastic search algorithms. In *Proceedings of the Fifteenth Conference on Uncertainty in Artificial Intelligence* (pp. 476-485). Morgan Kaufmann.

Pearl, J. (1988). *Probabilistic reasoning in intelligent systems: Networks of plausible inference.* San Mateo: Morgan Kaufman.

Tanese, R. (1989). Distributed genetic algorithms. In *Proceedings of the Third International Conference on Genetic Algorithms* (pp. 434-439).

Wong, M.L., Lam, W., & Leung, K.S. (1999). Using evolutionary programming and minimum description length principle for data mining of Bayesian Networks. *IEEE Transactions on Pattern Analysis and Machine Intelligence, 21*(2), 174-178.

This work was previously published in Bayesian Network Technologies: Applications and Graphical Models, edited by A. Mittal and A. Kassim, pp. 42-53, copyright 2007 by IGI Publishing, formerly known as Idea Group Publishing (an imprint of IGI Global).

Chapter 2.31
A Bayesian Framework for Improving Clusterng Accuracy of Protein Sequences Based on Association Rules

Peng-Yeng Yin
National Chi Nan University, Taiwan

Shyong-Jian Shyu
Ming Chuan University, Taiwan

Guan-Shieng Huang
National Chi Nan University, Taiwan

Shuang-Te Liao
Ming Chuan University, Taiwan

ABSTRACT

With the advent of new sequencing technology for biological data, the number of sequenced proteins stored in public databases has become an explosion. The structural, functional, and phylogenetic analyses of proteins would benefit from exploring databases by using data mining techniques. Clustering algorithms can assign proteins into clusters such that proteins in the same cluster are more similar in homology than those in different clusters. This procedure not only simplifies the analysis task but also enhances the accuracy of the results. Most of the existing protein-clustering algorithms compute the similarity between proteins based on one-to-one pairwise sequence alignment instead of multiple sequences alignment; the latter is prohibited due to expensive computation. Hence the accuracy of the clustering result is deteriorated. Further, the traditional clustering methods are ad-hoc and the resulting clustering often converges to local optima. This chapter presents a Bayesian framework for improving clustering accuracy of protein sequences based on association rules. The experimental results manifest that the proposed framework can significantly improve the performance of traditional clustering methods.

INTRODUCTION

One of the central problems of bioinformatics is to predict structural, functional, and phylogenetic features of proteins. A protein can be viewed as a sequence of amino acids with 20 letters (which is called the primary structure). The explosive growth of protein databases has made it possible to cluster proteins with similar properties into a family in order to understand their structural, functional, and phylogenetic relationships. For example, there are 181,821 protein sequences in the Swiss-Prot database (release 47.1) and 1,748,002 sequences in its supplement TrEMBL database (release 30.1) up to May 24, 2005. According to the secondary structural content and organization, proteins were originally classified into four classes: α, β, α+β, and α/β (Levitt & Chothia, 1976). Several others (including multi-domain, membrane and cell surface, and small proteins) have been added in the SCOP database (Lo Conte, Brenner, Hubbard, Chothia, & Murzin, 2002). Family is a group of proteins that share more than 50% identities when aligned, the SCOP database (release 1.67) reports 2630 families.

Pairwise comparisons between sequences provide good predictions of the biological similarity for related sequences. Alignment algorithms such as the Smith-Waterman algorithm and the Needleman-Wunsch algorithm and their variants are proved to be useful. Substitution matrices like PAMs and BLOSUMs are designed so that one can detect the similarity even between distant sequences. However, the statistical tests for distant homologous sequences are not usually significant (Hubbard, Lesk, & Tramontano, 1996). Pairwise alignment fails to represent shared similarities among three or more sequences because it leaves the problem of how to represent the similarities between the first and the third sequences after the first two sequences have been aligned. It is suggested in many literatures that multiple sequence alignment should be a better choice. While this sounds reasonable, it causes some problems we address here. The most critical issue is the time efficiency. The natural extension of the dynamic programming algorithm from the pairwise alignment to the multiple alignment requires exponential time (Carrillo & Lipman, 1988), and many problems related to finding the multiple alignment are known to be NP-hard (Wang & Jiang, 1994). The second issue is that calculating a distance matrix by pairwise-alignment algorithm is fundamental. ClustalW (Thompson, Higgins, & Gibson, 1994) is one of the most popular softwares for multiple-alignment problems. It implements the so-called progressive method, a heuristic that combines the sub-alignments into a big one under the guidance of a phylogenetic tree. In fact, the tree is built from a pre-computed distance matrix using pairwise alignment.

Many protein clustering techniques exist for sorting the proteins but the resulting clustering could be of low accuracy due to two reasons. First, these clustering techniques are conducted according to homology similarity, thus a preprocessing of sequence alignment should be applied to construct a homology proximity matrix (or similarity matrix). As we have mentioned, applying multiple sequence alignment among all proteins in a large data set is prohibited because of expensive computation. Instead, an all-against-all pairwise alignment is adopted for saving computation time but it may cause deterioration in accuracy. Second, most of the traditional clustering techniques, such as hierarchical merging, iterative partitioning, and graph-based clustering, often converge to local optima and are not established on statistical inference basis (Jain, Murty, & Flynn, 1999).

This chapter proposes a Bayesian framework for improving clustering accuracy of protein sequences based on association rules. With the initial clustering result obtained by using a traditional method based on the distance matrix, the strong association rules of protein subsequences for each cluster can be generated. These rules satisfying both minimum support and minimum confidence can serve as features to assign proteins to new clusters. We call the process to extract features from clusters the alignment-less alignment. Instead of merely comparing similarity from two protein sequences, these features capture important characteristics for a whole class from the majority, but ignore minor exceptions. These exceptions exist due to two reasons: the feature itself or the sequence itself. For the first reason, the feature being selected could be inappropriate and thus causes exceptions. Or, there does not exist a perfect feature that coincides for the whole class. The second reason is more important. The sequence causing the exception may be pre-classified into a wrong cluster; therefore, it should be re-assigned to the correct one. The Bayes classifier can provide optimal protein classifications by using the *a priori* feature information through statistical inference. As such, the accuracy of the protein clustering is improved.

The rest of this chapter is organized as follows. The background reviews existing methods relevant to protein clustering and the motivations of this chapter. The third section presents the ideas and the theory of the proposed method. The fourth section gives the experimental results with a dataset of protein sequences. The final section concludes this chapter.

BACKGROUND

Related Works

Many clustering techniques for protein sequences have been proposed. Among them, three main kinds of approaches exist, namely the hierarchical merging, iterative partitioning, and graph-based clustering. All of these methods use a pre-computed similarity matrix obtained by performing pairwise alignments on every pair of proteins. In the following we briefly review these approaches.

- **Hierarchical Merging:** The hierarchical merging clustering (Yona, Linial, & Linial, 1999, 2000; Sasson, Linial, & Linial, 2002) starts with a partition that takes each protein as a separate cluster, and then iteratively merges the two clusters that have the highest similarity from all pairs of current clusters. The similarity between two clusters is derived from the average similarity between the corresponding members. Thus, the hierarchical merging procedure forms a sequence of nested clusterings in which the number of clusters decreases as the number of iterations increases. A clustering result can be obtained by specifying an appropriate cutting-off threshold, in other words, the iterative merging procedure progresses until the maximal similarity score between any two clusters is less than the cutting-off threshold. The algorithm for hierarchical merging clustering is summarized in Figure 1.
- **Iterative Partitioning:** The iterative partitioning method (Jain & Dubes, 1988; Guralnik & Karypis, 2001; Sugiyama & Kotani, 2002) starts with an initial partition of *k* clusters. The initial partition can be obtained by arbitrarily specifying *k* proteins as cluster centers then assigning each protein to the closest cluster whose center has the most similarity with this protein. The next partition is obtained by computing the average similarity between each protein and all members of each cluster. The partitioning process is iterated until no protein

Figure 1. Summary of hierarchical merging clustering

```
For each protein p
   create.cluster(p);
Repeat
   Find clusters x, y such that similarity(x, y) is maximal;
   If similarity(x, y)>cutting_off_threshold then merge.
cluster(x, y);
   Otherwise, terminate;
```

Figure 2. Summary of iterative partitioning method

Select *k* proteins as initial cluster centers
Repeat
Compute the average similarity between each protein and each cluster;
Generate a new partition by assigning each protein to its closest cluster;
Until no proteins change assignment between successive iterations

Figure 3. Summary of graph-based clustering

Represent each input sequence as a vertex, and every pair of vertices are connected by an edge.
Label each edge with the similarity score between the two connected vertices.
Create a partition of the graph by cutting off the edges whose similarity scores are less than a specified threshold.

changes its assignment between successive iterations. Although the iterative partitioning method has some variants like ISODATA (Ball & Hall, 1964) and *K-means* (McQueen, 1967), their general principles can be described as shown in Figure 2. A post-processing stage can be added to refine the clusters obtained from the iterative partitioning by splitting or merging the clusters based on intra-cluster and inter-cluster similarity scores (Wise, 2002).

- **Graph-based Clustering:** These methods (Tatusov, Koonin, & Lipman, 1997; Enright & Ouzounis, 2000; Bolten, Schliep, Schneckener, Schomburg, & Schrader, 2001) represent each protein sequence as a graph vertex, and every pair of these vertices are connected by an edge with a label on it. The label denotes the similarity score between the two proteins represented by the vertices connecting to the corresponding edge. A partition of the graph can be generated by cutting off the edges whose labels are less than a specified similarity threshold, and each connected component of vertices corresponds to a cluster of proteins since the similarity scores between proteins from the same component are higher than those between proteins from different components. The general idea of the graph-based clustering is outlined in Figure 3. A post-refining process can be conducted by using the graph-based clustering result as the input to a cluster-merging algorithm which iteratively merges the nearest neighboring clusters if the relative entropy decreases with the merging of the clusters (Abascal & Valencia, 2002).

Motivations

The protein clustering result obtained by using the above mentioned approaches could be of low accuracy. This is partly due to the reason that these clustering methods use only pairwise sequence alignment information and partly because these clustering techniques often converge to local optima. Since it is computationally prohibitive to derive multiple sequence alignment information, an alternative is to calculate the statistics from the sequences directly by using data mining techniques. More precisely, the *association rules* between the amino acids in the protein sequences are mined within each cluster. Then the rules satisfying minimum confidence can serve as salient features to identify each cluster. Moreover, matching each protein sequence with the association rule provides a good estimate of the *a priori* probability that the protein satisfies the rule. The statistical inference can compensate the accuracy inadequacy of the clustering result due to the pairwise alignment and the local clustering technique. Therefore, we propose to improve the clustering accuracy by using the Bayes classifier with the conditional probabilities of association rules with each cluster and the alignment scores between protein sequences and these rules.

METHODS

With the assistance of association rule mining and Bayes classifier, our system improves the clustering accuracy of traditional protein-clustering methods. The system overview is shown in Figure 4. First, a traditional protein-clustering approach (either one of the hierarchical merging, iterative partitioning, or graph-based clustering methods) is performed to obtain an initial clustering result of the input protein sequences. In general, the traditional protein-clustering approach consists of three steps as shown in the upper grey box: (1) perform the local alignment (such as BLAST with scoring matrix of BLOSUM 62) between each pair of protein sequences, (2) construct a distance matrix from the raw distance scores (such as the E-values produced by BLAST) of the local alignment, and (3) apply the clustering method with the distance matrix to get the clustering result.

Second, the refining process (as shown in the lower grey box) is performed such that the clustering result is improved. The refining process also consists of three steps: (1) generate strong association rules which satisfy minimum confidence for each cluster, (2) perform sequence rule matching between each protein sequence and each association rule based on local alignment, and (3) adapt cluster membership of each protein sequence by using Bayes classifier and its matching score. The proposed refining process improves the clustering result based on statistical inference. The association rules used are statistically confident and the reassignment of protein sequences using Bayes classifier satisfies the maximum *a posteriori* criterion.

We now present the refining process in details.

Sequence Association Rule Generation

Association rule mining has been intensively used for finding correlation relationships among a large set of items and has delivered many successful applications, such as catalog design, cross marketing, and loss-leader analysis (Han & Kamber, 2001). Association rule mining finds significant associations or correlation relationships among a large repository of data items. These relationships, represented as rules, can assist the users to make their decisions. Traditionally, association rule mining works with unordered itemset, that is, the order of the items appearing in the itemset does not matter (Agrawal, Imielinski, & Swami, 1993). However, the order of the amino acids appearing in a protein sequence reserves important phylogeny information among homologies and should be taken into account. The *sequence Apriori algorithm* (Agrawal & Srikant, 1994, 1995) which adapts the classical association rule mining algorithm for sets of sequences can be used for this purpose.

Given a cluster C of protein sequences, the algorithm finds sets of sequences that have support above a given *minimum support*. The *support*

Figure 4. System overview

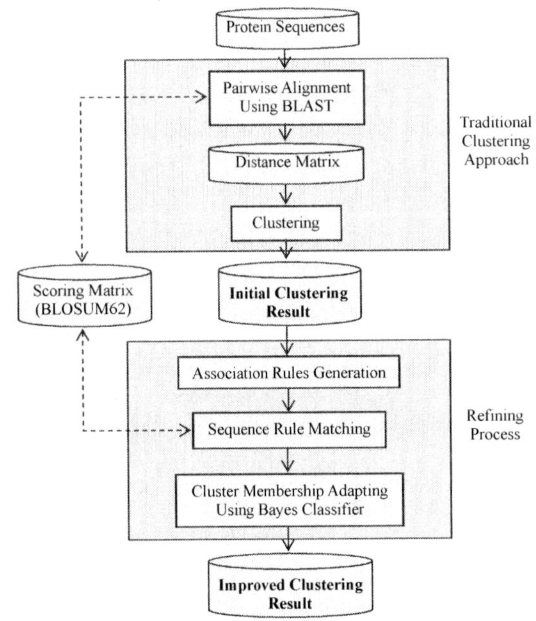

of a sequence X with respect to C, denoted by support$_C(X)$, is the number of sequences in C that contain X as subsequence. A sequence is called *frequent* if its support is greater than the given *minimum support*. To expedite the search for all frequent sequences, the algorithm enforces an iterative procedure based on the a priori property which states that any subsequence of a frequent sequence must be also frequent. It starts with the set of frequent 1-sequences which are the frequent sequences of length 1 and uses this set to find the set of frequent 2-sequences, then the set of frequent 2-sequences is used to find the set of frequent 3-sequences, and so on, until no more longer frequent sequences can be found. As the number of frequent sequences can be extremely large for a cluster of protein sequences having length of hundreds of amino acids, we retain the frequent sequences that are not contained in longer ones and restrict the search for frequent sequences of length 5 to 10. The reduced set of frequent sequences is sufficient for deriving similarity statistics in homology since longer frequent sequences of length more than 10 usually have lower support and a number of short frequent sequences within the specified range of lengths that the longer ones contain as subsequences can still be reserved.

With the set of found frequent sequences, we generate sequence association rules as follows. For a given frequent sequence X, it can be divided into two disjoint subsequences A and B with their position information attached. For example, let X be '*abcde*', one of the possible divisions could be A = '*a_c_e*' and B = '*_b_d_*'. A sequence association rule is of the form $A \Rightarrow B$ where both A and B contain at least one amino acid. A is called the rule *antecedent* and B is called the rule *consequent*. A sequence association rule is *strong* if it has a confidence value above a given *minimum confidence*. The confidence of a rule $A \Rightarrow B$ with respect to protein cluster C, denoted by confidence$_C(A \Rightarrow B)$, is defined as:

$$\text{confidence}_C\left(A \Rightarrow B\right) = \frac{\text{support}_C\left(A \cup B\right)}{\text{support}_C\left(A\right)} \quad (1)$$

where $A \cup B$ denotes the supersequence that is properly divided into A and B. We generate all strong sequence association rules for each cluster of protein sequences.

Sequence Association Rule Matching

In order to determine the cluster membership of each protein, we need to propose a measure which estimates the possibility that the evolution of a protein follows a particular association rule. Here we propose a rule matching scheme which is analogous to local sequence alignment without gaps. Given a sequence t and a strong association rule r, we compute the alignment score between t and r according to a substitution matrix with the constraint that no gap is allowed and the rule *antecedent* cannot be substituted in order to detect most similarities. For example, suppose we use the substitution matrix as shown in Figure 5 (a). Let the sequence be '*adabdacd*' and the association rule be '*da_d*' $\Rightarrow$ '*__d_*', there are two possible alignments as illustrated in Figure 5 (b) and the best alignment score without gaps is 11.

Assume that we obtain a set of strong association rules $\Re = \{r_1, r_2, ..., r_n\}$ from the procedure of sequence association rule generation, the probability that the evolution of a protein t follows a particular association rule r_i can b estimated by:

$$p(r_i \mid t) = \frac{w(r_i, t)}{\sum_{h=1}^{n} w(r_h, t)} \quad (2)$$

where $w(r_i, t)$ is the alignment score between r_i and t. As such, we can use te probabilities as the feature values of protein t, and determine its cluster membership by using the Bayes classifier.

Figure 5. An illustrative example of sequence association rule matching: (a) A substitution matrix, (b) Two possible alignments between the sequence and the association rule

	a	b	c	d
a	2	-1	2	0
b	-1	3	-2	1
c	2	-2	2	-1
d	0	1	-1	4

(a)

Sequence:	adabdacd	Score
Association rule:	dadd	4+2+1+4=11
Association rule:	dadd	4+2 – 1+4= 9

(b)

Cluster Membership Adapting by Using Bayes Classifier

The Bayes classifier is one of the most important techniques used in data mining for classification. It predicts the cluster membership probabilities based on statistical inference. Studies have shown that the Bayes classifier is comparable in performance with decision tree and neural network classifiers (Han & Kamber, 2001). Herein we propose to predict the cluster membership of each protein by using the Bayes classifier with association rules.

Let the initial clustering result consists of k clusters, denoted by $\Im = \{C_1, C_2, ..., C_k\}$, from which a set of n strong association rules is derived. The *a priori* probability $p(C_i)$ that a protein belongs to cluster C_i can be calculated by counting the ratio of C_i in size to the whole set of proteins. The condition probability $p(r_i|C_j)$ that a protein satisfies association rule r_i given that this protein is initially assigned to cluster C_j is estimated by the average probability $p(r_i|t)$ for any $t \in C_j$. The conditional probability $p(r_i|t)$ that the evolution of protein t follows association rule r_i is estimated by using Equation (2). We then use the naïve Bayes classifier to assign protein t to the most probable cluster C_{bayes} given by:

$$C_{Bayes} = \arg\max_{C_j \in \Im} \left\{ \prod_{i=1}^{n} \left(p(r_i \mid C_j) p(r_i \mid t) \right) p(C_j) \right\} \quad (3)$$

In theory, the naïve Bayes classifier makes classification with the minimum error rate.

EXPERIMENTAL RESULTS

We validate our method by using protein sequences selected from SCOP database (release 1.50, Murzin, Brenner, Hubbard, & Chothia, 1995; Lo Conte et al., 2002) which is a protein classification created manually. SCOP provides a hierarchy of known protein folds and their detailed structure information. We randomly select 1189 protein sequences from SCOP and these sequences compose 388 protein clusters according to manual annotations on structure domains. The mean length of these protein sequences is 188 amino acids. Our method is implemented in C++ programming language and the experiments are conducted on a personal computer with a 1.8 GHz CPU and 512 MB RAM.

Accuracy Measures

To define the accuracy measures of protein classification, some notations are first introduced. For every pair of protein sequences, the predicted classification and the annotated classification have four possible combinations. *TP* (true positive) is the number of pairs predicted in the same domain given that they are in the same SCOP domain, *TN* (true negative) is the number of pairs predicted with different domains given that they are in different SCOP domains, *FP* (false positive) is the number of different SCOP-domain pairs that are predicted in the same domain, and *FN* (false negative) is the number of SCOP-domain pairs that are predicted in different domains. Two accuracy measures, namely *sensitivity* (S_n) and *specificity* (S_p), are defined as follows:

$$Sn = \frac{TP}{TP + FN} \quad (4)$$

and.

$$Sp = \frac{TP}{TP + FP} \quad (5)$$

Sensitivity is the proportion of SCOP-domain pairs that have been correctly identified, and specificity is the proportion of pairs predicted in the same domain that are actually SCOP-domain pairs. Sensitivity and specificity cannot be used alone since perfect sensitivity can be obtained if all the pairs are predicted in the same domain, and specificity is not defined if all the pairs are predicted in different domains. To compare the performance between two competing methods, a sensitivity vs. specificity curve is usually used for evaluation. Or alternatively, the mean of sensitivity and specificity values can be used as a unified measure.

Performance Evaluation

We first evaluate the clustering performance using the traditional methods, in particular, we have implemented the hierarchical merging method (Sasson et al., 2002), the K-means algorithm (modified from Guralnik & Karypis, 2001, by changing the feature space to pair-wise E-value), and the graph-based clustering (Bolten et al., 2001). Various clustering thresholds are specified to obtain the performances at different specified numbers of clusters. Figures 6 (a) - 6 (c) show the variations of sensitivity and specificity as the number of clusters increases for hierarchical merging, K-means, and graph-based clustering, respectively. These curves are intuitive since the mean size of clusters is smaller if the partition with more clusters is obtained and, in general, the smaller the cluster-mean size is, the lower the sensitivity is, but the higher the specificity is. If we take the accuracy values obtained when the number of clusters is equal to the true number (388), the hierarchical merging method with S_n = 0.79 and S_p = 0.73 is superior to the K-means algorithm which produces S_n = 0.69 and S_p = 0.72, while the graph-based clustering has medium performance (S_n = 0.80 and S_p = 0.63).

Figure 6. Performances of traditional clustering methods, (a) hierarchical merging clustering, (b) K-means clustering, (c) graph-based clustering

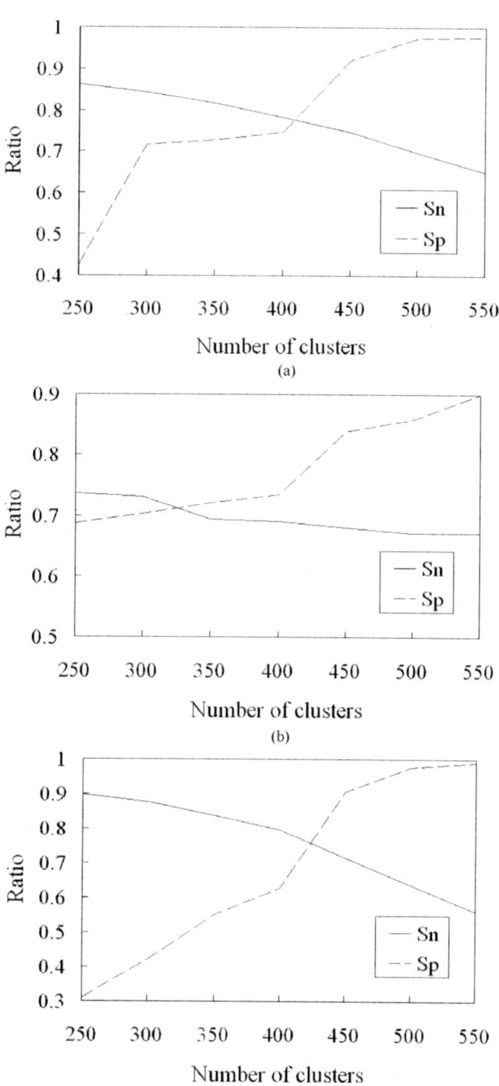

Next we evaluate the improvement for clustering accuracy due to our Bayesian framework. From the clustering results obtained by traditional clustering methods, we apply the sequence association rule mining to generate strong association rules. Each protein sequence is matched with these rules and updates its cluster membership by using the Bayes classifier in order to improve the accuracy. Because the sensitivity values obtained by hierarchical merging, K-means, and graph-based clusterings with the true cluster number (388) are 0.79, 0.69, and 0.80, respectively, we compute the specificity improvement for these methods within a range of sensitivity close to these values. Figures 7 (a) - 7 (c) show the sensitivity vs. specificity curve for illustrating the improvement achieved. It is observed that the proposed refining process can significantly improve the specificity of the traditional methods. The average improvements in Sp are 0.09, 0.11, and 0.06 for hierarchical merging, K-means, and graph-based clustering, respectively. We also compute the accuracy improvement by using the unified measure of $(S_n + S_p)/2$ as shown in Figs. 8 (a) - 8 (c). The average improvements over various specified numbers of clusters are 0.03, 0.04, and 0.03 for hierarchical merging, K-means, and graph-based clustering, manifesting the robustness of the proposed framework.

Table 1 shows the incurred computation time (in seconds) by each component of the proposed framework and the corresponding percentage to the whole for our collective database. We observe that the distance matrix computation using pairwise sequence alignment and the association rules generation using sequence Apriori algorithm are the most time-consuming components, and they consume 35% and 33% of the total computation time, respectively. Further, the computation time for the refining process involving the last two components of Table 1 is about half of the whole time needed. That is, for our collective database, the proposed framework provides a considerable amount of accuracy improvement (as shown in

Figure 7. Performance improvement for specificity vs. sensitivity, (a) hierarchical merging clustering, (b) K-means clustering, (c) graph-based clustering

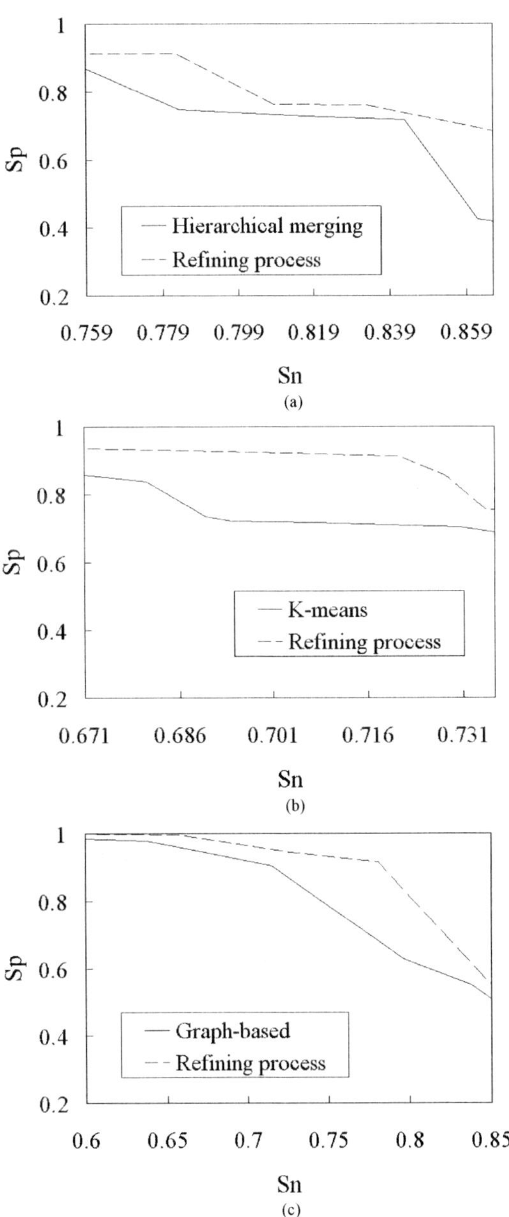

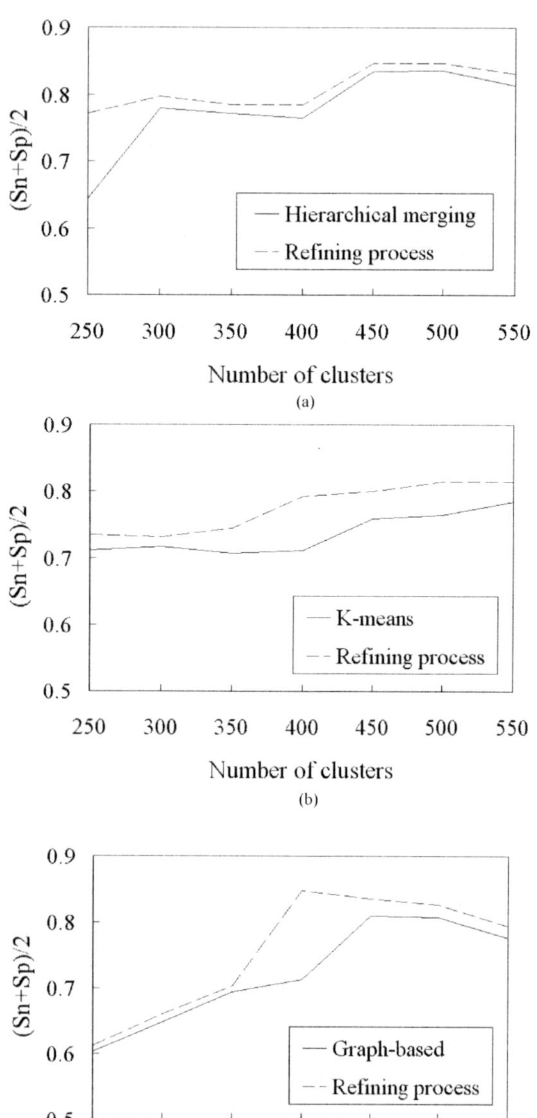

Figure 8. Performance improvement for the unified measure, (a) hierarchical merging clustering, (b) K-means clustering, (c) graph-based clustering

Figures 7-8) for the traditional protein-clustering algorithms by doubling the computation time. In general cases, the time proportion for the refining process diminishes when the number of sequences in the database increases. Suppose there are m sequences in the database, and for simplicity, assume that their average length is r. The time complexity for calculating a distance matrix is $\Theta(m^2r^2)$, for performing traditional clustering is $\Omega(m^2)$, and for generating the association rules of fixed length is $\Theta(mr)$. Rule matching and Bayesian classification depends on two factors: the number of features extracted and the number of clusters classified (and remember that each rule is within a fixed length). The former can be regarded as a constant once the lengths of rules and the alphabet (which is 20 for amino acids) are fixed to constants. Therefore a rough calculation asserts that it takes $\Theta(mrc)$ where c is the number of clusters. As a summary, the time complexity for each component is $\Theta(m^2r^2)$, $\Omega(m^2)$, $\Theta(mr)$, and $\Theta(mrc)$, respectively. Parameter r, the average length of peptide sequences, is around 300 and can be regarded as a constant. Parameter c grows mildly as the number of sequences increases. As a result, when the number of sequences in a database grows larger, the percentage of the computing time on the refining process becomes smaller.

SUMMARY

Protein sequence clustering is useful for structural, functional, and phylogenetic analyses. As the rapid growth in the number of sequenced proteins prohibits the analysis using multiple sequences alignment, most of the traditional protein-clustering methods derive the similarity among sequences from pairwise sequence alignment. In this chapter, we have proposed a Bayesian framework based on association rule mining for improving the clustering accuracy using existing methods. A selective dataset from SCOP has been experimented and the result manifests that the

Table 1. The computation time used by each component of the proposed method

	CPU Time (second)	Percentage
Distance Matrix Computation	162	35%
Traditional Clustering	64	14%
Association Rules Generation	157	33%
Rule Matching and Bayesian Classification	87	18%

proposed framework is feasible. The main features of the proposed framework include:

- The proposed framework improves the accuracy of a given initial clustering result which can be provided by any clustering methods. Therefore, the user is still able to choose a particular clustering method which is suited to his/her own analysis of the final result.
- From the initial clustering result, the sequence association rules among amino acids are mined. These rules represent important relationships for the sequences belonging to the same cluster.
- The association rules serve as features for classification of proteins. Using Bayes classifier, the classification error can be minimized based on statistical inference.

Future research is encouraged in expediting the computation for distance matrix and association rule generation in order to extend the application of the proposed framework to large protein databases such as SWISSPROT.

REFERENCES

Abascal, F., & Valencia, A. (2002). Clustering of proximal sequence space for the identification of protein families. *Bioinformatics, 18*(7), 908-921.

Agrawal, R., Imielinski, T., & Swami, A. (1993). Mining association rules between sets of items in large databases. *Proceedings of ACM-SIGMOD, International Conference on Management of Data*, Washington, DC (pp. 207-216).

Agrawal, R., & Srikant, R. (1994). Algorithms for mining association rules. *Proceedings of the 20th International Conference on Very Large Databases*, Santiago, Chile (pp. 487-499).

Agrawal, R., & Srikant, R. (1995). Mining sequential patterns. *Proceedings of the International Conference on Data Engineering (ICDE)*, Taipei, Taiwan.

Ball, G. H., & Hall, D. J. (1964). Some fundamental concepts and synthesis procedures for pattern recognition preprocessors. *Proceedings of International Conference on Microwaves, Circuit Theory, and Information Theory*, Tokyo.

Bolten, E., Schliep, A., Schneckener, S., Schomburg, D., & Schrader, R. (2001). Clustering protein sequences—structure prediction by transitive homology. *Bioinformatics, 17*(10), 935-941.

Carrillo, H., & Lipman, D. (1988). The multiple sequence alignment problem in biology. *SIAM Journal on Applied Mathematics, 48*, 1073-1082.

Enright, A. J., & Ouzounis, C. A. (2000). GeneRAGE a robust algorithm for sequence clustering and domain detection. *Bioinformatics, 16*(5), 451-457.

Guralnik, V., & Karypis, G., (2001, August 26). A scalable algorithm for clustering protein sequences. *Proceedings of Workshop on Data Mining in Bioinformatics,* San Francisco (pp. 73-80). ACM Press.

Han, J., & Kamber, M., (2001). *Data mining: Concepts and techniques.* San Francisco: Morgan Kaufmann Publishers.

Hubbard, T. J., Lesk, A. M., & Tramontano, A. (1996). Gathering them into the fold. *Nature Structure Biology, 4,* 313.

Jain, A. K., & Dubes, R. C. (1988). *Algorithms for clustering data.* Englewood Cliffs, NJ: Prentice Hall.

Jain, A. K., Murty, M. N., & Flynn, P. J. (1999). Data clustering: A review. *ACM Computing Surveys, 13*(3), 264-323.

Levitt, M., & Chothia, C. (1976). Structural patterns in globular proteins. *Nature, 261,* 552-558.

Lo Conte, L., Brenner, S. E., Hubbard, T. J. P., Chothia, C., & Murzin, A. G. (2002). SCOP database in 2002; refinements accommodate structural genomics. *Nucleic Acids Research, 30,* 264-267.

McQueen, J.B. (1967). Some methods of classification and analysis of multivariate observations. *Proceedings of Fifth Berkeleys Symposium on Mathematical Statistics and Probability* (pp. 281-297).

Murzin A. G., Brenner S. E., Hubbard T., & Chothia C. (1995). SCOP: A structure classification of proteins database for the investigation of sequences and structures. *Journal of Molecular Biology, 247,* 536-540.

Sasson, O., Linial, N., & Linial, M. (2002). The metric space of proteins—comparative study of clustering algorithms. *Bioinformatics, 18*(1), S14-S21.

Sugiyama, A., & Kotani, M. (2002). Analysis of gene expression data by self-organizing maps and k-means clustering. *Proceedings of IJCNN* (pp. 1342-1345).

Tatusov, R. L., Koonin, E. V., & Lipman, D. J. (1997). A genomic perspective on protein families. *Science, 278,* 631-637.

Thompson, J. D., Higgins, D. G., & Gibson, T. J. (1994). CLUSTAL W: Improving the sensitivity of progressive multiple sequence alignment through sequence weighting, position-specific gap penalties and weight matrix choice, *Nucleic Acids Research, 22,* 4673-4680.

Wang, L., & Jiang, T. (1994). On the complexity of multiple sequence alignment. *Journal of Computational Biology, 1,* 337-348.

Wise, M. J. (2002). The POPPs: clustering and searching using peptide probability profiles. *Bioinformatics, 18*(1), S38-S45.

Yona, G., Linial, N., & Linial, M. (1999). ProtoMap: Automatic classification of protein sequences, a hierarchy of protein families, and local maps of the protein space. *Proteins, 37,* 360-378.

Yona, G., Linial, N., & Linial, M. (2000). ProtoMap: Automatic classification of protein sequences, a hierarchy of protein families, and local maps of the protein space. *Nucleic Acids Research, 28,* 49-55.

Chapter 2.32
Improving Classification Accuracy of Decision Trees for Different Abstraction Levels of Data

Mina Jeong
Mokpo National University, Korea

Doheon Lee
Korea Advanced Institute of Science and Technology, Korea

ABSTRACT

Classification is an important problem in data mining. Given a database of records, each tagged with a class label, a classifier generates a concise and meaningful description for each class that can be used to classify subsequent records. Since the data is collected from disparate sources in many actual data mining environments, it is common to have data values in different abstraction levels. This article introduces the multiple abstraction level problem in decision tree classification, and proposes a method to deal with it. The proposed method adopts the notion of fuzzy relation for solving the multiple abstraction level problem. The experimental results show that the proposed method reduces classification error rates significantly when multiple abstraction levels of data are involved.

INTRODUCTION

Classification is one of the most widely used tasks in data mining. It consists of the construction phase and the assignment phase. The construction phase builds a classification model based on a given training set. The assignment phase utilizes the classification model to assign class labels to target records. Among many classification models proposed in the literature, decision trees are especially attractive in data mining environments due to their intuitive representation, which is easy for humans to understand (Gehrke et al., 2000; Gehrke et al., 1999; Berry & Linoff, 1997; Quinlan, 1993; Mehta et al., 1996; Shafer et al., 1996).

The training set is commonly collected from disparate sources in actual data mining environments. Information gathering agents access local sources, extract relevant data items, and deliver them to the central repository. Since the information gathering agents do not have any prior knowledge about global data distribution, the collected data in the training set is apt to be expressed in different abstraction levels. For example, a sales item description is expressed as "Coke" in a local table, while it is expressed as "Diet Coke 1.5 P.E.T." in another table. We call this the multiple abstraction level problem. Uncontrolled data entry by human users can also cause the same problem. Though sophisticated user interfaces of data entry software and employee training programs can help to standardize data entry, the problem cannot be eliminated due to mistakes and/or unavailable information (English, 1999).

The multiple abstraction levels can cause a severe problem in decision tree classification. Suppose that we have built a decision tree from multiple abstraction levels of training data as shown in Figure 1. Notice that the abstraction level of "Far-East" and "Mid-East," and that of "East" and "West" are different. Actually, "Far-East" and "Mid-East" are specializations of "East." Also suppose that a class label is assigned to a target record such that (Mid-East, 85k, Male). One choice is to follow the third branch of the root node, and assigns "E2" to the target record. However, since "Mid-East" belongs to "East," the other choice is to follow the first branch of the root node and in turn, the second branch of the "Income" node and assign "E1" to the target record. As a result, the decision tree yields two contradictory assignments.

As this sort of data quality problem has been a subject of long-standing discussions (English, 1999; Wang et al., 1995), data cleansing tools have been developed to improve the data quality on the market (Trilium Software systems, 1998; Vality Technology Inc., 1998; Williams, 1997). At first glance, it seems that there are two options to remedy the problem by using such tools. One option is to equalize abstraction levels by generalizing data values that are too specific. It is obvious that more specific data values yield more informative decision trees (Quinlan, 1993). Thus, this option results in the loss of useful information. The other option might be equalizing abstraction levels by specializing data values that are too general. Since it requires extracting additional information from the sources, it is hard or even impossible in actual data mining environments. Thus, the second option is also inapplicable due to the lack of information. Furthermore, it is also hard to determine how many levels we should consider generalizing or specializing data values in both options. Consequently, the existing data cleansing tools cannot provide satisfactory solutions for the problem.

There have been several fuzzy set-theoretic approaches for more flexible decision tree-based classification (Janikow, 1998; Dong & Kothari, 2001; Pelekis et al., 2005; Wang & Jiarong, 1998). They utilize fuzzy set-based measures for selecting the best-split attributes and usually deal with numerical values of attributes. In these researches more than one class labels can be assigned to each leaf node of the decision tree. Each of training data

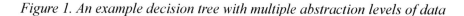

Figure 1. An example decision tree with multiple abstraction levels of data

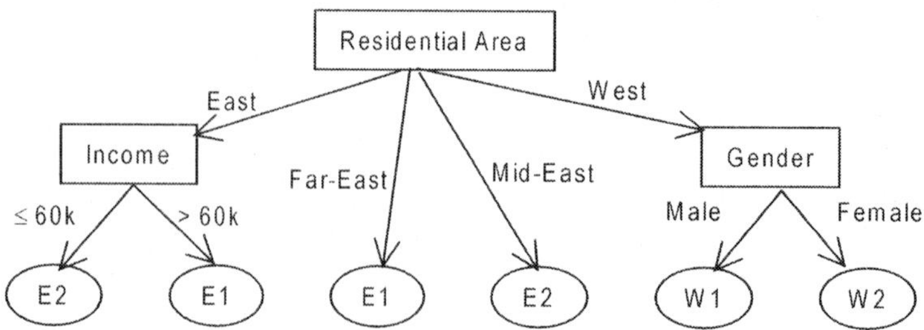

should be fuzzified; in other words, generalized to fuzzy sets, and their membership functions are used both for decision tree construction and classification phases. However, the assignment of membership values is assumed be carried out externally, and does not consider the notion of abstraction levels at all.

In this article, we explain that such multiple abstraction levels of data can cause undesirable effects in decision tree classification. There have been several approaches for handling multiple abstraction levels of data in data mining task (Kudoh et al., 2003; Lee et al., 2001). In one research, data value is replaced by generalized value. Since decision tree is constructed using generalized value, it is not more informative decision tree (Kudoh et al., 2003). However, in our research (Lee et al., 2001), the proposed method is rather than generalizing or specializing data values by force, we utilize the information as it is. The proposed ISA hierarchies simply present the relationship about generalization and specialization of data values. As a result, the proposed method adopts the notion of fuzzy relation for solving the multiple abstraction level problem. We utilize the information as it is, but accommodate the generalization/specialization relationship between data values in both the construction and assignment phases. The article is organized into five sections. The second section proposes modified versions of the best-split attribute selection and partitioning of training sets. We also extend the measure such as split info in C4.5 by introducing notion of fuzzy relation. The third section explains how to assign the class labels to target records when multiple abstraction levels of data exist. The fourth section presents experimental results to analyze classification accuracy improvement under more various conditions. The final section provides a summary of the findings and some closing remarks.

CONSTRUCTING DECISION TREES

To construct a decision tree, we need a training set. A training set is a data relation whose scheme consists of one or more predictor attributes and a class label. Briefly, a decision tree is a tree-structured representation of the distribution of the class label in terms of the predictor attributes (Gehrke et al., 2000; Gehrke et al., 1999; Quinlan, 1993). This section addresses how to construct a decision tree when a training set is given in multiple abstraction levels.

A decision tree is constructed through recursive partitioning of a training set. The first step is the selection of a predictor attribute that yields the best split of the training set. Once such a predictor attribute is selected, the training set is partitioned into subsets according to the attribute values. This selection-and-partitioning procedure is applied to each subset recursively. When the training set contains records expressed in multiple levels of abstraction, we revise partitioning the training set according to the best-split attribute and the selection of the best-split attribute.

Partitioning Training Sets With Multiple Abstraction Levels

The reason why partitioning the training set has to be modified can be explained through an example. Suppose that we have a training set of customer profiles as shown in Table 1. It consists of three predictor attributes, "Residential Area (RA)," "Gender," and "Income," and a class label "Frequently Visited Branch (FVB)." Though common training sets in actual data mining environment might contain at least tens of thousands of records, our example contains only ten records for simplicity.

For the moment, let us suppose that the predictor attribute "Residential Area" is selected as the best-split attribute with respect to the class label "FVB." The method of selecting the best-split attribute will be discussed precisely in the

Table 1. A training set of customer profiles

RID	RA	Gender	Income	FVB
t1	East	Male	90k	E2
t2	East	Male	70k	E1
t3	Far-East	Female	80k	E1
t4	Mid-East	Male	50k	E2
t5	Mid-East	Female	30k	E2
t6	West	Male	90k	W1
t7	West	Male	50k	W1
t8	West	Female	100k	W2
t9	West	Female	40k	W2
t10	West	Female	50k	W2

RA: Residential Area

FVB: Frequently Visited Branch (class attribute)

RID: Record ID (only for explanation)

following section. Now, we have to partition the training set according to "RA" values. Conventional partitioning results in $\{t_1, t_2\}$, $\{t_3\}$, $\{t_4, t_5\}$, and $\{t_6, t_7, t_8, t_9, t_{10}\}$, however, such partitioning misses the fact that "Far-East" and "Mid-East" belong to "East." Thus, we can include those records whose "RA" values are either "Far-East" or "Mid-East" in the first subset for "East." Then, the result becomes $\{t_1, t_2, t_3, t_4, t_5\}$, $\{t_3\}$, $\{t_4, t_5\}$, and $\{t_6, t_7, t_8, t_9, t_{10}\}$.

Unfortunately, it fails to address the fact that though the residential area of the first and second customers is registered as "East," their actual residential area may be either "Far-East" or "Mid-East." Since a proper training set reflects the value distribution of the entire domain, we can obtain the probability of the actual residential areas from the distribution of the training set. Three records, t_3, t_4, and t_5 have specialized "RA" values of "East." Among them, one record, t_3, has the value of "Far-East," while the other two records, t_4 and t_5, have the value of "Mid-East." From this distribution, we can say that the probability that the "RA" value of t_1 (or t_2) would actually be "Far-East" is 1/3 = 33%. Similarly, the probability that the "RA" value of t_1 (or t_2) would be actually "Mid-East" is 2/3 = 67%. As a result, we can partition the training set as $\{t_1, t_2, t_3, t_4, t_5\}$, $\{t_1/0.33, t_2/0.33, t_3\}$, $\{t_1/0.67, t_2/0.67, t_4, t_5\}$, and $\{t_6, t_7, t_8, t_9, t_{10}\}$, where t_i/μ denotes that t_i belongs to the set with the membership degree. Definitions 1 and 2 formalize this observation.

As seen in the last example, it is necessary to deal with the partial membership of a record in a set. Rather than inventing an ad-hoc representation for membership degrees, let us adopt the notion of fuzzy relations.

Definition 1. (Fuzzy Relation)

A fuzzy relation T is defined as follows:

$$T = \{(t, \mu_T(t)) \mid t \text{ is an ordinary record, } \mu_T(t) \text{ is the membership degree of } t \text{ in } T\}$$

A membership degree $\mu_T(t)$ is attached to each record to represent how completely the record belongs to the set T. If $\mu_T(t) = 1$, it implies a complete membership, while $\mu_T(t) < 1$ implies a partial membership. When the corresponding set T is obvious in the context, $\mu_T(t)$ is written simply as $\mu(t)$. In fact, an ordinary relation is a special case of a fuzzy relation where all the records have membership degrees of 1.0. Thus, a training set of records is regarded as a fuzzy relation from now on.

Definition 2. (Partitioning a Training Set)

When a fuzzy relation T (a training set) and an ISA hierarchy H on the domain of an attribute X are given, the selection from T with the condition such that "X is x," $SR(T, H, X, x)$ is defined as follows:

$$SR(T, H, X, x) = SR_{direct}(T, X, x) \cup SR_{descendent}(T, H, X, x) \cup SR_{antecedent}(T, H, X, x),$$

where

$SR_{direct}(T, X, x) = \{(t, \mu(t)) \mid t \in T, t.X = x, \mu(t) = \mu_T(t)\}$,

$SR_{descendent}(T, H, X, x) = \{(t, \mu(t)) \mid t \in T, t.X \in DESC(x, H), \mu(t) = \mu_T(t)\}$, and

$SR_{antecedent}(T, H, X, x) = \{(t, \mu(t)) \mid t \in T, t.X \in ANTE(x, H),$

$\mu(t) = \mu_T(t) \times (Card(\{(s, \mu_T(s)) \mid s \in T, s.X = x \text{ or } s.X \in DESC(x, H)\}) / Card(\{(s, \mu_T(s)) \mid s \in T, s.X \in DESC(t.X, H)\}))\}$.

ANTE(x) and *DESC(x)* denote sets of values appearing as antecedents and descendents of x, respectively, in the ISA hierarchy *H*. Card(T) = $\Sigma\ \mu_T(t)$. A membership degree $\mu_T(t)$ represents how completely the record belongs to the set.

Though the formulae in Definition 2 seem slightly complex, the idea is simple. The resulting set after selecting records whose *X* values are x from *T*, *(SR(T, H, T, x))*, consists of three parts. The first subset $SR_{direct}(T, X, x)$ is a set of records whose *X* values are literally identical to "x." Obviously, the μ(t) value is the membership degree of t in *T*, ($\mu_T(t)$). The second subset $SR_{descendent}(T, H, X, x)$ is a set of records whose "*X*" values are specializations of "x." In the previous example, "Far-East" represents a specialization of "East." In this case, the μ(t) values are still $\mu_T(t)$, since a specialization completely belongs to its generalized one. The third subset $SR_{antecedent}(T, H, X, x)$ is a set of records whose "*X*" values are generalizations of "x." In this case, μ(t) values become smaller than 1.0 (i.e., partial membership), since a generalized concept does not necessarily belong to its specialized one completely.

Figure 2 helps to improve understanding the membership degree assignment. Though actual fuzzy relations for decision tree construction would have several predictor attributes, the fuzzy relation in Figure 2 shows only one predictor attribute for simplicity. An ISA hierarchy on attribute *X* represents generalization/specialization relationships between values in the domain of X. Let us compute *SR(T, H, X, b)* as an example. (See Box 1.)

Extended Measures for Selecting the Best-Split Attribute

Now, let us return to the problem of selecting the best-split attribute. Recall that the best-split attribute is one that partitions the training set into the most homogeneous subsets. Though there are many measures proposed to evaluate the heterogeneity of a set, we adopt the information theoretic measure here, since it is commonly used in actual data mining systems. As we have regarded a fuzzy relation as a convenient representation of a training set, let us define how to measure the heterogeneity of a fuzzy relation by extending the measure of entropy (Quinlan, 1993).

Figure 2. An ISA hierarchy and a fuzzy relation as a training set

H: An ISA hierarchy on attribute X

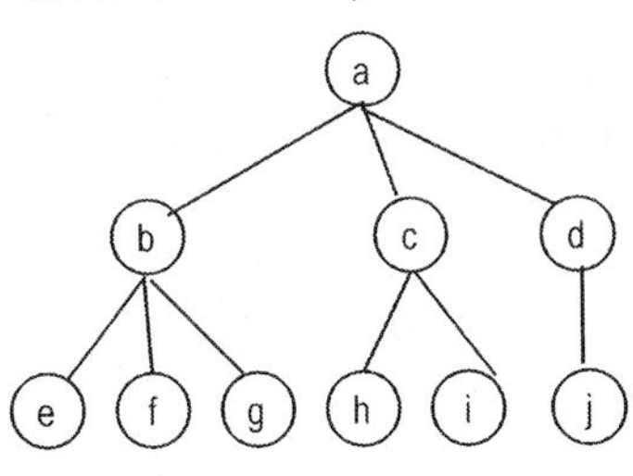

T: A training set

RID	X	μ(ti)	RID	X	μ(ti)
t1	a	1.0	t9	b	1.0
t2	c	1.0	t10	c	1.0
t3	d	1.0	t11	f	1.0
t4	h	0.7	t12	j	0.4
t5	a	0.9	t13	b	0.6
t6	c	1.0	t14	d	1.0
t7	e	0.9	t15	f	1.0
t8	i	1.0	t16	g	0.9

Box 1.

By Definition 2,

SR(T,H,X,b) = SRdirect(T,X,b) $\cup$ SRdescendent(T,H,X,b) $\cup$ SRantecedent(T,H,X,b). (1)

Since X values of t9 and t13 are literally identical to 'b,'

SRdirect(T,X,b) = {t9/1.0, t13/0.6} (2)

Since 'e,' 'f,' and 'g' are descendents of 'b,' and t7, t11, t15, and t16 have one of them. SRdescendent(T,H,X,b)

= {t7/0.9, t11/1.0, t15/1.0, t16/0.9} (3)

Since 'a' is the only antecedent of 'b,' and t1 and t5 have 'a' on X,

SRantecedent(T,H,X,b) = {t1/μ(t1), t5/μ(t5)} (4)

Since DESC(a) = {b, c, d, e, f, g, h, i, j} and DESC(b) = {e, f, g},

- μ(t1) = μT(t1) × Card({t7/0.9, t9/1.0, t11/1.0, t13/0.6, t15/1.0, t16/0.9}) / Card({t2/1.0, t3/1.0, t4/0.7,t6/1.0, t7/0.9, t8/1.0,t9/1.0, t10/1.0, t11/1.0, t12/0.4, t13/0.6, t14/1.0, t15/1.0, t16/0.9}) = 1.0 × 5.4/12.5 = 0.43 (5)
- μ(t5) = μT(t5) × Card({t7/0.9, t9/1.0, t11/1.0, t13/0.6, t15/1.0, t16/0.9}) / Card({t2/1.0, t3/1.0, t4/0.7, t6/1.0, t7/0.9, t8/1.0, t9/1.0, t10/1.0, t11/1.0, t12/0.4, t13/0.6, t14/1.0, t15/1.0, t16/0.9}) = 0.9 × 5.4/12.5 = 0.39 (6)

From (1), (2), (3), (4), (5), and (6),

SR(T,H,X,b) = {t9/1.0, t13/0.6} $\cup${t7/0.9, t11/1.0, t15/1.0, t16/0.9}$\cup${t1/0.43, t5/0.39}}

= {t1/0.43, t5/0.39, t7/0.9, t9/1.0, t11/1.0, t13/0.6, t15/1.0,t16/0.9}

Recall that only t9 and t13 are selected in conventional decision tree algorithms. However, our method also includes t1, t5, t7, t11, t15, and t16 with proper membership assignment.

Definition 3. (Entropy of a Fuzzy Relation)

Suppose that the domain of a class label C is $\{c_1, \ldots, c_m\}$, and a given a fuzzy relation T is partitioned into $T^{C1}, \ldots, T^{Cm}$ according to the values of C (i.e., $T^{Ci} = \{(t, \mu(t)) \mid t \in T, t.C = c_i, \mu(t) = \mu_T(t)\}$). Also, suppose that the domain of an attribute X is $\{x_1, \ldots, x_n\}$, and a given a fuzzy relation T is partitioned into $T^{X1}, \ldots, T^{Xn}$ according to the values of X (i.e., $T^{Xi} = \{(t, \mu(t)) \mid t \in T, t.X = x_i, \mu(t) = \mu_T(t)\}$). We call X the *discriminating attribute* of the partitioning. Then the entropy of T with respect to X or C is denoted as $info^C(T)$ or $info^X(T)$, and defined as follows.

$$info^C(T) = -\Sigma_{ci \in C} [Card(T^{ci})/Card(T) \times log_2(Card(T^{ci})/Card(T))],$$

$$info^X(T) = \Sigma_{xi \in X} [Card(T^{xi})/Card(T) \times info^S(T)]$$

Where S = $\{c_i \mid$ class label of the record which has x_i s as the attribute value$\}$,

$$Card(T^k) = \Sigma_{t \in T^k} \mu_T^k(t), k \in C \text{ or } k \in X$$

and

$$Card(T) = \Sigma_{t \in T} \mu_T(t).$$

Some algorithms such as C4.5 variations adopt an additional measure, such as *split info*, to eliminate a bias in favor of predictor attributes with more distinct values. In fact, the split info is also the entropy of a set, where the discriminating attribute is the predictor attribute — "RA" in the previous example — rather than the class label. By dividing the information gain value with split info, we can eliminate the unfair advantage that attributes with more distinct values may have. The measure of split info for a fuzzy relation can be computed with the formula in Definition 2.

Definition 4 (Split Info of a Fuzzy Relation)

Suppose that the domain of an attribute A is $\{a_1, \ldots, a_m\}$, and a given a fuzzy relation T is partitioned into $T^{a_1}, \ldots, T^{a_m}$ according to the attribute values of A, that is, $T^{a_j} = \{(t, \mu(t)) \mid t \in T, t.A = a_j, \mu(t) = \mu_T(t)\}$. We call A the *discriminating attribute* of the partitioning. Then the split *info(T)* is defined as follows.

$$split\ info(A) = -\Sigma_{j=1,\ldots,m} [Card(T^{a_j})/Card(T) \times log_2(Card(T^{a_j})/Card(T))],$$

where $Card(T^{a_j}) = \Sigma_{t \in T^{a_j}} \mu_T^{a_j}(t)$, and $Card(T) = \Sigma_{t \in T} \mu_T(t)$.

This represents the potential information generated by dividing T into m subsets. A membership degree $\mu_T^{a_j}(t)$ represents how completely the record containing to domain a_j belongs to the subset $_T^{a_j}$.

Let us compute the entropy of the training set in Table 1 with respect to "FVB." Since the table is an ordinary relation, the membership degree of each record is 1.0. Since the domain of the "FVB" attribute is $\{E_1, E_2, W_1, W_2\}$, the table is partitioned into four subsets as follows.

$T^{C1} = \{t_2/1.0, t_3/1.0\}$ (where "FVB" = "E_1"),
$T^{C2} = \{t_1/1.0, t_4/1.0, t_5/1.0\}$ (where "FVB" = "E_2"),
$T^{C3} = \{t_6/1.0, t_7/1.0\}$ (where "FVB" = "W_1"), and
$T^{C4} = \{t8/1.0, t9/1.0, t10/1.0\}$ (where "FVB" = "W_2").

Thus, $Card(T^{C1}) = 2.0$, $Card(T^{C2}) = 3.0$, $Card(T^{C3}) = 2.0$, and $Card(T^{C4}) = 3.0$. Since the training set itself has ten records whose membership degrees are 1.0, Card(T) = 10.0. As a result, $info^C(T) \approx [2.0/10.0 \times log_2(2.0/10.0) + 3.0/10.0 \times log_2(3.0/10.0) + 2.0/10.0 \times log_2(2.0/10.0) + 3.0/10.0 \times log_2(3.0/10.0)] = 1.97$.

We have to select the best-split attribute among three predictor attributes, "RA," "Gender," and "Income." First, let us consider "RA," By applying the partitioning method defined in Definition 2, we come to have four subsets if we partition the training set according to "RA" values as follows:

$T_1 = \{t_1/1.0, t_2/1.0, t_3/1.0, t_4/1.0, t_5/1.0\}$ (where "RA" is "East"),
$T_2 = \{t_1/0.33, t_2/0.33, t_3/1.0\}$ (where "RA" is "Far-East"),
$T_3 = \{t_1/0.67, t_2/0.67, t_4/1.0, t_5/1.0\}$ (where "RA" is "Mid-East"),
$T_4 = \{t_6/1.0, t_7/1.0, t_8/1.0, t_9/1.0, t_{10}/1.0\}$ (where "RA" is "West").

By applying the formula in Definition 3, we compute the entropy value of each subset with respect to the class label "FVB."

$info^{FVB}(T_1) = 5.0/15.9 \times [(2.0/5.0 \times log_2(2.0/5.0) + 3.0/5.0 \times log_2(3.0/5.0)]) = 0.42$,
$info^{FVB}(T_2) = 1.66/15.9 \times [(1.33/1.66 \times log_2(1.33/1.66) + 0.33/1.66 \times log_2(0.33/1.66)]) = 0.07$,
$info^{FVB}(T_3) = 3.34/15.9 \times [(0.67/3.34 \times log_2(0.67/3.34) + 2.67/3.34 \times log_2(2.67/3.34)] = 0.15$,
$info^{FVB}(T_4) = 5.0/15.9 \times ([2.0/5.0 \times log_2(2.0/5.0) + 3.0/5.0 \times log_2(3.0/5.0)] = 0.30$.

Then, by applying the formula in Definition 4, we compute the split info value of fuzzy relation with respect to the class label "FVB."

$splitinfo^{FVB}(T) \approx [5.0/15.9 \times log_2(5.0/15.9) + 1.66/15.9 \times log_2(1.66/15.9) + 3.34/15.9 \times log_2(3.34/15.9) + 5.0/15.9 \times log_2(5.0/15.9)] = 1.86$.

The normalized sum of the entropy values is 0.42 + 0.07 + 0.15 + 0.30 = 0.94. Thus, we can say that the entropy value is reduced by as much as 1.03 from 1.97 to 0.94. In other words, the information gain by partitioning the training set according to the attribute "RA" is 1.03 bits. If we apply the split info, the normalized sum of the entropy values is 1.03 (as before), and the resulting value is 1.03/1.86 = 0.55. Similarly, we can compute the information gain values of the attributes "Gender" and "Income." Among them, the attribute with the highest information gain is selected as the best-split attribute.

ASSIGNING CLASS LABELS WITH DECISION TREES

Suppose that we obtain a decision tree as shown in Figure 3 after applying the proposed method. The number attached to each terminal node represents the decision confidence (Quinlan, 1993). Also suppose that we have a target record such that (East, Male, 85k, Unknown). Notice that the last attribute "FVB," (the class label) is unknown. Our goal is to determine the "FVB" attribute of the record. Conventional methods simply follow the first branch from the root node and the second branch of the "Income" node to arrive at the assignment of "E_1" with a confidence of 67%.

However, we have to take account of the fact that "East" may be "Far-East." Suppose that 85% of records in the training set whose "Residential Area" attributes are specializations of "East" have "Mid-East" values. Then, we can follow the third branch of the root node, and assign "E_2" with a confidence of 85% × 98% = 83%. The assignment of "E_2" has a confidence of 83% while the assignment of "E_1" has a confidence of 67%. According to the principle of majority voting, we finally assign "E_2" with a confidence of 83%. The

Figure 3. An example decision tree with confidence levels

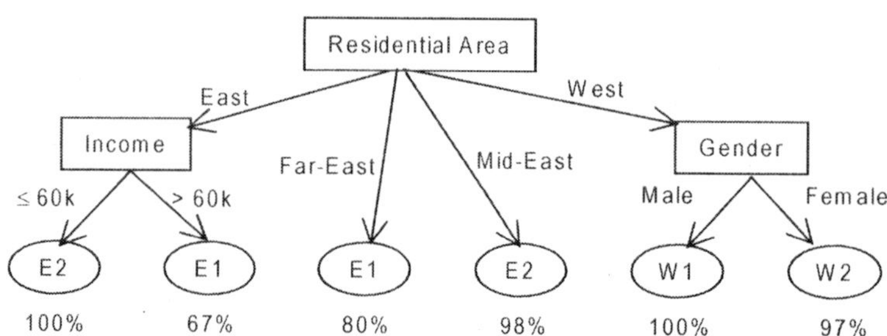

Figure 4. Class assignment with multiple abstraction levels of data

```
AssignClass (DecisionNode Attr, Record R) {
(1)   If Attr is a terminal node, return (Attr.Decision);
(2)   Child = the node followed by the branch with a label identical to R.Attr.
(3)   Answer = AssignClass (Child, R);
(4)   For each branch with a label
            that is a generalization of R.Attr {
(5)         Child = the node followed by this branch;
(6)         Temp = AssignClass (Child, R);
(7)         If Temp.Confidence > Answer.Confidence,
                then Answer = Temp;     }
(8)   For each branch with a label
            that is a specialization of R.Attr {
(9)             Child = the node followed by this branch;
(10)            Weight = the ratio of this specialization;
(11)            Temp = Weight × AssignClass (Child, R);
(12)            If Temp.Confidence > Answer.Confidence,
                    then Answer = Temp;   }
(13) return(Answer); }
```

algorithm in Figure 4 presents the assignment process precisely.

The conventional assignment process consists of Line (1) to Line (3). To accommodate generalizations and specializations, the loop from Line (4) to Line (7) and the loop from Line (8) to Line (12) are added, respectively. Among all the assignments, that with the maximum confidence is chosen for the final answer.

PERFORMANCE EVALUATION

We have conducted experiments with a benchmark data set to analyze classification accuracy of the proposed method. The experiment is performed on an Axil-Ultima 167 workstation with Solaris 7, 128 MB RAM, and a 10 GB Hard Disk.

To show that the proposed method outperforms the existing methods such as C4.5, it is necessary to have benchmark data sets containing multiple abstraction levels of data. However, handling of multiple abstraction levels of data in decision tree classification has been seldom studied so far. So it is hard to obtain such data sets from public sources. Instead, we deliberately distribute multiple abstraction levels of data across a public benchmark data set obtained from UCI Machine Learning Repository in a random and uniform way. The original benchmark data set contains 48,842 records with six continuous and eight categorical attributes. It has enough demographic information to classify high-income and low-income residents with the threshold income of $50,000. Since all the values in each attribute are in the same abstraction level, we select values in a random fashion, and substitute them with more general or specific abstraction levels. For example, an attribute value "federal government" in "workclass" attribute is substituted to a more general value "government." By controlling the ratio of data to substitute, we can generate two collections of data set with various heterogeneity degrees in their abstraction levels.

The algorithm proposed in this article can be regarded as an extension of the existing methods. We extend the ways of selecting best-split attribute and partitioning training data, which affect the effectiveness of the methods and their efficiency. It implies that most of the recent proposals for improving the efficiency including SLIQ (Mehta et al., 1996), SPRINT (Shafer et al., 1996), RainForest (Gehrke et al., 2000) and BOAT (Gehrke et al., 1999) are applicable to the proposed method without losing the efficiency benefit.

The experiment to follow presents the comparison and analysis for four methods. In C4.5, we can decide classes independently for each node in assigning the class, so application of the pruning is justifiable owing to the discussed advantages and disadvantages of the presence of the terminal nodes. However, in the proposed method, we compare and decide several terminal nodes to assign the classes. Consequently, we cannot discuss the advantages and disadvantages of the presence of a terminal node. The complexity arises from the combinatorial explosion resulting from any discussion of the advantages and disadvantages when considering the possible combinations of all terminal nodes in pruning. Nevertheless, our method has an effect on pruning in advance in the tree construction phase. Pruning is needed in splitting with more specific values. That our method has an effect on pruning in advance for constructing trees considering both general and specific values is proved through the experiment shown in Figures 5, 6, and 7.

Figure 5 shows how classification error accuracy is affected by the heterogeneity degrees of data sets. As the heterogeneity increases, the accuracy improvement of the proposed method increases higher up to 70%. We can see that our method has no variation for classification error rate over heterogeneity. However, notice that when the heterogeneity of the data set becomes larger than 1.0, the accuracy improvement decreases. This unexpected situation occurs because attributes with too diverse abstraction levels of data cannot

Figure 5. Relative error rates with respect to the heterogeneity in abstraction levels

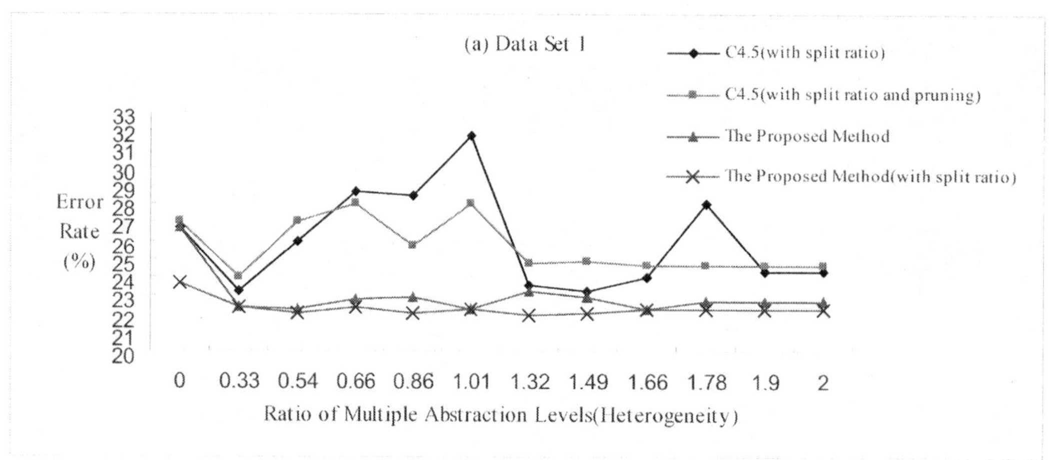

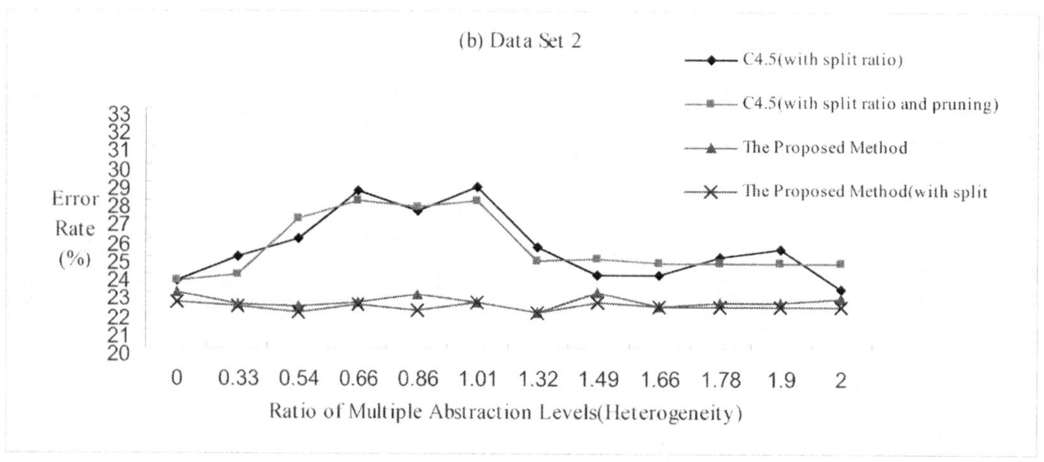

be selected as best-split attributes. In other words, the resulting decision trees do not include such attributes in their decision nodes. Consequently, handling multiple abstraction levels cannot take effect when the heterogeneity of the data set is too high.

To analyze the effect of the number of attributes with multiple abstraction levels, we also generate another collection of data sets by controlling data substitution to the original UCI data set. Other attributes with multiple abstraction levels become existent in data sets for this experiment. For example, because values ("10th," "11th," and "12th") of attribute "education" mean "high school" we added the attribute value "high school" as generalization value.

Figure 6 shows how classification error accuracy is affected as the number of attributes with multiple abstraction levels increases. As expected, the accuracy improvement becomes larger as the number attributes with multiple abstraction levels increases.

To analyze the effect of the depth of attributes with multiple abstraction levels, we also generate another collection of data sets by controlling data substitution to the original UCI data set as shown in Table 2. Figure 7 shows how classification error accuracy is affected as the depth of attributes with multiple abstraction levels increases. As expected, the improvement in accuracy increases as the depth of attributes with multiple abstraction levels increases.

Figure 6. Error rates with respect to increasing of the number of attributes with multiple abstraction levels of data

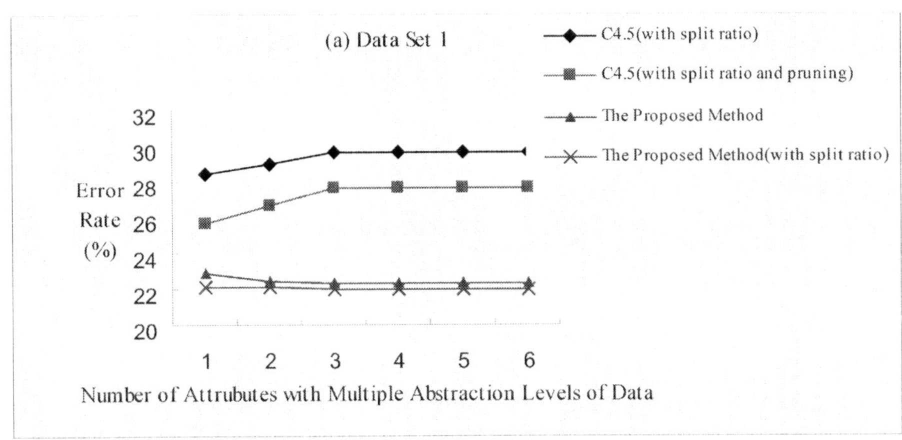

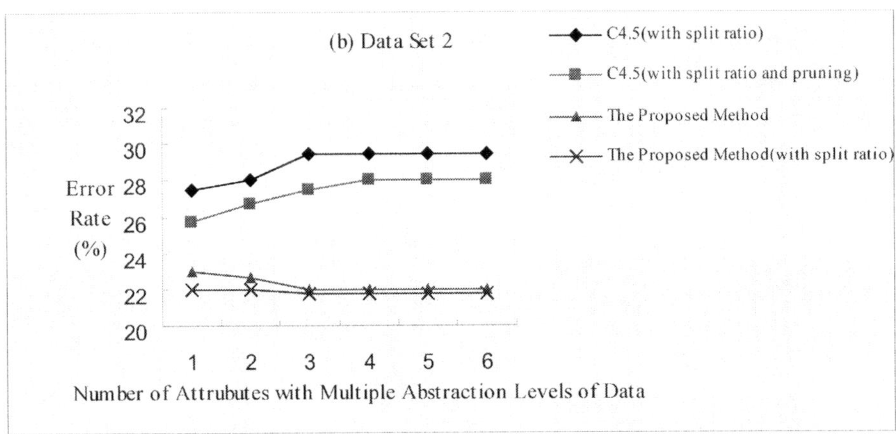

The asymptotic time complexity in constructing the decision tree is $O(m^2n)$ in the existing algorithm, where m is the number of attributes and n is the number of records. The time complexity in the proposed algorithm is also turned out to be $O(m^2n)$ when we assume that the depth of the given ISA hierarchy as the domain knowledge is independent of the input size. Furthermore, the complexity of classification given a decision tree of depth h is O(h) both for the existing algorithms and the proposed one. This analytical observation supports the idea that the proposed method outperforms the existing algorithms in the aspect of classification accuracy in the compatible computational costs.

Table 2. Depth of attributes with multiple abstraction levels

Depth of abstraction level	Depth of attributes with multiple abstraction levels
1	gov : State-gov, Local-gov, Federal-gov
2	State-gov : East-State-gov, West-State-gov
3	East-State-gov : Far-East-State-gov, Mid-East-State-gov

Figure 7. Error rates with respect to increasing of the depth of attributes with multiple abstraction levels of data

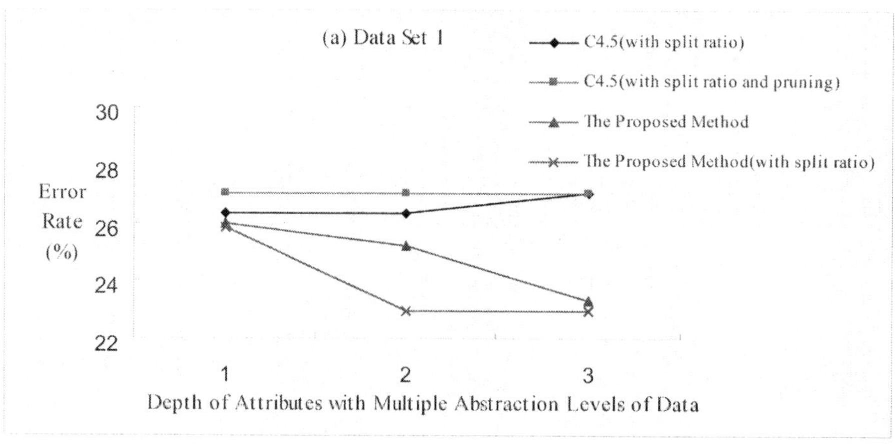

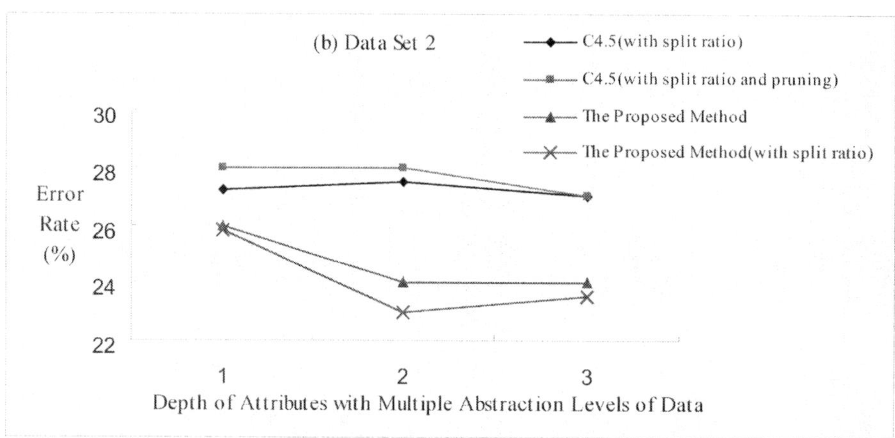

CONCLUDING REMARKS

This article has introduced the multiple abstraction level problem in decision tree classification, and shows that such multiple abstraction levels of data can cause undesirable effects in decision tree classification. It has explained that equalizing abstraction levels by force using data cleansing tools cannot solve the problem. We utilize the information as it is, but accommodate the generalization/specialization relationship between data values in both the construction and assignment phases. While a specialized value is compatible with its generalization completely, a generalized value is compatible with each of its specializations as much as the portion the specialization occupies among all existing specializations. In order to represent this partial compatibility, it adopts the notion of fuzzy relations. We also have an effect on pruning in advance for constructing trees considering both general and specific values. The experimental results conducted with a modified benchmark data set originating from UCI Machine Learning Repository have shown that the proposed method reduces classification error rates significantly when a data set includes multiple abstraction levels of data. For time complexity in constructing decision tree and time complexity in classifying experimental data by constructed decision tree, we used big-O notation for comparing C4.5 with the proposed algorithm.

REFERENCES

Berry, M., & Linoff, G. (1997). *Data mining techniques: For marketing, sales, and customer support*. New York: Wiley & Sons.

Dong, M., & Kothari, R. (2001). Look-ahead based fuzzy decision tree induction. *IEEE Transactions on Fuzzy Systems*, *9*(3), 461-468.

English, L. (1999). *Improving data warehouse and business information quality-method for reducing costs and increasing profits*. New York: Wiley & Sons.

Gehrke, J., Ramakrishinan, R., & Ganti, V. (2000). RainForest: A framework for fast decision tree construction of large datasets. *Data Mining and Knowledge Discovery*, *4*, 127-162.

Gehrke, J., Ganti, V., Ramakrishnan, R., & Loh, W.Y. (1999). BOAT: Optimistic decision tree construction. In *Proceedings of ACM SIGMOD Conference,* (pp. 169-180). Philadelphia, PA.

Janikow, C. (1998). Fuzzy decision trees: Issues and methods. *IEEE Transactions on, Systems, Man and Cybernetics, Part B*, *28*(1), 1 -14.

Kudoh, Y., Haraguchi, M., & Okubo, Y. (2003). Data abstractions for decision tree induction. *Theoretical Computer Science*, *292*(2), 387-416.

Lee, D., Jeong, M., & Won, Y. (2001). Decision trees for multiple abstraction levels of data.(LNAI 2182, 76-87).

Mehta, M., Agrawal, R., & Rissanen, J. (1996). SLIQ: A fast scalable classifier for data mining. In *Proceedings of the Fifth International Conference on Extending Database Technology (EDBT)*, Avignon, France.

Pelekis, N., Theodoulidis, B., Kopanakis, I., & Theodoulidis, Y. (2005). Fuzzy miner: Extracting fuzzy rules from numerical patterns. *International Journal of Data Warehousing and Mining, 1*(1), 57-81.

Quinlan, J. (1993). *C4.5: Programs for machine learning*. Morgan Kaufmann Publishers.

Shafer, J., Agrawal, R., & Mehta, M. (1996). SPRINT: A scalable parallel classifier for data mining. In *Proceedings of the 22nd International Conference on Very Large Databases*, Mumbai (Bombay), India.

Trillium Software System. (1998). *A practical guide to achieving enterprise data quality*, (White paper). Trillium Software.

Vality Technology Inc. (1998). *The five legacy data contaminants you will encounter in your warehouse migration*, (White paper). Vality Technology Inc.

Wang, R., Storey, V., & Firth, C. (1995). A framework for analysis of data quality research. *IEEE Transactions on Knowledge and Engineering,* 7(4), 623-640.

Williams, J. (1997). Tools for traveling data. *DBMS, 10*(7), 69-76.

Wang, X., & Jiarong, H. (1998). On the handling of fuzziness for continuous-valued attributes in decision tree generation. In *Fuzzy Sets and Systems,* (pp. 283-290).

This work was previously published in International Journal of Data Warehousing and Mining, Vol. 1, No. 3, edited by D. Taniar, pp. 1-14, copyright 2005 by IGI Publishing, formerly known as Idea Group Publishing (an imprint of IGI Global).

Chapter 2.33
Improving Similarity Search in Time Series Using Wavelets

Ioannis Liabotis
University of Manchester, UK

Babis Theodoulidis
University of Manchester, UK

Mohamad Saraee
University of Salford, UK

ABSTRACT

Sequences constitute a large portion of data stored in databases. Data mining applications require the ability to process similarity queries over a large amount of time series data. The query processing performance is an important factor that needs to be taken into consideration. This article proposes a similarity retrieval algorithm for time series. The proposed approach utilizes wavelet transformation in order to reduce the dimensionality of the time series. The transformed series are indexed using X-Trees, which is a spatial indexing technique able to efficiently index high-dimensional data. The article proves that this technique outperforms the usage of the Fourier transformation, since the wavelet transformation provides better approximation of the time series. Through the experiments, it can be concluded that the optimum performance is obtained using 16 to 20 wavelet coefficients. Furthermore, a novel mechanism for reducing the complexity of the calculation for the false alarms removal is proposed. Storing the approximation coefficients of the penultimate level of the decomposition tree, the Euclidean distance between the two sequences is calculated, thus reducing further the number of false alarms before calculating the actual Euclidean distance using the complete time series. The article concludes with a detailed performance evaluation of the proposed similarity retrieval algorithm using data from the Greek stock market and the temperature measurements from Athens. The comparison is done with techniques that use the Haar transform and the R*-Tree, and the proposed algorithm is shown to outperform them.

INTRODUCTION

Time series are a sequence of numerical values usually recorded at regular intervals (e.g., secondly, daily, weekly, monthly, or yearly). Regularity is known to be the major element considered in most of the times series. Although there are cases in which time series have no regularity (e.g., the history of stock splits), this work only considers time series with regularity. Time series appear in many database applications, from scientific to financial. Examples of such application domains include scientific experiments like temperature measurements over time; medical measurements, such as blood pressure or body temperature measurements taken in regular time intervals; business applications, including stock price indexes, such as opening, closing, minimum and maximum values or bank account histories, event sequences in automatic control, and musical or voice recordings.

Interesting features that can be retrieved from time series are similarities among them. There are various types of applications, depending on the type of sequence we are looking into. For example, a business analyst may desire to identify companies with similar patterns of growth to determine if they share other common characteristics. The identification of stocks with similar price movements could lead to predictions of their behavior. Also, similar patterns between two stocks that appear with a slight time difference can be identified. For example, the determination of time periods over the year that the daily temperature in Paris had the same fluctuation as January's daily temperature in Athens can help weather forecasting or prediction of unexpected weather conditions.

In this article, a similarity retrieval algorithm for time series is proposed. In the proposed algorithm, the wavelet transformation is used in order to reduce the dimensionality of the time series, and the transformed series are indexed using X-Trees (Berchtold et al., 1996). This indexing technique allows using a higher dimensionality for indexing without important loss in performance. From the evaluation process, it is shown that the wavelet transform reduces the false alarms significantly compared to the Fourier transform and provides better approximation of the time series. Also, using a number of dimensions between 16 and 20, the performance is optimized. Comparison has been done with techniques using the Haar transform and the R*-Tree (Chan & Fu, 1997; Wu et al., 2000; Chan et al., 2003), and the proposed algorithm has been shown to outperform them.

Furthermore, a novel mechanism for reducing the complexity of the false alarms removal calculation is proposed. Storing the approximation coefficients of the penultimate level of the decomposition tree, the Euclidean distance between the two sequences is calculated, thus reducing further the number of false alarms before calculating the actual Euclidean distance using the whole time series.

The organization of the rest of this article is as follows: the second section discusses the related work in relation to the work and assumptions reported in this article. The third section discusses the basics of the wavelets theory and the discrete wavelet transform. The similarity retrieval algorithm is discussed in the fourth section. The evaluation of the algorithm is presented in the fifth. Finally, the sixth section concludes the article.

RELATED WORK

In recent years, there has been a great deal of research in the similarity search area. There are several different similarity measurements that have been proposed in the literature as well as several algorithms for the efficient retrieval of these similarities. One approach to define the similarity of two time sequences is to use the Euclidean distance in an appropriate multidimensional space. In the work by Agrawal et al. (1993), two sequences are considered similar, if the Euclidean distance between them is less than a predefined threshold ε. The Euclidean distance between two sequences is defined as the square root of the sum of the squared differences: $D = (\Sigma\ (Xi\text{-}Yi)^2)^{1/2}$. The same distance metric is used by Faloutsos et al. (1994), Goldin and Kanellakis (1995), Rafiei and Mendelzon (1997), Kanth et al. (1998), Rafiei (1999), Chan and Fu (1999), Rafiei and Mendelzon (2000), Shahabi et al. (2000), Rafiei and Mendelzon (2002), and Kim and Park (2005). In particular, Agrawal et al. (1993) considered this metric to be one of the best measures to estimate signals corrupted by Gaussian additive noise.

Non-Euclidean metrics also have been used to define the similarity of two time sequences. Agrawal et al. (1995) uses the L_∞ distance metric, where two sequences are similar, if they have enough nonoverlapping time-ordered pairs of subsequences that are similar. Also, two sequences are considered similar, if one can be enclosed within an envelope of a specified width drawn around the other. Since time series databases contain outliers, the proposed method allows nonmatching gaps between the subsequences. The distance metric is defined as: $L_\infty = \max|p_i - q_i|$ for vectors p and q.

Agrawal et al. (1995) proposes a simple query language for retrieving objects based on shapes contained in historical sequences. The language is called SDL (Shape Definition Language) and provides a variety of different queries about the shapes that can be found in historical time series. A powerful characteristic of SDL is that it supports *blurry* matching. A blurry match is one where the user cares about the overall shape and does not care about specific details. This characteristic is similar to the threshold ε that other proposals provide when using an arithmetic distance metric.

Perng et al. (2000) proposes the Landmark model, where a time sequence is represented by a subset of its values, which are the peak points in the sequence. The distance between two sequences is defined as a tuple of values, one representing the time and the other the amplitude. Park et al. (2000) uses the idea of time warping distance and compares sequences of different lengths by stretching them. The idea of making two time series shift and scale invariant by transforming them into the shift eliminated plane is used in Chu and Wong (1999) and Kahveci and Singh (2004); the distance metric D_{norm} is defined by Lee et al. (2000), and it is used to compare multidimensional sequences. Finally, Vlachos et al. (2002) proposes the use of the Longest Common Subsequence (LCSS) technique and demonstrates that it is more robust to noise than the Euclidean distance and the time-warping distances.

Several proposals study the retrieval techniques that can be used in order to improve the searching performance. Using similarity metrics such as the Euclidean distance or the L_∞ norm each time series can be mapped to the multidimensional space, having n dimensions, where n equals the number of elements of each sequence. Having to index a sequence of 100 or 1,000 different values requires a large amount of space and leads to reduced performance. Reduction of the dimensionality can be the remedy to this problem.

Agrawal et al. (1993) proposed the use of the Discrete Fourier Transform (DFT) in order to translate the sequence from the time domain to the frequency domain. This is possible, because, according to Parseval's theorem, the Euclidean distance between two sequences in the time domain is the same as the Euclidean distance in the frequency domain. Given the fact that only a few Fourier coefficients are enough, it allows the building of an effective index with low dimensionality. The index is called F-index, because they were using only the first f_c Fourier coefficients, where f_c stands for cutoff frequency. Because Fourier coefficients are complex numbers, each sequence is mapped to the $2f_c$ dimensional space and indexed in an R*-Tree structure for fast retrieval. The reduction of the dimensionality introduces some *false alarms*. False alarms are considered the sequences that are selected as similar, but they are not. In order to avoid the false alarms, a measurement of the actual Euclidean distance is performed, and then only the qualifying time series are returned.

Extending the previous proposal, Faloutsos et al. (1994) allows the retrieval of subsequences that match a query sequence or a subsequence of it. The proposed method uses a sliding window over the data sequence. Then, using DFT, the sequence features are extracted, resulting in a trail in the feature space. Those trails are divided into subtrails that are represented by their Minimum Bounding Rectangles (MBRs) and are stored in R*-Trees for fast indexing. Specifically using a sliding window of size w and placing it at all the possible positions of the examined data sequences, the features of each subsequence are extracted. A query retrieves all MBRs that intersect the query region. The points that belong to each MBR are the candidate sequences, including some false alarms that are points that do not intersect the query region, but their MBRs do.

In order to make the indexing technique more efficient, Rafiei and Mendelzon (1997) use the last few coefficients of the DFT transform in the distance computation without storing them in the index. They prove that this method can accelerate the search time of the index, because every coefficient at the end is the complex conjugate of a coefficient at the beginning (Rafiei, 1999; Rafiei & Mendelzon, 2000; Rafiei & Mendelzon, 2002).

Korn et al. (1997) proposed the use of the Single Value Decomposition (SVD) of the sequences in order to compress very large time series and to pose the queries in the compressed data instead of the original. The method provides low loss of information during the compression so that the number of false alarms is minimized. SVD is used as a compression algorithm that permits quick reconstruction of arbitrary parts of the datasets. Withholding only the few most important principal components, a compressed version of the original sequence is obtained that can be used for reconstruction with only one disk access and with small average error in the reconstructed value.

Chan and Fu (1997) and Chan et al. (2003) used the Haar wavelet transform to reduce the number of dimensions and compared this method to DFT and proved that Haar wavelet transform performs better than DFT. Rafiei and Mendelzon (1998) and Wu et al. (2000) proved that the performance of DFT can be improved further using the symmetry of Fourier Transforms, in which case both methods give similar results. Finally, Wang and Wang (2000) proposed the use of B-spline wavelet transforms and the least square method to approximate time sequences. The drawback of this approach is that it may result in false dismissals.

DISCRETE WAVELET TRANSFORM

Wavelet Analysis

One of the most used tools in the field of signal processing is Fourier analysis. Fourier analysis breaks down a signal into constituent sinusoids of different frequencies. It is a mathematical technique for transforming the view of a signal from a time-based to a frequency-based one. Nevertheless, Fourier analysis has a serious drawback. When transforming to the frequency domain, information about time is lost. Most interesting signals, including historical stock prices or temperature measurements, contain a lot of non-stationary characteristics, such as drift, trend, and abrupt changes. Therefore, a new analysis tool is needed in order to detect them.

In order to overcome this problem, a new method has been introduced: the short-term Fourier analysis. This method uses the Fourier transform to analyze only one small section of the signal at a time. Putting a window over the signal and performing the Fourier transformation only for this window at a time provides some information about when and at what frequencies a specific event occurs. While this technique provides a solution to the problem of losing significant information about time, it has the drawback that once the window size is defined, it has to be constant for all frequencies. Wavelet analysis provides a solution by allowing the use of a variable-sized window over the sequence. A wavelet is a waveform of effectively limited duration with an average value of zero (Jaideva et al., 1999). Wavelet analysis is the breaking up of a signal into shifted and scaled versions of the original, or *mother*, wavelet.

The discrete wavelet transform is calculated by the following formula (Jaideva et al., 1999):

Equation 1. The discrete wavelet transform

$$W_{\psi}f(k2^{-S},2^{-S}) = 2^{S/2}\int_{-\infty}^{+\infty} f(t)\psi(2^{S}t-k)dt$$

where $k,S \in \mathbf{Z}$. The function is called *mother* wavelet. We assume that:

Equation 2. A property of a mother wavelet

$$\int \psi(t)dt = 0$$

The wavelet series representation is given by the formula:

Equation 3. Wavelet representation

$$f(t) = \sum_{S}\sum_{k} w_{k,S}\psi_{k,S}(t)$$

where

Equation 4. The mother wavelet

$$\psi_{k,S}(t) = 2^{S/2}\,\psi(2^{S}t-k)$$

It can be noticed that $\psi(2^{S}t-k)$ is obtained from a wavelet function $\psi(t)$ by a binary dilation (dilation by 2^{S}) and a dyadic translation. Using the previous, we can get the wavelet coefficients:

Equation 5. The wavelet coefficients

$$w_{k,S} = \langle f(t),\psi_{k,S}(t)\rangle nn = 2^{S/2}W_{\psi}f(\frac{k}{2^{2}},\frac{1}{2^{2}})$$

From the previous, it becomes clear that the discrete wavelet transform is a version of the wavelet transform, if scales and positions based on the power of two are chosen. An efficient way to implement this scheme using filters was developed by Mallat (1988). The filtering process, using a low-pass and a high-pass filter, separates the original signal into two different signals. The high-scale, low-frequency components constitute the approximation of the original signal, while the low-scale, high-frequency components constitute the details of the signal. The filtering process is shown in Figure 1.

This process, which is called decomposition, can be iterated. The new approximations created can be decomposed, resulting, in turn, in a lower resolution signal. The whole process results in the creation of a decomposition tree. Figure 2 illustrates the decomposition process up to level 3. The use of this discrete wavelet transform characteristic allows the representation of a sequence with length *l* only by its approximations, thus reducing its dimensionality.

Figure 1. The filtering process

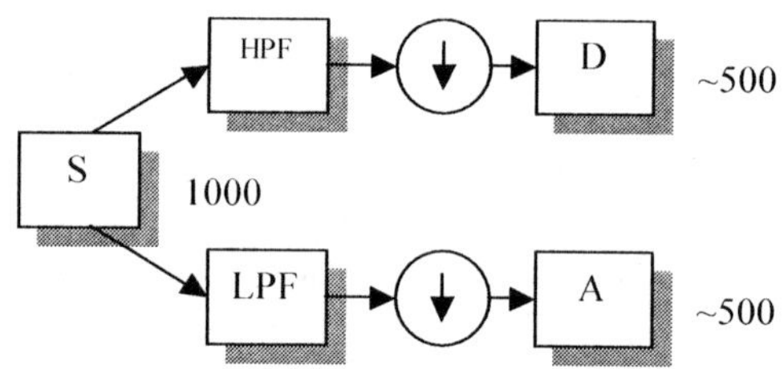

S: Sample
HPF: High pass Filter,
LPF: Low pass Filter.
D: Details,
A: Approximation.

Figure 2. The decomposition tree

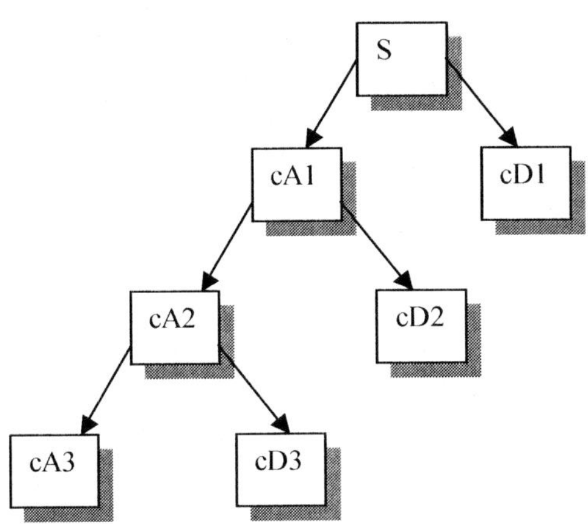

Wavelet Transform and Time Series

This section discusses the main idea behind the choice of the wavelet transform as an adequate one for approximation of time series, such as financial and temperature data sequences. A first look at some stock market fluctuations gives the impression that the values are not correlated in any way. The same happens with many other signals, like temperatures or sound recordings.

Figure 3 shows the fluctuation of the closing price of a stock for a period of 10 years. One can observe that the plot represents a random signal without any specific regularity.

In fact, it is proven that there is statistical self-similarity in many signals that represent natural phenomena (Wornell, 1996), such as:

- Geophysical time series such as variations in temperature and rainfall records.
- Financial time series such as the Dow Jones Industrial Average.
- Physiological sequences such as insulin uptake rate data for diabetics.
- Biological time series such as the voltage level across synthetic membranes and nerve.

These processes also are known as *1/f* processes. By definition, the *1/f* processes are processes that their power spectrum follows the power law relationship:

Equation 6. Power law relationship of a 1/f process

$$S_x(\omega) \approx \frac{\sigma_x^2}{|\omega|^\gamma}$$

Figure 3. A stock market fluctuation for 10 years

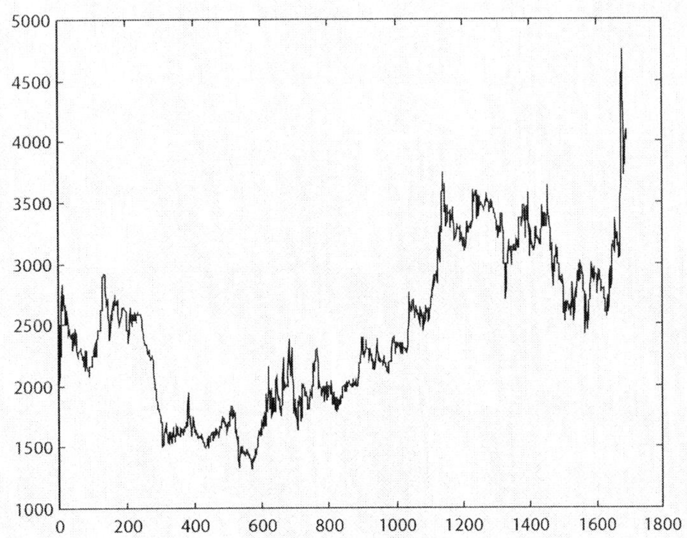

Figure 4. The Fourier transform amplitude plot (log-log) of the above sequence

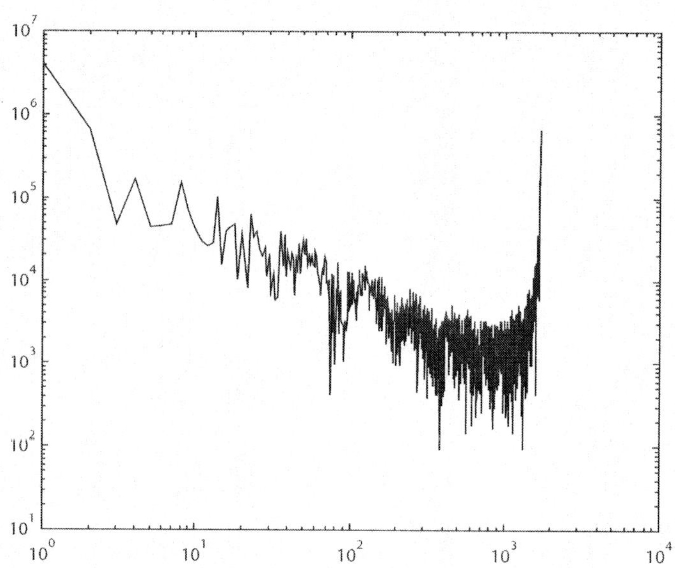

where $\gamma = 2H+1$ and H is a parameter related to the specific signal.

In practical terms, this property of a time series can be illustrated by plotting the amplitude of its Fourier Transform in a doubly logarithmic plot. Figures 3 and 4 illustrate this process.

The plotted amplitude in Figure 4 approximates closely the *1/f* line. Wornell (1996) shows that wavelet basis expansions are natural representations of *1/f* processes. Thus, it is efficient to approximate time series using the wavelet transform.

In addition to the previous mathematical argument, the theory of stock price states that the main objective is to detect trends that last between a month and a year. This agrees with the fact that the approximation coefficients of the discrete wavelet transform represent the high-scale, low-frequency components of a time series, making it an efficient tool for reduction of dimensionality for financial data.

One very interesting property of the Fourier and Discrete Fourier Transformation is Parseval's theorem, which ensures that the energy of a signal remains the same, if the signal is transformed from the time to the frequency domain. The same theorem also holds for the discrete wavelet transform (Ifeachor & Jervis, 2000). Preserving the energy of the signal also means that the Euclidean distance is the same between the two different domains. This allows the use of the discrete wavelet transform as a means to reduce the dimensionality, because it ensures no false dismissals. So, using the approximations of a sequence guarantees that no qualifying sequence will be rejected because of the transformation.

In the following discussion, we assume that the decomposition level is 1. The following results stand for any decomposition level. Let XA_0, YA_0, be the original time series. Performing the first-level decomposition, these two sequences are separated into details and approximations. That way we can calculate the Euclidean distance using the wavelet approximation and details coefficients.

Equation 7. The discrete wavelet transform preserves the Euclidean distance

$$\overrightarrow{XA_0} = \overrightarrow{XA_1} + \overrightarrow{XD_1}$$
$$\overrightarrow{YA_0} = \overrightarrow{YA_1} + \overrightarrow{YD_1}$$
$$\| \overrightarrow{XA_0} - \overrightarrow{YA_0} \|^2 = \| \overrightarrow{XA_1} + \overrightarrow{XD_1} - \overrightarrow{YA_1} - \overrightarrow{YD_1} \|^2$$

In order to ensure that there are no false dismissals due to the decrease of the dimensionality, we do the following. Suppose we want all sequences x that are similar to a query sequence q, within distance e:

$$D(\vec{x}, \vec{y}) \leq e$$

After the decomposition we have:

Equation 8. The discrete wavelet transform guaranties no false dismissals

$$\| \overrightarrow{XA_1} - \overrightarrow{YA_1} \| \leq \| \overrightarrow{XA_1} + \overrightarrow{XD_1} - \overrightarrow{YA_1} + \overrightarrow{YD_1} \| \leq e$$

We showed that after the decomposition, the Euclidean distance remains under the desired threshold *e*; thus, there are no false dismissals. This method, as with the DFT, guarantees no false dismissals but does not ensure no *false alarms*. A false alarm is raised when two sequences are considered similar, but they are not. This can happen because, as shown in Equation 8, when reducing the dimensionality of two time series, the calculated Euclidean distance is less that the actual distance calculated when using the original dimensions. So, as an extra step in the similarity retrieval algorithm, we need to check the original sequences and ensure that the similarity distance is under the desired threshold *e*.

Before we can implement the decomposition, we need to take two important decisions.

1. **Select the Wavelet Basis.** The wavelet transform does not have a single set of basic functions. The number of wavelet families is infinite, and each of them has wavelet subclasses distinguished by the number of coefficients (i.e., the filter length). The choice for the most effective wavelet function for the transformation is based on the given signal. Chan and Fu (1999) use the *Haar* wavelet. The *Haar mother* function is the simplest of all wavelets functions, making the implementation of the transformation straightforward, and it gives better approximation than the Fourier transform in all the possible dimensions. In our implementation, we choose to compare the Daubechies and the Coiflets families. The selected filter length is between four and six. Experimental results, presented in the fifth section, show that these two wavelet functions outperform the Discrete Fourier Transform for all the possible dimensions, and they also outperform the Haar transform in dimensions greater than eight.
2. **Determine the Decomposition Level.** We can continue the decomposition until the individual details consist of only one coefficient, but in practice, we need to select a maximum number of levels, so that we do not lose the necessary information about the sequence. The maximum suitable level of decomposition depends on the nature of the signal or on rules based on the *entropy* of the signal (Sutter, 1998). For our purposes, a very important factor that needs to be taken into consideration is the number of dimensions of the final time series. This depends on the indexing technique and on the trade-off between the search speed and the number of false alarms. Using an R*-Tree (Beckmann et al., 1990) for indexing performs better for dimensions between two to 10, while using the X-Tree (Berchtold et al., 1996), we can have very good performance even with 20

dimensional sequences. A comprehensive analysis of this trade-off and the decomposition level that gives better performance will be discussed in the fifth section.

The computation of the discrete wavelet transform for various decomposition levels is based on the pyramid algorithm by Cohen and Ryan (1995). This ensures that there is a Fast Wavelet Transform (FWT) with complexity N, while the Fast Fourier Transform (FFT) has complexity NlogN. This improves the efficiency of the similarity retrieval algorithm in the stage of the decomposition in order to reduce the dimensionality.

THE SIMILARITY RETRIEVAL ALGORITHM

Similarity and Time Series

Time Series data are sequences of real numbers representing measures at uniformly spaced temporal instances (Goldin & Kanellakis, 1995). The ith element of a sequence is $S[i]$, and a subsequence of S consisting of elements i through j is $S[i,j]$. The length l of the sequence $S[i,j]$ is equal to $j-i+1$. A typical example of a time series is the closing price of a stock over a working week (five days). This time series S can be denoted as $S = \{5,6,9,8,6\}$. The proposed similarity retrieval algorithm assumes different similarity problems and various similarity queries.

There are three major categories of similarity problems (Agrawal et al., 1993), as discussed next.

- **Whole Matching**, where two or more sequences with the same length n are compared.
- **Subsequence Matching.** In this case, the query sequence is smaller than the main sequence. We compare all the subsequences in the main large sequence with the query sequence in order to find those subsequences that are similar to the query sequence.
- **Multiple Subsequence Matching.** In this case, two large time series are defined. Two sets of query sequences are created by applying a moving window over the two large time series. Let the first set be the query set. We compare each of the subsequences of the query set with every subsequence of the other set in order to find the similar pairs of sequences.

As far as the similarity queries are concerned, we use the following concepts (Goldin & Kanellakis, 1995).

Approximate Matching of Time Series Data

Given a tolerance $e \geq 0$ and a distance metric D between sequences, sequences $S1$ and $S2$ match approximately within tolerance e when $D(S1,S2) \leq e$.

Similar Time-Series Data

In many cases, we are not interested in matching sequences that are only close to each other according to their Euclidean distance. For example, two companies may have exactly the same fluctuations in their closing price, but one could be worth twice the other. More formally, we can say that for the two sequences S1 and S2, we have S1(i) = a* S2(i) for 1 < i < N, where N is the length of the sequences. The difference between these two sequences is called *scale*. In another case, we may have two companies' ending prices that start at different values but then go up and down in exactly the same way. More formally, we can say that for the two sequences S1 and S2, we have S1(i) = b + S2(i)) for 1 < i < N, where N is the length of the sequences. The difference between these two sequences is called *shift*. There are also cases that we have combinations of scaling and shifting.

Figure 5. Four similar time sequences

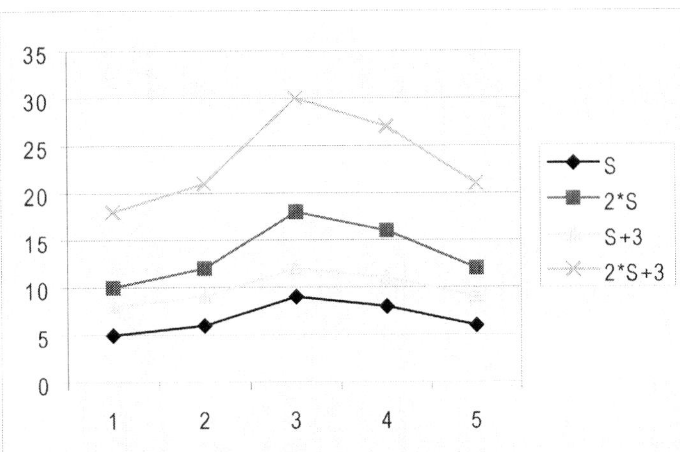

Figure 5 shows four different series S1 = *{5,6,9,8,6}*, *S2* = *2*S1* = *{10,12,18,16,12}*, *S3* = *S1+3* = *{8,9,12,11,9}* and *S4* = *2*S1* + *3* = *{18,21,30,27,21}*. These four sequences are considered to be similar according to the previous definition. In that notion, we define that:

Two sets of points are considered similar, if a transformation exists that maps one to the other.

Approximately Similar Time-Series Data

Having defined the notion of *similarity* and given a distance metric *D* between two sequences and a tolerance $e \geq 0$, two sequences, *Q* and *S*, are *approximately similar* within tolerance *e*, when there exists a similarity transformation *T* so that D(Q,T(S)) R e. If $e = 0$, then we have exact similarity like the sequences in Figure 5.

As described in the second section, there are various similarity distance metrics used in the literature. The choice of the appropriate similarity metric depends on the application domain. The most popular distance metric used in similarity retrieval in time series is the Euclidean distance between sequences. This is because it gives good results in most cases, and it also can be used with any other type of similarity measure, as long as this measure can be expressed as the Euclidean distance between vectors in a feature space. Another reason is that the property preserves its value under transformations like the Discrete Fourier Transformation (DFT) and the discrete wavelet transformation.

The Euclidean distance between two sequences, S1 and S2, that have the same length n, is defined as follows:

Equation 9. The Euclidean distance

$$D_E(S_1, S_2) = \sqrt{\sum_{1 \leq i \leq n} (S_1[i] - S_2[i])^2}$$

Having discussed the notion of approximate matching and similar time series, we give the following definition:

Given two sequences, S1 and S2, the similarity distance between them is the distance between their normal forms. (Goldin & Kanellakis, 1995)

$$D_S(S1,S2) = D_E(v(S1), v(S2))$$

To pose a similarity query between two sequences, *Q* and S, we define the following parameters:

- Distance threshold e, $e \geq 0$.
- Lower average and Upper average (l_a,u_a).
- Lower deviation and Upper deviation (l_σ,u_σ).

Using the above variables, we define the semantics of six different *similarity range queries* between a query sequence Q and a set of sequences S (whole matching):

1. **General Query:** find all *[S,x,y]* such that $D(Q,xS+y) \leq e$, $l_\alpha \leq x \leq u_\alpha$ and $l_\sigma \leq y \leq u_\sigma$.
2. **Unbounded Query:** find all *[S,x,y] such that* $D(Q,xS+y) \leq e$.
3. **Scaling:** find all [S,x] such that $D(Q,xS) < e$.
4. **Shifting:** find all [S,y] such that $D(Q,S+y) \leq e$.
5. **Approximate Match:** find all S such that $D(Q,S) \leq e$.
6. **Exact Similarity:** find all *[S,x,y]* such that $Q = xS+y$.

K nearest neighbor queries (whole matching) are defined using the number K of the nearest neighbor sequences we are interested in instead of the distance threshold *e*. More specifically, for a query sequence *Q* and a set if sequences *S*, we have:

7. **General Query:** find the first K *[S,x,y]* such that $l_\alpha \leq x \leq u_\alpha$ and $l_\sigma \leq y \leq u_\sigma$, *ordered by the value of D(Q,xS+y) in increasing order.*
8. **Unbounded Query:** find the first K *[S,x,y] ordered by the value of D(Q,xS+y).*
9. **Scaling:** find the first K [S,x] ordered by the value of *D(Q,xS).*
10. **Shifting:** find the first K [Si,y] ordered by the value of *D(Q,Si+y).*
11. **Approximate Match:** find the first K S ordered by the value of D(Q,S).
12. **Exact Similarity:** find the first K *[S,x,y]* such that $Q = xS+y$, order by search order. If there are more than *K exact similar* sequences, then there is loss of information. It there are less than *K exact similar* sequences, then only the exact similar ones are retrieved.

Figure 6. The similarity retrieval algorithm

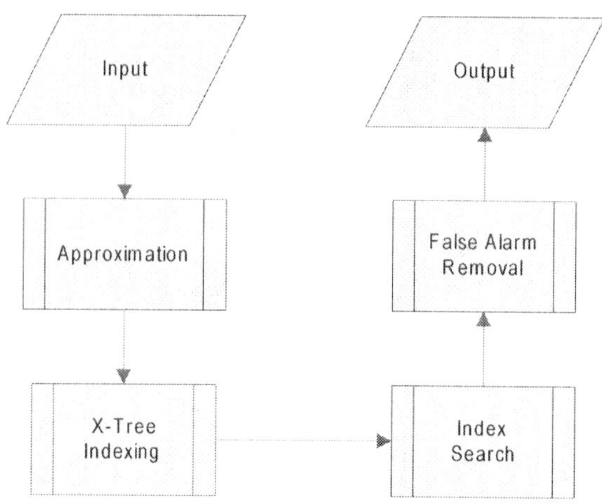

In the case of *subsequent matching*, we have the same classes of queries, but the set of sequences S is created from the sliding window that runs over the larger sequence. For the *multiple subsequence match*, both the query sequence Q and the sequence S are obtained by a moving windows over larger time series.

The Algorithm

The proposed similarity algorithm is shown diagrammatically in Figure 6. Regarding the input, for a whole match query, input data are all the sequences that are checked for identifying similarities. For a subsequence query, the input data are the sequences produced by applying a moving window of size equal to the size of the query sequence over the large time series. In both cases, the input data include also the query sequence.

Approximation

The first step of the algorithm is the *Approximation*. The objective of this step is to map each sequence to the multidimensional space, so that each time series is represented by a point that can be indexed using a spatial indexing structure. The approximation process is responsible for reducing the dimensionality of the input time series. Let the original time series be $S = \{s_1, s_2, \ldots, s_N\}$ and the original query time series be $Q = \{q_1, q_2, \ldots, q_N\}$, $N > 0$. For large N, the calculation of the Euclidean distance between time series S and Q is a complex process that becomes even more complex, if we consider that the similarity retrieval algorithm could involve a large number of time sequences.

As part of this step, the discrete wavelet transform is used in order to ensure no false dismissals and to reduce the false alarms introduced due to the dimensionality reduction. For the decomposition of the sequences, the Daubechies family of wavelets with length 4 is used. The algorithm used for the decomposition process is shown in Figures 7 and 8. Figure 7 demonstrates the use of the fast discrete wavelet transform for sequences of length $l = 2^s$, while Figure 8 presents the decomposition procedure in the general case.

Figure 7. Decomposition for a time sequence of length 2^s

```
Input: Sequence S0 of length l
Output: Sequence S'
[S1,S2,...,Sn] = FWT(S); // perform wavelet decomposition of S
S' = SAI;   // I the first approximation level that consists of / //
            less than 20 coefficients; it can be also S0
```

Figure 8. Decomposition for a time sequence of any length

```
Input: Sequence S0 of length l
Output: Sequence S'

While (l > 20)
        S = DWT(S);         //perform one level wavelet
                            //decomposition of S
S' = S;
```

X-Tree Indexing and Index Search

The next step of the algorithm includes the indexing of the low-dimensional sequences. The X-Tree indexing structure has been chosen because of its efficiency in indexing points of up to 20 dimensions (Berchtold et al., 1996). The X-Tree, in contrast with the R*-Tree or other structures of the spatial trees family, performs better for dimensions greater than six. Furthermore, the X-Tree has exactly the same behavior as the R*-Tree for dimensions less than six. An R-Tree, in general, is a height-balanced tree with index records in its leaf nodes containing pointers to data objects Guttman, 1984). The structure is designed so that a spatial search requires visiting only a small number of nodes. An example of some point in the two-dimensional space and the corresponding R-Tree is presented in Figure 9. The points into the two dimensional space are divided into groups that are covered by hyper-rectangles. These hyper-rectangles are divided into super-groups defined also by other hyper-rectangles, and so forth. Each hyper-rectangle is stored in an internal node of the tree, while the corresponding data points are stored into the leaf nodes.

The main problem in the design of the R-Trees is that by increasing the dimensionality, many overlapping hyper-rectangle areas are introduced. This reduces the performance of the query processing, because the overlap of directory nodes results in the necessity to follow multiple paths, even when searching between points (Berchtold et al., 1996). The X-Tree tries to avoid overlaps. When this is not possible, it uses the supernodes that are extended variable size directory nodes. As a result, the X-Tree has three different types of nodes: the data nodes, the normal directory nodes, and the supernodes. The supernodes avoid splits in a directory that could result in the creation of an inefficient structure. The structure of the X-Tree is presented in Figure 10, where the supernodes are shown in gray.

Figure 9. The R-Tree structure

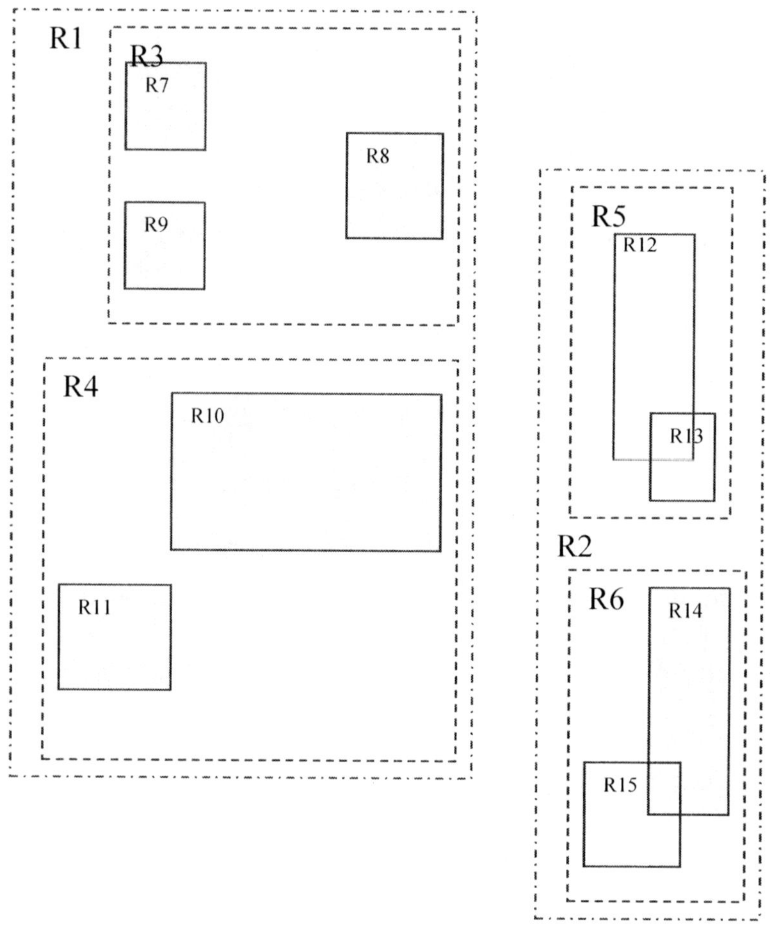

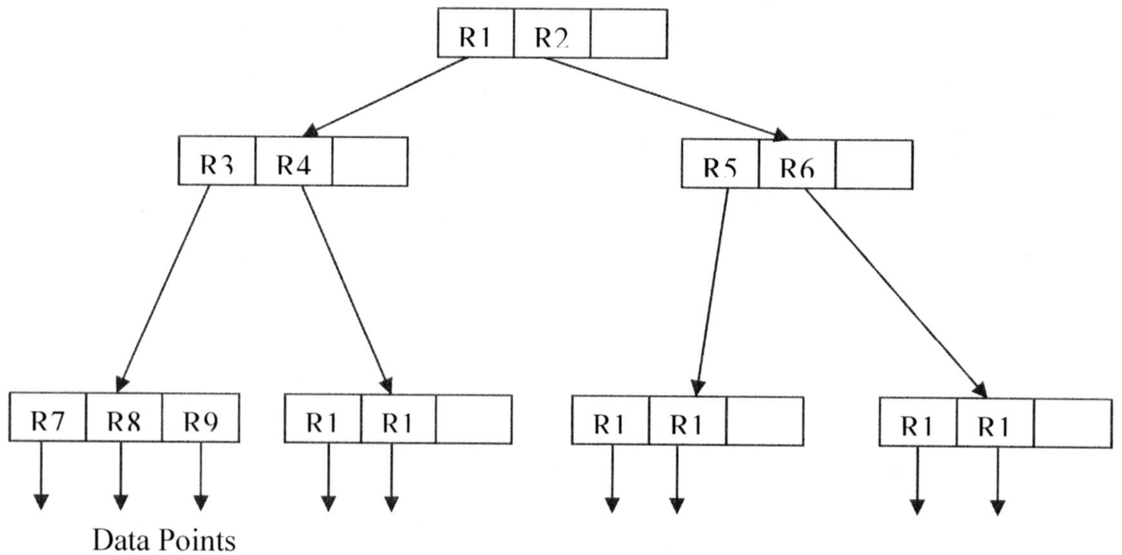

Figure 10. The X-Tree structure

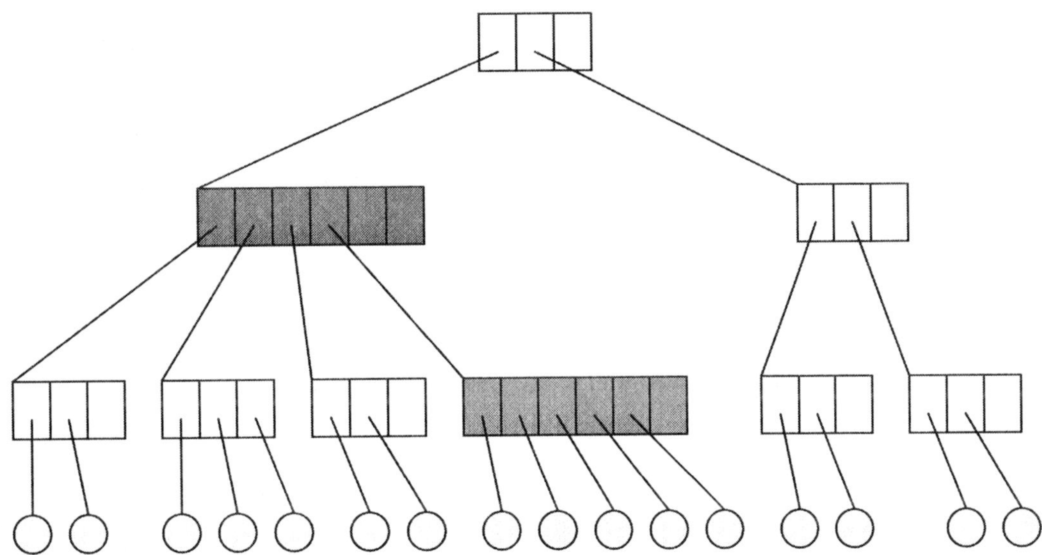

The process of indexing the time series and searching through the index is as follows. After the decomposition of the sequences to the desired level, an X-Tree with dimensionality equal to the number of coefficients of the wavelet approximation is constructed, and all the time series data are inserted into the tree. This process is shown in pseudo code in Figure 11.

Then, based on the query parameters, either a range query or a K nearest neighbor query is performed. This process is shown in pseudo code in Figure 12.

Figure 11. Create index algorithm

```
Input: The decomposed set of time Series S'i with length l.
Output: The X-Tree.

Xt = Create_Xtree(l);  // Create an X-Tree of dimension l.
for(all S'i) {
            Xt.insert(S'i);
}
```

False Alarm Removal

The next step of the algorithm is to remove the false alarms that are introduced because of the dimensionality reduction. This is done by evaluating the Euclidean distance between the original values of the query sequence and the sequences that qualify from the step of *Index Search.* This calculation can be time-consuming, especially for long-time series and for a larger number of qualified sequences. Again, the wavelet transform helps to reduce the complexity of this step. Let *s* be the

Figure 12. Search index algorithm

```
Input: The X-Tree xt, and the query parameters.
Output: The qualifying sequences [R'].

if(range_query) // range query
       [R'] = Xt.range_query(Q',range);        // Q' = the decomposed query
                                    // sequence
else          // K nearest neighbor query
       [R'] = Xt.knn_query(Q',k);
```

Figure 13. False alarms removal

```
Input: The qualifying time series [R'].
Output: The similar time series [R].

for(i =0; i < l; i++) {                          // for every time series R'i
        d = Euclidean_distance(Q,Ri);            // Q,Ri : Non decomposed T-S
        if(d < e)
                R.add(Ri);                       // add sequence Ri to the set of R
}
```

decomposition level. Before indexing the wavelet approximation coefficients of this level, the approximation coefficient of level *s-1* is stored. This step does not add any computational complexity, because the *s-1* level coefficients are calculated before the *s* level coefficients are retrieved. During the *False Alarms Removal* step, the Euclidean distance between the qualified sequences using the s-1 level coefficients is calculated. If the Euclidean distance is bigger than the predefined threshold, then the time series disqualifies, and it is removed. The removal is performed only in the case of the range queries, while for the K nearest neighbor queries, only the calculation of the real distance of the two sequences is used.

This helps to reduce further the number of false alarms and to reduce the complexity of the actual False Alarm Removal procedure. The algorithm for the removal of the false alarms is presented in Figure 13.

In the next section, the evaluation of the proposed sequence retrieval algorithm is discussed.

EVALUATION

This section describes the performance evaluation of the proposed similarity retrieval algorithm using two data sets. The evaluation of the performance of the proposed algorithm is based on three factors:

1. The quality of the approximation, which relates to the time spent to decompose the sequences. This consists of the time *t_Tranform* to decompose the sequences and the time *t_TreeBuild* to add the data to the index.
2. The time *t_Search*, which is needed to search through the index.
3. The time *t_FalseAlarmsReduction*, which is the time needed to reduce the false alarms.

The performance comparison is based on the total execution time, which is a function of the three factors and defined as:

t_Total = (t_Tranform + t_TreeBuild) + t_Search + t_FalseAlarmsReduction.

As part of the evaluation process, it was also possible to fine-tune the proposed algorithm in order to:

- Determine which mother wavelet function approximates better the time series.
- Identify the optimum number of coefficients, given that a large number of coefficients reduces the false alarms but decreases the efficiency of the indexing mechanism.
- Determine how the performance changes, given the size *n* of the sequences.
- Determine how the search time increases, given the number of time series.

For the evaluation, two datasets were used. The first dataset contains temperature measurements from Athens. These data include hourly temperature readings for the period from October 1, 1997 to September 30, 1998, and it contains 8,760 records. More specifically, the values for each record include the date, the hour in the day, and the temperature reading. The data set was provided by the Department of Civil Engineering of the National Technical University of Athens. (http://www.hydro.ntua.gr/).

The second dataset contains data from the Greek stock market provided by the Greek Stock Exchange. The data include daily stock prices taken for five days a week from November 26, 1990 to April 17, 1991, and it contains 4,455 records. More specifically, the values for each record include the share name, the transaction date, the minimum price of the share, the maximum price of the share, the closing price of the share, and the number of the exchanged shares. For simplicity reasons, only the closing price for each share is taken into consideration in this evaluation.

The experiments were performed in a SUN Spark Station5 with 512 Mbytes of main memory and enough free disk space running under the Solaris operating system. The implementation of the system was done in Visual C++ with its own data structures and was developed as a COM object so that it can be reused by other data mining software.

Quality of Approximation

The evaluation for the quality of the approximation was performed for the two datasets using a large time series of 1,900 data entries and varying the size of the query series from 32 to 512. A variety of wavelet functions also were used in order to investigate which one performed better.

In general, the quality of the approximation indicates which transformation estimates better the time series. A measurement of the approximation quality is the Precision, which is defined as:

$$Precision = Qtime/Qtransform$$

where *Qtime* is the number of sequences qualified when calculating the Euclidean distance in the time domain. *Qtransform* is the number of sequences qualified when calculating the Euclidean Distance in the transformed domain. The quality of the approximation was evaluated for the Fourier transform and the wavelet transform using as mother functions the Haar, the Coiflet 1, and the Daubechie 2. The experiments were carried out using the following pairs of values for the query size and the distance: (32,500), (64,600), (128, 1500), (256, 3000) for the Greek stock market dataset and (128,150) for the temperature dataset. The mean values of the results for the three different stocks have been used.

Figure 14 illustrates the precision of the approximation for the different query sizes in the case of the stock market data, and Figure 15 illustrates the precision of the approximation for the temperature data.

In Figure 14, we can see that the wavelet transformation outperforms the Fourier transformation in all cases. The range is between 10% for two to four coefficients and 75% for 14 to 20 coefficients. Another interesting observation is that the Fourier transform does not improve its precision from a specific number of coefficients upwards. This means that there is a limit to the utilization of the Fourier transform up to eight coefficients. Even in this low dimensionality, the wavelet transform gives better precision, so it is considered to be more effective in approximating the time series.

Some differences also can be identified between the various wavelet functions used for the wavelet decomposition. From Figures 14 and 15, it can be seen that for small number of dimensions (e.g., two to 12), the Haar transform outperforms all the others by an amount of 2% to 8%, while for dimensions 12 to 20, the Db2 transform outperforms all the others by an amount of 5% to 10%. For dimensions higher than 20, Coif1 and

Figure 14. Precision of wavelet approximation for stock market data

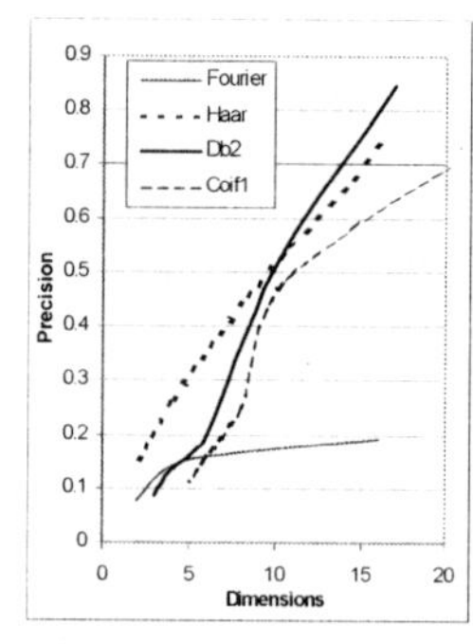

(a) Query Size = 32, Distance = 500

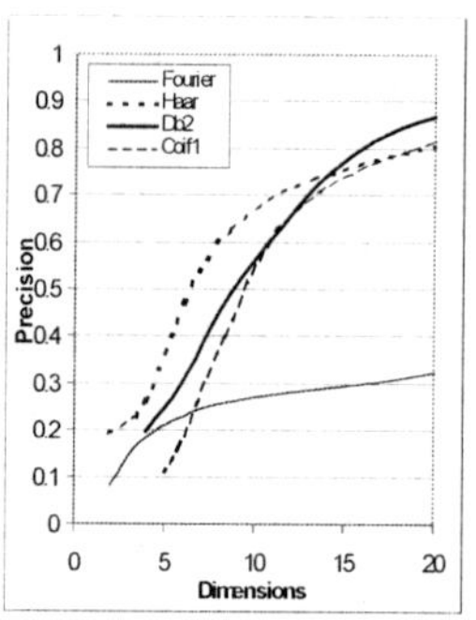

(b) Query Size = 128, Distance = 1500

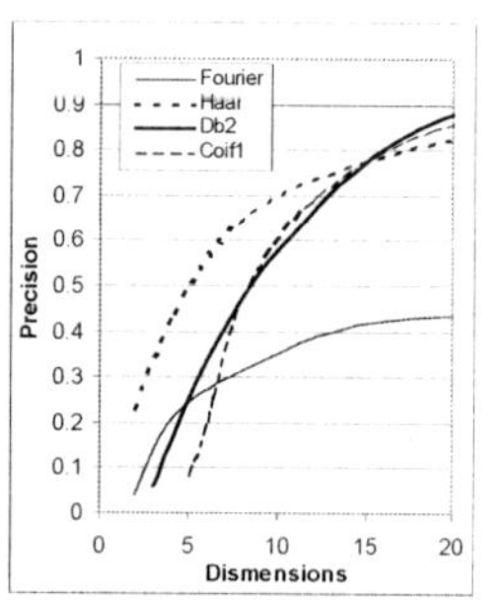

(c) Query Size = 64, Distance = 600

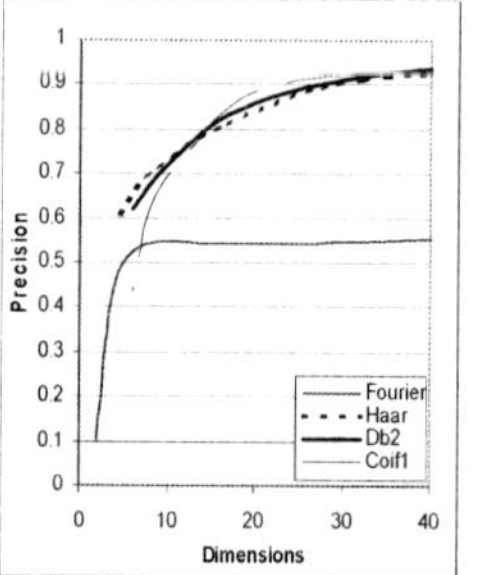

(d) Query Size = 256, Distance = 3000

Figure 15. Precision of wavelet approximation for temperature data

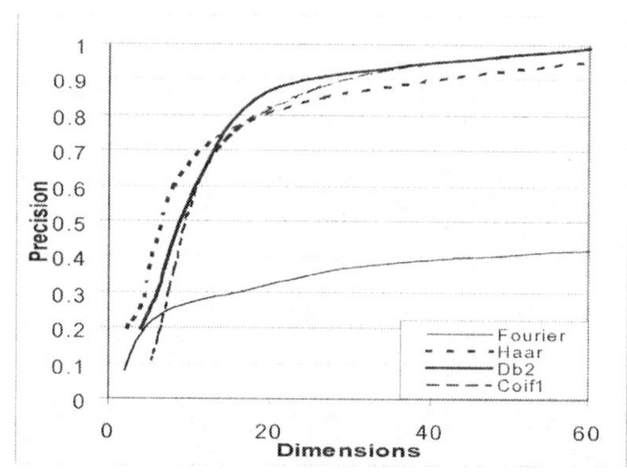

the other wavelet transforms give better precision, as shown in Figure 14d and Figure 15. If the dimensionality is around 40, then there is a 10% to 15% improvement in the approximation precision.

Based on the previous observations, it was decided to use the Db2 transform in the proposed algorithm, using a number of coefficients between two and 20. In this dimensionality, the Db2 transform gives 5% to 10% less false alarms than any other transformation. The same results are obtained by applying the wavelet transform to the temperature data, as shown in Figure 15. In this case, though the precision of the wavelet transform is lower, the size of the sequences becomes higher.

Searching the Index

The performance of the indexing technique depends on two time factors. The first is the time needed to build the index, and second is the time needed to search through the index structure. The first factor can be considered less important, because the index is built once, and this also includes the time of the transformation.

The main goal of the evaluation regarding indexing is to compare the performance of the X-Tree against the performance of the R*-Tree. In Berchtold et al. (1996), it is shown that the X-Tree outperforms the R*-Tree. More specifically, for dimensions from two to eight, the speedup factor is between 10 and 30, while for higher dimensions (e.g., for D = 16), the speedup factor is 270. The experiments carried out as part of this work used a synthetic database of uniformly distributed data, and the speedup obtained was even higher, reaching the factor of 430 for D = 16. Based on this, the X-Tree is preferred over the R*-Tree for any dimension. Even in lower dimensions between two to four, the performance gain is important.

False Alarms

Furthermore, it was investigated whether the increase of the dimensionality introduces additional delays. More specifically, it was investigated whether the increase from eight to 16 or 20 is going to affect the performance of the proposed algorithm.

Taking into consideration the discussion about the quality of the approximation earlier in this section, the Precision of the Db2 transform for dimensions 16 to 20 is from 20% to 25%, higher than the Precision of the Haar transform for dimensions six to eight. Thus, it is worth investigating the total execution time that is a factor of the time to search the index and the time for the false alarm detection and removal.

The experiments were performed in a SUN Spark Station with 512 Mbytes of main memory and adequate disk space. The page size for the X-Tree was set to 1024 Kbytes, which is considered a reasonable amount for achieving good performance results (Berchtold et al., 1996). The test data that were used during the evaluation process were real data retrieved from the approximation of the Greek stock market data. The number of dimensions used were D = 4 to D = 20.

The experiments performed for point queries used a data set of 2.5 Mbytes. For dimensions four to 14, the Haar approximations were used, while for dimensions 16 to 20, the Db2 approximations were used. Five different data sets were used, and the mean values were taken into consideration. Figure 16 presents the execution time of point queries for different dimensions. As it can be seen from Figure 16, the total execution time is reduced for dimensions 16 to 20. The time needed to reduce the false alarms affects significantly the total execution time. This is more true when the size of the sequences becomes higher, because the search time remains the same, while the time to reduce the false alarms increases. For dimensions higher than 20, the search time increases, while the time needed for the reduction of the false alarm decreases slightly, forcing the total execution time to increase.

Figure 16. Execution time for point queries

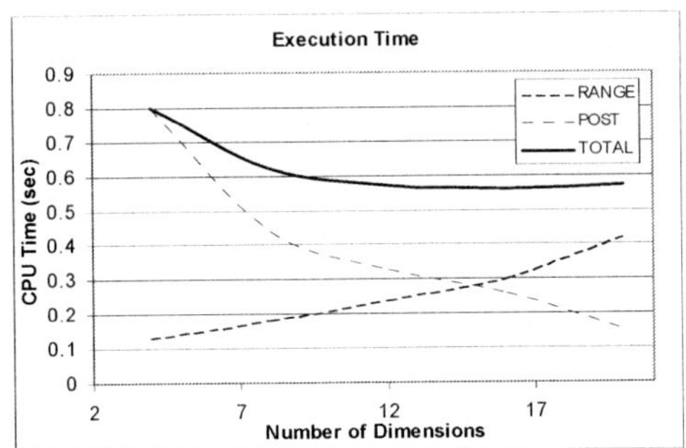

Figure 17. Execution time as a factor of the number of sequences

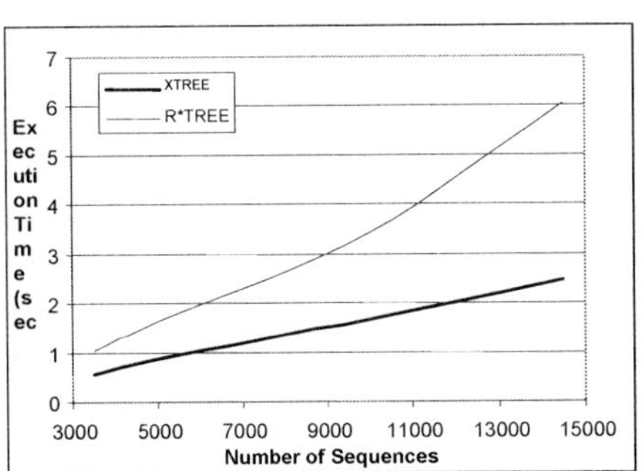

Scalability

Furthermore, the impact of the database size and the query size was evaluated by performing experiments and by measuring the performance using different number and varying the size of sequences.

The number of sequences evaluated was between 3,500 and 10,000, while the sequence size was varied from 64 to 1,024. All the tests have been performed for 16 dimensions. The performance measurements for the algorithm using the R*-Tree has been done using eight dimensions, because in that dimensionality, the R*-Tree performs better, as shown in Chan and Fu (1999) and, Chan et al. (2003). Figure 17 and Figure 18 show the results of these two experiments.

The performance evaluation experiments show that the use of the X-Tree for indexing time series outperforms the R-Tree. Moreover, the usage of the X-Tree allows to approximate the time series using 16 to 20 dimensions rather than using six to eight dimensions, which is the case using the R*-Tree as an indexing technique.

The fact that we are working with dimensions 16 to 20 suggests the use of the Db2 wavelet mother function for approximating the time series, because, in that dimensionality, it gives better approximation precision. Finally, the experiments showed that removing the false alarms using the s-1 level approximation coefficients optimizes the performance of the post-processing step of the proposed algorithm. Figure 19 shows that for query size equal to 256, a performance improvement of 7% is obtained. For bigger query sizes, the improvement increases up to 12% due to the fact that there is a higher difference in the approximation precision between the two levels of decomposition.

CONCLUSION

Time series comprise a large proportion of the data stored in today's databases. In scientific laboratories or financial departments of large companies, numerical values are measured and stored for further elaboration. One of the main problems that needs to be solved is the identification of similarities between time series. Due to the large amount of time series data, a fast algorithm to retrieve the similarities has to be developed.

Several techniques have been proposed to speed up the similarity identification process. Having to deal with gigabytes of data, even small improvements can be beneficial. In this article, a similarity retrieval algorithm for time series is proposed. In the proposed algorithm, the wavelet transformation is used in order to reduce the dimensionality of the time series, and the transformed series are indexed using X-Trees. This indexing technique allows using a

Figure 18. Execution time as a factor of the sequences size

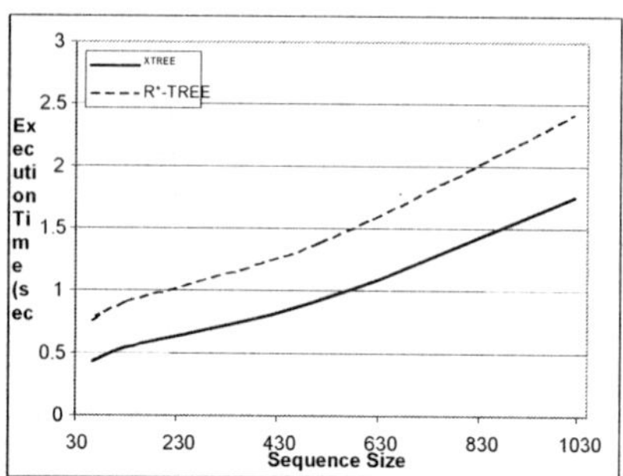

Figure 19. Removing false alarms using the s-1 level coefficients

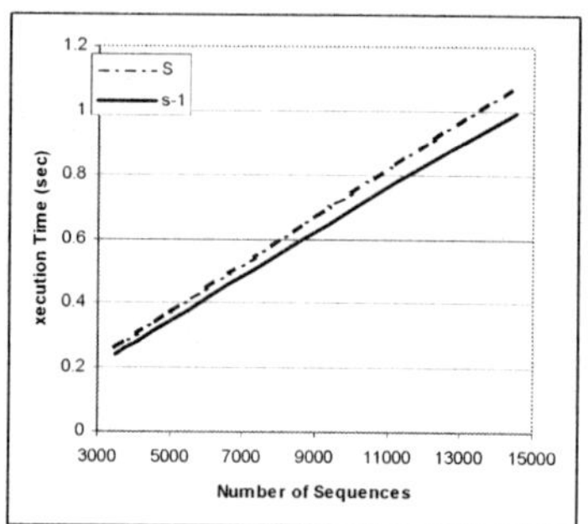

higher dimensionality for indexing without important loss in performance. Furthermore, it has been shown that the wavelet transform reduces the false alarms significantly compared to the Fourier transform and provides better approximation of the time series. Also, using a number of dimensions between 16 and 20, the performance is optimized. Comparison has been done with techniques using the Haar transform and the R*-Tree (Chan & Fu, 1997; Wu et al., 2000; Chan et al., 2003) and the proposed algorithm is shown to outperform them.

Furthermore, a novel mechanism for reducing the complexity of the false alarms removal calculation is proposed. Storing the approximation coefficients of the penultimate level of the decomposition tree, the Euclidean distance between the two sequences is calculated, thus reducing further the number of false alarms before calculating the actual Euclidean distance using the whole time series.

REFERENCES

Agrawal, R., Faloutses, C., & Swami, A. (1993, October). Efficient similarity search in sequence databases. In *Proceedings of the 4th International Conference on Foundations of Data Organization and Algorithms*, Chicago.

Agrawal, R., Giuseppe, P., Wimmers, E. L., & Zaït, M. (1995a, September). Querying shapes of histories. In *Proceedings of 21st International Conference on Very Large DataBases*, Zurich, Switzerland.

Agrawal, R., Lin, K., & Sawhney, H. S. (1995b, September). Fast similarity search in the presence of noise, scaling, and translation in time-series databases. In *Proceedings of the 21st International Conference on Very Large Databases*, Zurich, Switzerland.

Beckmann, N., Kriegel, H. P., Schneider, R., & Seeger, B. (1990). The R*-Tree: An efficient and robust access method for points and rectangles. In *Proceedings of the ACM SIGMOD International Conference on Management of Data.*

Berchtold, S., Keim, D. A., & Kriegel, H. P. (1996). The X-tree: An index structure for high-dimensional data. In *Proceedings of 22nd International Conference on Very Large DataBases.*

Chan, K. P., & Fu, W. C. (1999, March). Efficient time series matching by wavelets. In *Proceedings of the 15th International Conference on Data Engineering*, Sydney, Australia.

Chan, K. P., Fu, W. C., & Yu, C. (2003). Haar wavelets for efficient similarity search of time-series: With and without time warping. *IEEE Transactions on Knowledge and Data Engineering, 15*(3).

Chu, K. K. W., & Wong, M. H. (1999). Fast time-series searching with scaling and shifting. In *Proceedings of the Principles of Database Systems.*

Cohen, A., & Ryan, R. D. (1995). *Wavelets and multiscale signal processing.* Chapman and Hall.

Faloutsos, C., Ranganathan, M., & Manolopoulos, Y. (1994). Fast subsequence matching in time-series databases. In *Proceedings of the ACM SIGMOD International Conference on Management of Data*, Minneapolis, Minnesota.

Gelb, A. (1986). *Applied optimal estimation.* MIT Press.

Goldin, D. Q., & Kanellakis P. C. (1995). On similarity queries for time-series data: Constraint specification and implementation. In *Proceedings of the First International Conference on Constraint Programming*, Cassis, France.

Guttman, A. (1984). R-trees: A dynamic index structure for spatial searching. In *Proceedings of the SIGMOD Conference.*

Ifeachor, E., & Jervis, B. (2000). *Digital signal processing*. 2/E Prentice Hall.

Jaideva C., Goswami, A., & Chan, K. (1999). Fundamentals of wavelets, theory algorithms, and applications. In *Microwave and optical engineering*. Wiley.

Kahveci, T., & Singh, A. K. (2004). Optimising similarity search for arbitrary length time series queries. *IEEE Transactions on Knowledge and Data Engineering, 16*(4).

Kanth, K. V. R., Agrawal, D., & Singh, A. (1998). Dimensionality reduction for similarity searching in dynamic databases. In *Proceedings of the ACM SIGMOD Conference*.

Kim, J., & Park, S. (2005). Periodic streaming data reduction using flexible adjustment of time section size. *International Journal of Data Warehousing and Mining, 1*(1), 37-56.

Kopanakis, I., & Theodoulidis, B. (2003). Visual data mining modelling techniques for the visualisation of mining outcomes. *Journal of Visual Languages and Computing, 14*(6), 543-589.

Korn, F., Jagadish, H. V., & Faloutsos, C. (1997). Efficiently supporting ad hoc queries in large datasets of time sequences. In *Proceedings of the SIGMOD*, Tucson, Arizona.

Lee, S-L., Chun, S-J., Kim, D-H., Lee, J-H, & Chung, C-W. (2000). Similarity search for multidimensional data sequences. In *Proceedings of the International Conference on Data Engineering*.

Mallat, S. (1999). *A wavelet tour of signal processing*. Academic Press.

Park, S., Chu, W. W., Yoon, J., & Hsu, C. (2000). Efficient searches for similar subsequences of different lengths in sequence databases. In *Proceedings of the International Conference on Data Engineering*.

Pelekis, N. , Theodoulidis, B., Kopanakis, I., & Theodoridis, Y. (2005). Fuzzy miner: Extracting fuzzy rules from numerical patterns. *International Journal of Data Warehousing and Mining, 1*(1), 24-35.

Perng, C-S., Wang, H., Zhang, S. R., & Parker, D. S. (2000). Landmarks: A new model for similarity-based pattern querying in time series databases. In *Proceedings of the International Conference on Data Engineering*.

Rafiei, D. (1999, March). On similarity-based queries for time series data. In *Proceedings of the 15th International Conference on Data Engineering*, Sydney, Australia.

Rafiei, D., & Mendelzon, A. O. (1997). Similarity-based queries for time series data. In *Proceedings of the ACM SIGMOD International Conference on Management of Data*, Tucson, Arizona.

Rafiei, D., & Mendelzon, A. O. (1998). Efficient retrieval of similar time sequences using DFT. In *Proceedings of the International Conference on Foundations of Data Organization and Algorithms*.

Rafiei, D., & Mendelzon, A. O. (2000). Querying time series data based on similarity. *IEEE Transactions on Knowledge and Data Engineering, 12*(5).

Rafiei, D., & Mendelzon, A. O. (2002). Efficient retrieval of similar shapes. *VLDB Journal, 11*(1).

Shahabi, C., Tian, X., & Zhao, W. (2000). TSA-tree: A wavelet-based approach to improve efficiency of multi-level surprise and trent queries. In *Proceedings of 12th International Conference on Scientific and Statistical Database Management*.

Suter, B. W. (1998). *Multirate and wavelet signal processing*. Academic Press.

Taniar, D., Rahayu, J. W., & Tan. R. B-N. (2004, March). Parallel algorithms for selection query processing involving index in parallel database systems. *International Journal of Computer Systems Science & Engineering, 19*(2), 95-114.

Vlachos, M., Kollios, G., & Gunnopoulos, D. (2002, February). Discovering similar multidimensional trajectories. In *Proceedings of the International Conference on Data Engineering.*

Wang, C., & Wang, X. S. (2000). Supporting content-based searches on time series via approximation. In *Proceedings of the 12th International Conference on Scientific and Statistical Database Management.*

Wornell, G. W. (1996). *Signal processing with fractals. A wavelet-based approach.* Prentice Hall.

Wu, T-L., Agrawal, D., & El-abbadi, A. (2000). A comparison of DFT and DWT based similarity search in time-series databases. In *Proceedings of the International Conference on Information and Knowledge Management.*

This work was previously published in International Journal of Data Warehousing and Mining, Vol. 2, Issue 2, edited by D. Taniar, pp. 55-81, copyright 2006 by IGI Publishing, formerly known as Idea Group Publishing (an imprint of IGI Global).

Chapter 2.34
Cluster-Based Input Selection for Transparent Fuzzy Modeling[1]

Can Yang
Zhejiang University, China

Jun Meng
Zhejiang University, China

Shanan Zhu
Zhejiang University, China

ABSTRACT

Input selection is an important step in nonlinear regression modeling. By input selection, an interpretable model can be built with less computational cost. Input selection thus has drawn great attention in recent years. However, most available input selection methods are model-based. In this case, the input data selection is insensitive to changes. In this article, an effective model-free method is proposed for the input selection. This method is based on sensitivity analysis using Minimum Cluster Volume (MCV) algorithm. The advantage of our proposed method is that with no specific model needed to be built in advance for checking possible input combinations, the computational cost is reduced, and changes of data patterns can be captured automatically. The effectiveness of the proposed method is evaluated by using three well-known benchmark problems that show that the proposed method works effectively with small and medium-sized data collections. With an input selection procedure, a concise fuzzy model is constructed with high accuracy of prediction and better interpretation of data, which serves well the purpose of patterns discovery in data mining.

INTRODUCTION

Fuzzy Inference System (FIS) is one of the most important applications in fuzzy logic and fuzzy set theory (Zadeh, 1973). FIS is useful for description, prediction and control in many fields, such as data mining, diagnosis, decision support, system identification and control. The strength of FIS comes from two aspects: (1) it can handle linguistic concepts; (2) it is universal approximations (Wang & Mendel, 1992). Although expert knowledge can be easily incorporated in building a FIS, it has been seen that the expert knowledge would lead to an insufficient accuracy of FIS in modeling complex systems. So in recent years, many researchers have attempted to generate FIS from observed data (Wang & Mendel, 1992; Jang, 1993; Babuska, 1998; Sugeno & Yasukawa, 1993; Kosko, 1997).

In real-world applications such as system identification problems, it is common to have tens of potential inputs to a model under construction. The excessive inputs not only impair transparency of the underlying model, but also increase computational complexity in building the model. Therefore, the number of inputs actually used in modeling should be reduced to a sufficient minimum, especially when the model is nonlinear and has high dimensionality. Input selection is thus becoming a crucial step for the purposes of: (1) removing noises or irrelevant inputs that do not have any contribution to the output; (2) removing inputs that depend on other inputs; (3) making the underlying model more concise and transparent. A large array of input selection methods, like analysis of correlation, the principal component analysis (PCA) and the least squares method have been introduced in linear regression problems. However, they usually fail to discover significant inputs in real-world applications which often involve nonlinear modeling. Relatively a few methods are available for input selection in nonlinear modeling. These methods found in literature can generally be divided into two categories:

1. **Model-based methods** that use a specific model to search for significant inputs. Finding an optimal solution of input selection often requires examining different models for all possible combinations of inputs, which becomes computationally intractable even for a reasonable number of input attributes. In order to avoid this, heuristic criteria often were introduced. A relatively simple and fast method was proposed in Jang (1996) by using ANFIS. This method is based on an assumption that the ANFIS model with the smallest root mean squared error (RMSE) after one epoch of training has a greater potential to achieve a lower RMSE when given more epochs of training. The heuristic method, which generates ANFIS sequentially by involving increased number of inputs, is called forward selection. Although this method was developed for ANFIS, the same idea could be used for other types of FISs. Methods presented in Tanaka, Sano, and Watanabe (1995) and Chiu (1996) used a different heuristic method; namely, backward selection. Lin and Cunningham (1994) proposed a method based on fuzzy curves that represent the sensitivity of an output with respect to other inputs. Some other methods, based on criteria like individual discrimination power and entropy variation index, were proposed in Hong and Chen (1999) and Pal (1999). However, these methods are restricted to deal with classification problems and assumed that the input variables are independent. This assumption usually cannot be satisfied in real-world problems. Another algorithm was proposed in Wang (2003), based on mathematic analysis of approximation accuracy.
2. **Model-free methods** that do not need to develop models for measuring relevant inputs. For instance, the method proposed in He and Asada (1993) exploits the continuity property of nonlinear functions. The so-called Lipschitz coefficients are computed in order to find the optimal order of an input-output model. Emami, Turksen, and Goldenberg (1998) proposed a method based on geometric criterion, in which a non-significant index was developed in order to find the most important inputs. A

method for input selection was proposed in Hadjili and Wertz (2002), in which an input is judged to be irrelevant if removing it does not change fuzzy partition significantly. Another method that exploits both consistence of nonlinear model and fuzzy cluster was proposed in Sindelar and Babuska (2004). Compared with the model-based method, the model-free method is less time-consuming and not biased in choosing model types and structures.

A model-free method is proposed in this article. Unlike the model-based method, no specific model needs to be built in advance in order to check for irrelevant inputs. The proposed method is based on analysis of input sensitivity, in which minimum cluster volume (MCV) (Krishnapuram & Kim, 2000) algorithm is used to quantify the sensitivity of each input. The less the sensitivity is, the less relevant input would be. In this way, potential inputs can be ordered according to their relative importance, and irrelevant inputs then can be removed step-by-step until only a proper number of inputs remains.

FUZZY CLUSTER ALGORITHM

General Description of Fuzzy Cluster Algorithm

The purpose of clustering is to obtain natural groupings of data from a large data set in order to produce a concise representation of a system's behavior. Generally speaking, fuzzy clustering algorithms are less prone to local minima than crisp clustering algorithm, since they make soft decisions via memberships. Fuzzy c-Means clustering (FCM) (Bezdek, 1981) algorithm is effective only when all clusters are roughly spherical with similar volumes. Gustafson-Kessel (GK) algorithm (Gustafson & Kessel, 1979) tries to accommodate ellipsoidal clusters with fuzzy covariance matrix, but the main drawback of GK algorithm is that it cannot detect clusters that differ largely in their volumes. Although the Gath-Geva (GG) (Gath & Geva, 1989) cluster method can detect clusters that differ largely in their volumes, the GG method is sensitive to initialization. The shapes of clusters detected by some other clustering methods, such as mountain clustering (Yager & Filev, 1994) and subtractive clustering (Chiu, 1994), are not as good as GK and GG, since these two methods are developed based on density of data set.

Minimum Cluster Volume (MCV) Algorithm

In order to simplify the rule base and to make the model easily interpreted, we want to obtain a set of clusters that are able to cover full data sets by the smallest number of clusters. MCV algorithm is introduced for this purpose.

Let us denote data samples as a matrix z, which is formed by concatenating regression data matrix $\boldsymbol{X}$ and output vector y as:

$$\boldsymbol{X} = \begin{bmatrix} \boldsymbol{x}_1^T \\ \boldsymbol{x}_2^T \\ \vdots \\ \boldsymbol{x}_N^T \end{bmatrix}, \boldsymbol{y} = \begin{bmatrix} y_1 \\ y_2 \\ \vdots \\ y_N \end{bmatrix}, \boldsymbol{z}^T = [X^T \quad \boldsymbol{y}] \tag{1}$$

The clustering process is to partition a data set into a certain number (c) of clusters. The result of a fuzzy partition is represented by a matrix $\boldsymbol{U}=[u_{i,k}]_{c\times N}$, whose element $u_{i,k}$ represents the degree of the membership of observation z_k to cluster i. In this article, c as the number of clusters is assumed to be given, based on prior knowledge, or to be optimized, referring to Gath and Geva (1989).

The cost function of MCV is:

$$J_{fv} = \min_{u_{ij}, v_i} \sum_{i=1}^{c} |\boldsymbol{C}_{fi}|^{1/2} \tag{2}$$

where $|C_{fi}|$ denotes the determinant of the fuzzy covariance matrix C_{fi} of cluster i. It seeks c hyper-ellipsoids so that they cover all data in the data set with the smallest total volume. Similar to FCM, GK, and GG algorithms, MCV satisfies the following conditions:

$$U \in R^{c \times N} \text{ with } u_{i,k} \in [0,1], \forall i,k;$$

$$\sum_{i=1}^{c} u_{i,k} = 1, \forall k;\ 0 < \sum_{k=1}^{N} u_{ik} < N,\ \forall i; \tag{3}$$

The solution of (2) is sought by the optimization method given as follows:

Step 1: Initialization. Select $\varepsilon > 0$ (ε=0.01 in this article), which is the ending condition. Initialize centers, v_i, as the results of other simple clustering methods such as FCM and GK.

Step 2: Calculating fuzzy covariance matrix for each cluster:

$$C_{fi} = \frac{\sum_{j=1}^{N} u_{ij}^{m} (z_j - v_i)(z_j - v_i)^T}{\sum_{j=1}^{N} u_{ij}^{m}} \tag{4}$$

where, z_j is j-th data, v_i is i-th cluster center, u_{ij} is the membership of j-th data to i-th cluster, m is a fuzziness coefficient, $m \in [1\ \infty]$. The default value of m is set to be m=2.

Step 3: Calculating Mahalanobis Distance:

$$MD_{ij} = (z_j - v_i)^T C_{fi}^{-1} (z_j - v_i) \tag{5}$$

Step 4: Calculating distance D_{ij} defined by MCV

$$D_{ij} = \frac{|C_{fi}|^{1/2} (MD_{ij} - L)}{\sum_{j=1}^{N} u_{ij}^{m}} \tag{6}$$

where L is the dimension of data set z.

Step 5: Determining new membership u_{ij}. Update u_{ij} by:

$$u_{ij} = \begin{cases} \dfrac{(D_{ij})^{1/(1-m)}}{\sum_{s=1}^{c} (D_{sj})^{1/(1-m)}} & MD_{ij} \geq L, \forall i = 1,\dots,c; \\ 1 & i = \{k \mid \min_{l=1,\dots,c}(MD_{lj}) = MD_{jk}), and MD_{kj} < L; \\ 0 & i \neq k \text{ and } MD_{kj} < L \text{ for any } i = 1,\dots,c \end{cases} \tag{7}$$

Step 6: Updating cluster center $\mathbf{v}_i$:

$$v_i = \frac{\sum_{j=1}^{N} u_{ij}^{m} z_j}{\sum_{j=1}^{N} u_{ij}^{m}} \tag{8}$$

Step 7: Repeating Steps 2 through 5. Find changes of membership values, Δu_{ij}.

Step 8: If $\Delta u_{ij} > \varepsilon$, go back to Step 2; otherwise, stop.

The properties of previous MCV algorithm are discussed as follows:

1. The cost function is not to minimize the sum of distances between each data and cluster center but to minimize the sum of volumes of each cluster. MCV thus can detect clusters that are largely different in shapes and sizes.
2. MCV algorithm is a hybrid approach based on fuzzy and hard clustering algorithms, which can be seen from formula (7). When Mahalanobis Distance between data and cluster center is larger than data dimension L, the fuzzy update process ($MD_{ji} > L \forall i$) is performed using the following update law:

$$u_{ij} = \frac{(D_{ij})^{1/(1-m)}}{\sum_{r=1}^{C}(D_{rj})^{1/(1-m)}} \tag{9}$$

Otherwise, in a hard update process, membership can be set to either 1 or 0. With this update process, the obtained fuzzy partition matrix $\boldsymbol{U}=[u_{i,k}]_{c\times N}$ satisfies condition (3). Compared with other clustering methods, fuzzy partition matrix $\boldsymbol{U}$ obtained by MCV algorithm has two distinct properties: (1) membership functions obtained by fuzzy partition matrix $\boldsymbol{U}$ have larger core region and less overlap; (2) smoothness of the membership functions can be guaranteed due to the existence of fuzzy update process.

3. Comparing the distance defined by MCV of (6) and GG of:

$$D_{ij} = \frac{(2\pi)^{L/2}\,|\boldsymbol{C}_{fi}|^{1/2}\exp\left\{\frac{MD_{ij}}{2}\right\}}{\sum_{j=1}^{N} u_{ij}} \tag{10}$$

where L is dimension of the data set. It can be seen that they share a similar nature. Whereas GG method assumes that clusters are Gaussians, MCV algorithm does not make such an assumption. In this sense, it is a general criterion, and performance is expected to be improved when clusters cannot be approximated well by Gaussians. The exponent in formula (10) explains the difference of sensitivity to initialization of these algorithms, while term MD_{ij}-L in equation (6) contributes to the robustness of MCV, especially for data $\mathbf{z}$ in high dimension.

RANKING INPUT IMPORTANCE BASED ON COMMON SENSE

Before input selection is performed, it is necessary to understand the basic idea of importance ranking of input. The importance ranking of input can be performed based on common sense: (1) if output remains unchanged, even if one input attribute has changed largely, we say that the output is not sensitive to this attribute; (2) if output changes largely due to some small changes of one input attribute, we say that the output is sensitive to this attribute. Recall the main purpose of modeling: the model often is used to predict the changes of output and to test for underlying reasons of that change. Thus, it is natural to choose more sensitive attributes, since the model input for the changes of output can be reflected by these attributes. This way of common sense input selection can be called *sensitivity analysis* (SA).

SA can be formulated by calculus. Consider a well-defined function y = $f(\boldsymbol{x})$ (continuous, differentiable). It can be expanded within the neighborhood of a given point $\boldsymbol{x}_0$:

$$\Delta y \approx \frac{\partial f}{\partial x_1}\Delta x_1 + \cdots + \frac{\partial f}{\partial x_n}\Delta x_n \tag{11}$$

From equation (11), it can be seen that the calculation of partial differential coefficients can be performed, based on previous SA discussions. If the function $f(.)$ is a linear structure, then the input selection problem becomes fairly easy for partial differential coefficients of each input that is constant at any given point in a discourse domain. Thus, most linear methods work well, such as least square, ridge regression, lasso, PCA, and so forth. However, for a nonlinear function $f(.)$, traditional methods often fail due to the different partial differential coefficients in different regions.

Another challenge of developing a model-free method is the calculation of partial differential coefficients. Model-based methods can build a model in advance and then calculate the partial differential coefficients by using the model. In the next section, a model-free method is developed with partial differential coefficients obtained by applying MCV clustering.

Figure 1(a). Clusters approximate a nonlinear regression surface in 2D by partitioning it into linear subspaces (the dotted line) (the arrow indicates the eigenvector corresponding to the smallest eigenvalue of each cluster)

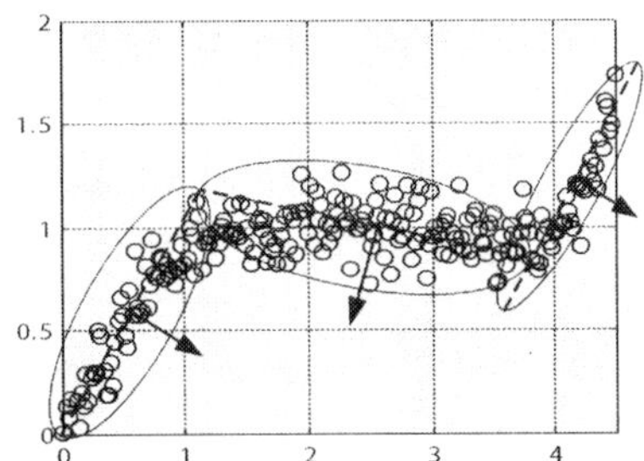

Figure 1(b). Clusters approximate a nonlinear regression surface in 3D (left) by partitioning it into linear subspaces (right) (the arrow indicates the eigenvector corresponding to the smallest eigenvalue of each cluster)

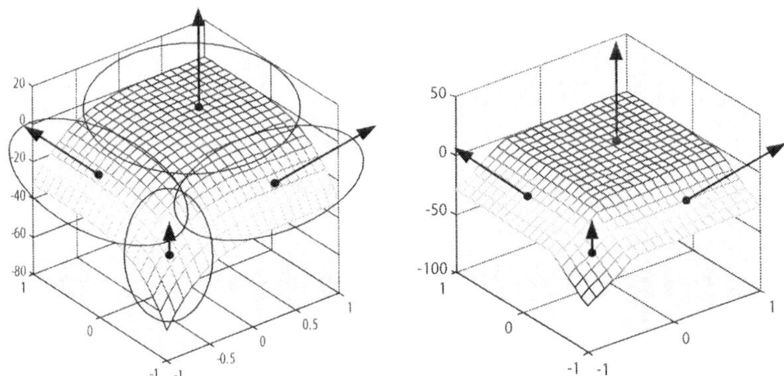

INPUT SELECTION BASED ON FUZZY CLUSTER

The basic idea of input selection through clustering is to partition a whole space into linear subspaces and then calculate partial differential coefficients within each linear subspace by MCV algorithm (we will show it in the following), and the intersection parts between linear spaces can be dealt with the membership values provided by MCV.

Clusters of MCV algorithm approximate the regression surface, which is illustrated in Figure 1. These clusters can be regarded as p-dimensional local linear subspaces, where p is the input number. This is reflected by the smallest eigenvalue $\lambda_{i,p+1}$ of cluster covariance matrices $\boldsymbol{C}_{fi}$ that typically are smaller than the remaining eigenvalues in magnitude.

The eigenvector corresponding to this smallest eigenvalue, $\boldsymbol{t}^i_{p+1}$, determines the normal vector to the hyperplane spanned by the remaining eigenvectors of the cluster (see Figure 2). In order to simplify the notation, $\boldsymbol{t}^i_{p+1}$ is denoted by $\boldsymbol{t}_i$. Recall that $z^{\mathrm{T}}=[\boldsymbol{x}^{\mathrm{T}}\ y]$ is the data matrix, and $\boldsymbol{v}_i$ is the cluster's prototype. The implicit normal form of the hyperplane can be given by:

$$(\boldsymbol{t}_i)^{\mathrm{T}}(z_k-\boldsymbol{v}_i)=0. \quad (12)$$

Figure 2. Eigenvectors for i-th cluster (local linear model)

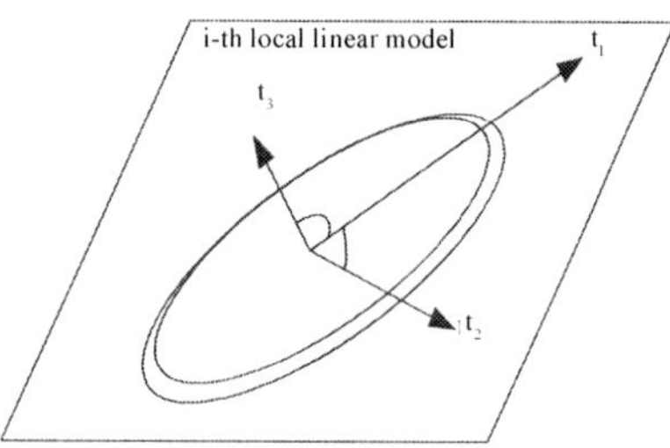

This expression indicates that inner product of normal vector $\boldsymbol{t}_i$ with any vector belonging to the hyperplane equals zero. For the following discussion, it is convenient to partition prototype $\mathbf{v}_i$ into a vector $\boldsymbol{v}_i^x$ corresponding to regressor $\boldsymbol{x}$ and a scalar $\boldsymbol{v}_i^y$ corresponding to regressand y: $\boldsymbol{v}_i^{\mathrm{T}}=[(\boldsymbol{v}_i^x)^{\mathrm{T}}\ v_i^y]$. The smallest eigenvector is partitioned in the same way as the cluster center (i.e., $\boldsymbol{t}_1^{\mathrm{T}}=[(\boldsymbol{t}_i^x)\ \boldsymbol{t}_i^y]$). Equation (11) now can be written as:

$$[(\boldsymbol{t}_i^x)^{\mathrm{T}}\ \boldsymbol{t}_i^y]([\boldsymbol{x}^{\mathrm{T}}\ y]-[(\boldsymbol{v}_i^x)^{\mathrm{T}}\ v_i^y])^{\mathrm{T}}=0. \tag{13}$$

Carrying out the inner product leads to the following equality:

$$(\boldsymbol{v}_i^x)^{\mathrm{T}}(x-\boldsymbol{v}_i^x)+\boldsymbol{t}_i^y(y-v_i^y)=0, \tag{14}$$

from which, by a simple algebraic manipulation, an explicit equation for the hyperplane is obtained:

$$y=\underbrace{-\frac{1}{t_i^y}(\boldsymbol{t}_i^x)^T}_{a_i^T}\boldsymbol{x}+\underbrace{\frac{1}{t_i^y}(\boldsymbol{t}_i)^T\boldsymbol{v}_i}_{b_i} \tag{15}$$

By comparing the previous expression with the linear expression $y=\boldsymbol{a}_i\boldsymbol{x}+b_1$, the calculation of $\boldsymbol{a}_i$ and b_i is directly performed by:

$$\boldsymbol{a}_1=-\frac{1}{\boldsymbol{t}_i^y}(\boldsymbol{t}_i^x)^T=-\frac{1}{t_{p+1}^i}(t_1^i,t_2^i,\ldots,t_p^i)^T$$
$$b_i=\frac{1}{t_i^y}(\boldsymbol{t}_i)^T\boldsymbol{v}_i\ . \tag{16}$$

Now let us investigate how to make use of the information obtained by MCV algorithm for input selection. Many nonlinear static or dynamic processes can be represented by the following regression model:

$$y_k=f(\boldsymbol{x}_k) \tag{17}$$

where $f(\cdot)$ is a nonlinear function, $\boldsymbol{x}$ represents its input vector, and k is the index of data number from 1 to N. Please note that equations (15, 16) provide us the gradient information of the input-output model. When the observation $\boldsymbol{x}_k$ is near the cluster i prototype, we have:

$$\frac{\partial y_k}{\partial x_k^l}=\frac{\partial f}{\partial x_k^l}=a_{ik}^l\approx-\frac{1}{t_{p+1}^i}t_l^i \tag{18}$$

where x_k^l is the l-th input. When the observation $\boldsymbol{x}_k$ belongs to cluster i and is not quite near the cluster i prototype, it is natural for us to use the degree of membership obtained by MCV in order to depict the distance between $\boldsymbol{x}_k$ and $\boldsymbol{v}_i$, and so we have:

$$\frac{\partial y_k}{\partial x_k^l}=\frac{\partial f}{\partial x_k^l}=a_{ik}^l\approx\sum_{i=1}^{i=c}u_{ik}(-\frac{1}{t_{p+1}^i}t_l^i) \tag{19}$$

In a general case, we also can have:

$$\frac{\partial y_k}{\partial x_k^l}=\frac{\partial f}{\partial x_k^l}=a_k^l\approx\sum_{i=1}^{i=c}u_k(-\frac{1}{t_{p+1}^i}t_l^i) \tag{20}$$

because u_{ik} will be one when the observation $\boldsymbol{x}_k$ is near the cluster i prototype. The sensitivity can be defined as the absolute value of the derivative. To evaluate the total sensitivity for all data, we define:

$$s_i^l=\sum_{k=1}^{N}|a_k^l| \tag{21}$$

However, since the input variables may have different ranges, normalized sensitivity Δ_i^l must be used:

$$\Delta_i^l = s_i^l \frac{\sigma_x^l}{\sigma_y} \quad (22)$$

where σ_x^l and σ_y are the standard deviations of l-th input and output, respectively. The most significant variables are selected using the maximum normalized sensitivity

$$\Delta_{\max}^l = \max_{1 \le j \le c} \Delta_j^k \quad (23)$$

The single input variable with the least maximum normalized sensitivity index (23) is a candidate for removal, as the output changes little even when this input changes greatly. The whole procedure can be summarized as follows:

Step 1: Collect all possible input variables and the output in the data set z, as shown in Equation (1). Determine the optimal cluster number c based on prior knowledge, or please refer to Gath and Geva, (1989).
Step 2: Run MCV algorithm to obtain $\boldsymbol{v}$, $\boldsymbol{U}$, and $\boldsymbol{t}^i$ (including $[t^i_1,\ldots,t^i_{p+1}]$):

$$[\boldsymbol{v}, \boldsymbol{U}, \mathbf{t}^i] = \mathrm{MCV}(z, c) \quad (24)$$

Step 3: Use equation (20)-(23) to calculate $\Delta^l_{\max}$ and then remove the most irrelevant input according to $\Delta^l_{\max}$.
Step 4: Form a new data set **z'** and redetermine the optimal cluster number c; then go to step 2 until the proper number of inputs is left for the modeling.

Note that the proper number of inputs should be chosen according to the specific modeling problems, such as available data size, required accuracy, and so forth.

CONSTRUCTION OF ANTECEDENT MEMBERSHIP FUNCTIONS AND PARAMETER IDENTIFICATION

TS fuzzy model is defined by:

$$\mathrm{R}_i\text{: if } \boldsymbol{x} \text{ is } \boldsymbol{A}_i(\boldsymbol{x}) \text{ then } y^* = \boldsymbol{a}_1^{\mathrm{T}}\boldsymbol{x} + b_i,\ i=1,\ldots,c \quad (25)$$

The antecedent proposition $x\ A_i(x)$ can be expressed usually in the following conjunctive form:

R_i: if x_1 is $A_{i,1}(x_1)$

and … and x_n is $A_{i,n}(x_n)$

then $y^* = \boldsymbol{a}_1^{\mathrm{T}}\boldsymbol{x} + b_i$. (26)

The rules are aggregated by using the fuzzy-mean formula:

$$y^* = \frac{\sum_{i=1}^{c} \beta_i(\boldsymbol{x})(\boldsymbol{a}_i^T \boldsymbol{x} + b_i)}{\sum_{i=1}^{c} \beta_i(\boldsymbol{x})} \quad (27)$$

where $\beta_i(\boldsymbol{x})$ is calculated as $\beta_i(\boldsymbol{x}) = \boldsymbol{A}_i(\boldsymbol{x}) = \prod \boldsymbol{A}_{i,j}(x_j)$.

Antecedent membership functions are obtained by projecting the multidimensional fuzzy sets defined point-wise in the rows of partition matrix $\boldsymbol{U}$ onto individual antecedent variables. Here, axis-orthogonal projection (see Figure 3) is used to inherit such good properties of MCV analyzed previously. The projected fuzzy set is not convex. In order to estimate the parameters of membership functions, the projected fuzzy set first should pass through a filter and then should be approximated by Gaussians-2 membership function of:

Figure 3. Axis-orthogonal projection of fuzzy set

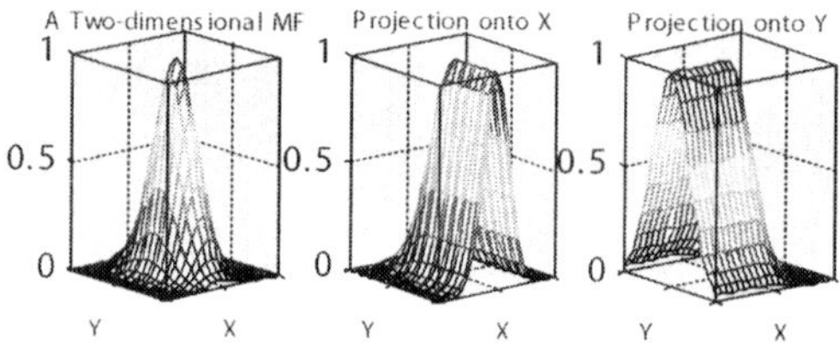

$$\mu(x;c_l,c_r,w_l,w_r)=\begin{cases}\exp\left(-\left(\dfrac{x-c_l}{w_l}\right)^2\right), & x<c_l\\ \exp\left(-\left(\dfrac{x-c_r}{w_r}\right)^2\right), & x>c_r\\ 1 & other\end{cases} \tag{28}$$

The identification for consequent parameters involves Least Square (LS) and Total Least Square (TLS):

1. LS: To minimize the criterion:

$$\min_{\theta_i}\frac{1}{N}(\boldsymbol{y}-\boldsymbol{X}_e\theta_i)^T\Phi_i(\boldsymbol{y}-\boldsymbol{X}_e\theta_i) \tag{29}$$

where $\boldsymbol{X}_e=[\boldsymbol{X}\boldsymbol{1}]$ is the regressor matrix and $\boldsymbol{\Phi}_i$ is a matrix having the membership degrees as:

$$\boldsymbol{\Phi}_i=\begin{bmatrix}\mu_{i,1} & 0 & \cdots & 0\\ 0 & \mu_{i,2} & \cdots & 0\\ \vdots & \vdots & \ddots & \vdots\\ 0 & 0 & \cdots & \mu_{i,N}\end{bmatrix} \tag{30}$$

The weighted least-squares estimate of the consequent parameters is given by $\boldsymbol{\theta}_i=[\boldsymbol{a}_i^T\ b_i]$ as

$$\boldsymbol{\theta}_i=(\boldsymbol{X}_e^T\boldsymbol{\Phi}_i\boldsymbol{X}_e)^{-1}\boldsymbol{X}_e\boldsymbol{\Phi}_i\boldsymbol{y} \tag{31}$$

2. **TLS:** Please refer to equation (15), (16).

The details of fuzzy modeling based on MCV can be found in Yang, Zhu, Meng, and Lu (2005) and Yang and Meng (2005), who also explained its advantages and gave many examples.

SIMULATIONS AND APPLICATIONS

In this part, three benchmark problems are used to illustrate efficiency of the proposed method. A simulated example is considered first in order to illustrate our basic idea. Second, an automobile MPG (miles-per-gallon) prediction is used as a case study, which is a nonlinear regression problem. The comparison between our modeling method and the method in Jang (1996) is given. Third, the proposed method is applied to a real dynamic process modeling problem — the well known Box and Jenkins gas furnace process.

A Simulation Example

This example is devoted to the estimation of a nonlinear relation based on measured data, which was also studied by Hadjili and Wertz (2002). The input-output data are generated via the following relation:

$$x=\frac{\sqrt{t^3}}{10}+e,\ for\quad t=0:170 \tag{32}$$

$$y=\begin{cases}10+e, & for\quad t=0:70\\ 0.05x+7.0087+e, & for\quad t=71:105\\ -0.03x+15.6623+e, & for\quad t=106:140\\ 10.6928+e, & for\quad t=141:170\end{cases} \tag{33}$$

where e is a sequence of random noise in the range of [0, 0.5]. In order to test the efficiency of input selection, we suppose that we have additional measured inputs generated as follows:

$$f=t^{(1/5)}+e, for\quad t=0:170 \tag{34}$$

$$g=\begin{cases}10\dfrac{\sin(t-40)}{t-40}+e, for & t=0:170, t\neq 40\\ 20+e \quad for & t=40\end{cases} \tag{35}$$

Now observe the noise-to-signal ratios (NSRs) of input-output signals. The NSR is defined as the ratio between the standard deviation of noise and the standard deviation of signal. We notice that signals f, g, and y are more corrupted by random noise than x (i.e., NSR(x) = 0.2%, NSR(f) = 34.7%, NSR(g) = 7.7%, and NSR(y) = 18.7%). Although output y depends only on input x, it highly correlates to the signals f and g because of the common noise sequence affecting f, g, and y but correlates less to the input x (see NSR). We check as follows whether the proposed method can effectively select the appropriate inputs. The candidate input data set used for modeling is $\boldsymbol{X}=\{x, f, g\}$ (see Figures 4(a), 4(b), and 4(c)). Using the fuzzy hypervolume criterion in Gath and Geva (1989), the appropriate number of fuzzy rules is taken as four, as shown in Figure 4(d). Next, we follow the steps presented in our algorithm.

It can be observed that g is the most irrelevant input from the sensitivity listed in Table 1, so input g is removed at the first round. We recomputed the sensitivity for only two inputs $\boldsymbol{X}=\{x, f\}$ and removed f at the second round (see Table 1). As a result, only input x is selected for modeling. Therefore, we construct a fuzzy model with the method in the last section. Figures 5(a), 5(b), and 5(c) show the cluster approximation, projected membership values, and parameterized membership functions (MFs), respectively.

The MFs can be given a linguistic concept, such as very small (VS), small (S), large (L), and very large (VL). Hence, with the consequent parameters identified by OLS, the rule base is obtained as follows:

Figure 4. Data set and cluster validation

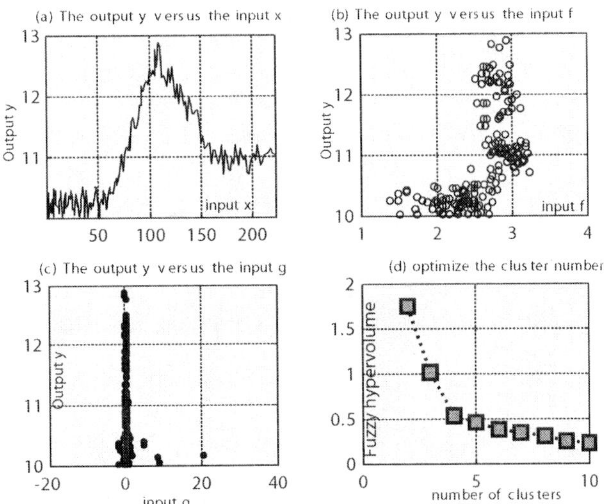

R$_1$: **If** x is VS, **then** y=0.0006x+10.2675;
R$_2$: **If** x is S, **then** y=0.0485x+7.3941;
R$_3$: **If** x is L, **then** y=-0.0256x+15.3529;
R$_4$: **If** x is VL, **then** y=0.0021x+10.5223.

The RMSE of the model is 0.1434, which is calculated by

$$RMSE = \sqrt{1/N\sum_{k=1}^{N}(y_k - \hat{y}_k)}.$$

From the rule base, we also can see that the consequent parameters are little different from equation (32). This indicates that the MCV-based fuzzy model gives less biased estimation than other methods (Babuska, 1998; Chiu, 1994). A careful discussion of this property and its reason was given in Yang, et al. (2005).

Automobile MPG Prediction

This section describes the use of a proposed method for modeling a well-known problem: automobile MPG prediction, which includes input selection and input space partition.

Table 1. Input selection procedure

Round	Input x	Input f	Input g
1	0.09319	0.02194	0.00693
	√	√	×
2	0.13388	0.00366	×
	√	×	×

Figure 5(a). Clusters approximation

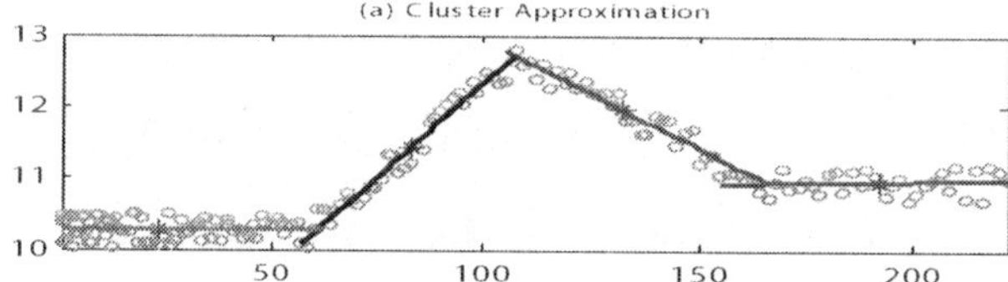

Figure 5(b). Projected membership values

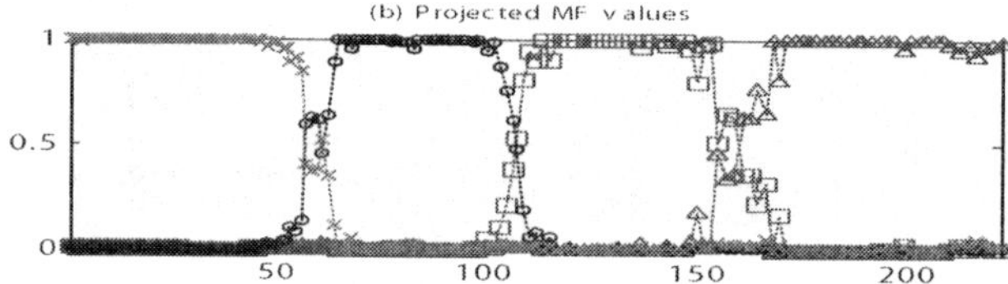

Figure 5(c). Parameterized MFs

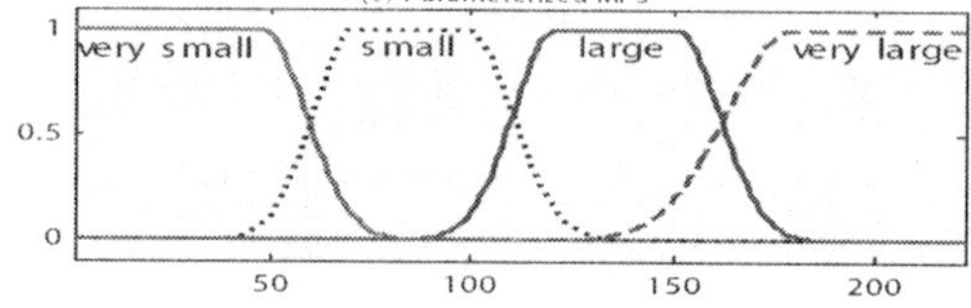

The automobile MPG prediction problem is a typical nonlinear regression problem in which several attributes (input variables) are used to predict another continuous attribute (output variable). In this case, the six input attributes include profile information about the automobiles (see Table 2).

In this typical nonlinear regression problem, six attributes (input variables) can be used to predict an automobile's fuel consumption. The data set is available from the UCI Repository of Machine Learning Databases and Domain Theories (ftp://ics.uci.edu/pub/machine-learning-databases/auto-mpg). After removing instances with missing values, the data set was reduced to 392 entries. Our objective then is to use the data set and the proposed method to construct a fuzzy model that could best predict the MPG of an automobile, given its six profile attributes.

Two issues should be taken into account before constructing a fuzzy model:

1. **Insufficient Data:** Generally speaking, for modeling a medium complex system, the principle is that at least 10 data values are required for a single input. Therefore, for the automobile MPG prediction problem in which there are six inputs, approximately 10^6 data values are needed, while we only have 392 data values. In this case, we actually have only $\sqrt[6]{392} \approx 2.5$ data values for each dimension, which is very insufficient. This problem is quite common in real-world problems. According to the principle, we

Table 2. Attributes for modeling and their data type

Attributes	**Data type**
No. of cylinders	multi-valued discrete
Displacement	continuous
Horsepower	continuous
Weight	continuous
Acceleration	continuous
Model year	Multi-valued discrete

Table 3. Input selection process for automobile MPG prediction problem

Round	**No. of Cylinders**	**Displacement**	**Horsepower**	**Weight**	**Acceleration**	**Model Year**
1	0.66894	1.9491	1.7295	4.3667	0.471	2.4805
	√	√	√	√	×	√
2	1.5799	2.5938	2.3708	3.5197	×	2.396
	×	√	√	√	×	√
3	×	0.8914	1.5914	3.5725	×	2.0002
	×	×	√	√	×	√
4	×	×	1.8455	4.5522	×	2.2491
	×	×	×	√	×	√
5	×	×	×	4.9203	×	1.9371
	×	×	×	√	×	×

should pick up two or three most important inputs for the modeling.

2. **Input Space Partition:** Grid partition is the most frequently used input partitioning method for fuzzy modeling. Here, we will use a clustering method for input space partition, which is an efficient method for input space partition and can be expected to have a better performance than grid partition.

The results of using the proposed method for input selection are in Table 3, in which the values is calculation result of Equation (23). We can see that the attribute Acceleration is removed at the first round, and then No. of cylinders, Displacement, Horsepower, and Model year are removed sequentially until only Weight is left. Now, let us turn to another well-known input selection methods proposed by Jang (1996), in which ANFIS was built 15 times in order to try all the combination of inputs to find out a good subset for modeling use, while our method needs to run only five times. The selection result is the same with our method (see Figure 6).

Now, we use two most important input variables—Weight and Year—in order to build a fuzzy model due to the previous discussion. First, this data set is divided into training and test sets with equal size (196) (see Figure 7(a)). The performance of the models is measured by RMSE. The result

Figure 6. Jang method for MPG input selection by ANFIS

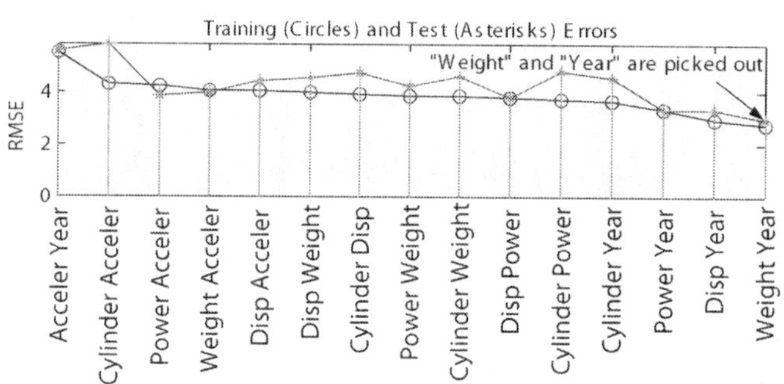

Figure 7. Data set and cluster partition

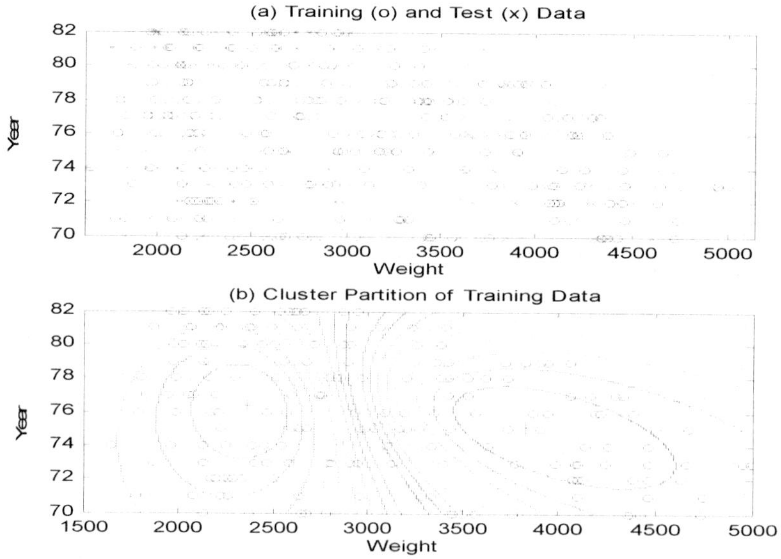

Figure 8. MFs for the two most important inputs

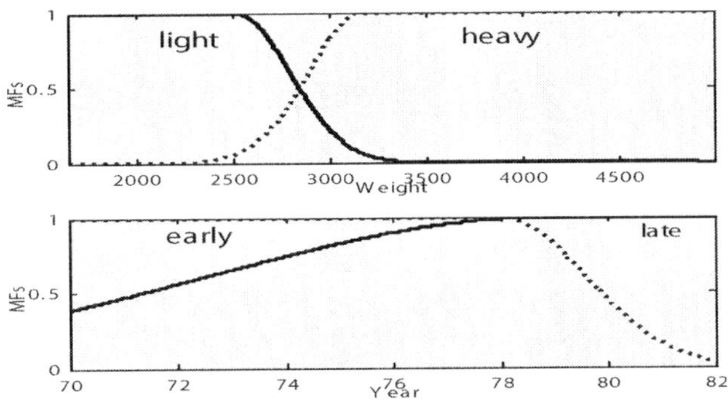

Figure 9(a). Regression surface obtained by proposed method

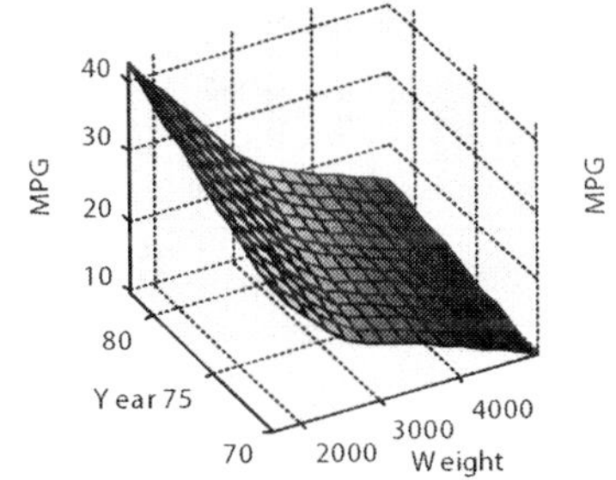

Figure 9(b). Regression surface obtained by ANFIS

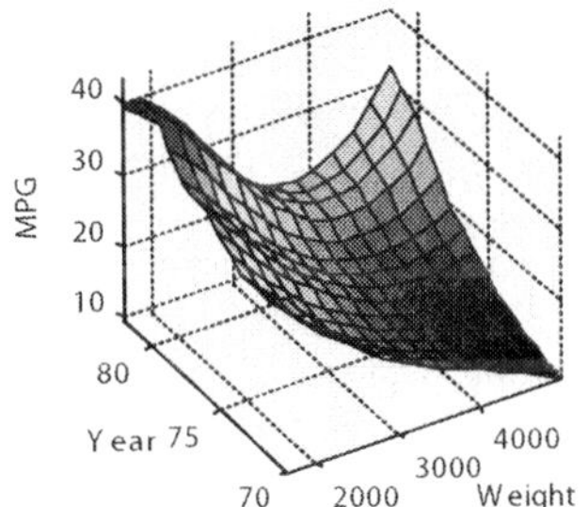

then is compared with some other technologies, like linear regression and ANFIS.

Applying the proposed method, the cluster partition of training data is shown in Figure 7(b). Projecting the fuzzy partition onto each dimension, MFs are obtained as shown in Figure 8. MFs on the attribute Year almost cover the entire domains to a high degree and, thus, possibly can be left out of the antecedents. This can be done according to the simplification algorithm in Babuska (1998).

Therefore, the obtained fuzzy model results in a compact fuzzy model, which gives a concise and accurate description of the problem:

$\mathbf{R_1}$: **If** *Weight* is *light*, **then** MPG = -0.01163 * Weight + 1.028 * Year - 22.93;

$\mathbf{R_2}$: **If** *Weight* is *heavy*, **then** MPG = -0.004196* Weight + 0.5011 * Year - 4.669.

The RMSE of the obtained model is 2.97 and 2.98 for training data and test data, respectively.

As a comparison, we now look at the result of linear regression, where the model is expressed as:

MPG = a0 + a1 *cyl +a2 *disp +a3 *hp +a4 *weight +a5 *accel +a6 *year

with a0, a1,..., a6 being seven modifiable linear parameters. The optimum values of these linear parameters are obtained directly by a least squares method, and the training and test RMSE are 3.45 and 3.44, which is much worse than our model. We can see from here that irrelative input might impair the modeling result. For further comparison, we also give the result of ANFIS. Its training and test RMSE are 2.61 and 2.99.

The model surfaces obtained by the proposed method and the ANFIS are shown in Figures 8(a) and 8(b), respectively. It can be seen that the surface increases toward the right upper corner, which is apparently a spurious result that heavy cars have high MPG ratings. Jang (1996) explained that this was because of luck of data (see Figure 7) due to the tendency of manufacturers to begin to build small compact cars during the mid 1970s, while the proposed approach does not suffer from this problem.

Nonlinear System Identification

In this section, the proposed method is applied to nonlinear system identification by means of the well-known Box and Jenkins (1970) gas furnace data as the modeling data set, which is a frequently used benchmark problem (Jang, 1996; Sugeno & Yasukawa, 1993). This is a time-series data set for a gas furnace process with gas flow rate $u(t)$ as the furnace input and CO_2 concentration $y(t)$ as the furnace output. For modeling purpose, the original data set containing 296 $[u(t)\ y(t)]$ data pairs is reorganized as $[y(t\text{-}1), \ldots, y(t\text{-}4), u(t\text{-}1), \ldots, u(t\text{-}6); y(t)]$. This reduces the number of the instances to 290, out of which the first 145 are used as training data, and the remaining 145 are used as test data. From the reorganized data set, one can see that there are 10 candidate input variables for modeling. It is reasonable and necessary to select input first in order to reduce the input dimension. For modeling dynamic process, the inputs selected must contain elements forming both the set of historical furnace outputs $\{y(t\text{-}1), \ldots, y(t\text{-}4)\}$ and the set of historical furnace inputs $\{u(t\text{-}1), \ldots, u(t\text{-}6)\}$. The proposed method is applied to the procedure until only two inputs are left (see Table 4). The input selection result indicates that the gas furnace process is a first order plus three sampling intervals time-delayed process. Now let us see another well-known input selection method

Table 4. Input selection process for Box and Jenkins (1970) gas furnace data modeling

Round	$y(t\text{-}1)$	$y(t\text{-}2)$	$y(t\text{-}3)$	$y(t\text{-}4)$	$u(t\text{-}1)$	$u(t\text{-}2)$	$u(t\text{-}3)$	$u(t\text{-}4)$	$u(t\text{-}5)$	$u(t\text{-}6)$
1	0.2869	0.4204	0.6226	0.3288	0.1419	0.2673	0.5404	0.4237	1.0865	0.0029
	√	√	√	√	√	√	√	√	√	×
2	1.0308	1.2424	0.2862	0.1590	0.3921	0.8703	0.7849	0.6747	0.0564	×
	√	√	√	√	√	√	√	√	×	×
3	0.7051	0.8056	0.0889	0.18234	0.35584	0.5632	0.33195	0.2871	×	×
	√	√	×	√	√	√	√	√	×	×
4	0.7005	0.7956	×	0.2557	0.0218	0.3446	0.8602	1.0188	×	×
	√	√	×	√	×	√	√	√	×	×
5	0.9407	1.0055	×	0.2794	×	0.0477	0.7271	1.0585	×	×
	√	√	×		×	×	√	√	×	×
6	1.4245	1.2743	×	0.1952	×	×	0.5897	0.2478	×	×
	√	√	×	×	×	×	√	√	×	×
7	0.3852	0.1005	×	×	×	×	0.3423	0.0410	×	×
	√	√	×	×	×	×	√	×	×	×
8	0.3944	0.1417	×	×	×	×	0.2205	×	×	×
	√	×	×	×	×	×	√	×	×	×
9	√	×	×	×	×	×	√	×	×	×

Figure 10. Jang's method for Box-Jenikin gas furnace input selection by ANFIS (exhaustive search)

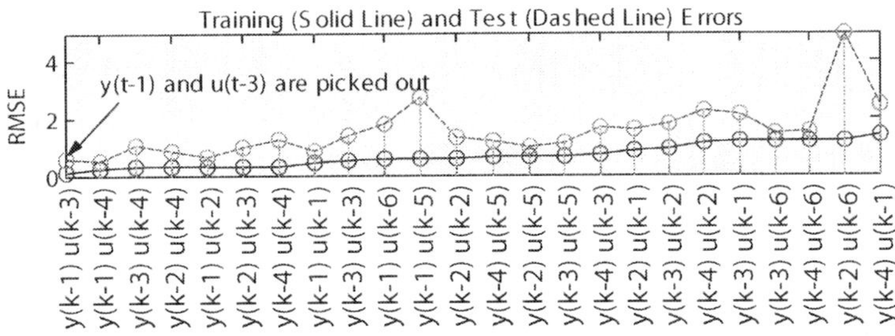

Figure 11(a). Cluster partition

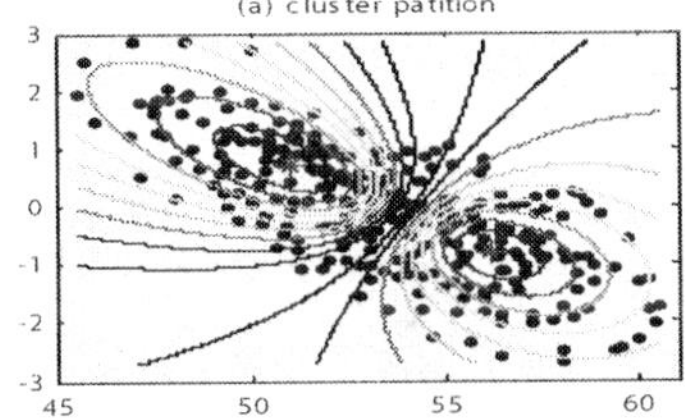

Figure 11(b). Data distribution

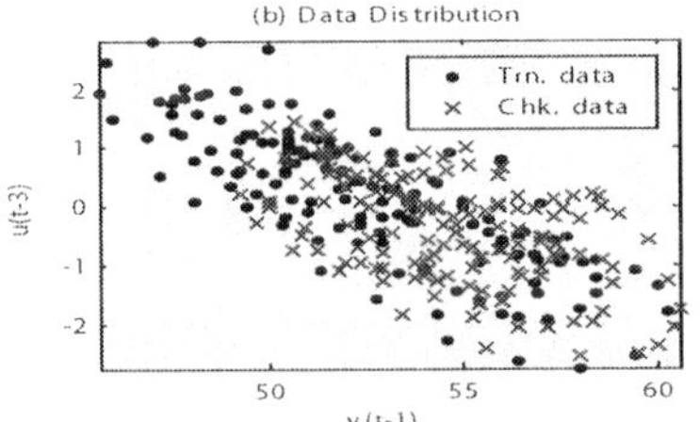

Figure 11(c). Model prediction

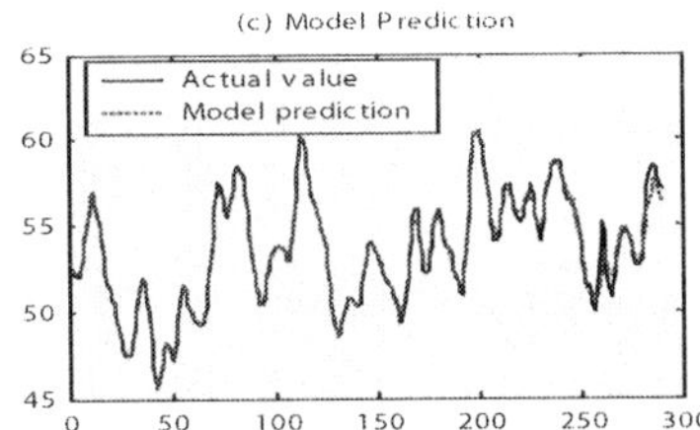

Figure 11(d). Model regression surface

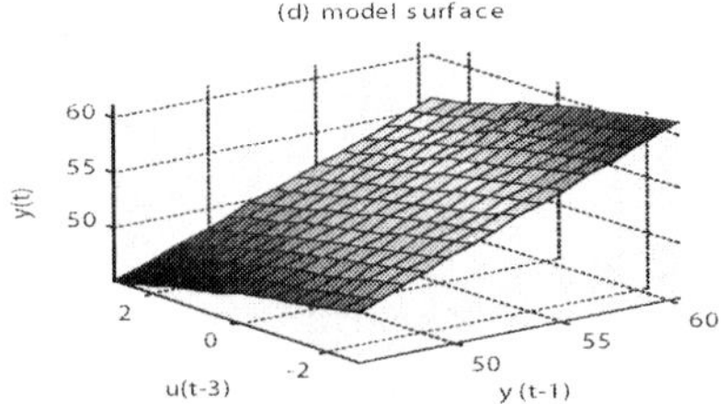

proposed by Jang (1996) in which ANFIS was built 24 times in order to try all the combination of inputs to find out a good subset for modeling use (see Figure 10). Our method needs to run only nine times and rank the importance of all inputs. The same selection result comes out, which is what we expected. Once the inputs are determined, the method proposed in this article can be employed for modeling. The rule base can be written in the following form:

R_i: **If** $y(t\text{-}1)$ is $\boldsymbol{A}$ and $u(t\text{-}3)$ is $\boldsymbol{B}$, **then** $y(t) = a_i^1 * y(t\text{-}1) + a_i^2 * u(t\text{-}3) + b_i$

With the proposed method presented in the previous section, the fuzzy model is obtained with high accuracy (see Figure 11). This result is compared with several other methods in Table 5.

The obtained MFs are shown in Figure 12, and its rule base is as follows:

R_1: **If** $y(t\text{-}1)$ is *small* and $u(t\text{-}3)$ is *positive*, **then** $y(t) = 0.7022 * y(t\text{-}1) - 1.014 * u(t\text{-}3) + 15.86$;

R_2: **If** $y(t\text{-}1)$ is *large* and $u(t\text{-}3)$ is *negative*, **then** $y(t) = 0.7195 * y(t\text{-}1) - 1.084 * u(t\text{-}3) + 14.72$.

From the rule base and the model surface, the process of Box-Jenkins (1970) gas furnace can be viewed as a linear process, which is a first-order

Figure 12. MFs for the two most important inputs

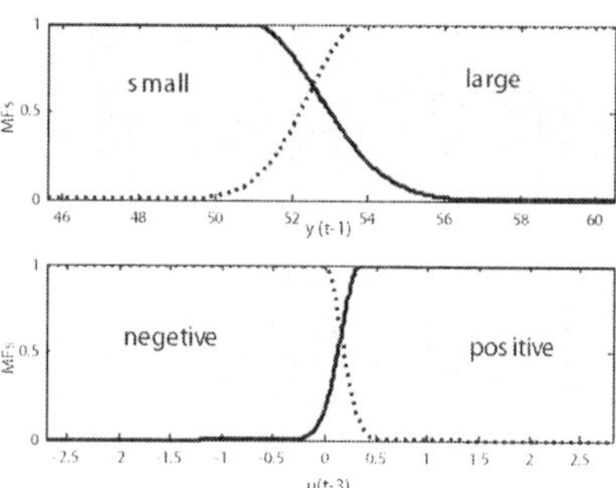

Figure 13. ANFIS surface of Box-Jenkins (1970) gas furnace

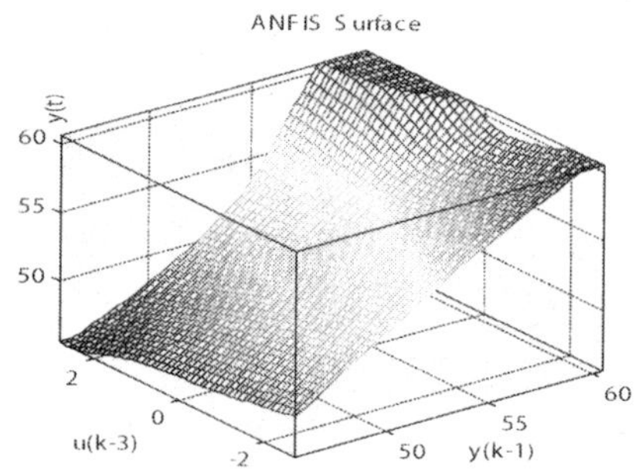

system with three sampling intervals time-delayed input. This information is very important for control purposes. The method of a controller designing for a linear system with input delay can be applied directly rather than looking for a nonlinear control scheme that improves real-time performance and saves a lot of costs. Now, we turn to consider the traditional method used for system identification. ARX model is built with n_a=2, n_b =5, n_k =4. Although it involves seven inputs, its training and test errors (RMSEs) are 0.6981 and 1.0877, respectively. So, the modeling result of ARX model is much worse than our model, and it only can be viewed as a black box with no additional information about the system. The modeling error (RMSE) of ANFIS is a little smaller than our proposed method, but our model structure is simpler, with only two rules, which is an easier interpretable model structure. ANFIS suffers the same problem in MPG modeling. From Figure 13, it can be seen that the surface of ANFIS increases toward the right upper corner, which is due to lack of data in that region (see Figure 11(a)). It may lead us to consider the Box-Jenkins (1970) gas furnace system as a nonlinear one and may encourage us to choose a nonlinear scheme for control purposes, which, in fact, is unnecessary.

Remarks

1. The numerical values in the column of RMSE are training error and test error, respectively.
2. ARX model structure is defined as:

$$A(q)y(t) = B(q)u(t - nk) + e(t)$$

where

$$A(q) = 1 + a_1 q^{-1} + \ldots + a_{n_a} q^{-n_a},$$

$$B(q) = b_1 + b_2 q^{-1} + \ldots + b_{n_b} q^{-n_b+1}$$

Both AIC and MDL (Ljung, 1999) select n_a =2, n_b =5, n_k =4.

CONCLUSION

In this article, we presented a model-free input selection method for nonlinear regression and fuzzy modeling. Our idea is based on the sensitivity analysis using MCV algorithm. The eigenvector associated with the smallest eigenvalue, which gives the gradient information, is used as a measurement of sensitivity. The efficiency of the proposed method is tested on three benchmark problems: (1) a noise nonlinear regression problem with two dummy inputs; (2) automobile MPG prediction problem with six inputs; and (3) Box and Jenkins (1970) gas furnace problem with 10 potential inputs. The case studies indicate that the proposed method works effectively for small and medium-sized problems (about 10 candidate inputs). With an input selection procedure, a model is reduced to a fairly concise structure, which only involves the most important inputs. This will contribute greatly in data mining, which places emphasis on high accuracy (for prediction) and easy interpretation (for description or visualization).

ACKNOWLEDGMENT

We are grateful to Prof. Dr. Robert Babuska at Delft University of Technology, The Netherlands, who gave us some suggestions when we were getting started on this project.

REFERENCES

Babuska, R. (1998). *Fuzzy modeling for control.* Boston: Kluwer Academic Publishers.

Bezdek, J.C. (1981). *Pattern recognition with fuzzy objective function.* New York: Plenum Press.

Box, G., & Jenkins, G. (1970). *Time series analysis, forecasting and control.* San Francisco: Holden Day.

Chiu, S. (1994). Fuzzy model identification based on cluster estimation. *Journal of Intelligent and Fuzzy Systems, 2*(3), 267–278.

Chiu, S. (1996). Selecting input variables for fuzzy models. *Journal of Intelligent and Fuzzy Systems, 4*, 243–256.

Emami, M.R., Turksen, I.B., & Goldenberg, A.A. (1998). Development of a systematic methodology of fuzzy logic modeling. *IEEE Transactions on Fuzzy Systems, 6*, 346–361.

Gath, I., & Geva, A. (1989). Unsupervised optimal fuzzy clustering. *IEEE Transactions on Pattern Analysis and Machine Intelligence, 7*, 773–781.

Gustafson, D.E., & Kessel, W.C. (1979). Fuzzy clustering with a fuzzy covariance matrix. *Proceedings of the IEEE CDC,* San Diego, California (pp. 761–766).

Hadjili, M.L., & Wertz, V. (2002). Takagi-Sugeno fuzzy modeling incorporating input variables selection. *IEEE Transactions on Fuzzy Systems, 10*(6), 728–742.

He, X., & Asada, H. (1993). A new method for identifying orders of input-output models for nonlinear dynamic system. *Proceedings of the American Control Conference*, San Francisco (pp. 2520–2523).

Hong, T.P., & Chen, J.B. (1999). Finding relevant attributes and membership functions. *Fuzzy Sets Systems, 103*, 389–404.

Jang, J.R. (1993). ANFIS: Adaptive-network-based fuzzy inference system. *IEEE Transactions on Systems, Man and Cybernetics, 23*, 665–685.

Jang, J.R. (1996). Input selection for ANFIS learning. *Proceeding of the IEEE International Conference on Fuzzy System*, New Orleans, Louisiana.

Kosko, B. (1997). *Fuzzy engineering.* Upper Saddle River, NJ: Prentice Hall.

Krishnapuram, R., & Kim, J. (2000). Clustering algorithms on volume criteria. *IEEE Transactions on Fuzzy Systems, 8*(2), 228–236.

Lin, Y., & Cunningham, G..A. (1994). A fuzzy approach to input variable identification. *Proceedings of the IEEE Conference on Fuzzy Systems*, Orlando, Florida (pp. 2031–2036).

Ljung, L. (1999). *System identification; theory for the user* (2nd ed.). Upper Saddle River, NJ: Prentice Hall.

Pal, N.R. (1999). Soft computing for feature analysis. *Fuzzy Sets Systems*, *103*, 201–221.

Sindelar, R., & Babuska, R. (2004). Input selection for nonlinear regression models. *IEEE Transactions on Fuzzy Systems*, *12*(5), 688–696.

Sugeno, M., & Yasukawa, T. (1993). A fuzzy logic based approach to qualitative modeling. *IEEE Transactions on Fuzzy Systems*, *1*(1), 7–31.

Tanaka, K., Sano, M., & Watanabe, H. (1995). Modeling and control of carbon monoxide concentration using a neuro-fuzzy technique. *IEEE Transactions on Fuzzy Systems*, *3*(3), 271–279.

Wang, L.X. (2003). The WM method completed: A flexible fuzzy system approach to data mining. *IEEE Transactions on Fuzzy Systems*, *11*(6), 768–782.

Wang, L.X., & Mendel, J.M. (1992a). Fuzzy basis functions, universal approximation, and orthogonal least squares learning. *IEEE Transactions on Neural Networks*, *3*(5), 807–814.

Wang, L.X., & Mendel, J.M. (1992b). Generating fuzzy rules by learning from examples. *IEEE Transactions on Systems, Man and Cybernetics*, *22*, 1414–1427.

Yager, R., & Filev, D. (1994). Generation of fuzzy rules by mountain clustering. *Journal of Intelligent and Fuzzy Systems*, *2*(3), 209–219.

Yang, C., & Meng, J. (2005). Optimal fuzzy modeling based on minimum cluster volume. *Proceedings of the First International Conference on Advanced Data Mining and Applications*, Wuhan, China.

Yang, C., Zhu, S.A., Meng, J., & Lu, L.M. (2005). Transparent fuzzy modeling based on minimum cluster volume. *Proceedings of the 5th International Conference on Control and Automation*, Budapest.

Zadeh, L.A. (1973). Outline of a new approach to the analysis of complex systems and decision processes. *IEEE Transactions on Systems, Man and Cybernetics*, *3*, 28–44.

ENDNOTE

1 The work was supported by National Natural Science Foundation of China (No. 60574079) and supported by Zhejiang Provincial Natural Science Foundation of China (No.601112).

This work was previously published in the International Journal of Data Warehousing and Mining, Vol. 2, Issue 3, edited by D. Taniar, pp. 57-75, copyright 2006 by IGI Publishing, formerly known as Idea Group Publishing (an imprint of IGI Global).

Chapter 2.35
Combinatorial Fusion Analysis:
Methods and Practices of Combining Multiple Scoring Systems

D. Frank Hsu
Fordham University, USA

Yun-Sheng Chung
National Tsing Hua University, Taiwan

Bruce S. Kristal
Burke Medical Research Institute and Weill Medical College of Cornell University, USA

ABSTRACT

Combination methods have been investigated as a possible means to improve performance in multi-variable (multi-criterion or multi-objective) classification, prediction, learning, and optimization problems. In addition, information collected from multi-sensor or multi-source environment also often needs to be combined to produce more accurate information, to derive better estimation, or to make more knowledgeable decisions. In this chapter, we present a method, called Combinatorial Fusion Analysis (CFA), for analyzing combination and fusion of multiple scoring. CFA characterizes each Scoring system as having included a Score function, a Rank function, and a Rank/score function. Both rank combination and score combination are explored as to their combinatorial complexity and computational efficiency. Information derived from the scoring characteristics of each scoring system is used to perform system selection and to decide method combination. In particular, the rank/score graph defined by Hsu, Shapiro and Taksa (Hsu et al., 2002; Hsu & Taksa, 2005) is used to measure the diversity between scoring systems. We illustrate various applications of the framework using examples in information retrieval and biomedical informatics.

INTRODUCTION

Many problems in a variety of applications domains such as information retrieval, social / welfare / preference assignments, internet/intranet search, pattern recognition, multi-sensor surveillance, drug design and discovery, and biomedical informatics can be formulated as multi-variable (multi-criterion or multi-objective) classification, prediction, learning, or optimization problems. To help obtain the maximum possible (or practical) accuracy in calculated solution(s) for these problems, many groups have considered the design and integrated use of multiple, (hopefully) complementary scoring schemes (algorithms or methods) under various names such as multiple classifier systems (Ho, 2002; Ho, Hull, & Srihari, 1992, 1994; Melnik, Vardi, & Zhang 2004; Xu, Krzyzak, & Suen, 1992; Kittler & Alkoot, 2003), social choice functions (Arrow, 1963; Young, 1975; Young & Levenglick, 1978), multiple evidences, Web page scoring systems or meta searches (Aslam, Pavlu, & Savell, 2003; Fagin, Kumar, & Sivakumar, 2003; Diligenti, Gori, & Maggini, 2004), multiple statistical analysis (Chuang, Liu, Brown, et al., 2004; Chuang, Liu, Chen, Kao, & Hsu, 2004; Kuriakose et al., 2004), cooperative multi-sensor surveillance systems (Collins, Lipton, Fujiyoshi, & Kanade, 2001; Hu, Tan, Wang, & Maybank, 2004), multi-criterion ranking (Patil & Taillie, 2004), hybrid systems (Duerr, Haettich, Tropf, & Winkler, 1980; Perrone & Cooper, 1992), and multiple scoring functions and molecular similarity measurements (Ginn, Willett, & Bradshaw, 2000; Shoichet, 2004; Yang, Chen, Shen, Kristal, & Hsu, 2005). For convenience and without loss of generality, we use the term **multiple scoring systems (MSS)** to denote all these aforementioned schemes, algorithms, or methods. We further note the need for the word "hopefully" above — there are limited practical means of predicting which combinations will be fruitful — the problem we address in the remainder of this report.

The main purpose in constructing multiple scoring systems is to combine those MSS's in order to improve the efficiency and effectiveness or increase the sensitivity and specificity of the results. This purpose has been met; it has been demonstrated that combining MSS's can improve the optimization results. Combination of multiple scoring systems has been studied under different names such as classification ensemble (Ho, 2002; Ho et al., 1992, 1994; Kittler & Alkoot, 2003; Tumer & Ghosh, 1999; Xu et al., 1992), evidence combination (Belkin, Kantor, Fox, & Shaw, 1995; Chuang, Liu, Brown, et al., 2004; Chuang, Liu, Chen, et al., 2004), data / information fusion (Dasarathy, 2000; Hsu & Palumbo, 2004; Hsu et al., 2002; Hsu & Taksa, 2005; Ibraev, Ng, & Kantor, 2001; Kantor, 1998; Kuriakose et al., 2004; Lee, 1997; Ng & Kantor, 1998, 2000), rank aggregation (Dwork, Kumar, Naor, & Sivakumar, 2001; Fagin et al., 2003), consensus scoring (Ginn et al., 2000; Shoichet, 2004; Yang et al., 2005), and cooperative surveillance (Collins et al., 2001; Hu et al., 2004). In addition, combination of MSS's has been also used in conjunction with other machine learning or evolutional computation approaches such as neural network and evolutional optimization (Garcia-Pedrajas, Hervas-Martinez, & Ortiz-Boyer, 2005; Jin & Branke, 2005). We use the term **combinatorial fusion** to denote all the aforementioned methods of combination.

Combination of multiple approaches (multiple query formulation, multiple retrieval schemes or systems) to solving a problem has been shown to be effective in data fusion in **information retrieval (IR)** and in internet meta search (Aslam et al., 2003; Belkin et al., 1995; Dwork et al., 2001; Fagin et al., 2003; Hsu et al., 2002; Hsu & Taksa, 2005; Lee, 1997; Ng & Kantor, 1998, 2000; Vogt & Cottrell, 1999). In performing classifier combination in the **pattern recognition (PR)** domain, rules are used to combine the output of multiple classifiers. The objective is to find methods (or rules) for building a hybrid classifier that would outperform each of the individual classifiers (Ho, 2002; Ho et al., 1992, 1994; Melnik et al., 2004). In **protein structure prediction (PSP)**, results from different features are combined to improve the accurate predictions of secondary classes or 3-D folding patterns (C.-Y. Lin et al., 2005; K.-L.

Lin et al., 2005a, 2005b). Biology may well represent a major area of future needs with respect to combinatorial fusion and related concepts. The last decade has seen an explosion in two data-driven concepts, so-called -omics level studies and *in silico* approaches to modeling.

Omics approaches are approaches that attempt to take snapshots of an organism at a specific level, for example, simultaneously measuring all the metabolites in a tissue and reconstructing pathways. The -omics levels studies range from the four major areas (e.g., genomics — the omics field of DNA analysis; transcriptomics — the omics field of RNA analysis; proteomics—the omics field of protein analysis; metabolomics — the omics field of metabolite analysis) to very specific subareas, such as glycomics (omics approaches to glycated proteins). Omics approaches represent a shift for two reasons: (1) the amount of data inherent either prevents or forces modifications in traditional data analysis approaches (e.g., *t*-tests being replaced by *t*-tests with false discovery rate calculations); and (2) these omics level approaches lead to data-driven and/or hypothesis-generating analyses, not (at least generally) to the testing of specific hypotheses. Most importantly, by offering greatly improved ability to consider systems as a whole and identify unexpected pieces of information, these approaches have opened new areas of investigation and are, perhaps too optimistically, expected to offer fundamentally new insights into biological mechanisms and fundamentally new approaches to issues such as diagnostics and medical classification.

In silico simulations are also becoming a major focus of some biological studies. There are at least two broad areas of simulation studies that are already playing major roles in biological investigations:

- *in silico* ligand-receptor (or drug-target) binding studies.
- *in silico* simulations of physiological systems.

In each of these cases, the great advantage lies not in the qualitative change empowered by technological advances, but in the quantitative savings of time and money. For example, obtaining and testing a small chemical library for binding or inhibition can readily cost between \$1 and \$100 and up (per compound), considering assay costs and obtaining (or synthesizing) the compounds. In contrast, once basic algorithms and binding site models are in place, *in silico* screening is limited only by computational costs, which are dropping exponentially. Similarly, *in silico* models of physiological systems can be somewhat costly to develop, but they are capable of identifying potential targets for intervention and to determine that other targets cannot work, saving tens or hundreds of millions of dollars in failed trials.

Fully utilizing the potential power of these omics and *in silico* based approaches, however, is heavily dependent on the quantity and quality of the computer resources available. The increases in hardware capacity, readily available software tools, database technology, and imaging and scanning techniques, have given the biomedical research community large scale and diversified data sets and the ability to begin to utilize these sets. The problem of how to manipulate, analyze, and interpret information from these biomedical data is a challenging and daunting task. Gradually biologists are discovering tools such as clustering, projection analyses, neural nets, genetic algorithms, genetic programs, and other machine learning and data mining techniques built in other fields, and these tools have found their way to the biomedical informatics domain.

A major issue here lies in the choice of appropriate tools; there is, for example, no clear "informatics pipeline" for omics level studies. One can readily pick up many of software tools, but their applicability for a given problem can be difficult to determine *a priori*. For example, clustering is often powerful for microarray experiments, but we have shown it is very limiting within some metabolomics experiments (Shi, et al., 2002a, 2002b). Similarly, principal components analyses and its supervised cousin, **Soft Independent Modeling of Class Analogy (SIMCA)** seem to work very well within defined cohorts, but they breakdown in complex multi-cohort studies (Shi et al., 2002a, 2002b, 2004), a problem apparently solvable using discriminant-based projection analyses (Paolucci, Vigneau-Callahan, Shi, Matson, & Kristal, 2004). Thus, the choice of an analysis method must often be determined empirically, in a slow, laborious step-wise manner. For the purpose of this article, we will break these biological studies into two broad areas, one of description (i.e., how do we best understand the data in front of us) and one of prediction (i.e., how can we use this information to make predictions about, for example, which ligand will bind or which person will become ill — questions which in many ways are mathematically equivalent).

The report focuses on mathematical issues related to the latter of these two broad issues, i.e., "can we use CFA to improve prediction accuracy?" The goal is a complex zero-sum game — we ideally want to further save time by reducing both false positives and false negatives, while simultaneously increasing accuracy on continuous measurements (e.g. binding strengths). In practice, it is almost certain that some trade-offs will have to be made. To be useful we must, at a minimum, identify an approach which enables some *a priori* decisions to be made about whether such fusion approaches are likely to succeed. Otherwise we have done nothing but to add a layer of complexity between where we are and where we need to be, without removing the time-consuming, laborious, and inherently limited stages of empirical validation. The system we choose to focus on is in **virtual screening (VS)**, the use of *in silico* approaches to identify potentially optimal binding ligands. VS is an area in which consensus scoring has been used in drug design and discovery and molecular similarity measurement for years, and in which data fusion approaches have recently been a major focus of efforts. (see Shoichet, 2004; Yang et al., 2005; and their references).

In this chapter, we present a method called combinatorial fusion analysis (CFA) which uses the **Cayley network Cay(S_n, T_n)** on the symmetric group S_n with generating set T_n (Biggs & White, 1979; Grammatikakis, Hsu, & Kraetzl, 2001; Heydemann, 1997; Marden, 1995). We study the fusion and combination process in the set $\mathbf{R}^n$, called score space, where **R** is the set of real numbers, and in the set S_n, called rank space. In the next section, we define rank and score functions and describe the concept of a rank/score graph defined by Hsu, Shapiro and Taksa (Hsu et al., 2002; Hsu & Taksa, 2005). The combination of two scoring systems (each with rank and score functions) is discussed and analyzed in the context of a Cayley network. The next section also entails the property of combined scoring system, performance evaluation and diversity issues, and various kinds of fusion algorithms on rank functions. The section "Data Mining Using Combinatorial Fusion" deals with mining and searching databases and the Web, and includes examples from application domains in biomedical informatics, in particular in virtual screening of chemical libraries and protein structure prediction. Then, in section "Conclusion and Discussion", we summarize our results with some discussion and future work.

COMBINATORIAL FUSION

Multiple Scoring Systems

Successfully turning raw data into useful information and then into valuable knowledge requires the application of scientific methods to the study of the storage, retrieval, extraction, transmission, diffusion, fusion/combination, manipulation, and interpretation at each stage of the process. Most scientific problems are multi-faceted and can be quantified in a variety of ways. Among many methodologies and approaches to solve complex scientific problems and deal with large datasets, we only mention three: (a) classification, (b) clustering, and (c) similarity measurement. Hybrid methods combining (a), (b), and (c) have been used.

Large data sets collected from multi-sensor devices or multi-sources or generated by experiments, surveys, recognition and judging systems are stored in a **data grid** $G(n, m, q)$ with n objects in $D = \{d_1, d_2, \ldots, d_n\}$, m features/attributes/indicators/cues in $G = \{a_1, a_2, \ldots, a_m\}$ and, possibly, q temporal traces in $T = \{t_1, t_2, \ldots, t_q\}$. We call this three-dimensional grid the **data space** (see Figure 1).

Since both m and q can be very big and the size of the datasets may limit the utility of single informatics approaches, it is difficult to use/design a single method/system because of the following reasons:

1. **Different methods/systems are appropriate for different features / attributes / indicators / cues and different temporal traces.** There have been a variety of different methods/systems used/proposed in the past decades such as statistical analysis and inference (e.g., t-test, non-parametric t-test, linear regression, analysis of variance, Bayesian systems), machine learning methods and systems (e.g., neural networks, self-organizing map, support vector machines), and evolutional computations (e.g., genetic algorithms, evolutionary programming).
2. **Different features / attributes / indicators / cues may use different kinds of measurements.** Many different measurements have been used such as variates, intervals, ratio-scales, and binary relations between objects. So in the data grid, each column on the D-G plane a_j can be an assignment of a score or a rank to each object d_i. For example, $M(i, j)$ is the score or rank assigned to object d_i by feature/attribute/indicator/cue a_j.
3. **Different methods/systems may be good for the same problem with different data sets generated from different information sources/experiments.** When different data sets generated from different information sources or experiments, different methods/systems should be used according to the style of the source and the nature of the experiment.
4. **Different methods/systems may be good for the same problem with the same data sets generated or collected from different devices/sources.** Even when the data sets are the same in the same experiments, different methods/systems should be adopted according to a variety of multi-sensor/multi-sources.

Figure 1. Data space $G(n, m, q)$

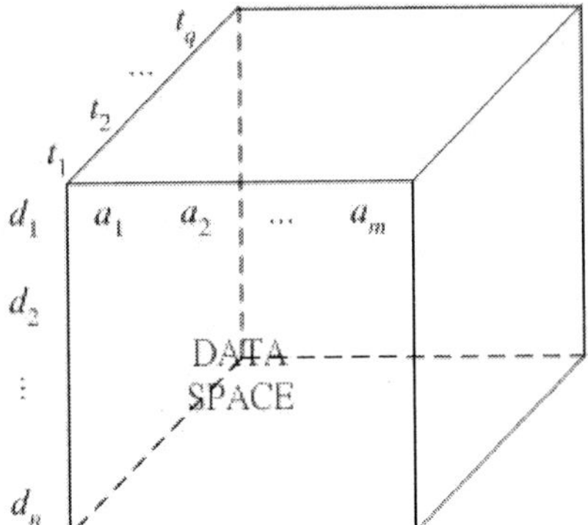

Due to the complexity of the problem involved, items (3) and (4) indicate that each single system/method, when applied to the problem, can be improved in performance to some extent, but it is difficult to become perfect. Item (1) indicates that performance of a single system / method may be optimal for some features/attributes/indicators/cues, but may downgrade its performance for other features/attributes/indicators/cues.

Recently, it has been demonstrated that combination of multiple systems/methods improves the performance of accuracy, precision, and true positive rate in several domains such as pattern recognition (Brown et al, 2005; Duerr et al., 1980; Freund, Iyer, Schapire, & Singer, 2003; Garcia-Pedrajas et al., 2005; Ho, 2002; Ho et al., 1992, 1994; Jin & Branke, 2005; Kittler & Alkoot, 2003; Perrone & Cooper, 1992; Triesch & von der Malsburg, 2001; Tumer & Ghosh, 1999; Xu et al., 1992), microarray gene expression analysis (Chuang, Liu, Brown, et al., 2004; Chuang, Liu, Chen, et al., 2004; Kuriakose et al., 2004), information retrieval (Aslam et al., 2003; Belkin et al., 1995; Diligenti et al., 2004; Dwork et al., 2001; Fagin et al., 2003; Hsu et al., 2002; Hsu & Taksa, 2005; Kantor, 1998; Lee, 1997; Ng & Kantor, 1998, 2000), virtual screening and drug discovery (Shoichet, 2004; Yang et al., 2005; and references in both), and protein structure prediction (C.-Y. Lin et al., 2005; K.-L. Lin et al., 2005a, 2005b).

There have been special meetings (such as the Workshop on Multiple Classifier Systems), conferences (such as International Conference on Information Fusion), societies (such as International Society of Information Fusion), and journals (e.g.: Information Fusion [Dasarathy, 2000]) dedicated to the scientific study of fusion/combination. The main title of the featured article "The Good of the Many Outweighs the Good of the One," by Corne, Deb, Fleming, & Knowles (2003) in IEEE Neural Networks Society typifies the scientific and philosophical merits of and motivations for fusion/combination.

Each system/method offers the ability to study different classes of outcomes, e.g., class assignment in a classification problem or similarity score assignment in the similarity measurement problem. In this chapter, we view the outcome of each system/method as a scoring system A which assigns (a) an object as a class among all objects in D, (b) a score to each object in D, and (c) a rank number to each object in D. These three outcomes were described as the abstract, score, and rank level respectively by Xu et al. (1992). We now construct the **system grid** $H(n, p, q)$ with n objects in $D = \{d_1, d_2, \ldots, d_n\}$, p systems in $H = \{A_1, A_2, \ldots, A_p\}$, and possibly, q temporal traces in $T = \{t_1, t_2, \ldots, t_q\}$. We call this three dimensional grid the **system space** for the multiple scoring systems (see Figure 2).

In the next section, we will define score function, rank function and the rank/score function for each scoring system A. In the section following the next, rank and score combination are defined and studied. Section "Method of Combinatorial Fusion" deals with performance evaluation criteria and diversity between and among different scoring systems.

Score Function, Rank Function and the Rank/Score Graph

Let $D = \{d_1, d_2, ..., d_n\}$ be a set of n objects and $N = \{1, 2, 3, \ldots, n\}$ be the set of all positive integers less than or equal to n. Let $\mathbf{R}$ be the set of real numbers. We now state the following three functions that were previously defined and studied by Hsu, Shapiro and Taksa (2002) and Hsu and Taksa (2005).

Definition 1

a. **Score Function:** A score function s is a function from D to $\mathbf{R}$ in which the function s assigns a score (a real number) to each object

Figure 2. System space H(n, p, q)

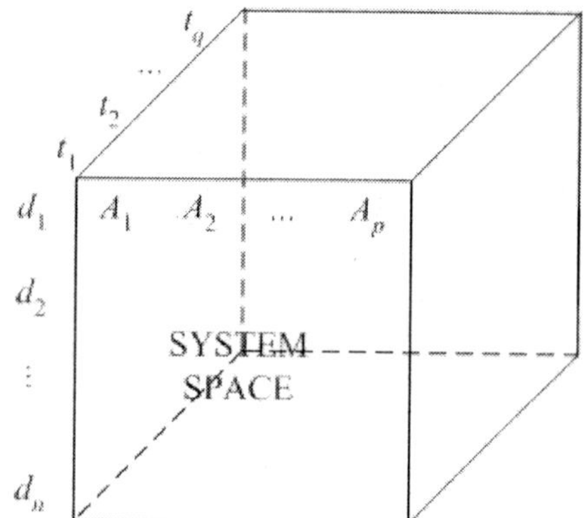

in D. In a more formal way, we write s: $D \to \mathbf{R}$ such that for every d_i in D, there exists a real number $s(d_i)$ in **R** corresponding to d_i.

b. **Rank Function:** A rank function r is a function from D to N such that the function $r : D \to N$ maps each element d_i in D to a natural number $r(d_i)$ in N. The number $r(d_i)$ stands for the rank number $r(d_i)$ assigned to the object d_i.

c. **Rank/Score Function:** Given r and s as rank and score function on the set D of objects respectively, the rank/score function f is defined to be $f : N \to \mathbf{R}$ such that $f(i) = (s \circ r^{-1})(i) = s(r^{-1}(i))$. In other words, the score function $s=$ is the composite function of the rank/score function and the rank function.

d. **Rank/Score Graph:** The graph representation of a rank/score function.

We note that in several application domains, one has to normalize the score function values before any combination can be performed. Hence it is quite natural to define the two functions s and f in the way that each of them has $[0, 1] = \{x \mid x$ in $\mathbf{R}, 0 \le x \le 1\}$ instead of **R** as their function range. Other intervals of real numbers can be also used depending on the situation and environment. We also note that since the rank function r' defined by Hsu, Shapiro and Taksa (see Hsu et al., 2002; Hsu & Taksa, 2005) is the inverse of r defined above, the rank/score function f would be such that $f = s^\circ r'$ instead of $f = s^\circ r'^{-1}$.

At this point, we would like to comment on some perspectives regarding rank vs. score function. Although these two functions both deal with the set of objects under study (in the case of PR, IR, VS and PSP, these would be classes, documents, chemical compounds, and classes or folding patterns, respectively), their emphases are different. The Score function deals more with the detailed data level while Rank function is more relevant for or related to the decision level. In theory, the Score function depends more on the variate data in the parametric domain while the Rank function depicts more on the ordinal data in the non-parametric fashion. The comparison can go on for a long time as score vs. rank, data level vs. decision level, variate data vs. ordinal data, and parametric vs. non-parametric. Historically and from the discipline perspective, scores are used in sciences, engineering, finance, and business, while ranks are used in social choices, ordinal data analysis and decision science. However, in biomedical informatics, since the data collected is large (and of multiple dimension) and the information we are seeking from biological and physiological systems is complex (and multi-variable), the information we find (or strive to find) from the relation between score and rank function would become valuable in biological, physiological, and pharmaceutical study.

The concept of a rank/score graph which depicts the graph representation of a rank/score function has at least three characteristics and advantages:

Remark 1

a. **Efficiency:** When a score function s_A is assigned resulting from scoring system A by either lab work or field study (conducted in vivo or in vitro), treating s_A as an array and sorting the array of scores into descending order would give rise to the rank function r_A. The rank/score function can be obtained accordingly. If there are n objects, this transformation takes $O(n \log n)$ steps.

b. **Neutrality:** Since the rank/score function f is defined from N to **R** or from N to [0, 1], it does not depend on the set of objects $D = \{d_1, d_2, \ldots, d_n\}$. Hence the rank/score function f_A of a scoring system A exhibits the behavior of the scoring system (scorer or ranker) A and is independent of who (or which object) has what rank or what score. The rank/score function f also fills the gap of relationship between the three sets D, N and **R**.

c. **Visualization:** The graph of the rank/score function f_A can be easily and clearly visualized. From the graph of f_A, it is readily concluded that the function f_A is a non-increasing monotonic function on *N*. The thrust of this easy-to-visualize property is that comparison (or difference) on two functions f_A and f_B can be recognized by drawing the graph of f_A and f_B on the same coordinate system.

Two examples of score functions, rank functions (derived from score function), rank/score functions with respect to scoring systems *A* and *B* are illustrated in Figure 3 and the rank / score graphs of f_A and f_B are included in Figure 4, where $D = \{d_i \mid i = 1 \text{ to } 10\}$ and $s(d_i)$ is in [0, 10].

Rank and Score Combination

As mentioned earlier, the combinations (or fusions) of multiple classifiers, multiple evidences, or multiple scoring functions in the PR, IR, VS, and PSP domain has gained tremendous momentum in the past decade. In this section, we deal with combinations of two functions with respect to both score and rank combinations. The following definitions were used by Hsu, Shapiro, and Taksa (2002) and Hsu and Taksa (2005).

Definition 2

a. **Score Combinations:** Given two score functions s_A and s_B, the score function of the score combined function s_F is defined as $s_F(d) = \frac{1}{2}[s_A(d) + s_B(d)]$ for every object *d* in *D*.

b. **Rank Combinations:** Given two rank functions r_A and r_B, the score function of the rank combined function s_E is defined as $s_E(d) = \frac{1}{2}[r_A(d) + r_B(d)]$ for every object *d* in *D*.

Since each of the scoring systems *A* and *B* has their score and rank function s_A, r_A and s_B, r_B respectively, each of the combined scoring systems *E* and *F* (by rank and by score combination) has s_E, r_E and s_F, r_F, respectively. These functions can be obtained as follows. The score function of the rank combination s_E is obtained from r_A and r_B using rank combination. Sorting s_E into ascending order gives rise to r_E. The score function of the score combination s_F is obtained from s_A and s_B using score combination. Sorting s_F into descending order gives rise to r_F. Hence for scoring systems *E* and *F*, we have the score function and rank function s_E, r_E and s_F, r_F respectively.

We note the difference between converting from s_E to r_E and that from s_F to r_F. Since the scoring

Figure 3. Score function, Rank function and Rank/Score function of A and B

D	d_1	d_2	d_3	d_4	d_5	d_6	d_7	d_8	d_9	d_{10}
$s_A(d_i)$	4	10	4.2	3	6.4	6.2	2	7	0	1
$r_A(d_i)$	6	1	5	7	3	4	8	2	10	9

(a) Score and Rank function for *A*

D	d_1	d_2	d_3	d_4	d_5	d_6	d_7	d_8	d_9	d_{10}
$s_B(d_i)$	4	7	3	1	10	8	5	6	9	2
$r_B(d_i)$	7	4	8	10	1	3	6	5	2	9

(b) Score and Rank function for *B*

N	1	2	3	4	5	6	7	8	9	10
$f_A(i)$	10	7	6.4	6.2	4.2	4	3	2	1	0
$f_B(i)$	10	9	8	7	6	5	4	3	2	1

(c) Rank/Score function for *A* and *B*

Figure 4. Rank/Score graphs of f_A and f_B

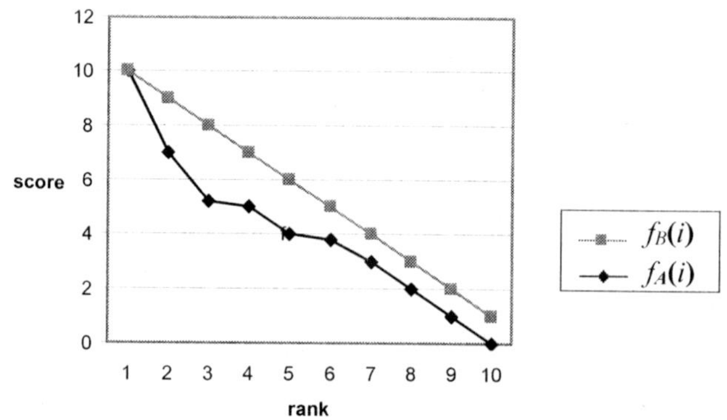

systems E and F are rank and score combination of A and B respectively, the transformation from s_E to r_E is by sorting into ascending order while that from s_F to r_F is by sorting into descending order. The fusion architecture given by Hsu, Shapiro and Taksa (Figure 5 in Hsu & Taksa, 2005) depicts the framework for the fusion method we use. Figure 5(a) and (b) show the rank and score function of the rank and score combination E and F (both related to the scoring systems A and B in Figure 3) respectively. The reader might have noticed that in the combination (rank or score), a simple average combination was used. Combination using weighted proportion on A and B is studied in Hsu and Palumbo (2004).

Recall that the main purpose of doing fusion/combination is whether or not the fused scoring system can outperform each individual scoring system in isolation. The concept of "performance" has to be defined for a given study (or, arguably, a given class of studies), so that it is meaningful to say "outperform". Recall also that our fusion framework is for application domains such as PR, IR, VS and PSP. In PR, each classifier produces a ranking of a set of possible classes. When given an image, each of the classifiers including the combined gives rise to a rank function and the class that was ranked at the top is predicted by the classifier as the identity of the image. So **the performance** (or **accuracy**) **of the classifier** A, written $P(A)$, is the percentage of times that this classification gives a correct prediction. From the perspective of biology and medicine, $P(A)$ may be defined as the reduction of false negatives (e.g., in cancer diagnostic) or the reduction of false positives (e.g., in screens of related compounds where follow-up is expensive).

In **IR**, a query Q is given. Then each of the ranking algorithms calculates similarity between the query and the documents in the database of n documents $D = \{d_1, d_2 \ldots, d_n\}$. A score function s_A for algorithm A is assigned and a rank function r_A is obtained. **The performance of the algorithm for the query**, written $P_q(A)$, is defined as the precision of A at q with respect to the query Q. More specifically, the following is defined and widely used in the information retrieval community.

Definition 3

a. Let $Rel(Q)$ be the set of all documents that are judged to be relevant with respect to the query Q. Let $|Rel(Q)| = q$ for some q, $0 \leq q \leq n$. On the other hand, let $A_{(k)} = \{d \mid d \text{ in } D \text{ and } r_A(d) \leq k\}$.
b. **Precision of A at q.** The performance of the scoring system A with respect to the query Q is defined to be $P_q(A) = |Rel(Q) \cap A_{(q)}| / q$, where $q = |Rel(Q)|$.

Figure 5. Score and rank function of E and F

D	d_1	d_2	d_3	d_4	d_5	d_6	d_7	d_8	d_9	d_{10}
$s_E(d_i)$	6.5	2.5	6.5	8.5	2	3.5	7	3.5	6	9
$r_E(d_i)$	6	2	4	9	1	3	8	7	5	10

(a) Score and rank function of E

D	d_1	d_2	d_3	d_4	d_5	d_6	d_7	d_8	d_9	d_{10}
$s_F(d_i)$	4.0	8.5	3.6	2.0	8.2	7.1	3.5	6.5	4.5	1.5
$r_F(d_i)$	6	1	7	9	2	3	8	4	5	10

(b) Score and rank functions of F

In **VS**, molecular compound libraries are searched for the discovery of novel lead compounds for drug development and/or therapeutical treatments. Let L be the total number of active ligands and T the total number of compounds in the database. Let L_h be the number of active ligands among the T_h highest ranking compounds (i.e., the hit list). Then the **goodness-of-hit (GH) score** for a scoring system A is defined as (see Yang et al., 2005):

$$GH(A) = \left(\frac{L_h (3L + T_h)}{4T_h L} \right) \left(1 - \frac{T_h - L_h}{T - L} \right).$$

The GH score ranges from 0.0 to 1.0, where 1.0 represents a perfect hit list. The GH score as defined in Yang et al. (2005) contains a coefficient to penalize excessive hit list size. We will come back to this topic in section "Virtual Screening and Drug Discovery".

In the **protein structure prediction problem (PSP)**, one wishes to extract structural information from the sequence databases as an alternative to determine the 3-D structure of a protein using the X-ray diffraction or NMR (resource and labor intensive, expensive, and, in practice, often difficult or impossible, particularly when one deals with variably modified proteins). Given a protein sequence, the objective is to predict its secondary structure (class) or its 3-D structures (folding patterns). The standard performance evaluation for the prediction of the n_i protein sequences $T_i = \{t_1, t_2, ..., t_{ni}\}$ for the ith class or ith folding pattern is the percentage accuracy rate $Q_i = Q(T_i) = (p_i / n_i) \times 100$, where n_i is the number of testing proteins in the ith class or ith folding pattern and p_i is the number of proteins within the n_i protein sequences correctly predicted. The overall prediction accuracy rate Q is defined as $Q = \sum_{i=1}^{k} q_i Q_i$, where $q_i = n_i / N$, N is the total number of proteins tested (i.e., $N = \sum_{i=1}^{k} n_i$) and k is the number of classes or folding patterns.

Now we come back to the fundamental issue of when the combined scoring system E or F outperforms its individual scoring system A and B. The following two central questions regarding combination and fusion were asked by Hsu, Shapiro, and Taksa (2002) and Hsu and Taksa (2005).

Remark 2

a. For what scoring systems A and B and with what combination or fusion algorithm, $P(E)$ (or $P(F)) \geq \max\{P(A), P(B)\}$, and
b. For what A and B, $P(E) \geq P(F)$?

Four important issues are central to CFA: (1) What is the best fusion algorithm / method to use? (2) Does the performance of E or F depend very much (or how much) on the relationship between A and B? (3) Given a limited number of primary scoring systems, can we optimize the specifics of the fusion? and (4) Can we answer any or all of the previous three issues without resorting to empirical validation? The general issue of combination algorithm / method will be discussed in the next section. In this section, we simply use the average combination regardless of rank or score combination.

Arguably, issues (1) and (2) may be considered primary issues, and issues (3) and (4) secondary or derivative. We propose that issue (2) is as, if not more, important as issue (1). It has been observed and reported extensively and intensively that the combined scoring system E or F performs better than each individual scoring system A and B when A and B are "different", "diverse", or "orthogonal". In particular, Vogt and Cottrell (1999) studied the problem of predicting the performance of a linearly combined system and stated that the linear combination should only be used when the individual systems involved have high performance, a large overlap of relevant documents, and a small overlap of non-relevant documents. Ng and Kantor (2000) identified two predictive variables for the effectiveness of the combination: (a) the output dissimilarity of A and B, and (b) the ratio of the performance of A and B. Then Hsu, Shapiro and Taksa (2002) and Hsu and Taksa (2005) suggested using the difference between the two rank/score functions f_A and f_B as the diversity measurement to predict the effectiveness of the combination. This diversity measurement has been used in microarray gene expression analysis (Chuang, Liu, Brown, et al., 2004; Chuang, Liu, Chen, et al., 2004) and in virtual screening (Yang et al., 2005). We will discuss in more details the use of graphs for rank/score functions f_A and f_B in the diversity measurement between A and B in section "Virtual Screening and Drug Discovery".

Method of Combinatorial Fusion

As defined in section "Score Function, Rank Function and the Rank/Score Graph", a rank function r_A of a scoring system A is an one-one function from the set $D = \{d_1, d_2, \ldots, d_n\}$ of n objects to the set N of positive integers less than or equal to n. When considering r_A as a permutation of N, we have $r_A(D) = [r_A(d_1), r_A(d_2), \ldots, r_A(d_n)]$. Without loss of clarity, we write r instead of r_A.

Since the set of all permutations of the set N is a group with the composition $\alpha^{\circ}\beta$ as the binary operation, called symmetric group S_n, $r(D)$ as a permutation of N is an element in the group S_n. Each element a in S_n has two different ways of representation as $[\alpha(1), \alpha(2), ..., \alpha(n)]=[\alpha_1, \alpha_2, ... ,\alpha_n]=[\alpha_1\ \alpha_2\ \alpha_3\ ...\ \alpha_n]$ called **standard representation** and as the product of disjoint cycles each consisting of elements from N called **cycle representation**. The general concept of a Cayley graph (or network) can be found in the book and article (Grammatikakis et al., 2001; Heydemann, 1997).

Definition 4

a. **Example of Permutations:** The permutation $a(i) = 2i$ for $i = 1, 2, 3$ and $2i - 7$ for $i = 4, 5, 6$ in S_6 can be written as [2 4 6 1 3 5] and (124) (365). The permutation $b(i) = i$ for $i = 1, 4, 5, 6$ and $b(2) = 3$, $b(3) = 2$ can be written as $b = [1\ 3\ 2\ 4\ 5\ 6] = (23)$. Note that in the cycle representation, we ignore the cycles that are singletons.
b. In the group S_n, the set of $n - 1$ adjacent transpositions such as b= (23) in S_6 is denoted as T_n. In other words, T_n consists of all cycles of length 2 which are adjacent transpositions and $T_n = \{(1\ 2), (2\ 3), \ldots, (n-1\ n)\}$ is a subset of S_n. With this in mind, we can define a Cayley network based on S_n and T_n:

Cayley network Cay (S_n, T_n)**:** The Cayley network Cay(S_n, T_n) is a graph $G(V, E)$ with the node set $V = S_n$ and arc set $E = \{(\alpha, \alpha \circ t) \mid \alpha$ in S_n and t in $T_n\}$.

The concept of a Cayley network extends the group structure in S_n to the graph structure in Cay(S_n, T_n). By doing so, a distance measure between any two permutations (and hence any two rank functions) is well defined in the context of applications that will prove to be very useful in biomedical informatics. In fact, it has been mentioned by Hsu, Shapiro and Taksa (2002) and Hsu and Taksa (2005) that the graph distance in Cay(S_n, T_n) is the same as Kendall's tau distance in the rank correlation analysis (RCA) (see e.g., Kendall &Gibbons, 1990; Marden, 1995). This striking coincidence supports the importance and usefulness of using Cayley networks as a framework for fusion and combination. Moreover, we point out that the combinatorial fusion we proposed and the rank correlation studied by many researchers in the past bear similarity but have differences. They are very similar because they all study ranks although one treats ranks as a function and the other treats them as ordinal data or the order of the values of a random variable. On the other hand, they are quite different. The CFA (combinatorial fusion analysis) views the set S_n as a rank space aiming to produce a dynamic process and reasonable algorithms to reach a better combined rank function (or in general, combined scoring system). The RCA (rank correlation analysis) views the set S_n as a population space aiming to calculate the static correlation and significant *P*-value to reach a hypothesis testing result.

Suppose we are given the following p rank functions A_j obtained from the data set $D = \{d_1, d_2, \ldots, d_n\}$, $A_j, j = 1, 2, \ldots, p$:

$$A_j = (a_{1j}, a_{2j}, a_{3j}, \ldots, a_{nj})^t,$$

where V^t is the transpose of the vector V. Let M_r be the matrix, called **rank matrix**, with dimension $n \times p$ such that $M_r(i, j) = M(i, j) =$ the rank assigned to the object d_i by scoring system A_j. Now, we describe the rank version of the combinatorial fusion problem.

Definition 5

a. **Combinatorial Fusion Problem (rank version):** Given p nodes $A_j, j = 1, 2, \ldots, p$ in the Cayley network $\mathrm{Cay}(S_n, T_n)$ with respect to n objects $D = \{d_1, d_2, \ldots, d_n\}$, find a node A^* in S_n which "performs" as good as or better than the best of A_j's in the sense of performance as defined as accuracy, precision or goodness-of-hit in IR, VS and PSP described previously.

There are several ways to find the candidates for the node A^* when given the p nodes $A_j, j = 1, 2, \ldots, p$ in S_n. We briefly describe, in Definition 6, the following six types of methods / algorithms to fuse the given p nodes and generate the candidate node. All of these approaches aim to construct a score function which, when sorted, would lead to a rank function.

Definition 6

a. **Voting:** Scoring function $s^*(d)$, d in D. The score of the object d_i, $s^*(d_i)$, is obtained by a voting scheme among the p values $M(i, j)$, $j = 1, 2, \ldots, p$. These include max, min, and median.

b. **Linear Combination:** These are the cases that $s^*(d_i)$ is a weighted linear combination of the $M(i,j)$'s, i.e., $s^*(d_i) = \sum_{j=1}^{p} w_j \cdot M(i, j)$ for some weighted function so that $\sum_{j=1}^{p} w_j = 1$. When $w_j = 1/p$, $s^*(d_i)$ is the average of the ranks $M(i, j)$'s, $j = 1, 2, \ldots, p$.

c. **Probability Method:** Two examples are the Bayes rule that uses the information from the given p nodes $A_j, j = 1, 2, \ldots, p$ to predict the node A^*, and the Markov Chain method that calculates a stochastic transition matrix.

d. **Rank Statistics:** Suppose that the p rank functions are obtained by the p scoring systems or observers who are ranking the n objects. We may ask: what is the true ranking of each of the n objects? Since the real (or true) ranking is difficult to come by, we may ask another question: What is the best estimate of the true ranking when we are given the p observations? Rank correlation among the p rank functions can be calculated as $W = 12S / [p^2(n^3 - n)]$, where $S = \sum_{i=1}^{n} R_i^2 - np^2(n+1)^2/4$ and $R_i = \sum_{j=1}^{p} M(i, j)$. The significance of an observed value of W is then tested in the $(n!)^p$ possible sets of rank functions.

e. **Combinatorial Algorithm:** For each of the n objects and its set of p elements $\{M(i, j) \mid j = 1, 2, \ldots, p\} = C$ as the candidate set for $s^*(d_i)$, one combinatorial algorithm considers the power set 2^C and explores all the possible combinatorial combinations. Another algorithm treats the n objects as n vectors $d_i = (a_{i1}, a_{i2}, \ldots, a_{ip})$ where $a_{ij} = M(i, j)$, $i = 1, 2, \ldots, n$ and $1 \leq a_{ij} \leq n$. It then places these n vectors in the context of a **partially ordered set (Poset)** L consisting of all the n^p vectors $(a_{i1}, a_{i2}, \ldots, a_{ip})$, $i = 1, 2, \ldots, n$ and a_{ij} in N. The scores $s^*(d_i)$, $i = 1, 2, \ldots, n$ is then calculated based on the relative position of the vector d_i in the Poset L.

f. **Evolutional Approaches:** Genetic algorithms and other machine learning techniques such as neural networks and support vector machines can be used on the p rank functions to process a (large) number of iterations in order to produce a rank function that is closest to the node A^*.

The voting schemes in Definition 6 (a) have been used in social choice functions (Arrow, 1963; Young & Levenglick, 1978). Linear combination and average linear combination in 6(b), due to their simplicity, have been widely used in many application domains (Kendall & Gibbons, 1990; Kuriakose et al., 2004; Hsu & Palumbo, 2004; Hsu et al., 2002; Hsu & Taksa, 2005; Vogt & Cottrell, 1999). In fact, the concept of Borda count, used by Jean-Charles de Borda of the L'Academie Royale des Sciences in 1770, is equivalent to the average linear combination. Dwork et al. (2001) used Markov chain method to aggregate the rank functions for the Web. As described in Definition 6(d), the significance of S depends on the distribution of S in the $(n!)^p$ possible set of rank functions. Due to the manner that S is defined, it may be shown that the average linear combination gives a "best" estimate in the sense of Spearman's rho distance (see Kendall & Gibbons, 1990, Chapter 6). Combinatorial algorithms stated in Definition 6(e) have been used in Mixed Group Ranks and in Rank and Combine method by researchers (Chuang, Liu, Brown, et al., 2004; Chuang, Liu, Chen, et al., 2004; Melnik et al., 2004). Although genetic algorithms such as GemDOCK and GOLD were used to study the docking of ligands into a protein, the authors in Yang et al. (2005) use linear combination and the rank/score graph as a diversity measurement. We will discuss the application in more details in next section.

Definition 6 lists six different groups of methods/algorithms/approaches for performing combination. Here we return to the second issue raised by Remark 2. That is: What are the predictive variables / parameters / criteria for effective combination? In accordance with Remark 2, we focus on two functions A and B (i.e. $p = 2$) at this moment although the methods / algorithms / approaches in Definition 6, are able to deal with the multiple functions ($p \geq 2$). We summarize, in Definition 7, the two variables for the prediction of effective combination among two scoring systems A and B (Chuang, Liu, Brown, et al., 2004; Chuang, Liu, Chen, et al., 2004; Hsu et al., 2002; Hsu & Taksa, 2005; Ng & Kantor, 2000; Vogt & Cottrell, 1999; Yang et al., 2005).

Definition 7

a. **The performance ratio,** P_l / P_h, measures the relative performance of A and B where P_l and P_h are the lower performance and higher performance among $\{P(A), P(B)\}$ respectively.
b. **The bi-diversity between** A **and** B, $d_2(A, B)$, measures the "difference / dissimilarity / diversity" between the two scoring systems A and B.

We note that in order to properly use diversity $d_2(A, B)$ as a predictive parameter for effective combination of functions A and B, $d_2(A, B)$ might be defined to reflect different combination algorithms and different domain applications. However, for the diversity measurement to be effective, it has to be universal at least among a variety of data sets in applications domain. Diversity measurement between two scoring systems A and B, $d_2(A, B)$ have been defined and used (see Chuang, Liu, Brown, et al., 2004; Chuang, Liu, Chen, et al., 2004; Ng & Kantor, 2000; Yang et al., 2005). We summarize in Definition 8.

Definition 8

Diversity Measure: The bi-diversity (or 2-diversity) measure $d_2(A, B)$ between two scoring systems A and B can be defined as one of the following:

a. $d_2(A, B) = d(s_A, s_B)$, the distance between score functions s_A and s_B. One example of $d(s_A, s_B)$ is the covariance of s_A and s_B, $\mathrm{Cov}(s_A, s_B)$, when s_A and s_B are viewed as two random variables.
b. $d_2(A, B) = d(r_A, r_B)$, the distance between rank functions r_A and r_B. One example of $d(r_A, r_B)$ is the Kendall's tau distance as we defined in S_n.
c. $d_2(A, B) = d(f_A, f_B)$, the distance between rank/score functions f_A and f_B.

We note that diversity measure for multiple classifier systems in pattern recognition and classification has been studied extensively (Kuncheva, 2005).

Definition 9

In the data space $G(n, m, q)$, $m = 2$, defined in Figure 1, given a temporal step t_i in $T = \{t_1, t_2, \ldots, t_q\}$ and the two scoring systems A and B, we define:

a. $d_{t_i}(A, B) = \sum_j |f_A(j) - f_B(j)|$, where j is in $N = \{1, 2, \ldots, n\}$, as the function value of the **diversity score function** $d_x(A, B)$ for t_i.
b. If we let i vary and fix the system pair A and B, then $s_{(A,B)}(x)$ is the diversity score function, defined as $s_{(A,B)}(t_i) =$, from $T = \{t_1, t_2, \ldots, t_q\}$ to **R**.
c. Sorting $s_{(A,B)}(x)$ into ascending order leads to the **diversity rank function** $r_{(A,B)}(x)$ from T to $\{1, 2, \ldots, q\}$.
d. The **diversity rank/score function** $f_{(A,B)}(j)$ can be obtained as $f_{(A,B)}(j) = (s_{(A,B)} \circ r^{-1}_{(A,B)})(j) = s_{(A,B)}(r^{-1}_{(A,B)}(j))$, where j is in $\{1, 2, \ldots, q\}$.
e. The **diversity rank/score graph** (or diversity graph) is the graph representation of the diversity rank/score function $f_{(A,B)}(j)$ from $\{1, 2, \ldots, q\}$ to **R**.

We note the difference between the rank/score function and the diversity rank/score function. In the definition of rank/score function $f_A: N \rightarrow [0, 1]$ (see Definition 1(c)), the set N is different from the set D which is in turn the set of objects (classes, documents, ligands, and classes or folding patterns). The set N is used as the index set for the rank function values. The rank/score function f_A so defined describes the scoring (or ranking) behavior of the scoring system A and is independent of the objects under consideration. The diversity rank/score function (see Definition 9(d)) $f_{(A,B)}(j)$ is defined from $Q = \{1, 2, \ldots, q\}$ to **R** (or [0, 1]). The set Q is different from the set $T = \{t_1, t_2, \ldots, t_q\}$ which is the set of temporal steps under study. The set Q is used as the index set for the diversity rank function values. The diversity rank/score function $f_{(A,B)}$ so defined describes the diversity trend of the pair of scoring systems A and B and is independent of the specific temporal step t_i for some i under study.

DATA MINING USING COMBINATORIAL FUSION

In this section, we present three examples of data mining using combinatorial fusion as defined in the previous section. These three examples are from applications in information retrieval (IR) (Hsu et al, 2002; Hsu & Taksa, 2005; Ng & Kantor, 1998, 2000), consensus scoring in virtual screening (VS) (Yang et al., 2005), and protein structure prediction (PSP) (C.-Y. Lin et al., 2005; K.-L. Lin et al., 2005a, 2005b). But before we concentrate on special cases, we will further discuss the relation between rank and score functions as defined in the previous section.

Rank/Score Transfer

Let $M_r(i, j)$ be the **rank matrix** defined before, where M_{ij} in $M_r(i, j)$ is the rank assigned to the object d_i by scoring system A_j. Let $M_s(i,j)$ be the $n \times p$ **score matrix** defined similarly with respect to the p score functions so that M_{ij} (without ambiguity) in $M_s(i, j)$ is the score value assigned to the object d_i by scoring system A_j. The algorithms and approaches described in Definition 6 can be applied to the rank matrix M_r. However, some of these algorithms have been also applied to the score matrix M_s. Up to this point in this chapter, we have emphasized rank combination algorithms and considered ranks of objects as the basic data of a given situation / experiment / work, regardless of the manner in which they were obtained. However, in many situations, the ranking takes place according to the score values of a variable or variate. It is, therefore, of considerable interest to study the relationship between $M_r(i, j)$ and $M_s(i, j)$. As we mentioned before, $M_r(i, j)$ can be derived from $M_s(i, j)$ by sorting each column, A_j, into descending order and assigning higher value with higher rank (i.e., smaller number). One of the interesting questions is that: Is there any difference between the information represented by $M_s(i, j)$ and that by $M_r(i, j)$? In 1954, A. Stuart showed the following (see Kendall & Gibbons, 1990, Chapters 9, 10):

Remark 3 (Correlation Between Scores and Ranks)

With $n = 25$, the correlation between scores and ranks for a scoring system (ranker/scorer) A is as high as 0.94 under the assumption of normal distribution and 0.96 for the uniform distribution among the score values. These values increase when the sample size (i.e., the number n) increases and reach the limits of 0.98 and 1, for normal and uniform distribution, respectively.

In light of this close relationship between ranks $M_r(i, j)$ and scores $M_s(i, j)$, we might expect that operating on $M_r(i, j)$ and on $M_s(i, j)$ would draw the same conclusion. This appears to be so in a number of special cases. But in general, it has to be approached with certain amount of caution and care. It is clear, for example, that a few points with comparatively high residuals (i.e. poor correlations) would not have major effects on the overall correlation and correlation structure of the dataset, but these outliers may well be the key target of the investigation. We list the special features of transforming from $M_s(i, j)$ to $M_r(i, j)$.

Remark 4 (Rank/Score Transfer)

When transforming from a score function s_A to a rank function r_A on n objects, we have:

a. The dimension of sample space is reduced from $\mathbf{R}^n$ for the score space to N^n (and then $N!$ because of permutation) for the rank space, and
b. The score function values have been standardized to the scale while the mean is fixed for every rank function.

Remark 4(a) states that dimension reduction is obtained by a rank/score transform process that gives rise to the concept of a rank/score function, and the rank/score graph has at least three advantages: efficiency, neutrality, and visualization (see Remark 1). Remark 4(b) states that in $M_r(i, j)$, the mean of each column (a rank function on the n objects) is fixed to be $(n + 1)/2$. The same phenomenon is also true for some score data under certain cases of specific functions and study objectives. However, we note that when non-parametric and ordinal rank data is used, emphasis is not on the mean of the data. Rather it is on the discrete order and position each of the rank data is placed.

We recall that fully ranked data on n objects are considered as rank functions on the symmetric group of n elements S_n. Since S_n does not have a natural linear ordering, graphical methods such as histograms and bar graphs may not be appropriate for displaying ranked data in S_n. However, a natural partial ordering on S_n is induced in the Cayley network Cay(S_n, T_n). Moreover, since a polytope is the convex hull of a finite set of points in $\mathbf{R}^{n-1}$, the $n!$ nodes in Cay(S_n, T_n) constitute a permutation polytope when regarded as vectors in $\mathbf{R}^n$ (see Marden, 1995; McCullagh, 1992; Thompson, 1992). In fact, the $n!$ nodes of Cay(S_n, T_n) lie on the surface of a sphere in $\mathbf{R}^{n-1}$. The six nodes and twenty four nodes of Cay(S_3, T_3) and Cay(S_4, T_4) are exhibited in Figure 6(a) and 6(b), respectively.

Information Retrieval

We now turn to the application of these data mining techniques to the information retrieval domain. We use as an example the study by Ng and Kantor (1998, 2000) (We call this the NK-study). Their exploratory analysis considered data from TREC competition with 26 systems and 50 queries for each system on a large but fixed database consisting of about 1000 documents. The results from these 26 systems are then fused in a paired manner. As such, there are [(26 × 25) / 2] × 50 = 16, 250 cases of data fusion in the training data set. In 3,623 of these cases, the performance measures, as P_{100}, of the combined system is better than the best of the two original systems. We refer to these as **positive cases**. There are 9,171 **negative cases** where the performance of the combined system is worse than the best of the two original systems. In order to understand these two outcomes, two predictive variables are used. The first is the ratio of P_{100}, $r = P_l / P_h$ (see Definition 7(a)). The second variable is the normalized dissimilarity $z = d(r_A, r_B)$ (see Definition 8(b)). We summarize the results of the NK-study as follows. See Figure 7 for an illustration.

Figure 6. Nodes in Cay(S3, T3) and Cay(S4, T4), (a) six nodes in Cay(S3, T3), and (b) 24 nodes in Cay(S4, T4)

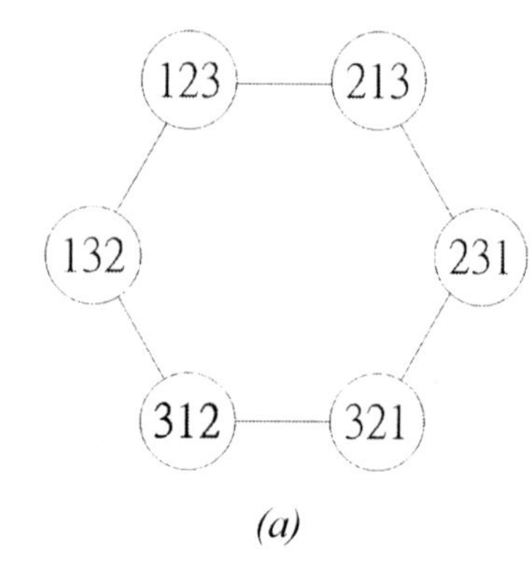

(a)

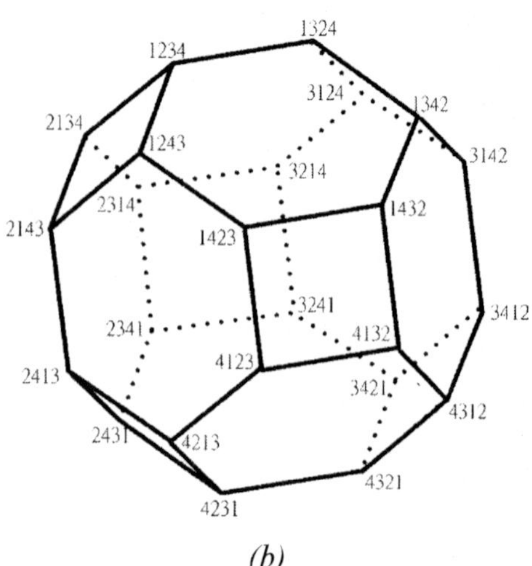

(b)

Remark 5 (NK-study)

The results of the NK-study shows that (a) the positive cases tend to lie above the diagonal line $r + z = 1$, and (b) the negative cases are more likely to scatter around the line $r + z = 1$.

Remark 5 gives the general trend as to where the positive cases and negative cases should fall. There are very few negative cases with small r and small z and comparatively very few cases with high r and high z. Since the negative cases all spread around the line $r + z = 1$, z approaches 0 as r approaches 1 and vice versa. This means that for the negative cases, when the performances P_{100} of the two IR systems are about the same, their rank functions are similar to each other. For the positive cases, it was found that there are very few cases with small r and z and comparatively few cases with large r and z. But as Remark 5(a) indicated, the positive cases are more likely to lie above the diagonal $r + z = 1$. This indicates that systems with dissimilar (i.e., diverse) rank functions but comparable performance are more likely to lead to effective fusion.

Virtual Screening and Drug Discovery

We now turn to the application of the data mining technique to biomedical informatics. In particular, we discuss in more details the study by Yang et al. (2005) (we call this paper the YCSKH-study). The study explores criteria for a recently developed virtual screening technique called "Consensus Scoring" (CS). It also provides a CS procedure for improving the enrichment factor in CS using combinatorial fusion analysis (CFA) techniques and explores diversity measures on scoring characteristics between individual scoring functions.

In structure-based virtual screening, a docking algorithm and a scoring function are involved (see Shoichet, 2004, and its references). The primary purpose for a docking program is to find out the most favorable combination of orientation and conformation (**Pose**). It also requires a comparison of the best pose (or top few poses) of a given ligand with those of the other ligands in chemical data base such that a final ranking or ordering can be obtained. In essence, VS uses computer-based methods to discover new ligands on the basis of biological structure. Although it was once popular in the 1970s and 1980s, it has since struggled to meet its initial promise. Drug discovery remains dominated by empirical screening in the past three decades. Recent successes in predicting new ligands and their receptor-bound structure have re-invigorated interest in VS, which is now widely used in drug discovery.

Although VS of molecular compound libraries has emerged as a powerful and inexpensive method for the discovery of novel lead compounds, its major weakness — the inability to consistently identify true positive (leads) — is likely due to a lack of understanding of the chemistry involved in ligand binding and the subsequently imprecise scoring algorithms. It has been demonstrated that consensus scoring (CS), which combines multiple scoring functions, improves enrichment of true positions. Results of VS using CS have largely focused on empirical study. The YCSKH-study is one attempt to provide theoretical analysis using combinatorial fusion (Yang et al., 2005).

The YCSKH-study developed a novel CS system that was tested for five scoring systems (A, B, C, D, and E) with two evolutionary dock-

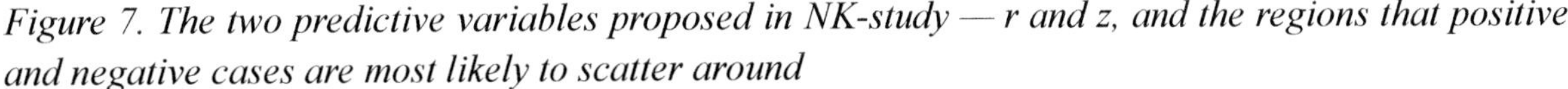
Figure 7. The two predictive variables proposed in NK-study — r and z, and the regions that positive and negative cases are most likely to scatter around

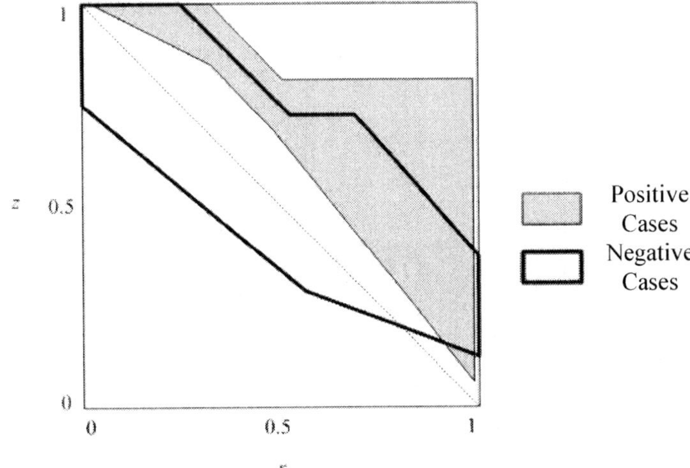

ing algorithms (GemDOCK and GOLD) on four targets: thymidine kinase (TK), human dihydrofolate reductase (DHFR), and estrogen receptors (ER) of antagonists and agonists (ERA). Their scoring systems consist of both rank-based and score-based CS systems (RCS and SCS). They used the GH (goodness-of-hit) score to evaluate the performance of each individual and combined systems. That is, $P(A)$ = GH score of the system A. Two predicative variables are used: (a) the performance ratio $PR(A, B) = P_l/P_h$ and (b) the diversity measure $d_2(f_A, f_B)$ as defined in Definitions 7 and 8 (see Yang et al., 2005).

$PR(A, B) = P_l / P_h = \min\{P(A), P(B)\} / \max\{P(A), P(B)\}$,

and

$$d_2(f_A, f_B) = \left\{ \sum_{j=1}^{n} \left(f_A(j) - f_B(j)\right)^2 / n \right\}^{1/2},$$

where $P(A)$ and $P(B)$ are the performances of the two scoring systems A and B to be combined.

Let $g(x)$ denote the normalization function for the two predictive variables $PR(A, B)$ and $d_2(f_A, f_B)$ so that $g(x)$ is in [0, 1]. Their results regarding bi-diversity and combination of two scoring systems are summarized in the following remark where we use x and y as the coordinates for $g(P_l / P_h)$ and $g(d_2(f_A, f_B))$ respectively (see Figure 8).

Remark 6 (YCSKH-study)

The YCSKH-study shows that numbers of positive and negative cases split into roughly half and (a) most of the positive cases are located above the line $x + y = 1$ while none of the few cases below the line have both $x \leq 0.30$ and $y \leq 0.30$., and (b) most of the negative cases tend to be located below the line $x + y = 1$ while only one of the few cases above the line have both $x \geq 0.50$ and $y \geq 0.50$. The exceptional case has $g(d_2(f_A, f_B)) \approx 0.60$ and $g(PR(A, B)) \approx 0.95$ but both P_l and P_h are very small.

Remark 6 reconfirms that combining two scoring systems (rank function or score function) improves performance only if (a) each of the individual scoring systems has relatively high performance and (b) the scoring characteristics of each of the individual scoring systems are quite different. This suggests that the two systems to be combined have to be fairly diverse so that they can complement each other, and the performance of each system has to be good, although we cannot yet quantitate/constrain the quality of "good" outside of our specific study.

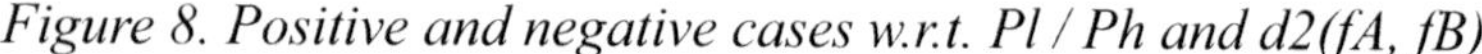

Figure 8. Positive and negative cases w.r.t. Pl / Ph and d2(fA, fB)

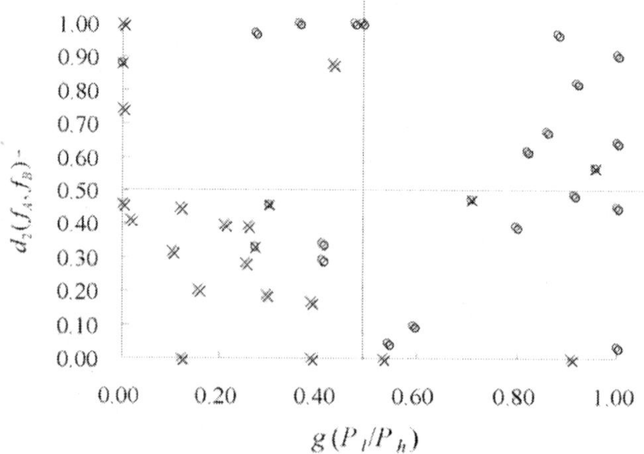

Protein Structure Prediction

Following their previous work establishing a hierarchical learning architecture (HLA), two indirect coding features, and a gate function to differentiate proteins according to their classes and folding patterns, C.-Y. Lin et al. (2005) and K.-L. Lin et al. (2005a, 2005b) have used combinatorial fusion to improve their prediction accuracy on the secondary class structure and/or 3-D folding patterns. Using 8 and 11 features respectively (i.e., scoring systems) and neural networks as a multi-class classifier to build HLA, they adopted the radical basis function network (RBFN) model to predict folding pattern for 384 proteins. In other words, in the system space $M(n, p, q)$ defined in Figure 2, the number of scoring systems $p = 8$ or 11 and the number of temporal steps $q = 384$.

The work by C.-Y. Lin et al. (2005) using 8 features has an overall prediction rate of 69.6% for the 27 folding categories, which improves previous results by Ding and Dubchak (2001) of 56% and by Huang et al. (2003) of 65.5%. The work by K.-L. Lin et al. (2005a, 2005b), using CFA to facilitate feature selection using 11 features, achieves a prediction accuracy of 70.9% for 27 folding patterns.

Both works utilize the concept of diversity rank/score graph (Definition 9(e)) to select features to combine. In C.-Y. Lin et al. (2005), the scoring systems (features in this case) E, F, G, H are selected from the eight features {A, B, C, D, E, F, G, H}. Then the diversity rank/score functions for the six pairs of the four scoring systems E, F, G, H, are calculated (see Figure 9), where the diversity rank/score graph for the pair (E, H) is found to have the highest overall value across the $q = 384$ protein sequences tested. Hence E and H are considered to have the highest diversity among the six pairs of scoring systems. The best result was obtained by combining these two scoring systems E and H.

The work by K.-L. Lin et al. (2005a, 2005b) uses 11 features and has selected H, I, K out of 11 features {A, B, C, D, E, F, G, H, I, J, K} because of their higher individual performance. The diversity rank/score function is then calculated for any pair among H, I, K (see Figure 10). From the graphs of these three functions, they conclude that the pair H, I has the highest diversity across the $q = 384$ protein sequences tested. This pair of scoring systems H and I are then used to perform combination to achieve the desired result.

Figure 9. Diversity graphs for the six pairs of scoring systems E, F, G and H

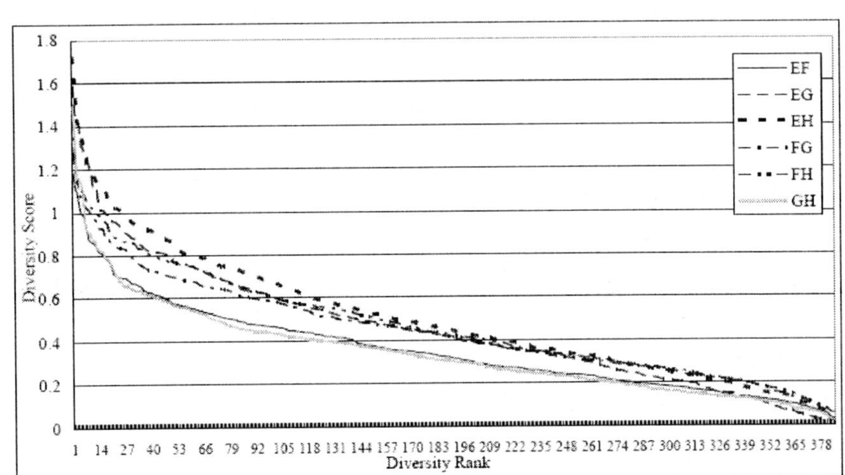

Figure 10. Diversity graphs for the three pairs of scoring systems H, I and J.

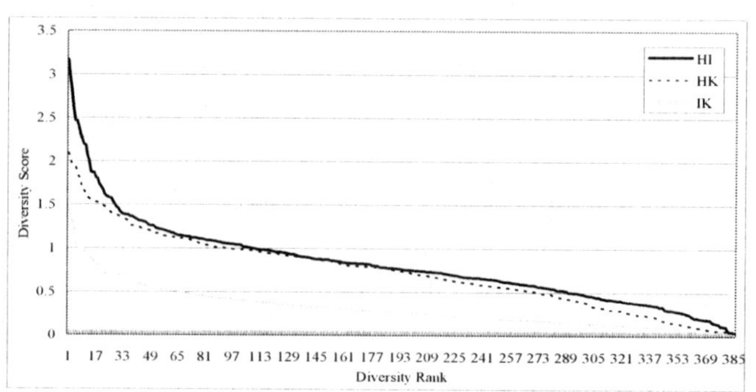

CONCLUSION AND DISCUSSION

In this chapter, we have described a method, called Combinatorial Fusion Analysis (CFA), for combining multiple scoring systems (MSS) each of which is obtained from a set of homogeneous or heterogeneous features/attributes/indicators/cues. The method (CFA) is based on information obtained from multi-variable (multi-criterion/multi-objective) classification, prediction, learning or optimization problems, or collected from multi-sensor / multi-source environments or experiments. We distinguished between the data space (see Figure 1) which consists of using features/attributes/indicators/cues to describe objects (e.g., pattern classes / documents / molecules / folding patterns), and the system space (see Figure 2), which consists of using different scoring systems (statistical methods, learning systems, combinatorial techniques, or computational algorithms) to assign a score and a rank to each of the objects under study. In the extreme case when each scoring system is a feature / attribute / indicator / cue, the system space coincides with the data space.

We use the concepts of score functions, rank functions and rank/score functions as defined by Hsu, Shapiro and Taksa (2002) and Hsu and Taksa (2005) to represent a scoring system. Rank / score transfer and correlation between scores and ranks are described and examined. We also described various performance measurements with respect to different application domains. Various combination/fusion methods/algorithms of combining multiple scoring systems have been explored. Theoretical analysis that gives insights into system selection and methods of combination/fusion has been provided. In particular, we observe that combining multiple scoring systems improves the performance only if (a) each of the individual scoring systems has relatively high performance, and (b) the individual scoring systems are distinctive (different, diverse or orthogonal). We have initiated study on these two issues (a) and (b) for the special case of two scoring systems A and B. Two predictive parameters are used. The first parameter is the performance ratio, P_l/P_h, which measures the relative performance of A and B where P_l and P_h are the lower performance and higher performance respectively. The second parameter deals with the bi-diversity between A and B, $d_2(A, B)$, which measures the degree of difference/dissimilarity/diversity between the two scoring systems A and B.

The bi-diversity (or 2-diversity) measure $d_2(A, B)$ between two scoring systems A and B is defined as one of the following three possibilities: (1) $d_2(A, B) = d(s_A, s_B)$, the distance between the score functions s_A and s_B of scoring system A and B; (2) $d_2(A, B) = d(r_A, r_B)$, the distance between the rank functions r_A and r_B; and (3) $d_2(A, B) = d(f_A, f_B)$, the distance between rank/score functions f_A

and f_B. Diversity measures have been studied extensively in pattern recognition and classification (Kuncheva, 2005; Brown et al., 2005). Diversity measure defined in the form of rank function was used in information retrieval (Ng & Kantor, 1998, 2000). The work of Hsu and Taksa (2005) and Yang et al. (2005) used rank/score functions to measure diversity between two scoring systems in their study of comparing rank vs. score combination and consensus scoring criteria for improving enrichment in virtual screening and drug discovery. For the protein structure prediction problem, the rank/score functions of A and B, f_A and f_B, are used for each protein sequence p_i, where $d_{p_i}(A,B)=\sum_j |f_A(j)-f_B(j)|$, j in $N=\{1, 2, \ldots, n\}$, is the diversity score function for p_i (C.-Y. Lin et al., 2005; K.-L. Lin et al., 2005a, 2005b).

Considering all protein sequences p_i in $P=\{p_1, p_2, \ldots, p_q\}$, the diversity score function, $s_{(A,B)}(x)$ written as $s_{(A,B)}(p_i) = d_{p_i}(A,B)$, is a function from P to $\mathbf{R}$. Consequently, sorting $s_{(A,B)}(x)$ into descending order leads to a diversity rank function $r_{(A,B)}(x)$ from P to $Q = \{1, 2, \ldots, q\}$. Hence the diversity rank/score function (or diversity function) $f_{(A,B)}(j)$ defined as:

$$f_{(A,B)}(j)=(s_{(A,B)}\circ r^{-1}_{(A,B)})(j)=s_{(A,B)}(r^{-1}_{(A,B)}(j))$$

where j is in Q, is a function from Q to $\mathbf{R}$. The diversity function and its graph play important roles in system selection and method combination in the PSP problem. We note that the diversity function so defined for PSP problem can be applied to other classification or prediction problems as well.

We illustrate the method of combinatorial fusion using examples from three different domains IR, VS, and PSP. In all three applications, multiple scoring systems are used. The issue of bi-diversity was discussed. The diversity rank/score function was calculated in the protein structure prediction problem.

In summary, we have discussed the method of combinatorial fusion analysis developed and used recently in pattern recognition (PR), information retrieval (IR), virtual screening (VS), and protein structure prediction (PSP). Our current work has generated several issues and topics worthy of further investigation. Among them, we list four:

1. We have so far emphasized more on bi-diversity (i.e., 2-diversity). How about tri-diversity (i.e., 3-diversity)? How about higher level diversity measurement? Can this be (or is this best) treated as a single optimization or a sequential series of bi-diversity problems?
2. Our diversity score function for the feature pair or scoring systems A and B with respect to temporal step t_i is defined using the variation of the rank/score functions between A and B (i.e., $d(f_A, f_B)$). In general, the variation of the score functions $d(s_A, s_B)$ or the rank functions $d(r_A, r_B)$ could be used to define the diversity score function.
3. There have been several results concerning the combination of multiple scoring systems in the past few years. Three important issues we would like to study are: (a) selecting the most appropriate scoring systems to combine, (b) finding the best way to combine (Definition 6), and (c) establishing the predictive variables for effective combination. But due to page limitation and chapter constraint, we have included only three examples in IR, VS and PSP with respect to all three issues (a), (b), and (c) described above.
4. We will explore more examples combining multiple heterogeneous scoring systems in the future. For example, we are working on multiple scoring systems which are hybrid of classifier systems, prediction systems, and learning systems (such as neural networks, support vector machines, or evolutionary algorithms).

ACKNOWLEDGMENT

D.F. Hsu thanks Fordham University, National Tsing Hua University, and Ministry of Education of Taiwan, for a Faculty Fellowship and for support and hospitality during his visit to NTHU in Spring 2004. B.S. Kristal acknowledges support from NIH and a NY State SCORE grant.

REFERENCES

Arrow, K.J. (1963). *Social choices and individual values*. New York: John Wiley.

Aslam, J.A, Pavlu, V., & Savell, R. (2003). A unified model for metasearch, pooling, and system evaluation. In O. Frieder (Ed.), *Proceedings of the Twelfth International Conference on Information and Knowledge Management* (pp. 484-491). New York: ACM Press.

Belkin, N.J., Kantor, P.B., Fox, E.A., & Shaw, J.A. (1995). Combining evidence of multiple query representation for information retrieval. *Information Processing & Management, 31*(3), 431-448.

Biggs, N.L., & White, T. (1979). *Permutation groups and combinatorial structures* (LMS Lecture Note Series, Vol. 33). Cambridge: Cambridge University Press.

Brown, G., Wyatt, J., Harris, R., & Yao, X. (2005). Diversity creation methods: a survey and categorization. *Information Fusion, 6*, 5-20.

Chuang, H.Y., Liu, H.F., Brown, S., McMunn-Coffran, C., Kao, C.Y., & Hsu, D.F. (2004). Identifying significant genes from microarray data. In *Proceedings of IEEE BIBE'04* (pp. 358-365). IEEE Computer Society.

Chuang, H.Y., Liu, H.F., Chen, F.A., Kao, C.Y., & Hsu, D.F. (2004). Combination method in microarray analysis. In D.F. Hsu et al. (Ed.), *Proceedings of the 7th International Symposium on Parallel Architectures, Algorithms and Networks (I-SPAN'04)* (pp. 625-630). IEEE Computer Society.

Collins, R.T., Lipton, A.J., Fujiyoshi, H., & Kanade, T. (2001). Algorithms for cooperative multisensor surveillance. *Proceedings of the IEEE, 89* (10), 1456-1477.

Corne, D.W., Deb, K., Fleming, P.J., & Knowles, J.D. (2003). The good of the many outweighs the good of the one: evolutional multi-objective optimization [Featured article]. *IEEE Neural Networks Society*, 9-13.

Dasarathy, B.V. (2000). Elucidative fusion systems—an exposition. *Information Fusion, 1*, 5-15.

Diligenti, M., Gori, M., & Maggini, M. (2004). A unified probabilistic framework for web page scoring systems, *IEEE Trans. on Knowledge and Data Engineering, 16*(1), 4-16.

Ding, C.H.Q., & Dubchak, I. (2001). Multi-class protein fold recognition using support vector machines and neural networks. *Bioinformatics, 17* (4), 349-358.

Duerr, B., Haettich, W., Tropf, H., & Winkler, G. (1980). A combination of statistical and syntactical pattern recognition applied to classification of unconstrained handwritten numerals. *Pattern Recognition, 12*(3), 189-199.

Dwork, C., Kumar, R., Naor, M., & Sivakumar, D. (2001). Rank aggregation methods for the web. In *Proceedings of the Tenth International World Wide Web Conference, WWW 10* (pp. 613-622). New York: ACM Press.

Fagin, R., Kumar, R., & Sivakumar, D. (2003). Comparing top *k*-lists. *SIAM Journal on Discrete Mathematics, 17*, 134-160.

Freund, Y., Iyer, R., Schapire, R.E., & Singer, Y. (2003). An efficient boosting algorithm for combining preferences. *Journal of Machine Learning Research, 4*, 933-969.

Garcia-Pedrajas, N., Hervas-Martinez, C., & Ortiz-Boyer, D. (2005). Cooperative coevolution of artificial neural network ensembles for pattern classification. *IEEE Trans. on Evolutional Computation, 9*(3), 271-302.

Ginn, C. M. R., Willett, P., & Bradshaw, J. (2000). Combination of molecular similarity measures using data fusion [Perspectives]. *Drug Discovery and Design, 20,* 1-15.

Grammatikakis, M.D., Hsu, D.F., & Kraetzl, M. (2001). *Parallel system interconnections and communications.* Boca Raton, FL: CRC Press.

Heydemann, M.C. (1997). Cayley graphs and interconnection networks. In G. Hahn & G. Sabidussi (Eds.), *Graph symmetry* (pp. 161-224). Norwell, MA: Kluwer Academic Publishers.

Ho, T.K. (2002). Multiple classifier combination: Lessons and next steps. In H. Bunke & A. Kandel (Ed.), *Hybrid methods in pattern recognition* (pp. 171-198). Singapore: World Scientific.

Ho, T.K., Hull, J.J., & Srihari, S.N. (1992). Combination of decisions by multiple classifiers. In H.S. Baird, H. Burke, & K. Yamulmoto (Eds.), *Structured document image analysis* (pp. 188-202). Berlin: Springer-Verlag.

Ho, T.K., Hull, J.J., & Srihari, S.N. (1994). Decision combination in multiple classifier system. *IEEE Trans. on Pattern Analysis and Machine Intelligence, 16*(1), 66-75.

Hsu, D.F., & Palumbo, A. (2004). A study of data fusion in Cayley graphs $G(S_n, P_n)$. In: D.F. Hsu et al. (Ed.), *Proceedings of the 7th International Symposium on Parallel Architectures, Algorithms and Networks (I-SPAN'04)* (pp. 557-562). IEEE Computer Society.

Hsu, D.F., Shapiro, J., & Taksa, I. (2002). *Methods of data fusion in information retrieval: Rank vs. score combination* (Tech. Rep. 2002-58). Piscataway, NJ: DIMACS Center.

Hsu, D.F., & Taksa, I. (2005). Comparing rank and score combination methods for data fusion in information retrieval. *Information Retrieval, 8*(3), 449-480.

Hu, W., Tan, T., Wang, L., & Maybank, S. (2004). A survey on visual surveillance of object motion and behaviors. *IEEE Trans. on Systems, Man, and Cybernetics—Part C: Applications and Review, 34*(3), 334-352.

Huang, C.D., Lin, C.T., & Pal, N.R. (2003). Hierarchical learning architecture with automatic feature selection for multi-class protein fold classification. *IEEE Trans. on NanoBioscience, 2*(4), 503-517.

Ibraev, U., Ng, K.B., & Kantor, P.B. (2001). *Counter intuitive cases of data fusion in information retrieval* (Tech. Rep.). Rutgers University.

Jin, X., & Branke, J. (2005). Evolutional optimization in uncertain environments—a survey. *IEEE Trans. on Evolutional Computation, 9*(3), 303-317.

Kantor, P.B. (1998, Jan). *Semantic dimension: On the effectiveness of naïve data fusion methods in certain learning and detection problems.* Paper presented at the meeting of the Fifth International Symposium on Artificial Intelligence and Mathematics, Ft. Lauderdalee, FL.

Kendall, M., & Gibbons, J.D. (1990). *Rank correlation methods.* London: Edward Arnold.

Kittler, J., & Alkoot, F.M. (2003). Sum versus vote fusion in multiple classifier systems. *IEEE Transactions on Pattern Analysis and Machine Intelligence, 25*(1), 110-115.

Kuncheva, L.I. (2005). Diversity in multiple classifier systems [Guest editorial]. *Information Fusion, 6,* 3-4.

Kuriakose, M.A., Chen, W.T., He, Z.M., Sikora, A.G., Zhang, P., Zhang, Z.Y., Qiu, W.L., Hsu, D.F., McMunn-Coffran, C., Brown, S.M., Elango, E.M.,

Delacure, M.D., & Chen, F.A., (2004). Selection and Validation of differentially expressed genes in head and neck cancer. *Cellular and Molecular Life Sciences, 61*, 1372-1383.

Lee, J.H. (1997). Analyses of multiple evidence combination. In N.J. Belkin, A.D. Narasimhalu, P. Willett, W. Hersh (Ed.), *Proceedings of the 20th Annual International ACM SIGIR Conference on Research and Development in Information Retrieval* (pp. 267-276). New York: ACM Press.

Lin, C.-Y., Lin, K.-L., Huang, C.-D., Chang, H.-M., Yang, C.-Y., Lin, C.-T., & Hsu, D.F. (2005). Feature selection and combination criteria for improving predictive accuracy in protein structure classification. In *Proceedings of IEEE BIBE'05* (pp. 311-315). IEEE Computer Society.

Lin, K.-L., Lin, C.-Y., Huang, C.-D., Chang, H.-M., Lin, C.-T., Tang, C.Y., & Hsu, D.F. (2005a). Improving prediction accuracy for protein structure classification by neural networks using feature combination. *Proceedings of the 5th WSEAS International Conference on Applied Informatics and Communications (AIC'05)*(pp. 313-318).

Lin, K.-L., Lin, C.Y., Huang, C.-D., Chang, H.-M., Lin, C.-T., Tang, C.Y., & Hsu, D.F. (2005b). Methods of improving protein structure prediction based on HLA neural networks and combinatorial fusion analysis. *WSEAS Trans. on Information Science and Application, 2*, 2146-2153.

Marden, J.I. (1995). *Analyzing and modeling rank data.* (Monographs on Statistics and Applied Probability, No. 64). London: Chapman & Hall.

McCullagh, P. (1992). Models on spheres and models for permutations. In *Probabality Models and Statistical Analysis for Ranking Data* (M.A. Fligner, & J.S. Verducci, Ed., pp. 278-283). (Lecture Notes in Statistics, No. 80). Berlin: Springer-Verlag.

Melnik, D., Vardi, Y., & Zhang, C.U. (2004). Mixed group ranks: preference and confidence in classifier combination. *IEEE Trans. on Pattern Analysis and Machine Intelligence, 26*(8), 973-981.

Ng, K.B., & Kantor, P.B. (1998). An investigation of the preconditions for effective data fusion in information retrieval: A pilot study. In C.M. Preston (Ed.), *Proceedings of the 61st Annual Meeting of the American Society for Information Science* (pp. 166-178). Medford, NJ: Information Today.

Ng, K.B., & Kantor, P.B. (2000). Predicting the effectiveness of naïve data fusion on the basis of system characteristics. *Journal of the American Society for Information Science. 51*(13), 1177-1189.

Paolucci, U., Vigneau-Callahan, K. E., Shi, H., Matson, W. R., & Kristal, B. S. (2004). Development of biomarkers based on diet-dependent metabolic serotypes: Characteristics of component-based models of metabolic serotype. *OMICS, 8*, 221-238.

Patil, G.P., & Taillie, C. (2004). Multiple indicators, partially ordered sets, and linear extensions: multi-criterion ranking and prioritization. *Environmental and Ecological Statistics, 11*, 199-288.

Perrone, M.P., & Cooper, L.N. (1992). *When networks disagree: Ensemble methods for hybrid neural networks* (Report AF-S260 045). U.S. Dept. of Commerce.

Shi, H., Paolucci, U., Vigneau-Callahan, K. E., Milbury, P. E., Matson, W. R., & Kristal, B. S. (2004). Development of biomarkers based on diet-dependent metabolic serotypes: Practical issues in development of expert system-based classification models in metabolomic studies. *OMICS, 8*, 197-208.

Shi, H., Vigneau-Callahan, K., Shestopalov, I., Milbury, P.E., Matson, W.R., & Kristal B.S. (2002a). Characterization of diet-dependent metabolic serotypes: Proof of principle in female and male rats. *The Journal of Nutrition, 132*, 1031-1038.

Shi, H., Vigneau-Callahan, K., Shestopalov, I., Milbury, P.E., Matson, W.R., & Kristal, B.S. (2002b). Characterization of diet-dependent metabolic serotypes: Primary validation of male and female serotypes in independent cohorts of rats. *The Journal of Nutrition*, *132*, 1039-1046.

Shoichet, B.K (2004). Virtual screening of chemical libraries. *Nature*, *432*, 862-865.

Thompson, G. L. (1992). Graphical techniques for ranked data. In *Probabality Models and Statistical Analysis for Ranking Data* (M.A. Fligner, & J.S. Verducci, Ed., pp. 294-298). (Lecture Notes in Statistics, No. 80). Berlin: Springer-Verlag.

Triesch, J., & von der Malsburg, C. (2001). Democratic integration: self-organized integration of adaptive cues. *Neural Computation*, *13*, 2049-2074.

Tumer, K., & Ghosh, J. (1999). Linear and order statistics combinations for pattern classification. In: Amanda Sharkey (Ed.), *Combining artificial neural networks* (pp. 127-162). Berlin: Springer-Verlag.

Vogt, C. C., & Cottrell, G. W. (1999). Fusion via a linear combination of scores. *Information Retrieval*, *1*(3), 151-173.

Xu, L., Krzyzak, A., & Suen, C.Y. (1992). Methods of combining multiple classifiers and their applications to handwriting recognition. *IEEE Transactions on Systems, Man, and Cybernetics*, *22*(3), 418-435.

Yang, J.M., Chen, Y.F., Shen, T.W., Kristal, B.S., & Hsu, D.F. (2005). Consensus scoring for improving enrichment in virtual screening. *Journal of Chemical Information and Modeling*, *45*, 1134-1146.

Young, H.P. (1975). Social choice scoring functions. *SIAM Journal on Applied Mathematics*, *28*(4), 824-838.

Young, H. P., & Levenglick, A. (1978). A consistent extension of Condorcet's election principle. *SIAM Journal on Applied Mathematics*, *35*(2), 285-300.

This work was previously published in Advanced Data Mining Technologies in Bioinformatics, edited by H. Hsu, pp. 32-62, copyright 2006 by IGI Publishing, formerly known as Idea Group Publishing (an imprint of IGI Global).

Chapter 2.36
Databases Modeling of Engineering Information

Z. M. Ma
Northeastern University, China

ABSTRACT

Information systems have become the nerve center of current computer-based engineering applications, which hereby put the requirements on engineering information modeling. Databases are designed to support data storage, processing, and retrieval activities related to data management, and database systems are the key to implementing engineering information modeling. It should be noted that, however, the current mainstream databases are mainly used for business applications. Some new engineering requirements challenge today's database technologies and promote their evolvement. Database modeling can be classified into two levels: conceptual data modeling and logical database modeling. In this chapter, we try to identify the requirements for engineering information modeling and then investigate the satisfactions of current database models to these requirements at two levels: conceptual data models and logical database models. In addition, the relationships among the conceptual data models and the logical database models for engineering information modeling are presented in the chapter viewed from database conceptual design.

INTRODUCTION

To increase product competitiveness, current manufacturing enterprises have to deliver their products at reduced cost and high quality in a short time. The change from sellers' market to buyers' market results in a steady decrease in the product life cycle time and the demands for tailor-made and small-batch products. All these changes require that manufacturing enterprises quickly respond to market changes. Traditional production patterns and manufacturing technologies may find it difficult to satisfy the requirements of current product development. Many types of advanced manufacturing techniques, such as Computer Integrated Manufacturing (CIM), Agile Manufacturing (AM), Concurrent Engineering (CE), and Virtual Enterprise (VE) based on global manufacturing have been proposed to meet these requirements. One of the foundational supporting strategies is the computer-based information technology. Information systems have become the nerve center of current manufacturing systems. So some new requirements on information modeling are introduced.

Database systems are the key to implementing information modeling. Engineering information modeling requires database support. Engineering applications, however, are data- and knowledge-intensive applications. Some unique characteristics and usage of new technologies have put many potential requirements on engineering information modeling, which challenge today's database systems and promote their evolvement. Database systems have gone through the development from hierarchical and network databases to relational databases. But in non-transaction processing such as CAD/CAPP/CAM (computer-aided design/computer-aided process planning/computer-aided manufacturing), knowledge-based system, multimedia and Internet systems, most of these data-intensive application systems suffer from the same limitations of relational databases. Therefore, some non-traditional data models have been proposed. These data models are fundamental tools for modeling databases or the potential database models. Incorporation between additional semantics and data models has been a major goal for database research and development.

Focusing on engineering applications of databases, in this chapter, we identify the requirements for engineering information modeling and investigate the satisfactions of current database models to these requirements. Here we differentiate two levels of database models: conceptual data models and logical database models. Constructions of database models for engineering information modeling are hereby proposed.

The remainder of the chapter is organized as follows: The next section identifies the generic requirements of engineering information modeling. The issues that current databases satisfy these requirements are then investigated in the third section. The fourth section proposes the constructions of database models. The final section concludes this chapter.

NEEDS FOR ENGINEERING INFORMATION MODELING

Complex Objects and Relationships

Engineering data have complex structures and are usually large in volume. But engineering design objects and their components are not independent. In particular, they are generally organized into taxonomical hierarchies. The specialization association is the well-known association. Also the part-whole association, which relates components to the compound of which they are part, is another key association in engineering settings.

In addition, the position relationships between the components of design objects and the configuration information are typically multi-dimensional. Also, the information of version evolution is obviously time-related. All these kinds of information should be stored. It is clear that spatio-temporal data modeling is essential in engineering design (Manwaring, Jones, & Glagowski, 1996).

Typically, product modeling for product family and product variants has resulted in product data models, which define the form and content of product data generated through the product lifecycle from specification through design to manufacturing. Products are generally complex (see Figure 1, which shows a simple example of product structure) and product data models should hereby have advanced modeling abilities for unstructured objects, relationships, abstractions, and so on (Shaw, Bloor, & de Pennington, 1989).

Data Exchange and Share

Engineering activities are generally performed across departmental and organization boundaries. Product development based on virtual enterprises, for example, is generally performed by several independent member companies that are physically located at different places. Information exchange and share among them is necessary. It is also

Figure 1. An example illustration of product structure

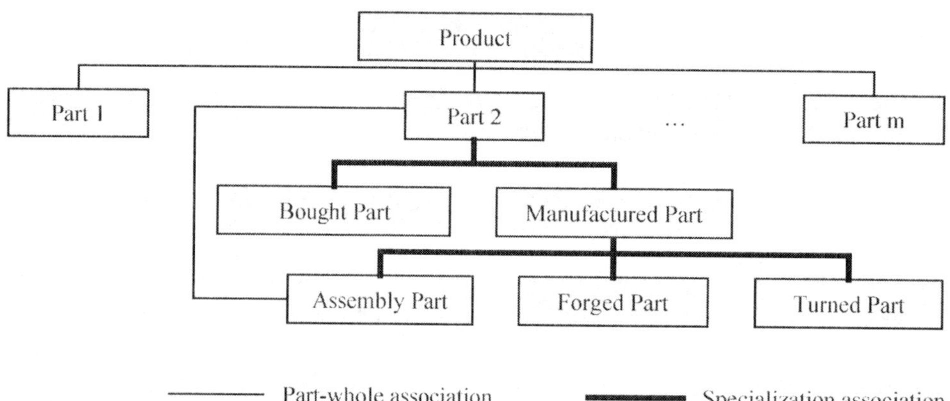

true in different departments or even in different groups within a member company. Enterprise information systems (EISs) in manufacturing industry, for example, typically consist of supply chain management (SCM), enterprise resource planning (ERP) (Ho, Wu, & Tai, 2004), and CAD/CAPP/CAM. These individual software systems need to share and exchange product and production information in order to effectively organize production activities of enterprise. However, they are generally developed independently. In such an environment of distributed and heterogeneous computer-based systems, exchanging and sharing data across units are very difficult. An effective means must be provided so that the data can be exchanged and shared among deferent applications and enterprises. Recently, the PDM (product data management) system (CIMdata, 1997) is being extensively used to integrate both the engineering data and the product development process throughout the product lifecycle, although the PDM system also has the problem of exchanging data with ERP.

Web-Based Applications

Information systems in today's manufacturing enterprises are distributed. Data exchange and share can be performed by computer network systems. The Internet is a large and connected network of computers, and the World Wide Web (WWW) is the fastest growing segment of the Internet. Enterprise operations go increasingly global, and Web-based manufacturing enterprises can not only obtain online information but also organize production activities. Web technology facilitates cross-enterprise information sharing through interconnectivity and integration, which can connect enterprises to their strategic partners as well as to their customers. So Web-based virtual enterprises (Zhang, Zhang, & Wang, 2000), Web-based PDM (Chu & Fan, 1999; Liu & Xu, 2001), Web-based concurrent engineering (Xue & Xu, 2003), Web-based supply chain management, and Web-based B2B e-commerce for manufacturing (Fensel et al., 2001; Shaw, 2000a, 2000b; Soliman & Youssef, 2003; Tan, Shaw, & Fulkerson, 2000) are emerging. A comprehensive review was given of recent research on developing Web-based manufacturing systems in Yang and Xue (2003).

The data resources stored on the Web are very rich. In addition to common types of data, there are many special types of data such as multimedia data and hypertext link, which are referred to as semi-structured data. With the recent popularity of the WWW and informative manufacturing enterprises, how to model and manipulate semi-structured data coming from various sources in manufacturing databases is becoming more and more important. Web-based applications, including Web-based supply chain management, B2B e-commerce, and PDM systems, have been evolved from information publication to information share and exchange. HTML-based Web application cannot satisfy such requirements.

Intelligence for Engineering

Artificial intelligence and expert systems have extensively been used in many engineering activities such as product design, manufacturing, assembly, fault diagnosis, and production management. Five artificial intelligence tools that are most applicable to engineering problems were reviewed in Pham and Pham (1999), which are *knowledge-based systems*, *fuzzy logic*, *inductive learning*, *neural networks*, and *genetic algorithms*. Each of these tools was outlined in the paper together with examples of their use in different branches of engineering. In Issa, Shen, and Chew (1994), an expert system that applies analogical reasoning to mechanism design was developed. Based on fuzzy logic, an integration of financial and strategic justification approaches was proposed for manufacturing in Chiadamrong (1999).

Imprecision and Uncertainty

Imprecision is most notable in the early phase of the design process and has been defined as the choice between alternatives (Antonsoon & Otto, 1995). Four sources of imprecision found in engineering design were classified as *relationship imprecision*, *data imprecision*, *linguistic imprecision*, and *inconsistency imprecision* in Giachetti et al. (1997). In addition to engineering design, imprecise and uncertain information can be found in many engineering activities. The imprecision and uncertainty in activity control for product development was investigated in Grabot and Geneste (1998). To manage the uncertainty occurring in industrial firms, the various types of buffers were provided in Caputo (1996) according to different types of uncertainty faced and to the characteristics of the production system. Buffers are used as alternative and complementary factors to attain technological flexibility when a firm is unable to achieve the desired level of flexibility and faces uncertainty. Nine types of flexibility (*machine*, *routing*, *material handling system*, *product*, *operation*, *process*, *volume*, *expansion*, and *labor*) in manufacturing were summarized in Tsourveloudis and Phillis (1998).

Concerning the representation of imprecision and uncertainty, attempts have been made to address the issue of imprecision and inconsistency in design by way of intervals (Kim et al., 1995). Other approaches to representing imprecision in design include using utility theory, implicit representations using optimization methods, matrix methods such as Quality Function Deployment, probability methods, and necessity methods. An extensive review of these approaches was provided in Antonsoon and Otto (1995). These methods have all had limited success in solving design problems with imprecision. It is believed that fuzzy reorientation of imprecision will play an increasingly important role in design systems (Zimmermann, 1999).

Fuzzy set theory (Zadeh, 1965) is a generalization of classical set theory. In normal set theory, an object may or may not be a member of a set. There are only two states. Fuzzy sets contain elements to a certain degree. Thus, it is possible to represent an object that has partial membership in a set. The membership value of element u in a fuzzy set is represented by $\mu(u)$ and is normalized such that $\mu(u)$ is in [0, 1]. Formally, let F be a fuzzy set in a universe of discourse U and $\mu_F: U \rightarrow [0, 1]$ be the membership function for the fuzzy set F. Then the fuzzy set F is described as:

$F = \{\mu(u_1)/u_1, \mu(u_2)/u_2, ..., \mu(u_n)/u_n\}$, where $u_i \in U$(i = 1, 2, …, n).

Fuzzy sets can represent linguistic terms and imprecise quantities and make systems more flexible and robust. So fuzzy set theory has been used in some engineering applications (e.g., engineering/product design and manufacturing, production management, manufacturing flexibility, e-manufacturing, etc.), where, either crisp information is not available or information flexible processing is necessary.

1. Concerning engineering/product design and manufacturing, the needs for fuzzy logic in the development of CAD systems were identified and how fuzzy logic could be used to model aesthetic factors was discussed in Pham (1998). The development of an expert system with production rules and the integration of fuzzy techniques (fuzzy rules and fuzzy data calculus) was described for the preliminary design in Francois and Bigeon (1995). Integrating knowledge-based methods with multi-criteria decision-making and fuzzy logic, an approach to engineering design and configuration problems was developed in order to enrich existing design and configuration support systems with more intelligent abilities in Muller and Sebastian (1997). A methodology for making the transition from imprecise goals and requirements to the precise specifications needed to manufacture the product was introduced using fuzzy set theory in Giachetti et al. (1997). In Jones and Hua (1998), an approach to engineering design in which fuzzy sets were used to represent the range of variants on existing mechanisms was described so that novel requirements of engineering design could be met. A method for design candidate evaluation and identification using neural network-based fuzzy reasoning was presented in Sun, Kalenchuk, Xue, and Gu (2000).
2. In production management, the potential applications of fuzzy set theory to new product development; facility location and layout; production scheduling and control; inventory management; and quality and cost-benefit analysis were identified in Karwowski and Evans (1986). A comprehensive literature survey on fuzzy set applications in product management research was given in Guiffrida and Nagi (1998). A classification scheme for fuzzy applications in product management research was defined in their paper, including job shop scheduling; quality management; project scheduling; facilities location and layout; aggregate planning; production and inventory planning; and forecasting.
3. In manufacturing domain, flexibility is an inherently vague notion. So fuzzy logic was introduced and a fuzzy knowledge-based approach was used to measure manufacturing flexibility (Tsourveloudis & Phillis, 1998).
4. More recently, the research on supply chain management and electronic commerce have also shown that fuzzy set can be used in customer demand, supply deliveries along the supply chain, external or market supply, targeted marketing, and product category description (Petrovic, Roy, & Petrovic, 1998, 1999; Yager, 2000; Yager & Pasi, 2001).

It is believed that fuzzy set theory has considerable potential for intelligent manufacturing systems and will be employed in more and more engineering applications.

Knowledge Management

Engineering application is a knowledge-intensive application. Knowledge-based managements have covered the whole activities of current enterprises (O'Leary, 1998; Maedche et al., 2003; Wong, 2005), including manufacturing enterprises (Michael & Khemani, 2002). In Tan and Platts (2004), the use of the connectance concept for managing manufacturing knowledge was proposed. A software tool called Tool for Action Plan Selection (TAPS) has been developed based on the connectance concept, which enables managers to sketch and visualize their knowledge of how variables interact in a connectance network. Based on the computer-integrated manufacturing open-system architecture reference model (CIMOSA), a formalism was presented in de Souza, Ying, and Yang (1998) to specify the business processes and enterprise activities at the knowledge level. The formalism used an integration of multiple types of knowledge, including precise, muddy, and random symbolic and numerical knowledge to systematically represent enterprise behavior and functionality. Instead of focusing on individual human knowledge, as in Thannhuber, Tseng, and Bullinger (2001), the ability of an enterprise to dynamically derive processes to meet the external needs and internal stability was identified as the organizational knowledge. On the basis, a knowledge management system has been developed.

The management of engineering knowledge entails its modeling, maintenance, integration, and use (Ma & Mili, 2003; Mili et al., 2001). Knowledge modeling consists of representing the knowledge in some selected language or notation. Knowledge maintenance encompasses all activities related to the validation, growth, and evolution of the knowledge. Knowledge integration is the synthesis of knowledge from related sources. The use of the knowledge requires bridging the gap between the objective expressed by the knowledge and the directives needed to support engineering activities.

It should be noticed that Web-based engineering knowledge management has emerged because of Web-based engineering applications (Caldwell et al., 2000). In addition, engineering knowledge is closely related to engineering data, although they are different. Engineering knowledge is generally embedded in engineering data. So it is necessary to synthetically manage engineering knowledge and data in bases (Xue, Yadav, & Norrie, 1999; Zhang & Xue, 2002). Finally, the field of artificial intelligence (AI) is usually concerned with the problems caused by imprecise and uncertain information (Parsons, 1996). Knowledge representation is one of the most basic and active research areas of AI. The conventional approaches to knowledge representation, however, only support exact rather than approximate reasoning, and fuzzy logic is apt for knowledge representation (Zadeh, 1989). Fuzzy rules (Dubois & Prade, 1996) and fuzzy constraints (Dubois, Fargier, & Prade, 1996) have been advocated and employed as a key tool for expressing pieces of knowledge in fuzzy logic. In particular, fuzzy constraint satisfaction problem (FCSP) has been used in many engineering activities such as design and optimization (Dzbor, 1999; Kapadia & Fromherz, 1997; Young, Giachetti, & Ress, 1996) as well as planning and scheduling (Dubois, Fargier, & Prade, 1995; Fargier & Thierry, 1999; Johtela et al., 1999).

Data Mining and Knowledge Discovery

Engineering knowledge plays a crucial role in engineering activities. But engineering knowledge is not always represented explicitly. Data mining and knowledge discovery from databases (KDD) can extract information characterized as "knowledge" from data that can be very complex and in large quantities. So the field of data mining and knowledge discovery from databases has emerged as a new discipline in engineering (Gertosio & Dussauchoy, 2004) and now is extensively studied and applied in many industrial processes. In Ben-Arieh, Chopra, and Bleyberg (1998), data mining application for real-time distributed shop-floor control was presented. With a data mining approach, the prediction problem encountered in engineering design was solved in Kusiak and Tseng (2000). Furthermore, the data mining issues and requirements within an enterprise were examined in Kleissner (1998).

With the huge amount of information available online, the World Wide Web is a fertile area for data mining research. The Web mining research is at the crossroads of research from several research communities such as database, information retrieval, and within AI, especially the sub-areas of machine learning and natural language processing (Kosala & Blockeel, 2000). In addition, soft computing methodologies (involving fuzzy sets, neural networks, genetic algorithms, and rough sets) are most widely applied in the data mining step of the overall KDD process (Mitra, Pal, & Mitra, 2002). Fuzzy sets provide a natural framework for the process in dealing with uncertainty. Neural networks and rough sets are widely used for classification and rule generation. Genetic algorithms (GAs) are involved in various optimization and search processes, like query optimization and template selection. Particularly, a review of Web Mining in Soft Computing Framework was given in Pal, Talwar, and Mitra (2002).

CURRENT DATABASE MODELS

Engineering information modeling in databases can be carried out at two different levels: conceptual data modeling and logical database modeling. Therefore, we have conceptual data models and logical database models for engineering information modeling, respectively. In this chapter, database models for engineering information modeling refer to conceptual data models and logical database models simultaneously. Table 1 gives some conceptual data models and logical database models that may be applied for engineering information modeling. The following two sub-sections give the more detailed explanations about these models.

Conceptual Data Models

Much attention has been directed at conceptual data modeling of engineering information (Mannisto et al., 2001; McKay, Bloor, & de Pennington,

Table 1. Database models for engineering information modeling

Database Models					
Conceptual Data Models L		***ogical Database Models***			
Generic Conceptual Data Models	Specific Conceptual Data Models for Engineering	Classical Logical Database Models	XML Databases	Specific & Hybrid Database Models	Extended Database Models
• ER data model • EER data model • UML data model • XML data model	• IDEF1X data model • EXPRESS data model	• Relational databases • Nested relational databases • Object-oriented databases • Object-relational databases	• Classical logical databases • Native XML databases	• Active databases • Deductive databases • Constraint databases • Spatio-temporal databases • Object-oriented active databases • Deductive object-relational databases ...	• Fuzzy relational databases • Fuzzy nested relational databases • Fuzzy object-oriented databases • Deductive fuzzy relational databases ...

1996). Product data models, for example, can be viewed as a class of semantic data models (i.e., conceptual data models) that take into account the needs of engineering data (Shaw, Bloor, & de Pennington, 1989). Recently, conceptual information modeling of enterprises such as virtual enterprises has received increasing attention (Zhang & Li, 1999). Generally speaking, traditional ER (entity-relationship) and EER (extended entity-relationship) can be used for engineering information modeling at conceptual level (Chen, 1976). But limited by their power in engineering modeling, some improved conceptual data models have been developed.

IDEF1X is a method for designing relational databases with a syntax designed to support the semantic constructs necessary in developing a conceptual schema. Some research has focused on the IDEF1X methodology. A thorough treatment of the IDEF1X method can be found in Wizdom Systems Inc. (1985). The use of the IDEF1X methodology to build a database for multiple applications was addressed in Kusiak, Letsche, and Zakarian (1997).

In order to share and exchange product data, the Standard for the Exchange of Product Model Data (STEP) is being developed by the International Organization for Standardization (ISO). STEP provides a means to describe a product model throughout its life cycle and to exchange data between different units. STEP consists of four major categories, which are *description methods*, *implementation methods*, *conformance testing methodology and framework*, and *standardized application data models/schemata*, respectively. EXPRESS (Schenck & Wilson, 1994), as the description methods of STEP and a conceptual schema language, can model product design, manufacturing, and production data. EXPRESS model hereby becomes one of the major conceptual data models for engineering information modeling.

With regard to CAD/CAM development for product modeling, a review was conducted in Eastman and Fereshetian (1994), and five information models used in product modeling, namely, ER, NAIM, IDEF1X, EXPRESS and EDM, were studied. Compared with IDEF1X, EXPRESS can model complex semantics in engineering application, including engineering objects and their relationships. Based on EXPRESS model, it is easy to implement share and exchange engineering information.

It should be noted that ER/EER, IDEF1X and EXPRESS could model neither knowledge nor fuzzy information. The first effort was done in Zvieli and Chen (1996) to extend ER model to represent three levels of fuzziness. The first level refers to the set of semantic objects, resulting in fuzzy entity sets, fuzzy relationship sets and fuzzy attribute sets. The second level concerns the occurrences of entities and relationships. The third level is related to the fuzziness in attribute values of entities and relationships. Consequently, ER algebra was fuzzily extended to manipulate fuzzy data. In Chen and Kerre (1998), several major notions in EER model were extended, including fuzzy extension to generalization/specialization, and shared subclass/category as well as fuzzy multiple inheritance, fuzzy selective inheritance, and fuzzy inheritance for derived attributes. More recently, using fuzzy sets and possibility distribution (Zadeh, 1978), fuzzy extensions to IDEF1X and EXPRESS were proposed in Ma, Zhang, and Ma (2002) and Ma (in press), respectively.

UML (Unified Modeling Language) (Booch, Rumbaugh, & Jacobson, 1998; OMG, 2003), being standardized by the Object Management Group (OMG), is a set of OO modeling notations. UML provides a collection of models to capture many aspects of a software system. From the information modeling point of view, the most relevant model is the class model. The building blocks in this class model are those of classes and relationships. The class model of UML encompasses the concepts used in ER, as well as other OO concepts. In addition, it also presents the advantage of being open and extensible, allowing its adaptation to the specific needs of the application such as workflow modeling of e-commerce (Chang et al., 2000) and product structure mapping (Oh, Hana, & Suhb, 2001). In particular, the class model of UML is extended for the representation of class constraints and the introduction of stereotype associations (Mili et al., 2001).

With the popularity of Web-based design, manufacturing, and business activities, the requirement has been put on the exchange and share of engineering information over the Web. XML (eXtensible Markup Language), created by the World Wide Web Consortium, lets information publishers invent their own tags for particular applications or work with other organizations to define shared sets of tags that promote interoperability and that clearly separate content and presentation. XML provides a Web-friendly and well-understood syntax for the exchange of data. Because XML impacts on data definition and share on the Web (Seligman & Rosenthal, 2001), XML technology has been increasingly studied, and more and more Web tools and Web servers are capable of supporting XML. In Bourret (2004), product data markup language, the XML for product data exchange and integration, has been developed. As to XML modeling at concept level, UML was used for designing XML DTD (document- type definition) in Conrad, Scheffner, and Freytag (2000). In Xiao et al. (2001), an object-oriented conceptual model was developed to design XML schema. ER model was used for conceptual design of semi-structured databases in Lee et al. (2001). But XML does not support imprecise and uncertain information modeling and knowledge modeling. Introducing imprecision and uncertainty into XML has increasingly become a topic of research (Abiteboul, Segoufin, & Vianu, 2001; Damiani, Oliboni, & Tanca, 2001; Ma, 2005).

Logical Database Models

Classical Logical Database Models

As to engineering information modeling in database systems, the generic logical database models such relational databases, nested relational databases, and object-oriented databases can be used. Also, some hybrid logical database models such as object-relational databases are very useful for this purpose.

In Ahmed (2004), the KSS (Kraftwerk Kennzeichen System) identification and classification system was used to develop database system for plant maintenance and management. On top of a relational DBMS, an EXPRESS-oriented information system was built in Arnalte and Scala (1997) for supporting information integration in a computer-integrated manufacturing environment. In this case, the conceptual model of the information was built in EXPRESS and then parsed and translated to the corresponding relational constructs. Relational databases for STEP/EXPRESS were also discussed in Krebs and Lührsen (1995). In addition, an object-oriented layer was developed in Barsalou and Wiederhold (1990) to model complex entities on top of a relational database. This domain-independent architecture permits object-oriented access to information stored in relational format-information that can be shared among applications.

Object-oriented databases provide an approach for expressing and manipulating complex objects. A prototype object-oriented database system, called ORION, was thus designed and implemented to support CAD (Kim et al., 1990). Object-oriented databases for STEP/EXPRESS have been studied in Goh et al. (1994, 1997). In addition, an object-oriented active database was also designed for STEP/EXPRESS models in Dong, Y. et al. (1997). According to the characteristics of engineering design, a framework for the classification of queries in object-oriented engineering databases was provided in Samaras,

Spooner, and Hardwick (1994), where the strategy for query evaluation is different from traditional relational databases. Based on the comparison with relational databases, the selections and characteristics of the object-oriented database and database management systems (OODBMS) in manufacturing were discussed in Zhang (2001). The current studies and applications were also summarized.

XML Databases

It is crucial for Web-based applications to model, store, manipulate, and manage XML data documents. XML documents can be classified into data-centric documents and document-centric documents (Bourret, 2004). Data-centric documents are characterized by fairly regular structure, fine-grained data (i.e., the smallest independent unit of data is at the level of a PCDATA-only element or an attribute), and little or no mixed content. The order in which sibling elements and PCDATA occurs is generally not significant, except when validating the document. Data-centric documents are documents that use XML as a data transport. They are designed for machine consumption and the fact that XML is used at all is usually superfluous. That is, it is not important to the application or the database that the data is, for some length of time, stored in an XML document. As a general rule, the data in data-centric documents is stored in a traditional database, such as a relational, object-oriented, or hierarchical database. The data can also be transferred from a database to a XML document. For the transfers between XML documents and databases, the mapping relationships between their architectures as well as their data should be created (Lee & Chu, 2000; Surjanto, Ritter, & Loeser, 2000). Note that it is possible to discard some information such as the document and its physical structure when transferring data between them. It must be pointed out, however, that the data in data-centric documents such as semi-structured data can also be stored in a native XML database, in which a document-centric document is usually stored. Document-centric documents are characterized by less regular or irregular structure, larger-grained data (that is, the smallest independent unit of data might be at the level of an element with mixed content or the entire document itself), and lots of mixed content. The order in which sibling elements and PCDATA occurs is almost always significant. Document-centric documents are usually documents that are designed for human consumption. As a general rule, the documents in document-centric documents are stored in a native XML database or a content management system (an application designed to manage documents and built on top of a native XML database). Native XML databases are databases designed especially for storing XML documents. The only difference of native XML databases from other databases is that their internal model is based on XML and not something else, such as the relational model.

In practice, however, the distinction between data-centric and document-centric documents is not always clear. So the previously-mentioned rules are not of a certainty. Data, especially semi-structured data, can be stored in native XML databases, and documents can be stored in traditional databases when few XML-specific features are needed. Furthermore, the boundaries between traditional databases and native XML databases are beginning to blur, as traditional databases add native XML capabilities and native XML databases support the storage of document fragments in external databases.

In Seng, Lin, Wang, and Yu (2003), a technical review of XML and XML database technology, including storage method, mapping technique, and transformation paradigm, was provided and an analytic and comparative framework was developed. By collecting and compiling the IBM, Oracle, Sybase, and Microsoft XML database products, the framework was used and each of these XML database techniques was analyzed.

Special, Hybrid, and Extended Logical Database Models

It should be pointed out that, however, the generic logical database models such as relational databases, nested relational databases, and object-oriented databases do not always satisfy the requirements of engineering modeling. As pointed out in Liu (1999), relational databases do not describe the complex structure relationship of data naturally, and separate relations may result in data inconsistencies when updating the data. In addition, the problem of inconsistent data still exists in nested relational databases, and the mechanism of sharing and reusing CAD objects is not fully effective in object-oriented databases. In particular, these database models cannot handle engineering knowledge. Some special databases based on relational or object-oriented models are hereby introduced. In Dong and Goh (1998), an object-oriented active database for engineering application was developed to support intelligent activities in engineering applications. In Liu (1999), deductive databases were considered as the preferable database models for CAD databases, and deductive object-relational databases for CAD were introduced in Liu and Katragadda (2001). Constraint databases based on the generic logical database models are used to represent large or even infinite sets in a compact way and are suitable hereby for modeling spatial and temporal data (Belussi, Bertino, & Catania, 1998; Kuper, Libkin, & Paredaens, 2000). Also, it is well established that engineering design is a constraint-based activity (Dzbor, 1999; Guiffrida, & Nagi, 1998; Young, Giachetti, & Ress, 1996). So constraint databases are promising as a technology for modeling engineering information that can be characterized by large data in volume, complex relationships (structure, spatial and/or temporal semantics), intensive knowledge and so forth. In Posselt and Hillebrand (2002), the issue about constraint database support for evolving data in product design was investigated.

It should be noted that fuzzy databases have been proposed to capture fuzzy information in engineering (Sebastian & Antonsson, 1996; Zimmermann, 1999). Fuzzy databases may be based on the generic logical database models such as relational databases (Buckles & Petry, 1982; Prade & Testemale, 1984), nested relational databases (Yazici et al., 1999), and object-oriented databases (Bordogna, Pasi, & Lucarella, 1999; George et al., 1996; van Gyseghem & de Caluwe, 1998). Also, some special databases are extended for fuzzy information handling. In Medina et al. (1997), the architecture for deductive fuzzy relational database was presented, and a fuzzy deductive object-oriented data model was proposed in Bostan and Yazici (1998). More recently, how to construct fuzzy event sets automatically and apply it to active databases was investigated in Saygin and Ulusoy (2001).

CONSTRUCTIONS OF DATABASE MODELS

Depending on data abstract levels and actual applications, different database models have their advantages and disadvantages. This is the reason why there exist a lot of database models, conceptual ones and logical ones. It is not appropriate to state that one database model is always better than the others. Conceptual data models are generally used for engineering information modeling at a high level of abstraction. However, engineering information systems are constructed based on logical database models. So at the level of data manipulation, that is, a low level of abstraction, the logical database model is used for engineering information modeling. Here, logical database models are often created through mapping conceptual data models into logical database models. This conversion is called *conceptual design of databases*. The relationships among conceptual data models, logical database models, and engineering information systems are shown in Figure 2.

Figure 2. Relationships among conceptual data model, logical database model, and engineering information systems

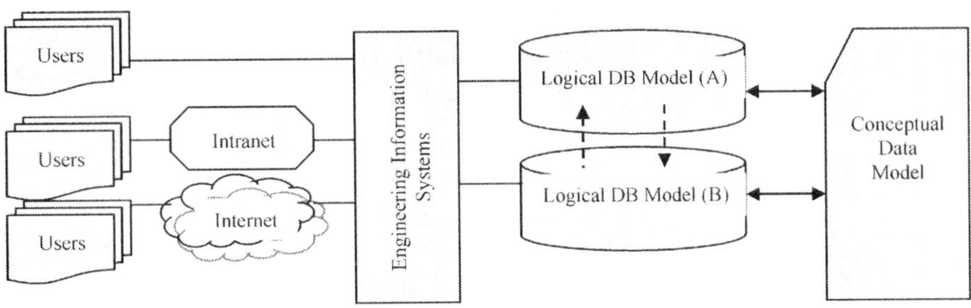

In this figure, *Logical DB Model (A)* and *Logical DB Model (B)* are different database systems. That means that they may have different logical database models, say relational database and object-oriented database, or they may be different database products, say *Oracle*™ and *DB2*, although they have the same logical database model. It can be seen from the figure that a developed conceptual data model can be mapped into different logical database models. Besides, it can also be seen that a logical database model can be mapped into a conceptual data model. This conversion is called *database reverse engineering*. It is clear that it is possible that different logical database models can be converted one another through database reverse engineering.

Development of Conceptual Data Models

It has been shown that database modeling of engineering information generally starts from conceptual data models, and then the developed conceptual data models are mapped into logical database models. First of all, let us focus on the choice, design, conversion, and extension of conceptual data models in database modeling of engineering information.

Generally speaking, ER and IDEF1X data models are good candidates for business process in engineering applications. But for design and manufacturing, object-oriented conceptual data models such EER, UML, and EXPRESS are powerful. Being the description methods of STEP and a conceptual schema language, EXPRESS is extensively accepted in industrial applications. However, EXPRESS is not a graphical schema language, unlike EER and UML. In order to construct EXPRESS data model at a higher level of abstract, EXPRESS-G, being the graphical representation of EXPRESS, is introduced. Note that EXPRESS-G can only express a subset of the full language of EXPRESS. EXPESS-G provides supports for the notions of entity, type, relationship, cardinality, and schema. The functions, procedures, and rules in EXPRESS language are not supported by EXPRESS-G. So EER and UML should be used to design EXPRESS data model conceptually, and then such EER and UML data models can be translated into EXPRESS data model.

It should be pointed out that, however, for Web-based engineering applications, XML should be used for conceptual data modeling. Just like EXPRESS, XML is not a graphical schema language, either. EER and UML can be used to design XML data model conceptually, and then such EER and UML data models can be translated into XML data model.

That multiple graphical data models can be employed facilitates the designers with different background to design their conceptual models

Figure 3. Relationships among conceptual data models

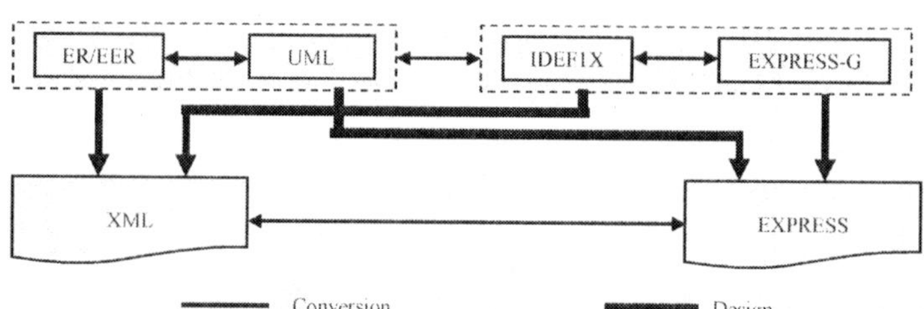

easily by using one of the graphical data models with which they are familiar. However, a complex conceptual data model is generally completed cooperatively by a design group, in which each member may use a different graphical data model. All these graphical data models, designed by different members, should be converted into a union data model finally. Furthermore, the EXPRESS schema can be turned into XML DTD. So far, the data model conversions among EXPRESS-G, IDEF1X, ER/EER, and UML only receive few attentions although such conversions are crucial in engineering information modeling. In (Cherfi, Akoka, and Comyn-Wattiau, 2002), the conceptual modeling quality between EER and UML was investigated. In Arnold and Podehl (1999), a mapping from EXPRESS-G to UML was introduced in order to define a linking bridge and bring the best of the worlds of product data technology and software engineering together. Also, the formal transformation of EER and EXPRESS-G was developed in Ma et al. (2003). In addition, the comparison of UML and IDEF was given in Noran (2000).

Figure 3 shows the design and conversion relationships among conceptual data models.

In order to model fuzzy engineering information in a conceptual data model, it is necessary to extend its modeling capability. As we know, most database models make use of three levels of abstraction, namely, the data dictionary, the database schema, and the database contents (Erens, McKay, & Bloor, 1994). The fuzzy extensions of conceptual data models should be conducted at all three levels of abstraction. Of course, the constructs of conceptual data models should accordingly be extended to support fuzzy information modeling at these three levels of abstraction. In Zvieli and Chen (1996), for example, three levels of fuzziness were captured in the extended ER model. The first level is concerned with the schema and refers to the set of semantic objects, resulting in fuzzy entity sets, fuzzy relationship sets and fuzzy attribute sets. The second level is concerned with the schema/instance and refers to the set of instances, resulting in fuzzy occurrences of entities and relationships. The third level is concerned with the content and refers to the set of values, resulting in fuzzy attribute values of entities and relationships.

EXPRESS permits null values in array data types and role names by utilizing the keyword *Optional* and used three-valued logic (*False*, *Unknown*, and *True*). In addition, the select data type in EXPRESS defines one kind of imprecise and uncertain data type which actual type is unknown at present. So EXPRESS indeed supports imprecise information modeling but very weakly. Further fuzzy extension to EXPRESS is needed. Just like fuzzy ER, fuzzy EXPRESS should capture three levels of fuzziness and its constructs such as the basic elements (reserved words and literals), the data types, the entities, the expressions and so on, should hereby be extended.

Development of Logical Database Models

It should be noticed that there might be semantic incompatibility between conceptual data models and logical database models. So when a conceptual data model is mapped into a logical database model, we should adopt such a logical database model which expressive power is close to the conceptual data model so that the original information and semantics in the conceptual data model can be preserved and supported furthest. Table 2 shows how relational and object-oriented databases fair against various conceptual data models. Here, *CDM* and *LDBM* denote conceptual data model and logical database model, respectively.

It is clear from the table that relational databases support ER and IDEF1X well. So, when an ER or IDEF1X data model is converted, relational databases should be used. Of course, the target relational databases should be fuzzy ones if ER or IDEF1X data model is a fuzzy one. It is also seen that EER, UML, or EXPRESS data model should be mapped into object-oriented databases. EXPRESS is extensively accepted in industrial application area. EER and UML, being graphical conceptual data models, can be used to design EXPRESS data model conceptually, and then EER and UML data models can be translated into EXPRESS data model (Oh, Hana, & Suhb, 2001). In addition, the EXPRESS schema can be turned into XML DTD (Burkett, 2001). So, in the following, we focus on logical database implementation of EXPRESS data model.

In order to construct a logical database around an EXPRESS data model, the following tasks must be performed: (1) defining the database structures from EXPRESS data model and (2) providing SDAI (STEP Standard Data Access Interface) access to the database. Users define their databases using EXPRESS, manipulate the databases using SDAI, and exchange data with other applications through the database systems.

Relational and Object-Oriented Database Support for EXPRESS Data Model

In EXPRESS data models, entity instances are identified by their unique identifiers. Entity instances can be represented as tuples in relational databases, where the tuples are identified by their keys. To manipulate the data of entity instances in relational databases, the problem that entity instances are identified in relational databases must be resolved. As we know, in EXPRESS, there are attributes with UNIQUE constraints. When an entity type is mapped into a relation and each entity instance is mapped into a tuple, it is clear that such attributes can be viewed as the key of the tuples to identify instances. So an EXPRESS data model must contain such an attribute with UNIQUE constraints at least when relational databases are used to model EXPRESS data model. In addition, inverse clause and where clause can be implemented in relational databases as the constraints of foreign key and domain, respectively. Complex entities and subtype/superclass in EXPRESS data models can be implemented in

Table 2. Match of logical database models to conceptual data models

CDM \ LDBM	Relational Databases	Object-Oriented Databases
ER	*good*	*bad*
IDEF1X	*good*	*bad*
EER	*fair*	*good*
UML	*fair*	*good*
EXPRESS	*fair*	*good*

relational databases via the reference relationships between relations. Such organizations, however, do not naturally represent the structural relationships among the objects described. When users make a query, some join operations must be used. Therefore, object-oriented databases should be used for the EXPRESS data model.

Unlike the relational databases, there is no widely accepted definition as to what constitutes an object-oriented database, although object-oriented database standards have been released by ODMG (2000). Not only is it true that not all features in one object-oriented database can be found in another, but the interpretation of similar features may also differ. But some features are in common with object-oriented databases, including object identity, complex objects, encapsulation, types, and inheritance. EXPRESS is object-oriented in nature, which supports these common features in object-oriented databases. Therefore, there should be a more direct way to mapping EXPRESS data model into object-oriented databases. It should be noted that there is incompatibility between the EXPRESS data model and object-oriented databases. No widely accepted definition of object-oriented database model results in the fact that there is not a common set of incompatibilities between EXPRESS and object-oriented databases. Some possible incompatibilities can be found in Goh et al. (1997).

Now let us focus on fuzzy relational and object-oriented databases. As mentioned previously, the fuzzy EXPRESS should capture three levels of fuzziness: the schema level, the schema/instance, and the content. Depending on the modeling capability, however, fuzzy relational databases only support the last two levels of fuzziness, namely, the schema/instance and the content. It is possible that object-oriented databases are extended to support all three levels of fuzziness in fuzzy EXPRESS.

Requirements and Implementation of SDAI Functions

The goal of SDAI is to provide the users with uniform manipulation interfaces and reduce the cost of integrated product databases. When EXPRESS data models are mapped into databases, users will face databases. As a data access interface, SDAI falls into the category of the application users who access and manipulate the data. So the requirements of SDAI functions are decided by the requirements of the application users of databases. However, SDAI itself is in a state of evolution. Considering the enormity of the task and the difficulty for achieving agreement as to what functions are to be included and the viability of implementing the suggestions, only some basic requirements such as data query, data update, structure query, and validation are catered for. Furthermore, under fuzzy information environment, the requirements of SDAI functions needed for manipulating the fuzzy EXPRESS data model must consider the fuzzy information processing such as flexible data query.

Using SDAI operations, the SDAI applications can access EXPRESS data model. However, only the specifications of SDAI operations are given in STEP Part 23 and Part 24. The implementation of these operations is empty, which should be developed utilizing the special binding language according to database systems. One will meet two difficulties when implementing SDAI in the databases. First, the SDAI specifications are still in a state of evolution. Second, the implementation of SDAI functions is product-related. In addition, object-oriented databases are not standardized. It is extremely true for the database implementation of the SDAI functions needed for manipulating the fuzzy EXPRESS data model, because there are no commercial fuzzy relational database management systems, and little research is done on fuzzy object-oriented databases so far.

It should be pointed out that, however, there exists a higher-level implementation of EXPRESS data model than database implementation, which is knowledge-based. Knowledge-based implementation has the features of database implementations, plus full support for EXPRESS constraint validation. A knowledge-based system should read and write exchange files, make product data available to applications in structures defined by EXPRESS, work on data stored in a central database, and should be able to reason about the contents of the database. Knowledge-based systems encode rules using techniques such as frames, semantic nets, and various logic systems, and then use inference techniques such as forward and backward chaining to reason about the contents of a database. Although some interesting preliminary work was done, knowledge-based implementations do not exist. Deductive databases and constraint databases based on relational and/or object-oriented database models are useful in knowledge-intensive engineering applications for this purpose. In deductive databases, rules can be modeled and knowledge bases are hereby constituted. In constraint databases, complex spatial and/or temporal data can be modeled. In particular, constraint databases can handle a wealth of constraints in engineering design.

CONCLUSION

Manufacturing enterprises obtain increasing product varieties and products with lower price, high quality and shorter lead time by using enterprise information systems. The enterprise information systems have become the nerve center of current computer-based manufacturing enterprises. Manufacturing engineering is typically a data- and knowledge-intensive application area and engineering information modeling is hereby one of the crucial tasks to implement engineering information systems. Databases are designed to support data storage, processing, and retrieval activities related to data management, and database systems are the key to implementing engineering information modeling. But the current mainstream databases are mainly designed for business applications. There are some unique requirements from engineering information modeling, which impose a challenge to databases technologies and promote their evolvement. It is especially true for contemporary engineering applications, where some new techniques have been increasingly applied and their operational patterns are hereby evolved (e.g., e-manufacturing, Web-based PDM, etc.). One can find many researches in literature that focus on using database techniques for engineering information modeling to support various engineering activities. It should be noted that, however, most of these papers only discuss some of the issues according to the different viewpoints and application requirements. Engineering information modeling is complex because it should cover product life cycle times. On the other hand, databases cover wide variety of topics and evolve quickly. Currently, few papers provide comprehensive discussions about how current engineering information modeling can be supported by database technologies. This chapter tries to fill this gap.

In this chapter, we first identify some requirements for engineering information modeling, which include complex objects and relationships, data exchange and share, Web-based applications, imprecision and uncertainty, and knowledge management. Since the current mainstream databases are mainly designed for business applications, and the database models can be classified into conceptual data models and logical database models, we then investigate how current conceptual data models and logical database models satisfy the requirements of engineering information modeling in databases. The purpose of engineering information modeling in databases is to construct the logical database models, which are the foundation of the engineering information systems. Generally the constructions of logical database models start from the constructions of conceptual data models and then the developed conceptual data models are converted into the logical database models. So the chapter presents not only the development of some conceptual data models for engineering information modeling, but also the development of the relational and object-oriented databases which are used to implement EXPRESS/STEP. The contribution of the chapter is to identify the direction of database study viewed from engineering applications and provide a guidance of information modeling for engineering design, manufacturing, and production management. It can be believed that some more powerful database models will be developed to satisfy engineering information modeling.

REFERENCES

Abiteboul, S., Segoufin, L., & Vianu, V. (2001). Representing and querying XML with incomplete information, In *Proceedings of the 12th ACM SIGACT-SIGMOD-SIGART Symposium on Principles of Database Systems,* California (pp. 150-161).

Ahmed, S. (2004). Classification standard in large process plants for integration with robust database. *Industrial Management & Data Systems, 104*(8), 667-673.

Antonsoon, E. K., & Otto, K. N. (1995). Imprecision in engineering design. *ASME Journal of Mechanical Design, 117*(B), 25-32.

Arnalte, S., & Scala, R. M. (1997). An information system for computer-integrated manufacturing systems. *Robotics and Computer-Integrated Manufacturing, 13*(3), 217-228.

Arnold, F., & Podehl, G. (1999). Best of both worlds — A mapping from EXPRESS-G to UML. *Lecture Notes in Computer Science, Vol. 1618*, 49-63.

Barsalou, T., & Wiederhold, G. (1990). Complex objects for relational databases. *Computer-Aided Design, 22*(8), 458-468.

Belussi, A., Bertino, E., & Catania, B. (1998). An extended algebra for constraint databases. *IEEE Transactions on Knowledge and Data Engineering, 10*(5), 686-705.

Ben-Arieh, D., Chopra, M., & Bleyberg, M. Z. (1998). Data mining application for real-time distributed shop floor control. In *Proceedings of 1998 IEEE International Conference on Systems, Man, and Cybernetics,* San Diego, California (pp. 2738-2743).

Booch, G., Rumbaugh, J., & Jacobson, I. (1998). *The unified modeling language user guide*. Reading, MA: Addison-Welsley Longman.

Bordogna, G., Pasi, G., & Lucarella, D. (1999). A fuzzy object-oriented data model for managing vague and uncertain information. *International Journal of Intelligent Systems, 14*, 623-651.

Bostan, B., & Yazici, A. (1998). A fuzzy deductive object-oriented data model. In *Proceedings of the IEEE International Conference on Fuzzy Systems,* Alaska (vol. 2, pp. 1361-1366). IEEE.

Bourret, R. (2004). *XML and databases*. Retrieved October 2004, from http://www.rpbourret.com/xml/XMLAndDatabases.htm

Buckles, B. P., & Petry, F. E. (1982). A fuzzy representation of data for relational database. *Fuzzy Sets and Systems, 7*(3), 213-226.

Burkett, W. C. (2001). Product data markup language: A new paradigm for product data exchange and integration. *Computer Aided Design, 33*(7), 489-500.

Caldwell, N. H. M., Clarkson, Rodgers, & Huxor (2000). Web-based knowledge management for distributed design. *IEEE Intelligent Systems, 15*(3), 40-47.

Caputo, M. (1996). Uncertainty, flexibility and buffers in the management of the firm operating system. *Production Planning & Control, 7*(5), 518-528.

Chang, Y. L., et al. (2000). Workflow process definition and their applications in e-commerce. In *Proceedings of International Symposium on Multimedia Software Engineering,* Taiwan (pp. 193-200). IEEE Computer Society.

Chen, G. Q., & Kerre, E. E. (1998). Extending ER/EER concepts towards fuzzy conceptual data modeling. In *Proceedings of the 1998 IEEE International Conference on Fuzzy Systems,* Alaska (vol. 2, pp. 1320-1325). IEEE.

Chen, P. P. (1976). The entity-relationship model: Toward a unified view of data. *ACM Transactions on Database Systems, 1*(1), 9-36.

Cherfi, S. S. S., Akoka, J., & Comyn-Wattiau, I. (2002). Conceptual modeling quality — From EER to UML schemas evaluation. *Lecture Notes in Computer Science, Vol. 250*, 414-428.

Chiadamrong, N. (1999). An integrated fuzzy multi-criteria decision-making method for manufacturing strategy selection. *Computers and Industrial Engineering, 37*, 433-436.

Chu, X. J., & Fan, Y. Q. (1999). Product data management based on Web technology. *Integrated Manufacturing Systems, 10*(2), 84-88.

CIMdata. (1997). *Product data management: The definition.* Retrieved September 1997, from http://www.cimdata.com

Conrad, R., Scheffner, D., & Freytag, J. C. (2000). XML conceptual modeling using UML. *Lecture Notes in Computer Science, Vol. 1920*, 558-571.

Damiani, E., Oliboni, B., & Tanca, L. (2001). Fuzzy techniques for XML data smushing. *Lecture Notes in Computer Science, Vol. 2206*, 637-652.

de Souza, R., Ying, Z. Z., & Yang, L. C. (1998). Modeling business processes and enterprise activities at the knowledge level. *Artificial Intelligence for Engineering Design, Analysis and Manufacturing: AIEDAM, 12*(1), 29-42.

Dong, Y., & Goh, A. (1998). An intelligent database for engineering applications. *Artificial Intelligence in Engineering, 12*, 1-14.

Dong, Y., et al. (1997). Active database support for STEP/EXPRESS models. *Journal of Intelligent Manufacturing, 8*(4), 251-261.

Dubois, D., Fargier, H., & Prade, H. (1995). Fuzzy constraints in job-shop scheduling. *Journal of Intelligent Manufacturing, 6*(4), 215-234.

Dubois, D., Fargier, H., & Prade, H. (1996). Possibility theory in constraint satisfaction problems: Handling priority, preference, and uncertainty. *Applied Intelligence, 6*, 287-309.

Dubois, D., & Prade, H. (1996). What are fuzzy rules and how to use them. *Fuzzy Sets and Systems, 84*, 169-185.

Dzbor, M. (1999). Intelligent support for problem formalization in design. In *Proceedings of the 3rd IEEE Conference on Intelligent Engineering Systems,* Stara Lesna, Slovakia (pp. 279-284). IEEE.

Eastman, C. M., & Fereshetian, N. (1994). Information models for use in product design: A comparison. *Computer-Aide Design, 26*(7), 551-572.

Erens, F., McKay, A., & Bloor, S. (1994). Product modeling using multiple levels of abstract: Instances as types. *Computers in Industry, 24*, 17-28.

Fargier, H., & Thierry, C. (1999). The use of qualitative decision theory in manufacturing planning and control: Recent results in fuzzy master production scheduling. In R. Slowinski, & M. Hapke (Eds.), *Advances in scheduling and sequencing under fuzziness* (pp. 45-59). Heidelberg: Physica-Verlag.

Fensel, D., Ding, & Omelayenko (2001). Product data integration in B2B e-commerce. *IEEE Intelligent Systems and Their Applications, 16*(4), 54-59.

Francois, F., & Bigeon, J. (1995). Integration of fuzzy techniques in a CAD-CAM system. *IEEE Transactions on Magnetics, 31*(3), 1996-1999.

George, R., et al. (1996). Uncertainty management issues in the object-oriented data model. *IEEE Transactions on Fuzzy Systems, 4*(2), 179-192.

Gertosio, C., & Dussauchoy, A. (2004). Knowledge discovery form industrial databases. *Journal of Intelligent Manufacturing, 15*, 29-37.

Giachetti, R. E., Young, Roggatz, Eversheim, & Perrone (1997). A methodology for the reduction of imprecision in the engineering design process. *European Journal of Operations Research, 100*(2), 277-292.

Goh, A., et al. (1994). A study of SDAI implementation on object-oriented databases. *Computer Standards & Interfaces, 16*, 33-43.

Goh, A., et al. (1997). A STEP/EXPRESS to object-oriented databases translator. *International Journal of Computer Applications in Technology, 10*(1-2), 90-96.

Grabot, B., & Geneste, L. (1998). Management of imprecision and uncertainty for production activity control. *Journal of Intelligent Manufacturing, 9*, 431-446.

Guiffrida, A., & Nagi, R. (1998). Fuzzy set theory applications in production management research: A literature survey. *Journal of Intelligent Manufacturing, 9*, 39-56.

Ho, C. F., Wu, W. H., & Tai, Y. M. (2004). Strategies for the adaptation of ERP systems. *Industrial Management & Data Systems, 104*(3), 234-251.

Issa, G., Shen, S., & Chew, M. S. (1994). Using analogical reasoning for mechanism design. *IEEE Expert, 9*(3), 60-69.

Johtela, T., Smed, Johnsson, & Nevalainen (1999) A fuzzy approach for modeling multiple criteria in the job grouping problem. In *Proceedings of the 25th International Conference on Computers & Industrial Engineering,* New Orleans, LA (pp. 447-50).

Jones, J. D., & Hua, Y. (1998). A fuzzy knowledge base to support routine engineering design. *Fuzzy Sets and Systems, 98*, 267-278.

Kapadia, R., & Fromherz, M. P. J. (1997). Design optimization with uncertain application knowledge. In *Proceedings of the 10th International Conference on Industrial and Engineering Application of Artificial Intelligence and Expert Systems,* Atlanta, GA (pp. 421-430). Gordon and Breach Science Publishers.

Karwowski, W., & Evans, G. W. (1986). Fuzzy concepts in production management research: A review. *International Journal of Production Research, 24*(1), 129-147.

Kim, K., Cormier, O'Grady, & Young (1995). A system for design and concurrent engineering under imprecision. *Journal of Intelligent Manufacturing, 6*(1), 11-27.

Kim, W., et al. (1990). Object-oriented database support for CAD. *Computer-Aided Design, 22*(8), 521-550.

Kleissner, C. (1998). Data mining for the enterprise. In *Proceedings of the Thirty-First Annual Hawaii International Conference on System Sciences,* Hawaii (vol. 7, pp. 295-304). IEEE Computer Society.

Kosala, R., & Blockeel, H. (2000). Web mining research: A survey. *SIGKDD Explorations, 2*(1), 1-15.

Krebs, T., & Lührsen, H. (1995). STEP databases as integration platform for concurrent engineering. In *Proceedings of the 2nd International Conference on Concurrent Engineering,* Virginia (pp. 131-142). Johnstown, PA: Concurrent Technologies.

Kuper, G., Libkin, L., & Paredaens, J. (2000). *Constraint databases*. Springer Verlag.

Kusiak, A., Letsche, T., & Zakarian, A. (1997). Data modeling with IDEF1X. *International Journal of Computer Integrated Manufacturing, 10*(6), 470-486.

Kusiak, A., & Tseng, T. L. (2000). Data mining in engineering design: A case study. In *Proceedings of the IEEE Conference on Robotics and Automation,* San Francisco (pp. 206-211). IEEE.

Lee, D. W., & Chu, W. W. (2000). Constraints-preserving transformation from XML document type definition to relational schema. *Lecture Notes in Computer Science,* Utah (vol. 1920, pp. 323-338).

Lee, M. L., et al. (2001). Designing semi-structured databases: A conceptual approach. *Lecture Notes in Computer Science, Vol. 2113* (pp. 12-21).

Liu, D. T., & Xu, X. W. (2001). A review of Web-based product data management systems. *Computers in Industry, 44*(3), 251-262.

Liu, M. C. (1999). On CAD databases. In *Proceedings of the 1999 IEEE Canadian Conference on Electrical and Computer Engineering,* Edmonton, Canada (pp. 325-330). IEEE Computer Society.

Liu, M. C., & Katragadda, S. (2001). DrawCAD: Using deductive object-relational databases in CAD. *Lecture Notes in Artificial Intelligence* (vol. 2113, pp. 481-490). Munich, Germany: Springer.

Ma, Z. M. (2005). *Fuzzy database modeling with XML*. Springer.

Ma, Z. M. (in press). Extending EXPRESS for imprecise and uncertain engineering information modeling. *Journal of Intelligent Manufacturing.*

Ma, Z. M., & Mili, F. (2003). Knowledge comparison in design repositories. *Engineering Applications of Artificial Intelligence, 16*(3), 203-211.

Ma, Z. M., et al. (2003). Conceptual data models for engineering information modeling and formal transformation of EER and EXPRESS-G. *Lecture Notes in Computer Science, Vol. 2813* (pp. 573-575). Springer Verlag.

Ma, Z. M., Zhang, W. J., & Ma, W. Y. (2002). Extending IDEF1X to model fuzzy data. *Journal of Intelligent Manufacturing, 13*(4), 295-307.

Maedche, A., Motik, Stojanovic, Studer, & Volz (2003). Ontologies for enterprise knowledge management. *IEEE Intelligent Systems, 18*(2), 2-9.

Mannisto, T., Peltonen, Soininen, & Sulonen (2001). Multiple abstraction levels in modeling product structures. *Date and Knowledge Engineering, 36*(1), 55-78.

Manwaring, M. L., Jones, K. L., & Glagowski, T. G. (1996). An engineering design process supported by knowledge retrieval from a spatial database. In *Proceedings of Second IEEE International Conference on Engineering of Complex Computer Systems,* Montreal, Canada (pp. 395-398). IEEE Computer Society.

McKay, A., Bloor, M. S., & de Pennington, A. (1996). A framework for product data. *IEEE Transactions on Knowledge and Data Engineering, 8*(5), 825-837.

Medina, J. M., et al. (1997). FREDDI: A fuzzy relational deductive database interface. *International Journal of Intelligent Systems, 12*(8), 597-613.

Michael, S. M., & Khemani, D. (2002). Knowledge management in manufacturing technology: An A.I. application in the industry. In *Proceedings of the 2002 International Conference on Enterprise Information Systems,* Ciudad Real, Spain (pp. 506-511).

Mili, F., Shen, Martinez, Noel, Ram, & Zouras (2001). Knowledge modeling for design decisions. *Artificial Intelligence in Engineering, 15,* 153-164.

Mitra, S., Pal, S. K., & Mitra, P. (2002). Data mining in soft computing framework: A survey. *IEEE Transactions on Neural Networks, 13*(1), 3-14.

Muller, K., & Sebastian, H. J. (1997). Intelligent systems for engineering design and configuration problems. *European Journal of Operational Research, 100*, 315-326.

Noran, O. (2000). *Business Modeling: UML vs. IDEF* (Report/Slides). Griffith University, School of CIT. Retrieved February 2000, from http://www.cit.gu.edu.au/~noran

O'Leary, D. E. (1998). Enterprise knowledge management. *IEEE Computer, 31*(3), 54-61.

ODMG. (2000). *Object Data Management Group*. Retrieved November 2000, from http://www.odmg.org/

Oh, Y., Hana, S. H., & Suhb, H. (2001). Mapping product structures between CAD and PDM systems using UML. *Computer-Aided Design, 33*, 521-529.

OMG. (2003). *Unified Modeling Language (UML)*. Retrieved December 2003, from http://www.omg.org/technology/documents/formal/uml.htm

Pal, S. K., Talwar, V., & Mitra, P. (2002). Web mining in soft computing framework: Relevance, state of the art and future directions. *IEEE Transactions on Neural Networks, 13*(5), 1163-1177.

Parsons, S. (1996). Current approaches to handling imperfect information in data and knowledge bases. *IEEE Transactions on Knowledge and Data Engineering, 8*(2), 353-372.

Petrovic, D., Roy, R., & Petrovic, R. (1998). Modeling and simulation of a supply chain in an uncertain environment. *European Journal of Operational Research, 109*, 299-309.

Petrovic, D., Roy, R., & Petrovic, R. (1999). Supply chain modeling using fuzzy sets. *International Journal of Production Economics, 59*, 443-453.

Pham, B. (1998). Fuzzy logic applications in computer-aided design. *Fuzzy Systems Design*. In L. Reznik, V. Dimitrov, & J. Kacprzyk (Eds.), *Studies in Fuzziness and Soft Computing, 17*, 73-85.

Pham, D. T., & Pham, P. T. N. (1999). Artificial intelligence in engineering. *International Journal of Machine Tools & Manufacture, 39*(6), 937-949.

Posselt, D., & Hillebrand, G. (2002). Database support for evolving data in product design. *Computers in Industry, 48*(1), 59-69.

Prade, H., & Testemale, C. (1984). Generalizing database relational algebra for the treatment of incomplete or uncertain information. *Information Sciences, 34*, 115-143.

Samaras, G., Spooner, D., & Hardwick, M. (1994). Query classification in object-oriented engineering design systems. *Computer-Aided Design, 26*(2), 127-136.

Saygin, Y., & Ulusoy, O. (2001). Automated construction of fuzzy event sets and its application to active databases. *IEEE Transactions on Fuzzy Systems, 9*(3), 450-460.

Schenck, D. A., & Wilson, P. R. (1994). *Information modeling: The EXPRESS way*. Oxford University Press.

Sebastian, H. J., & Antonsson, E. K. (1996). *Fuzzy sets in engineering design and configuration*. Boston: Kluwer Academic Publishers.

Seligman, L., & Rosenthal, A. (2001). XML's impact on databases and data sharing. *IEEE Computer, 34*(6), 59-67.

Seng, J. L., Lin, Y., Wang, J., & Yu, J. (2003). An analytic study of XML database techniques. *Industrial Management & Data Systems, 103*(2), 111-120.

Shaw, M. J. (2000a). Information-based manufacturing with the Web. *The International Journal of Flexible Manufacturing Systems, 12*, 115-129.

Shaw, M. J. (2000b). Building an e-business from enterprise systems. *Information Systems Frontiers, 2*(1), 7-17.

Shaw, N. K., Bloor, M. S., & de Pennington, A. (1989). Product data models. *Research in Engineering Design, 1*, 43-50.

Soliman, F., & Youssef, M. A. (2003). Internet-based e-commerce and its impact on manufacturing and business operations. *Industrial Management & Data Systems, 103*(8), 546-552.

Sun, J., Kalenchuk, D. K., Xue, D., & Gu, P. (2000). Design candidate identification using neural network-based fuzzy reasoning. *Robotics and Computer Integrated Manufacturing, 16*, 383-396.

Surjanto, B., Ritter, N., & Loeser, H. (2000). XML content management based on object-relational database technology. In *Proceedings of the First International Conference on Web Information Systems Engineering,* Hong Kong (vol. 1, pp. 70-79).

Tan, G. W., Shaw, M. J., & Fulkerson, B. (2000). Web-based supply chain management. *Information Systems Frontiers, 2*(1), 41-55.

Tan, K. H., & Platts, K. (2004). A connectance-based approach for managing manufacturing knowledge. *Industrial Management & Data Systems, 104*(2), 158-168.

Thannhuber, M., Tseng, M. M., & Bullinger, H. J. (2001). An autopoietic approach for building knowledge management systems in manufacturing enterprises. *CIRP Annals — Manufacturing Technology, 50*(1), 313-318.

Tsourveloudis, N. G., & Phillis, Y. A. (1998). Manufacturing flexibility measurement: A fuzzy logic framework. *IEEE Transactions on Robotics and Automation, 14*(4), 513-524.

van Gyseghem, N., & de Caluwe, R. (1998). Imprecision and uncertainty in UFO database model. *Journal of the American Society for Information Science, 49*(3), 236-252.

Wizdom Systems Inc. (1985). *U.S. Air Force ICAM Manual: IDEF1X.* Naperville, IL.

Wong, K. Y. (2005). Critical success factors for implementing knowledge management in small and medium enterprises. *Industrial Management & Data Systems, 105*(3), 261-279.

Xiao, R. G., et al. (2001). Modeling and transformation of object-oriented conceptual models into XML schema. *Lecture Notes in Computer Science, Vol. 2113* (pp. 795-804).

Xue, D., & Xu, Y. (2003). Web-based distributed systems and database modeling for concurrent design. *Computer-Aided Design, 35*, 433-452.

Xue, D., Yadav, S., & Norrie, D. H. (1999). Knowledge base and database representation for intelligent concurrent design. *Computer-Aided Design, 31*, 131-145.

Yager, R. R. (2000). Targeted e-commerce marketing using fuzzy intelligent agents. *IEEE Intelligent Systems, 15*(6), 42-45.

Yager, R. R., & Pasi, G. (2001). Product category description for Web-shopping in e-commerce. *International Journal of Intelligent Systems, 16*, 1009-1021.

Yang, H., & Xue, D. (2003). Recent research on developing Web-based manufacturing systems: A review. *International Journal of Product Research, 41*(15), 3601-3629.

Yazici, A., et al. (1999). Uncertainty in a nested relational database model. *Data & Knowledge Engineering, 30*, 275-301.

Young, R. E., Giachetti, R., & Ress, D. A. (1996). A fuzzy constraint satisfaction system for design and manufacturing. In *Proceedings of the Fifth IEEE International Conference on Fuzzy Systems,* New Orleans, LA (vol. 2, pp. 1106-1112).

Zadeh, L. A. (1965). Fuzzy sets. *Information and Control, 8*(3), 338-353.

Zadeh, L. A. (1978). Fuzzy sets as a basis for a theory of possibility. *Fuzzy Sets and Systems, 1*(1), 3-28.

Zadeh, L. A. (1989). Knowledge representation in fuzzy logic. *IEEE Transactions on Knowledge and Data Engineering, 1*(1), 89-100.

Zhang, F., & Xue, D. (2002). Distributed database and knowledge base modeling for intelligent concurrent design. *Computer-Aided Design, 34*, 27-40.

Zhang, Q. Y. (2001). Object-oriented database systems in manufacturing: selection and applications. *Industrial Management & Data Systems, 101*(3), 97-105.

Zhang, W. J., & Li, Q. (1999). Information modeling for made-to-order virtual enterprise manufacturing systems. *Computer-Aided Design, 31*(10), 611-619.

Zhang, Y. P., Zhang, C. C., and Wang, H. P. B. (2000). An Internet-based STEP data exchange framework for virtual enterprises. *Computers in Industry, 41*, 51-63.

Zimmermann, H. J. (1999). *Practical applications of fuzzy technologies*. Boston: Kluwer Academic Publishers.

Zvieli, A., & Chen, P. P. (1996). Entity-relationship modeling and fuzzy databases. In *Proceedings of the 1986 IEEE International Conference on Data Engineering* (pp. 320-327).

This work was previously published in Database Modeling for Industrial Data Management: Emerging Technologies and Applications, edited by Z. M. Ma, pp. 1-34, copyright 2006 by Information Science Publishing (an imprint of IGI Global).

Index

Symbols

%SkinPixels 1971
1084
10th Pacific-Asia Conference on Knowledge Discovery and Data Mining (PAKDD) 84–92
2-component property 230, 231
2-component property of patterns 246
3G mobile telecommunications network 2558–2565
3M 2750

A

a metric incremental clustering algorithm (AMICA) 863
a posteriori criterion 1095
a priori 1093
ab initio 101, 104
absolute deviation proportion (ADP) 2214
abstract data mining service (ADMS) 924
abstract data types (ADT) 3595
abstract information 1938, 1950, 1954
abstract syntax 3464–3465
accounting information systems 398
accuracy measures 1097
acquisition 439
acquisitions management 2674
active data warehouses (ADWHs) 757
active databases (ADBs) 757
active intelligence 1899
activity monitoring (BAM) 1852
ad hoc networks 2711
AdaBoost 359
adaptive estimator (AE) 2209, 2212
adaptive genetic algorithm (AGA) 946
adaptive threshold 3229, 3232
administration and management 1
adult-related queries 1929
affinity analysis 97
affordances 2321
agent migration 712
aggregate information (AI) 2397
aggregate usage profiles 3563
aggregation logic 247
agile manufacturing (AM) 1182
akaike information criterion 366
algorithm RulExSVM 1271
algorithm selection 2307
algorithmic approach 2966

algorithms 95
alignment 104
alignment-less alignment 1093
AlltheWeb 1926
alphabetic tokens 1366, 1368
AltaVista 1437
AltaVista 1926
ambiguous motifs 1725
amino acids (AAs) 1722
analysis manpower 2567
analysis of the variance (ANOVA) 946
analytical hierarchy process (AHP) 2966
analytical information technologies 2618
analytics 1825, 1827, 1832, 1835, 1838, 1840, 1842, 1844, 1849, 1850, 1853
anomaly detection 2881
anti money laundering (AML) 1837
application domains 2273
application integration (EAI) 1852
application server logging 2893
approximate query answering (AQA) 975
apriori algorithm 259, 303, 309, 312, 324, 325
apriori property techniques 2902
apriori-based approach 3525
apriori-inverse 3222–3233
AR mining task formulation 515
ARF method 3675
ARMA models 944
ART algorithm 3496
Artificial Immune Recognition System (AIRS) 3441
artificial intelligence (AI) 99, 1800, 2289, 2445, 2586, 2659, 2965
artificial intelligence techniques 1689
artificial neural network (ANN) 138, 154, 2234, 2916, 1405
artificial recognition ball (ARB) 3441
ASIA networks 1084, 1085, 1088
asset liability management (ALM) 1837
asset mentality 3422
association mining 2516
association rule 97, 309, 310, 313, 314
association rule generation 2664
association rule mining 2993, 3349
association rule mining in XML data 509–529
association rule semantics 3466–3467
association rules 645, 1303, 2004, 2009, 2523, 2524, 2533, 2535, 2536, 2537, 2549, 2550
association rules mining 336
association rules mining (ARM) 231
association rules mining paradigm 2994
association-rule mining system (ARMS) 2803
atomic change 248
atomic process 922
attribute generalization algorithm 2121
attribute value decomposition (AVD) 1611
attribute-oriented induction 2122
attributes and class labels through linguistic rules. The extracted rules are in 1271
auction-bots 1489
Australian and New Zealand Bank (ANZ) 2775
Australian banking industry 2773
autocorrelation 1749
automated teller machines (ATMs) 2445
automatic machine-learning-based pornographic Website classification 1958
automatic semantics extraction 3545
automatic Website categorization 1962
automating actions using plans 2653
automation 2857
AVC-group 1206

B

B2B e-commerce 1185
bagging 358
Bank for International Settlements (BIS) 1825
banking industry 2774
bankruptcies (BK) 1869
banks 1825, 1826, 1828, 1830, 1840, 1841, 1845, 1850
Basel II 1836, 1837, 1839
basic bitmap index 1593
Bayes theorem 99
Bayesian approach 2946
Bayesian classifier 99
Bayesian criteria 366
Bayesian Information Criterion 366
Bayesian method 1405, 2376
Bayesian network (BN) 1081–1090
Bayesian network tools in Java (BNJ) 1085
Bayesian neural network 1320
BBC (see byte-aligned bitmap code) 1597
behavioral model 1856
belief networks 963, 969, 970, 973
belief systems 2304
benchmarking 2754
benefits of data mining 449
bibliomining 1816, 2674
big bang approach 2434
binary classification 64–74
binary large objects (BLOBs) 261
binning 1595

bioanalytics 1759
BioCarta 1790
bioinformatics 1643, 1714, 1816, 2607
bioinformatics data 2273
biomarker identifier (BMI) 1766
bit transposed file model 3695
bitmap index tuning 1598
bit-sliced index (BSI) 1611
Blocking-Based Techniques 56
BOAT 240
BOAT algorithm 1206
body mass index (BMI) 1823
body of evidence (BOE) 2945–2946
boosting 359
bootstrapping 1206
bootstrapping techniques 1975
bounded rationality problem 3664
Box-Muller algorithm
brain-derived neurotrophic factor (BDNF) 38
branch-and-bound 104
breadth-first frequent pattern mining 1288
brief psychiatric rating scale (BPRS) 3683
budgeting 2376
building XML data warehouses 530–555
business ethics 2291
business event logging 2893
business intelligence (BI) 756, 2421, 2688
business intelligence and knowledge management pha 2723
business process execution language for Web service (BPEL4WS) 2031
business support 2754
business to business (B2B) 810
business-to-business negotiation 2900
byte-aligned bitmap code (BBC) 1597

C

caBIO 1790
cache insertion algorithm 626
cache replacement policy 1541
CAD/CAM 2638
CAD/CAPP/CAM 1183
CAESAR 3615
call detail records (CDRs) 760
candidate rules 3379
capillary electrophoresis (CE) 1760
capital adequacy ratio (CAR) 1825
CaRBS technique 2955
cart 2559–2565
CART (classification and regression tree analysis) 12
CART algorithm 2499
CART model confusion matrix 1526
CART tree 3681
cascading strategy 1974
case-based reasoning (CBR) 2659
CASP 116
causal connection matrix 1426
causal loop diagrams 2690
caWorkBench2 1790
Cayley network 1160, 1167
CCMine 1337
central nervous system (CNS) 1650
centralized data 2739
centroid 240
CFIs 231
change detector mechanism 237
channel dimension 799
charge-off (CO) 1869
Chisholm, Roderick 3199
Chi-square 298
Chubb & Son 1888
CIMOSA 1187
circulating fluidized boiler (CFB) 2234
CiteSeer 3562
classic decision tree (CDT) 2982
classical genetic algorithm (CGA) 1087–1088
classical mutation operator (CM) 1087
classification 336, 356
classification algorithm 137
classification algorithms 3213
classification and ranking belief simplex (CaRBS) 2943, 2960
classification error (CE) 243
classification rules 1780
classification systems 1962
classification, associative 69
classification, binary 64–74
classification, multi-label 64–74
classifier, item-based 3215
classifier, pairwise-based 3216
CLDS, definition of 3
clickstream data 2553
client physical schema (CPS) 570
clinical study report (CSR) 3676
CLIQUE algorithm 233
closed association rule mining (CHARM) 1661
closed pattern discovery by transposing tables that are extremely long (CARPENTER) 1661
CLOSET 1661
cluster analysis 152
cluster analysis 3288–3289

cluster identification via connectivity kernels (CLICK) 1652
cluster-based input selection 1138–1156
clustering 97, 336, 643, 1301, 1816, 2008, 2282, 2516, 2735, 2858, 2889, 3455–3457, 3459–3467, 3665
clustering algorithms 1091
clustering and classification 2373
clustering genetic algorithm (CGA) 1270
clustering method 1858
COBWEB algorithm 233
COG-mode 1791
cognitive map 1421
collaborative filtering (CF) 3558
collaborative knowledge production 439
collaborative virtual environments 1624
collection management 2674
collective data mining 2597
co-location patterns 3482
Combinatorial Algorithm 1168
combinatorial fusion 1159
combinatorial fusion analysis (CFA) 1173
CombineDims 326–327
combining row and column enumeration (COBBLER) 1662
combustion dataset 1600
committee methods 356
common practice notation view (C.P.N. View) 3590
Commonwealth Bank of Australia (CBA) 2775
communication and interaction between partners 2709
communication requirements 2390
communities, informal 1456
comparative genome annotation 1785
comparative genomics 1784
compensation planning 2371
competitive advantage 2290, 2773
complete linkage 240
complex networks, supply chain management 2468–2475
component diagram 567
component rating 1894
composite process 922
compression 1596
computation independent model (CIM) 560
computational intelligence 146
computational intelligence 2088
computational intelligence techniques 1438
computational time complexity 3245
computer terminal network (CTN) 2763
computer vision 3621
computer-based training 2424
concept formation 2058
concept hierarchies 2128
concept-effect relationships, data mining for 2931
conceptive landmark 51
conceptual clustering, and video data 1634
conceptual construction 3028
conceptual data model 563
conceptual data modeling (CDM) 280, 281, 282, 300
conceptual data models 188
conceptual defined sequence 3393
conceptual modeling 209
concrete motifs 1725
concurrent engineering (CE) 1182
conditional market-basket probability (CMBP) 236
conference publication system (CPSys), case study 493
confusion matrix 367, 1524
connectivity matrix 1082
connectivity phenomena 1378
consensus scoring (CS) 1173
consistent fuzzy concept hierarchy 2130
constraint databases 1192
consumer value 2843
contact map overlap (CMO) 115
content-based music information retrieval (CBMIR) 3587
continuous evolution 1938
continuous replacement planning 2627
continuous value discretisation (CVD) 3013
contractual charge-offs (CCO) 1869
control access 478
control signatures 2226, 2227, 2233
convenience users 1856
conventional music notation (CMN) 3590
cookies 2007
core region 109
corporate analysis 2603
corporate information factory (CIF) 4
corporate lending 2438
corporate libraries 2675
correlation-based feature (CSF) 870
corridor-based heuristic 248
cost of turnover 2372
cost planning 2024
cost-effective revenue growth 1889
counter-terrorism 693
counts per second (cps) 1763
coupled two-way clustering (CTWC) 1654
coupling type 247

cube presentation model (CPM) 1007
creating a data warehouse 552
creating dimensions 547
creating intermediate XML documents 541
credit card users' data mining 2464–2467
credit risk 2466
credit risk models 2441
credit scoring 2449, 2451
crisp concept hierarchy 2129
criteria based on score functions 365
CRM (customer relationship management) 1488
CRM data 787
CRM factor 787
CRM models 787
CRM quadrant 798
cross industry standard process for data mining (CRISP-DM) 149, 918, 1843, 2294
crossover operator 1082
cross-sell 1890
cross-selling 2447
cross-validation 367, 1975
cubeDT 1206, 1207, 1211, 1212, 1213
customer acquisition 1889
customer differentiation 1893
customer information management (CIM) 383
customer purchase support 2660
customer recognition 3070
customer relationship management (CRM) 788, 810, 2588, 2660, 2749, 3068
customer requirement analysis 2798
customer value index (CVI) 1893
customization transportation diagram (CTD) 570
CVI calculation 1893
cyber attacks 2880
cyber communities 2339
cyber security 3640
cyberpatrol 1979
cybersitter 1979
cycle representation 1167
cytoscape 1790

D

D&B WorldBase 2759
DARPA Agent Markup Language (DAML) 2025
DARPA Agent Markup Language Semantic Markup extension for Web Services (DAML-S) 2031
DAS XML 1789
data acquisition 2834
data allocation scheme 3327
data analysis process engineering 449
data cleaning 32, 533, 544, 1802, 3637
data clustering 2508
data collection 95, 1950
data constraints 2602
data conversion (DC) 2714
data cube 2123, 2202, 2203, 2204, 2205, 2206, 2207, 2209, 2210, 2212, 2213, 2224, 2225
data cube 387
data cube generation 3177, 3182, 3185, 3190
data definitions 3428
data dictionary 387, 581, 919
data formatting 2907
data fusion 1950
data integration 2292
data interception by remote transmission (DIRT) 2864
data level conflicts 536
data marts 382, 430
data mining 119, 356
data mining
data mining 93, 95, 228, 350, 881, 942, 944, 959, 963, 966, 972, 1131, 1136, 1689, 1877, 1878, 1880, 1883, 1884, 1885, 1886, 1887, 2226, 2227 2772, 2850, 3346
data mining applications 2688
Data Mining Benchmarking Association 2375
data mining by using database statistics 352
data mining generations 2588
data mining learning curve 2780
data mining medical information 2915–2927
data mining of dynamic data 2611
data mining ontology (DMO) 914, 923
data mining practitioner 1855
data mining process 95
data mining query language (DMQL) 151
data mining solutions 1896
data mining tasks 96
data mining techniques 336, 1747, 2508, 2566, 2674, 2900
data mining textbooks, and privacy 2872–2879
data mining tools 2290
data mining, and ethical dilemmas 2841–2849
data mining, and ethics 2834–2840
data mining, data models 150–151
data mining, privacy concerns 693
data mining, process 149
data modeling levels 562
data modelling languages 3194
data modification 54
data normalization 2907
data overload problem 3612
data overview 1522

data ownership issues 2750
data partitioning 53
data preparation 1969
data preparation 919
data preparation techniques 1802
data preprocessing 95
data quality 1104, 1115
data quality 3633
data requirements 2750
data restriction techniques 55
data security 2851
data selection 882
data sourcing 1545
data stream management 2644–2658
data stream, memory 1231
data transformation 32
data treatment. 882
data understanding 918
data visualization 1639
data visualization 1718
data warehouse 18, 661, 679, 1820, 1823, 1831, 2627, 2750, 3117
data warehouse administrator (DWA) 3
data warehouse design 208
data warehouse development 2
data warehouse integration 3386
data warehouse loader 3375
data warehouse maintenance 1907
data warehouse metrics 408–428
data warehouse network 382
data warehouse operational processes 3049
data warehouse physical design 569
data warehouse physical schema 576
data warehouse refreshment 2627
data warehouse synergies 3411–3415
data warehouse system 3116
data warehouse team 2
data warehouse, physical modeling of 591–621
data warehousing 1, 99, 2585, 2735, 3346
data warehousing (DW 1005
data warehousing (DW) 486
data warehousing applications 2566
data warehousing architecture 2428
data warehousing success 2432
data warehousing, seismological 3645–3661
data, streaming 1231
database management system (DBMS) 4, 169, 660, 680, 697, 3365
database marketing 1849
database marketing 2824–2833
database modeling 188
databases 94
data-intensive business 1799
datamart 2373
data-mining algorithms 1983
data-mining applications 2273
data-mining based learning 1969
data-mining query languages (DMQL) 2588
data-mining technique 1965, 1969, 2273
datasets selection 32
data-stream management, 3 problems 2647
dataveillance 3630
days past due (DPD) 1848
days sales outstanding (DSO) 2460
DB2 1193
DDID-PD framework 832
DDID-PD process model 835
decision intelligence 2427
decision support systems (DSS) 1, 2618, 2641, 2749, 2965, 3086
decision theoretic framework 1320
decision tree (DT) 11, 100, 231, 1301, 2454, 2497, 2516, 2523, 2524, 2532, 2539, 2540, 2541, 2547, 2548, 2978
decision tree algorithms 1205, 1211
decision tree analysis 2374
decision tree mining 2508
decisional model 222
decision-making 3420
decision-making processes 2674
decision-tree approach 2494
default management 2446
defense software collaborators (DACS) 2376
DEMON framework 248
Dempster-Shafer theory (DST) 2943, 2945, 2960
density based spatial clustering of applications with noise (DBSCAN) 2382
density-based algorithm 2901
density-based methods 233
deoxyribonucleic acid molecule (DNA) 1644
Department of Human Services (DHS) 2434
Department of Revenue and Finance (DRF) 2434
dependency modeling 2736, 2737
deployment diagram 568
deployment landmark 51
depth-first frequent pattern mining method 1289
deviation analysis 96
deviation detection 2736
DGX distribution 1337
diabetes 1820
dialysis 2506, 2516
dice 3357

differential association rules 1747
digital forensics 3642
digital library collections 2674
digital nervous system 2723
digital revolution 343
dimension/level constraints 2602
dimensional fact model (DFM) 210
dimension-reduction 2036, 2037, 2039
dimensions as UML packages 494
diplotype analysis 1685
direct acyclic graph (DAG) structure 127
directed acyclic graph (DAG) 120, 1082–1083
discovery services via search engine model (DSSEM) 710–717
discovery-driven data mining 2967
discrete wavelet transform 1119
discretionary access control (DAC) 686
discriminant analysis 12
discriminant analysis (DA) 2454
disease management data mining 1805
disease management programs 1801
distance function 232
distance-based association rule mining (DARM) 1660
distributed annotation system (DAS) 1787
distributed data mining 2597
distribution channel 1890
distribution channel access 2781
diversity rank function 1170
diversity score function 1170
divide-and-conquer 104
DM technology 229
DNA chip 1706
domain knowledge 2318
domain knowledge representation 3567
domain ontology 1078
domain ontology acquisition 3560
domain-driven data mining, key components of 836
domain-driven in-depth pattern discovery (DDID-PD) framework 832
DOS group 42
double dynamic programming 108
drill-down 3356
DS-Web 1446
Dublin Core 3533
DWFIST approach 3143
dynamic health strategies (DHS) 2376
dynamic hierarchies 218
dynamic media language 1472
dynamic personalization 1488
dynamic programming 105
dynamic time warping (DTW) 947

E

earthquake data 3645–3661
eavesdropping 2711
e-business 1436
echelon and transient electromagnetic monitoring P 2864
e-commerce 1486
e-commerce solutions 1889
e-commerce transactions 343
economic/business cycles 2438
E-Government Act of 2002 2430
eigenfaces 3624
electronic banking 343
electronic business 1436
electronic commerce 2273
electronic customer relationship management (e-CRM 2292
electronic data interchange (EDI) 2763
electronic numerical integrator and computer (ENIAC) 343
ellite data acquisition (ETL) 1846
e-mail filing 1454
e-mail mining 1454
e-mail worms 2036, 2037, 2038, 2039, 2049
emantic associations 3526
e-manufacturing 1197
EMBL nucleotide sequence database 95
emerging patterns (EPs) 1658
e-metrics 2555
employee retention 2371
employment practices liability 1889
empowerment and collaboration phase 2723
encoded bitmap index (EBI) 1610
encoding 1593
encryption 2711
enhanced data mining 3029
ensemble 356
enterprise data warehouse 561
enterprise decision-making 2290
enterprise information systems (EISs) 1184
enterprise resource planning (ERP) 810, 1184
enterprise-wide information systems 399
entity resolution 3068
entity-relationship (ER) model 283, 300
EnviroFacts 2430
equality encoded bit-sliced index (EEBSI) 1611
equality-encoded bitmap 1602
ER modeling 181
estimation error, measuring of 1237

ethical standards 2708
ethics, and data mining 2835–2840
ETL workflow, optimization of 3059
Euclidean distance 1649
Euclidean distance 952
evaluation function 232
exception rules mining 336
execution agents 471
executive information systems (EIS) 2423, 2749
exogenous variables 1965
expectation maximization (EM) 2069, 2386
expectation maximization (EM) algorithms 99
experiential learning 2900
expert systems 1185, 1438, 1800
explicit profiling 1495
exploratory data analysis (EDA) 80, 1893, 3675
exploratory data mining 2567
exposure at default (EAD) 1837
EXPRESS 1189
expression component 230
extensible Markov models 1409
extensible markup language (XML) 485, 510
external data 2263, 2725
external information sources 400
extraction 3545
extraction transformation loading (ETL) 2, 391–396, 3050
extraction, transformation, loading (ETL) processes 556
extreme programming (XP) 1844

F

Fair Isaac Corporation (FICO) 2451
false trails 3372
fast update (FUP) algorithm 238
FASTA 1789
feature-selection 2036, 2037, 2039, 2250, 2251, 2253, 2256, 2257, 2258
Federal Aviation Administration 2421
Federal Trade Commission (FTC) 2861
file archiving 2653
file transfer protocol (FTP) 1730
filtering system 1958
filtering unification module 1387
filters 1765
financial applications 1446
financial forecasting 2480
finding interesting association rule groups by enumeration of rows (FARMER) 1660
first generation data-mining System 2588
first-class citizens 1043
FIs 231
fitness function 1082
five forces model 2773
fixed threshold 3229, 3232
flat files 2892
flat hierarchy 2136
flexible gap motifs 1725
flood prediction 1402
flow injection analysis (FIA) 1763
flow measures 220
FLP classification 36
FLP method 31
FlyBase 1788
FOCUS measure 235
folklore term 451
Food and Drug Administration 3676
forecasting 2088–2090, 2103–2104
Forecasting 3424
forest cover 2088–2100
four levels architecture 2728
franchise "family" relationship 2723
franchising 2722
fraud detection 2423
fraud detection 2447, 2603
French Ministry of Education 1959
frequent episodes 2004, 2012
frequent itemset mining (FIM) 230
frequent wayfinding-sequence (FWS) 1573
frozen approximation 108
fum, filtering unification module 1387
functional magnetic resonance imaging (fMRI) 943
fuzzy adaptive genetic algorithms (FAGA) 947
fuzzy class hierarchy 2125
fuzzy class schema 2127
fuzzy clustering algorithm 2123
fuzzy c-means (FCM) 942
fuzzy c-means (FCM) algorithm 944
Fuzzy concept hierarchy 2130
fuzzy conceptual data modeling 188
fuzzy data mining 2123
fuzzy deixis 1471
fuzzy inference system (FIS) 1138, 1139
fuzzy learning vector quantization (FLVQ) 1652
fuzzy logic 187, 1185, 1438, 1439, 2143, 2145
fuzzy logical database modeling 188
fuzzy modeling 1146, 1150, 1154, 1155, 1156
fuzzy object-oriented model 2122, 2125
fuzzy observations 3029
fuzzy patterns 3029
fuzzy rule-based system (FRBS) 2455
fuzzy set 2123

fuzzy set theory 1185, 2944
fuzzy sets 189
fuzzy systems 100
fuzzy transformation 3030

G

GA mechanism 2227
gap symbols 105
GCDM 1077
GEDW DataExpress 2759
GemDOCK 1174
gene annotation 101
gene chip 1706
gene discovery 2250
gene expression analysis 1698
gene fusion analysis 1793
gene neighborhood analysis 1793
gene ontology 128, 138
gene selection methods 1707
general environment 2303
general purpose techniques (GPT) 57
generalized regression neural network 2481
generate generalized metapatterns 518
generative topographic mapping (GTM) 2069
generic framework 128
generic rule model (GRM) 237
genetic algorithm (GA) 100, 138, 942, 944, 959, 1270, 1717, 1858, 2226, 2227
genetic algorithm technique 1082–1084
genetic clustering 943, 944, 948, 960
genetic-algorithm(GA)-based methods 942
GenNLI 322–327
genome annotation 1784
genome reviews 95
genome selection flexibility 1785
genomics 1698
GeoCache 622–641
geographic data mining 2600
geographic identifier 881
geographic information system (GIS) 622–641, 660, 2618, 3478
geometric techniques 2106
geo-referenced data sets 881
getensemble algorithm 2817
global as view (GAV) 3542
global data warehouse (GDW) 1846
global error rate 1961, 1975
global information systems 398
global pricing 2753
global sales 2753
global sequence patterns 3483
globalization 3477
globus Toolkit 3 (GT3) 755
goodness-of-hit (GH) 1166
G-protein-coupled-receptor (GPCR) 140
gradually expanded trees 2816–2823
graph-based clustering 1094, 1098
graph-based data mining 3525
graphical user interface (GUI) 2765
graphical user interface (GUI) tools 2641
greatest common refinement (GCR) 234, 246
grid-based methods 233
GridMiner assistant (GMA) 913, 923
group health management 2376
growing neural gas (GNG) 1651
GT 2639
guaranteed-error estimator (GEE) 2209, 2211, 2212

H

Hadden-Kelly methodology 382
haplotype analysis, and diseases 1674–1688
haplotype, multi-locus 1681
hash token 3068
HAvg 303–335
health and wellness initiatives 2376
health information technology industry 1800
Health Insurance Portability and Accountability Act 2376
health plan benefits 2376
health plan employer data and information set 2376
healthcare organizations 2494
hearst patterns 1988, 1989
heterogeneous meta-ensemble 1320
heuristic, linear 3170
didden Markov models 1408, 3594
hierarchical clustering 1649
hierarchical clustering 97, 3600
hierarchical merging 1093
hierarchical merging method 1098
hierarchical methods 232
hierarchical partitioning 2374
hierarchical techniques 2106
hierarchy-driven indexed quad-tree summary (H-IQTS) 976
high business value creation and implementation ph 2723
high performance computing (HPC) 1321
high-energy physics dataset 1600
HIPAA 3663
HIV-1 associated dementia (HAD) 27, 46
homeland security 3639
homogeneous composition 489

homology modeling 104
hop extraction process 963, 965
Hopfield neural network 3030
horizontal partitioning (HP) 22
human lung cancer 2088, 2091, 2093, 2100
human resource information systems (HRIS) 2371
human-computer interaction 1638
human-computer interaction 471
humanities data warehousing 2364
humanities research 2364
human-machine interactions 2051
hybrid association rule 309, 310
hybrid evolutionary programming (HEP) algorithm 1082
hybrid knowledge base 2665
hybrid recommendation 2666
hybrid technique 3366
hybrid ordering algorithm 1252
hypermedia data mining 2598
Hyper-Node Based Compression 2345
hypersoap 3622
hypertext markup language (HTML) 1464
hyper-view 2628, 2632, 2634
hyperwalk 2342
hypotheses 1388
hypothesis-driven approach 2305

I

i-agents 1437
IBL paradigm 125
iconographic techniques 2106
identity theft 2375
image reconstruction 3493
image retrieval 262, 268, 275
IMAGO 712–717
imperfect information 188
impersonation 2711
implementation in Oracle 580
implicit profiling 1495
inclusion dependency mining 351
indexed XML tree (IX-tree), construction 513
indexing techniques 18
inductive database 3525
inductive logic processing (ILP) 3525
industrial computed tomography (ICT) 3494
industry cycles 2438
InferenceEngine 1388
information age 2290, 2294
information disseminators 1436
information ethics 2874
information extraction 1811
information extraction 470, 1816
information files (CIFs) 1850
information filtering 1438, 1439, 2594
information fusion 1158
information gain (IG) 1766
information paralysis 2289, 2290
information quality schema (IQS) 2397
information retrieval 1811, 1816
information retrieval 95
information retrieval system 1425
informational privacy 2836
information-gain 2476
information-retrieval 2263
instance matching 350
instance-based learning (IBL) 124
institutional environment 2303
interface agents 471
integer program (IP) 111
integrated data mining systems 2610
integrated spatial reasoning 888
integrating data selection and AR mining 521
integration of data sources 350
integration transportation diagram (ITD) 569, 578
intelligent agent 1437, 1489
intelligent cache management 1539–1556
intelligent discovery assistant (IDA) 917
intelligent interfaces 2517
intelligent search agents 2594
interactive and parallel mining supports 840
interactive association rule mining 2110
internal node 231
internal ratings-based (IRB) 1837
International Organization for Standardization 1189
Internet data-mining 1446
internet learning agent (ILA) 2594
Internet shopping 1486
internetworked 2705
interoperability 3637
interpretations of rules 2057
inter-quartile range (IQR) 2027
interrelated two-way clustering 1654
interval-based heuristic 248
interviews 2737
introns 101
intrusion detection 3642
Intrusion Detection System 2881
Inventory Management 2792
Invisible Data Mining 2611
issues in mining digital libraries 1812
itemset 3144

itemset problem 3229
iterative partitioning 1093
IX-tree efficiency 525

J

Java data mining API (JDM) 921
Java Server Pages (JSP) 1469
just-in-time (JIT) 1903

K

k nearest neighbor (kNN) classification 3212
KBNMiner 1422
KEGG database 1787
kernel function approach 3525
kernel width selection 3308–3323
k-fold cross-validation 1427
Killer-Victim Scoreboard (KVS) method 952
K-means 1094
K-means algorithm 97, 233, 1098
K-medoids 233
K-medoids algorithm 233
k-medoids-based genetic algorithms 944
knowledge age 2294
knowledge and organizational networks 2705
knowledge application system 1426
knowledge base (KB) 916
knowledge base construction 3561
knowledge capturing 1489
knowledge consolidation 440
knowledge creation 438
knowledge discovery 280, 2121, 2494, 2697, 2802, 2859
knowledge discovery (KD) 3451, 3452–3454
knowledge discovery from data (KDD) 93, 228, 449, 923, 1418, 1558, 1741, 2302, 2660, 3348,
knowledge domains 1907
knowledge elicitation 1489
knowledge generation 439
knowledge management (KM) 438, 1186, 2291, 2421, 3423
knowledge marts 440
knowledge structure (KS) 10
knowledge warehouses 441
knowledge-discovery and data-mining 343
knowledge-producing agents 440
knowledge-type constraints 2602
known-worms set 2043
Kolmogorov-Smirnov (K-S) statistic 1862

L

large AR rules 520
large-itemsets growth tree (LIS-Growth Tree) 1662
lattice structure concept 1543
LD block 1680
lead qualification process 1890
leaf node 231
learning classifier system (LCS) 1320
learning diagnosis system, development 2935
learning, multiple-instance 65
least absolute difference (LAD) algorithm 1868
least squares (OLS) algorithm 1868
legacy data 2726
lexicographic order 1304
library and information scientists 2674
library and information services 2673
library data records 2674
library managers 2674
library workflow 2675
lifecyle events 2652
lifetime value (CLV) 1837
linear combination 1168
linear format (LINEAR) 11
linear programming 104
linear regression 2476
linear regression 99
linguistics computing 2365
link analysis 97, 2698, 2703, 2967, 3639
link mining 3527
liquid chromatography (LC) 1760
local as view (LAV) 3542
local sequence patterns 3483
location-based service pattern 3477
location-sensitive sequence patterns 3483
logical data model 563
logical database models 188, 1183, 1190
logistic regression 12
logistic regression (LR) 13
logistics support system 2421
logo image discrimination 1980
loop-closed iterative refinement 840
loss control specialist 1889
loss function based criteria 367
loss given default (LGD) 1837

M

machine intelligence 2967
machine learning 94, 99, 119, 356, 1401, 1816, 1877, 1878, 1885, 2122, 2254, 2260, 2587, 2620

management information systems 398
mandatory access control (MAC) 686
mandatory leaf-node prediction 128
map cube operator 664
mapping process 1042
marine underwriting business 1888
market basket analysis 97
market basket association rules 2106
market competition 1963
marketing campaign, design of 2825
marketing managers 2660
Markov Chain Monte Carlo 1405
Markov model 1406
Markov process 1406
mars 2560–2565
mass spectrometry (MS) 1760, 1761, 1780
massively parallel processing (MPP) 2756
materialized view selection (MVS) 2201, 2202, 2203, 2204, 2205, 2206, 2203, 2206, 2207, 2208, 2209, 2212, 2215, 2217, 2222
maximum difference subset (MDSS) 1658
maximum sub-array 1310
MCP 28
MD modeling 558, 559
mean-squared error 2481
medical data mining 2517
medical digital libraries 1810
medical discovery 2517
medical field 3662
medoid 240
mental healthcare 2496
metabolic pathway analysis 1794
metabolic pathways 101
metabolomics 1759, 1779, 1780
Metadata 19, 2265, 3418
meta-learning 3308
metaqueries 3453–3454, 3455
metarule 311, 313, 315
methods for gene selection 1708
Metis methodology 382
MFIs 231
microarray 1706
microarray data sets 1707
middle agents 471
MIDI 3590
minimum description length (MDL) 1083, 1659
minimum error ordering in the wavelet domain (MEOW) 1255
mining biomedical literature 1814
mining minimum support 3150
mining schema 920
mining sentences 1457
mining systems, and human involvement 838
Minnesota Mining and Manufacturing (3M) 2752
mise en place problem 2645
missing imputation 2559–2565
misuse detection 2881
mixed decision trees 3510
mixed decision trees, and evolutionary induction 3509–3523
mixed decision trees, global induction of 3511
MLEM2 algorithm 963
multimedia message service (MMS) 2027
mobile (m-) business 2697
mobile business data 2697
mobile data warehouse systems 1542
mobile equipment 1504
mobile phone customer type discrimination 1519–1538
mobile phone usage, classifying of 3440–3450
mobile phones 1504
mobile technology 2699
mobile user data mining 1502–1518, 1503
model combiners 356
model formulation 2307
model formulation and classification 32
model identification 2281
model performance 1863
model verification 920
model-based algorithm 2901
modeling context 1520
monitoring agents 2651
monitoring and discovery system (MDS) 916
Monte Carlo simulation 1877, 1885
most valuable customers (MVCs) 1834
MP3 3611
MPEG 3611
multi-agent approach 439, 469
multi-agent approach advantages 472
multi-agent systems 439
multi-dimension expression (MDX) 2715
multidimensional clustering (MDCLUST) 1654
multidimensional database 2
multidimensional database design 2183
multidimensional expressions (MDX) 682
multidimensional modeling 594
multidimensional OLAP 174
multidimensional security constraint language (MDS 682
multi-label classification 64
multilevel object-modeling technique (MOMT) 681
multi-level spatial measure (MuSD) 666

multimedia data mining 2600
multimedia database mining, with virtual reality 1557–1572
multi-paradigm architecture 2341
multiple criteria linear programming (MCLP) 29, 38
multiple criteria programming (MCP) 26, 27
multiple criteria quadratic programming (MCQP) 29
multiple discriminant analysis (MDA) 13
multiple genome comparison 1793
multiple reaction monitoring (MRM) 1761
multiple scoring systems (MSS) 1158
multirelational databases (Multi-DB) 509
multi-relational databases of XML data, construction 513
multivariate adaptive regression splines (MARS) 1858
music data mining 3586
music information retrieval (MIR) 3589
mutation operator 1082
MYL learning dataset 1967, 1974
MYL test dataset 1960, 1976

N

Naïve Bayes (NB) 1321, 2036, 2037
Naïve Bayes approach 1658
narrow fan-beam scan mode 3494
national identification card 2868
national security 693, 3528
natural language processing (NLP) 98, 1816
navigation assistance 1573, 1588
nearest neighbor 2073
nearest neighbor method 1718
nearest neighbor prediction 2858
nearest neighbour (NN) 3602
nearest triangulation 2073
negative event information 1422
negative predictive value (NPV) 2921
negotiation 2903
nesting level 1042
NetNanny 1979
neural data mining system (ndms) 2713
neural networks 9, 13, 100, 1419, 1438, 1717, 1800, 1824, 2454, 2475, 2714
neutral loss (NL) 1761
new drug application (NDA) 3676
new entrants 2776
Newton's method 1276
NFK measure 415
narrative knowledge representation language (NKRL) 1376
node partitioned data warehouse (NPDW) 719
noise addition techniques 55
nominal attribute 232
nonlinear dynamics 2620
nonlinear regression 2481
nonperturbative algorithms 2385
nonpornographic Websites 1961
non-spatial data dominated generalization 3479
Non-text media 1461
non-volatile 1829, 2735
normalization 2365
North West Shelf (NWS) 1920
Norton Internet Security 1979
novel data analysis methods 1714
novel detection accuracy 2041, 2042, 2044, 2046, 2047, 2048, 2049
NP-hard 107
nucleotide sequence database 95
numeric tokens 1365
numerical attribute 232, 1304

O

objective function (OB) 2949
object-modeling technique (OMT) 681
object-oriented databases 1192
object-oriented databases (OODBs) 2121
observation 2737
occurrence-based grouping heuristic 248
ODMG 1196
OLAM system 76
OLAP 181, 1591, 3234, 3236, 3249, 3251
OLAP analysis 2727
OLAP dimension flattening process 984
OLAP hierarchies 974
OLAP query 1544
OLAP technology 2164–2184
OLAP visualization of multidimensional data cubes 974
OnePass algorithm 3280
OnePass-AllSI algorithm 3281
one-way association 2055
one-way clustering 1649
online analytical processing 170, 179, 208, 255, 382, 430, 451, 557, 564, 682, 718, 1005, 1335, 1539, 1591, 2585, 2749, 2767, 2714, 2964, 3176
online analytical processing (OLAP) technologies 2641
online mining 76

online transaction processing (OLTP) 20, 208, 1591, 2767
online transaction processing transactions 179
online transactional processing 169
ontology 3525, 3537, 3559
ontology classes 916
ontology definition language 917
ontology engineering 3560
ontology mapping tools 1057
ontology systematic shared 1904
ontology Web ontology language services (OWL-S) 921
ontology-based data warehouse 1915
Ontology-Based Personalization 3574
OO concepts 487
OO requirement model 492
open grid service infrastructure (OGSI) 755
open grid services architecture (OGSA) 758
operational data 2725
operational data store (ODS) 757, 1829, 1831
operational scale requirements 2781
optimal alignment 105
optimization/ant colony optimization (PSO/ACO) algorithm 139
order-entry systems 429
ordinal attribute 232
organizational culture 2291
organizational data analysis 449
organizational data cognition 2302
organizational data mining (ODM) 2289, 2296, 2856
organizational data mining project factors 2292
organizational decision support 1689
organizational framework 450
organizational learning and behavior 2291
organizational performance 2292
organizational politics 2291
organizational theory (OT) 2289
outlier treatment 2559–2565
OWL 3532

P

Pacific-Asia Conference on Knowledge Discovery and Data Mining (PAKDD) 1320
packet sniffer 2893
Padrão 880
Padrão System 897
page crawlers 1469
page specification languages 1468
pairwise comparison database (PCDB) 1793
PANDA 246
PANDA framework 236, 247
PANDA project 3145
pan-organism classification 128
parallel database systems (PDBS) 3324
parallel implementation scheme of relational tables 3324–3345
parallel processing 95
Parthasarathy-Ogihara 235
partially ordered set (Poset) 1168
partially supervised classification (PSC) 1216–1230
partition around medoids (PAM) 943
partition tree 2130
partitioning and parallel processing 18
partitioning methods 232
passive intelligence 1899
path analysis 3424
patient diagnosis 2517
Patriot Act 2863
pattern analysis 2005
pattern discovery 2005
pattern finding 1632
pattern mining 2004
pattern monitor (PAM) 237, 248
pattern recognition 2122, 3621
product data management (PDM) system 1184
peer-to-peer (P2P) infrastructure 1786
percentage of total skin color pixels 1974
periodic data 740
permanent rules 248
perpetual inventory (PI) system 2768
personalized Web agents 2594
personnel hiring records 2372
petroleum resources description framework (PRDF) 1920, 1921
phenomenal data mining 2602
phylogenetic tree 101
physical data model 563
piece-wise affine (PWA) identification 2281
pixel-based techniques 2107
PLATCOM 1792
point-of-sales (POS) data 2764
pooled information 2376
pornographic Website classification 1962
position weight matrices (PWM) 1724
positive and negative syndrome scale (PANSS) 3683
positive predictive value (PPV) 2921
possibility distribution 189
posteriori error rate 1975
p-problems 10
precursor (PS) 1761

prediction by collective likelihoods (PCL) 1658
prediction tools 1401
predictive accuracy 3013
predictive attributes 1965
predictive model markup language (PMML) 151, 919, 2295
preference-based frequent pattern mining, general framework 1296
preference-based mining, efficiency of 1294
preprocessing 2005
primary structure 100, 1092
principal component (PC) 1766
principal component analysis (PCA) 1763, 2036, 3624
priori error rate 1961, 1975
privacy 1515, 2004, 2835, 2856
privacy breach 2386
privacy constraints 695
privacy issues 2375
privacy laws 2412, 2415
privacy preserving data mining (PPDM) 2380, 2403, 2419, 2851
privacy preserving data mining techniques 2380
privacy violation 52
privacy, in data mining text books 2872–2879
privacy, relational 2836
privacy-preserving data mining 3642
probabilistic neural network 2482
probabilistic principal surfaces (PPS) 2067
probability method 1168
probability of default (PD) 1836, 2457
process management 2
process-based data mining 343–349
Products on the Web! (PoW!) 2759
professional stage of growth 2729
profile initialization 1497
proliferation 1959, 1964
promoter recognition 2248, 2250, 2251, 2252, 2258, 2259, 2260, 2261
property map (PMap) 1612
PROSPECT 109
prospective landmark 51
Protection of privacy 1515
protein 1091
protein data bank (PDB) 95
protein interaction 1747
protein interaction network 101
protein structure 103
protein structure alignment 104
Protein structure prediction 103
protein structure prediction (PSP) 1158
protein threading 104
protein threading with constraints 104
protein-clustering 1091
proteomics 1698
proto-collection 1995, 1998
prototype system 1351
pseudo-C code 1303
public access to data warehouses 2429

Q

quadrupole time of flight (qTOF) 1761
qualifying leads 1891
qualitative spatial reasoning 885
quality of care 2376
quality of service open shortest path first (QOSPF) 3033
quaternary structure 100, 1738
query mechanism 1543
query modifiers 174

R

radical basis function network (RBFN) 1175, 2641, 2769
ragged (or incomplete) hierarchy 217
random error rate technique 1975
range-encoded bitmap indices 1601
ranking algorithms 1437
rapid application development (RAD) 382
RAPTOR 109
RBF kernel functions 1277
rank correlation analysis (RCA) 1168
receiver operating characteristic (ROC) 368, 1520
reconfigurability 1785
recursive partitioning 1211
redundant array of inexpensive disk (RAID) system 561
redundant information 2892
Regression 96, 99, 2374
regression techniques 2967
regression trees 1301
relational database management systems (RDBMS) 1831, 2587
relational database systems (RDBMS) 20, 2371
relational modeling 181, 2364
relational OLAP 174
relational transformation of XML data, advantages 515
relationship-mining algorithm 2932
relative strength 1428
renal failure 2516

representative function 1082
reproduction operator 1082
requent wayfinding-sequence (FWS) 1574
resources monitoring 915
retaining strategy 2061
retention time (rt) 1761
return on investment (ROI) 1843, 2432
return on investment (ROI) analysis 2765
revelation time 740
RFID 2762
right-time enterprise (RTE) 1851
rigid gap motifs 1725
risk exposure 2376
risk model 1859
risk models 2376
RNA secondary structures 104
robust split 3679
ROCK algorithm 240
ROLAP 181, 3176, 3177, 3182, 3190
root paths 3253
root subtrees 3254
root-mean-square error (RMSE) 1086
rough set methodology 3038
rough set theory (RST) 3006
rough sets 1824
rule constraints 2602
rule generation 2283
rule induction 963, 966, 968, 970, 1718, 2907, 3164–3175
rule induction, divide-and-conquer 3167
rule induction, exhaustive search 3167
rule induction, separate-and-conquer 3167

S

salient variables 2281
sanitization-based techniques 56
schema level conflicts 534
schema matching 350
science 2619
score function 105
score matrix 1170
score-based greedy algorithm 1082
search algorithm 2967
search engines 1462, 2273
search heuristic, forming of a 3169
search patterns 1387
search-bots 1489
seasonal decomposition 3274
second generation data-mining system 2588
secondary structure 100
secondary structure elements (SSEs) 1738
Secondary structure prediction 104
secure multiparty computation (SMC) 2387
secure multi-party computation (SMC) 54
Securities and Exchange Commission (SEC) 2440
security labels 684
security management 2
SEED 1786
segmentation 97
seismic data management and mining system (SD-MMS) 3645–3661
seismological data warehousing 3645–3661
selective superiority problem 652
self-insured 2376
self-organising map (SOM) 260, 1649, 2069, 2097, 3030
self-organization latent lattice (SOLL) 1650
semantic context 3527
semantic data mining 3524
semantic genetic algorithm (SGA) 1084–1090
semantic metadata 3526
semantic mutation operator (SM) 1084
semantic relations 1987, 1988
semantic similarities measurement 3571
Semantic Web 470
Semantic Web Mining 3531
Semantic Web rule language (SWRL) 923
semiosis 2321
sentence completion 1457
sequence apriori algorithm 1095
sequence clustering 1794
sequence data 3480
sequence data mining 2601
sequence pattern 3481
sequence to-model 943
sequential pattern 647, 2004, 2011, 3365
sequential patterns mining 336
server session 2005
service discovery protocols (SDPs) 710
sessionization 2004, 2553
sexology 1962
shared hierarchies 217
Shlosser's estimator (SE) 2211, 2212
significant interval 3275
significant interval discovery (SID) 3272
similar time-series data 1124
similarity and time series 1124
similarity match ratio (SMR) 1369
similarity relationship 2122
similarity retrieval algorithm 1124
simple object access protocol (SOAP) 2023
simple process 922

simple quadratic programming (SQP) 45
simple spatial features 663
Simpson's paradox 3234, 3236, 3237, 3242, 3244, 3247
simultaneous iterative reconstruction technique (SIRT) 3496
single linkage 240
single nucleotide polymorphisms (SNPs) 1674
single pass algorithm, and time-series data 3272–3284
single point cyclic crossover operator (SPCC) 1087
single point semantic crossover operator (SPSC) 1084
single tree CART models 1526
singular value decomposition (SVD) 1774, 2029
skin colour related visual content-based analysis 1968
skin-colour model 1969
sk-navigation 1350
sk-navigation system 1336
sliced average variance estimation (SAVE) 1658
sliced inverse regression (SIR) 1658
slowly-changing dimensions 740
small business scoring service (SBSS) 2453
smaller error variance 2792
short message service (SMS) 2027
snapshot fact 740
snapshot model 740
social network 2338, 2339, 2340, 2341, 2342, 2343, 2344, 2346, 2349, 2351, 2352, 2353, 2357, 2358, 2359, 2360, 2361, 2362
social network system 2341
social sciences 2296
societal value 2847
soft independent modeling of class analogy (SIM-CA) 1160
software agents 439
software maintenance 472
source physical schema 576
space transformation techniques 55
spam filtering 1454
sparse statistics trees (SST) 1207, 1208, 1209, 1211, 1212
spatial association patterns 3479
spatial association rule 3479
spatial characteristic rule 3479
spatial data 660
spatial data dominant generalization 3479
spatial data mining 2600
spatial data warehousing 659
spatial databases 3480
spatial fact 666
spatial measure levels 669
spatial mining 3478
spatial multidimensional data model 662
spatial OLAP (SOLAP) 661
spatial patterns 3480
spatio-temporal database 3477
spatiotemporal prediction 1401
spectral bioclustering 1655
s-problems 10
SQL 2715
stability index (SI) 1865
standard ARM 1750
standard deviation 1802
stanford stream data manager (STREAM) 1233
star schema 182, 209, 307, 308, 309, 314, 327, 3390
state-oriented data 740
static approach 3085, 3090, 3112
static view selection 3085, 3105, 3111, 3112
statistical algorithms 1717
statistical hypothesis testing 364
statistical process control (SPC) 164
statistical techniques 2906
statistics tree (ST) 1206, 1207
StatSoft 162
steganography 2711
STING 233
stochastic gradient boosting 1519–1538
stochastic models 1404
stock market returns 2477
stock-and-flow diagrams 2690
storage limit m, flexible adjustment of 1237
storage management mechanism 1549
stored knowledge 469
strategic decision-making process 2304
strategic HR decisions 2371
strategic utilization 1689
strategy formulation (SF) phase 2304
strategy of homogeneity 1971
stream mining, data reduction-based 1234
streaming data, summarizing 1234
structural motif 1738
structured query language (SQL) 1820
student learning problems in science courses, data mining approach 2928–2942
subject-oriented 1829
substitute products 2776
aubtree mining 3255, 3259
subtrees 3254, 3255, 3261
successive decision tree (SDT) 2978–2979, 2983

summarization 98
super computer data mining (SCDM) 1320
superparamagnetic clustering algorithm (SPC) 1652
supervised learning 96, 100, 1816
supply chain 2637
supply chain management
supply chain management (SCM) 810, 1184, 2468, 2641
supply chain processes 2763
support vector machine (SVM) 28, 44, 100, 1230, 1269, 1271, 2036, 2454
surveillance tools 2864
survival analysis (SA) 2454
suspicious activity report (SAR) 2324
switchboard user interface 2651
synergy 3412
syntax development 2326
system dynamics, limitations 2690
systematic querying approach for virtual dimension/s (VDim/s) 498
systems analysis 2620
systems development life cycle (SDLC) 3

T

table-based techniques 2107
tablespaces creation 581
task agents 471
task environment 2303
technology hype 2433
technology value 2775
telephony data mining 2646
temporal association patterns 3481
temporal database mining 3480
temporal hierarchy 217
temporarily rules 248
Teradata Corporation 2764
termination reports 2372
terrorist information awareness 2422
tertiary structure 100
tertiary structure 1738
text categorization 2595
text classification 1454
text data mining 2594
text mining 98, 1810, 2674
The Data Warehouse Institute (TDWI) 1842
third generation (3G) 1322
third generation data-mining systems 2588
time complexity 1601
time sequences 1117, 1118, 1119, 1127, 1136
time series 942, 943, 944, 953
time series data mining 2601
time-series data 3480
topic maps 3534
topological spatial relations 892
TOPSIS 1877, 1878, 1879, 1881, 1882, 1877, 1884, 1885, 1886, 1887
total information awareness 2422
traditional online transaction (OLTP) 163
traditional ontologies 1376
train set vs. prediction set data 1523
training data set 1205, 1208, 1209, 1210
transactional data 2638, 2892
transactional database 2641
transactional fact 740
transactional model 740
transactional supply chain data 2638
transaction-based data 2641
transient data 740
transparent fuzzy modeling 1138–1156
trauma audit 2917
trauma audit & research network (TARN) 2917
tree sketch 3679, 3680
TreeMiner 3266, 3267, 3269
TreeNet confusion matrix 1528
trigonometric differential evolution (TDE) method 2948
true rule 3379
true traversals 3379
trust building 2707
trust maintenance 2709
two-phase selection 2036, 2049
two-way association 2055

U

ubiquitous data mining (UDM) 2597
UCR time series data mining archive 951
UML diagrams 566
UML extensibility mechanism 562
UML extensibility mechanism package 595
unbalanced (or recursive) hierarchy 218
underpinnings 1992
unified modeling language (UML) 283, 286, 288, 301, 558, 561, 563, 570, 572, 576, 584, 1189
unified modeling language (UML), and data warehouse modeling 591–621
unified process (UP) 558, 560, 564
uniform resource identifiers (URI) 3534
universal description, discovery, and integration (UDDI) 2023
unsupervised feature selection via two-way ordering (UFS-2way) 1655

unsupervised learning 1816
unsupervised learning 97, 100
update with early pruning (UWEP) algorithm 238
up-sell 1890
URI (universal resource identifier) 2005
user identification 2892
user identification issues 1516
user sessions 3365
user agent interface 478
user agents 471
user-centered three-layer framework 2051, 2054
uses for mining medical digital libraries 1814

V

variable precision rough set theory (VPRS) 3005
variation of information (VI) 242
verification-driven data mining 2967
vertical database design, for scalable data mining 3694–3699
VidaMine 3458–3472
video association mining 1635
video classification 1632
video clustering 1632
video data mining 1632–1637
view selection problem (VSP) 21
virtual dimensions 494
virtual enterprise (VE) 1182
virtual environments (VEs) 1573, 1587
virtual reality 1624
virtual screening (VS) 1160
visual data mining (VDM) 1558, 1623–1630, 2105, 2599
visual shop 2889
visualization 96, 2107
visualization techniques 2968
visualization, and data mining 1623–1630
visualization, information 1623
visualization, three-dimensional 1627
VRMiner 1557–1572

W

WAH code (see word-aligned-hybrid code) 1597
Waikato environment for knowledge analysis (WEKA) 167
warehouse user requirements 491
wavelet coefficients, ordering signifigance of 1253
wavelet datasets, ordering coefficients 1253
wavelet transform and time series 1121
wavelets 1116
wavelets, summarization of 1235
wayfinding 1573, 1575, 1576
Web agents 472
Web characterization activity (WCA) 2005
Web communities 2339
Web content mining 98, 2590, 3364
Web crawler 1969
Web data 2888
Web data mining 2661
Web data techniques (WDT) 57
Web log 3365
Web log analysis 2890
Web log data 2659
Web log data set 2008
web log mining 1490
web logs 2004
Web mining 2888
Web mining 3364, 3539
Web ontology language (OWL) for services (OWL-S) 921
Web ontology language for services (OWL-S) 2031
Web page captions 1463
Web personalization 2604, 3286
Web robot 2554
Web service log mining 1953
Web service modeling toolkit (WSMT) 1057
Web services description language (WSDL) 2023
Web services usage data 1938
Web site traffic 3411
Web structure mining 2591, 2662, 3364
Web technology 3411–3415
Web usage 3293
Web usage mining 642–658, 2004, 2551, 2663, 3364, 3559
Web-based multi-agent system 469
Web-based warehouse 3428
WebGuard 1958
WebGuard-TS 1973
WebGuard-V 1973
WebGurad-TS 1967
WebParser 470
WebParser Architecture 473
WebParser Integration 479
Website topology 653
wiretapping 2867
WiseNut 1926
WMAvg 303–335
word-aligned hybrid (WAH) code 1597
workflow composer 915
workflow engine 916
workflow management (WFM) 810
workflow, description 915

working knowledge 2723
working smarter 2328
wrapper approach 470
wrappers 1765

X

XAR-miner in AR mining, efficiency 526
XML 3533, 3622
XML 469
XML (extensible markup language) 1190
XML (extensible markup language) 2023
XML data extraction 512
XML data warehouse 492
XML document warehouse (XDW), proposed model 491
XML document warehouses 485–508
XML document warehouses (XDW) 485
XML FACT repository (xFACT) 493
XML schema 487
X-tree indexing 1127
XTREEM-SA 1987, 1988, 1990–2001

Y

YCSKH-study 1173, 1174

Z

zero latency enterprise (ZLE) framework 757
zero-latency data warehouse (ZLDWH) 756